Advances in polymer nanocomposites

## Related titles:

*Physical properties and applications of polymer nanocomposites*
(ISBN 978-1-84569-672-6)
Polymer nanocomposites are polymer matrices reinforced with nano-scale fillers. Understanding the physical properties of polymer nanocomposites is a key factor in gaining wider uptake of the materials in new applications. The book is divided into sections covering polymer/nanoparticle composites, polymer/nanoplatelet composites and polymer/nanotube composites. It finishes by reviewing the range of applications for these important materials.

*Creep and fatigue in polymer matrix composites*
(ISBN 978-1-84569-656-6)
*Creep and fatigue in polymer matrix composites* reviews the latest research in modelling and predicting creep and fatigue in polymer matrix composites. The first part of the book reviews the modelling of viscoelastic and viscoplastic behaviour as a way of predicting performance and service life. Part II discusses techniques for modelling creep rupture and failure. The final part of the book discusses ways of testing and predicting long-term creep and fatigue in polymer matrix composites.

*Failure analysis and fractography of polymer composites*
(ISBN 978-1-84569-217-9)
The growing use of polymer composites is leading to an increasing demand for fractographic expertise. Fractography is the study of fracture surface morphology and is an essential tool for advancing the understanding of composite structural behaviour. *Failure analysis and fractography of polymer composites* gives an insight into damage mechanisms and underpins the development of physically based failure criteria. It provides a vital link between predictive models and experimental observations. Finally, it is essential for post-mortem analysis of failed or crashed structures, the findings of which are used to optimise future designs. This authoritative work provides both an overview of the field and a reference text for engineers.

Details of these and other Woodhead Publishing books can be obtained by:

- visiting our web site at www.woodheadpublishing.com
- contacting Customer Services (e-mail: sales@woodheadpublishing.com; fax: +44 (0) 1223 832819; tel.: +44 (0) 1223 499140 ext. 130; address: Woodhead Publishing Limited, 80 High Street, Sawston, Cambridge CB22 3HJ, UK)
- contacting our US office (e-mail: usmarketing@woodheadpublishing.com; tel: (215) 928 9112; address: Woodhead Publishing, 1518 Walnut Street, Suite 1100, Philadelphia, PA 19102-3406, USA)

If you would like e-versions of our content, please visit our online platform: www.woodheadpublishingonline.com. Please recommend it to your librarian so that everyone in your institution can benefit from the wealth of content on the site.

# Advances in polymer nanocomposites

## Types and applications

Edited by
Fengge Gao

Oxford   Cambridge   Philadelphia   New Delhi

Published by Woodhead Publishing Limited,
80 High Street, Sawston, Cambridge CB22 3HJ, UK
www.woodheadpublishing.com
www.woodheadpublishingonline.com

Woodhead Publishing, 1518 Walnut Street, Suite 1100, Philadelphia,
PA 19102-3406, USA

Woodhead Publishing India Private Limited, G-2, Vardaan House,
7/28 Ansari Road, Daryaganj, New Delhi – 110002, India
www.woodheadpublishingindia.com

First published 2012, Woodhead Publishing Limited


British Library Cataloguing in Publication Data
A catalogue record for this book is available from the British Library.

Library of Congress Control Number: 2012945570

ISBN 978-1-84569-940-6 (print)
ISBN 978-0-85709-624-1 (online)

The publisher's policy is to use permanent paper from mills that operate a sustainable forestry policy, and which has been manufactured from pulp which is processed using acid-free and elemental chlorine-free practices. Furthermore, the publisher ensures that the text paper and cover board used have met acceptable environmental accreditation standards.

Typeset by RefineCatch Limited, Bungay, Suffolk
Printed by Lightning Source

# Contents

# Contributor contact details

(* = main contact)

## Editor

Dr Fengge Gao
Reader in Nanotechnology
School of Science and Technology
Nottingham Trent University
Clifton Lane
Nottingham
NG11 8NS
UK

E-mail: Fengge.gao@ntu.ac.uk

## Chapter 1

V. M. Karbhari*
University of Alabama in Huntsville
366 Shelbie King Hall
Huntsville
AL 35899
USA

E-mail: vistasp.karbhari@uah.edu

C. T. Love
U.S. Naval Research Laboratory
Alternate Energy Section
4555 Overlook Avenue SW
Washington
DC 20375
USA

E-mail: corey.love@nrl.navy.mil

## Chapter 2

Vikas Khanna*
Department of Civil and Environmental Engineering
University of Pittsburgh
Benedum Hall
3700 O'Hara Street
Pittsburgh
PA 15261
USA

E-mail: khannav@pitt.edu

Laura Merugula and Bhavik R. Bakshi
William G. Lowrie Department of Chemical and Biomolecular Engineering
The Ohio State University, Koffolt Laboratories
140 W, 19th Ave
Columbus
OH 43210
USA

E-mail: bakshi.2@osu.edu

## Chapter 3

Dr Yong Lin and Professor Chi-Ming Chan*
Department of Chemical and Biomolecular Engineering
The Hong Kong University of Science and Technology
Clear Water Bay
Hong Kong

E-mail: kecmchan@ust.hk

## Chapter 4

Professor Dr José Ignacio Velasco*
University Professor
Centre Català del Plàstic
Departament de Ciència dels Materials i Enginyeria Metal-lúrgica
Universitat Politècnica de Catalunya
C/ Colom 114
E-08222 Terrassa, Barcelona
Spain

E-mail: jose.ignacio.velasco@upc.edu

Dr Mónica Ardanuy
Assistant Professor
Centre Català del Plàstic
Departament d'Enginyeria Textil i Paperera
Universitat Politècnica de Catalunya
C/ Colom 11
E-08222 Terrassa, Barcelona
Spain

E-mail: monica.ardanuy@upc.edu

Dr Marcelo Antunes
Assistant Professor
Centre Català del Plàstic
Departament de Ciència dels Materials i Enginyeria Metal-lúrgica
Universitat Politècnica de Catalunya
C/ Colom 114
E-08222 Terrassa, Barcelona
Spain

E-mail: marcelo.antunes@upc.edu

## Chapter 5

Elaine C. Ramires
Instituto de Química de São Carlos (IQSC)
Universidade de São Paulo (USP)
C.P. 780
13560-970 São Carlos
Brazil

Alain Dufresne*
The International School of Paper, Print Media and Biomaterials – Pagora
Grenoble Institute of Technology, BP 65
38402 Saint Martin d'Hères Cedex
France

E-mail: Alain.Dufresne@pagora.grenoble-inp.fr

## Chapter 6

Dr Greg Heness
Senior Lecturer in Materials Science & Engineering
School of Physics and Advanced Materials and the Institute of Nanoscale Technology
University of Technology Sydney
PO Box 123
Broadway, NSW, 2007
Australia

E-mail: g.heness@uts.edu.au

## Chapter 7

Tapas Kuila* and Joong Hee Lee
Department of BIN Fusion Technology
Chonbuk National University
Jeonju, Jeonbuk
561-756
Republic of Korea

E-mail: tkuila@gmail.com

Tridib Tripathy
Department of Chemistry
Midnapore College
Midnapore
West Bengal
721101
India

## Chapter 8

Professor Marianne Gilbert
Department of Materials
Loughborough University
Loughborough
Leicestershire
LE11 3TU
UK

E-mail: M.Gilbert@lboro.ac.uk

## Chapter 9

Szu-Hui Lim
Singapore Institute of Manufacturing Technology (Simtech)
71 Nanyang Drive
638075 Singapore

Aravind Dasari*
School of Materials Science & Engineering
Blk. N4.1
50 Nanyang Avenue
Nanyang Technological University (NTU)
Singapore 639798

E-mail: aravind@ntu.edu.sg

and

Madrid Institute for Advanced Studies of Materials (IMDEA Materials Institute)
C/ Profesor Aranguren s/n
28040 Madrid
Spain

E-mail: aravind.dasari@imdea.org

## Chapter 10

Jayita Bandyopadhyay
DST/CSIR Nanotechnology Innovation Centre
National Centre for Nanostructured Materials
Council for Scientific and Industrial Research
Pretoria 0001
Republic of South Africa

Suprakas Sinha Ray*
DST/CSIR Nanotechnology Innovation Centre
National Centre for Nanostructured Materials
Council for Scientific and Industrial Research
Pretoria 0001
Republic of South Africa

E-mail: rsuprakas@csir.co.za

and

Department of Chemical Technology
University of Johannesburg
Doornfontein 2018
Johannesburg
Republic of South Africa

## Chapter 11

Professor D. J. Martin*
Group Leader, The Australian Institute for Bioengineering and Nanotechnology (AIBN)
Professor, School of Chemical Engineering
Building 75, Cr College and Cooper Rd
University of Queensland
Brisbane, 4072
Australia

E-mail: darren.martin@uq.edu.au

A. F. Osman and Y. Andriani
The Australian Institute for Bioengineering and Nanotechnology (AIBN)
Building 75, Cr College and Cooper Rd
University of Queensland
Brisbane, 4072
Australia

Dr G. A. Edwards
The Australian Institute for Bioengineering and Nanotechnology (AIBN)
Building 75, Cr College and Cooper Rd
University of Queensland
Brisbane
4072
Australia

## Chapter 12

Dr Kazutoshi Haraguchi
Material Chemistry Laboratory
Kawamura Institute of Chemical Research
631 Sakado
Sakura
Chiba 285-0078
Japan

E-mail: hara@kicr.or.jp

## Chapter 13

Dr Andy McLauchlin*
Exeter Advanced Technologies
University of Exeter
North Park Road
Exeter
Devon
EX4 4QF
UK

E-mail: A.McLauchlin@exeter.ac.uk

Dr Noreen Thomas
Department of Materials
Loughborough University
Loughborough
Leicestershire
LE11 3TU
UK

E-mail: n.l.thomas@lboro.ac.uk

## Chapter 14

Professor Hee-Woo Rhee* and Lee-Jin Ghil
Department of Chemical and Biomolecular Engineering
Sogang University
1 Shinsu-dong, Mapo-Gu
Seoul 121-742
South Korea

E-mail: hwrhee@sogang.ac.kr

## Chapter 15

James Njuguna*, Krzysztof Pielichowski and Jiying Fan
Cranfield University
Bedfordshire
MK43 0AL
UK

E-mail: j.njuguna@cranfield.ac.uk

and

Cracow University of Technology, ul.
Warszawska 24
31-155 Kraków
Poland

## Chapter 16

José-Marie Lopez-Cuesta
Centre des Matériaux (CMGD)
Ecole des Mines d'Alès
6 Avenue de Clavières
30319 Alès Cedex
France

E-mail: Jose-Marie.Lopez-Cuesta@mines-ales.fr

## Chapter 17

Dr Dorothée Vinga Szabó*
Karlsruhe Institute of Technology
Institute for Applied Materials – Material Process Technology
Hermann-von-Helmholtz-Platz 1
76344 Eggenstein-Leopoldshafen
Germany

E-mail: dorothee.szabo@kit.edu

Professor Dr Thomas Hanemann
Karlsruhe Institute of Technology
Institute for Applied Materials – Material Process Technology
Hermann-von-Helmholtz-Platz 1
76344 Eggenstein-Leopoldshafen
Germany

E-mail: thomas.hanemann@kit.edu

and

Albert-Ludwigs-University of Freiburg
Department of Microsystems Engineering
Georges-Koehler-Allee 102
79110 Freiburg
Germany

## Chapter 18

Tsao-Cheng Huang and Jui-Ming Yeh*
Department of Chemistry and Center for Nanotechnology
Chung Yuan Christian University
Taiwan

E-mail: juiming@cycu.edu.tw

Cheng-Yuan Lai
Department of Material Development
Sipix Technology, Inc.
Taiwan

# Preface

Filler-enhanced polymer nanocomposite technology is one of the driving forces in stimulating and promoting nanotechnology development. At present, nanotechnology has expanded into almost every aspect of science and technology. It may be appropriate to summarise today's nanotechnology as the science and technology developed within the 'nanometre range'. This summary appears to be too generous and somewhat confusing but may truly represent what has happened in this field. The original initiative for nanotechnology was to develop new and much better materials within the nano-size range, which had been ignored before. For filler-enhanced materials, innovation has led to greatly improved properties and functions, which cannot be obtained using traditional macro- and micro-technology within normal filler loading levels. For devices, innovation has led to novel functions and applications, which were not previously possible. However, this aim has only been partially achieved over the past one and half decades. There are two reasons why there has been a move away from the initial aim. First, the commonly accepted size range in the original definition for nanotechnology, 1 to 100 nm, is fairly random and during the early stages of development there was a lack of scientific evidence to show any difference between materials in this range and in the micro-range. In fact, it is has become more and more clear that many types of particle larger than 50 nm have similar properties to microparticles. Therefore, most particles between 50 nm and 100 nm, and also larger particles, do not show a dramatic enhancement of material properties as originally expected. The second reason was that some scientists took advantage of the better opportunities for funding and the publications in nanotechnology, to rename well-developed scientific fields with some relevance to the 1–100 nm range as nanotechnology. As a consequence, today's nanotechnology is a mixture of both innovative new developments and less competitive traditional science and technology. Filler-enhanced nanocomposite technology is one of those truly innovative areas, whose results could not have been achieved before. In the early days of this technology, the Toyota team demonstrated that exfoliating smectite clay to create nanoparticles as polymer fillers, could lead to a dramatic enhancement of a wide range of physical and engineering properties using a very small fraction of filler loading compared with traditional micro-dispersed clay/polymer composites. Since then clay/polymer nanocomposite technology has been used with almost every type of polymer. More and more nanofillers have been identified and used for different polymeric products. Whilst other areas of nanotechnology were still in the concept stage, filler-enhanced polymer nanocomposite technology was used in various commercial products from the very beginning. These developments were a catalyst in promoting innovative nanotechnology in other polymer fields.

This book was originally designed to be a handbook covering fundamental knowledge, major scientific progress and applications of filler-enhanced polymer nanocomposites for each filler and polymer. However, the rapid progress and expansion of this field and the busy lives of the experts forced us to change the format of the book. We have started with selected nanocomposite research fields, which have the potential to form a series of chapters to cover the progress made in this subject on a regular basis systematically. This first book follows the original design and covers research on polymer systems, types of filler and applications. Part II of this book is on polymer systems and includes relatively mature developments in nylon and difficult polymer systems such as polyolefins, PVC and PET. It will also highlight new developments in soft nanocomposites and gels, thermoplastic polyurethanes and biodegradable polymer nanocomposites, which generate much public interest. In Part I, on the types of filler, there are no chapters on fillers such as clay since these are discussed extensively in most of the chapters in Part II on polymer systems and in Part III on applications. Part I covers those filler systems that have not been extensively discussed elsewhere, such as cellulose, metal and calcium carbonate nanofillers, together with the relatively popular fillers, carbon nanotubes and layered double hydroxides. We felt that it was important to address the health and safety aspects of nanoparticles and their composites and the life cycle performance of this new class of material. Unfortunately we could not include a chapter on the toxicity of nanocomposites in this current volume, but there is a chapter on life cycle assessment in Part I. Although nanocomposite technology has only been under development for a relatively short time, its applications have spread to many fields. Part III of this volume includes only a few selected applications in fuel cells, aerospace products, optical products, fire retardancy and coatings. More applications will be included in subsequent volumes in this series.

Finally I would like to thank all the authors who squeezed valuable time out of their busy lives to contribute to this book and those reviewers who gave invaluable critical comments on the manuscripts. The editors and staff at Woodhead Publishing Limited played a major role in editing this book. Many thanks for your hard work.

*Fengge Gao*

# Part I

## Types of polymer nanocomposites according to fillers

# 1
# Processing of nanotube-based nanocomposites

V. M. KARBHARI, University of Alabama in Huntsville, USA
and C. T. LOVE, U.S. Naval Research Laboratory, USA

**Abstract:** With the continuing developments in materials synthesis and characterization at the nanoscale the potential of true materials tailoring has been enhanced substantially. In general this class of materials involves structures that have at least one dimension at the nanometer scale (usually taken to be up to 100 nanometers). The utilization of nanoscale fibers in polymer composites not only enables the development of uniquely created structures but also provides a means for the development of unique properties and functionalities at levels not possible with conventional fiber reinforced composites. It is the goal of this chapter to introduce the advances in carbon nanotube (CNT) nanocomposite research with specific emphasis on processing routes used to disperse, align and fabricate carbon nanotube reinforced polymer nanocomposites for enhanced physical and mechanical properties. Acknowledging that the potential for CNTs as reinforcement and conductive media has not yet been fully realized, a discussion on future trends is also given.

**Key words:** carbon nanotubes, composites, alignment, processing, performance attributes.

**Note:** This chapter is a revised and updated version of Chapter 19 'Processing of carbon nanotubes and carbon nanotube based nanocomposites' by V. M. Karbhari and C. T. Love, also published in *Advances in polymer processing: From macro to nano scales*, ed. S. Thomas and Y. Weimin, Woodhead Publishing Limited, published 2009, ISBN: 978-1-84569-396-1.

## 1.1 Introduction

### 1.1.1 Nanoscience and composites

With the continuing developments in materials synthesis and characterization at the nanoscale the potential of true materials tailoring has been enhanced substantially. In general this class of materials involves structures that have at least one dimension at the nanometer scale (usually taken to be up to 100 nanometers). Remarkable progress has been made regarding the use of carbon nanotubes (CNTs) to reinforce polymer matrices since the helical tube geometry of carbon nanotubes was first discovered by Iijima in 1991 (Iijima, 1991). As the name implies, CNTs are cylindrical tube structures of varying lengths made of carbon atoms. They exist as either single-walled carbon nanotubes (SWCNT) or multi-walled carbon nanotubes (MWCNT) where individual tubes are nested concentrically inside one another like tree rings. These unique nanostructures are considered one-dimensional due to their high aspect ratio (length-to-diameter) leading to superior mechanical and electrical properties. The diameter and chirality of the CNTs produce either metallic (conducting) or semi-conducting nanotubes. Additionally, their anisotropic nature makes CNTs interesting reinforcing fibers for multi-functional ultra-light, high strength and stiffness composite materials and devices. Compared with carbon fibers, the very high modulus versions of which have elastic moduli of over 750 GPa, the elastic moduli of CNTs have been measured in the range of 1–4.7 TPa (Cooper *et al.*, 2002; Lau *et al.*, 2006). The tensile strength is also approximately two orders of magnitude higher than that shown by currently available carbon fibers (Cooper *et al.*, 2002). In addition, the change in diameter of these materials also results in a significant increase in surface area for the same volume fraction in a composite, leading to a variety of very interesting characteristics. For example, a 30 nm diameter nanotube has 150 times more surface area than a 5 μm diameter carbon fiber for the same filler volume fraction (Eitan *et al.*, 2006).

The utilization of these fibers in polymer composites not only enables the development of uniquely created structures but also provides a means for the development of unique properties and functionalities at levels not possible with conventional fiber reinforced composites. Several comprehensive reviews of carbon nanotube reinforced polymer composites have been published over the last few years (Bal and Samal, 2007; Coleman *et al.*, 2006; Hussain *et al.*, 2006; Thostenson *et al.*, 2001, 2005; Gibson *et al.*, 2007; Xie *et al.*, 2005; Wang *et al.*, 2004). It is the goal of this chapter to introduce the advances in CNT nanocomposite research since the publication of these reviews with specific emphasis on processing routes used to disperse, align and fabricate carbon nanotube reinforced polymer nanocomposites for enhanced physical and mechanical properties.

### 1.1.2 Aspects for consideration

Just as on the macroscale, the properties of the nanocomposite are dictated by the distribution, orientation and fiber/matrix interactions. Because nanotubes tend to form clusters and bundles, the biggest challenges on the nanoscale are to fully disperse individual nanotubes in the matrices and achieve good interfacial adhesion between them and polymer for load transfer capabilities. The tendency for the reinforcement to agglomerate persists unless high shear forces are applied by vigorous mixing of the polymer. However, the mixing intensity must be controlled since overmixing often damages CNT structures, compromising their properties. Another issue is that the polymer–nanotube mixtures are highly viscous (due to the large surface area of nanotubes). This creates process-related problems, because the composites do not flow easily and are hence more difficult to mold. Viscosities of nanotube-filled polymers are known to show abrupt increases above fairly low loading thresholds following a Schulz–Blaschke type response. Processing is also hindered by the poor compatibility of nanotubes with most solvents and polymers. Nevertheless, several approaches have been successfully adopted to obtain intimate mixing of nanotubes with polymer phases, including dry powder mixing, melt mixing, polymerization of monomers onto and surrounding CNT surfaces, and surfactant-assisted mixing. More creative processing techniques are still needed. It should be noted that the potential of the mechanical, electrical, and thermal properties offered by nanotubes has not fully been realized and is mostly limited by processing methods. It is in this context that the current chapter provides a state-of-the-art review of the topic, highlighting important advances in a variety of processing routes, and ending with a brief identification of potential future directions.

## 1.2 Structure of carbon nanotubes

### 1.2.1 Carbon nanotube morphology

There are two primary types of CNTs available. SWCNTs consist of a single graphene sheet seamlessly wrapped into a cylindrical tube. The one-dimensional nature of the CNTs means that they exhibit electrical conductivity as high as copper, thermal conductivity as good as diamond, and strength levels as much as 100 times greater than steel at a fraction of the weight. The structure of MWCNTs can be thought of as concentrically nested SWCNTs where dimensions such as inner and outer tube diameter are important for strength and conduction. In most cases tube diameter is linearly proportional to tube thickness (due to concentric tube layering). MWCNTs offer higher stiffness than SWCNTs, especially in compression, due to the reinforcing efforts of centrically aligned tubes.

Single-walled carbon nanotubes can exist in three distinct structures: armchair, zigzag and chiral. The packing of the carbon hexagons in the graphitic sheets defines a chiral vector $(m,n)$ and angle. The indices of the vector determine the morphology of the CNT (Hussain *et al.*, 2006). Zigzag structures are identified by $(n,0)$ type indices, while armchair are $(m,n)$, and chiral are $(n,m)$ where $n \neq m$. When $(m-n)/3$ is an integer, the resulting structure is conductive and is termed 'metallic'. Other variations have semiconductive properties. Therefore all armchair and one-third of zigzag CNTs are metallic, having a continuous conduction band (Hussain *et al.*, 2006). The remaining two-thirds of the zigzag and chiral CNTs are semiconducting, having an energy gap in the conduction band (Hussain *et al.*, 2006).

### 1.2.2 Synthesis of carbon nanotubes

The mechanisms by which CNTs may be synthesized can be grouped into two categories: ablation of graphite or decomposition of carbon-containing compounds. The main methods of CNT synthesis using graphite sublimation are direct-current arc discharge and laser ablation. Many versions of the arc discharge method have been reported. The main disadvantage of this process is the concurrent formation of amorphous carbon soot, metal clusters coated with carbon and non-tubular fullerenes along with CNTs (Rakov, 2006). Such inefficiencies limit the production of SWCNTs to a 20–40% yield. Additionally, the SWCNTs produced this way are generally entangled, with poor linear orientation. SWCNT synthesis by laser ablation was first achieved by Smalley using a pulsed laser that strikes a target of graphite and a metal catalyst such as Co or Ni (Scott *et al.*, 2001). The laser vaporization plume condenses on a collector outside the reaction chamber to form the SWCNT structure. Another common technique for the production of SWCNTs is electric arc discharge in which carbon is vaporized between two graphite electrodes acting as the anode and the cathode. When a direct current (dc) voltage is applied, large quantities of electrons from the arc-discharge move to the anode and collide onto the anodic rod. Upon cooling, the electrode deposits condense on the cathode. The contents contain carbon nanotubes, nanoparticles, and clusters. Therefore, when comparing laser ablation and arc-discharge synthesis methods, the nanotubes produced by laser ablation are purer (up to about 90% pure) than those produced in the arc process (Scott *et al.*, 2001).

Chemical vapor deposition (CVD) is based upon the mechanism of decomposition of carbonaceous compounds to grow vertically aligned CNT arrays of controlled length. In this process a continuous supply of carbon-containing vapor is introduced into a reaction furnace at elevated temperature (Deck and Vecchio, 2006). Carbon is deposited onto conductive seed particles which can be patterned onto a silicon wafer to produce uniquely designed CNT arrays. Fang *et al.* (2005) used a CVD process to grow vertically aligned MWCNTs on a thin polysilicon film deposited on top of a silicon wafer. An iron film is then deposited and patterned onto polysilicon by means of a lift-off technique. Aligned MWCNTs were grown by CVD using acetylene as the precursor at 800°C in an $Ar/H_2$ flow for 10 minutes. This method has been shown to produce more intricate patterns of aligned MWCNTs depending upon the seed particle (Fe) patterning.

Plasma enhanced CVD (PECVD) results in further CNT uniformity within the array. Figure 1.1 shows the controlled length and diameter of vertically aligned MWCNTs grown from the PECVD of acetylene on a silicon wafer. Of all the synthesis techniques, variations of the CVD process show the most promise for economically viable, large-scale synthesis of high quality CNTs. Even so, there is no clear understanding of the influence of key parameters (temperature, pressure, carbon source) on CNT properties (diameter, length, and morphology), which shows that further research is necessary to optimize this process (See and Harris, 2007). Recent work has focused on scaling-up production of SWCNTs to supply large quantities to meet the demands of implementing CNT reinforced composite devices in a variety of application fields. A scaled-up pulsed laser vaporization process also shows promise for high production of SWCNTs to produce up to 10 g/day SWCNT (Rinzler *et al.*, 1998).

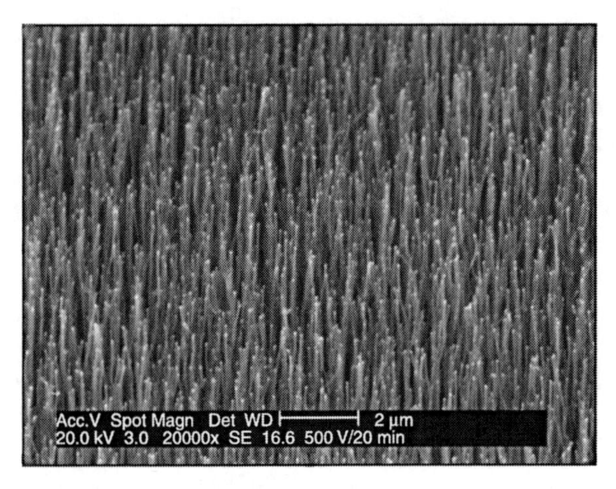


*1.1* SEM image of vertically grown MWCNTs on silicon substrate produced by PECVD process.

### 1.2.3 Carbon nanotube surface treatments

Surface treatments such as oxidation and chemical functionalization of CNT sidewalls and endcaps assist in the separation of bundles and entanglements. Acid etch pretreatments are often used as a means of purifying the CNTs and removing catalyst 'seed' particles that result from the CVD synthesis process. Just as macroscale fiber sizing promotes interfacial adhesion, chemical functionalization on the surface of the CNTs improves interaction with the polymer matrix at the nanoscale of the composite.

Functionalization along CNT sidewalls offers more coverage area than on endcaps and thus a higher capacity for property enhancement. Possible surface treatments include the attachment of functional groups or the grafting of low molecular weight polymers to the CNTs by covalent interactions or by adsorption where weaker secondary bonding interactions take place. Covalent bonding to CNT surfaces can be achieved through the attachment of chemical functionality to carboxylic acid groups on the CNT surface as a result of the acid etch treatment, or through direct bonding to surface carbon double bonds.

*Oxidation*

Oxidation of CNT surfaces is typically the starting point for various functionalization techniques. Oxygen-containing carbonyl, carboxylic acid and hydroxyl moieties can be easily introduced via an acid treatment which also helps to purify the CNTs by removing elemental impurities. Concentrated $HNO_3$ and mixtures of $H_2SO_4$ with $HNO_3$, $H_2O_2$, or $KMnO_4$ have been used for attaching oxygenated functionality to CNTs. However, excessive acid treatments may result in a decrease in physical properties. Functionalization by vigorous oxidation processes followed by covalent bonding has been shown to introduce defect sites where $sp^2$ hybridization is converted to $sp^3$ hybridization, which may affect the properties of the nanotube itself (Odegard *et al.*, 2005).

As a precursor for CNT functionalization, the oxidation conditions were explored by Hong *et al.* (2007), who indicated that MWCNTs were damaged during the oxidation process and became shorter. As the temperature of the oxidation step increased the CNT length decreased; however, with increasing oxidation time, the length distribution becomes comparatively narrow with an optimal oxidation time around 10 min for acid oxidation at 80°C. While physical properties of CNTs may be diminished with decreasing CNT length, the shorter lengths have been shown to aid in the dispersion of MWCNT in polypropylene (PP) composites (Hong *et al.*, 2007). Jin *et al.* (2007) also found that nitric/sulfuric acid treated CNTs showed a lower degree of entanglement due to shorter length. Another technique showed that the exposure of CNTs to $CO_2$/Ar plasma optimized with respect to time, pressure, power and gas concentration resulted in 14.5 at% in the first atomic layers forming hydroxyls, carbonyls, and carboxyl groups as detected by X-ray photoelectron spectroscopy (XPS) (Andrews and Weisenberger, 2004). Bubert *et al.* (2003) used plasma to introduce oxygen-containing functional groups to the surface of carbon nanotubes, resulting in polar groups being formed on the surface and a modified surface morphology which was shown to increase the adhesion between CNTs and a polymer matrix. Optimal treatment conditions were found to include the use of carbon dioxide as the plasma gas, a low pressure of approximately 0.1 hPa, a low plasma power of approximately 80 W or less and a treatment time of not more than 3–5 min.

*Polymer grafting*

Functionalization not only increases dispersibility in various solvents and polymers but also increases the strength of the interface between CNTs and the polymer (Andrews and Weisenberger, 2004). Poly(vinyl alcohol) (PVA) has been used to functionalize SWCNTs to render them dispersible and stable in water (Vandervorst *et al.*, 2006) and to create a dissipative interface for introduction into a poly(butyl acrylate) latex dispersion (Wang *et al.*, 2006). Hwang *et al.* (2004) showed that emulsion polymerization, when used to graft polymethyl methacrylate (PMMA) onto the surface of MWCNTs prior to their addition to PMMA at loadings up to 20%, would result in a significant increase in elastic modulus when processed using solution casting in a chloroform solvent. The *in situ* emulsion polymerization of monomer butyl acrylate (BA) or methyl methacrylate (MMA) has also been shown to proceed while CNTs are redispersed, consequently coating the CNTs, resulting in yield strength increases of about 31% and elastic modulus increases of about 35% at only 1.0 wt% polymer encapsulated MWCNTs dispersed in polymide 6 (PA6) (Xia *et al.*, 2003). Graft polymerization via plasma chemical vapor deposition has also been shown to produce PMMA coating on MWCNT, which can then be processed via melt mixing and drawing, causing a significant improvement in mechanical properties compared with pure PMMA (Gorga *et al.*, 2006). Functionalized MWCNTs with PMMA via *in situ* polymerization have also been demonstrated to result in improvement of 94% in elastic modulus, 360% for ultimate tensile strength, 373% for breaking strength, 1282% for toughness, and 526% for elongation at break with less than 0.5 wt% MWCNTs (Blond *et al.*, 2006).

Amine-functionalized CNTs have received a lot of research attention because of the potential for covalent bonding with epoxy resins. A comprehensive review of fluorinated trichlorosilane functionalization of MWCNTs was provided by Vast *et al.* (2004) and is hence not repeated herein. Amine terminated CNTs have been shown to increase the modulus, tensile strength, and elongation at break when introduced into natural rubber due to high polymer/CNT interaction between the silanized CNT and natural rubber vulcanizates (Shanmugharaj *et al.*, 2007). Amino functionalization makes the CNTs very effective crosslinking agents. Zhu *et al.* (2004) demonstrated that the functionalized CNTs form strong covalent bonds during the cure of epoxy matrices and hence become an integral part of the crosslinking system. Ramanathan *et al.* (2005) explored two techniques for introducing amine functionalization to the surface of SWCNTs. In the first approach, carboxylated SWCNTs were treated with ethylenediamine in a controlled reaction to link the free amine group to the SWCNTs to form amide functionality, while the second approach involved the reduction of the carboxyl group to hydroxymethyl, followed by transformation into aminomethyl groups. In both approaches, amine groups are tethered to the CNT surface, which is available for reaction with a variety of reactive polymers. Diamine-MWCNTs can be obtained from the reaction between acid-treated MWCNTs and 1,10-diaminodecan (DA10) with the lower degree of aggregation in functionalized MWCNTs being attributed not only to the functional drops, such as carboxyl and diamine, but also to their shorter lengths (Jin *et al.*, 2007).

*Gamma-radiation*

Guo *et al.* used $\gamma$-irradiation to prepare various functionalized modifications on CNTs, through the formation of an amine bond wherein the concentration of functional groups bound to the CNT increased due to the number of defect sites created by $\gamma$-photons (Guo *et al.*, 2005).

*Mechanochemical milling*

The large-scale production of functionalized short CNTs using mechanochemical ball milling has also been shown to introduce functional groups such as thiol, amine, amide, carbonyl and chloride onto the surface of MWCNTs (Konya *et al.*, 2002), providing a cheaper, albeit less uniform, mechanism of surface treatment.

## 1.3 Processing methods for nanotube-based polymer nanocomposites

### 1.3.1 Solution blending

Perhaps the most commonly applied processing technique for carbon nanotube polymer composites is the solution mixing or solution blending technique. Solution processing of CNT/polymer composites involves the dispersion of nanotubes in a polymer solution by energetic agitation, controlled evaporation of solvent and finally composite film casting. Agitation can be done by magnetic stirring, shear mixing, reflux or, most commonly, sonication. Sonication or ultrasonication excitation of filler/resin mixtures breaks up CNT clusters through cavitation (nucleation and collapse of bubbles) in the liquid and/or exciting resonant vibrations of the clusters. Sonication can be achieved either by immersing a beaker of solution of CNT in solvent in an ultrasonic bath or by partially submerging an ultrasonic wand (tip or horn) directly into the CNT/solvent solution. The processing parameters, such as CNT weight percent, bath and tip sonication, sonication times, surfactant type and solvent type, have an effect on the dispersion of SWCNTs in the bulk polymer.

Solution-based processes make use of low viscosities to facilitate mixing and dispersion. Many studies have used this method for processing of both thermosetting and thermoplastic polymers. The appropriate selection of a solvent is crucial to the efficacy of the process since it must not only provide complete dissolution of the polymer matrix material but also ensure the attainment of an appropriate viscosity to evenly disperse CNTs with the aid of sonication. Khan *et al.* (2007) studied the effect of solvent choice on the mechanical properties of CNT/polymer composites via a solution drop-casting technique and demonstrated that the removal of solvent from the CNT/polymer system also plays a critical role in the physical properties of the composite, with the $T_g$ decreasing if there is residual solvent due to the build-up of solvent at the CNT/polymer interface.

*Thermoplastic polymers*

Solution blending of thermoplastics begins with the dissolution of the polymer in a solvent. Esawi and Farag (2007) observed that the selection of a solvent can influence the mechanical performance of the composite, as shown in Table 1.1.

Solution mixing may be done at room temperature or with the application of external heat to accelerate the thermodynamic dissolution process. The nanotubes are then added to the liquid polymer solution and mixed via ultrasonication. To assist in the complete dispersion of nanotubes, a suspension may be created with

*Table 1.1* Mechanical properties of PVA and double-walled CNTs/PVA composites as a function of solvent selection and drying time

| Solvent | Drying time (days) | $E_{polymer}$ (GPa) | $E_{composite}$ (GPa) | Strain at break (%) | Tensile strength (MPa) |
|---|---|---|---|---|---|
| $H_2O$ | 1 | 1.28 ± 0.1 | 1.85 ± 0.01 | 9 | 78 ± 0.0 |
| | 5 | 1.18 | 1.43 ± 0.03 | 19 | 49 ± 15 |
| DMSO | 1 | 0.41 ± 0.01 | 0.50 ± 0.05 | 63 | 35 ± 2 |
| | 5 | 0.65 ± 0.05 | 0.90 ± 0.0 | 110 | 38 ± 1 |
| NMP | 1 | 0.35 ± 0.05 | 0.24 ± 0.02 | 85 | 25 ± 1 |
| | 5 | 0.78 ± 0.02 | 0.46 ± 0.01 | 72 | 37 ± 4 |

Source: Adapted from Esawi and Farag (2007).

CNTs in solvent prior to being combined with the polymer solution. Once ultrasonication is completed for the polymer solution with nanotubes, the mixture can be cast in a mold. Film casting may be done in a vacuum oven to assist in the solvent removal; however, it has been observed by the authors that slow evaporation of solvent (no vacuum oven) yields a composite with higher polymer crystallinity than a more rapid solvent evaporation method (Love and Karbhari, 2007).

Li *et al.* (2004) used ultrasonication to successfully disperse 10 wt% MWCNTs in ethylene-co-vinyl acetate (EVA) by using solution processing technique in a mixture of solvent oil and ethanol wherein the CNTs were observed to act as nucleating agents and hence noticeably increased the crystallization temperature of the EVA. Foster *et al.* (2005) explored the effect of solvent choice on SWCNT dispersion in polystyrene (PS) and thermoplastic polyurethane (TPU) matrices using a dip-coating method and reported that, while toluene and chloroform should be disregarded due to poor dispersion characteristics, 1-methyl-2-pyrrolidinone (NMP), which dissolved both PS and TPU, provided adequate dispersion quality and stable suspension even after several months of storage. It is noted that the solution casting method is capable of producing high CNT loading up to 50 wt% with reasonably good dispersion (Coleman *et al.*, 2006), with MWCNT dispersed in water and blended with poly(vinyl alcohol) in water yielding even as high as 60 wt% nanotube composites.

Both the choice of solvent and sonication time have a profound effect on the distribution of CNTs in composites. Safadi *et al.* (2002), for example, were able to dissolve PS in toluene with MWCNTs and use sonication to cast thin film composite of random orientation CNTs with thickness 100–400 μm, whereas Qian *et al.* (2000) found the optimum sonication time needed for complete dispersion increased with CNT length (30 min for 15 μm length and 1 h for 50 μm length) for the case of MWCNT solvent cast in PS with ultrasonic wand at 150 W. The variable distribution quality of MWCNT in a thermoplastic matrix process via solution casting is shown in Fig. 1.2. Insufficient sonication time results in the presence of MWCNT bundles.

Peng *et al.* (2006) used an electrostatic-assembly solution technique to evenly disperse and distribute CNTs and polymer particles in an aqueous suspension. MWCNTs underwent an acid etch process to introduce oxygenated surface functionality to the CNT walls, making them negatively charged. Polymer particles of PMMA prepared by soap-free emulsion polymerization were dispersed in a cationic surfactant in water. Electrostatic coupling between the negatively

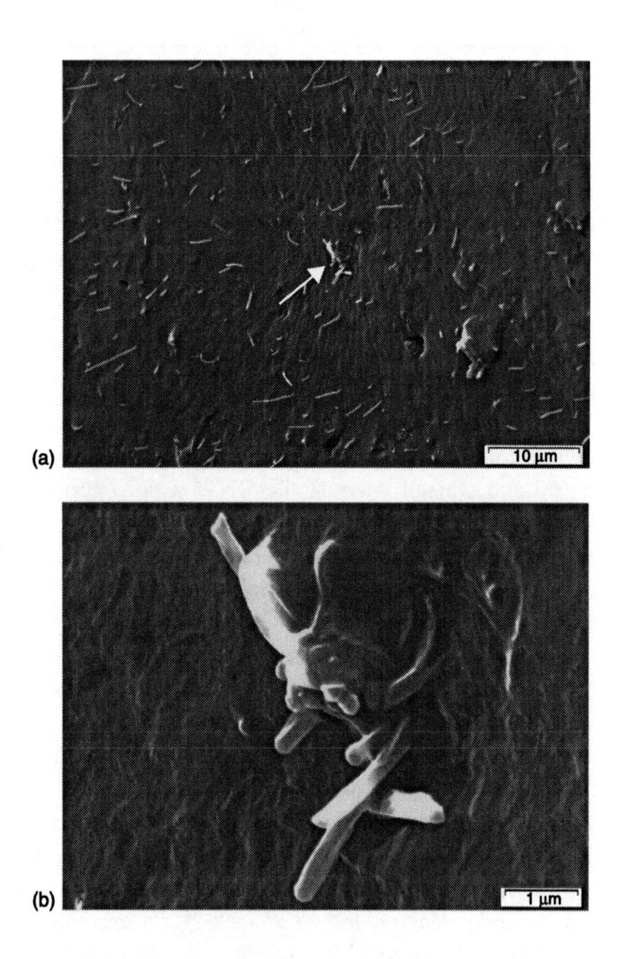


*1.2* SEM fracture surface of a solvent cast MWCNT/reactive ethylene terpolymer composite illustrating (a) dispersion of individual MWCNTs with some agglomerations, (b) close-up of MWCNT agglomeration identified by arrow. (Copyright, Dr Corey Love, University of California-San Diego, 2007.)

charged CNTs and positively charged PMMA particles facilitates the formation of dispersed and distributed filler networks. The water is removed by vacuum heating and the CNT/polymer particles are fused at elevated temperature to create composite films.

The solvent choice is generally made based upon the solubility of the polymer. Some pure CNTs may not be easily dispersed in specific solvents. To overcome this, a surfactant may be used to break up CNT bundles and entanglements. The most commonly used surfactant is sodium dodecylsulfate to (SDS), which has been demonstrated to assist the dispersion of MWCNTs in epoxy at levels between 0.01 and 0.5 wt% CNT loading (Fidelus *et al.*, 2005). The use of an anionic surfactant such as polyoxyethylene 8 lauryl ($C_{12}EO_8$), as a wetting/dispersing agent, in water has also been shown to stabilize CNTs, leading to good dispersion of MWCNTs (Dalmas *et al.*, 2005) and to improve the $T_g$ from 63°C to 88°C and increase the elastic modulus more than 30% over an unreinforced epoxy (Gong *et al.*, 2000). The non-ionic surfactant contains a hydrophilic oxyethylenated segment, which essentially interacts with the resin via hydrogen bonding, while the hydrocarbon hydrophobic segment acts as a dispersing agent for carbon where steric repulsive forces overcome the van der Waals bundling forces and facilitate CNT dispersion.

*Thermosetting polymers*

Epoxy systems are the most commonly used matrices in thermosetting resin composites. Typically, CNTs are dispersed in the resin via sonication. Solvents are added to lower the solution viscosity. A catalyzing agent or hardener is then added to initiate the formation of chemical crosslinking under an exothermic reaction. While composites have been successfully formed by pouring into a cast (after dispersing the nanotubes in a solution of a block copolymer in ethanol solvent, followed by the addition of the liquid epoxy, and removal of the ethanol by evaporation), CNTs have also been efficiently dispersed ultrasonically for as long as 48 hours in dichloroethane or *N*-*N*-dimethylformamide to form a stable solution into which the epoxy was then dissolved and the suspension placed under vacuum to remove trapped air for 1 h at 130°C, followed by the addition of a curing agent to cure at room temperature for 2–4 days before a 2-h post-cure bake at 120°C (Biercuk *et al.*, 2002).

Gojny and Schulte (2004) in contrast dispersed functionalized and non-functionalized MWCNTs in a polyetheramine curing agent rather than the epoxy resin prior to curing and attributed the high dispersion quality to the long chain polar solvent interacting with functionalized CNTs, causing stabilization of the nanotube suspension. For epoxy resins with UV curing no hardener is needed to initiate crosslinking and curing since excitation by UV light initiates the cure. CNT composites have been formed in this medium by dispersing them in chloroform solution before pouring into a photoresist epoxy resin, followed by further curing carried out by baking (Coleman *et al.*, 2006).

While solution casting offers many significant advantages for processing nanocomposite materials, there are several issues which must be addressed, chief among which are the large quantity of highly toxic solvents needed for polymer dissolution and the eventual environmental health and safety hazard posed by the CNT dispersion (Barraza *et al.*, 2002).

### 1.3.2 Melt mixing

Melt blending of thermoplastic polymers as a processing technique for the incorporation of fillers and fibers has existed for some time. The reversible thermo-physical nature of thermoplastics makes melt blending possible with carbon nanotube reinforcements. The combination of thermal stimulation (melting) and high shear mixing forces provides the necessary energy to disperse nanotubes in thermoplastic matrices. However, relative to solution blending methods, melt blending is generally less effective at dispersing nanotubes in polymers and is limited to lower concentrations due to high viscosities of the composite at higher nanotube loadings (Moniruzzaman and Winey, 2006). For polymers which are not readily soluble, such as most thermoplastics, melt processing is an acceptable alternative. As a general rule, amorphous polymers may be processed above their glass transition temperature while semi-crystalline polymers must be heated above their melting temperature. The application of heat for melting polymers creates a viscous 'gel' to which CNTs may be added and mixed using standard mixing techniques using mall laboratory benchtop mixers. The use of a larger standard Haake internal mixer by Thermo-Electron Corp. has obtained good dispersions and distributions (Lozano and Barrera, 2001).

As the mixing energy is increased, the dispersion quality increases. This can be done by increasing the mixing time or speed. However, energy input from shear mixing can break nanotubes into shorter segments with lengths decreasing up to 75% of their original, decreasing their aspect ratio in the final composite while simultaneously improving their dispersibility (Andrews and Weisenberger, 2004). MWCNTs have been fairly uniformly dispersed in HDPE by the shearing action of a twin-screw extruder where small aggregates of MWCNTs are evenly dispersed on a size scale of 10 μm or larger (Tang *et al.*, 2003).

Good dispersion using a dilution technique was confirmed via AFM and TEM investigations for MWCNTs melt mix compounded and extruded with polycarbonate (PC) (Potschke *et al.*, 2004). Starting from a high concentration of MWCNTs in PC masterbatch, subsequent PC was added to dilute to the desired composition. A percolation threshold for electrical conductivity was observed at 1.5 wt%. In another work, the optimal conditions for MWCNTs in PS, PP and aminobenzene sulfonic acid (ABS) were found to be 20 rpm for 15 min, which was adequate for complete dispersion (Andrews *et al.*, 2002a).

Often a combination of melt processing techniques is employed to produce a high degree of CNT separation. Melt mixing prior to extrusion is often important for producing accurate CNT loading levels, since the CNTs can stick to the walls of the hopper, making the final composition less than the desired weight fraction. There are several advantages of using melt mixing methods of process CNT/polymer composites, most notably the speed, simplicity, and compatibility with industrial techniques. Important parameters to consider for shear mixing of nanocomposites include temperature, mixing time, rotor speed, and CNT concentration.

*Extrusion*

The extrusion processing parameters on the dispersion quality of MWCNTs in HDPE have been studied by Zou *et al.* (2004), who concluded that at low screw speeds large agglomerations of loosely bound MWCNTs occur in a bundle form. This occurs since, with an increase in MWCNT loading, the average energy the screw can transfer to the filler agglomerate is decreased such that the lower the screw rate, the poorer is the dispersion and the bigger the agglomeration of MWCNTs. Dispersion quality was found to be higher at 150 rpm than at 80 rpm, as shear thinning effects at high screw speeds prevent the formation of large agglomerations by breaking them into smaller particles. Similar findings by Li and Shimizu (2007) showed improved dispersion with increasing screw rotation speed, with the optimal screw speed for MWCNTs in a thermoplastic elastomer as high as 2000 rpm, as shown in Table 1.2.

*Injection molding*

While extrusion has been shown to result in the best dispersion quality, injection molding of MWCNT composites is often resorted to as a second step in fabricating individual components using extruded pellets as a means of enhancing overall shape complexity and speed.

*Table 1.2* Mechanical properties of neat SBBS and its composites processed at various screw shear rates containing 3 wt% MWCNT

| Screw shear rate (rpm) | Modulus (MPa) | Elongation at break (%) | Strength (MPa) | Residual strain (%) |
|---|---|---|---|---|
| Neat SBBS | 12.5 | 1108 | 19.59 | 20.34 |
| 300 | 14.31 | 628 | 9.45 | 34.41 |
| 1000 | 23.74 | 870 | 14.99 | 30.22 |
| 2000 | 25.31 | 917 | 24.08 | 22.63 |

Source: Adapted from Li and Shimizu (2007).

*Dynamic packing injection molding (DPIM)*

Dynamic packing injection molding has been shown to control polymer morphology and mechanical properties of filled thermoplastics. The main feature of this technology is that the specimen is forced to move repeatedly by two pistons that move reversibly with the same frequency during cooling, resulting in preferential orientation of the dispersed phase or filler as well as the matrix. Polypropylene/MWNT composites with good dispersion have also been prepared by the DPIM with enhancement in tensile strength and impact strength at very small loadings of MWNTs (only 0.6 wt%). The repeated shear force offered by DPIM is powerful enough for the dispersion of MWNTs in polymer matrix.

*Compression molding*

Compression molding or hot press molding has been successfully demonstrated by Haggenmueller *et al.* (2000), who cut a masterbatch solvent cast composite of 10 wt% SWCNT PMMA into segments of 1–1.5 $cm^2$, which were then stacked together and hot pressed between 180°C platens for 3 min under 3000 lb pressure. The resulting 50–100 μm thick films were subsequently cut into pieces again and re-pressed, with the procedure being repeated as many as 25 times to achieve complete dispersion of SWCNTs in PMMA, where the dispersion increased with each additional melt cycle.

### 1.3.3 *In situ* polymerization

The *in situ* polymerization of nanocomposites can be described as a 'bottom up' approach to composite manufacturing, wherein nanotubes are dispersed in a monomer solution which is then polymerized to form a solid-state composite. The advantage of bottom-up fabrication is that interaction at the molecular level allows high nanotube loadings and good miscibility with almost any polymer type. This technique is particularly important for insoluble and thermally unstable polymers which cannot be processed by solution or processed through melting techniques. Depending upon the required molecular weight and molecular weight distribution of polymers, chain transfer, radical, anionic, and ring-opening metathesis polymerizations can be used for *in situ* polymerization. The introduction of hydroxyl, carboxyl and isocyanate groups on the surface of MWCNTs has been shown to catalyze the ring-opening reaction of benzoxazine (Chen *et al.*, 2006), with the reaction of the isocyanate group with the phenolic hydroxyl groups generated by the ring-opening of benzoxazine resulting in significantly improved polybenzoxazine (PBZ) MWCNT interfaces.

Jia *et al.* (1999) used *in situ* radical polymerization in conjunction with PMMA using a 2,2′-azobisisobutyronitrile (AIBN) radical initiator such that the $\pi$-bonds in carbon nanotubes were initiated by AIBN and therefore the nanotubes could participate in PMMA polymerization to form a strong interface between the MWCNTs and the PMMA matrix. Addition polymerization was used to produce PMMA/MWCNT composites. In this process, 0.08 wt% of free radical initiator-AIBN (2,2′-azobisisobutyronitrile) was added into MMA at the reaction temperature of 358–363 K, at which an AIBN molecule yields $N_2$ and forms two free radicals, which causes the double-bonded MMA to open and polymerize. Their findings indicated that the CNTs may take part in the polymerization process, as more AIBN is consumed in the presence of CNTs, causing an opening of the *p*-bonds linking the CNTs to PMMA. This bond resists the growth of PMMA but may produce a single C–C bond, creating a strong interface between CNT and PMMA. Velasco-Santos *et al.* (2003a) later used the same initiator to incorporate non-functionalized and carboxyl functionalized MWCNTs into a PMMA matrix.

*In situ* hydrolytic polymerization of ε-caprolactam can be completed in the presence of pristine and carboxylated nanotubes since the monomer of ε-caprolactam forms electron-transfer complexes with the MWCNTs, giving a homogeneous polymerizable master solution which facilitates the formation of composites with homogeneously dispersed nanotubes. Gao *et al.* (2005) improved upon the technology by adding a continuous spinning technique to form PA6 fibers by ring-opening polymerization of caprolactam in the presence of SWCNTs, resulting in the formation of a new hybrid material with excellent compatibility between SWCNTs and PA6.

Acid purified MWCNTs have also been added to PMMA via an *in situ* polymerization technique (Park *et al.*, 2003) in which the radicals induced on the MWCNT by AIBN were found to trigger the grafting of PMMA. During polymerization MWCNTs consumed the initiator AIBN by opening π-bonds on the MWCNT surface to form radicals, as a result of which the molecular weight of PMMA increased with the MWCNT content.

Xia *et al.* (2006) reported on the use of ultrasonication as a means of initiating polymerization of monomers without a chemical initiator. The rigorous nature of the sonication process elevates temperature and pressure to the level where radicals can be generated due to the decomposition of water or monomer surfactant, or through the rupture of polymer chains to initiate polymerization of the monomer.

### *Electropolymerization*

Wanekaya *et al.* (2006) studied the electropolymerization and characterization of polypyrrole films doped with poly(*m*-aminobenzene sulfonic acid) (PABS) functionalized SWCNTs (PPy/SWCNT-PABS). This functionalization makes the SWCNTs water-soluble and negatively charged, resulting in the SWCNT-PAB acting as anodic dopants during the electropolymerization of pyrrole monomer between an applied potential of −0.65 V and 1.05 V with a galvanostatic sweep rate of 50 mV/s to synthesize the composite. Scanning electron microscopy (SEM) and Atomic Force Microscopy (AFM) studies indicate that the use of doped SWCNT-PAB significantly changes the morphology of polypyrrole and the composite film forms a porous and cross network (Wanekaya et al 2006). The addition of PBS-functionalized SWCNTs improved the electronic performance when compared with pure PPy.

*Inverse microemulsion*

An inverse microemulsion, which consists of oil, surfactant, and water molecules, is a thermodynamically stable and isotropic transparent solution and is used to form layers on substrates through trapping of microdrops of an aqueous phase within assemblies of surfactant molecules dispersed in a continuous oil phase. Yu *et al.* (2005) demonstrated an approach to the preparation of MWCNT/polypyrrole (PPy) core-shell nanowires by *in situ* inverse microemulsion wherein the CNTs were evenly coated with a thin coating of PPy of the order of tens of nanometers through the precipitation of inorganic nanoparticles onto the nanotube surface. Strong interaction between the $\pi$-bonded surface of the carbon nanotubes and the conjugated structure of the PPy shell layer was deduced from FTIR spectra. This approach provides a simple and reproducible procedure in which the thickness and adherence of the coating are fairly easily controlled by the monomer concentration and the reaction conditions.

## 1.3.4 Mechanochemical pulverization

Mechanochemistry deals with stress-induced chemical reactions and structural changes of materials, and mechanochemical processes such as high-energy ball milling, vibromilling, pan milling and jet milling are typically used to prepare ultrafine metal, ceramic or polymer powders. The benefit of mechanochemical pulverization is the production of CNT/polymer composite powders which can then be processed using a number of melt processing techniques such as extrusion, injection molding, compression molding, and melt spraying. Repetitive milling cycles reduce the particle size even further. Xia *et al.* (2004) used mechanochemical pulverization (pan milling) to prepare 3% MWCNT/PP composite powders which were subsequently melt-mixed with a twin-roll masticator to obtain composite sheets. During the pan milling process, high shear radial and tangential forces cause a reduction in particle size and chain scission of polymers. The ground CNTs are attached to the milled polymer particles via adsorption. After 20 milling cycles the average particle size was reduced to a few microns. It should, however, be noted that milling does have a detrimental effect on CNT geometry as the high shear forces cause the average CNT length range to decrease from 1–10 μm to 0.4–0.5 μm, although the CNTs do become straighter and are more uniform in size (Xia *et al.*, 2004).

## 1.3.5 Electrostatic assembly

The electrostatic interaction between cationic polystyrene (PS) latex and anionic carbon nanofibers provides a 'bottom-up' synthesis method for a composite in which a cationic polystyrene latex was synthesized by emulsion polymerization and mixed with an aqueous suspension of oxidized carbon nanofibers, resulting in a material form obtained through hetero-coagulation. The final product is formed through a subsequent compression molding step. Xu *et al.* (2005) found that the percolation threshold was below 2 wt% (1 vol%) carbon nanofibers in PS in this case.

### 1.3.6 Alternative processing techniques

In addition to the more conventional techniques described earlier, other alternate processing routes also exist for nanofiber based composites. Hassanien *et al.* (2001) showed the potential of electrochemical deposition of conducting polymer films onto individual nanotubes. A novel spray-coating technique was developed by Zhang *et al.* (2006) to cover ultrahigh molecular weight polyethylene (UHMWPE) particles with SWCNTs, which were then processed through dissolution of the SWCNT/UHMWPE particles in xylene solvent, forming a composite with a three-dimensional network having a threshold composition of 0.6% SWCNTs in UHMWPE, which improved electrical conductivity by nine orders of magnitude. Safadi *et al.* (2002) used a spin casting technique to form thin PS/MWCNTs composite films of 30–50 μm. In this case the nanocomposites were constituted of an amorphous poly(styrene-*co*-butyl acrylate) latex as a matrix material with MWCNTs as fillers. Regev *et al.* (2004) demonstrated that SWCNTs can be easily integrated into a PS or PMMA matrix by dispersing SWCNTs in a surfactant which is added to a latex nanoparticle suspension obtained by emulsion polymerization. When the surfactant and water are removed, the remaining polymer nanoparticles and absorbed SWCNTs can be melt processed by compression molding.

## 1.4 Nanotube alignment

Typical solution casting techniques result in nanocomposites with random orientation of CNTs. However, in addition to structural rationale, CNT alignment is desired for anisotropic applications such as in conductivity and thermal management devices. A comprehensive review on the development of aligned and micro-patterned carbon nanotubes was published by Dai *et al.* (2003). The alignment of carbon nanotubes for reinforcing polymer materials to form composites will be discussed in this section with a focus on in-plane and transversely aligned systems.

### 1.4.1 In-plane alignment

*Melt drawing*

Thostenson and Chou (2002) successfully dispersed and aligned prefabricated MWCNT/PS solution cast composites by post-processing with a twin-screw extruder though a rectangular die (13 mm × 0.35 mm) with a 100 rpm screw speed and drawing the film prior to cooling at various take-up rates. The drawing alignment of CNTs in PS caused a significant improvement in elastic modulus compared with extrusion compounded composites without the drawing step. Additionally, increases in yield strength and ultimate strength indicated an improvement in load transfer capability between PS and CNTs reinforcements. Ruan *et al.* (2003) observed the toughening performance of ultrahigh molecular weight polyethylene using MWCNTs aligned via hot-drawing. The final anisotropic films with different draw ratios were obtained by tensile drawing at 120°C using an Instron tensile tester. The composites showed a ductile stress–strain response with the maximum strain at break decreasing with increasing draw ratio. It should be noted that tensile toughness and modulus are also enhanced due to the presence of CNT inclusions. Melt mixing via extrusion coupled with melt drawing has been reported to yield unique modulus, strength and toughness behavior for MWCNT reinforced PP as shown in Table 1.3 (Dondero and Gorga, 2006).

Drawing fibers not only makes use of CNT anisotropies, but also causes stress fields unique to the processing method, which, even in unfilled polymers, leads to unique morphologies such as the 'shish-kebab' morphology (Probst *et al.*, 2004).

*Table 1.3* Mechanical properties for PP and MWCNT/PP composites as a function of CNT concentration and draw ratio

| MWCNT loading (wt%) | Draw ratio | Tensile toughness (MJ/m$^3$) | Modulus (MPa) | Yield stress (MPa) |
|---|---|---|---|---|
| 0 | 0 | 0.8 | 257 | 15.8 |
| 0 | 12:1 | 457 | 551 | 31.8 |
| 0 | 23:1 | 397 | 585 | 29.9 |
| 0 | 24:1 | 523 | 611 | 33.3 |
| 0.25 | 0 | 1.1 | 528 | 21.2 |
| 0.25 | 12:1 | 608 | 1312 | 34.6 |
| 0.25 | 23:1 | 549 | 1167 | 33.4 |
| 0.25 | 24:1 | 512 | 1250 | 40.3 |
| 0.5 | 0 | 1.7 | 254 | 19.4 |
| 0.5 | 12:1 | 620 | 852 | 38.0 |
| 0.5 | 23:1 | 474 | 877 | 37.7 |
| 0.5 | 24:1 | 237 | 967 | 31.7 |
| 1.0 | 0 | 0.7 | 266 | 15.8 |
| 1.0 | 12:1 | 500 | 782 | 38.8 |
| 1.0 | 23:1 | 355 | 1064 | 38.8 |
| 1.0 | 24:1 | 511 | 949 | 40.0 |
| 3.0 | 0 | 0.9 | 238 | 16.4 |
| 3.0 | 12:1 | 254 | 957 | 34.2 |
| 3.0 | 23:1 | 286 | 662 | 32.6 |

Source: Adapted from Dondero and Gorga (2006).

The stress involved in producing fibers can also create polymer chain orientation along the fiber axis if the stress levels are high enough, and, additionally, CNT alignment could play a role in the strength.

*Magnetic field induced alignment*

The application of magnetic fields has been demonstrated to effectively cause alignment of SWCNTs with the anisotropic magnetic susceptibility of CNTs, causing them to align along the direction of an applied magnetic field. Gonnet *et al.* (2006) applied a 17.3 T magnetic field to a SWCNT suspension to align the nano-reinforcement into 'buckypapers' which were then infiltrated with an epoxy resin to form an encapsulated composite yielding fiber loading up to 50 vol%. A solution approach was taken by Kimura *et al.* (2002) by adding styrene to lower the viscosity of a mixture of MWCNTs dispersed in an unsaturated polyester monomer resin. The CNTs became mobile upon application of a magnetic field of 10 T, after which a radical initiator was added to polymerize the polymer surrounding the aligned MWCNTs into a composite film. This mode of alignment is possible because of the diamagnetic susceptibilities parallel and perpendicular to the tube axis. If the perpendicular diamagnetic susceptibility is larger than the parallel one, a MWCNT tends to align parallel to the magnetic field by overcoming thermal energy. This is consistent with earlier reported theoretical analysis in which the diamagnetism perpendicular to the tube axis is predicted to be three orders higher than that in the parallel direction. In the parallel direction, the room temperature electrical conductivity is comparable to individual SWCNT ropes.

Shi *et al.* (2005) used NiO and CoO coatings on carbon nanofibers to further help align SWCNTs in the presence of a magnetic field.

*Electric field induced alignment*

Alignment of SWCNTs by an electric field provides a simple yet efficient method. SWCNTs are highly aligned along the direction of the electric flux, which is strongly dependent upon the magnitude and frequency of the electric field, with electrostatic forces determining the aligning speed. When the magnitude of the electric field is decreased, fewer CNTs are attracted. Because of the anisotropic nature of the CNTs, the dipole moment in the direction parallel to the tube axis is stronger than that in the perpendicular direction; therefore CNTs can be aligned along the direction of the electric field if the movement is not restricted by its environment (Chen *et al.*, 2001).

Martin *et al.* (2005) applied both dc and ac electric fields to align carbon nanotubes to form a conductive network in an epoxy resin and reported that the use of ac fields resulted in more uniformly aligned CNTs, although the initial onset time was slightly longer (10 minutes). However, even the improved conductivity with the electric field alignment process is still low compared with pure MWCNTs and indicates the presence of polymer barriers preventing direct contact between individual particles. Excellent alignment of CNT bundles between conducting electrodes has been shown to occur at frequencies above 1 kHz for alternating electric fields, with the mobility of CNTs aligned parallel to the electric field estimated to be as high as $5 \times 10^{-5}\,cm^2/V{\cdot}s$ (Yamamoto *et al.*, 1996).

*Doctor blade technique*

The doctor blade technique involves the application of shear forces in the direction parallel to the blading direction to align CNTs in a polymer composite. Wood *et al.* (2001) used a doctor blade flow orientation method to align SWCNTs in UV-curable urethane acrylate. A solution of polymer/SWCNTs was spread on a glass surface and sheared twice with a doctor blade to induce flow orientation before being exposed to a UV source to cure. The blade acts as a 'squeegee' where the flow orientation upon shearing of the polymer encourages polymer molecular orientation and therefore alignment within the polymer matrix. Prior to curing, the polymer is a workable viscous gel, but it can be cured in a matter of seconds after excitation from a UV source, minimizing the time for relaxation of the SWCNTs.

*Layer-by-layer assembly*

The use of layer-by-layer (LBL) assembly techniques has been demonstrated to have exceptional uniformity and versatility for constructing nanostructure composites with SWCNTs. Shim and Kotov (2005) introduced a fusion method for aligning SWCNTs during LBL assembly by taking advantage of air–water interfacial force, as shown in Fig. 1.3. Pressurized air flows over a wet surface with randomly absorbed SWCNTs so that the surface topography changes from random orientation to a unidirectional orientation. The ability of LBL to deposit one layer of SWCNT at a time can also be used to prepare criss-crossed SWCNT composites that are expected to have excellent mechanical properties with respect to several key parameters essential for both biomedical and space applications.

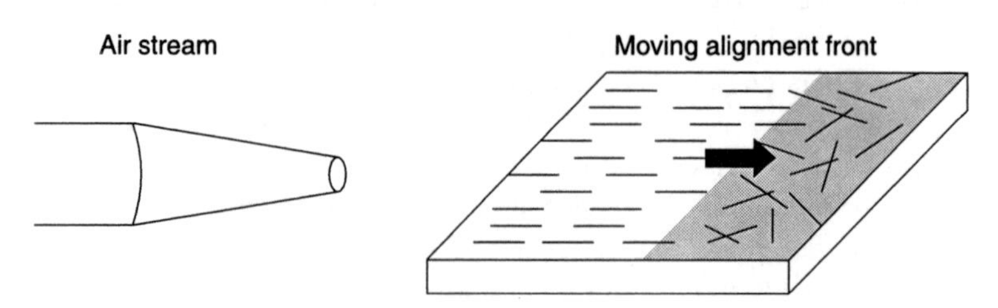


*1.3* Schematic of SWCNT alignment via layer-by-layer assembly. Adapted from Shim and Kotov (2005).

*Polar solvent exposure*

Chen and Tao (2005) demonstrated the viability of using polar solvent exposure as a means of aligning CNTs. The polar solvent used, tetrahydrofuran, penetrates inside thermoplastic polyurethane, affecting the three-dimensional hydrogen-bonded structure in the hard segments and causing the arrested chain segments to relax and align during the swelling and moisture curing stage. This serves as the driving force for the self-assembly of the oriented nanotubes in the polymer. The solvent is then removed during the casting step.

*Solution spinning*

Solution spinning consists of dispersing nanotubes in a surfactant solution, recondensing the CNTs in the flow of the polymer solution to form a nanotube mesh and then collating this mesh to a nanotube fiber (Poulin *et al.*, 2002). The flow-induced alignment leads to the formation of long ribbons that remain stable in the absence of flow. Control of the process parameters, such an injection rate, flow conditions and dimensions of the capillary tube, can be used to vary the diameter of the fabricated composite from a few microns up to 100 microns.

### 1.4.2 Transverse alignment

Transversely aligned CNT films currently focus mainly on electronic field emission (Fan *et al.*, 1999; Fang *et al.*, 2005). Self-oriented arrays of CNTs have been successfully grown by using CVD or PECVD methods. Integration of polymers to bind vertically aligned CNTs for microelectromechanical systems applications was explored by Fang *et al.* (2005), who used the technique to synthesize vertically aligned MWCNTs on silicon substrates. The small gaps between nanotubes were filled with Parylene dimers which were first vaporized at 150°C in a stainless steel chamber. The polymer vapor was mixed with methylene and introduced into a furnace at 680°C to yield a stable, monomeric diradical *para*-xylylene. Monomers were subsequently redirected into a room-temperature deposition chamber where they were simultaneously polymerized and absorbed on the MWCNTs. A simple etch step removes the polymer-reinforced nanotube forest from the underlying silicon substrate, forming a free-standing composite film.

## 1.5 Properties and characteristics

### 1.5.1 Mechanical properties – elastic modulus

In most CNT/polymer composites, the increase in CNT composition enhances the mechanical properties. Ryan *et al.* (2007) reported that 0.1 vol% SWCNTs added to polyvinyl alcohol (PVOH) increased the elastic modulus over pure PVOH 300% and increased the tensile strength 200%. The same increase in modulus was observed for MWCNTs; however, the concentration of CNTs had to be increased to a higher level. The stiffness of the composites also increased due to the crystallization effects of PVOH around the CNT surface. For crystallization stiffening, smaller diameter CNTs provide enhanced crystallinity of PVOH. The variation in CNT wall structures, length, diameter, and chirality along with a large variety of host polymers makes correlating mechanical properties challenging. Additionally, as stated previously, the variety of processing techniques will also have a significant influence on the mechanical properties of the CNT/polymer nanocomposite. The elastic modulus resulting from a range of CNT/polymer composites categorized by processing technique is given in Table 1.4 (solution casting), Table 1.5 (melt mixing), and Table 1.6 (*in situ* polymerization). Several researchers have used theoretical models to predict CNT/polymer composite properties with increased CNT loading. Kanagaraj *et al.* (2007) used the Halpin–Tsai model and a modified series approach to predict the Young's modulus of

*Table 1.4* Elastic modulus of polymers and nanocomposites processed via solvent-casting

| Nanotube type | Polymer matrix | CNT loading | $E_{polymer}$ (GPa) | $E_{composite}$ (GPa) | Reference |
|---|---|---|---|---|---|
| CVD-MWCNT | PVA | 60 wt% | 6.3 | 12.6 | Shaffer and Windle, 1999 |
| CVD-MWCNT | PS | 1.0 wt% | 1.2 | 1.69 | Qian *et al.*, 2000 |
| Arc-MWCNT | PVA | 0.6 vol% | 7 | 12.6 | Cadek *et al.*, 2002 |
| Arc-MWCNT | PVK | 4.8 vol% | 2 | 5.6 | Cadek *et al.*, 2002 |
| CVD-MWCNT | PSBA | 8.3 vol% | $0.52 \times 10^{-3}$ | $0.52 \times 10^{-3}$ | Dufresne *et al.*, 2002 |
| CVD-MWCNT | PS | 2.5 vol% | 1.53 | 3.4 | Safadi *et al.*, 2002 |
| Arc-MWCNT | MEMA | 1.0 wt% | 0.71 | 2.34 | Velasco-Santos *et al.*, 2003a |
| CVD-MWCNT | | 0.6 vol% | | 4.2 | Cadek *et al.*, 2004 |
| CVD-MWCNT | | 0.6 vol% | | 3.2 | Cadek *et al.*, 2004 |
| CVD-MWCNT | | 0.6 vol% | | 3.0 | Cadek *et al.*, 2004 |
| Arc-MWCNT | | 1.5 vol% | | 3.6 | Cadek *et al.*, 2004 |
| SWCNT | | 2.0 vol% | | 3.2 | Cadek *et al.*, 2004 |
| CVD-MWCNT | PVA | 0.6 vol% | 1.9 | 7.04 | Coleman *et al.*, 2004 |
| SWCNT | Epoxy | 1.0 wt% | 2.03 | 2.65 | Zhu *et al.*, 2004 |
| SWCNT | Epoxy | 4.0 wt% | 2.03 | 3.4 | Zhu *et al.*, 2004 |
| CVD-SWCNT | TPU | 0.5 wt% | 7.7 | 14.5 | Chen and Tao, 2005 |

Note: PS = polystyrene; PVA = polyvinyl acetate; PSBA = polystyrene butylacrylate; TPU = thermoplastic polyurethane.

*Table 1.5* Elastic modulus of polymers and nanocomposites processed via melt mixing

| Nanotube type | Polymer matrix | CNT loading | $E_{polymer}$ (GPa) | $E_{composite}$ (GPa) | Reference |
|---|---|---|---|---|---|
| Arc-MWCNT | PMMA | 17 wt% | 0.73 | 1.63 | Jin *et al.*, 2001 |
| CVD-MWCNT | PS | 25 vol% | 2 | 4.5 | Andrews *et al.*, 2002b |
| CVD-MWCNT | PS | 5 wt% | 2.3 | 2.6 | Thostenson and Chou, 2002 |
| CVD-MWCNT | PC | 15 wt% | 0.8 | 1.04 | Potschke *et al.*, 2004 |
| MWCNT | UHMWPE | 1.0 wt% | 0.98 | 1.35 | Ruan *et al.*, 2003 |
| CVD-MWCNT | PMMA | 10 wt% | 2.7 | 3.7 | Gorga and Cohen, 2004 |
| CVD-MWCNT | PA6 | 12.5 wt% | 2.6 | 4.2 | Meincke *et al.*, 2004 |
| CVD-MWCNT | PA | 2 wt% | 0.4 | 1.24 | Liu *et al.*, 2004 |
| Arc-SWCNT | PP | 0.75 wt% | 0.85 | 1.19 | Manchado *et al.*, 2005 |
| CVD-MWCNT | PC | 2.0 wt% | 2.0 | 2.6 | Eitan *et al.*, 2006 |
| CVD-MWCNT | PC | 5 wt% | 2.0 | 3.3 | Eitan *et al.*, 2006 |
| CVD-MWCNT | EVA (12% VA) | 3.0 wt% | $8.06 \times 10^{-2}$ | $1.17 \times 10^{-1}$ | Peeterbroeck *et al.*, 2007 |
| CVD-MWCNT | EVA (19% VA) | 3.0 wt% | $2.83 \times 10^{-2}$ | $4.49 \times 10^{-2}$ | Peeterbroeck *et al.*, 2007 |
| CVD-MWCNT | EVA (27% VA) | 3.0 wt% | $1.17 \times 10^{-2}$ | $2.43 \times 10^{-2}$ | Peeterbroeck *et al.*, 2007 |
| MWCNT | HDPE | 44 vol% | 1.095 | 1.338 | Kanagaraj *et al.*, 2007 |

Note: PMMA = polymethyl methacrylate; PC = polycarbonate; UHMWPE = ultra-high molecular weight polyethylene; PA6 = polyamide 6; PP = polypropylene; EVA = ethylene vinyl acetate; HDPE = high density polyethylene).

*Table 1.6* Elastic modulus of polymers and nanocomposites processed via *insitu* polymerization

| Nanotube type | Polymer matrix | CNT loading | $E_{polymer}$ (GPa) | $E_{composite}$ (GPa) | Reference |
|---|---|---|---|---|---|
| SWCNT | PI | 1 vol% | 1.0 | 1.6 | Park *et al.*, 2002 |
| SWCNT | PBO | 10 wt% | 138 | 167 | Kumar *et al.*, 2002 |
| Arc-MWCNT | PMMA | 1 wt% | 1.5 | 2.5 | Velasco-Santos *et al.*, 2003b |
| SWCNT | PMMA | 0.01 wt% | 0.3 | 0.38 | Putz *et al.*, 2004 |
| CVD-MWCNT | waterborne PU | 1.5 wt% | 0.0102 | 0.0149 | Kwon and Kim, 2005 |

Note: PI = polyimide; PU = polyurethane.

HDPE with increasing volume fraction of MWCNTs with good agreement. A comprehensive computational micromechanics analysis of high stiffness hollow fiber nanocomposites was performed by Hammerand *et al.* (2007) using finite element analysis wherein CNTs were modeled as isotropic hollow tubes or equivalent traversely isotropic effective solid cylinders with properties computed using a micromechanics-based composite cylinders method.

## 1.5.2 Thermal properties

The thermal conductivity of carbon nanotubes can be as high as 3000 W/mK, about seven times higher than copper (385 W/mK) (Tang *et al.*, 2003). However, the theoretical estimates for room-temperature thermal conductivity of an isolated SWCNT are significantly higher (6000 W/mK) (Biercuk *et al.*, 2002). The addition of 1.0 wt% unpurified SWCNTs to epoxy resin showed a 70% increase in thermal conductivity at 40 K and an increase of 125% at room temperature, with a percolation threshold evident by a sharp onset in conductivity observed between just 0.1 and 0.2 wt% (Biercuk *et al.*, 2002). Gonnet *et al.* (2006) determined the thermal conductivity for the aligned SWCNTs alone to be 42 W/mK at room temperature; however, upon introduction into an epoxy matrix, the thermal conductivity was seen to drop by nearly an order of magnitude, or equal to the level of a randomly oriented SWCNT in the epoxy due to molecular interactions between the SWCNTs and the epoxy which cause phonon scattering.

## 1.5.3 Electrical properties

The addition of CNTs to polymers significantly influences electrical properties of the polymer with additions as low as 0.1 vol% resulting in an 8–10 order of magnitude increase in conductivity, such as from $5 \times 10^{-8}$ S/m for an unreinforced epoxy to 0.01 S/m for the CNT reinforced epoxy. In most cases the addition of CNTs results in the formation of a percolation network due to the presence of a three-dimensional path through the bulk resin. The addition of MWCNTs to polyethylene (PE) via twin-screw extrusion has been reported to improve the electrical conductivity from $10^{-20}$ to $10^{-4}$ S/cm with a percolation threshold of about 7.5 wt%, which is higher than that of most similar systems (McNally *et al.*, 2005). A high concentration of CNJs is needed since PE encapsulates and isolates the MWCNTs, rendering them less effective at conduction. In this instance extrusion-induced CNT alignment reduces contact points between CNTs and it disrupts the three-dimensional percolation network. The addition of 1 wt% MWCNT to polypyrrole (PPy) increases the electrical conductivity by an order of magnitude, with the MWCNT/PPy composite exhibiting metallic electrical characteristics (current-voltage curve) (Yu *et al.*, 2005). It should be noted that Park *et al.* (2006) reported that the conductivity can be controlled over six orders of magnitude by controlling the strength of the alignment electric field, with conductivities ranging from insulating levels of $10^{-13}$ S/cm to conductive levels of $10^{-6}$ S/cm being attained in the alignment direction with 0.3 wt% SWCNT loading, which is lower than the percolation threshold for isotropic dispersion. Remarkably, the conductivity of the composite also increases perpendicular to the electric field alignment direction by about two orders of magnitude.

## 1.6 Future trends

The potential for CNTs as reinforcement and conductive media has not yet been fully realized. Mechanical properties of nanocomposites, for example, are still significantly lower than the theoretical estimates. Among the major challenges are the attainment of high quality dispersion, good alignment, and achieving a good interfacial bond between the CNT and the polymer. In addition there are still a number of challenges associated with maintaining the original length of the CNT through the process steps.

The establishment of strong interfacial adhesion between the nano-reinforcement and polymer matrix or an ordered interphase is important for maximizing the interfacial stress transfer capability of the nanocomposite. Nanotubes have been shown to nucleate polymer crystallinity, creating an ordered CNT coating maximizing CNT/polymer binding (Coleman *et al.*, 2004). Strength increases due to morphological changes in the polymer matrix can be further explored through new processing techniques. Changes within the polymer structure, such as the addition of reactive functionalities or through the altering of solution pH, may be attractive options for further study. Peeterbroeck *et al.* (2004) found that increasing the content of polar vinyl acetate (VA) in ethylene vinyl acetate tends to slightly favor the dispersion of CNTs, whereas Grunlan *et al.* (2006) have proposed using a pH sensitive polymer to separate and distribute CNTs through the polymer matrix, such that at low pH the polymer is uncharged with a coiled structure, and as the pH is increased the polymer loses a proton and becomes negatively charged, causing extension of the polymer chains and repulsion, leading to dispersion of the graft CNTs.

Recent work has focused on using CNTs to improve the cohesive energy of polymer adhesives (Love *et al.*, 2007). Property enhancements such as increased elongation and strain to failure with CNT loading are desired for adhesive applications. Changing the composition of CNTs in adhesive epoxy between composite laminates can also favorably influence the debonding and shear characteristics of the interface due to the large surface area and interlock between the CNTs and epoxy adhesive (Meguid and Sun, 2004). In the case of aerospace applications where specifically high levels of electrical conductivity are essential to avoid collection of electrostatic charges on control surfaces, the use of CNTs in adhesives can be a major advantage, even over the use of lower-cost carbon black, since the addition of CNTs actually enhances adhesive response.

The addition of CNTs to polymeric materials has been shown thus far to improve mechanical, thermal, and electrical properties. In most instances only a small amount of CNTs are needed to greatly increase properties such as modulus for elastomers and thermoplastics, electrical properties of polymers such as HDPE, etc. The addition of CNTs into natural rubber consumes less energy than incorporating carbon black into natural rubber (Sui *et al.*, 2007), indicating advantages even in energy consumption. The use of conductive CNTs in combination with conductive polymer matrices such as polyaniline and polypyrrole could lead to the development of materials that integrate functionalities such as structural response with sensor function. Small amounts of CNTs added to PA6 are known to enhance the fire retardance of polymeric systems through an increased melt viscosity that prevents dripping and flowing at or near fire conditions. In addition, the ability of the CNTs to form a percolation network allows faster and more efficient dissipation of heat, thereby also enhancing flame retardancy. Alternative energy sources, relying on high conductivity and charge capacitance capabilities, open the door to new-generation composite electrode materials, proton exchange membranes for fuel cells, and solar cells for photovoltaic devices. There is significant promise that the incorporation of CNTs into traditional composites would further enhance functionality, compression and shear response, and even through-thickness characteristics. However, the primary challenges are still those related to processing, initiating at the level of processing of the CNTs themselves at scale, through their dispersion in melts and solvents and then their alignment in actual products at levels and with the functionality desired. The solution of the challenges posed by these issues will undoubtedly open new vistas for the development of new tailored materials designed for specific applications.

## 1.7 References

Andrews, R. and Weisenberger M. C., *Current Opinion in Solid State and Materials Science* 8 (2004) 31.

Andrews R., Jacques D., Minot M. and Rantell T., *Macromolecular Materials and Engineering* 287 (2002a) 395.

Andrews R., Jacques D., Qian D. and Rantell T., *Acc. Chem. Res.* 35 (2002b) 1008.

Bal S. and Samal S. S., *Bulletin of Materials Science* 30 (2007) 379.

Barraza H. J., Pompeo F., O'Rear E. A. and Resasco D. E., *Nano Lett.* 2 (2002) 797.

Biercuk M. J., Llaguno M. C., Radosavljevic M., Hyun J. K., Johnson A. T. *et al.*, *Applied Physics Letters* 80 (2002) 2767.

Blond D., Barron V., Ruether M., Ryan K. P., Nicolosi V., *et al.*, *Advanced Functional Materials* 16 (2006) 1608.

Bubert H., Haiber S., Brandl W., Marginean G., Heintze M., *et al.*, *Diamond and Related Materials* 12 (2003) 811.

Cadek M., Coleman J. N., Barron V., Hedicke K. and Blau W. J., *Applied Physics Letters* 81 (2002) 5123.

Cadek M., Coleman J. N., Ryan K. P., Nicolosi V., Bister G., *et al.*, *Nano Lett.* 4 (2004) 353.

Chen Q., Xu R. and Yu D., *Polymer* 47 (2006) 7711.

Chen W. and Tao X., *Macromolecular Rapid Communications* 26 (2005) 1763.

Chen X. Q., Saito T., Yamada H. and Matsushige K., *Applied Physics Letters* 78 (2001) 3714.

Coleman J. N., Cadek M., Blake R., Nicolosi V., Ryan K. P., *et al.*, *Advanced Functional Materials* 14 (2004) 791.

Coleman J. N., Khan U., Blau W. J. and Gun'ko Y. K., *Carbon* 44 (2006) 1624.

Cooper C. A., Ravich D., Lips D., Mayer J. and Wagner H., *Composites Science and Technology* 62 (2002) 1105.

Dai L., Patil A., Gong X., Guo Z., Liu L., *et al.*, *ChemPhysChem* 4 (2003) 1150.

Dalmas F., Chazeau L., Gauthier C., Masenelli-Varlot K., Dendievel R., *et al.*, *Journal of Polymer Science Part B: Polymer Physics* 43 (2005) 1186.

Deck C. P. and Vecchio K., *Carbon* 44 (2006) 267.

Dondero W. E. and Gorga R. E., *Journal of Polymer Science Part B: Polymer Physics* 44 (2006) 864.

Dufresne A., Paillet M., Putaux J. L., Canet R., Carmona F., *et al.*, *Journal of Materials Science* 37 (2002) 3915.

Eitan A., Fisher F. T., Andrews R., Brinson L. C. and Schadler L. S., *Composites Science and Technology* 66 (2006) 1162.

Esawi A. M. K. and Farag M. M., *Materials & Design* 28 (2007) 2394.

Fan S., Chapline M. G., Franklin N. R., Tombler T. W., Cassell A. M. and Dai H., *Science* 283 (1999) 512.

Fang W., Chu H. Y., Hsu W. K., Cheng T. W. and Tai N. H., *Advanced Materials* 17 (2005) 2987.

Fidelus J. D., Wiesel E., Gojny F. H., Schulte K. and Wagner H. D., *Composites Part A: Applied Science and Manufacturing* 36 (2005) 1555.

Foster J., Singamaneni S., Kattumenu R. and Bliznyuk V., *Journal of Colloid and Interface Science* 287 (2005) 167.

Gao J., Itkis M. E., Yu A., Bekyarova E., Zhao B. *et al.*, *J. Am. Chem. Soc.* 127 (2005) 3847.

Gibson R. F., Ayorinde E. O. and Wen Y.-F., *Composites Science and Technology* 67 (2007) 1.

Gojny F. H. and Schulte K., *Composites Science and Technology* 64 (2004) 2303.

Gong X., Liu J., Baskaran S., Voise R. D. and Young J. S., *Chem. Mater.* 12 (2000) 1049.

Gonnet P., Liang Z., Choi E. S., Kadambala R. S., Zhang C., *et al.*, *Current Applied Physics* 6 (2006) 119.

Gorga R. E. and Cohen R. E., *Journal of Polymer Science Part B: Polymer Physics* 42 (2004) 2690.
Gorga R. E., Lau K. K. S., Gleason K. K. and Cohen R. E., *Journal of Applied Polymer Science* 102 (2006) 1413.
Grunlan J. C., Liu L. and Kim Y. S., *Nano Lett.* 6 (2006) 911.
Guo J., Li Y., Wu S. and Li W., *Nanotechnology* 16 (2005) 2385.
Haggenmueller R., Gommans H. H., Rinzler A. G., Fischer J. E. and Winey K. I., *Chemical Physics Letters* 330 (2000) 219.
Hammerand D. C., Seidel G. D. and Lagoudas D. C., *Mechanics of Advanced Materials and Structures* 14 (2007) 277.
Hassanien A., Gao M., Tokumoto M. and Dai L., *Chemical Physics Letters* 342 (2001) 479.
Hong C.-E., Lee J.-H., Kalappa P. and Advani S. G., *Composites Science and Technology* 67 (2007) 1027.
Hussain F., Hojjati M., Okamoto M. and Gorga R. E., *Journal of Composite Materials* 40 (2006) 1511.
Hwang G. L., Shieh K. T. and Hwang K. C., *Advanced Functional Materials* 14 (2004) 487.
Iijima S., *Nature* 354 (1991) 56.
Jia Z., Wang Z., Xu C., Liang J., Wei B., *et al.*, *Materials Science and Engineering: A* 271 (1999) 395.
Jin S. H., Park Y.-B. and Yoon K. H., *Composites Science and Technology* 67 (2007) 3434.
Jin Z., Pramoda K. P., Xu G. and Goh S. H., *Chemical Physics Letters* 337 (2001) 43.
Kanagaraj S., Varanda F. R., Zhil'tsova T. V., Oliveira M. S. A. and Simoes J. A. O., *Composites Science and Technology* 67 (2007) 3071.
Khan U., Ryan K. P., Blau W. J. and Coleman J. N., *Composites Science and Technology* 67 (2007) 3158.
Kimura T., Ago H., Tobita M., Ohshima S., Kyotani M. *et al.*, *Advanced Materials* 14 (2002) 1380.
Konya Z., Vesselenyi I., Niesz K., Kukovecz A., Demortier A., *et al.*, *Chemical Physics Letters* 360 (2002) 429.
Kumar S., Dang T. D., Arnold F. E., Bhattacharyya A. R., Min B. G., *et al.*, *Macromolecules* 35 (2002) 9039.
Kwon J. and Kim H., *Journal of Polymer Science Part A: Polymer Chemistry* 43 (2005) 3973.
Lau K.-T., Gu C. and Hui D., *Composites Part B: Engineering* 37 (2006) 425.
Li S.-N., Li Z.-M., Yang M.-B., Hu Z.-Q., Xu X.-B. *et al.*, *Materials Letters* 58 (2004) 3967.
Li Y. and Shimizu H., *Polymer* 48 (2007) 2203.
Liu T. X., Phang I. Y., Shen L., Chow S. Y. and W. D. Zhang, *Macromolecules* 37 (2004) 7214.
Love C. T. and Karbhari V. M., *unpublished results* (2007).
Love C. T., Gapin A. and Karbhari V. M., in *Nanopolymers 2007* (Rapra, Berlin, Germany, 2007).
Lozano K. and Barrera E. V., *Journal of Applied Polymer Science* 79 (2001) 125.
Manchado M. A. L., Valentini L., Biagiotti J. and Kenny J. M., *Carbon* 43 (2005) 1499.
Martin C. A., Sandler J. K. W., Windle A. H., Schwarz M. K., Bauhofer W., *et al.*, *Polymer* 46 (2005) 877.
McNally T., Potschke P., Halley P., Murphy M., Martin D., *et al.*, *Polymer* 46 (2005) 8222.
Meguid S. A. and Sun Y., *Materials & Design* 25 (2004) 289.

Meincke O., Kaempfer D., Weickmann H., Friedrich C., Vathauer M. *et al.*, *Polymer* 45 (2004) 739.

Moniruzzaman M. and Winey K. I., *Macromolecules* 39 (2006) 5194.

Odegard M., Frankland S. J. V. and Gates T. S., *American Institute of Aeronautics and Astronautics Journal* 43 (2005) 1828.

Park C., Ounaies Z., Watson K. A., Crooks R. E., Smith J., *et al.*, *Chemical Physics Letters* 364 (2002) 303.

Park C., Wilkinson J., Banda S., Ounaies Z., Wise K. E., *et al.*, *Journal of Polymer Science Part B: Polymer Physics* 44 (2006) 1751.

Park S. J., Cho M. S., Lim S. T., Choi H. J. and Jhon M. S., *Macromolecular Rapid Communications* 24 (2003) 1070.

Peeterbroeck S., Breugelmans L., Alexandre M., BNagyc J., Viville P., *et al.*, The influence of the matrix polarity on the morphology and properties of ethylene vinyl acetate copolymers–carbon nanotube nanocomposites, *Composites Science and Technology*, 67(7) (2004).

Peeterbroeck S., Breugelmans L., Alexandre M., Nagy J. B., Viville P., *et al.*, *Composites Science and Technology* 67 (2007) 1659.

Peng M., Li D., Chen Y. and Zheng Q., *Macromolecular Rapid Communications* 27 (2006) 859.

Potschke P., Bhattacharyya A. R. and Janke A., *European Polymer Journal* 40 (2004) 137.

Poulin P., Vigolo B. and Launois P., *Carbon* 40 (2002) 1741.

Probst O., Moore E. M., Resasco D. E. and Grady B. P., *Polymer* 45 (2004) 4437.

Putz K. W., Mitchell C. A., Krishnamoorti R. and Green P. F., *Journal of Polymer Science Part B: Polymer Physics* 42 (2004) 2286.

Qian D., Dickey E. C., Andrews R. and Rantell T., *Applied Physics Letters* 76 (2000) 2868.

Rakov E. G., in *Carbon Nanomaterials*, edited by Gogotsi Y., Taylor & Francis Group, Boca Raton, FL (2006) p. 77.

Ramanathan T., Fisher F. T., Ruoff R. S. and L. C. Brinson, *Chem. Mater.* 17 (2005) 1290.

Regev O., ElKati P. N. B., Loos J. and Koning C. E., *Advanced Materials* 16 (2004) 248.

Rinzler A. G., Liu J., Dai H., Nikolaev P., Huffman C. B., *et al.*, *Applied Physics A: Materials Science & Processing* 67 (1998) 29.

Ruan S. L., Gao P., Yang X. G. and Yu T. X., *Polymer* 44 (2003) 5643.

Ryan K. P., Cadek M., Nicolosi V., Blond D., Ruether M., *et al.*, *Composites Science and Technology* 67 (2007) 1640.

Safadi B., Andrews R. and Grulke E. A., *Journal of Applied Polymer Science* 84 (2002) 2660.

Scott C. D., Arepalli S., Nikolaev P. and Smalley R. E., *Applied Physics A: Materials Science & Processing* 72 (2001) 573.

See C. H. and Harris A. T., *Ind. Eng. Chem. Res.* 46 (2007) 997.

Shaffer M. S. P. and Windle A. H., *Advanced Materials* 11 (1999) 937.

Shanmugharaj A. M., Bae J. H., Lee K. Y., Noh W. H., Lee S. H., *et al.*, *Composites Science and Technology* 67 (2007) 1813.

Shi D., He P., Lian J., Chaud X., Bud'ko S. L., *et al.*, *Journal of Applied Physics* 97 (2005) 064312.

Shim B. and Kotov N. A., *Langmuir* 21 (2005) 9381.

Sui G., Zhong W., Yang X. and Zhao S., *Macromolecular Materials and Engineering* 292 (2007) 1020.

Tang W., Santare M. H. and Advani S. G., *Carbon* 41 (2003) 2779.

Thostenson E. T. and Chou T. W., *Journal of Physics D: Applied Physics* (2002) L77.

Thostenson E. T., Ren Z. F. and Chou T. W., *Composites Science and Technology* 61 (2001) 1899.
Thostenson E. T., Li C. and Chou T. W., *Composites Science and Technology* 65 (2005) 491.
Vandervorst P., Lei C. H., Lin Y., Dupont O., Dalton A. B., *et al.*, *Progress in Organic Coatings* 57 (2006) 91.
Vast L., Philippin G., Destrée A., Moreau N., Fonseca A., *et al.*, *Nanotechnology* 15 (2004) 781.
Velasco-Santos C., Martinez-Hernandez A. L., Fisher F., Ruoff R. and Castano V. M., *Journal of Physics D: Applied Physics* 36 (2003a) 1423.
Velasco-Santos C., Martinez-Hernandez A. L., Fisher F. T., Ruoff R. and Castano V. M., *Chem. Mater.* 15 (2003b) 4470.
Wanekaya A. K., Lei Y., Bekyarova E., Chen W., Haddon R., *et al.*, *Electroanalysis* 18 (2006) 1047.
Wang C. G., Guo Z. X., Fu S., Wu W. and Zhu D., *Progress in Polymer Science* 29 (2004) 1079.
Wang T., Lei C. H., Dalton A. B., Creton C., Lin Y., *et al.*, *Advanced Materials* 18 (2006) 2730.
Wood J. R., Zhao Q. and Wagner H. D., *Composites Part A: Applied Science and Manufacturing* 32 (2001) 391.
Xia H., Wang Q. and Qiu G., *Chem. Mater.* 15 (2003) 3879.
Xia H., Wang Q., Li K. and Hu G. H., *Journal of Applied Polymer Science* 93 (2004) 378.
Xia H., Qiu G. and Wang Q., *Journal of Applied Polymer Science* 100 (2006) 3123.
Xie X. L., Mai Y. W. and Zhou X. P., *Materials Science and Engineering: R: Reports* 49 (2005) 89.
Xu Y., Higgins B. and Brittain W. J., *Polymer* 46 (2005) 799.
Yamamoto K., Akita S. and Nakayama Y., *Japanese Journal of Applied Physics* 35 (1996) 917.
Yu Y., Ouyang C., Gao Y., Si Z., Chen W., *et al.*, *Journal of Polymer Science Part A: Polymer Chemistry* 43 (2005) 6105.
Zhang Q., Lippits D. R. and Rastogi S., *Macromolecules* 39 (2006) 658.
Zhu J., Peng H., Rodriguez-Macias F., Margrave J. L., Khabashesku V. N., *et al.*, *Advanced Functional Materials* 14 (2004) 643.
Zou Y., Feng Y., Wang L. and Liu X., *Carbon* 42 (2004) 271.

# 2
# Environmental life-cycle assessment of polymer nanocomposites

V. KHANNA, University of Pittsburgh, USA and
L. MERUGULA and B. R. BAKSHI, The Ohio
State University, USA

**Abstract:** This chapter discusses the needs, challenges, and benefits of life-cycle assessment (LCA) in the environmental evaluation of polymer nanocomposites (PNCs). The chapter provides a detailed description of the LCA approach and describes the unique challenges surrounding the LCA of nanotechnology, specifically PNCs. The benefits of LCA and an overview of the major studies on the LCA of nanotechnology are also presented. Case studies on the LCA of carbon nanofiber (CNF)-reinforced polymer nanocomposites are presented along with a general discussion about the LCA of specific nanoproducts. The role of predictive approaches in the evaluation of emerging nanoproducts is also discussed.

**Key words:** life-cycle assessment, nanomaterial, carbon nanofiber, greenhouse gas, cumulative energy demand.

## 2.1 Introduction

No emerging technology has evoked more interest in recent years than the rapidly evolving field of nanotechnology. Given the enormous promise that nanotechnology offers in shaping the future global economy, it is almost a necessity to investigate the large-scale impacts of this new technology. While studies of the health impacts of nanomaterials and nanoproducts are essential, a holistic understanding of the effect of engineered nanomaterials and nanoproducts using systems analysis is central for developing safe nanoproducts. Systems analysis is a good tool for evaluating the broader effects of emerging technologies. However, systems analysis poses unique challenges to researchers and designers, especially for technologies in an early stage of research. Life-cycle assessment (LCA) is one such approach that considers a holistic view and offers great promise in the environmental evaluation of emerging nanotechnologies.

The aim of this chapter is to discuss the needs, challenges, and benefits of life-cycle assessment in the environmental evaluation of polymer nanocomposites (PNCs). The rest of the chapter is organized as follows. Section 2.2 is a detailed description of the LCA approach and describes the unique challenges surrounding the LCA of nanotechnology, specifically polymer nanocomposites. The benefits of LCA and an overview of the major studies on the LCA of nanotechnology are also given in Section 2.2. A brief discussion along with guidelines for compiling a life-cycle inventory of nanomanufacturing is given in Section 2.3. Case studies of the life-cycle assessment of carbon nanofiber (CNF)-reinforced polymer nanocomposites is also given in Section 2.3 along with a general discussion about the LCA of specific nanoproducts. The role of predictive approaches in the evaluation of emerging nanoproducts is discussed in Section 2.4. Finally, Section 2.5 summarizes the overall discussion.

## 2.2 The life-cycle assessment (LCA) approach to nanotechnology

Life-cycle assessment is a methodology to quantify the environmental impact of a product or a process over its entire life cycle. The methodology is standardized via ISO 14000. In its simplest form, the LCA methodology has four basic steps. Figure 2.1 is a schematic description of these four phases, which are briefly described below.

### 2.2.1 Goal definition and scope

The scope defines the boundaries of the life-cycle study and specifies which major and minor processes are to be considered. In general, the scope of an LCA comprises the extraction of resources, the synthesis of inputs, product manufacture, waste treatment, packaging, use, and ends when the product is finally disposed or recycled. Such an analysis is 'cradle-to-grave.' The purpose of the LCA should be stated clearly in this phase. The purpose could be a comparative assessment of nanoproducts vs. alternatives, the preparation of a nanoproduct using different synthesis routes, or the use of a nanoproduct for a new application. This is especially important to aid in defining the appropriate functional unit for the study, so as to give a fair comparison. As an example, consider a life-cycle environmental comparison of plastic vs. paper grocery bags. One paper bag might hold the same amount of groceries as two plastic bags. Thus the appropriate functional unit in this case should be one paper bag vs. two plastic bags. Once the functional unit is fixed, the process boundary should be specified clearly to define

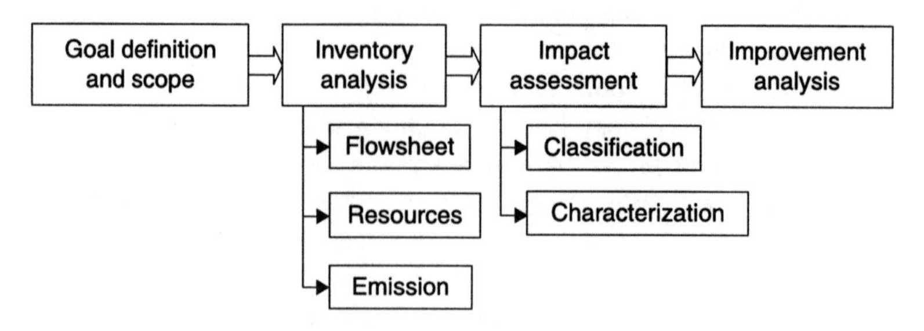


*2.1* Phases of an LCA based on ISO 14040.

the scope of the LCA. Since data collection and analysis can be very time consuming and expensive, it is tempting to narrow the inputs considered or to make the boundary too narrow, leading to invalid or flawed results. Also, since a process or product is connected to the rest of the economy through innumerable links, any attempt to get each and every input may make the study too expensive or time consuming. For these reasons, proper care and vigilance are crucial during this stage to yield meaningful and interpretable results. In addition, performing multiple studies with different boundaries to determine the effect of boundary limits can be helpful in certain situations. The life-cycle practitioner should always state the purpose of the study, which usually helps in defining the boundary.

### 2.2.2 Inventory analysis

This is the most resource-intensive phase of an LCA. It involves identifying and collecting input and output data for each of the processes that are included in the goal definition and scope phase. The quantity of data needed depends on the boundaries of the study. There are many ways of obtaining data for life-cycle assessment. Sources include published databases, data from papers, and data from manufacturers. Input data typically includes material and energy resources, labor, capital, and equipment. Output data consists of products, byproducts, and emissions of substances to air, water, and soil.

### 2.2.3 Impact assessment

This phase involves classifying and characterizing emissions into various impact categories to provide indicators for analyzing the potential contributions of resource extraction, and the waste and emissions in the inventory for a number of potential impacts.[1] The individual results for the different impact categories can be further normalized and weighted based on valuation techniques. Since normalization and valuation techniques are subjective, these are likely to introduce uncertainty into the results; thus only the classification and characterization steps are performed in this study. Several life-cycle impact assessment (LCIA) approaches exist and these have been described and critically assessed in detail in the literature.[2] The choice of which impact categories and hence impact assessment approach to use is a subjective one and at the discretion of the LCA practitioner. More often, the LCIA methodology to be used for the problem at hand is selected based on the goal of the study, i.e. whether the goal is to make comparisons between alternative products with the same purpose, or identifying areas for process improvement. Emissions can be aggregated. They are classified and characterized based on their impact into various impact categories, which are often described as midpoint indicators. Common impact categories include the potential for global warming, acidification, eutrophication, human toxicity, ozone layer depletion, and photochemical smog formation. Different chemicals can have a variety of different impacts. For example, carbon dioxide causes global warming and hydrochlorofluorocarbons (HCFCs) have an adverse effect on the ozone layer. Classification combines different chemical effects into a common metric. For example, 1 kg of methane has a global warming impact equivalent to 23 kg of carbon dioxide; similarly 1 kg of methyl chloride has a global warming impact equivalent to 16 kg of carbon dioxide. These impacts are then expressed in terms of total equivalents of carbon dioxide to quantify the potential for global warming. A similar approach is followed for the other impact categories. Characterization factors are available in the literature for a large number of chemicals and their impacts in various categories.

Mathematically this is expressed as:

$$\text{Impact potential} = \sum_{i=1}^{n} \left(\text{Characterization factor}\right)_{\text{i}} \times \left(\text{Chemical flow}\right)_{\text{i}}$$

Midpoint indicators can be further aggregated to obtain endpoint indicators. Endpoints are entities that are valuable to society, which include buildings, natural resources, humans, and plant and animal species. Endpoint indicators are often combined together to provide more meaningful and interpretable damage indicators.

### 2.2.4 Interpretation

Interpretation is the step in the LCA where the results of the impact assessment and the other phases are put together into a form that can be used directly to draw conclusions and make recommendations. Before starting the interpretation, the LCA practitioner will have to verify the validity of the results using sensitivity or uncertainty analysis. If enough data is available, statistical methods are of great utility at this stage.

### 2.2.5 Nanotechnology LCA: challenges

LCA is a data-intensive approach, which has found wide use in industry and is standardized via the ISO 14000 series. Many companies, like Dow, Proctor and Gamble, United Solar, etc., have found it advantageous to employ LCA techniques as a way of moving beyond compliance. They use the techniques to create win-win opportunities by improving the quality of their products while minimizing the environmental impact. LCA encompasses detailed information about resource consumption and emissions across the entire life cycle of a product. Despite the fact that LCA methodology is standardized via the ISO series, and pertinent software and inventory databases are available, the methodology faces several challenges. These include getting high-quality life-cycle inventory data, combining data in disparate units and at multiple spatial and temporal scales, dealing with high-dimensionality data involving varying degrees of uncertainty, and dealing with processes having a range of emissions.[1] These challenges make it even more difficult to apply LCA to emerging technologies like nanotechnology.

Researchers and various agencies have identified the need to use LCA for potential nanoproducts.[3,4] However, the LCA of nanotechnology poses several formidable challenges. Some of these are:

- Existing life-cycle inventory databases such as the National Renewable Energy Laboratory (NREL) life-cycle inventory database, SimaPro,[5] etc. are limited in scope and are useful for evaluating only common products and processes.
- Little is known about the inputs and outputs of nanomanufacturing processes since most of the data is proprietary or only available at the laboratory scale.
- There is very little quantifiable data available on the effects on human health and ecosystems of products and byproducts of nanomanufacturing.
- Besides these, forecasting nanotechnology life-cycle processes and activities is difficult since the technology is in its infancy and evolving rapidly.

These are severe problems for obtaining high-quality inventory data for nanomaterials and nanoprocesses. In the light of these challenges, the LCA of emerging nanotechnologies has received very little attention and there are only a handful of studies available addressing the life-cycle issues and complexities of nanoproducts and nanoprocesses. Lloyd and Lave studied the life-cycle implications of replacing automobile body panels made of steel with those of PNCs and aluminum.[6] The study employed an economic input-output LCA framework for quantifying savings in petroleum use and production and hence reductions in $CO_2$ emission. They also studied the use of nanotechnology to stabilize platinum group metals in automobile catalytic converters.[7] Pietrini and coworkers evaluated the environmental performance of biocomposites in two different end products, namely, cathode ray tube monitors and the internal panels of a car.[8] Nanoscale clay and sugar cane bagasse were used as the fillers. Osterwalder and coworkers[9] compared the wet and dry synthesis methods for oxide nanoparticle production with respect to energy consumption. Not all life-cycle stages could be included in these studies, especially the end-of-life and disposal phase of nanoproducts, due to the lack of data. Further, equipment-level information, i.e. both energy and emissions, for the synthesis of nanoparticles is missing. Although these studies represent the first step towards creating LCAs for nanomanufacturing, their accuracy is limited for guiding selection among alternatives. In the literature, there are only a handful of nanoparticle-specific LCAs that use detailed process-level information and combine it with details from other steps in the life cycle to quantify the energy and environmental burden associated with the production of nanomaterials. Such studies can be especially useful since they will lay the basis for compiling the life-cycle inventory for nanoparticle synthesis and hence pave the way for LCA of specific nanoproducts.

### 2.2.6 Nanotechnology LCA: expected benefits

The need for LCA of potential nanoproducts cannot be overemphasized. Although not a panacea, especially for emerging technologies, LCA can help address many critical concerns like material use, energy use, and the environmental impact intensity of products, which, complemented with the toxicological information for nanomaterials, will have the potential to inform, so that the emerging field of nanotechnology can develop in a sustainable manner. Some of the expected benefits of nanotechnology LCA are:

- Identify the phases in a product's life cycle that use the most energy and have the most environmental impact and thus the potential for improvement.
- Quantify how much of the energy savings during the use-phase of nanoproducts are offset by energy expenses during the production-phase of the nanomaterials.
- Identification of end-of-life scenarios that might be specific to nanoproducts.
- LCA results can help in the selection of various process alternatives during product design, which may positively influence the downstream phase of the product life cycle.
- Evaluation of economic and environmental trade-offs of nanoproducts vs. conventional ones.
- Identify and address public concerns about the emerging field of nanotechnology.
- LCA results can be used as a screening tool for evaluating competing technologies during the preliminary phases of product design.

Any LCA of nanoproducts will most likely suffer from high uncertainty at the early stages of nanotechnology research. At the early stages, efforts should be directed towards compiling a life-cycle inventory of engineered nanomaterials, which will help pave the way for LCA of specific nanoproducts. LCA practitioners need to work in close association with toxicologists to fill in the missing gaps created by unknowns such as human health and environmental hazards of nanoscale materials. As toxicological data on engineered nanomaterials becomes available, it should be incorporated into the LCA framework.

The next section illustrates the life-cycle approach in the environmental evaluation of polymer nanocomposite-based products.

## 2.3 The environmental LCA of polymer nanocomposites

### 2.3.1 Life-cycle inventory for nanotechnology LCA: an overview

The ISO methodology as outlined in ISO series 14040 and 14044, in general, is applicable for evaluating potential nanoproducts and nanoprocesses. This section describes the general approach that can be followed for compiling the LCA of potential nanoproducts in the light of the challenges mentioned in the preceding sections.

Compiling a life-cycle inventory for nanomaterials and nanoproducts can be an excruciating and resource-intensive task. This is because nanoprocesses are evolving rapidly and most of the industry data is proprietary. As a starting step, LCA practitioners can compile the life-cycle inventory for nanoproducts based on laboratory experience and data available in the open literature. Wherever possible, plant-specific data should be used. In the absence of any information, parallels can be drawn with similar technologies to get data on material and energy consumption. For example, the chemical vapor deposition (CVD) process used for the synthesis of carbon-based nanomaterials, i.e. single-walled carbon nanotubes and carbon nanofibers, is similar in nature to the CVD process used for chip manufacturing in the semiconductor industry. Missing data can be reconciled using the conservation laws for mass and energy. Unlike traditional products and processes, additional information about the possible quantity and mode of release of nanoparticles is crucial for estimating the extent of the human and ecological impact of nanoproducts. Various assumptions related to the boundary and selected processes should also be clearly stated. Since nanoproducts have only recently started penetrating the consumer market, information regarding the end-of-life of nanoproducts is missing. Sensitivity analyses based on different scenarios can provide useful insight and extreme bounds for the end-of-life impact of nanoproducts.

### 2.3.2 Case study: carbon nanofiber/polymer composites for automotive applications: life-cycle energetic evaluation and greenhouse gas impact

Traditionally, conventional fiber composites consisting of glass or micron-sized carbon fibers have been most commonly used for composite applications. Although traditional fiber-reinforced plastics have good in-plane mechanical properties that are governed by conventional fibers, they have poor properties in the transverse and thickness directions characterized mainly by the properties of the polymer matrix. Failure of the polymer matrix between the fibers can take place under impact or shear. Carbon nanofibers can directly reinforce the polymer matrix between the long fibers, thereby enhancing the strength in the transverse and thickness directions. The result is a high-strength composite, which combines the advantages of conventional long fibers and carbon nanofibers. Besides enhancing mechanical strength, CNFs can also impart a desirable level of electrical conductivity to the polymer composites. Electrically conductive polypropylene composites (with an electrical resistivity of 10 000 ohm.cm) have been prepared with CNF loading levels of 3 vol% in the polymer matrix.[10] Electrical resistivity values as low as 100 ohm.cm have been reported for CNF-reinforced epoxy composites at CNF loading content of 10 wt%.[11]

This section presents and discusses the life-cycle energetic and greenhouse gas (GHG) impact associated with the production and use of carbon nanofiber PNCs. The use of PNCs in automotive body panels as a substitute for traditional materials is also evaluated and compared with traditional materials. PNCs have enhanced mechanical properties, high strength-to-weight ratios, and are capable of offering specific functionalities such as a desired level of electrical conductivity. These combinations of properties are making PNCs one of the fastest-growing plastic segments and an attractive alternative to conventional materials like steel and aluminum.

A typical life cycle of a PNC product is shown in Fig. 2.2. Several alternatives exist at each step in the complete life cycle. The selection of the polymeric resin depends on the application and hence the desired properties. Polypropylene (PP) and unsaturated polyester resin (UPR) are used as the thermoplastic and thermosetting resins, respectively, as these have been widely studied in several nanocomposite experimental studies with respect to their mechanical and electrical properties. Specifically, both simple CNF and carbon nanofiber-glass fiber (CNF-GF) hybrid PNCs are evaluated and compared with steel for designs with the same stiffness. The life-cycle inventory was developed based on published literature and the best available engineering information.

The analysis is performed at two levels. In the first level, the functional unit is the cradle-to-gate life-cycle comparison of a standard steel plate, 48″ × 96″ × 0.1875″ (122 × 244 × 0.5 cm), with plates of CNF- and CNF-GF-reinforced polymer composites designed to have the same stiffness. The mass and the corresponding thickness of the PNC material were estimated using Ashby's

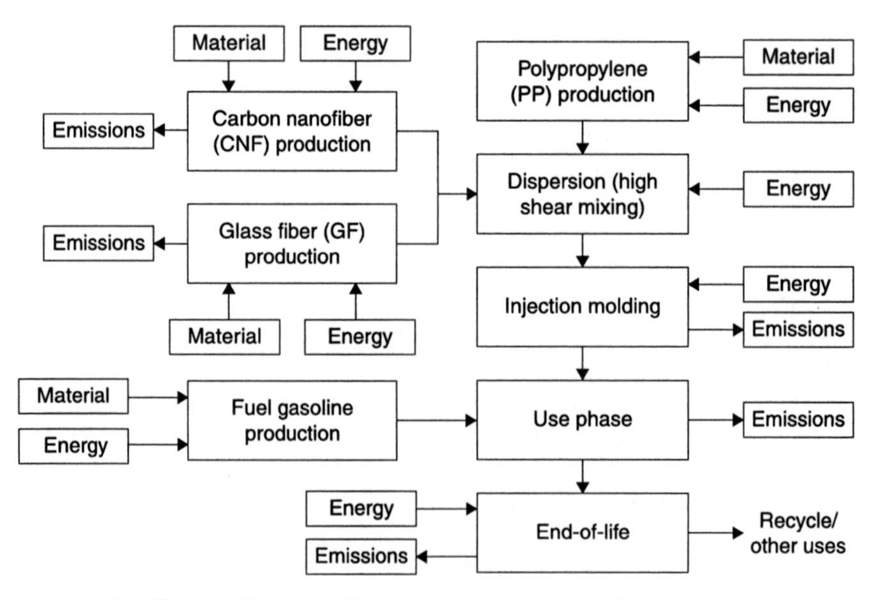


*2.2* Typical life-cycle of a polymer nanocomposite product.

approach for equal stiffness design.[12,13] The use- and end-phases were not modeled in this analysis. In the second level of analysis, the life-cycle energetic impact associated with the use of CNF- and CNF-GF-reinforced polymer composites vs. steel in the body panels of a midsize car weighing 3300 lbs (1497 kg) is estimated with the functional unit being 150 000 vehicle miles (241 402 km) traveled. The end-of-life phase of the automobile is not included in this analysis.

This study is a cradle-to-gate study as it does not include the end-of-life issues specific to CNF-reinforced polymer composite materials, primarily because of the lack of quantifiable information about the recovery of CNF and glass fiber (GF) from thermoplastic and thermoset polymer composite materials. Comparison of a standard steel plate with PNC materials for equal stiffness design evaluates the life-cycle energetic impact of the different materials without including the use-phase. The case study involving the use of CNF- and CNF-GF-reinforced PNCs vs. steel in the body panels of a midsize car goes a step further by including the vehicle use-phase but still does not consider the end-of-life-phase. Including end-of-life considerations may not change the overall conclusions of this study.

The basis for comparison is the equal stiffness design of the components. This is justified on the basis that, for structural and automotive applications, the component is assumed to have sufficient strength once it meets the stiffness criteria. Although this criterion may be true for automotive applications, other considerations such as impact properties may also influence the selection of the final material. Impact energy, an indicator of energy-absorbing capacity, decreases with an increasing content of CNFs in long-fiber thermoplastic materials.[10]

The primary weight savings for the automotive body panels were calculated directly using Ashby's approach. Secondary weight savings of 0.5 kg/kg of primary weight savings were considered resulting from downsizing the chassis and auxiliaries. Based on the different values previously used in comparing aluminum-intensive vehicles with steel, a sensitivity analysis was performed by considering a range of 0.5–1 kg of secondary weight savings per kilogram of primary weight reduction.[6,14,15] An improvement in fuel economy relative to the vehicle with steel body panels was calculated using the sedan equivalent estimation proposed and used previously:[6,16,17]

$$\mathrm{FE}_2 = \mathrm{FE}_1\left(\frac{m_1}{m_2}\right)^{0.72}$$

where $\mathrm{FE}_i$ is the fuel economy of vehicle $i$ and $m_i$ is the mass of vehicle $i$.

The results of a cradle-to-gate (excluding the use-phase) life-cycle comparison of PNCs vs. steel for equal stiffness of the components are presented in Fig. 2.3. A standard steel plate was considered as the basis for comparison. Figure 2.3 reveals that, on a cradle-to-gate basis, CNF-based polymer composite materials are 1.6–12 times more energy intensive compared with steel. However, products based on nanomaterials may be environmentally superior to alternatives for a given

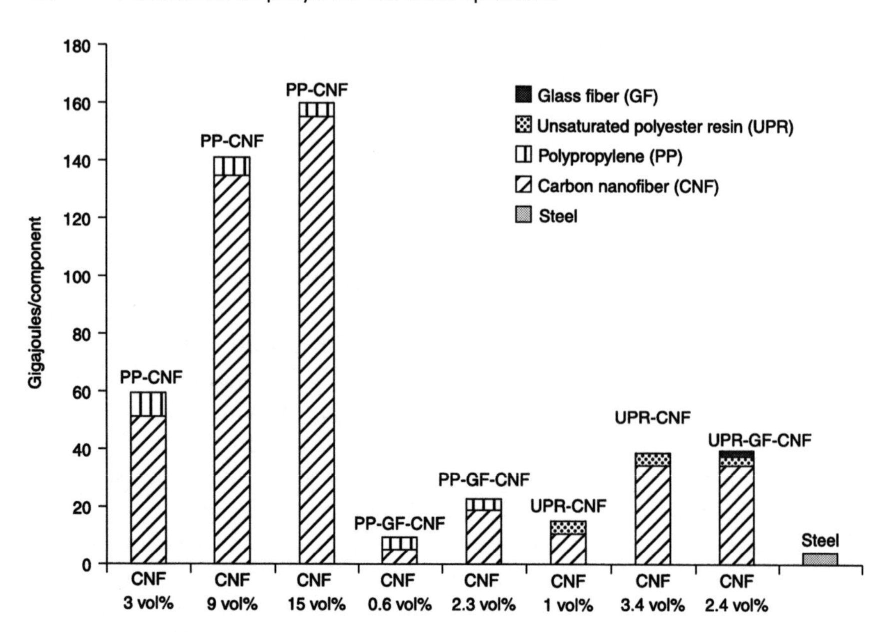


*2.3* Life-cycle energy comparison of polymer nanocomposites with steel for equal stiffness design.

application. The amount of engineered nanomaterials used and the resource savings due to the use of these products could be the deciding factors. It is essential that specific nanoproducts are evaluated on a functional unit basis for direct comparison with conventional products. This is particularly important as product use-phase might govern whether the high upstream manufacturing energy for certain nanomaterials can be offset during the use-phase to realize any life-cycle energy savings.

In light of the results and discussion in the previous paragraphs, the use of CNF-reinforced nanocomposites in body panels of light-duty vehicles was further evaluated. The life cycle of a midsize automobile with CNF-reinforced nanocomposite body panels was evaluated and compared with one with conventional steel panels. The results are shown in Fig. 2.4. Overall, for the range of CNF compositions and scenarios investigated, lifetime use-phase fuel gasoline savings of 1.4–10% were observed for PNC-based body panels relative to steel. Other factors such as cost, toxicity impact of CNF, and end-of-life issues specific to CNFs need to be considered to evaluate the final economic and environmental performance of CNF-reinforced PNC materials.

Finally, the net change in the cradle-to-gate life-cycle energy consumption resulting from the substitution of steel with PNCs in the body panels of a midsize car is presented in Fig. 2.5. Inclusion of the vehicle use-phase, as depicted in Fig. 2.5, indicates that, on a life-cycle basis, the use of PNCs in automotive body panels as a potential replacement for steel offers net savings in life-cycle energy consumption relative to steel. The higher upstream manufacturing energy for

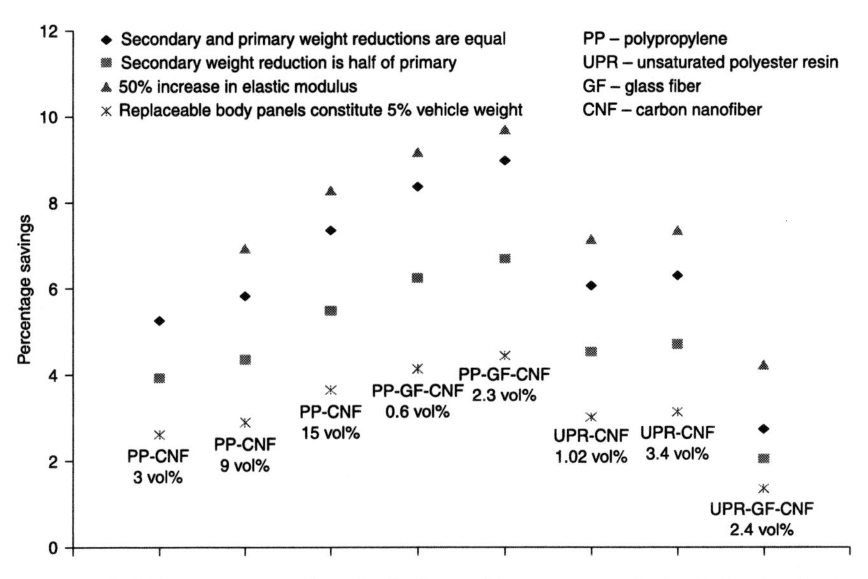


*2.4* Percentage savings in fuel gasoline consumption relative to steel for use-phase in auto body panels.

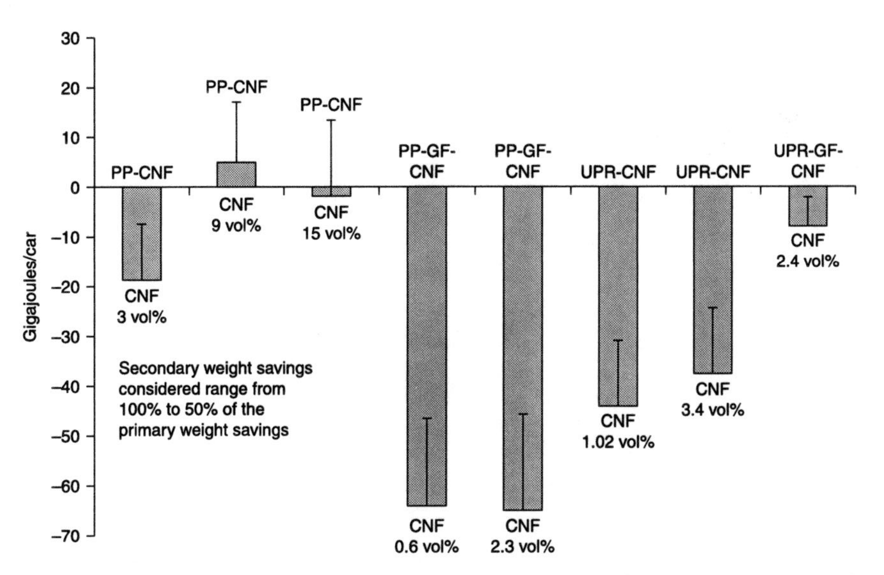


*2.5* Net cradle-to-gate life-cycle energy savings for the substitution of steel with CNF-reinforced PNCs in automobile body panels. Values less than zero indicate net cradle-to-gate life-cycle energy savings relative to steel.

PNC synthesis, as shown in Fig. 2.3, is offset by the energy savings over the use-phase of the vehicle lifetime. These savings in the use-phase energy consumption are a result of the improved vehicle fuel economy that arises because of the weight reduction of the automotive body panels.

The use of PNC-based automotive body panels with higher CNF loading ratios, in particular CNF contents of 9 and 15 vol%, might actually result in an overall increase in the fossil energy consumed over the entire life cycle of the car. This is because the fuel savings in the use-phase are outweighed by increased upstream energy consumption for CNF synthesis. However, these might represent unnecessarily higher loading ratios for commercial product applications. Further, for polypropylene-based PNCs, the life-cycle fossil-energy saving is highest for GF-CNF hybrid PNCs, around 65 GJ/car compared with simple CNF-based PNCs with a lifetime energy saving of 18 GJ/car. The corresponding numbers for UPR-based composites range from 8 to 44 GJ/ car over the entire life cycle. A sensitivity analysis was also conducted to study the effect of achievable secondary weight reduction on the lifetime fossil-energy savings and the results are shown using the negative error bars in Fig. 2.5. A range of secondary weight savings from 0.5 to 1 kg secondary weight saved/kg of primary weight saving is considered.[9] The negative bars indicate that the net life-cycle energy savings relative to steel decrease with a decrease in the secondary weight savings.

Figure 2.6 depicts the cradle-to-gate GHG impact comparison of CNF-based polymer composite materials vs. steel for equal stiffness design of the components.

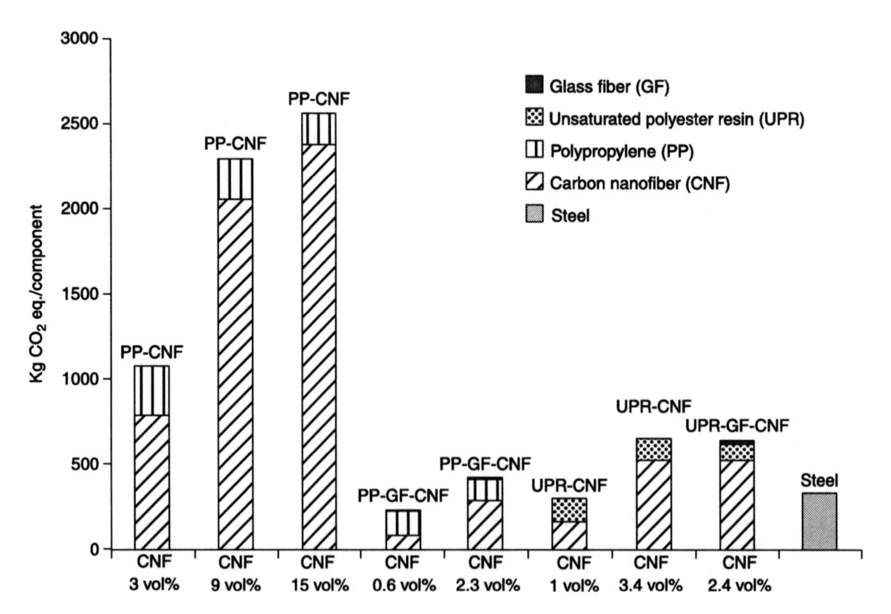


*2.6* Greenhouse-gas impact comparison of CNF/polymer composites with steel for equal component stiffness. The basis for comparison is a standard steel plate (48″ × 96″ × 0.1875″).

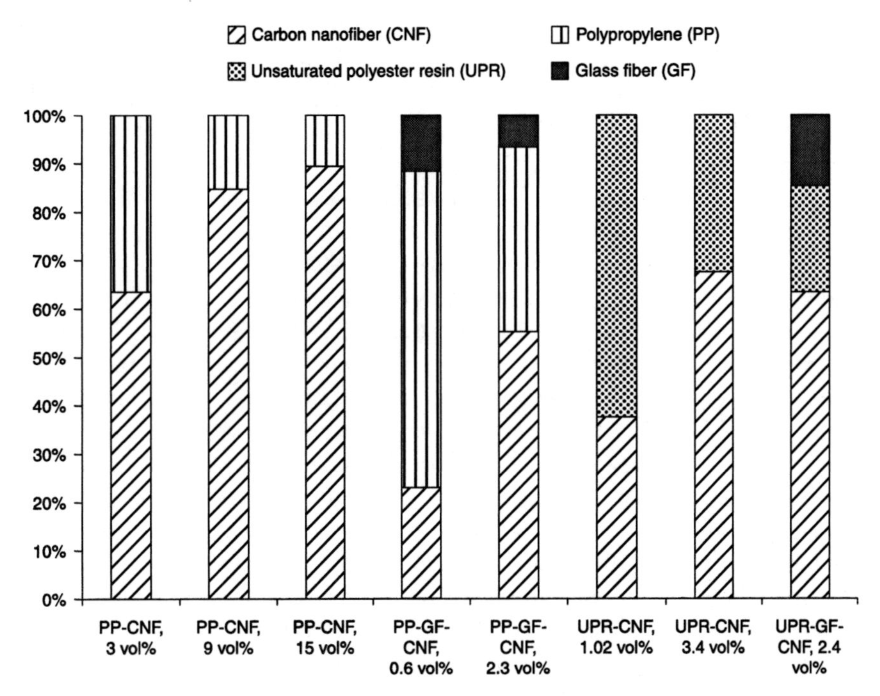


*2.7* Distribution of life-cycle energy consumption for the manufacture of CNF-based polymer composite automotive body panels. Vehicle use-phase is not included.

As expected, the trend for GHG impact comparison is similar to that observed for the life-cycle energy consumption.

The distribution of the life-cycle energy consumed during manufacture for constituents for a CNF-reinforced polymer composite automotive body panel is depicted in Fig. 2.7. Figure 2.7 does not include the use-phase of the vehicle. It is observed that the majority of the life-cycle energy used results from the life cycle for carbon nanofiber production, especially for polymer composites with higher loading ratios of CNFs. For the thermoset-based PNCs, CNFs constitute 38–68% of the cradle-to-gate life-cycle energy consumption. The next highest impact results from the production of polymeric resins, followed by glass fibers.

### 2.3.3 Case study: polymer nanocomposites for windmill applications

Not much long-term data is available for multimegawatt wind energy converters (WECs) that are currently being deployed. Whereas lifespan ratings of WECs are typically 20 years, the first installations of 2-MW systems occurred as recently as 1999.[18] Units of this size are representative of current offshore deployments. Vestas has released summaries of life-cycle assessments for its V80, V82, and V90 models, which have ratings of 2.0 MW, 1.65 MW, and 3.0 MW, respectively.[19–21] Elsam Engineering A/S has also released assessments of the V80 system for both onshore and offshore use.[22] The life-cycle inventory database Ecoinvent has data on a 2-MW offshore system in the Middelgrund installation outside Copenhagen; the availability of the inventory has led to its use as surrogate data in further LCAs.[23–27] Few assessments have been published for systems larger than 2 MW.[21,27–29]

Few published LCAs of nanoproducts exist.[30] Khanna *et al.*[31] developed a life-cycle inventory for the manufacture of carbon nanofibers. These nanofibers may be integrated into polymer resin systems forming PNCs, which could be used to manufacture larger, lighter and more resilient wind turbine blades with longer lifespans. These would maximize electricity production and, hence, increase the penetration of WECs. Khanna *et al.*'s CNF inventory has been extended to a PNC system formed by a novel prebinding process described in Movva.[32] Prebinding of glass fiber mats is followed by vacuum-assisted resin transfer molding with epoxy, resulting in reinforcement of the glass–epoxy interface with carbon nanofibers and improved manufacturability of large pieces. This extension of the CNF manufacturing inventory to the PNC system has been used to determine the implications of scaling production to large deployments, both in size and market penetration, of the material for multimegawatt WECs.

Considering 5 wt% as an upper limit on the amount of CNFs that could make up a wind turbine blade, a linear model was developed to assess the life-cycle impact of incorporating this new material using the prebinding method. Potential differences in processing energy for molding a blade with prebound CNF-glass fiber mats were not considered; unlike premixing, prebinding does not increase resin viscosity and simply alters permeability of the fiber mat, reducing the demand for pumping energy. The assessment incorporated mass loadings of the nanofibers and relative amounts of acetone required for dispersion, including a waste scenario for solvent reuse, recovery, and incidental evaporation. Analysis boundaries do not include capital and utility investments for reuse and recovery but focus on the effects of acetone feed and emissions. End-of-life scenarios for the nanofibers were not extended due to the lack of toxicity, transport, and fate data, and it was assumed that the disposal of the fiber-reinforced plastics would not change from current practices. It was hypothesized that CNF exposure is greatest during the manufacturing phase; whereas, in use, the nanofibers would be locked in a solid matrix, making ecosystem exposure unlikely until incineration and deposition of the blades in local landfills.

The analysis includes a possible decrease in the mass of glass fiber and epoxy, an increase in fiber loading for better reinforcement, and a reduction in the dispersing solvent from the R&D ratio. Furthermore, the cumulative energy demand for carbon nanofiber production was varied to consider current manufacturing yields and predictions for maturation of the nanomanufacturing process.

Figure 2.8 shows the cumulative energy demand (CEnD) for the addition of a PNC without other system improvements. Assuming that there is a reduction

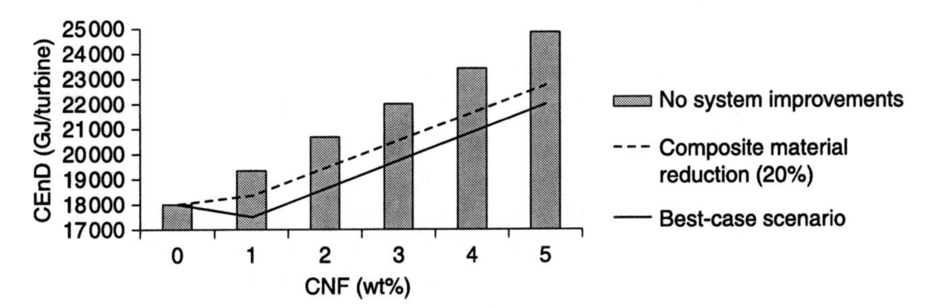


*2.8* Cumulative energy demand of a 2-MW windmill system assuming maturation of CNF manufacturing.

of solvent use by 90% from the laboratory scale because of maturation and best engineering practices, then the figure shows that there would be a 38% increase in CEnD. If the addition of the CNFs yields improved mechanical properties and leads to a reduction in material use or increased fiber loading, the CEnD increase would be lessened to 22–32%. While CEnD is shown in units of gigajoules per turbine, it does not take into account the operation of the WEC.

Figure 2.9 demonstrates how the energy return on investment (EROI) is affected by system improvements projected as increases in capacity factor or lifespan as well as material effects. The columns show the effect of adding the nanocomposite with no system changes and assume a 20-year lifespan and a capacity factor of 30. From the columns, it can be seen that just adding the PNC material results in a decrease in the life-cycle EROI of nearly 30%. Much of the impact is because of the large amount of solvent required for dispersion and application of the nanofibers; the magnitude of this effect is implicit for Fig. 2.8 and Fig. 2.9, which assume a 90% reduction of solvent use. Reductions beyond 90% do not substantially change EROI results. However, since current methods emit 100% of the acetone to the atmosphere, further reductions would be aimed more at reducing emissions-based impacts. The 'best-case scenario'

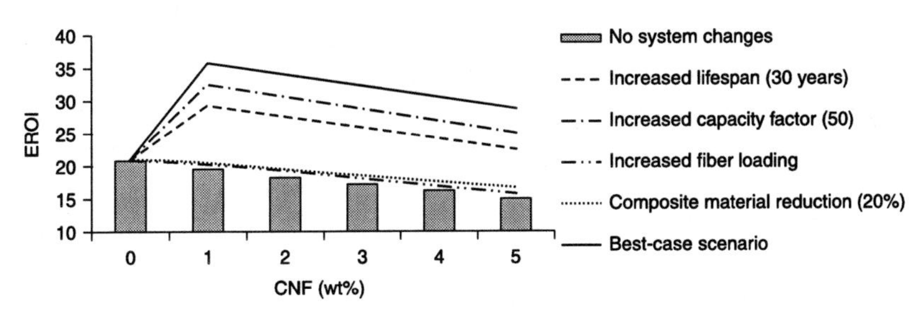


*2.9* Energy return on investment (EROI) of the 2-MW windmill system assuming maturation of CNF manufacturing.

is defined as a coupling of an increased capacity factor of 50, a material reduction by 20%, and increased fiber loading. It is recognized that overall system changes are not likely to occur in isolation from other changes to WEC system components such as load-mitigating designs.

The results show a high dependency of the cumulative energy demand and hence life-cycle EROI on the weight percentage of carbon nanofibers, despite their low mass loading, which would be less than 0.1% of the WEC mass. An assessment of available LCAs for WECs made from traditional materials shows stable CEnD trends with improvements being made as the technology has matured over the last decade. However, the incorporation of CNFs and the accompanying solvent use force a large deviation from the base case despite the small amounts used, which can only be justified if there are significant technological and economic improvements, such as increased penetration of electricity-generating technology that outperforms thermoelectric power generation in terms of emissions while remaining competitive in life-cycle EROI.

## 2.4 Future trends and alternate approaches for evaluating emerging technologies

Although a LCA of emerging nanotechnology-based products such as PNCs can be very informative, the traditional LCA approach described in Section 2.2 and illustrated in Section 2.3 suffers from limitations. A traditional LCA is mainly an 'output-side' method due to its focus on detailed data of emissions and their impact throughout the life cycle. Such data is often not available for nanotechnology and other emerging technologies due to a sheer lack of information or the slow pace of toxicological studies. In contrast to the lack of output information, data about material and energy inputs for even emerging technologies is more readily available from the scientific literature and laboratory or pilot plant studies. Since the use of materials and energy affects the emissions from a process, it is likely that input-side metrics may be a proxy indicator of environmental impact. Such a relation, if discovered, would serve as a way of doing a quick or streamlined LCA of a new technology or at the early stages of decision-making.

A variety of input-side metrics can be formulated based on mass, energy, and exergy. Total mass or cumulative mass consumption (CMC) is the overall material consumption in the life cycle of a product or service, which is studied by material flow analysis (MFA). Another popular approach is net energy analysis, which determines cumulative energy consumption (CEnC) – the amount of energy consumed in all activities necessary to provide the goods or services. Traditionally, energy analysis only focused on fossil fuels; likewise MFA only focuses on minerals because of concerns about the depletion of non-renewable resources. Since determining the renewability of a resource can be challenging due to the dependence on the renewal time and consumption rates, more recent studies tend to include renewable energy and material resources too, such as solar energy and wood. Often, non-renewable and renewable resources are simply added up or compared without accounting for the difference in their qualities.[33]

The objective of studying mass and energy flow is to reduce resource consumption and enhance process efficiency, which potentially have inherent connections with the environmental impact. Researchers have considered whether input-side metrics based on mass and energy can predict environmental impacts. Two independent studies[34,35] used large life-cycle inventory modules and found a strong correlation between cumulative energy demand (CEnD) and impact. The correlation was higher than 90% for air pollution categories.

Although input-side metrics based on mass and energy show a certain correlation with impact, a thermodynamic view indicates that they may not be the best tools. Total mass only captures the property of weight. It is clear that a tonne of iron is not equivalent to a tonne of mercury in terms of environmental impact. Similarly, energy content lacks the ability to distinguish between the qualities of the energy. For instance, a joule of solar energy cannot do the same amount of work as a joule of electricity. Moreover, each of the methods considers only part of the total inputs. MFA includes materials while energy analysis includes energy resources. The different units, gram vs. joule, make it difficult to combine them into one unitary indicator.

The laws of thermodynamics indicate that the second law may be better at capturing the relation with environmental impact. This is because all industrial and ecological processes and their life cycles are networks of energy flows. Manufacturing involves the reduction of entropy to make the desired products, which, according to the second law, must result in an increase of entropy or disorder in the surroundings. This implies that, among alternatives with similar utility, the process with a higher life-cycle thermodynamic efficiency would have a smaller portion of useful work converted to entropy, and so a smaller life-cycle environmental impact. Therefore, it may be true that life-cycle thermodynamic efficiency is inversely proportional to life-cycle environmental impact.

Available energy or exergy satisfies the first and second laws. Exergy is defined as the maximum amount of work that can be extracted when a material or energy stream is brought to equilibrium with its environment. It captures information from mass, energy, concentration, velocity, location, and other properties of a material. Its unit is the joule, the same as energy, but with a different meaning. Besides the ability to indicate entropy formation by exergy loss, exergy can also compare and combine diverse material and energy flows, making it more comprehensive than mass or energy. When only one unitary measurement is required, exergy is a more natural choice for measuring resource consumption than mass or energy. Therefore, exergy is also expected to be more attractive than mass or energy in connecting life-cycle inputs with life-cycle impacts.[36]

Exergy analysis is a method that measures exergy consumption and efficiency needed to make a product or service. Traditionally, exergy analysis focuses on a single process to increase exergy efficiency. But this type of analysis ignores life-cycle information, and it is not able to help environmental decision-making on a broader scale. Cumulative exergy consumption (CEC) analysis is an extension of exergy analysis. Industrial CEC (ICEC) analysis only includes industrial processes.[37,38] ICEC takes a life-cycle view, taking into account exergy consumed in all steps from natural resource extraction to the final product. ICEC ignores ecological stages outside the economic system. Different natural resources are created by different ecological processes, which vary considerably in their thermodynamic efficiencies. For instance, solar radiation is obtained freely from the sun, whereas coal is made available to the economic system by the geological cycle and has a much lower efficiency than solar energy. Accounting for the exergy consumption in the relevant ecosystem processes gives a better representation of the quality differences between different types of natural resources.

Ecological CEC (ECEC) analysis extends ICEC analysis by including ecological stages of the production network, such as natural processes for producing, transporting, and concentrating natural resources, such as coal and rain; various pollution dissipation functions in ecological systems are also part of ECEC analysis. Consider power generation from solar energy and coal as an example to compare ICEC and ECEC.[37] Without considering ecological processes, coal is more efficient for generating electricity than solar energy because coal is a higher quality natural resource. This result contradicts traditional LCA, which shows that sunlight is a cleaner energy resource than coal. By including ecological stages, the ECEC of coal is much larger than solar energy, which reduces the life-cycle thermodynamic efficiency of coal compared with solar energy. This simple illustration indicates that life-cycle thermodynamic efficiency should be defined with respect to ECEC, and not ICEC or cumulative energy.

Zhang and coworkers developed a model for an ecologically based LCA that augments the 1997 US Economic Input–Output model with information on the flow of resources from nature to various industrial sectors and emissions.[39–41] This model provides data for almost 500 sectors and is well suited for statistical analysis. Resources are evaluated in different units: mass, energy, and exergy. CMC, CEnC, and ICEC include the flows in the economic system. ECEC expands the boundary to ecological systems. Disability adjusted life years (DALY) represents the environmental impact. Figure 2.10 shows the preliminary results from these studies. It shows that specific emissions do have a reasonably high correlation with the life-cycle consumption of some resources. For example, $CO_2$ emissions are most correlated with the use of fossil fuels, the emission of particulate matter (PM10) is most correlated with the use of stone, and methanol emissions are most correlated with the use of wood. If a single aggregated metric is desired, then ECEC seems to have the best correlation with the life-cycle environmental impact. This may indicate that ECEC, which accounts for the exergy consumed in industrial and ecological systems, may be the most appropriate

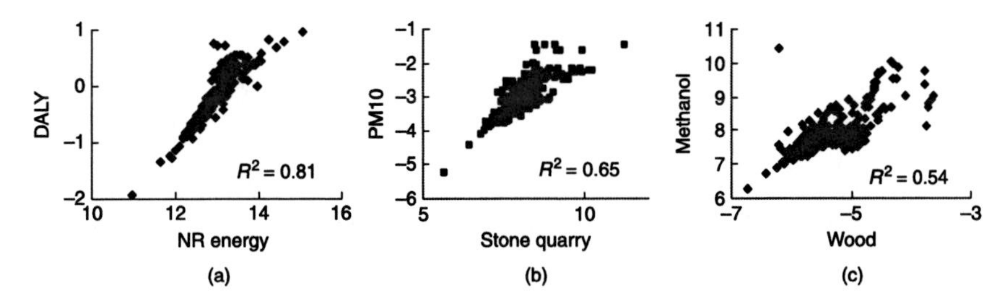


*2.10* (a) Cumulative non-renewable energy consumption vs. overall impact (DALY). (b) Stone quarry vs. PM10 emission. (c) Wood vs. methanol emission.

way of combining different types of resources. Although such a study is not comprehensive enough for validating the relation, this method seems to be worth further study with advanced statistical models. The results of such statistical analysis may be useful for guiding the development of nanotechnology and other emerging technologies until detailed LCAs can be completed.

## 2.5 Conclusions

Nanotechnology is a fast-emerging technology and is expected to play a pivotal role in influencing future global markets. Although, on the one hand, altered physicochemical properties of engineered nanomaterials enable novel applications, they also raise a multitude of concerns about the human and ecosystem impact of these materials. 'Is nanotechnology a leap towards sustainability?' is a question that is being asked repeatedly as research in this burgeoning field progresses at an increasing pace. While studies regarding the environmental and health impacts of nanomaterials are a necessary evil, it is also critical to take a holistic view for developing environmentally sustainable nanoproducts.

The need for LCA of potential PNC products cannot be overemphasized. Any LCA of nanoproducts will most likely suffer from high uncertainty at the early stages of nanotechnology research. At the early stages, efforts should be directed towards compiling a life-cycle inventory of engineered nanomaterials, which will help pave the way for LCAs of specific PNC-based products. LCAs can be used to quantify how much of the savings in energy and material consumption are offset by increases in the manufacturing phases of the life cycle. LCA practitioners need to work in close association with toxicologists to fill in the missing gaps created by unknowns such as the human health and environmental hazards of nanoscale materials. As toxicological data on engineered nanomaterials becomes available, it can be incorporated into the LCA framework. LCA, with its powerful toolbox for addressing the material, energy, and environmental impact intensity of products, has the potential to inform, so that the emerging nanotechnologies can develop in a sustainable manner.

## 2.6 References


1. Rebitzer, G., *et al.*, Life cycle assessment Part 1: Framework, goal and scope definition, inventory analysis, and applications. *Environment International*, 2004. 30(5): pp. 701–720.
2. Bare, J.C. and T.P. Gloria, Critical analysis of the mathematical relationships and comprehensiveness of life cycle impact assessment approaches. *Environ Sci Technol*, 2006. 40(4): pp. 1104–1113.
3. Royal Academy of Engineering and the Royal Society. Nanoscience and nanotechnologies: Opportunities and uncertainties. Accessed 20 November 2010. Available from: www.nanotec.org.uk/finalReport.htm.
4. Initiative, TNN Environmental, *Health, and Safety Research Needs for Engineered Nanoscale Materials*. Accessed 20 November 2010. Available from: www.nano.gov.
5. Product Ecology Consultants, SimaPro 7 software. Accessed 20 September 2010. Available from: http://www.pre.nl/simapro/.
6. Lloyd, S.M. and L.B. Lave, Life cycle economic and environmental implications of using nanocomposites in automobiles. *Environ Sci Technol*, 2003. 37(15): pp. 3458–3466.
7. Lloyd, S., L. Lave, and H. Matthews, Life cycle benefits of using nanotechnology to stabilize platinum-group metal particles in automotive catalysts. *Environ Sci Technol*, 2005. 39(5): pp. 1384–1392.
8. Pietrini, M., Roes, A.L., Patel, M.K., and Chiellini, E., Comparative life cycle studies on poly (3-hydroxybutyrate)-based composites as potential replacement for conventional petrochemical plastics. *Biomacromolecules*, 2007. 8(7): pp. 2210–2218.
9. Osterwalder, N., *et al.*, Energy consumption during nanoparticle production: How economic is dry synthesis? *Journal of Nanoparticle Research*, 2006. 8(1): pp. 1–9.
10. van Hattum, F., *et al.*, Conductive long fibre reinforced thermoplastics by using carbon nanofibres. *Plastics, Rubber and Composites*, 2006. 6(7): pp. 247–252.
11. Choi, Y., *et al.*, Mechanical and physical properties of epoxy composites reinforced by vapor grown carbon nanofibers. *Carbon*, 2005. 43(10): pp. 2199–2208.
12. Ashby, M., *Materials Selection in Mechanical Design*. 2005: Butterworth-Heinemann.
13. Ashby, M. and D. Jones, *Engineering Materials. 1. An Introduction to their Properties and Applications*. 1980: Pergamon Press, Oxford, England.
14. Stodolsky, F., *et al.*, Life-cycle energy savings potential from aluminum-intensive vehicles. 1995. SAE Technical Paper 951837, 1995, doi:10.4271/951837.
15. Das, S., The life-cycle impacts of aluminum body-in-white automotive material. *Journal of the Minerals, Metals and Materials Society*, 2000. 52(8): pp. 41–44.
16. MacLean, H. and L. Lave, Environmental implications of alternative-fueled automobiles: Air quality and greenhouse gas tradeoffs. *Environ Sci Technol*, 2000. 34(2): pp. 225–231.
17. Lave, L., Conflicting objectives in regulating the automobile. *Science*, 1981. 212(4497): p. 893.
18. Energy, D. Final Report on Offshore Wind Technology. NEEDS: *New Energy Externalities Development for Sustainability*. Accessed 20 November 2010. Available from: http://www.needs-project.org/RS1a/WP10%20Final%20report%20on%20offshore%20wind%20technology.pdf.
19. Vestas. An environmentally friendly investment: Lifecycle assessment of a V80-2.0 MW onshore wind turbine. Available from: http://www.vestas.com/en/about-vestas/sustainability/wind-turbines-and-the-environment/life-cycle-assessment-(lca).aspx

20. Vestas. Life cycle assessment of electricity produced from onshore sited wind power plants based on Vestas V82-1.65 MW turbines. 29 December 2006. Available from: http://www.vestas.com/en/about-vestas/sustainability/wind-turbines-and-the-environment/life-cycle-assessment-(lca).aspx.
21. Vestas. Life cycle assessment of offshore and onshore sited wind power plants based on Vestas V90-3.0 MW turbines. 21 June 2006. Available from: http://www.vestas.com/en/about-vestas/principles.
22. Elsam Engineering A/S life cycle assessment of offshore and onshore sited wind farms. 20 October 2004. Available from: http://www.apere.org/manager/docnum/doc/doc1252_LCA_V80_2004_uk[1].fiche%2042.pdf.
23. Swiss Center for Life Cycle Inventories. Ecoinvent v2.2. 2010.
24. Jungbluth, N., Bauer, C., Dones, R., and Frischknecht, R., Life cycle assessment for emerging technologies: Case studies for photovoltaic and wind power. *International Journal of Life Cycle Assessment*, 2005. 10(1): pp. 24–34.
25. Martinez, E., Sanz, F., Pellegrini, S., Jimenez, E., and Blanco, J., Life cycle assessment of a multi-megawatt wind turbine. *Renewable Energy*, 2009. 34(3): pp. 667–673.
26. Martínez, E., Sanz, F., Pellegrini, S., Jimenez, E., and Blanco, J., Life-cycle assessment of a 2-MW rated power wind turbine: CML method. *International Journal of Life Cycle Assessment*, 2009. 14(1): pp. 52–63.
27. Tryfonidou, R. and H. Wagner, Multi-megawatt wind turbines for offshore use: Aspects of life cycle assessment. *International Journal of Global Energy Issues*, 2004. 21(3): pp. 255–262.
28. Tremeac, B. and F. Meunier, Life cycle analysis of 4.5 MW and 250 W wind turbines. *Renewable and Sustainable Energy Reviews*, 2009. 13(8): pp. 2104–2110.
29. Weinzettel, J., *et al.*, Life cycle assessment of a floating offshore wind turbine. *Renewable Energy*, 2009. 34(3): pp. 742–747.
30. Meyer, D., M. Curran, and M. Gonzalez, An examination of existing data for the industrial manufacture and use of nanocomponents and their role in the life cycle impact of nanoproducts. *Environmental Science & Technology*, 2009. 43(5): pp. 1256–1263.
31. Khanna, V., B.R. Bakshi, and L.J. Lee, Carbon nanofiber production: Life cycle energy consumption and environmental impact. *Journal of Industrial Ecology*, 2008. 12(3): pp. 394–410.
32. Movva, S., Effect of carbon nanofibers on mold filling in a vacuum assisted resin transfer molding system. *Journal of Composite Materials*, 2009. 43(6): p. 611.
33. Haberl, H., Weisz, H., Amann, C., Bondeau, A., Eisenmenger, N., *et al.*, The energetic metabolism of the European Union and the United States: Decadal energy input time series with an emphasis on biomass. *Journal of Industrial Ecology*, 2006. 10(4): pp. 151–171.
34. Haes, H., Life cycle assessment and the use of broad indicators. *Journal of Industrial Ecology*, 2006. 10(3): pp. 5–7.
35. Huijbregts, M., Rombouts, L., Hellweg, S., Frischknecht, R., Hendricks, A., *et al.*, Is cumulative fossil energy demand a useful indicator for the environmental performance of products? *Environ Sci Technol*, 2006. 40(3): pp. 641–648.
36. Ukidwe, N. and B. Bakshi, Thermodynamic accounting of ecosystem contribution to economic sectors with application to 1992 US economy. *Environ Sci Technol*, 2004. 38(18): pp. 4810–4827.
37. Hau, J. and B. Bakshi, Expanding exergy analysis to account for ecosystem products and services. *Environ Sci Technol*, 2004. 38(13): pp. 3768–3777.

38. Szargut, J., D. Morris, and F. Steward, *Energy Analysis of Thermal, Chemical, and Metallurgical Processes*. 1988: Hemisphere Publ. Corp., New York.
39. Center for Resilience, TOSU. Ecologically based life cycle assessment. Accessed 20 September 2010. Available from: http://resilience.eng.ohio-state.edu/ecolca-cv/.
40. Zhang, Y., A. Baral, and B.R. Bakshi, Accounting for ecosystem services in life cycle assessment, Part II: Toward an ecologically based LCA. *Environ Sci Technol*, 2010. 44(7): pp. 2624–2631.
41. Zhang, Y., S. Singh, and B.R. Bakshi, Accounting for ecosystem services in life cycle assessment, Part I: a critical review. *Environ Sci Technol*, 2010. 44(7): pp. 2232–2242.

# 3
# Calcium carbonate nanocomposites

Y. LIN and C.-M. CHAN, The Hong Kong University of Science and Technology, PR China

**Abstract:** This chapter describes the use of calcium carbonate nanoparticles in polymers. The chapter first reviews the production of calcium carbonate nanoparticles and their surface treatments. The chapter then reviews various polymer nanocomposites. Finally, it presents the toughening mechanisms of polymers filled with calcium carbonate nanoparticles.

**Key words:** calcium carbonate nanoparticle, surface treatment, nanocomposite, toughening mechanism, precipitated calcium carbonate.

## 3.1 Introduction: applications of calcium carbonate nanoparticles

Calcium carbonate particles have been used in the plastics industry for many years. The original purpose of adding ground calcium carbonate (GCC) particles as filler material for plastics was to reduce material costs. With the development of precipitated calcium carbonate (PCC) particles, which are smaller than GCC particles, many more industrial applications for this type of nanocomposite have emerged. Many recent studies have shown that PCC nanoparticles can be used as fillers not only to reduce the cost of materials but also to improve the mechanical properties of the polymers. The purpose of this paper is to review the uses of PCC particles in polymer nanocomposites and the toughening mechanisms of the nanocomposite materials.

## 3.2 Calcium carbonate as filler material

Calcium carbonate is a versatile mineral. It is an essential ingredient in paper, paints and coatings, rubbers, adhesives, sealants, and plastics. Usually, calcium carbonate particles are added to plastics to reduce material costs. In certain applications, the thermal, mechanical, and rheological properties of the resulting plastics are also improved. Due to the low cost and abundance of calcium carbonate, it has become one of the most popular fillers for plastics.

### 3.2.1 The manufacture of calcium carbonate (GCC) and precipitated calcium carbonate (PCC)

Calcium carbonate is one of the most common minerals on earth, making up about 4% of the earth's outer crust. Natural calcium carbonate, which occurs as

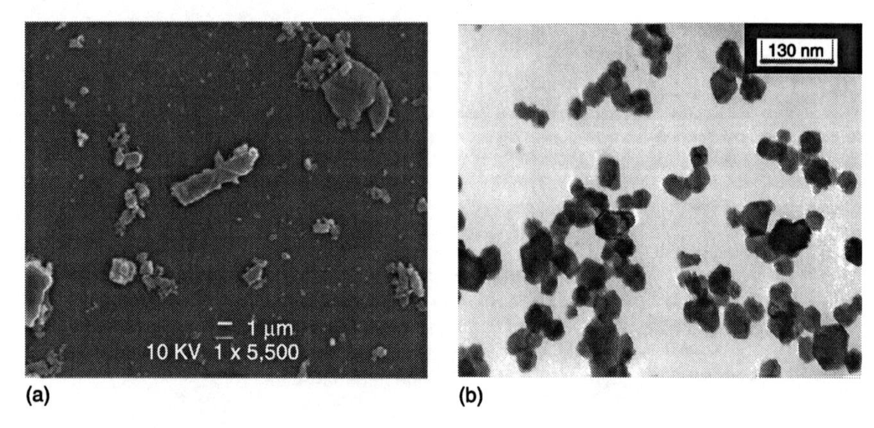


*3.1* (a) SEM micrograph of GCC particles. (b) TEM micrograph of PCC nanoparticles.

limestone, chalk, and marble, differs in the degree of whiteness, particle-size distribution, abrasivity, and opacity. All natural calcium carbonate varieties are suitable for the production of calcium carbonate fillers. GCC and PCC are the two primary types of fillers for industrial applications. Figures 3.1(a) and (b) show a scanning electron micrograph (SEM) of GCC particles and a transmission electron micrograph (TEM) of PCC particles, respectively.

GCC particles are produced by crushing and grinding calcium carbonate deposits. Ground particles are classified according to size and size distribution. Irregular shapes and wide particle-size distributions are common. In addition, the particle-size distribution is heavily dependent on the raw material. For example, GCC particles from chalk have a broader particle-size distribution than from limestone or marble (Huwald, 2001). In general, GCC particles are bigger than 1 μm in size. In addition to particle size, the degree of whiteness is another important parameter and is strongly dependent on the purity of the starting $CaCO_3$. Any traces of iron or manganese can reduce the whiteness of the particles.

PCC, which is a precipitate produced by bubbling $CO_2$ through a solution of calcium hydroxide, has high purity and a very fine and controlled particle size of the order of 0.02 to 2 μm in diameter. In general, particles with diameters less than 100 nm are referred to as nanoparticles. The Solvay process is the most widely used and most economical re-carbonation process. First, limestone is decomposed into calcium oxide and carbon dioxide at a temperature above 900 °C. After calcination, the lime is slaked with water and the resulting solution is purified and carbonized with the carbon dioxide generated from the calcination process:

| | | |
|---|---|---|
| Burning of limestone | $CaCO_3 \longrightarrow$ | $CaO + CO_2$ |
| Slaking of quicklime | $CaO + H_2O \longrightarrow$ | $Ca(OH)_2$ |
| Precipitation | $Ca(OH)_2 + CO_2 \longrightarrow$ | $CaCO_3 + H_2O$ (PCC) |

A suspension of $CaCO_3$ is produced by the re-carbonization process. The particle size and the crystal forms of the PCC particles are controlled by temperature, the concentration of the reactants, and the reaction time. The suspension is subsequently filtered, surface treated and dried. PCC particles have different crystal forms (calcite or aragonite) and shapes (cubic-like or needle-like), as shown in Fig. 3.2. Surface modification of PCC particles, which is needed in many industrial applications, changes the surface of the particles from hydrophilic to hydrophobic. Fatty acid is a commonly used surfactant and is usually reacted with the synthesized $CaCO_3$ prior to filtration.

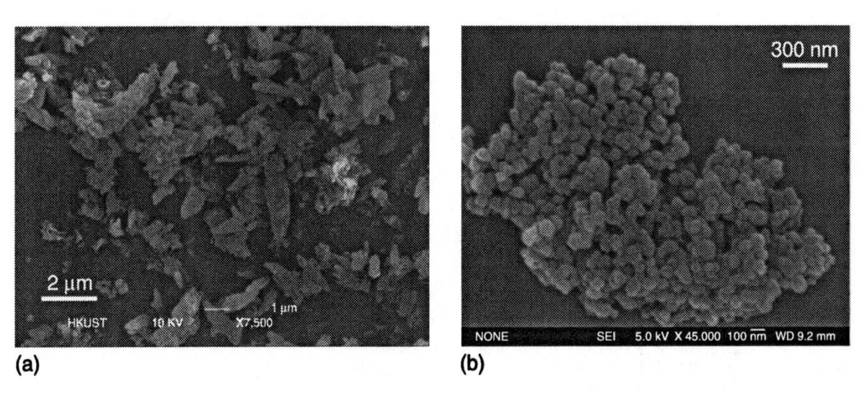


*3.2* (a) Needle-like PCC particles. (b) Cubic-like PCC particles.

## 3.2.2 Suppliers of $CaCO_3$ particles

As the demand for PCC particles has been increasing constantly because of their many industrial applications, the supply of PCC particles has also been growing noticeably. In addition to many well-known companies, many new companies have emerged and are competing in the global market. Table 3.1 lists some of the suppliers of PCC particles and some of their selected grades. The important attributes of PCC particles, such as particle size, whiteness, and surface nature, as well as the recommended applications are summarized.

## 3.2.3 The effects of particle size

Particle size has a dramatic effect on the mechanical properties of nanocomposites because the interparticle distance and interfacial area decreases and increases,

*Table 3.1* Suppliers and selected grades of PCC particles

| Supplier | Selected grades | Particle size | Surface treatment | Shape | Specific surface area ($m^2/g$) | Whiteness (%) | Recommended application by the suppliers |
|---|---|---|---|---|---|---|---|
| Imerys Performance Minerals (www.imerys-perfmins.com/) | Carbilux | 99 wt% below 2 microns | | | | | Dextrin adhesives, polyvinyl acetate adhesives, starch adhesives |
| | Carbital 110 | Below 10 microns | Uncoated | Ground | 5 | 95 ± 1.5 | Paint, filler for PVC, polyolefin film |
| | Carbital 110S | Below 10 microns | Stearate 0.9 wt% | Ground | 5 | 95 ± 1.5 | PVC products, adhesives and sealants |
| | Carbital 95 | 97 wt% below 2 microns | | Ground | | | Adhesives |
| | Carbital SL | 90 ± 3 wt% below 2 microns | Stearate 1.15 wt% | Ground | 10 | 94.5 ± 1.5 | PVC products, injection molding, adhesives and sealants |
| Solvay Precipitated Calcium Carbonate (www.solvaypcc.com/) | WINNOFIL S, SPM, SPT | 130–140 nm | Coated | | 16–30 | 93–98 | N/A |
| | SOCAL 31 | 70 nm | Uncoated | Cube-like | 17 | 97 | N/A |
| | SOCAL P3, P2, N2R | 500–180 nm | Uncoated | Cigar-like | 6–11 | 95–99 | N/A |
| Minerals Technology Inc. (www.mineralstech.com/) | Ultra-Pflex | 0.07 μm | Coated | | 19 | 98 | PVC plastisol, polysulfide, polyurethane system, rigid PVC |

| | | | | | | | |
|---|---|---|---|---|---|---|---|
| | Thixo-Carb 300 | 0.15 μm | Surface treated | Cubic | 12 | N/A | Polyurethanes, polysulfide, PVC plastisol, epoxy and butyl systems |
| | SUPER-PFLEX 200 | 0.7 μm | Surface treatment of 2% | Ground | 7 | 97 | Rigid PVC applications |
| | Calofort SM | 0.07 μm | 2.7–3.3 wt% stearate coating | Cubic | 19–25 | N/A | Polyurethane, polysulfide and other polymer systems that need viscosity control |
| Shanghai Yaohua Nano-Tech Co., Ltd (www.caco3-yhnano.com/) | XM-301 | <25 nm | Hydrophilic | Cubic | >55 | >90 | Hydrophilic paint |
| | XM-302 | 60–90 | Compound activated agent | Cubic | >21 | >95 | PVC anti-stone coating |
| | XM-303 | 70–90 nm | N/A | Cubic | 19–21 | >93 | PVC, PP, PE, ABS |
| | XM-202 | <1.5 μm | Compound activated agent | Cubic | N/A | >90 | UPVC, PVC, PE, PP, ABS, HIPS |
| | YH-304 | 60–90 nm | Fatty acid | Cubic | 19–21 | >90 | Rubber product |
| | YH-303 | 70–90 nm | | Cubic | 15–26 | >90 | PVC, PP, PE, ABS, HIPS |
| | YH-306 | 60–100 nm | Special agent | Cubic | | >90 | Adhesive and sealant |
| Shanghai Perfection Nanometre New Material Co. Ltd (www.nano-metre.com/index.html) | SPT/SP | 60–90 nm | Fatty acid | Cubic | 23 ±3 | >90 | Paint |
| | SPA-100 | 60–90 nm | Fatty acid | Cubic | N/A | >90 | PVC sealant |
| | SPA-200 | 30–60 nm | Resin acid | Cubic | N/A | N/A | PVC plastisol |
| | SP-100 | <100 nm | Fatty acid | Cubic | 23± 3 | >90 | PVC, PP, PE, ABS, HIPS |
| | SP-200 | <100 nm | Fatty acid | Cubic | 23 ± 3 | >90 | Filler for thermoplastic |

respectively, as the size of the particle decreases at fixed particle content. The large interfacial areas of polymer nanocomposites play an important role in determining the tensile modulus and toughness of the polymers. A strong interface increases the tensile modulus of the polymer if the interfacial strength is strong while a weak interface allows debonding of the nanoparticles from the matrix, which has proven to be a precondition for improved toughness. Assuming all the particles are spherical, Wu's model (Wu, 1985) can be used to estimate the interparticle distance, $d_{IP}$, or critical ligament thickness:

$$d_{IP} = d_c\left[k\left(\frac{\pi}{6\phi_r}\right)^{1/3} - 1\right], \quad [3.1]$$

where $d_c$ is the particle diameter, $\phi_r$ is the particle volume fraction, and $k$ is a geometric constant (i.e. $k=1$ for a cubic lattice, $k=2^{1/3}$ for a body-centered lattice and $k=4^{1/3}$ for face-centered lattice). It is clear from Eq. 3.1 that the interparticle distance is determined by the size and volume fraction of the particles. A small interparticle distance or ligament thickness, which allows large plastic deformation of the matrix, is a key parameter that promotes the toughness of a polymer. A toughness window showing the relation between toughness, interfacial adhesion and interparticle distance was proposed by Kausch and Michler (2007) (Fig. 3.3). This can be used as a rough guideline in the development of nanocomposites.

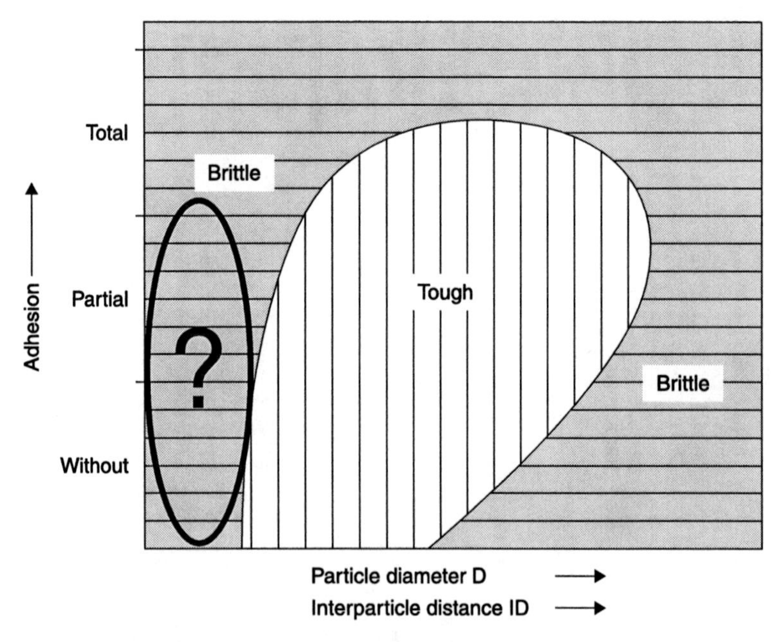


*3.3* Relation between toughness, interparticle distance and interfacial adhesion (Kausch and Michler, 2007).

### 3.2.4 The effects of surfactant coating on dispersion

Small PCC particles, especially nanoparticles, have a great tendency to form agglomerates because they have a large specific surface area and high surface tension. The formation of agglomerates is especially undesirable and even detrimental when PCC particles are used as reinforcing filler in plastics. Applying a coating of fatty acid on the surface of PCC particles is a cost-effective and common industrial practice to reduce the surface tension and to improve the dispersibility of the particles in various polymers. It should be emphasized that the surface coverage of the surfactant is an important parameter, which characterizes the surface tension of the particles and ultimately determines their dispersibility (Kiss *et al.*, 2007). The surface coverage can be classified into three categories: partial coating, monolayer coating, and excessive coating. They are shown in Fig. 3.4. In a monolayer coating, the particle surface is saturated with chemically bonded surfactant molecules. A certain amount of bare calcium carbonate surface is still present on partially coated particles. Excessive coating forms physically absorbed layers in addition to the complete chemically bonded layer.

Móczó *et al.* (2004) showed that the surface tension of $CaCO_3$ particles with a monolayer coverage of fatty acid is at a minimum value, as shown in Fig. 3.5. Under this condition, the interactions among the monolayer-coated PCC particles are minimized and, therefore, they are expected to have the best dispersibility in polymers. The amount of surfactant required for a monolayer coating is a function of the specific surface area of the PCC particles. Lin *et al.* (2010, 2011b) investigated the effects of the amount of coating on the dispersibility of PCC nanoparticles in a polypropylene matrix. Figures 3.6(a) and (b) are SEM micrographs of a polypropylene (PP) containing 20 wt% $CaCO_3$ (70 nm) with partial and monolayer coatings, respectively. Clearly, the monolayer-coated $CaCO_3$ nanoparticles have smaller clusters and better dispersion in the PP than the

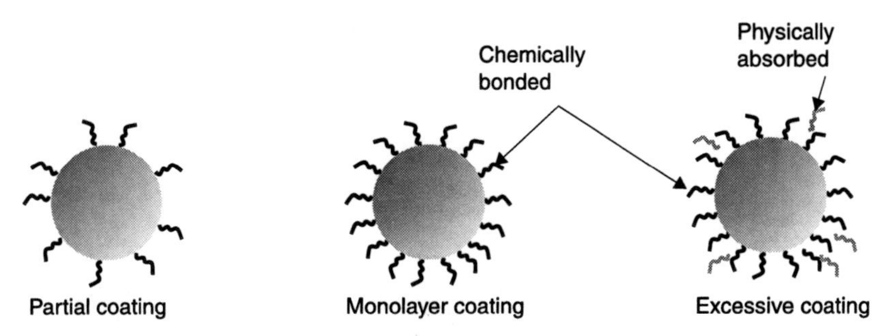


*3.4* Particles with different surface coverage: partial coating, monolayer coating, and excessive coating.

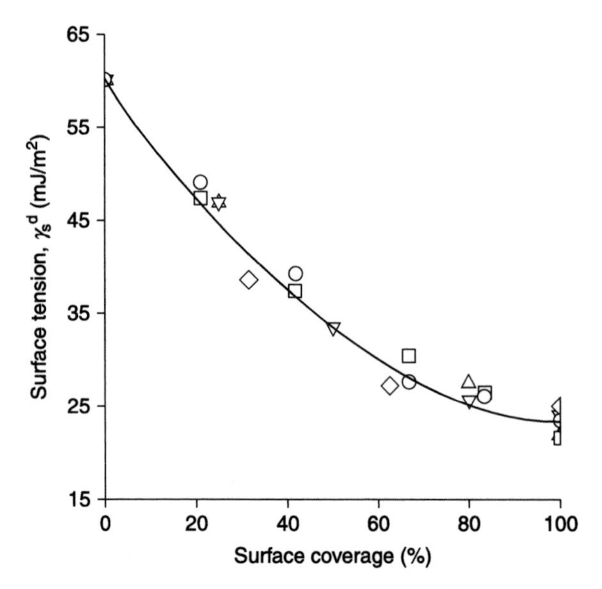


*3.5* Relation between surface tension ($\gamma_s^d$) and surface coverage. ▽: oleic acid; △: lauric acid; □: stearic acid; ○: behenic acid; ◇: 2-ethylhexanoic acid (Móczó *et al.*, 2004).

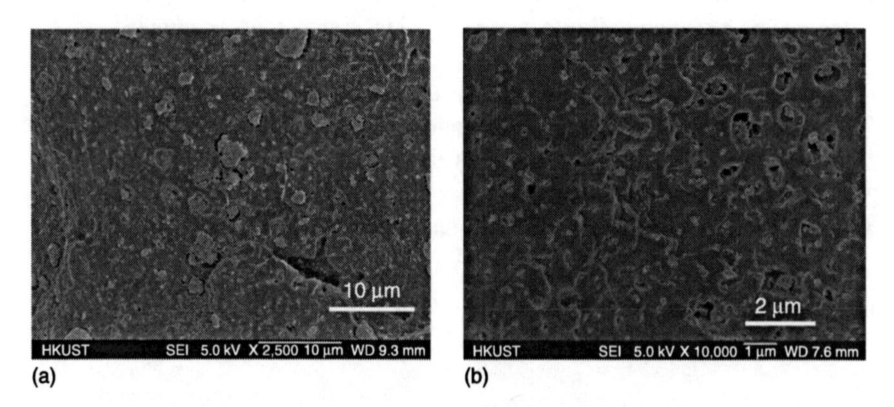


*3.6* SEM micrographs of PP/20 wt% $CaCO_3$ (70 nm) nanocomposites: (a) $CaCO_3$ with partial coating (2.3 wt% stearic acid). (b) $CaCO_3$ with monolayer coating (5.3 wt% stearic acid).

partially coated $CaCO_3$ nanoparticles. An observation made at a much higher resolution with transmission electron microscopy on the same sample revealed that the nanoparticles were well dispersed (Fig. 3.7). Even so, some small clusters ranging from tens of nanometers to hundreds of nanometers could still be found.

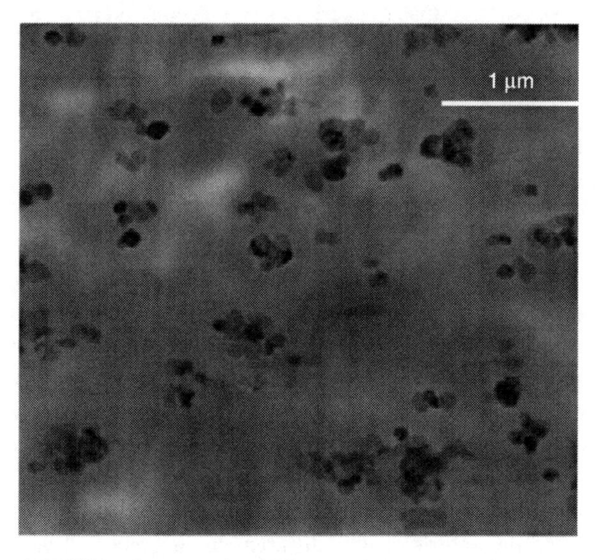


*3.7* TEM micrograph of PP containing 20 wt% monolayer-coated $CaCO_3$ nanoparticles (70 nm).

## 3.2.5 Characterization of the monolayer coating

To achieve the best dispersion of nanoparticles in a polymer, a monolayer coating on the particles is desirable. Methods that can determine the amount of surfactant needed to produce monolayer-coated $CaCO_3$ particles are therefore required. Several techniques have been established to characterize the surface coverage of coated $CaCO_3$ particles, including sedimentation methods (Liauw *et al.*, 1995; Schofield *et al.*, 1998), the dissolution method (Fekete *et al.*, 1990), X-ray photoelectron spectroscopy (XPS) (Sutherland *et al.*, 1998), inverse gas chromatography (IGC) (Fekete *et al.*, 2004), diffuse reflectance infrared Fourier-transform spectroscopy (DRIFTS), differential scanning calorimetry (DSC), thermogravimetric analysis (TGA), and derivative thermogravimetric (DTG) analysis (Osman and Suter, 2002). For example, TGA, DTG, DRIFTS, and DSC were used to characterize $CaCO_3$ nanoparticles (SPT from Solvay Chemical Company) with an average diameter of 70 nm and an original coating of 2 wt% stearic acid (SA). These $CaCO_3$ nanoparticles were further coated with different amounts of stearic acid to investigate the effects of the amount of coating on the dispersibility of the nanoparticles (Lin *et al.*, 2011a). Table 3.2 lists different samples of $CaCO_3$ nanoparticles with different amounts of stearic acid coating.

TGA was used to measure the amount of coating on the surface-treated $CaCO_3$ nanoparticles and DTG was used to identify the amount of surfactant needed for a monolayer coating. A DTG curve can reveal physically and chemically absorbed

*Table 3.2* Modified nanoparticles and their corresponding amount of stearic acid coating

| Sample | SA3 | SA4 | SA5 | SA6 | SA7 | SA8 |
|---|---|---|---|---|---|---|
| Total coating amount based on nanoparticles (wt%) | 2.8 | 3.7 | 5.3 | 5.6 | 6.5 | 7.5 |

components of a surfactant, which decompose at different temperatures due to their different thermal stabilities. If there is only one layer of chemisorbed coating, then only one decomposition peak appears in the DTG curve. However, if the coating is thicker than a monolayer, physisorbed layers will appear. As a result, a shoulder or even a new peak, depending on the amount of coating, will emerge at a lower temperature. The critical amount of coating that is associated with the emergence of a shoulder can be regarded as the amount needed for a monolayer. DTG traces of $CaCO_3$ nanoparticles with different coating amounts are displayed in Fig. 3.8. SA3, SA4, and SA5 have only a single peak at about 330 °C. The shifting of the peak at 330 °C to lower temperatures started from SA6. Therefore, 5.3 wt% (SA5) was regarded as the amount needed for a complete monolayer coating. The amount of stearic acid needed for a monolayer coating can also be estimated theoretically: the surface area of $CaCO_3$ nanoparticles is 22.2 $m^2/g$ as reported by the supplier and each $Ca^{2+}$ ion occupies an area of 0.2 $nm^2$ (Padden and Keith, 1966). If each stearic acid molecule connects to one $Ca^{2+}$ ion and each stearic acid molecule is oriented vertically to the particle surface (Binsbergen and

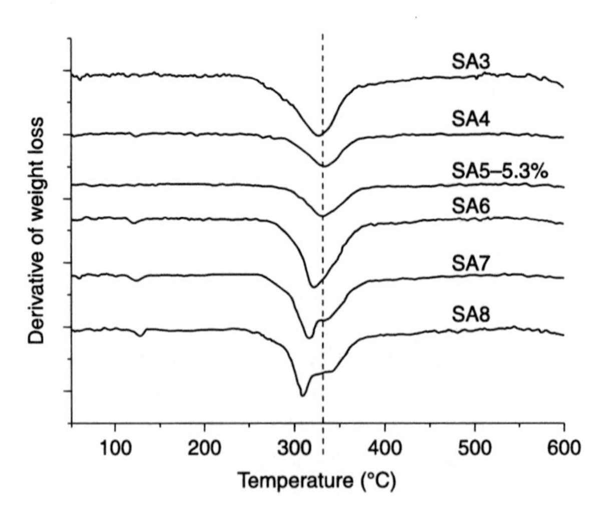


*3.8* DTG curves of PCC nanoparticles coated with different amounts of stearic acid (Lin, 2009).

de Lange, 1968), the calculated amount of stearic acid that is needed to produce a complete monolayer coating is about 5.3 wt%. The calculated and DTG results are in good agreement.

Fourier transform infrared (FTIR) spectra can provide fingerprint references about the nature of the coating on the surface of $CaCO_3$ nanoparticles. Successful coating with stearic acid is evidenced by the presence of C–H vibration bands in the range 2800 to 3300 $cm^{-1}$. In addition, the conformation and packing density of the alkyl chains on the particle surface revealed in FTIR spectra can be used to distinguish a monolayer coating from other types of coating.

It is known that methylene symmetric, $\nu_s$-$CH_2$, and anti-symmetric stretching bands, $\nu_a$-$CH_2$, which are very sensitive to conformational changes of the alkyl chains, shift to lower frequencies when the alkyl chains adopt an all-trans conformation (Kawai *et al.*, 1985; Sakai and Umemura, 1993, 2008; Kew and Hall, 2006). In this conformation, all the alkyl chains are fully extended and densely packed. In addition, the $\nu_s$-$CH_2$ and $\nu_a$-$CH_2$ peaks shift to higher frequencies at high temperatures because of the increased mobility of the alkyl chains (Singh *et al.*, 2002). DRIFTS spectra of the coated nanoparticles at 50 °C are shown in Fig. 3.9 (Lin *et al.*, 2011a). The $\nu_s$-$CH_2$ and $\nu_a$-$CH_2$ peaks of SA2, which were at 2850 and 2917 $cm^{-1}$, respectively, were at the same frequencies as those of pure stearic acid (Osman and Suter, 2002). The $\nu_s$-$CH_2$ and $\nu_a$-$CH_2$ peaks of SA5, on the other hand, were at 2847 and 2912 $cm^{-1}$, respectively, which were lower than the frequencies for pure stearic acid, indicating that the alkyl chains of SA5 had a much more compact packing structure, implying the presence of a monolayer structure. When the coating was in excess of a monolayer, for

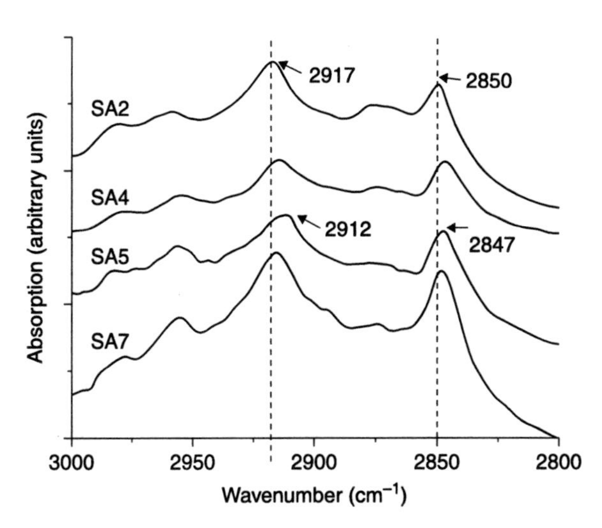


*3.9* DRIFTS spectra of nanoparticles with different amounts of coating coverage at 50 °C (Lin *et al.*, 2011a).

SA7, the $\nu_s$-$CH_2$ and $\nu_a$-$CH_2$ peaks were shifted back to higher frequencies, possibly because the physically absorbed stearic acid molecules were more mobile than the chemically bonded calcium stearate.

DSC is commonly used in calorimetric studies of the coating layer on these PCC nanoparticles. A monolayer coating can be identified due to its unique calorimetric features. Hydrated calcium stearate is a common form of the product when coating is performed in an aqueous medium (Shi *et al.*, 2010; Garnier *et al.*, 1988; Vold *et al.*, 1949). Monohydrated calcium stearate has a reported dehydration peak at 115 °C and one major peak at 130 °C, corresponding to a crystalline-to-smectic transition (Vold *et al.*, 1949). DSC curves of the nanoparticles with varied amounts of coverage are shown in Fig. 3.10. SA4 and SA7 have two endothermic peaks. The lower temperature peak between 115 and 120 °C (Fig. 3.10) was determined as the dehydration peak and the higher temperature peak at 131 °C (Fig. 3.10) was attributed to the crystalline-to-smectic transition. In contrast, the DSC curves of SA2 and SA5 were horizontal straight lines, implying that there was no structural change in their coatings at temperatures below 200 °C. In the case of SA2, the coating amount was small. It is possible that the alkyl chains were so far apart from each other that they did not interact with each other, leading to the absence of an endothermic peak, while for SA5, which had a monolayer coating, the stearate molecules were so closely packed that steric hindrance prohibited the aliphatic chains from any motion even at high temperatures. As a result, no crystalline-to-smectic transition over the scanned temperature range was observed, leading to the absence of endothermic peaks.

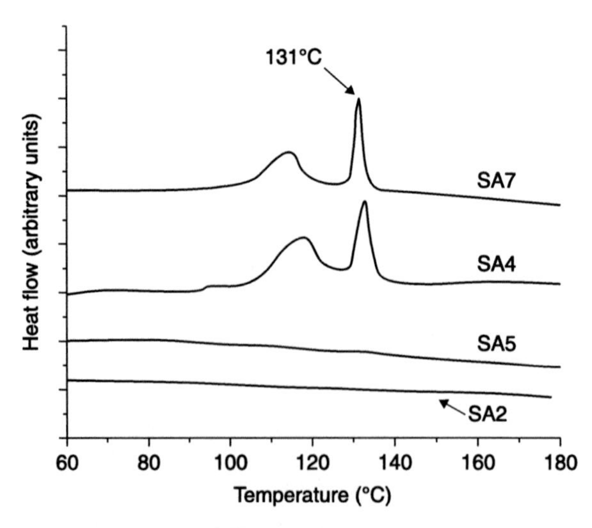


*3.10* DSC temperature scans of $CaCO_3$ nanoparticles with different amounts of calcium stearate coating (Lin *et al.*, 2011a).

### 3.2.6 $CaCO_3$ polymer nanocomposites

*Poly(vinyl chloride)/$CaCO_3$ nanocomposites*

Adding PCC nanoparticles to poly(vinyl chloride) (PVC) has the following advantages:

- shorter fusion and gelation times
- enhanced impact strength, particularly at low temperatures
- improved surface finish and gloss
- elimination of plate-out
- improved weatherability
- efficient hydrogen chloride gas absorption.

PVC, a commodity plastic, has many desirable properties such as non-flammability, relatively low cost and formulating versatility. Depending on the amount of plasticizer added, PVC can be classified as flexible (or plasticized) or rigid (unplasticized). Flexible PVC, which is soft and bendable, is commonly used for flexible hoses and tubing, wire coating materials, and floor coverings. Due to its very nature, rigid PVC is very durable and long-lasting, and it is widely used as a construction material for profiles, pipes, conduits, and fittings. However, the impact strength of PVC is low, especially at low temperatures. Impact modifiers, for example, acrylic, chlorinated polyethylene (CPE), and methacrylate butadiene styrene (MBS), are commonly used to improve the impact strength of PVC.

GCC particles have been used as fillers in thermoplastics and especially in PVC. The purpose of adding GCC particles is to lower the material costs of the finished products. The loadings of particles in PVC are usually lower than 10 parts per hundred (pph). However, GCC microparticles may generally impair certain performance parameters of the compounds, including the tensile strength and impact strength. Moreover, compounds with high loadings of GCC microparticles larger than 10 μm in diameter are not desirable because these particles, which are highly abrasive, damage the barrels and screws of the extruders.

Due to the ultra-fine particle size, surface-modified PCC nanoparticles are well recognized as a good processing aid and an impact modifier for rigid PVC (Cornwell, 2001). Surface-modified PCC, which is more easily dispersed uniformly in a polymer, provides various benefits to PVC compounds. The compounding of PVC with surface-modified PCC nanoparticles significantly improves the surface gloss and the processability of the PVC during injection molding. Also, the damage to extruder barrels and screws is minimized if the PCC nanoparticles are surface-coated with a fatty acid. Furthermore, surface-modified PCC nanoparticles can reduce the gelation time of PVC because of better compatibility between the surface-modified PCC nanoparticles and PVC primary particles. Another benefit of the addition of surface-modified PCC nanoparticles is that they can stabilize the dispersion components of PVC compounds and eliminate the plate-out problem. More importantly, surface-modified PCC nanoparticles can increase both the impact strength and stiffness of PVC.

Particle size is a very important parameter: it can significantly affect the performance of composites (Fernando and Thomas, 2007; Sun *et al.*, 2006; Bryant and Wiebking, 2002). In general, the Young's modulus of PVC increases with increasing PCC nanoparticle (60 nm) concentration, as shown in Fig. 3.11 (Kemal *et al.*, 2009), whereas the tensile strength and yield strength slightly decrease or show no effects with the addition of PCC particles. A plot showing the relation between the particle size and the impact strength of PVC is shown in Fig. 3.12 (Bryant and Wiebking, 2002). Generally, the smaller the particle size, the higher the impact strength of the finished compounds. For particles of 1 μm or larger, the toughening effect was moderate. As the particle size decreased below 1 μm, the impact strength of PVC increased sharply.

The impact strength of PVC increases with the loading level of PCC (Xie *et al.*, 2004; Kemal *et al.*, 2009; Wu *et al.*, 2004). Figure 3.13 shows the positive linear relation between PVC impact strength and concentration of 44-nm PCC nanoparticles (Wu *et al.*, 2004). In addition to particle size, the surface treatment of PCC nanoparticles has a great influence on the impact strength of PVC compounds. As shown in Fig. 3.14 (Sun *et al.*, 2006), PVC filled with untreated PCC nanoparticles had a lower impact strength than one filled with treated PCC

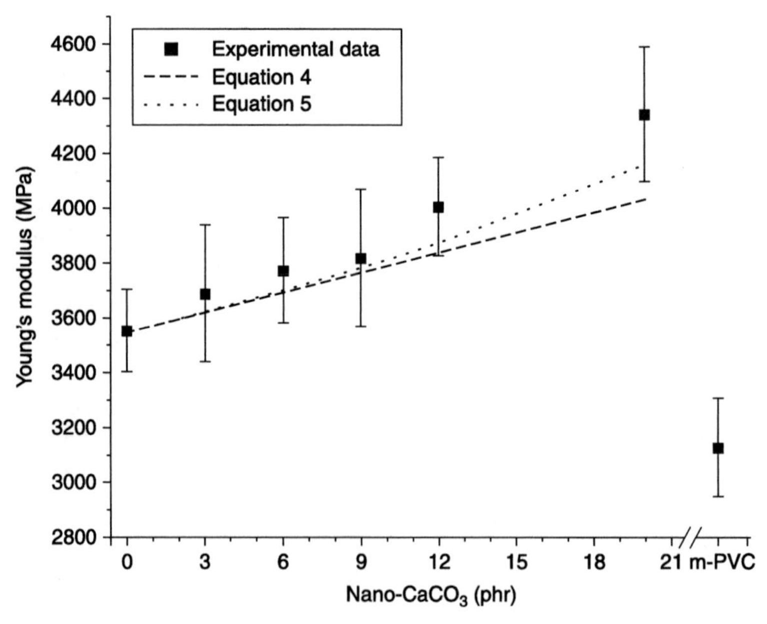


*3.11* Effect of nano-$CaCO_3$ content on the Young's modulus of PVC nanocomposites (Kemal *et al.*, 2009).

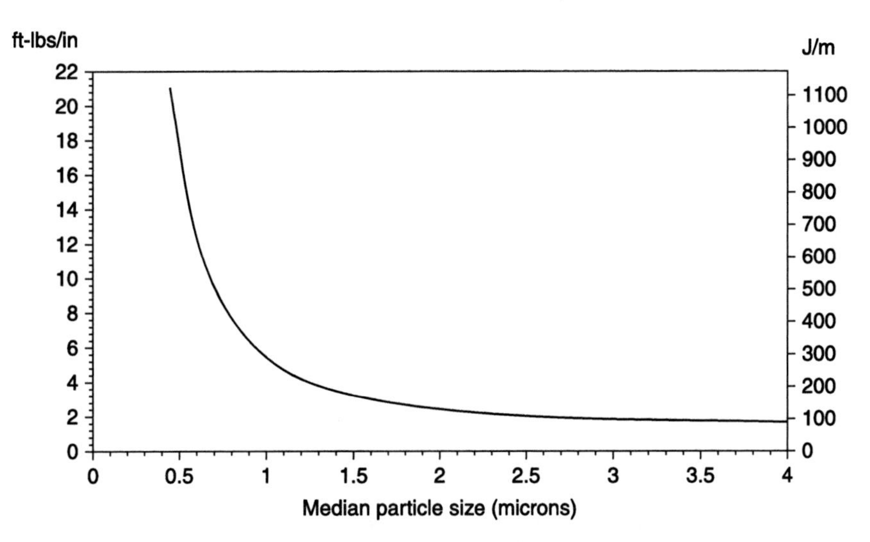


*3.12* Notched Izod impact strength of PVC filled with calcium carbonate particles (20 pph) (Bryant and Wiebking, 2002).

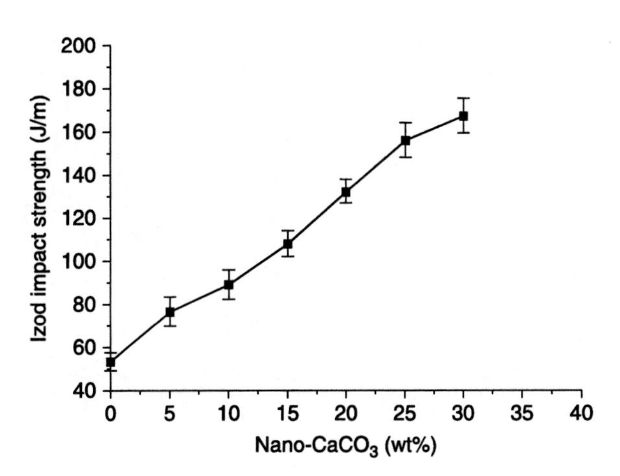


*3.13* Notched Izod impact strength of PVC/nano-$CaCO_3$ composites vs. nano-$CaCO_3$ content (Wu *et al.*, 2004).

nanoparticles. In addition, the choice of surfactant also affected the impact strength of PVC compounds, as shown in Fig. 3.14.

Kemal *et al.* (2009) studied experimentally the toughening mechanism of PVC filled with PCC nanoparticles (60 nm) and performed microstructure-based finite element modeling. Many nanovoids were observed in the plastic deformation zone in the impact-fractured samples, as shown in Fig. 3.15 (Kemal *et al.*, 2009).

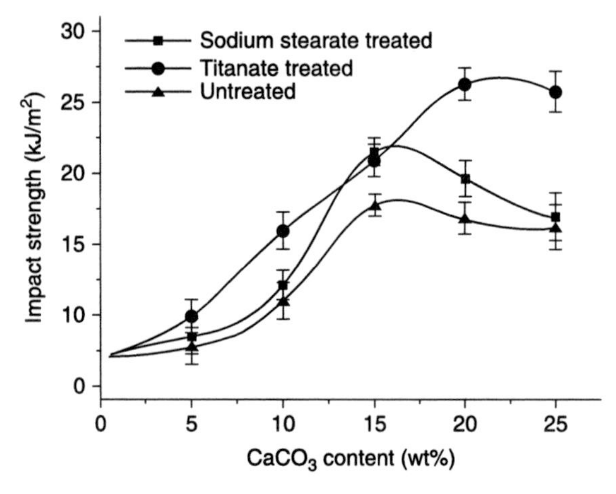


*3.14* Impact strength of PVC composites filled with untreated, sodium-stearate-treated, and titanate-treated $CaCO_3$ particles vs. $CaCO_3$ particle content (Sun *et al.*, 2006).

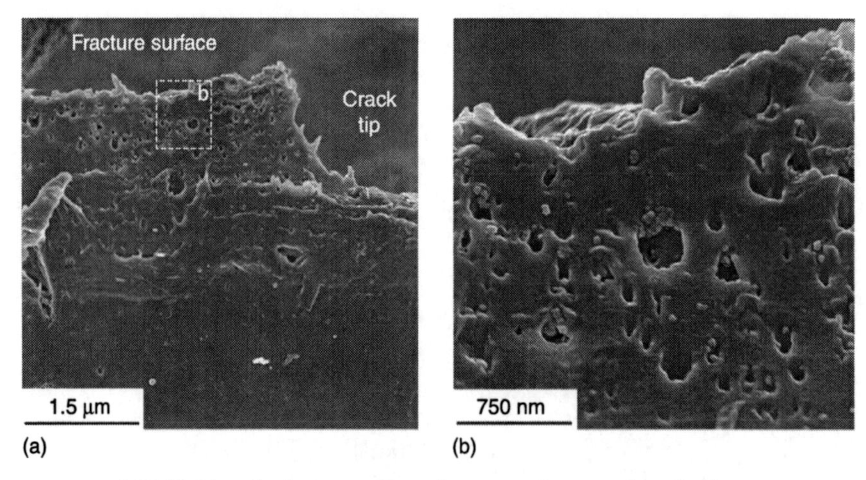


*3.15* Field emission scanning electron micrographs of microtome-cut cross sections of the process zone of impact-fractured PVC nanocomposites with 9 phr nano-$CaCO_3$ content: (a) Low magnification. (b) High magnification (Kemal, *et al.*, 2009).

The authors attributed the increased impact strength to the enhanced formation of cavities initiated by the nanoparticles, leading to improved toughness.

A substantial increase in the impact strength of PVC can be achieved when both PCC nanoparticles and an impact modifier are added, due to their synergetic effects. Figures 3.16 and 3.17 show the loading effects of PCC nanoparticles

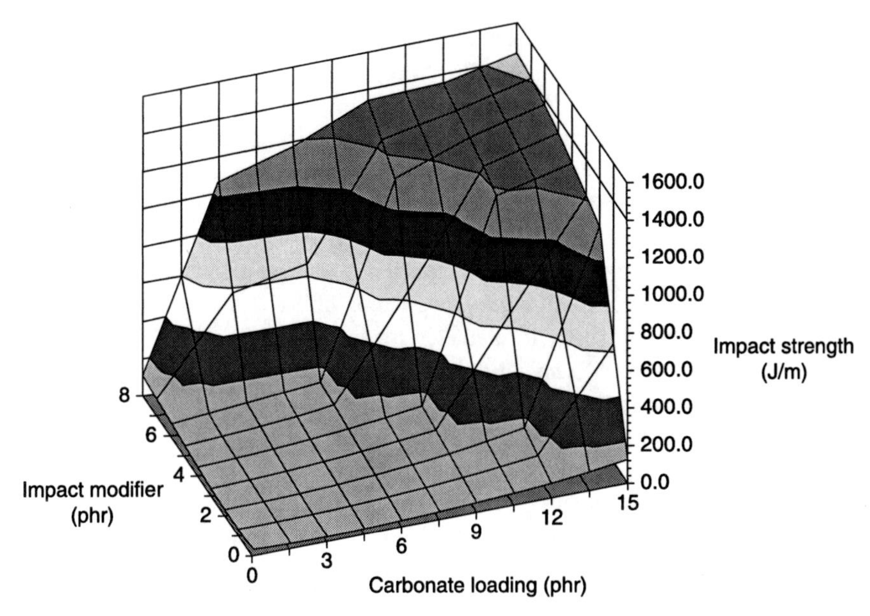


*3.16* Notched Izod impact strength of PVC filled with 0.07-μm PCC and an acrylic impact modifier (Bryant and Wiebking, 2002).

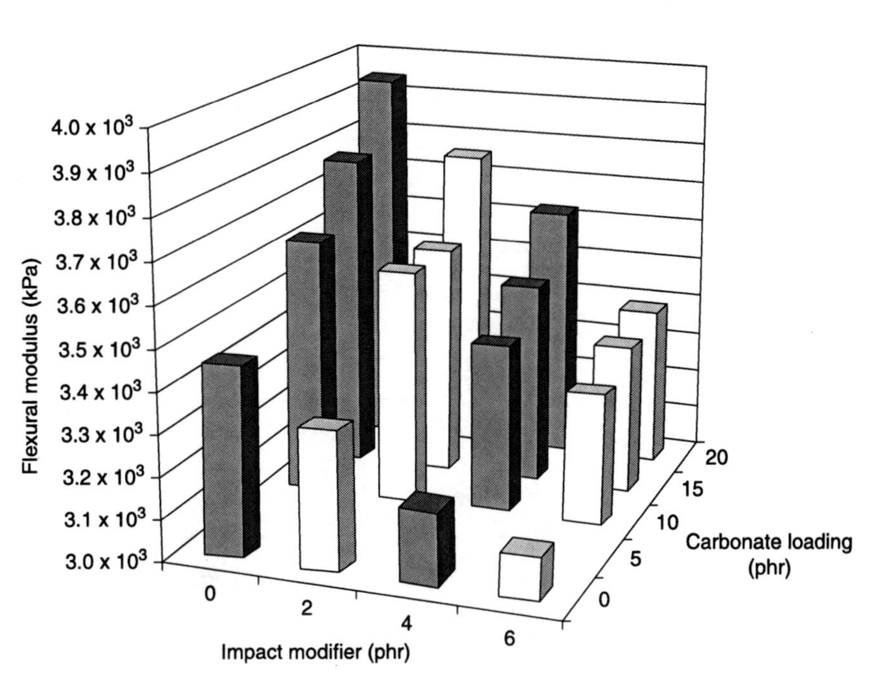


*3.17* Flexural modulus (kPa × 1000) of PVC filled with 0.07-μm PCC and an acrylic impact modifier (Bryant and Wiebking, 2002).

(Ultra-Pflex, 70 nm) and an impact modifier on the impact strength and flexural modulus of PVC (Bryant and Wiebking, 2002), respectively. Optimized formulations can be obtained by varying the loading level of the PCC nanoparticles and the impact modifier to achieve balanced mechanical properties, i.e. a balance between stiffness and impact strength. In this example, the nanoparticles played two important roles in the PVC compounds. First, the stiffness of the compounds, which could have been reduced due to the presence of the impact modifier, was maintained. Second, the costs of the finished products were reduced.

Wu *et al.* (2004) reported that the toughening effect was absent if PVC, PCC nanoparticles (average particle diameter of 40 nm and coated with a titanate coupling agent), and chlorinated polyethylene were melt blended at the same time. However, the ternary blends exhibited excellent toughness if PVC was subsequently blended with a CPE/PCC master batch. As revealed in TEM micrographs of the PVC ternary blends prepared using the two-step blending procedure, the PCC nanoparticles were encapsulated by CPE. Figure 3.18 (Wu *et al.*, 2004) shows the micro-morphology of the ternary system: the PCC nanoparticles are encapsulated by a CPE layer and are well dispersed in the PVC matrix. The toughening effect was attributed to the presence of the core-shell particles, which triggered the rubber-toughening mechanism, leading to the remarkable improvement in impact toughness.

### *Polyethylene (PE)/$CaCO_3$ nanocomposites*

Adding PCC particles to PE has the following advantages:

- increased thermal conductivity
- reduced cycle time of injection molding
- increased stiffness
- enhanced dimensional stability
- lower shrinkage
- relaxation of internal stress.

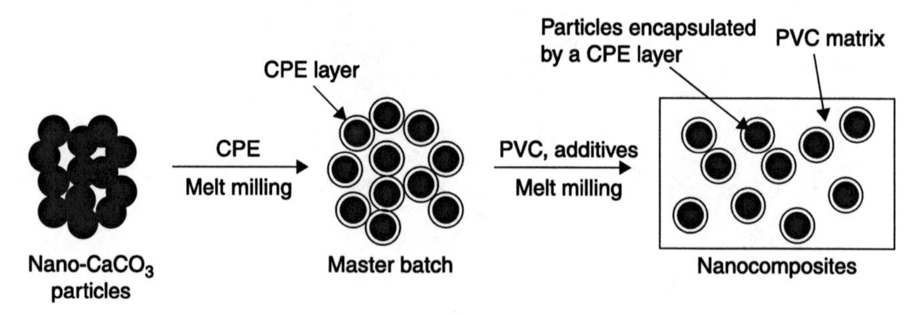


*3.18* Preparation of PVC/CPE/nano-$CaCO_3$ nanocomposites (Wu *et al.*, 2004).

Low-density polyethylene (LDPE) and high-density polyethylene (HDPE) are versatile thermoplastics because of their low cost, recyclability, availability, and ease of processing. The addition of calcium carbonate nanoparticles leads to increased output, self-cleaning of the tools, and improved printing properties of the films with reduced doses of corona treatment. More importantly, the addition of calcium carbonate nanoparticles to LDPE can improve the anti-blocking and slip properties of the resulting films, which are used widely for plastic bags (Hess, 2001). Bottles, containers, and pipes are the major applications of HDPE. These products require materials with good mechanical strength. An increase in the modulus and yield stress of HDPE without sacrificing its impact strength can be achieved by incorporating calcium carbonate nanoparticles. To improve the compatibility between calcium carbonate nanoparticles and HDPE, either the calcium carbonate particles are surface-treated or the HDPE is chemically modified, for example, by grafting with maleic acid (Liu *et al.*, 2002; Yuan *et al.*, 2009).

GCC particles are effective reinforcement fillers for HDPE. Fu *et al.* (1993a, 1993b) and Liu *et al.* (1998) systematically investigated the effects of GCC particles on the impact strength of HDPE. Their studies revealed that the surface treatment of GCC particles, the volume percentage of GCC particles, and the intrinsic toughness of the HDPE matrix have a great influence on the impact toughness of HDPE/$CaCO_3$ composites. Surface treatment is essential for good dispersion of GCC particles in a HDPE matrix and therefore constitutes an effective improvement for the toughness of HDPE (Fu and Wang, 1993). A critical matrix ligament thickness, regardless of particle size, was observed, corresponding to the brittle-to-ductile transition of HDPE (Fig. 3.19). In addition, Fig. 3.19 also

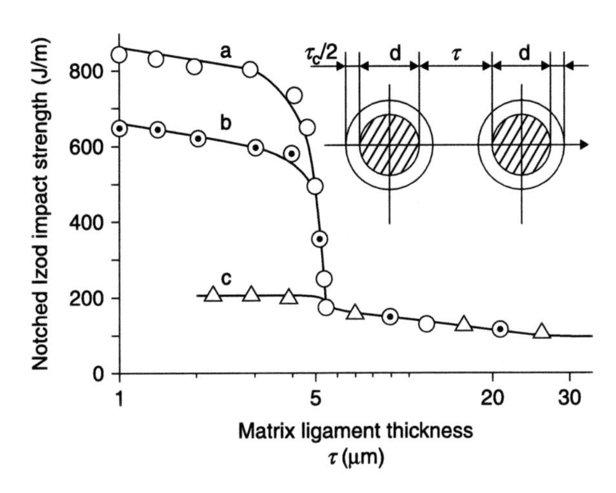


*3.19* Notched Izod impact strength vs. matrix ligament thickness for HDPE/$CaCO_3$ blends, with an average $CaCO_3$ particle diameter of (a) 6.66 μm; (b) 7.44 μm; (c) 15.9 μm (Fu and Wang, 1993).

*Table 3.3* Effect of matrix impact strength on toughness of HDPE/$CaCO_3$ (50/50) blends

| Matrix notched Izod impact strength (J/m) | Notched Izod impact strength of HDPE/$CaCO_3$ (50/50) blends (J/m) | Increase (%) |
|---|---|---|
| 180.0 | 850.0 | 372 |
| 162.7 | 488.3 | 200 |
| 138.8 | 309.9 | 123 |
| 71.4 | 224.2 | 214 |
| 44.5 | 146.7 | 230 |
| 42.7 | 58.1 | 36 |
| 38.9 | 50.0 | 29 |
| 29.7 | 31.7 | 7 |

Source: Fu and Wang (1993).

reveals that the smaller the GCC particles, the higher the impact toughness of the HDPE/$CaCO_3$ blends. Furthermore, the toughening effect of calcium carbonate microparticles strongly depended on the intrinsic toughness of the HDPE matrix. As shown in Table 3.3, the enhancement in toughness of the HDPE/$CaCO_3$ composites, compared with the corresponding pure HDPE, increased with the intrinsic toughness of HDPE.

Bartczak *et al.* (1999) compared the tensile and impact properties of HDPE with PCC and GCC particles. The GCC microparticles (CC1) and PCC particles (CC2), which had weight-average particle sizes of 3.5 and 0.7 μm, respectively, were surface-treated with calcium stearate. Table 3.4 and Fig. 3.20 show the tensile properties and impact properties of the filled HDPE composites, respectively. Young's modulus increased but the yield stress decreased with increasing calcium carbonate particle concentration for both GCC- and PCC-filled HDPE composites, whereas the differences in stiffness and yield stress between HDPE/CC1 and HDPE/CC2 were not significant. However, the PCC particles had a more noticeable toughening effect on HDPE than GCC particles, as shown in Fig. 3.20. For the HDPE/CC2 blend (i.e. HDPE filled with PCC particles), a steep rise in the impact energy occurred at the filler concentration between 5 and 10 vol%. For HDPE/CC1 (i.e. HDPE filled with GCC particles), the impact toughness was almost constant at low particle concentrations and showed a slight improvement at the particle concentration of 25 vol%.

As shown in Fig. 3.21, the HDPE/CC2 blend maintained its high impact strength at low temperatures whereas a rubber-filled HDPE lost most of its ductility at temperatures below −20 °C. The high impact resistance of HDPE/PCC composites at low temperatures is very desirable and attractive for many outdoor applications of HDPE. For low-temperature applications, HDPE filled with PCC particles is superior to HDPE filled with rubber particles.

*Table 3.4* Tensile properties of HDPE/CC blends at room temperature and the initial deformation rate of $1.67 \times 10^{-2}$ $S^{-1}$

| Blend code | Volumetric composition (HDPE:CC) | Young's modulus (MPa) | Yield stress (MPa) | Yield strain (%) | Stress at break (MPa)[a] | Elongation at break (%) |
|---|---|---|---|---|---|---|
| HDPE (control) | 100:0 | 756.1 | 24.9 | 11.5 | 14.5 | 730 |
| HDPE/CC1 | 90:10 | 896.8 | 21.9 | 10.3 | | 319 |
| | 80:20 | 1235.0 | 19.6 | 7.6 | | 36 |
| | 75:25 | 1274.1 | 18.8 | 7.8 | 11.7 | 64 |
| | 70:30 | 1806.4 | 15.6 | 3.3 | 11.8 | 10 |
| HDPE/CC2 | 95:5 | 818.2 | 23.8 | 10.5 | 13.8 | 696 |
| | 90:10 | 961.9 | 22.9 | 9.1 | 13.5 | 200 |
| | 85:15 | 1136.5 | 21.3 | 7.5 | | 37 |
| | 80:20 | 1209.8 | 20.2 | 6.3 | | 33 |
| | 75:25 | 1207.8 | 20.0 | 5.6 | 13.8 | 205 |
| | 70:30 | 1604.8 | 19.2 | – | 19.2 | 5 |
| HDPE/CC3 | 75:25 | 1197.90 | 20.45 | 11.50 | 10.50 | 186 |

Note: [a]Fracture load divided by initial cross-section area.

Source: Bartczak *et al.* (1999).

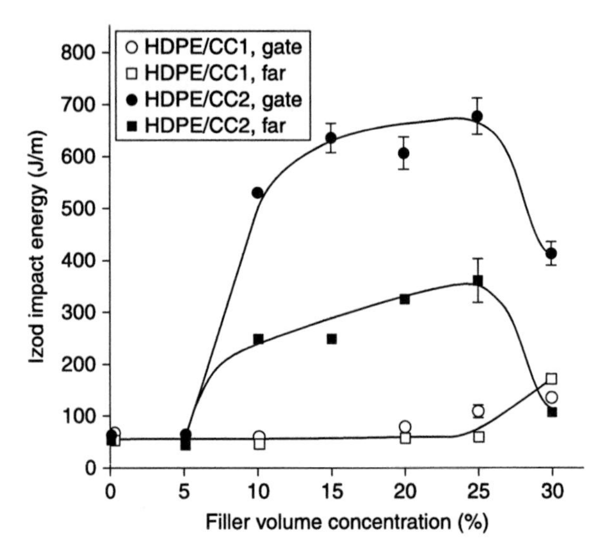


*3.20* Dependence of notched Izod impact energy on the concentration of filler, determined for samples of HDPE/CC1 and HDPE/CC2. Data for specimens from the gate end and the far end are shown separately (Bartczak *et al.*, 1999).

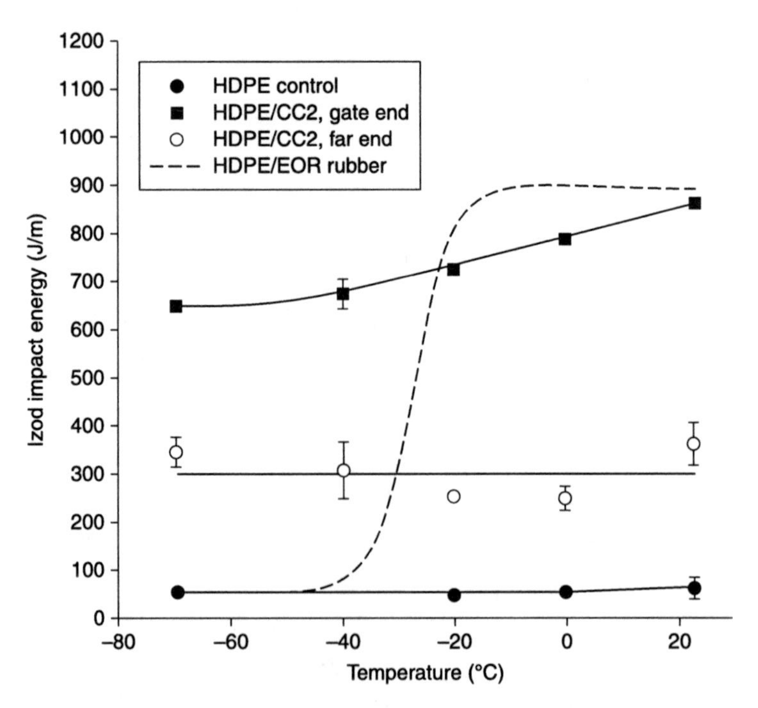


*3.21* Dependence of notched Izod impact energy on temperature for HDPE/CC2 blends. For comparison, a curve for a blend of HDPE with EOR rubber (78:22 by volume) is shown (Bartczak *et al.*, 1999).

However, several studies have shown that using PCC particles that are too small might have an adverse effect on the impact strength of HDPE (Lazzeri *et al.*, 2005; Sahebian *et al.*, 2007). Sub-micrometer (~ 0.7 μm) PCC particles are beneficial to the impact toughness of HDPE while PCC nanoparticles (less than 100 nm) cause a decrease in the impact toughness of HDPE. Lazzeri *et al.* (2005) and Sahebian *et al.* (2007) studied the effects of coated and uncoated PCC nanoparticles (70 nm) on the impact strength of HDPE. The addition of uncoated PCC nanoparticles caused a sharp drop in the impact toughness of HDPE (Fig. 3.22). Increasing the amount of the stearic acid (SA) coating progressively increased the impact toughness of HDPE/PCC nanocomposites. However, even if the coating amount was more than a monolayer, which was approximately 2.5 $mg/m^2$, the impact strength of HDPE/PCC nanocomposites was still lower than that of pure HDPE (Fig. 3.22). The authors concluded that the stearic acid coating could not completely prevent the formation of agglomeration, although it was able to reduce the sizes of the agglomerates.

The results shown in Fig. 3.20 and Fig. 3.22 raised questions about the ability of PCC nanoparticles to toughen HDPE. Sub-micron PCC particles can improve the toughness of HDPE (Bartczak *et al.*, 1999) whereas PCC nanoparticles (less than 100 nm) were detrimental to the impact toughness of HDPE (Lazzeri *et al.*, 2005). However, as suggested by the authors (Lazzeri *et al.*, 2005), poor dispersion of PCC nanoparticles was possibly responsible for the decreased impact strength of HDPE. Justified and solid conclusions about the toughening effect of PCC nanoparticles can only be made after the successful preparation of HDPE/PCC nanocomposite samples with well-dispersed PCC nanoparticles.

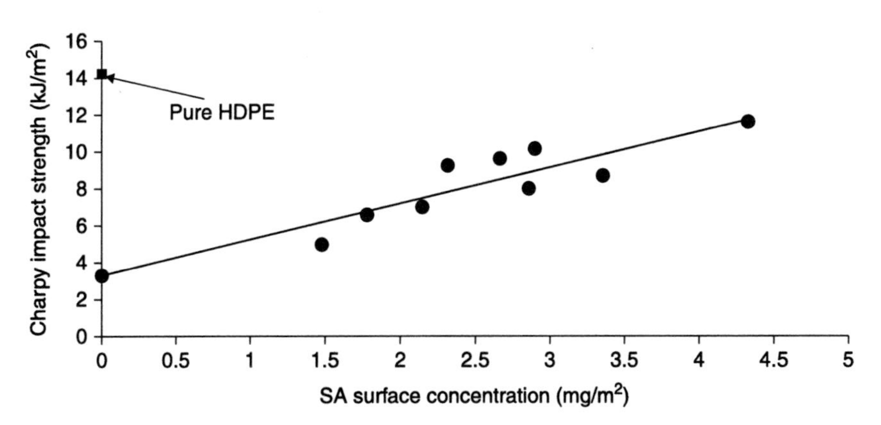


*3.22* Dependence of the impact strength of PCC/HDPE nanocomposites against SA concentration (Lazzeri *et al.*, 2005).

*Polypropylene nanocomposites*

The addition of PCC nanoparticles to polypropylene (PP) provides the following benefits:

- increased thermal conductivity
- increased crystallization temperature and shorter cycle times for injection molding
- increased stiffness
- increased impact strength
- enhanced dimensional stability
- lower shrinkage
- improved creep resistance.

Isotactic polypropylene (iPP), which is one of the most versatile polymers, has excellent mechanical properties, good chemical resistance, high deflection temperature, good fatigue resistance, and good processability. Isotactic PP has a long list of applications, such as in household appliances, flexible and rigid packaging, pipes, and automotive interiors. However, iPP is brittle at low temperatures and in the presence of a notch. The brittle temperature of iPP varies between 5 and 15 °C. The use of PP block-copolymers with 5 to 15% ethylene is one way to improve the toughness by extending its service temperature below −20 °C. However, this approach is at the expense of stiffness.

It is a common industrial practice to add PCC particles to iPP to reduce the material cost. Other benefits include improved impact toughness, better surface finishes, higher strength-to-weight ratio, and improved thermal conductivity. Table 3.5 (Yang *et al.*, 2006) shows the influence of particle concentration on the

*Table 3.5* Mechanical properties of PP1 nanocomposites

| Composition (weight ratio) PP1 | CC0.07 | Yield strength (MPa) | Flexural strength (MPa) | Flexural modulus (MPa) | Izod impact strength (kJ/m$^2$) |
|---|---|---|---|---|---|
| 100 | 0 | 32.4 | 35.4 | 1386.7 | 2.5 |
| 98 | 2 | 34.8 | 35.4 | 1420.8 | 2.5 |
| 96 | 4 | 34.2 | 35.6 | 1451.4 | 2.5 |
| 94 | 6 | 33.6 | 35.8 | 1469.4 | 2.8 |
| 92 | 8 | 33.2 | 36.1 | 1485.1 | 2.9 |
| 90 | 10 | 33.1 | 36.4 | 1536.3 | 3.2 |
| 80 | 20 | 26.7 | 37.5 | 1722.4 | 4.7 |
| 70 | 30 | 26.0 | 38.3 | 1868.5 | 5.7 |
| 60 | 40 | 24.2 | 39.2 | 2085.9 | 6.0 |

Source: Yang *et al.* (2006).

tensile properties of iPP/$CaCO_3$ nanocomposites. The $CaCO_3$ nanoparticles were supplied by Solvay (Winnofil S) with an average particle size of 0.07 μm and further coated with a 1 wt% liquid silane coupling agent. Young's modulus increased but the yield stress decreased with increasing particle concentration. Generally speaking, if the PCC nanoparticles are not coupled to the polymer matrix using a coupling agent, the interaction between the particles and the polymer matrix is weak. Debonding takes place at the iPP/$CaCO_3$ interface at the very beginning of plastic deformation, leading to a reduced yield stress.

Many studies have shown that the impact toughness of iPP/$CaCO_3$ compounds is controlled by the surface treatment and loading level of PCC nanoparticles, as well as the matrix properties of the iPP (Zhang *et al.*, 2004; Wang *et al.*, 2002; Thio *et al.*, 2002; Lin *et al.*, 2011b). Figure 3.23 (Lin *et al.*, 2011b) shows the effects of the amount of stearic acid coating on the impact strength of iPP/20 wt% $CaCO_3$ nanocomposites. A stearic acid coating of 5.3 wt% was determined to be the required quantity for a monolayer for the 70-nm nanoparticles. When the amount of coating was less than a monolayer, the dispersion of the nanoparticles was usually poor, leading to the deteriorated impact strengths of the nanocomposites compared with neat iPP. An appreciable increase in the impact strength of iPP/$CaCO_3$ nanocomposites was achieved when monolayer-coated nanoparticles were used. Coatings of more than a monolayer were also detrimental to the impact strength of iPP/$CaCO_3$ nanocomposites, possibly because the excessive stearic acid might segregate at the boundaries of spherulites, leading to a deteriorated physical network of polymer chains.

The toughening effect strongly depends on the physical properties of the iPP matrix (Lin *et al.*, 2010). In general, the higher the molecular weight (MW) of the

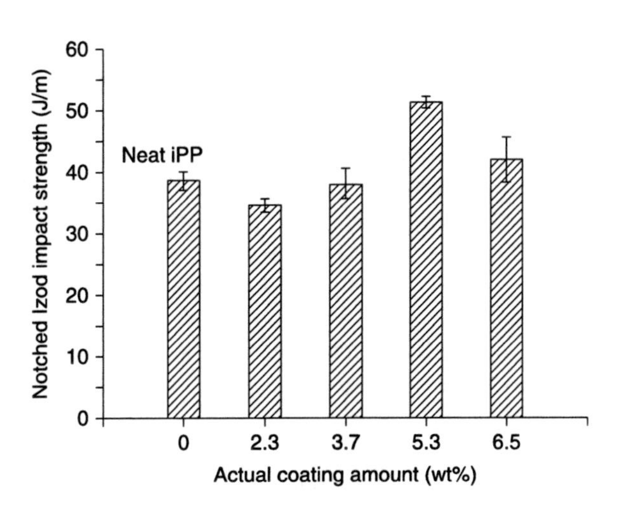


*3.23* Effect of the amount of stearic acid coating on the impact strength of PP/20 wt% $CaCO_3$ nanocomposites (Lin *et al.*, 2011b).

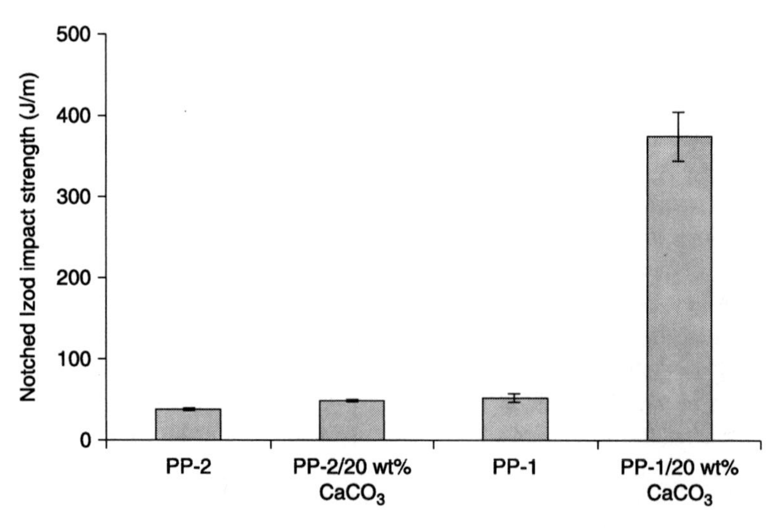


*3.24* Notched Izod impact strength of iPPs and their nanocomposites.

iPP matrix, the better the toughening effect. Figure 3.24 shows the impact strength of a high-MW ($\bar{M}_w$ = 472 000 g/mol) iPP, PP-1, and a low-MW ($\bar{M}_w$ = 346 000 g/mol), iPP, PP-2, nanocomposite containing 20 wt% monolayer-coated $CaCO_3$ nanoparticles (70 nm). The intrinsic toughness of PP-1 was 25% higher than PP-2. With the addition of nanoparticles, the impact strength of PP-1 increased by 740% whereas the impact strength of PP-2 increased only by 20%. This indicates that the high-MW matrix and the monolayer-coated nanoparticles act in synergy to improve the impact strength of the nanocomposites. Isotactic PP with a high MW has a higher strength, which amplifies the toughening effect of the well-dispersed nanoparticles. A similar phenomenon was reported for HDPE filled with GCC microparticles (Fu *et al.*, 1993). The degree of enhancement in the impact toughness of the HDPE increased as the impact strength of the HDPE matrix increased.

The influence of particle concentration on the impact toughness of iPP/$CaCO_3$ nanocomposites is also dependent on the polymer matrix (Lin *et al.*, 2010). Figure 3.25(a) and (b) are plots showing the impact strength of PP-1 and PP-2 nanocomposites, respectively, as a function of particle concentration. A sharp brittle-to-ductile transition for PP-2 occurred at the particle loading of 25 wt%. At particle loadings above 25 wt%, the impact strength dropped slightly and gradually. Interestingly, the brittle-to-ductile transition was not distinct for PP-1. It would be considered to be between 5 and 10 wt%, if indeed there was one. The impact strength of PP-1 increased steadily and significantly with the particle concentration from 10 to 30 wt%.

Annealing was reported to have a profound effect on the impact strength of PP/$CaCO_3$ nanocomposites, as shown in Fig. 3.26 (Lin *et al.*, 2008, 2011c). The impact bars were annealed at 155 °C for 3 h prior to preparation of the notch. The impact

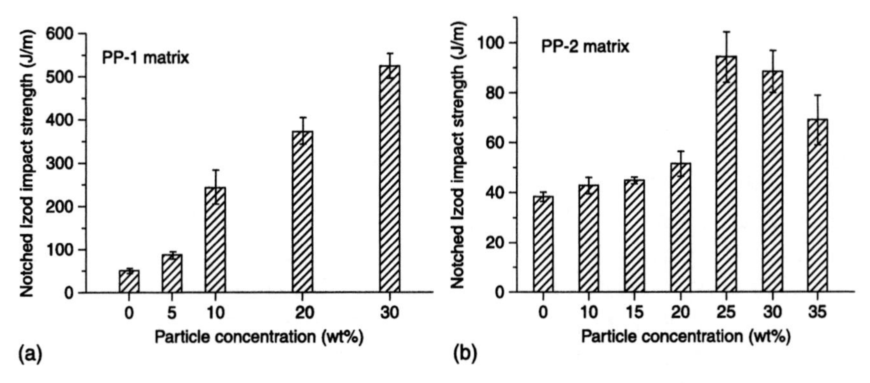


*3.25* Effect of particle concentration on the impact strength of PP/$CaCO_3$ nanocomposites with different PP matrices: (a) PP-1 with $\bar{M}_w = 472\,000$ g/mol. (b) PP-2 with $\bar{M}_w = 346\,000$ g/mol.

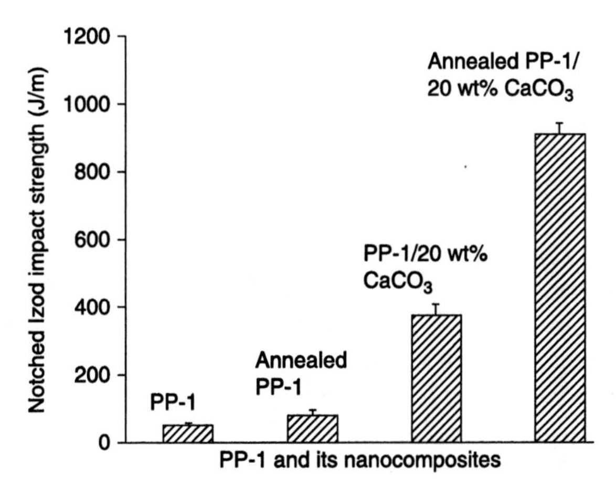


*3.26* Effect of annealing on the impact toughness of PP-1/$CaCO_3$ nanocomposites (Lin *et al.*, 2011c).

strength of PP increased only moderately on annealing. In contrast, annealing remarkably increased the impact strength of the PP/$CaCO_3$ nanocomposite from 370 to 900 J/m. The annealing-induced toughening mechanism is due to the large cavities present around the nanoparticles at the beginning of plastic deformation.

## 3.3 The toughening mechanisms of polymers filled with precipitated calcium carbonate (PCC) nanoparticles

PCC particles have been successfully used to improve the impact strength of polymers, such as PVC, PP, and HDPE. Many researchers have investigated the toughening mechanisms of rigid-particulate-toughened polymers (Thio *et al.*, 2002; Galeski, 2003; Chan *et al.*, 2002; Zuiderduin *et al.*, 2003; Fu *et al.*, 2008). The brittle-to-ductile transition of polymers is susceptible to high temperature and a low strain rate. Both the brittle and ductile responses, i.e., the toughness of polymers, are controlled by the competition between plastic resistance (or tensile yield stress) and brittle fracture stress. The tensile yield stress has a strong dependence on temperature and strain rate, while the brittle fracture stress, which is governed by microstructural flaws or extrinsic imperfections, is basically independent of temperature and strain rate. While it is not always technologically feasible or profitable to produce polymers with a number of flaws and a low degree of imperfection, one of the most practical and feasible ways to increase the toughness of a polymer is to decrease the plastic resistance (i.e. the tensile yield stress). Once the plastic response is initiated, amelioration of some of the microstructural imperfections will often result and lead eventually to molecular alignment or texture development, which can significantly elevate the strength of the polymers along the extension direction (Argon and Cohen, 2003).

The most effective way to reduce plastic resistance is through the incorporation of rubber particles, which produce cavitation in a polymer, or the incorporation of rigid particles, which can debond from the polymer matrix prior to plastic flow. In both approaches, the continuous homopolymer is transformed into thin ligaments, which are much more likely to undergo large strain plastic flow, and are less prone to catastrophic crack propagation. Rigid particulates outperform rubber particles in that rigid particles can simultaneously increase the stiffness of the polymer and extend its service to low temperatures. The size of rigid fillers is critical to the toughening effect. GCC microparticles usually cannot increase the impact toughness of polymer composites because large particles will trigger sharp macro-cracks and consequently initiate brittle-crack propagation. Large aggregates behave similarly to GCC microparticles and result in brittle failure. Therefore, good dispersion of PCC nanoparticles is of primary importance.

The fracture process and toughening mechanisms of PVC, HDPE, and iPP filled with PCC nanoparticles are basically the same. We will use iPP/$CaCO_3$ nanocomposites as an example to illustrate the toughening mechanisms of nanocomposites. PCC nanoparticles, which are often coated with a fatty acid, usually have weak interaction with the polymer matrix, resulting in easy debonding between the PCC nanoparticles and the matrix at the very beginning of deformation. These nanovoids transform the continuous matrix into thin ligaments and therefore release the plastic constraint and change the stress state of the polymer matrix (from the plane-strain to the plane-stress condition), allowing a large plastic deformation of the matrix. Zuiderduin *et al.* (2003) presented the deformation mechanism of iPP/$CaCO_3$ nanocomposites using the concept of three-stage deformation proposed by Kim and Michler (1998a, 1998b). The concept is shown schematically in Fig. 3.27.

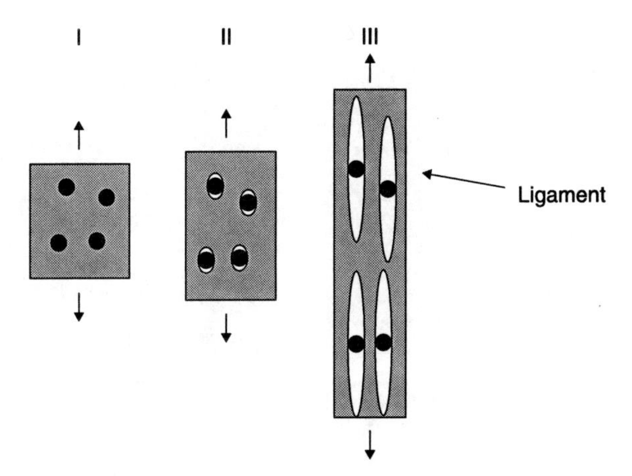


*3.27* Toughening mechanisms of polymers with rigid particles (Zuiderduin *et al.*, 2003).

The deformation process consists of three steps:

I Stress concentration: Because the inorganic particles have high rigidity, they cannot deform compliantly with the matrix and act as stress concentrators.

II Debonding: The concentration of stress causes a build-up of triaxial stresses around the filler particles, leading to debonding at the particle–polymer interface. Ligaments are formed as a result of the debonding.

III Shear yielding: The voids resulting from the debonding alter the stress state of the host matrix polymer surrounding the voids. This reduces the sensitivity towards crazing since the volume strain is released. The shear yielding deformation becomes operative and the material is able to absorb large quantities of energy on fracture.

This toughening mechanism is strongly supported by SEM observations of impact-fractured PP-1/$CaCO_3$ nanocomposites (Lin *et al.*, 2010), which show an extensive plastic deformation zone. The SEM micrographs (Fig. 3.28) are cross sections underneath the fracture surfaces of iPP/$CaCO_3$ nanocomposites. A schematic picture showing the location of sampling with respect to the impact bar is at the left of Fig. 3.28(a). Letters b, c, and d in Fig. 3.28(a) show the locations from where Figures 3.28(b), (c) and (d) were taken, respectively. The arrows indicate the crack-propagation direction. The morphology of Fig. 3.28(b) appeared to be the result of melt flow of the polymer, possibly due to the adiabatic impact process. As revealed by the highly fibrillated ligaments (Fig. 3.28(c)), extensive plastic deformation was clear in the iPP/$CaCO_3$ nanocomposites. Debonding was also found in the deeper layers (Fig. 3.28(d)). The plastic deformation zone of this fractured sample was at least 40 μm beneath the fracture surface and the length of the plastic deformation zone was over 180 μm, as indicated in Fig. 3.28(a).

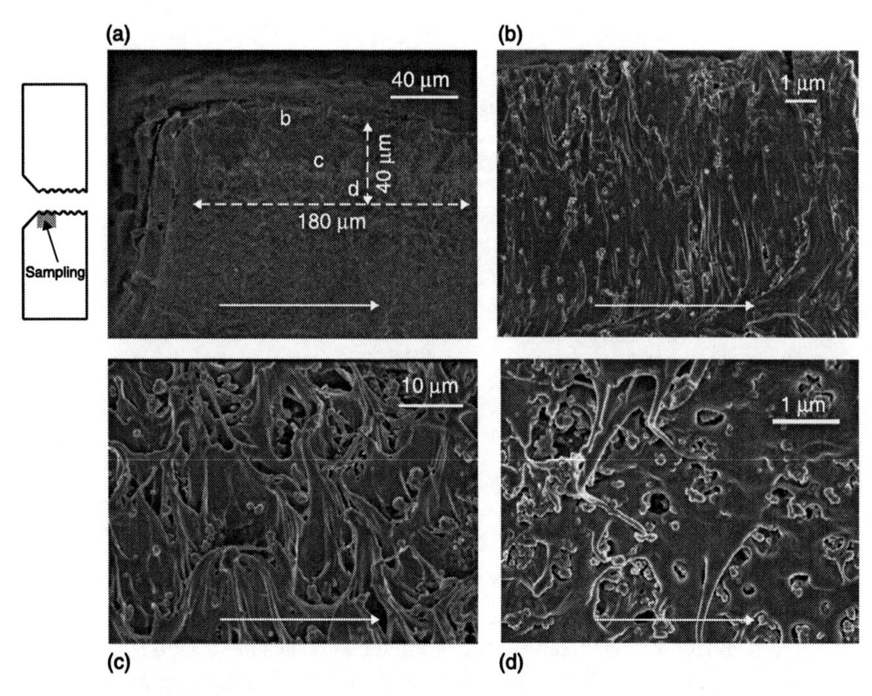


*3.28* SEM micrographs of cross sections underneath the fracture surfaces of impact-fractured PP-1/20 wt% $CaCO_3$ nanocomposites; (b), (c) and (d) are close-up views of the areas indicated in (a).

The micro-deformation of an iPP/$CaCO_3$ nanocomposite during impact loading was deduced from observations made on an arrested crack tip, the morphology of which is well preserved at the crack-initiation stage, as shown in Fig. 3.29 (Lin *et al.*, 2008). Debonding first occurred in response to an impact load, followed by shearing and stretching of the ligaments when their plastic constraint was sufficiently released. A collective failure of ligaments at the notch root was responsible for the formation of a macro-crack, which was immediately followed by the catastrophic fracture of the nanocomposite. Strong ligaments, which have high fracture stresses, were suspected to be good for stabilizing the crack-initiation process and hence promoting energy dissipation.

In another study, Lin *et al.* (2010) reported that strong matrix ligaments could be achieved using a high-MW PP matrix. As shown in Fig. 3.24, given the same particle concentration, the toughening effect is more profound for a high-MW iPP matrix. As discussed earlier, the toughness of a polymer can be improved by either decreasing the plastic resistance or increasing the brittle fracture stress, which is controlled by the microstructure of the polymer matrix. An increase in the molecular weight increases the brittle fracture stress. Ishikawa *et al.* (1996) showed that the strength of oriented iPP fibrils increased with molecular weight.

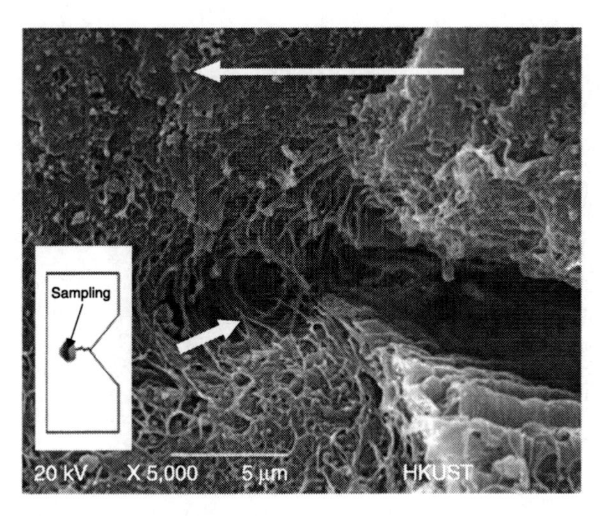


*3.29* SEM micrograph of an arrested crack tip in an annealed nanocomposite (Lin *et al.*, 2008).

They attributed the high fracture stress of a high-MW iPP to the high concentration of tie molecules.

In iPP/$CaCO_3$ nanocomposites, the plastic constraint is reduced by the addition of PCC nanoparticles. However, the extent to which the reduction of the plastic constraint is dependent on the thickness of the polymer-matrix ligaments, i.e., the interparticle distance, can be estimated from the particle concentration using the Wu model (Wu, 1988). A threshold value of particle concentration was observed for the effective toughening of iPP (Lin *et al.*, 2011b), indicating a critical ligament thickness below which the tensile yield stress is reduced to below the fracture stress of the ligaments, resulting in plastic deformation. The critical ligament thickness actually corresponds to the plane-strain to plane-stress transition of polymer-matrix ligaments under the impact testing conditions, leading to significantly enhanced toughness (Fernando and Williams, 1980). It was noted that the critical ligament thickness depends on the polymer matrix properties. The critical particle concentration occurred at a lower particle concentration with a higher-MW PP matrix. This can be explained by the competition between the tensile yield stress and brittle stress.

Another feasible way to reduce the plastic constraint is to introduce large nanocavities in the polymer matrix (Dasari *et al.*, 2010). Submicron voids can be created by directly adding polyoxyethylene nonylphenol (Dasari *et al.*, 2010) or can be initiated by annealing. The extraordinary effect of annealing on the impact toughness of iPP/$CaCO_3$ nanocomposites is demonstrated in Fig. 3.26. Annealing reinforces crystals, leading to stronger mechanical contrast between the amorphous and crystalline domains. Because stresses are more concentrated in the amorphous regions, cavitation is mostly likely to occur in annealed nanocomposites. A large

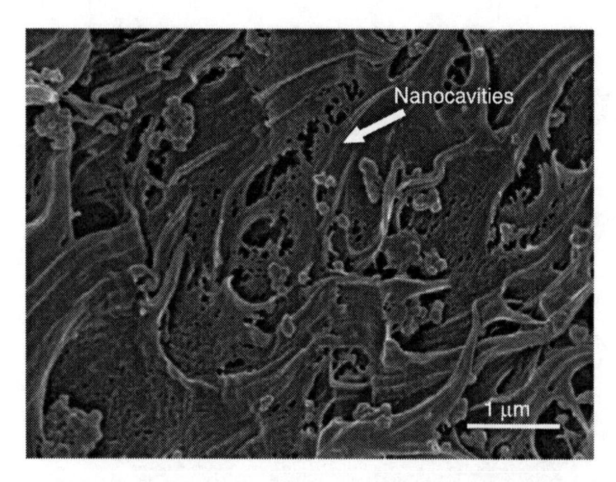


*3.30* SEM micrograph of a cross section of a broken Izod 155°C-annealed PP-1/$CaCO_3$ sample. The sampling location was just underneath the fracture surface (Lin *et al.*, 2011c).

number of nanocavities were found to have formed immediately after debonding but prior to intensive ligament stretching, as shown in Fig. 3.30 (Lin *et al.*, 2011c). The creation of a large number of nanocavities was beneficial to the release of the plastic constraint, which was rather severe during impact loading. As a result, there was considerable plastic deformation due to the combined effects of debonding and cavitation. Consequently, the impact toughness of annealed iPP/$CaCO_3$ nanocomposites was greatly improved.

Plastic deformation takes place only when the fracture stress of a polymer matrix is sufficiently higher than its tensile yield stress. Any modifications introduced into the iPP matrix have to satisfy the condition that the plastic constraint is low enough to ensure the activation of shearing of ligaments before ultimate fracture. The plastic constraint can be reduced by transforming the polymer matrix into thin ligaments through the debonding of nanoparticles or the creation of a large number of nanocavities. A high fracture stress of the ligaments helps to stabilize the crack-initiation stage and allows for the expansion of the plastic deformation zone. Ligaments with high fracture stress can be realized with a high-MW polymer matrix. The combination of high fracture stress and low tensile yield stress helps to enlarge the plastic deformation zone and therefore improve the impact toughness of the polymer.

## 3.4 Conclusions

This paper presented an overview of the current developments in calcium carbonate polymer nanocomposites. PCC particles and nanoparticles have been shown to be very useful for reducing the costs of the finished products and for reinforcing polymers, such as PVC, LDPE, HDPE, and iPP. The surface coatings of commercially available PCC nanoparticles are usually less than a monolayer. This significantly limits the use of PCC nanoparticles as reinforcing fillers in polymers because it is difficult to achieve good dispersion even with dedicated mixing equipment. Nanocomposites with a high impact toughness and tensile modulus have been successfully prepared using monolayer-coated PCC nanoparticles. New coating materials, which can control the bonding strength between the nanoparticles and various polymers, and which can be dispersed uniformly in various polymers, should be developed.

## 3.5 References

Argon, A.S. and Cohen, R.E. (2003) Toughenability of polymers. *Polymer* 44 (19), pp. 6013–6032.

Bartczak, Z., Argon, A.S., Cohen, R.E. and Weinberg, M. (1999) Toughness mechanism in semi-crystalline polymer blends: II. High-density polyethylene toughened with calcium carbonate filler particles. *Polymer* 40 (9), pp. 2347–2365.

Binsbergen, F. L. and de Lange, B.G.M. (1968) Morphology of polypropylene crystallized from the melt. *Polymer* 9, pp. 23–40.

Bryant, W.S. and Wiebking, H.E. (2002) The effect of calcium carbonate size and loading level on the impact performance of rigid PVC compounds containing varying amounts of acrylic impact modifier, *ANTEC conference proceeding* Vol. 3.

Chan, C.-M., Wu, J., Li, J.-X. and Cheung, Y.-K. (2002) Polypropylene/calcium carbonate nanocomposites. *Polymer* 43 (10), pp. 2981–2992.

Cornwell, D.W. (2001) Plastic performance: Benefits of PCC as a PVC additive. *Industrial Minerals*, pp. 35–37.

Dasari, A., Zhang, Q.-X., Yu, Z.-Z. and Mai, Y.-W. (2010) Toughening polypropylene and its nanocomposites with submicrometer voids. *Macromolecules* 43 (13), pp. 5734–5739.

Fekete, E., Pukanszky, B., Toth, A. and Bertoti, I. (1990) Surface modification and characterization of particulate mineral fillers. *Journal of Colloid Interface Science* 135 (1), pp. 200–208.

Fekete, E., Móczó, J. and Pukánszky, B. (2004) Determination of the surface characteristics of particulate fillers by inverse gas chromatography at infinite dilution: A critical approach. *Journal of Colloid and Interface Science* 269 (1), pp. 143–152.

Fernando, N.A.S. and Thomas, N.L. (2007) Effect of precipitated calcium carbonate on the mechanical properties of poly (vinyl chloride). *Journal of Vinyl and Additive Technology* 13 (2), pp. 98–102.

Fernando, P.L. and Williams, J.G. (1980) Plane stress and plane strain fractures in polypropylene. *Polymer Engineering and Science* 20 (3), pp. 215–220.

Fu, Q. and Wang, G. (1993) Effect of morphology on brittle-ductile transition of HDPE/$CaCO_3$ blends. *Journal of Applied Polymer Science* 49 (11), pp. 1985–1988.

Fu, Q., Wang, G. and Shen, J. (1993) Polyethylene toughened by $CaCO_3$ particle: Brittle-ductile transition of $CaCO_3$-toughened HDPE. *Journal of Applied Polymer Science* 49 (4), pp. 673–677.

Fu, S.-Y., Feng, X.-Q., Lauke, B. and Mai, Y.-W. (2008) Effects of particle size, particle/matrix interface adhesion and particle loading on mechanical properties of particulate-polymer composites. *Composites Part B: Engineering* 39 (6), pp. 933–961.

Galeski, A. (2003) Strength and toughness of crystalline polymer systems. *Progress in Polymer Science* (Oxford) 28 (12), pp. 1643–1699.

Garnier, P., Gregoire, P., Montmitonnet, P. and Delamare, F. (1988) Polymorphism of crystalline phases of calcium stearate. *Journal of Materials Science* 23 (9), pp. 3225–3231.

Hess, P. (2001) Plastics. In: Tegethoff, F.W. ed. *Calcium Carbonate from the Cretaceous Period into the 21st Century*. Basel, Boston, Berlin: Birkhäuser Verlag, pp. 238–259.

Huwald, E. (2001) Calcium carbonate – Pigment and filler. In: Tegethoff, F.W. ed. *Calcium Carbonate from the Cretaceous Period into the 21st Century*. Basel, Boston, Berlin: Birkhäuser Verlag, pp. 160–193.

Ishikawa, M., Ushui, K., Kondo, Y., Katada, K. and Gima, S. (1996) Effect of tie molecules on the craze strength of polypropylene. *Polymer* 37 (24), pp. 5375–5379.

Kausch, H.H. and Michler, G.H. (2007) Effects of nanoparticle size and size-distribution on mechanical behavior of filled amorphous thermoplastic polymers. *Journal of Applied Polymer Science* 105 (5), pp. 2577–2587.

Kawai, T., Umemura, J., Takenaka, T., Kodama, M. and Seki, S. (1985) Fourier transform infrared study on the phase transitions of an octadecyltrimethylammonium chloride-water system. *Journal of Colloid and Interface Science* 103 (1), pp. 56–61.

Kemal, I., Whittle, A., Burford, R., Vodenitcharova, T. and Hoffman, M. (2009) Toughening of unmodified polyvinylchloride through the addition of nanoparticulate calcium carbonate. *Polymer* 50 (16), pp. 4066–4079.

Kew, S.J. and Hall, E.A.H. (2006) Structural effect of polymerisation and dehydration on bolaamphiphilic polydiacetylene assemblies. *Journal of Materials Chemistry* 16 (21), pp. 2039–2047.

Kim, G.M., Michler, G.H., Gahleitner, M. and Fiebig, J. (1996) Relationship between morphology and micromechanical toughening mechanisms in modified polypropylenes. *Journal of Applied Polymer Science* 60 (9), pp. 1391–1403.

Kim, G.M. and Michler, G.H. (1998a) Micromechanical deformation processes in toughened and particle-filled semicrystalline polymers: Part 1. Characterization of deformation processes in dependence on phase morphology. *Polymer* 39 (23), pp. 5689–5697.

Kim, G.M. and Michler, G.H. (1998b) Micromechanical deformation processes in toughened and particle filled semicrystalline polymers. Part 2: Model representation for micromechanical deformation processes. *Polymer* 39 (23), pp. 5699–5703.

Kiss, A., Fekete, E. and Pukánszky, B. (2007) Aggregation of $CaCO_3$ particles in PP composites: Effect of surface coating. *Composites Science and Technology* 67 (7–8), pp. 1574–1583.

Lazzeri, A., Zebarjad, S.M., Pracella, M., Cavalier, K. and Rosa, R. (2005) Filler toughening of plastics. Part 1 – The effect of surface interactions on physico-mechanical properties and rheological behaviour of ultrafine $CaCO_3$/HDPE nanocomposites. *Polymer* 46 (3), pp. 827–844.

Liauw, C.M., Lees, G.C., Hurst, S.J., Rothon, R.N. and Dobson, D.C. (1995) Investigation of the surface modification of aluminium hydroxide filler and optimum modifier dosage level. *Plastic Rubber and Composites Processing and Applications* 24 (4), pp. 211–219.

Lin, Y., Chen, H.B., Chan, C.-M. and Wu, J.S. (2008) High impact toughness polypropylene/$CaCO_3$ nanocomposites and the toughening mechanism. *Macromolecules* 41 (23), pp. 9204–9213.

Lin, Y. (2009) *Toughening mechanism of polypropylene/Calcium carbonate nanocomposites*. Thesis (PhD), the Hong Kong University of Science and Technology.

Lin, Y., Chen, H.B., Chan, C.-M. and Wu, J.S. (2010) The toughening mechanism of polypropylene/calcium carbonate nanocomposites. *Polymer* 51 (14), pp. 3277–3284.

Lin, Y., Chen, H.B., Chan, C.-M. and Wu, J.S. (2011a) Nucleating effect of calcium stearate coated $CaCO_3$ nanoparticles on polypropylene. *Journal of Colloid and Interface Science* 345, pp. 570–576.

Lin, Y., Chen, H.B., Chan, C.-M. and Wu, J.S. (2011b) Effects of coating amount and particle concentration on the impact toughness of polypropylene/$CaCO_3$ nanocomposites. *European Polymer Journal* 47, pp. 294–304.

Lin, Y., Chen, H.B., Chan, C.-M. and Wu, J.S. (2011c) Annealing-induced toughening mechanism of polypropylene/$CaCO_3$ nanocomposites. *Journal Applied Polymer Science* (in print).

Liu, Z.H., Kwok, K.W., Li, R.K.Y. and Choy, C.L. (2002) Effects of coupling agent and morphology on the impact strength of high density polyethylene/$CaCO_3$ composites. *Polymer* 43 (8), pp. 2501–2506.

Liu, Z.H., Zhu, X.G., Li, Q., Qi, Z.N. and Wang, F.S. (1998) Effect of morphology on the brittle ductile transition of polymer blends: 5. The role of $CaCO_3$ particle size distribution in high density polyethylene/$CaCO_3$ composites. *Polymer* 39 (10), pp. 1863–1868.

Móczó, J., Fekete, E. and Pukánszky, B. (2004) Adsorption of surfactants on $CaCO_3$ and its effect on surface free energy. *Progress in Colloid and Polymer Science* 125, pp. 134–141.

Osman, M.A. and Suter, U.W. (2002) Surface treatment of calcite with fatty acids: Structure and properties of the organic monolayer. *Chemistry of Materials* 14 (10), pp. 4408–4415.

Padden, F. J. and Keith, H.D. (1966) Crystallization in thin films of isotactic polypropylene. *Journal of Applied Physics* 37(11), pp. 4013–4020.

Sahebian, S., Zebarjad, S.M., Sajjadi, S.A., Sherafat, Z. and Lazzeri, A. (2007) Effect of both uncoated and coated calcium carbonate on fracture toughness of HDPE/$CaCO_3$ nanocomposites. *Journal of Applied Polymer Science* 104 (6), pp. 3688–3694.

Sakai, H. and Umemura, J. (1993) Infrared external reflection spectra of Langmuir films of stearic-acid and cadmium stearate. *Chemistry Letters* 12, pp. 2167–2170.

Sakai, H. and Umemura, J. (2008) Evaluation of structural change during surface pressure relaxation in Langmuir monolayer of zinc stearate by infrared external reflection spectroscopy. *Colloid and Polymer Science* 286 (14–15), pp. 1637–1641.

Schofield, W.C.E., Hurst, S.J., Lees, G.C., Liauw, C.M. and Rothon, R.N. (1998) Influence of surface modification of magnesium hydroxide on the processing and mechanical properties of composites of magnesium hydroxide and an ethylene vinyl acetate copolymer. *Composite Interfaces* 5 (6), pp. 515–528.

Shanghai Perfection Nanometre New Material Co., Ltd (2010) Available from: http://www.nano-metre.com/EN/cpjj.htm [Accessed 10 October, 2010].

Shanghai Yaohua Nano-tech Co., Ltd (2010) Available from: http://www.caco3-yhnano.com/html/products-application.html [Accessed 10 October, 2010].

Shi, X.T., Rosa, R. and Lazzeri, A. (2010) On the coating of precipitated calcium carbonate with stearic acid in aqueous medium. *Langmuir* 26 (11), pp. 8474–8482.

Singh, S., Wegmann, J., Albert, K. and Muller, K. (2002) Variable temperature FT-IR studies of n-alkyl modified silica gels. *Journal of Physical Chemistry B* 106 (4), pp. 878–888.

Sun S., Li, C.Z., Zhang, L., Du, H.L. and Burnell-Gray, J.S. (2006) Interfacial structures and mechanical properties of PVC composites reinforced by $CaCO_3$ with different particle size and surface treatments. *Polymer International* 55 (2), 158–164.

Sutherland, I., Maton, D. and Harrison, D.L. (1998) Filler surfaces and composite properties. *Composite Interfaces* 5 (6), pp. 493–502.

Thio, Y.S., Argon, A.S., Cohen, R.E. and Weinberg, M. (2002) Toughening of isotactic polypropylene with $CaCO_3$ particles. *Polymer* 43 (13), pp. 3661–3674.

Vold, M.J., Hattiangdi, G.S., and Vold, R.D. (1949) Crystal forms of anhydrous calcium stearate derivable from calcium stearate monohydrate. *Journal of Colloid Science* 4 (2), pp. 93–101.

Wang, G., Chen, X.Y., Huang, R. and Zhang, L. (2002) Nano-$CaCO_3$/polypropylene composites made with ultra-high-speed mixer. *Journal of Materials Science Letters* 21 (13), pp. 985–986.

Wu, D., Wang, X., Song, Y. and Jin, R. (2004) Nanocomposites of poly(vinyl chloride) and nanometric calcium carbonate particles: Effects of chlorinated polyethylene on mechanical properties, morphology, and rheology. *Journal of Applied Polymer Science* 92 (4), pp. 2714–2723.

Wu, S. (1985) Phase structure and adhesion in polymer blends: A criterion for rubber toughening. *Polymer* 26, pp. 1855–1863.

Wu, S.H. (1988) A generalized criterion for rubber toughening: The critical matrix ligament thickness. *Journal of Applied Polymer Science* 35(2), pp. 549–561.

Xie, X.L., Liu, Q.X., Li, R.K.Y., Zhou, X.P., Zhang, Q.X. *et al.* (2004) Rheological and mechanical properties of PVC/$CaCO_3$ nanocomposites prepared by *in situ* polymerization. *Polymer* 45 (19), pp. 6665–6673.

Yang, K., Yang, Q., Li, G., Sun, Y. and Feng, D. (2006) Morphology and mechanical properties of polypropylene/calcium carbonate nanocomposites. *Materials Letters* 60 (6), pp. 805–809.

Yuan, Q., Shah, J.S., Bertrand, K.J. and Misra, R.D.K. (2009) On processing and impact deformation behavior of high density polyethylene (HDPE)-calcium carbonate nanocomposites. *Macromolecular Materials and Engineering* 294 (2), pp. 141–151.

Zhang, Q.X., Yu, Z.Z., Xie, X.L. and Mai, Y.W. (2004) Crystallization and impact energy of polypropylene/$CaCO_3$ nanocomposites with nonionic modifier. *Polymer* 45 (17), pp. 5985–5994.

Zuiderduin, W.C.J., Westzaan, C., Huetink, J. and Gaymans, R.J. (2003) Toughening of polypropylene with calcium carbonate particles. *Polymer* 44 (1), pp. 261–275.

# 4

# Layered double hydroxides (LDHs) as functional fillers in polymer nanocomposites

J. I. VELASCO, M. ARDANUY and M. ANTUNES,
Universitat Politècnica de Catalunya, Spain

**Abstract:** Layered double hydroxides (LDHs) are lamellar inorganic solids with a brucite-like structure similar to hydrotalcite, where the partial substitution of trivalent for divalent cations results in a positive sheet charge, compensated by anions situated in the interlayer galleries. Their layered structure enables the intercalation of polymer macromolecules, in some cases resulting in the exfoliation of their sheets, and hence in the formation of polymer nanocomposites. This chapter reviews the state-of-the-art preparation of hybrid LDHs and incorporation into polymers, the structure of these nanocomposites and the resulting properties: rheological, mechanical, and thermal. Particular attention is given to their possible applications and future trends.

**Key words:** layered double hydroxides, hybrid LDH, nanocomposites.

## 4.1 Introduction: the role of layered double hydroxides (LDHs) as reinforcements

Layered double hydroxides (LDHs) are a family of lamellar compounds containing exchangeable anions in the interlayer space. For this reason, they are also known as anionic clays. Their structure consists of brucite-like sheets with a typical thickness of 0.5 nm, in which partial substitution of trivalent for divalent metallic ions results in a positive charge, compensated by anions situated within the interlayer space (see Fig. 4.1). The general formula of LDHs is $(M^{2+}_{1-x} M^{3+}_{x} (OH)_2)(A^{n-}_{x/n} \cdot mH_2O)$, where $M^{2+}$ and $M^{3+}$ are divalent and trivalent metal cations, respectively, which occupy octahedral positions in the hydroxide layers, and $A^{n-}$ is an interlayer anion. The layers are based on octahedral $M(OH)_6$ units, where the metal (M) is coordinated by six hydroxyl groups (OH), hence forming $M(OH)_2$ brucite-like sheets. As shown in Fig. 4.1, the centers of the octahedral units are occupied by divalent and trivalent metal cations and the vertices by the hydroxyl anions (Rives, 2001, 2002).

Typically, the metal ions have ionic radii similar to $Mg^{2+}$ (0.65 Å). Combinations of divalent cations like $Mg^{2+}$, $Ni^{2+}$, $Zn^{2+}$, $Co^{2+}$, $Cu^{2+}$, or $Cd^{2+}$, and trivalent cations such as $Al^{3+}$, $Cr^{3+}$, $Fe^{3+}$, $Mn^{3+}$, or $Ga^{3+}$, are usual. The most common interlayer anions are inorganic in nature, such as carbonates, chlorides, nitrates, and sulfates. The most used LDH is hydrotalcite, where the constituents of the layers are magnesium and aluminum and the interlayer anions are mainly carbonates. Its general formula is $(Mg^{2+}_{1-x} Al^{3+}_{x} (OH)_2)(CO_3^{2-}{}_{x/n} \cdot mH_2O)$ $(0.2 \leq x \leq 0.33)$ (Rives, 2002).

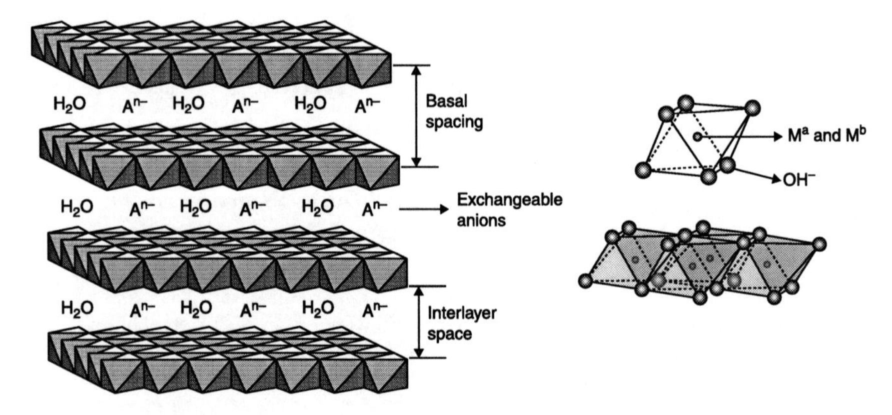


*4.1* General structure of LDHs (left) and detail of the octahedral sheets (right).

In order to prepare hybrid LDHs, the interlayer anions are exchanged by organic ones with a larger volume. In the literature, several works have considered the preparation of a great variety of LDH hybrids. Generally speaking, hydrocarbonated sulfate and sulfonate anions lead to a high expanded crystalline structure along the *c* axis, i.e., with a higher interlayer space, as illustrated in Fig. 4.2. Specifically, LDH hybrids have been made with expanded interlayer spaces from 21.1 Å ($C_8$) to 32.6 Å ($C_{18}$) (Newman and Jones, 1998; Trujillano *et al.*, 2002).

The role of LDHs as functional or structural fillers in polymers lies in combining, in an adequate manner, different aspects of their composition, dispersion efficiency, distribution, and orientation.

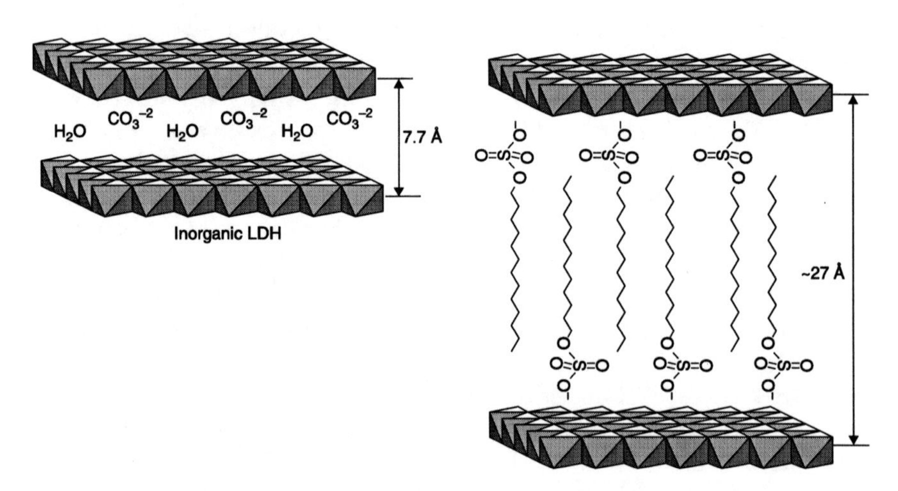


*4.2* Formation of an LDH hybrid by anion exchange between carbonate and dodecyl sulfate anions.

As expected when preparing a nanocomposite, LDH concentrations should be rather low. Typical LDH weight concentrations are below 10 wt.%, of which an important part (around 50%) corresponds to the organic part of the prepared hybrid.

Concerning the nature of the metal cations, although most of the studies use LDHs containing MgAl or ZnAl, other cations can promote interesting properties. For instance, Manzi-Nshuti *et al.* (2008b) studied the effect of NiAl–LDH ZnAl and CoAl–LDH on the mechanical properties and flame retardancy of poly(methyl methacrylate) (PMMA), and found high efficiency with these LDHs; Peng *et al.* (2009) used CoAl–LDH in polyamide 6 (PA6) to prepare PA6 nanocomposites. Wang *et al.* (2008) found that cobalt chelate or the respective compound showed better flame retardancy on polyethylene than other types of metal chelates or metal chelate compounds.

The interlayer anions play a double role. On one hand, they contribute in expanding the layered structure of the LDHs, increasing their interlayer distance; on the other hand, they promote compatibilization with the polymer matrix. A careful selection of the interlayer anion may allow the preparation of hybrids, which, depending on the nature of the selected anion, may add functionality to the nanocomposite.

The dispersion of the LDH nanoparticles in the polymer matrix is mainly done through dissolution or melt-blending. In order to achieve a high dispersion and hence a high efficiency of the LDH nanoparticles, the local strains that have to be applied during this step must exceed the cohesive forces that keep the hybrid platelets together. Of these two methods, melt-blending is the less effective due to the high viscosity of the matrix at typical processing temperatures. High dispersive mixing devices are required, although the degree of dispersion is not optimal. Generally speaking, the morphology is a combination of crystalline aggregates (known as *tactoids*), groups of a couple of platelets, and individual platelets. Due to the expanded interlayer structure of hybrid LDHs, as well as its compatibility, polymer molecular intercalation is favored, although full exfoliation is rather difficult to attain. LDH layer morphology induces a high anisotropy in the prepared nanocomposites. During molding, the platelets or aggregates tend to orientate preferentially along the flow direction. Likewise, a certain inhomogeneity is commonly observed in terms of particle distribution.

The combination of all these factors plays a key role in the use of LDHs as functional or structural fillers in polymer nanocomposites.

## 4.2 Preparation of hybrid LDHs for polymer nanocomposites

The preparation of hybrid LDHs is commonly made by one of the following three routes (Newman and Jones, 1998): direct synthesis, anionic exchange, and reconstitution (rehydration).

### 4.2.1 Direct synthesis

Direct synthesis by co-precipitation is the most common method used to prepare LDHs with inorganic anions and it can be used to synthesize hybrid LDHs. This method consists of the addition of the desired anion to a solution containing metal salts of the ions that will form the layers (usually chloride or nitrate salts). Control of the pH prevents co-precipitation of other phases, such as oxide impurities from the metals. To avoid the incorporation of metal salt anions used in the synthesis, it is important that the organic anions have a good affinity for the hydroxide layers. The main advantages of this method are the control of the charge density of the layers ($M^{2+}$ to $M^{3+}$ ratio) and the high purity of the synthesized LDHs.

As can be seen from Table 4.1, a wide variety of surfactant anions have been used in this method, leading to variable results in terms of structural modification efficiency. For instance, in the case of MgAl–LDHs, the best results concerning the interlayer distance have been found for hybrids with stearate anions (Liu *et al.*, 2008) (3.37 nm interlayer distance), dodecylbenzenesulfonate (Zammarano *et al.*, 2006; Wang *et al.*, 2009) (interlayer distances from 2.76 to 3.15 nm) and dodecyl sulfate (Du *et al.*, 2007; Kuila *et al.*, 2007; Zhang *et al.*, 2007; Martínez-Gallegos *et al.*, 2008; Bao *et al.*, 2006, 2008b; Liu *et al.*, 2007, 2008; Kotal *et al.*, 2009) (interlayer distances between 2.59 and 2.75 nm). In contrast, no significant effects were observed when using borate-like anions, dimethyl 5-sulfoisophthalate, benzoate, or adipic acid, with interlayer distances below 1.6 nm.

With ZnAl–LDH, significant increases of the interlayer distance were observed using oleate (Manzi-Nshuti *et al.*, 2009b) (3.96 nm), 4-(12-(methacryloylamino) dodecanoylamino)benzenesulfonate (Illaik *et al.*, 2008; Leroux *et al.*, 2009) (3.75 nm), and dodecyl sulfate (Zhang *et al.*, 2008a; Chen *et al.*, 2004b) (between 2.54 and 2.85 nm). Borate (Nyambo *et al.*, 2009a) and benzoate (He *et al.*, 2006) anions produced the least remarkable results of all works presented in Table 4.1.

### 4.2.2 Anionic exchange

This procedure is very simple and consists of the dispersion of an LDH precursor in a solution containing an excess of the organic anions to be incorporated. Prior to choosing the LDH precursor, it is important to take into account the affinity of the anions with the LDH structure. Hence, the biggest difficulty lies in the carbonate anions, due to the high affinity of the layers for small divalent anions, followed by sulfate anions. On the other hand, among the most frequent monovalent anions present in LDHs, hydroxide shows the highest interchange difficulty, followed by chloride, fluoride, bromine, nitrate, and iodide (Miyata and Kumura, 1973).

*Table 4.1* Basal spacing of various LDH hybrids prepared by direct synthesis

| Organic anion | LDH type | Basal spacing (nm) | Reference |
|---|---|---|---|
| Adipic acid | MgAl–LDH | 1.40 | Herrero *et al.*, 2010 |
| Dodecyl sulfate | MgAl–LDH | 2.75 | Du *et al.*, 2007 |
| | | 2.61 | Kuila *et al.*, 2007 |
| | | 2.75 | Zhang *et al.*, 2007 |
| | | 2.60 | Martínez-Gallegos *et al.*, 2008 |
| | | 2.73 | Bao *et al.*, 2006 |
| | | 2.72 | Liu *et al.*, 2007, 2008 |
| | | 2.73 | Bao *et al.*, 2008b |
| | | 2.59 | Kotal *et al.*, 2009 |
| | ZnAl–LDH | 2.54 | Chen *et al.*, 2004b |
| | | 2.85 | Zhang *et al.*, 2008a |
| 4-dodecylbenzenesulfonate | MgAl–LDH | 2.76–3.15 | Zammarano *et al.*, 2006 |
| | | 2.90 | Wang *et al.*, 2009 |
| | CoAl–LDH | 2.91 | Wang *et al.*, 2010 |
| Borate | MgAl–LDH | 1.08 | Nyambo *et al.*, 2009a |
| | ZnAl–LDH | 1.05 | |
| Benzoate | ZnAl–LDH | 1.55 | He *et al.*, 2006 |
| Amino benzoate | MgAl–LDH | 1.50 | Tseng *et al.*, 2007 |
| | | 1.50 | Hsueh and Chen, 2003b |
| Dimethyl 5-sulfoisophthalate | MgAl–LDH | 1.20 | Lee and Im, 2007 |
| 4,4′-azobis(4-cyano-valerate) | ZnAl–LDH | 1.60 | Manzi-Nshuti *et al.*, 2009a |
| | MgAl–LDH | 1.60 | |
| Dye blue molecules | ZnAl–LDH | 2.04 | Marangoni *et al.*, 2008 |
| | | 2.06 | |
| | | 2.34 | |
| 4-(12-(Methacryloylamino) dodecanoylamino) benzenesulfonate | ZnAl–LDH | 3.75 | Illaik *et al.*, 2008<br>Leroux *et al.*, 2009 |
| Dye direct yellow | ZnAl–LDH | 2.65 | Taviot-Gueho *et al.*, 2007 |
| 10-undecanoate | ZnAl–LDH | 2.74–3.10 | Manzi-Nshuti *et al.*, 2008a |
| | CoAl–LDH | 2.70 | Manzi-Nshuti *et al.*, 2008b |
| | CuAl–LDH | 2.76 | |
| | NiAl–LDH | 2.70 | |
| | MgAl–LDH | 2.33 | Wang *et al.*, 2005 |
| Bis(2-ethylhexyl) phosphate | MgAl–LDH | 2.34 | Wang *et al.*, 2009 |
| 2-ethylhexyl sulfate | MgAl–LDH | 2.14 | Wang *et al.*, 2009 |
| Stearate | MgAl–LDH | 3.37 | Liu *et al.*, 2008 |
| Oleate | ZnAl–LDH | 3.96 | Manzi-Nshuti *et al.*, 2009b |
| Laurate | MgAl–LDH | 2.10 | Hsueh and Chen, 2003a |

This procedure was first reported by Miyata and Kumura (1973) and used to prepare an LDH hybrid with acetate anions. Meyn *et al.* (1990) also used ionic exchange to prepare hybrids using different organic anions.

Since then, many researchers have prepared hybrids with a wide variety of organic molecules when making LDH polymer nanocomposites, as shown in Table 4.2.

*Table 4.2* Basal spacing of various LDH hybrids prepared using anionic exchange

| Organic anion | LDH type | Anion precursor | Basal spacing (nm) | Reference |
|---|---|---|---|---|
| Adipic acid | MgAl-LDH | Nitrate | 1.30 | Zhu *et al.*, 2008 |
| Dodecyl sulfate | MgAl-LDH | Nitrate | 2.72 | Chen *et al.*, 2004a<br>Chen and Qu, 2005 |
| | | | 2.75 | Lv *et al.*, 2009b |
| | | Chloride | 2.57 | Lonkar *et al.*, 2009, 2010 |
| | | | 2.49 | Qiu *et al.*, 2006 |
| | ZnAl-LDH | Chloride | 2.57 | Lonkar *et al.*, 2010 |
| | | | 2.58 | Ding and Qu, 2006 |
| | | | 2.75 | Du and Qu, 2006 |
| | | Nitrate | 2.53 | Peng *et al.*, 2010 |
| | MgFe-LDH | Nitrate | 3.11 | Ding *et al.*, 2008 |
| 2-hydroxydodecanoate | MgAl-LDH | Nitrate | 2.27 | Sorrentino *et al.*, 2005<br>Pucciariello *et al.*, 2007<br>Romeo *et al.*, 2007 |
| 2-acrylamido-2-methyl-1-propane sulfonate | ZnAl-LDH<br>CoAl-LDH<br>CoFe-LDH | Nitrate | 1.80<br>1.79<br>1.76 | Leroux *et al.*, 2010 |
| $K_2S_2O_8$ | MgAl-LDH | Nitrate | 0.92 | Qiu and Qu, 2006 |
| 10-undecanoate | MgAl-LDH | Nitrate | 2.20 | Nyambo *et al.*, 2008 |
| Stearate | ZnAl-LDH | Nitrate | 3.05 | Costantino *et al.*, 2005 |
| 3-amino benzeno-sulfonate | | | 1.59 | Zammarano *et al.*, 2005 |
| 3-toluene sulfonate | MgAl-LDH | Nitrate | 1.71 | |
| 4-hydroxy benzeno-sulfonate | | | 1.52 | |
| Polyoxyethylene alkyl propenyl ether ammonium sulfate | LiAl-LDH | Nitrate | 2.42 | Tsai *et al.*, 2008 |
| Poly(oxo-propylene)-amines | MgAl-LDH | Nitrate | 2.59 | Chan *et al.*, 2008 |

### 4.2.3 Hydration and reconstitution from oxides

Miyata (1980) reported that layered double hydroxides, after being transformed to oxides by calcination at temperatures between 500 and 800 °C, are able to rehydrate to their original form in the presence of anions and water, i.e., that LDHs are able to regenerate after calcination. This ability, known as a 'memory effect,' may be used to prepare hybrids using different anions in a rather easy way.

A wide variety of LDHs have been prepared with inorganic anions such as carbonates (Hibino and Tsunashima, 1998), organic-like naphthalene carboxylates (Tagaya *et al.*, 1993), or carboxylates (Dimotakis and Pinnavaia, 1990) using this method. Regardless of the preparation method, the main problem when preparing high-purity hybrids lies in their high affinity for carbonates, which requires specific conditions in order to avoid the presence of this anion. For this reason, it is essential to use water free of carbonates (bi-distilled water is normally used) and an inert atmosphere (nitrogen or other inert gases are conventionally used).

As shown in Table 4.3, this method has been also used for the preparation of hybrids with a wide variety of organic molecules.

There are a few publications that compare the procedures described previously or in which a combination of more than one method is used to prepare a specific hybrid. One of the groundbreaking studies was Ulibarri *et al.* (1994), which

*Table 4.3* Basal spacing of various LDH hybrids prepared using rehydration

| Organic anion | LDH type | Basal spacing (nm) | Reference |
|---|---|---|---|
| Dodecyl sulfate | MgAl-LDH | 2.65 | Ardanuy *et al.*, 2009, 2010 |
| | | 2.75 | Chen and Qu, 2003 |
| | | 2.65 | Acharya *et al.*, 2007a, 2007b |
| | | 2.80 | Zubitur *et al.*, 2009 |
| | | 2.75 | Du and Qu, 2006 |
| | ZnAl-LDH | 2.85 | Zhang *et al.*, 2008a |
| | | 2.75 | Zhang *et al.*, 2008b |
| 4-dodecyl benzene sulfonate | MgAl-LDH | 2.80 | Lee *et al.*, 2006 |
| | | 2.96 | Costa *et al.*, 2005, 2006a, 2006b, 2007 |
| | | 2.99 | Schonhals *et al.*, 2009 |
| | | | Zubitur *et al.*, 2009 |
| Octyl sulfate | MgAl-LDH | 2.26 | Lee *et al.*, 2006 |
| Phenyl phosphonate | MgAl-LDH | 1.38 | Nyambo *et al.*, 2009b |
| Palmitate | MgAl-LDH | 2.85 | Nyambo *et al.*, 2009a |
| Stibnite ($Sb_2S_3$) | MgAl-LDH | 0.86 | Chen *et al.*, 2010 |
| 1-decanesulfonate | MgAl-LDH | 2.27 | Pradhan *et al.*, 2008 |

compared anionic exchange and reconstitution after calcination for preparing LDH hybrids using vanadate anions. Trujillano *et al.* (2002) compared anionic exchange and direct synthesis for preparing nickel and aluminum hybrids with alkyl sulfate or sulfonate anions. Finally, a recent work (Dupin *et al.*, 2004) compared the three methods previously mentioned (anionic exchange, direct synthesis, and reconstitution after calcination) in the preparation of magnesium and aluminum LDHs using dichlorophenyl amino phenyl acetate.

## 4.3 Nanocomposite preparation routes

The successful preparation of nanocomposites with lamellar nano-sized particles depends on the rupture of their primary structure (exfoliation of the layers) or the intercalation of polymer molecules between the platelets (intercalation of the layers), as well as on their homogeneous dispersion in the polymer matrix. Moreover, it is important to achieve good interaction (compatibilization) between the polymer matrix and the layers (Pinnavaia and Beall, 2001).

Several strategies may be used to prepare polymer-layered double hydroxide nanocomposites. The most commonly used are *in situ* polymerization, melt-mixing, and solution blending (Alexandre, 2002).

### 4.3.1 *In situ* polymerization

With *in situ* polymerization, the LDH hybrid is first mixed with a solution of the monomer or with the liquid monomer to allow intercalation into the interlayer gallery. Second, a catalyst is added to the solution and the monomer is polymerized *in situ*. The polymerization of the monomer in the interlayer gallery leads to rupture of the lamellar structure and promotes the homogeneous dispersion of the layers (see Fig. 4.3).

This method was the first reported in the literature to prepare a polymer-clay nanocomposite (Okada *et al.*, 1990). It is conventionally used to prepare

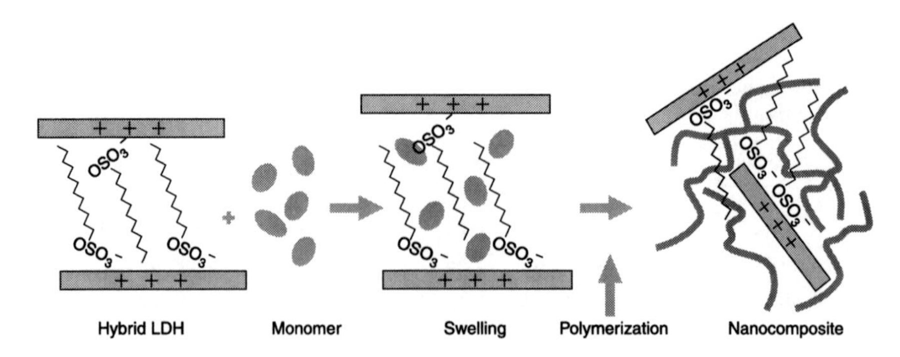


*4.3 In situ* polymerization (adapted from Kornmann, 2001).

nanocomposites based on polar polymers or thermoset resins. In the latter, the resins are cured in the presence of the layered particles. Nonetheless, it is not the usual method for preparing nanocomposites based on commodity thermoplastics such as polyolefins.

The main advantage of this method is that it promotes high exfoliation of the layered particles. However, it also has drawbacks, as it is a somewhat complex process and is rather industrially unfeasible for commodity polymers.

This method has been used successfully to prepare LDH nanocomposites based on PA6.6 (Herrero *et al.*, 2010; Zhu *et al.*, 2008), poly(ethylene terephthalate) (Lee and Im, 2007; Martínez-Gallegos *et al.*, 2008), polystyrene (PS) (Leroux *et al.*, 2005, 2010; Manzi-Nshuti *et al.*, 2008b, 2009b; Qiu and Qu, 2006; Ding and Qu, 2006; Marangoni *et al.*, 2008; Illaik *et al.*, 2008; Taviot-Gueho *et al.*, 2007; Matusinovic *et al.*, 2009), poly(methyl methacrylate) (Nyambo *et al.*, 2008; Wang *et al.*, 2005, 2006, 2009; Chen *et al.*, 2004a; Chen and Qu, 2005; Ding *et al.*, 2008), poly(vinyl chloride) (Bao *et al.*, 2006, 2008b), epoxy (Hsueh and Chen, 2003a; Zammarano *et al.*, 2005; Tseng *et al.*, 2007; Tsai *et al.*, 2008; Chan *et al.*, 2008; Lv *et al.*, 2009a), and polyimide (Hsueh and Chen, 2003b), among others.

## 4.3.2 Melt-mixing

The melt-mixing preparation of polymer–LDH nanocomposites consists of the dispersion of a LDH hybrid in a polymer matrix by applying high local shear stresses and using rather high temperatures in a melt-mixer dispositive, such as a twin-screw extruder or internal mixer (Fig. 4.4).

With this procedure it is possible to prepare a wide range of nanocomposites with intercalated or partially exfoliated structures, depending on the degree of intercalation of the polymer molecules in the interlayer gallery of the layered

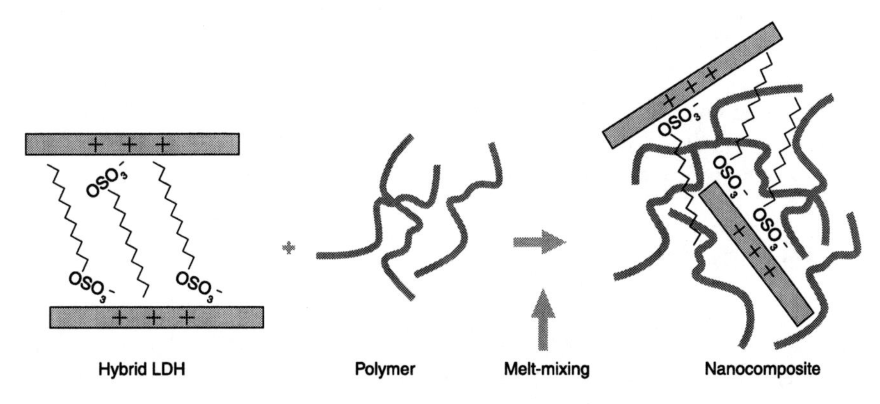


*4.4* Melt-mixing (adapted from Kornmann, 2001).

particles. It also has important advantages over other methods. First of all, from an environmental point of view, it does not require the use of organic solvents, and second, it is compatible with common plastic-processing technologies such as extrusion or injection.

The first report of this method to prepare polymer nanocomposites with cationic clays was by Giannelis (1996). Since then, it has been widely used due to its versatility. Gopakumar *et al.* (2002) suggest that, in the case of non-polar polymers, in order to attain good dispersion and partial exfoliation of the platelets, it is necessary, besides prior organophilization of the layered particles, to add a third component to act as a compatibilizer between the polymer and the particles. This compatibilizer needs polar groups, which should interact with the surface of the inorganic particles, promoting, alongside the high shear stresses generated during mixing, their exfoliation. In the case of polypropylene (PP), the most commonly used compatibilizer is a polypropylene copolymer with grafted groups of maleic anhydride (PP-g-MAH).

One of the first studies published describing melt-mixing was the work of Costa *et al.* (2005), which described the preparation and characterization of nanocomposites of low-density polyethylene (LDPE) modified with a polyethylene copolymer with methacrylic acid and hydrotalcite hybrids, prepared using dodecylbenzene sulfonate anions. Though total exfoliation of the LDH layers in the polymer was not achieved, significant changes were observed in its viscoelastic response compared with the pure polymer, even at very low LDH concentrations. It is the most widely used method nowadays. Alongside the polymers used to prepare LDH nanocomposites by means of *in situ* polymerization, melt-mixing has been employed with different results for many more polymers. For instance, different nanocomposites have been prepared with PA6 (Du *et al.*, 2007; Zammarano *et al.*, 2006), ethylene vinyl acetate (EVA) copolymer (Du *et al.*, 2006; Du and Qu, 2007; Zhang *et al.*, 2007, 2008a; Nyambo and Wilkie, 2009; Nyambo *et al.*, 2009b), poly($\varepsilon$-caprolactone) (Pucciariello *et al.*, 2007), poly(ethylene terephthalate) (Lee *et al.*, 2006), poly(methyl methacrylate) (Manzi-Nshuti *et al.*, 2008b; Nyambo *et al.*, 2008, 2009a; Wang *et al.*, 2009), poly(vinyl chloride) (Bao *et al.*, 2006; Chen *et al.*, 2010; Xu *et al.*, 2006), polypropylene (Ardanuy *et al.*, 2008, 2010; Lonkar *et al.*, 2009; Zhang *et al.*, 2008b; Wang *et al.*, 2010; Ding and Qu, 2006; Shi *et al.*, 2010), polyethylene (Manzi-Nshuti *et al.*, 2009b; Costa *et al.*, 2005, 2006a, 2006b, 2007; Schonhals *et al.*, 2009; Ardanuy *et al.*, 2009; Costantino *et al.*, 2005; Du and Qu, 2006), among others.

## 4.3.3 Solution blending

With solution, the pre-expanded layered particles are dispersed in a polymer solution to promote the entry of the polymer molecules into the interlayer gallery. The remaining solvent is evaporated afterwards, resulting in the precipitation of the polymer incorporated between the inorganic composite platelets (see Fig. 4.5).

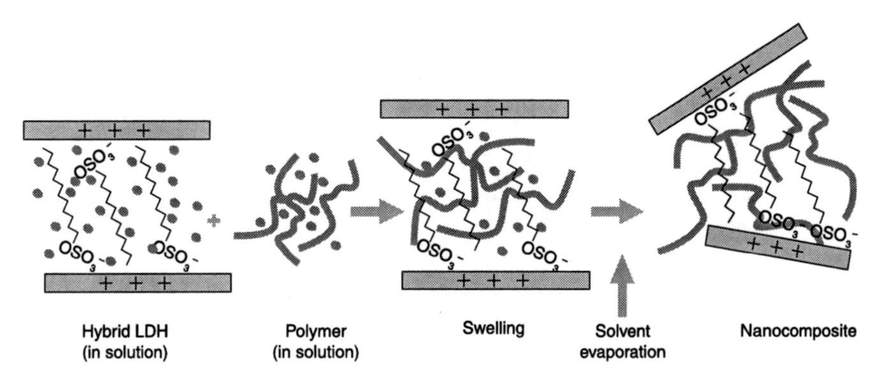


*4.5* Solution blending (adapted from Kornmann, 2001).

Though this procedure is quite simple, in practice it is rather complicated to find a solvent that is able to completely dissolve the polymer and fully disperse the layered particles. This method is suitable for the intercalation of polymers with a low or null polarity, facilitating the preparation of thin films with a high orientation of the inorganic particles. However, from an environmental point of view, this procedure has the inconvenience of using organic solvents, which may also be quite expensive.

In the literature, this technique has been described in the preparation of nanocomposites from water-soluble polymers such as poly(ethylene oxide) or polyvinyl alcohol (Ramaraj *et al.*, 2010), as well as the preparation of nanocomposites in non-polar solvents like toluene (EVA: Kuila *et al.*, 2007; PS: He *et al.*, 2006), xylene (EVA: Ramaraj and Yoon, 2008; Zhang *et al.*, 2008a; Kuila *et al.*, 2008; PS: Qiu *et al.*, 2005), linear low-density polyethylene (LLDPE): Qiu *et al.*, 2006; Chen and Qu, 2003; Chen *et al.*, 2004b), tetrahydrofuran (poly(vinyl chloride) (PVC): (Liu *et al.*, 2007, 2008), or dimethylformamide (poly(vinylidene fluoride): Bao *et al.*, 2008a), among others.

## 4.3.4 Others

A new procedure for preparing poly(ε-caprolactone)–LDH nanocomposites, high-energy ball milling (HEBM), was proposed by Sorrentino *et al.* (2005) and successfully used by other authors like Bugatti *et al.* (2010) and Romeo *et al.* (2007). This method consists in mixing the LDH particles (in this specific case a hydroxydecanoate-hybrid) with the polymer, both in powder form. Through an intensive mixing it is possible to break down the powder particles, causing the diffusion of the atoms and creating an intimate mix that results in the exfoliation of the LDH particles in the polymer. In this work, the dynamic–mechanical–thermal properties of the nanocomposites were better over the whole temperature range, with an increase of the storage modulus and tan $\delta$ decrease.

## 4.4 Structure of polymer–LDH nanocomposites

In order to obtain a higher efficiency of LDH, the process used to prepare the nanocomposites should ideally result in the rupture of the pilled structure of the particles, leading to a good dispersion of the layers (Fig. 4.6). Thus, there will be an increase in the contact surface with the matrix, improving the possible mechanical reinforcement efficiency of the particles by going from micro-sized particles to nano-sized ones. Nonetheless, it is not always possible to achieve full exfoliation of the particles in the polymer matrix. A microcomposite morphology such as the one displayed in Fig. 4.6 could occur, in which the layered particles keep their primary structure while mixed with the polymer; also possible is an intercalated morphology, in which the polymer molecules are between the platelets, thus increasing the contact surface between the particles and the matrix; finally, it is possible to obtain hybrid structures where fully exfoliated particle layers coexist with intercalated polymer-particle structures.

Among the several techniques used to elucidate the structure of silicate layered polymer nanocomposites, X-ray diffraction (XRD) and transmission electron microscopy (TEM) are the most extensive for evaluating the degree of exfoliation (Morgan and Gilman, 2003). XRD is used to determine the position, shape, and

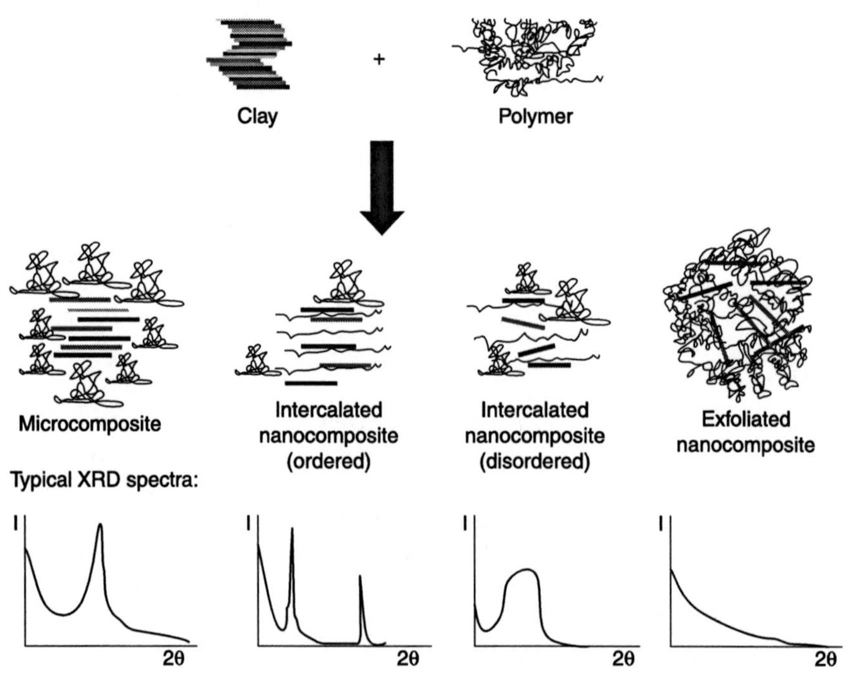


*4.6* Possible polymer nanocomposite microstructures and typical XRD spectra.

intensity of the peaks corresponding to the diffraction planes of the layered particles. More specifically, the first diffraction peak is commonly taken as a reference, due to its high intensity and to the fact that it is the one used to determine the interlayer distance. In the case where the primary structure of the layered composite is unaltered (immiscible mixtures, leading to the formation of *microcomposites*), so does the first diffraction peak (having the same position as in the pure particle) (see Fig. 4.6). For *intercalated structures*, the peak corresponding to the first diffraction plane appears at a lower angle because of an increment of the interlayer distance as a consequence of the presence of polymer molecules in the interlayer gallery. Last but not least, the total *exfoliation* of the particles leads to the disappearance of this diffraction peak (Vaia, 2001). However, XRD only detects a periodic order of the layers; thus it is important to complement the XRD measurements with TEM observations, which give the distribution of the particles in the matrix.

The exfoliation of the LDH particles in the matrix depends on several factors, such as the chemical nature and concentration of the cations in the inorganic layers, the interlayer distance between the layers, as well as the nature of the interlayer anions, the particle concentration in the composites, the preparation method, the nature of the polymer matrix, and the use of compatibilizing agents, among others.

The ionic exchange capacity of the layered particles is determined by the density charge of its layers. Comparing the residual charge of layered double hydroxides with that of cationic clays such as montmorillonite, the first have a much higher ionic interchange capacity and hence a higher density charge per unit area. Thus, using an LDH with a lower ionic exchange capacity should promote the formation of a nanocomposite, since particle exfoliation should be promoted due to lower attraction between the layers. For this reason, Zammarano *et al.* (2006) analyzed the effect of the density charge of LDHs in composites with a 5 wt.% of MgAl–LDH hybrids prepared with two different Mg/Al ratios (3:1 and 6:1). These authors found that, for complete exfoliation of the particles, it was necessary to use low anionic-exchange-capacity (AEC) LDHs combined with a mixing system with a minimum applied shear stress. Exfoliated PA6–LDH nanocomposites can be prepared by melt-mixing when suitable modified LDHs with appropriate AEC values have been synthesized. Shear, together with the inner ionic exchange capacity of LDHs, seems to be a key factor for the delamination of LDH in PA6.

On the other hand, it is also possible to regulate exfoliation through the type of cations present in the layers. For instance, Lonkar *et al.* (2010) have compared different PP composites with a concentration of 5 and 10 wt.% of LDHs with layers formed by combinations of MgAl, MgZnAl, and ZnAl cations, and DS as the interlayer anion. The PP–LDH composites were prepared in two steps using a co-rotating twin-screw extruder and PP-g-MAH was used as compatibilizer. The authors found that partial exfoliation was achieved for all combinations, though the rates of oxidation of the composites were influenced by the composition of the layers.

Another way to facilitate and promote the exfoliation of the particles without modifying the density charge of the layers consists in separating them until the attraction forces decrease through the incorporation of a highly voluminous organic anion. For example, Romeo *et al.* (2007) reported that for MgAl/LDH–poly($\varepsilon$-caprolactone) composites a partial exfoliation is only reached when using an LDH hybrid with 12-hydroxydecanoate. Tseng *et al.* (2007) compared the structure of epoxy nanocomposites with LDH–amino benzoate and LDH–carbonate particles. In the former, no diffraction peak was observed for contents between 1 and 7 wt.%. This is because the LDHs nanolayers can be sufficiently exfoliated in the epoxy matrix to form exfoliated nanocomposites. In contrast, characteristic diffraction peaks were observed for all LDH–carbonate/epoxy nanocomposites. For these reasons, LDH–amino benzoate/epoxy nanocomposites exhibit important thermal and mechanical improvements compared with the pristine epoxy resin. In contrast, LDH–carbonate/epoxy nanocomposites show only a small enhancement because of their intercalated morphology.

On the other hand, it is also important to take into account the type of organic anion used. Bugatti *et al.* (2010) compared poly($\varepsilon$-caprolactone) composites with ZnAl–LDHs prepared using different anions (benzoate (Bz), 2,4-dichlorobenzoate (BzDC), para-hydroxybenzoate (p-BzOH), and ortho-hydroxybenzoate (o-BzOH) anions) by the high-energy ball milling. They found that both the nature and position of the aromatic ring substituent affect the value of the interlayer distance and the hydrogen bonds of the hybrids. X-ray diffraction analysis of all the composites indicated that LDHs containing the BzDC anion were mainly exfoliated in the polymer matrix, whereas those containing p-BzOH remained almost unchanged, resulting in microcomposites. A partially exfoliated/partially intercalated structure was found for LDH modified with Bz and o-BzOH anions.

Lee *et al.* (2006) compared the structure of polyethylene terephthalate (PET) composites with LDHs modified using different anions, dodecyl sulfate (DS), dodecylbenzene sulfonate (DBS), and octyl sulfate (OS), prepared by melt-mixing. XRD analysis combined with TEM observations showed exfoliation for all cases, though to a higher extent in the MgAl hybrid prepared with DS. This composite also displayed better thermal and mechanical properties, thus supporting the higher exfoliation degree and interaction of the particles. Moreover, these authors (Lee and Im, 2007) used dimethyl 5-sulfoisophthalate (DMSI) anions for *in situ* polymerized LDH nanocomposites (LDH–DMSI), in order to enhance the compatibility between the PET matrix and LDH, resulting in full exfoliation of the LDH particles. The morphology of the nanocomposites was studied by TEM and XRD, clearly showing that LDH–DMSI was exfoliated in the PET matrix.

Nyambo *et al.* (2008) compared the structure of PMMA composites prepared with modified MgAl LDHs using decanoate (MgAl–C10), undecanoate (MgAl–C11), laurate (MgAl–C12), myristate (MgAl–C14), palmitate (MgAl–C16), stearate (MgAl–C18), and behenate (C22) as anions. From these, C10 produced an exfoliated morphology, C12 and C14 in mixed intercalated–exfoliated, and C18 and C22 in intercalated.

Shi *et al.* (2010) compared the structure of PP composites with LDHs modified with sodium dodecyl sulfate (SDS), itaconic acid (IA), or a combination of both, prepared by melt-mixing. The highest degree of exfoliation was obtained when combining both anions.

Another factor that affects the formation of exfoliated particle morphology is the concentration of particles in the polymer matrix. For instance, Herrero *et al.* (2010) and Zhu *et al.* (2008) showed that, for *in situ* polymerized PA6.6 composites prepared with hybrid LDHs, the best dispersion was achieved for the nanocomposites with a low LDH concentration (with an exfoliated structure for LDH concentrations below 0.5 wt.%). Similarly, Du *et al.* (2007) observed an exfoliated structure in PA6 nanocomposites with SDS-modified LDH particles prepared by melt-mixing for LDH concentrations below 10 wt.%, while Zhang *et al.* (2007) observed this exfoliated morphology in EVA for SDS-modified LDHs concentrations < 5 wt.%.

Martínez-Gallegos *et al.* (2008) demonstrated that, for *in situ* polymerized PET composites, exfoliation and dispersion seems to be complete for LDH contents up to 5 wt.%. Higher concentrations lead to the formation of LDH aggregates in the composite.

Kuila *et al.* (2008) found that, for EVA/LDPE/DS–LDH composites prepared by solution blending, there was full delamination of the DS–LDH layers at low DS–LDH contents, whereas partially exfoliated structures were observed for higher filler loadings. Similarly, Zhang *et al.* (2008a) compared the structure of EVA composites prepared by melt-mixing and solution blending and found an exfoliated particle morphology for particle contents < 10 wt.%.

Regarding preparation conditions, Qiu *et al.* (2005) showed that the exfoliation of PS/ZnAl–LDH nanocomposites prepared by solution blending involved two steps during refluxing: solvent swelling and layer breaking. Chen *et al.* (2004b) described a method, based on refluxing the LDH particles in a non-polar xylene solution of the base polymer, which allowed the preparation of LDH nanocomposites with a high degree of exfoliation even for high concentrations.

Another important factor in achieving exfoliated nanocomposite morphology is the nature of the polymer matrix. Hence, Wang *et al.* (2009) compared different anions, to give different LDHs (2-ethylhexyl sulfate (SEHS), bis(2-ethylhexyl) phosphate (HDEHP), and dodecylbenzenesulfonate (SDBS)), and also different preparation methods (melt-mixing and *in situ* polymerization) and different polymers (PMMA and PS). They found that it was much easier to disperse the particles in PMMA than in PS, although SEHS did not disperse well in either polymer.

## 4.5 Properties

### 4.5.1 Rheological properties

To understand the role of the LDH particles as fillers, rheological measurements are used to reveal possible interactions and relate them with the degree of dispersion of LDH in the polymer. The linear viscoelastic regime provides important information on LDH dispersion, as well as possible particle/particle and particle/polymer interactions and, therefore, may be used to relate structure and properties.

The rheological properties of PET nanocomposites have been studied by Lee *et al.* (2006). PET nanocomposites containing 2.0 wt.% of MgAl–LDH with carbonate ($MgAlCO_3$), dodecylbenzene sulfonate (MgAlDBS), and octyl sulfate (MgAlOS) showed similar $G'$ vs. $G''$ curves, obtained from dynamic parallel plate rheometry measurements. All these nanocomposites displayed lower slopes than pure PET (1.27 vs. 2), with higher values of the storage modulus at a low loss modulus, indicative that the system was heterogeneous in terms of filler/filler interactions. However, PET nanocomposites with 2.0 wt.% dodecyl sulfate (MgAlDS) displayed a shallower slope (around 0.87) because of network structures with filler/filler and filler/matrix interactions due to the increased hydrophobic nature and exfoliation of the MgAlDS system. However, for $G'' > 10^3$, the $G'$ vs. $G''$ slopes were steeper for all nanocomposites and approached that of the PET homopolymer. The authors argued that these results reflected the fact that some network structures in filler/filler or filler/matrix interactions were ruptured by the applied shear force, the system becoming isotropic and homogeneous.

The viscosity vs. frequency curve of PET/$MgAlCO_3$ indicated a drastic shear-thinning behavior due to slip between the polymer and filler caused by a low filler/matrix interaction. Additionally, although the viscosity of PET with MgAlDBS, MgAlDS, and MgAlOS has shear thinning in a low-frequency range, PET with MgAlDBS and MgAlOS showed continuous-like shear-thinning behavior, due to the breakdown of the network structure and slip between PET and the filler. PET/MgAlDS retained the viscosity at high frequencies, much like the PET homopolymer, due to improved filler/matrix interactions.

The rheological behavior of PS nanocomposites with a $Zn_2Al$–LDH dye (Chicago sky blue (CB), Evans blue (EB), and Niagara blue (NB)) hybrid filler was studied by Marangoni *et al.* (2008). The linear viscoelastic properties were studied under dynamic oscillatory shearing. The elastic $G'(\omega)$ and loss $G''(\omega)$ moduli of the unfilled PS showed in the low-$\omega$ region the usual Newtonian behavior with the expected relaxation exponent $G' \propto \omega^{1.92}$ and $G'' \propto \omega^{0.98}$, close to the ideal slopes $G' \propto \omega^2$ and $G'' \propto \omega^1$, characteristic of the frequency-dependent liquid-like flow behavior (with a plateau in the complex viscosity master curve). For a PS nanocomposite containing either $Zn_2Al$/CB or $Zn_2Al$/EB, both the curvature of $G'(\omega)$ and $G''(\omega)$ and the relaxation parameters in the low-$\omega$ region remained similar to PS. In contrast, PS–$Zn_2Al$/NB displayed a change in slope of both moduli in the terminal zone.

This was even more pronounced in the $|\eta^*|(\omega)$ curve, where a plateau was not observed (see Table 4.4). This effect was considered by the authors to be an incomplete relaxation of PS chains. Although this is consistent with the shift of $T_g$, it was, however, reported as unusual for an immiscible PS nanocomposite structure. The pseudosolid-like behavior of PS–$Zn_2Al$/NB was explained by attrition phenomena between PS and the filler. The changes occurring in $T_g$ and the modulus in the low-$\omega$ domain were explained by interfacial interactions. The pseudosolid behavior of PS–$Zn_2Al$/NB was considered to be a rupture of the filler's lamellar structure, in strong contradiction to the trend usually observed, where solid-like behavior is seen in intercalated or exfoliated nanocomposite structures. The authors surmised that the attrition phenomena largely developed at the interface of $Zn_2Al$/NB platelets and the PS chain should be attributed to the molecular arrangement of NB at the surface of the platelets.

Finally, it was noted that the molecular similarity between EB and CB was the reason for the super imposable rheological behavior, while the difference from NB was easily seen in the $|\eta^*|(\omega)$ curve.

Leroux *et al.* (2009) proved that the shear-thinning exponent did not allow a quantitative discrimination between intercalated and exfoliated nanocomposite structures. The change in the shear-thinning parameter, directly correlated to the relaxation parameter, was interpreted as a modification in the interfacial interaction between the LDH filler and PS chains. In order to see if this effect a rose from the alkyl chain alone, a PS–$Zn_2Al$/DBS nanocomposite was studied. As with the $Zn_2Al$/MADABS hybrid phase, $Zn_2Al$/DBS presented a well-defined lamellar structure, with several basal peaks. When incorporated into PS, the basal spacing distance remained unaltered (2.93 nm), hence defining a non-miscible PS nanocomposite structure, apparently comparable to PS–$Zn_2Al$/MADABS. The

*Table 4.4* Rheological properties obtained under dynamic oscillatory shearing of PS nanocomposites with $Zn_2Al$–LDH modified with different dyes

| Nanocomposite | $\Delta d$[a] (nm) | n[b] $G \propto \omega^n$ | n[b] $G'' \propto \omega^n$ | n[c] $\lvert\eta^*\rvert \propto \omega^n$ |
|---|---|---|---|---|
| PS | – | 1.92 | 0.98 | −0.03 |
| PS–$Zn_2Al$/CB | −0.017 | 1.61 | 0.96 | −0.05 |
| PS–$Zn_2Al$/EB | −0.021 | 1.43 | 0.95 | −0.04 |
| PS–$Zn_2Al$/NB | −0.352 | 0.89 | 0.84 | −0.13 |

Notes:[a] Change in the basal spacing compared with the blue LDH filler before their dispersion into styrene. A negative value is interpreted as a reduction.

[b] Relaxation parameters in the terminal zone.

[c] Shear-thinning parameters in the terminal zone.

Source: Adapted from Marangoni *et al.*, 2008.

glass transition temperature was similar to that of unfilled PS, as were the relaxation and shear-thinning exponents. Indeed, the presence of $Zn_2Al$/DBS strongly decreased the molecular weight of PS. Also, the amplitude of the elastic modulus $G'$ was drastically reduced in the low-$\omega$ region. Consequently, the attrition phenomena cannot be explained solely by the length of the surfactant molecule.

The rheological behavior of hot-pressed films of LDH–PS nanocomposites was also studied. A filler percentage of 5 wt.% was used. The glass transition temperature shifted in comparison to unfilled PS, although not for the same magnitude as for a 10 wt.% content sample. PS films with 5 wt.% had intermediate rheological behavior, i.e. incomplete PS chain relaxation in the low-$\omega$ region with a shear-thinning factor different from zero. Once again, the lowering observed in the relaxation can be considered to be tethering of a polymer chain on a particle's surface and particle–particle interactions. The authors concluded that, after checking the consistency between the rheological behavior of LDH–PS nanocomposites and the state of dispersion observed by TEM, one can infer that XRD is not the most appropriate technique for elucidating the structure of polymer nanocomposites, as their morphology is much more complex than previously thought. The general trend can be explained by the existence of an LDH-percolated network. The use of lower filler loadings should induce a smaller interface, which would cause smaller attrition phenomena. These observations are to some extent similar to those for cationic clays, which form three-dimensional deformable networks in PS. Nonetheless, the effect is much more pronounced in the case of the present filler, hence the interest in the use of LDH-based materials as inorganic platelet-like fillers. Filler preparation was also reported to be important, since polymerization of the compatibilizer at the platelets' surface may give rise to aggregation when performed concomitantly with styrene bulk polymerization. Bulk polymerization acts as mechanical grinding when the platelets are pushed together in advance after a thermal pre-treatment, thus leading to a concatenated filler structure with solid-like rheological behavior.

Taviot-Gueho *et al.* (2007) also studied the rheological properties of several LDH–PS nanocomposites. The linear viscoelastic dynamic storage and loss modulus (G′ and G″, respectively) for PS–$Zn_2Al$/yellow were good matches with those of PS, whereas for PS–$Zn_2Al$/DBS there was a large decrease over the entire frequency domain, considerably more so for $G'$ than for $G''$, hence explaining the increase in tan $\delta$. The increase in tan $\delta$ has to be explained by greater molecular mobility. It appears that the presence of $Zn_2Al$/yellow in PS, even as an intercalated structure and in contrast to $Zn_2Al$/DBS, does not result in worse rheological properties. The rheological behavior, expressed as $\eta''$ vs. $\eta'$, was found to be different for the three samples. At 200 °C, PS and PS–$Zn_2Al$/yellow curves matched well. The viscosity at zero shear ($\eta_0$), similar for the two systems, suggested similar molecular weights, which were confirmed by gel permeation chromatography (GPC). For PS–$Zn_2Al$/DBS, the intercept was at a much lower value, once again consistent with the GPC results. Finally, the size of the hybrid LDH aggregates was argued as a possible disadvantage during bulk polymerization. Both fillers had an intercalated structure, though it was ill-defined for PS–$Zn_2Al$/yellow and well-defined for PS–$Zn_2Al$/DBS. The presence of large $Zn_2Al$/DBS filler particles prevented the formation of extended polymer chains and consequently impoverished the rheological properties.

The rheological properties of LDH–polyethylene nanocomposites were studied by Costa *et al.* (2006a). The evolution of the complex viscosity of LDH–LDPE nanocomposites at low frequency was studied from 180 to 260 °C. The dependence of the rheological properties on temperature was weaker for the nanocomposites with an unfilled polymer, especially beyond a critical LDH content, representing a deviation of the liquid-like low-frequency flow behavior towards a pseudosolid one.

On the other hand, HDPE-g-MAH nanocomposites (HDPE is high-density polyethylene) showed the shifting behavior at much lower LDH concentrations (around 5 phr) compared with the LDPE/compatibilizer composites, where changes in viscosity with temperature were still observed at a 10 phr LDH concentration. This means that the nature of the polymer matrix strongly influences the effect of increasing the amount of LDH in terms of the rheological behavior of LDH nanocomposites. With increasing LDH nanoparticle concentration, the low-frequency response of the complex viscosity vs. frequency shifted to a shear-thinning mode, with similar behavior for both nanocomposites. Two factors could explain this situation: first of all, the polarity difference between the two matrices, and second, the presence of a low molecular weight compatibilizer in one of the composites.

It is well known that both unfilled and particulate polyolefin microcomposites conventionally display a Newtonian liquid-like behavior at low frequency. Nevertheless, LDH–LDPE nanocomposites display shear-thinning behavior with a higher shear-thinning exponent and lower relaxation exponent. Both exponents become almost independent of LDH concentration at higher contents. The shear-thinning exponent has been used as a semi-quantitative parameter of the degree of clay delamination in polymer nanocomposites (Wagner and Reisinger, 2003). Although higher negative values of this exponent are normally considered to be due to higher clay exfoliation, the higher shear-thinning exponent values of the LDH–LDPE nanocomposites were due to polymer molecule confinement, delaying the relaxation process. Higher LDH loadings enhanced this effect, with $G'$ increasing and being less frequency dependent. Additionally, as the average distance between the dispersed particles decreased, the tendency to interlock, which acts as an energy barrier for molecular relaxation, increased.

## 4.5.2 Mechanical properties

Taking into account the large surface area of layered double hydroxides, in some cases reaching 800 $m^2/g$, a large interaction with polymer molecules is possible, and thus a large mechanical reinforcement effect is to be expected. This situation could partly explain why these layered materials may dramatically improve the modulus of polymers even when present in small amounts.

Most research into the mechanical properties of polymer–LDH nanocomposites has analyzed their tensile properties, particularly the modulus and tensile strength, as a function of LDH content. Generally speaking, the addition of low amounts of LDH increases the Young's modulus and tensile strength, mainly due to the ease of attaining better particle dispersion in the polymer matrix at low LDH contents. Higher LDH concentrations increase the probability of particle cluster formation, leading to deterioration of the mechanical properties.

Zhang *et al.* (2008a) reported a gradual increase in Young's modulus from 42.8 to 72.1 MPa for EVA composites containing up to 20 wt.% of ZnAl–LDH particles and an increase in tensile strength up to 5 wt.% LDH, decreasing for higher loadings. Another study (Sorrentino *et al.*, 2005) showed how small amounts of LDH particles (up to 2.8 wt.%) significantly improved the mechanical properties of composites, with an increase of both Young's modulus and yield stress of about 100%. However, for concentrations up to 6 wt.%, the mechanical properties, although still higher than for the pure polymer, decreased. A similar phenomenon was observed in PU/DS–LDH nanocomposites (PU is polyurethane) (Kotal *et al.*, 2009). Kotal *et al.* studied the mechanical properties of nanocomposites containing 1, 3, 5, and 8 wt.%, finding that the maximum enhancement of tensile strength and elongation at break corresponded to the nanocomposite containing 3 wt.% DS–LDH. These authors suggested that the decrease observed beyond this DS–LDH content was probably due to the increasing tendency of DS–LDH to aggregate in the PU matrix at high loadings. Du *et al.* (2006) also studied the effect of DS/MgAl–LDH particle content (5 and 10 wt.% loadings were used) in poly(propylene carbonate) (PPC). They found that the nanocomposites displayed enhanced tensile strengths and Young's moduli compared with pure PPC. The nanocomposite containing 5 wt.% showed the highest tensile strength and Young's modulus (72% and 57% higher than those of pure PPC, respectively). This improvement was due to the structure found in the nanocomposites: exfoliated for the 5 wt.% LDH nanocomposite and partly exfoliated for the 10 wt.% one. A similar phenomenon was observed in aminobenzoate/MgAl–LDH/polyimide nanocomposites (Hsueh and Chen, 2003b). When LDH particle contents exceeded 5 wt.%, a decrease in the tensile strength at break was found (although still higher than for pure polyimide). Once again, this was due to the aggregation of the LDH layers.

The results of tensile tests performed by the same authors (Hsueh and Chen, 2003a) on laurate-modified MgAl–LDH epoxy nanocomposites showed how both the tensile strength and modulus increased gradually when the amount of LDH was increased up to 7 wt.%, the maximum concentration analyzed. However, these authors also found that elongation at break decreased slightly for LDH contents higher than 3 wt.%. Wang *et al.* (2006) also found interesting results for PMMA nanocomposites reinforced with different amounts of undecanoate-modified LDH (LDH-U). The authors found that the tensile modulus was enhanced with increasing LDH-U content (38% and around 80%, respectively, for LDH contents of 3 and 5 wt.%). This was due to good dispersion of the LDH layers and a strong interfacial adhesion between the layers and the PMMA matrix. A similar trend was observed for tensile strength. Nevertheless, a slight decrease was noticed for the elongation at break with increasing LDH content, due to the high resistance of the LDH layers to deformation of PMMA.

Acharya *et al.* (2007b) also reported an increase of both the tensile modulus and strength when increasing the amount of LDH up to 8 wt.% in EPDM/DS–LDH nanocomposites.

Other studies reported the effect of the interlamellar anion and the composition of the layers in the mechanical properties of LDH nanocomposites. For example, Manzi-Nshuti *et al.* (2008b) studied the effect in PMMA systems of LDH particles prepared with different combinations of NiAl, CoAl, and ZnAl layers, as well as different Zn/Al ratios (2:1 and 3:1). The authors did not find significant effects of the LDH particles on the mechanical properties, regardless of the composition and ratios of the divalent to trivalent metal cations. In another study, Bugatti *et al.* (2010) analyzed the effect of LDH hybrids prepared using different interlayer anions (benzoate, dichlorobenzoate, p-hydroxybenzoate, and o-hydroxybenzoate), as well as their concentration on the mechanical properties of polycaprolactone (PCL) films. Slight improvements were observed for all mechanical parameters, with some interesting differences between the differently dispersed nanohybrids. These authors showed that for a given LDH concentration, for instance 3 wt.%, the tensile modulus may vary between 218 and 300 MPa, depending on the type of interlayer anion (in this case the higher value corresponded to dichlorobenzoate-modified LDH). Wang *et al.* (2009) also analyzed the effect of LDH concentration and the use of different interlayer anions on the mechanical properties of PMMA and PS nanocomposites. No significant differences were found when varying the type of interlayer anion.

An interesting study was conducted by Bao *et al.* (2006), which analyzed the effect of LDH particles on the tensile strength and Young's modulus of PVC composites prepared by both melt-mixing and *in situ* polymerization. The authors found that, although the tensile strength and Young's modulus increased for both PVC/LDH–DS composites when increasing the amount of LDH–DS, this increase was higher for the ones prepared by *in situ* polymerization for a given LDH concentration. On the other hand, Pradhan *et al.* (2008) showed that, when well dispersed, LDH acted as a mechanical reinforcement in both ethylene propylene diene monomer (EPDM) and acrylonitrile butadiene carboxy monomer (XNBR) elastomers. In particular, the tensile strength increased from 1.54 and 1.94 MPa to values as high as 3.25 and 17.83 MPa for a 10 wt.% LDH content, respectively, for the EPDM and XNBR elastomers, clearly demonstrating the mechanical reinforcement effect, especially in the case of the XNBR elastomer.

Regarding some of our previous results, we studied the mechanical properties of PS and styrene-acrylonitrile resin (SAN) composites with 5 and 10 wt.% of dodecyl-sulfate-modified MgAl layered double hydroxides (o-HT) (Realinho *et al.*, 2009). The tensile modulus and strength of PS remained almost unaltered when incorporating the o-HT particles, with only a slight increase being observed for Young's modulus. In contrast, the addition of o-HT to SAN, although decreasing its tensile strength, increased the stiffness to a higher extent than in PS and noticeably increased the ductility of the material in terms of strain at break. These results were due to better o-HT dispersion in SAN compared with PS. The decrease in tensile strength observed when incorporating o-HT into SAN was explained by the combination of a relatively weak particle–matrix interface adhesion and inherent high deformability of SAN, which promoted o-HT particle debonding at low stress levels and subsequent plastic flow.

For semi-crystalline polypropylene (Ardanuy *et al.*, 2010), we showed that the incorporation of 10 wt.% of unmodified (HT) and dodecyl-sulfate-modified hydrotalcite (HTDS) to a PP matrix with and without PP-g-MAH and PP-g-SEBS compatibilizers resulted in nanocomposites with slightly higher Young's moduli. However, this increase was considerably lower than that observed for similar PP–montmorillonite (PP-MMT) nanocomposites, due to a higher inherent flexibility of HT layers compared with cationic clays such as MMT. Comparing HT- and HTDS-reinforced nanocomposites, the latter had higher values of Young's modulus, probably due to a higher exfoliation of the HTDS particles in PP, especially when using the PP-g-MAH compatibilizer (there were more interactions between the polymer and HT platelets). A low degree of adhesion was considered to be the reason for the slight decrease observed regarding the tensile strength for both the HT- and HTDS-reinforced PP nanocomposites, compared with the higher values obtained for the nanocomposites with compatibilizer.

### 4.5.3 Thermal stability

Generally speaking, the incorporation of layered particles into polymer matrices results in an enhancement of their thermal stability (Pavlidoua and Papaspyrides, 2008). The increase in thermal stability may be due to the layers hindering the diffusion of oxygen and volatile products throughout the polymer, as well as to the formation of a char layer after thermal decomposition of the organic matrix. Additionally, endothermic decomposition of the host metal hydroxide layer can provide a cooling effect, which could delay the combustion process of the organic species (surfactant anion, polymer chain segments, etc.) constrained within the interlayer gallery of the LDH particles.

As can be seen in Table 4.5, except for some specific polymer matrices like polyamide 6 (PA6) or PCL, the addition of LDH hybrids to different polymers results in an increase in the decomposition temperatures corresponding to a 50% weight loss.

*Table 4.5* Decomposition temperatures for a 50wt.% loss ($T_{0.5}$) of various polymer nanocomposites as a function of LDH content

| Nanocomposite | LDH content (wt.%) | $T_{0.5}$ (°C) | | Reference |
|---|---|---|---|---|
| EVA/SDS ZnAl–LDH (18% vinyl acetate) | 0 | 435 | | Zhang *et al.*, 2008a |
| | | Melt-mixing | Solution blending | |
| | 2 | 448 | 451 | |
| | 5 | 448 | 464 | |
| | 10 | 461 | 470 | |
| EVA/borate-modified–LDH | 0 | 449 | | |
| | | MgAl–LDH | ZnAl–LDH | Nyambo and Wilkie, 2009 |
| | 3 | 456 | 460 | |
| | 5 | 462 | 464 | |
| | 10 | 463 | 463 | |
| | 20 | 466 | 470 | |
| | 40 | 473 | 476 | |
| EVA/LDPE/DS MgAl–LDH | 0 | 431 | | Kuila *et al.*, 2008 |
| | 1 | 437 | | |
| | 3 | 442 | | |
| | 5 | 442 | | |
| | 8 | 445 | | |
| PET/DMSI MgAl–LDH | 0 | 386 (5wt.% loss) | | Lee and Im, 2007 |
| | 0.5 | 392 | | |
| | 1.0 | 400 | | |
| | 2.0 | 395 | | |
| PS/DS ZnAl–LDH | 0 | 383.8 | | Ding and Qu, 2006 |
| | 1 | 396.9 | | |
| | 5 | 411.8 | | |
| | 10 | 385.8 | | |
| PS-co-MMA/Benzoate CaAl–LDH | 0 | 445 | | Matusinovic *et al.*, 2009 |
| | 1 | 455 | | |
| | 2.5 | 459 | | |
| | 5 | 466 | | |
| | 7.5 | 471 | | |
| PMMA/undecanoate | 0 | 365 | | Manzi-Nshuti *et al.*, 2008a |
| | | NiAl–LDH | CuAl–LDH | |
| | 5 | 380 | 294 | |
| | 10 | 385 | 306 | |
| | | CoAl–LDH | ZnAl–LDH | |
| | 5 | 386 | 383 | |
| | 10 | 388 | 391 | |
| PE-g-MAH/DS MgAl–LDH | 0 | 394 | | Chen and Qu, 2003 |
| | 2 | 446 | | |
| | 5 | 453 | | |
| LLDPE/DS ZnAl–LDH | 0 | 416 (40wt.% loss) | | Du and Qu, 2006 |
| | 2 | 444 | | |
| | 5 | 447 | | |
| | 10 | 459 | | |

(*Continued*)

*Table 4.5* Continued

| Nanocomposite | LDH content (wt.%) | $T_{0.5}$ (°C) | Reference |
|---|---|---|---|
| PCL/ hydroxydodecanoate MgAl–LDH | 0 | 402 | Pucciariello *et al.*, 2007 |
| | 1 | 387 | |
| | 2 | 361 | |
| | 3 | 342 | |
| PCL/DS CoAl–LDH | 0 | 402 | Peng *et al.*, 2010 |
| | 1 | 384 | |
| | 2 | 374 | |
| | 4 | 370 | |
| PA6/SDS MgAl–LDH | 0 | 447.8 | Du *et al.*, 2007 |
| | 5 | 428.1 | |
| | 10 | 413.1 | |
| | 20 | 420.7 | |
| PA6/DBS MgAl–LDH | 0 | 420.1 | Zammarano *et al.*, 2006 |
| | 5 | 457.6 | |

Zhang *et al.* (2008a) studied the thermal stability of DS/ZnAl–LDH/EVA nanocomposites as a function of LDH concentration and preparation procedure. These authors observed that the thermal degradation temperatures of the nanocomposites were between 13 and 35 °C higher than that of pure EVA, increasing with increasing LDH concentration. The authors explained this effect based on the fact that at a higher loading the LDH layers favored the formation of an obstructive char layer at the surface of the polymer, increasing the thermal stability of the nanocomposites. Moreover, samples prepared by solution blending displayed higher degradation temperatures than those prepared via melt-mixing for identical LDH contents. The authors attributed this effect to a better dispersion of the LDH particles in the nanocomposites prepared by solution blending.

Nyambo and Wilkie (2009) reported similar results for EVA-based polymer nanocomposites reinforced with borate-modified LDHs with a different composition of the metal cations. The authors found that the addition of MgAl or ZnAl-borate LDHs improved the thermal stability of EVA, more markedly with increasing LDH concentration. In particular, the best thermal stability effect was obtained for a 40 wt.% LDH loading regardless of the nature of the layer metal cations. Using a similar approach, Kuila *et al.* (2008) studied the thermal decomposition of EVA/LDPE-based LDH nanocomposites. They found that, although the initial weight loss for the nanocomposites was accelerated due to the early degradation of DS molecules, the thermal decomposition temperatures corresponding to a 50% weight loss for nanocomposites with 1, 3, 5, and 8 wt.% DS–LDH were higher than for pure EVA/LDPE, increasing with increasing LDH content. This increase in thermal stability was attributed to the homogeneous dispersion of DS–LDH in the EVA/LDPE matrix. In another study, Qiu *et al.* (2005) analyzed the effects of LDH concentration on the thermal decomposition of PS nanocomposites containing 5, 10, and 20 wt.% ZnAlDS–LDH. They showed that the thermal decomposition temperature of the nanocomposites was between 16 and 39 °C higher than that of pure PS, with optimal concentration being 10 wt.%. Higher LDH loadings (20 wt.%) resulted in nanocomposites with lower thermal stability. These authors also analyzed the thermal behavior of the same nanocomposite systems prepared using different methods. They found that the samples prepared by rapid precipitation and solvent evaporation at 140 °C, although displaying similar thermal behavior when varying the amount of LDH, showed better thermal stability compared with samples prepared by solvent evaporation at 60 °C. This was explained by the different proportions of exfoliated and intercalated structures in these nanocomposites.

Different effects of LDH content on the thermal stability of polyolefin-based nanocomposites have been reported in the literature. For example, for PP–LDH systems, Shi *et al.* (2010) reported that the thermal decomposition temperatures of a series of PP–LDH nanocomposites with various LDH concentrations were higher than for pure PP (from 7 to 30 °C). Moreover, the authors found that the nanocomposites containing organo-modified MgAl–LDH particles exhibited higher decomposition temperatures than the nanocomposites containing unmodified MgAl–LDH particles. In the same way, Ardanuy and Velasco (2011) reported that the decomposition temperature of pure PP increased with organophilized LDH particles, while the thermal stability decreased when using unmodified LDH particles. These authors also found that the increase of thermal stability was more noticeable when PP-g-MAH was used as a compatibilizer. They noted that the exfoliated layers of organophilized LDH hindered the diffusion of oxygen after combustion ignition to a greater extent. Nevertheless, Wang *et al.* (2010) did not find any differences between the thermal decomposition of pure PP and nanocomposites with organic modified CaAl–LDH for particle loadings between 1 and 6 wt.%. These authors concluded that the particles did not improve the thermal stability of PP.

Concerning polyethylene–LDH nanocomposites, Qiu *et al.* (2006) compared the thermal stability of LLDPE systems with different MgAl, ZnAl–LDH, and MMT contents. They showed that the thermal stability of the nanocomposites was enhanced compared with virgin LLDPE, although the enhancement was quite different depending on the composition of the layered particles. The $T_{0.2}$ values (decomposition temperature corresponding to a 20 wt.% loss) for the LLDPE/MMT and LLDPE/MgAl–LDH samples increased gradually to 399 and 429 °C, respectively, with increasing MMT and $Mg_3Al(DS)$ contents from 0 to 10 wt%. However, LLDPE/ZnAl–LDH with only 2.5 wt% of $Zn_3Al(DS)$ showed important improvements, $T_{0.2}$ increasing till 428 °C. It reached a maximum value of 434 °C for 5 wt.% $Zn_3Al(DS)$, and decreased to 420 °C for 10 wt.% $Zn_3Al(DS)$. The possible reason behind these differences could be the relative extent of exfoliation in the different nanocomposites.

Costa *et al.* (2007) also studied LDPE–LDH nanocomposites, reporting that the onset of the decomposition temperature was delayed significantly with the addition of only 2.4 wt.% of LDH particles. Once again, the decomposition temperature increased with increasing LDH concentration, with a maximum increase of 70 °C compared with pure LDPE for a 9 wt.% LDH content. Ardanuy *et al.* (2009) reported similar results for HDPE nanocomposites. The onset of decomposition of HDPE was significantly delayed by the effect of the organo-modified LDH particles and, similarly, $T_{0.5}$ increased because of these particles.

However, there are results in the literature in which the incorporation of LDH particles resulted in a decrease in the decomposition temperature of the polymer. For example, Du *et al.* (2007) reported that the thermal degradation temperature of PA6/DS MgAl–LDH nanocomposites was lower than that of pure PA6. This effect was highlighted with increasing LDH content. These authors considered that this effect was due to a catalytic degradation phenomenon caused by the organically modified LDH layers. Herrero *et al.* (2010) reported similar results for PA6.6–LDH systems with filler contents higher than 1 wt.%. This behavior is because the high content of Ad–LDH could catalyze the alkaline degradation of PA6.6. Zammarano *et al.* (2006) also reported a reduction in the onset of the thermal decomposition temperature for LDH–PA6 nanocomposites. Again, it was suggested that this effect was due to a nucleophilic attack mechanism.

### 4.5.4 Other properties

Besides the few works that analyze properties such as arc resistance and dielectric strength (Ramaraj and Yoon, 2008), barrier properties (Bugatti *et al.*, 2010; Sorrentino *et al.*, 2005), dielectric properties (Leroux *et al.*, 2010; Schonhals *et al.*, 2009), and optical properties (Marangoni *et al.*, 2008; Zhang *et al.*, 2008b; Acharya *et al.*, 2007b), one of the most extensively studied properties in polymer-layered double hydroxide nanocomposites is fire resistance, mainly using cone calorimetry and limiting oxygen index (LOI) determination.

For instance, Du *et al.* (2007) reported the combustion characterization of PA6/MgAl–LDH nanocomposites. They found that the heat release rate (HRR) decreased considerably when increasing the amount of LDH in the polymer matrix. Moreover, they found that the ignition time was delayed compared with pure PA6 for low LDH contents. However, ignition times increased when the content of MgAl(H-DS) exceeded 10 wt.%. The authors suggested that this phenomenon could be due to a catalytic degradation effect of PA6 promoted by the organo-modified particles, which decreased the molecular weight of the polymer and thus caused the material to burn more easily.

Ramaraj and Yoon (2008) reported the LOI estimation and flammability rating of EVA–LDH nanocomposite films. Results showed a significant improvement in the flame retardancy of EVA nanocomposites due to the LDH particles. They showed that the minimum oxygen concentration required for a flame to appear increased from 19 to 26% and the flammability rating improved from no rating to V1 rating with increasing LDH concentration. They reported that the flame-retardant characteristics of LDH were due to their $Mg(OH)_2$-like behavior, which involved endothermic decompositions with the formation of water vapor and a metal oxide residue. This residue prevented the burning process from proceeding by reducing the oxygen supply to the underlying combustible polymer.

Another study by Manzi-Nshuti *et al.* (2009a) gave the cone calorimetric results of PS–LDH systems. The best reduction in peak heat release rate (PHRR) (35%) was obtained for PS filled with 10 wt.% of an organo-modified ZnAl-MgAl–LDH. Lower LDH contents did not give significant fire retardancy improvements. Nyambo *et al.* (2009a) also studied the fire behavior of PMMA/MgAl–LDH nanocomposites using cone calorimetry. A significant reduction was observed regarding PHRR, due to improved physical interaction and good nano-dispersion of the organically modified MgAl–LDH particles in the PMMA matrix. A lower reduction was observed when incorporating unmodified MgAl–LDH particles.

Similar improvements in fire resistance with LDH particles were found by other researchers, such as Costa *et al.* (2007). These authors found that the incorporation of MgAl–LDH particles resulted in a very efficient reduction of the heat release rate and total heat released during combustion. MgAl–LDH particles were also found to facilitate the formation of a carbonaceous char on the burning surface, causing a lower emission of carbon monoxide during the initial phase of burning. For LDH concentrations above 10 wt.%, the nanocomposites not only burnt at extremely slow rates, but also showed low dripping tendencies. However, they reported that LDH by itself, even at these concentrations, was not enough to obtain a high LOI value or V0 burning rate.

All of these results prove the effectiveness of LDH particles as flame retardants in polymer systems.

## 4.6 Applications and future trends

LDH nanocomposites are promising functional materials, as they combine the characteristics and properties of both a polymer and an inorganic phase. This is a particularly interesting field for academics, since it offers the possibility of merging different scientific fields such as organic, macromolecular or inorganic chemistries, as well as for materials engineers, with potential applications in different industrial sectors such as energy production and storage (photovoltaics, batteries, fuel cells, . . .), medicine (drug delivery, biomaterials, imaging, . . .), functional coatings, environment and safety (membranes, fire resistance, . . .), nano- and microelectronics, optics, etc.

The use of LDH in polymer nanocomposites is an emerging research field with several potential applications. LDH–polymer nanocomposites have advantages compared with montmorillonite-based ones, due to the versatility of LDHs regarding chemical composition, as well as tunable charge density, allowing multiple interactions with the polymer. This makes them attractive for enhancing thermal stability and improving fire resistance in different polymeric materials (Camino *et al.*, 2001; Nyambo *et al.*, 2009b; Nyambo and Wilkie, 2009; Manzi-Nshuti *et al.*, 2009b). In particular, for flame-retardant polyolefins, different products such as cables and construction panels can be produced using LDH nanocomposites to meet the increasingly more stringent ratings of the low-smoke zero-halogen (LSZH) classification with reduced contents of mineral flame retardants such as aluminum trihydroxide or magnesium dihydroxide.

The fire behavior of LDH–polymer nanocomposites has been investigated by microcombustion calorimetry (Wang *et al.*, 2010), which revealed O-CoAl–LDH as a very efficient system in reducing both HRR and PHRR during burning. The heat release capacity and total heat released during burning were also reduced significantly with increasing nanofiller concentration. These results emphasize the fact that organo-LDH's flame-retardancy mechanisms are different from those of organo-layered silicates.

*Nanofire* products, developed by CIMTECLAB (Italy), based on LDH nanocomposites, have been reported as a solution for considerably reducing the content of toxic substances in polymeric formulations whilst maintaining excellent fire-retardant properties. *Nanofire* additives, applied to thermoplastic polymers, combine nanotechnology and a synergistic effect with a liquid plasticizer in various thermoplastic materials, such as PVC, polyamides, and polycarbonate/acrylonitrile–butadiene–styrene copolymer blends, hence enabling the reduction of toxic additives such as antimony and halogen without altering fire performance.

Other applications of polyolefin–LDH nanocomposites include water-storage systems, automotive injection-molded parts, and even foams. Moreover, in polyolefin packaging, films, bottles, and cups can be produced with improved mechanical, barrier, and rheological properties, resulting in better thermoforming properties.

Over the past decade, significant interest has been given to the synthesis of LDHs with new compositions, which would allow them to be used in different application areas. Das *et al.* (2011) prepared and characterized transparent rubber nanocomposites using LDHs, which are a more environmentally friendly rubber composite vulcanized without the use of ZnO. In this material, LDH nanoparticles deliver zinc ions during vulcanization as accelerators and stearate anions as activators, while the mineral platelets act as a nanofiller reinforcement.

Martínez *et al.* (2011) prepared and characterized microcellular foams from nanocomposites of PS, SAN, and PMMA with LDH, displaying fine uniform cellular structures with sub-micrometric cells. The thermo-mechanical properties and gas impermeability of these foamed nanocomposites were enhanced compared with pure polymer foams.

Kotal *et al.* (2011) produced thermoplastic polyurethane (TPU) nanocomposites with improved mechanical strength and ductility due to stearate-intercalated LDH particles. These nanocomposites showed maximum improvements in tensile strength (45%) and elongation at break (53%) for 1 and 3 wt.% LDH contents. Maximum improvements in storage and loss moduli (20%), with a shift of the glass transition temperature (15 °C) and an increase in thermal stability (32 °C) at 50% weight loss, were observed for an 8 wt.% LDH loading.

Guo *et al.* (2011) reported a slower thermo-oxidative rate in polyurethane/CoAl–LDH nanocomposites compared with neat PU, from 160 to 340 °C, probably due to the barrier effect of the exfoliated LDH layers. These results suggest potential applications of CoAl–LDH as flame retardants in PU.

Zhao *et al.* (2011) described transparent ethylene–vinyl alcohol copolymer (EVOH) nanocomposite films containing partially exfoliated LDHs intercalated with UV absorbers, prepared using DMSO as a solvent. The composite film obtained had a visible light transmittance of 90%, comparable to that of the pure matrix, was flexible and exhibited excellent UV-shielding capability, as well as improved thermal stability.

Bugatti *et al.* (2011) prepared films of microcomposites and exfoliated composites of PCL containing sodium 2,4-dichlorobenzoate or p-hydroxybenzoate simply dispersed in the polymer, which released antimicrobial species when in contact with physiologic solutions. The release of antimicrobial moieties 'freely dispersed' in the polymer happened much faster than for molecular anions bonded to the inorganic compound, and occurred in one step. In contrast, the release from the nanohybrids occurred in two stages: the first, a fast 'burst,' occurring during the first few days; and the second, very slow, extending over many months. The diffusion of the active molecules out of the microcomposite seemed to be slower than for the exfoliated nanohybrid. Moreover, the composite showed a chemical absorption hysteresis; hence this parameter has to be taken into account when tuning a release.

Lv *et al.* (2009b) developed UV-cured coatings consisting of urethane acrylates as oligomers and a diacrylate monomer reinforced with layered double hydroxides, resulting in coatings with enhanced thermal and mechanical properties. Also, a novel UV-cured polymer–LDH nanocomposite was prepared (Yuan and Shi, 2011) by modifying LDH with sodium dodecyl sulfate and (3-(methyl-acroloxy) propyl)trimethoxysilane (KH570), followed by UV irradiation after blending with the acrylate system. The storage modulus and glass transition temperature of the nanocomposite containing 5 wt.% LDH–KH increased to 47.5 MPa and 67.8 °C, respectively, compared with 39.7 MPa and 66 °C for the pure polymer. The tensile strength and Persoz hardness increased to 10.6 MPa and 111 s, respectively, from the 7.7 MPa and 85 s for the pure polymer. Similarly, Hu *et al.* (2011) prepared exfoliated polymer–LDH nanocomposites by UV-initiated photo-polymerization of acrylate systems using 5 wt.% of an Irgacure 2959-modified LDH precursor (LDH-2959) as a photo-initiator complex. The glass transition temperature of the

UV-cured exfoliated nanocomposites increased from the 55 °C of the pure polymer to 64 °C. The tensile strength increased from 10.1 MPa to 25.2 MPa; the Persoz hardness also increased, while the elongation at break remained at an acceptable level.

Marangoni *et al.* (2011) showed the possibilities for a new range of applications for layered double hydroxides intercalated with dyes in the preparation of polymer composite multifunctional materials. They prepared transparent, homogeneous, and colored nanocomposite films by casting after dispersing dye-intercalated LDHs (pigments) into commercial polyvinyl alcohol (PVA). They used ZnAl–LDH intercalated with anions of the dyes orange G, orange II, and methyl orange.

Jaymand (2011) synthesized and characterized an exfoliated modified syndiotactic polystyrene/MgAl-layered double-hydroxide nanocomposite. Compared with pure sPS-g-(PS-b-PMS), the nanocomposite showed a much higher decomposition temperature and higher glass transition temperature.

Intercalation, an important step for pillaring, means that layered double hydroxides can be used as carrier systems. The micro-porous properties can be used for controlled drug delivery. Cao *et al.* (2011) evaluated the potential use of an LDH–polymer nanocomposite as a drug delivery system for ocular delivery. Diclofenac was successfully intercalated into $ZnAlNO_3$–LDH using co-precipitation. *In vivo* pre-corneal retention studies were conducted with diclofenac sodium saline, diclofenac–LDH nanocomposite dispersion, 2% poly(vinyl pyrrolidone) (PVP) K30-diclofenac–LDH nanohybrid dispersion, and 10% PVP K30-diclofenac–LDH nanohybrid dispersion, separately. Compared with diclofenac sodium saline, all the dispersions extended the detectable time from 3 to 6 h; $C_{max}$ and AUC0-t of diclofenac–LDH nanocomposite dispersion showed 3.1-fold and 4.0-fold increases, respectively; $C_{max}$ and AUC0-t of 2% PVP K30–LDH nanohybrid dispersion were enhanced by about 5.3 and 6.0-fold, respectively. Additionally, no eye irritation was demonstrated in rabbits after single and repeated administration. These results show that this novel ocular drug delivery system is promising for improving the bioavailability of drugs in ophthalmic applications.

Wang *et al.* (2011) studied the combustion behavior via microcombustion calorimetry of polypropylene/organo-modified MgAl–LDH (PP/o-MgAl–LDH) composites prepared by melt-mixing. The results showed that the specific heat release rate (HRR), the heat release capacity (HRC), and the total heat release (THR) were reduced compared with the pure polymer. Further enhancements were observed by increasing the concentration of MgAl–LDH.

Kakati *et al.* (2011) prepared PP/NiAl–LDH nanocomposites with enhanced thermal stability due to a barrier effect induced by the LDH lamellar layers. For 10 wt.% loss, the decomposition temperature of a PP–LDH (5 wt.%) nanocomposite was 15 °C higher than that of neat PP. The thermal stability of the nanocomposites also increased with increasing LDH loading.

Hintze-Bruening *et al.* (2011) described an economically scalable approach to realize impact-resistant coatings built on substantially ordered platelets. Unlike other artificial structured nanocomposites, the parallel alignment of the LDH particles is obtainable with a single coating process. The alignment originated from an aqueous intermediate formed by a lyotropic liquid-crystal phase of polymer-stabilized LDH particles.

## 4.7 Conclusions

In conclusion, LDH–polymer nanocomposites are quite promising materials in several application areas. Although advances have been made in the development of LDHs for novel applications, further research is still necessary in order to develop feasible synthetic procedures for mass production to be used in full-scale anion exchange applications, as well as to improve selective absorption from multi-anionic systems. The great versatility of LDHs, that is the ability to modify their structures, provides many possibilities in using them as multifunctional materials for applications in materials science.

## 4.8 Sources of further information and advice

*Patents*

Although many patents discuss LDH–polymer nanocomposites, most of them include these materials merely as a particular case of clay–polymer nanocomposites. Therefore, not many patents are specifically about LDH–polymer nanocomposites. Some of the most recent ones consider the application of LDH nanocomposites in preventing hair loss.

Choy *et al.* (2010a, 2010b, 2010c) described an organic/inorganic nanohybrid complex containing eicosapentaenoic acid (EPA), vitamin C, or an indole-3-acetic acid, respectively, to maximize hair growth, having stability in heat, light, air, and oxygen. The nanohybrid complex for promoting hair growth and preventing hair loss was inserted between the LDH with the structural formula $(M^{2+}{}_{(1-x)}N^{3+}{}_{x}(OH)_2)A_{xy}H_2O$. A specific composition for preventing hair loss and promoting hair growth contained 0.01 to 20 wt.% of nanocomposite with EPA. Another composition used the nanohybrid complex as an active ingredient in which vitamin C was inserted between LDH layers. Other compositions used indole-3-acetic acid or a nanohybrid complex as the active ingredient and were applied as a lotion, emulsion, cream, essence, or spray.

Gook *et al.* (2008) patented a polyelectrolyte nanocomposite and a manufacturing method capable of improving the tensile modulus and ion conductance to obtain a high photoelectric effect and to improve productivity while maintaining uniform physical properties. The polyelectrolyte nanocomposite consisted of LDH dispersed in a polyethylene glycol acrylate (PEDGA) polymer matrix.

Ferrara *et al.* (2010) prepared a polyolefin nanocomposite material based on a crystalline or semi-crystalline polyolefin matrix and 0.02 to 6 wt.% of modified hydrotalcite. In particular, these authors reported the development of fibers, films, blisters, thermoformed, blow-molded, and injection-molded products based on these materials all having improved properties (good barrier properties, good stiffness and optical properties and, in some cases, improved thermo-mechanical and processing properties) compared to similar systems reinforced with conventional clays.

Yongzhong *et al.* (2005) described a method for preparing poly(vinyl chloride)/ hydrotalcite nanocomposites including steps of reacting the hydrotalcite with fatty acids or anion surface activators, and obtained modified hydrotalcites through filtering and drying. The addition of these modified hydrotalcites, the initiator, and the dispersing agent to vinyl chloride and deionized water inside a reactor at room temperature produced PVC composites by increasing the temperature to between 35 and 65 °C.

*LDH producers*

- Süd-Chemie (www.sud-chemie.com/scmcms/web/page_en_5375.htm) produces the hydrotalcite grades SORBACID and HYCITE.
- Sasol (www.sasolgermany.de/index.php?id=47) produces boehmite, high-purity alumina, and hydrotalcite (PURAL MG grades).
- INEOS Silicas (www.ineossilicas.com) produces hydrotalcite grades MACROSORB.
- AkzoNobel (www.akzonobel.com/brands_products) produces aluminum magnesium layered double hydroxide modified with hydrogenated fatty acids (PERKALITE grades).
- CIMTECLAB (www.cimteclab.net) produces modified hydrotalcite for fire-retardancy applications (Nanofire grades).

## 4.9 References

Acharya H, Srivastava S K and Bhowmick A K (2007a), 'A solution blending route to ethylene propylene diene terpolymer/layered double hydroxide nanocomposites', *Nanoscale Res Lett*, 2, 1–5.

Acharya H, Srivastava S K and Bhowmick A K (2007b), 'Synthesis of partially exfoliated EPDM/LDH nanocomposites by solution intercalation: Structural characterization and properties', *Compos Sci Tech*, 67, 2807–2816.

Alexandre M, Dubois P, Sun T, Garces J M and Jérome R (2002), 'Polyethylene-layered silicate nanocomposites prepared by the polymerization-filling technique: synthesis and mechanical properties', *Polymer*, 43, 2123–2132.

Ardanuy M and Velasco J I (2011), 'Mg–Al Layered double hydroxide nanoparticles. Evaluation of the thermal stability in polypropylene matrix', *Appl Clay Sci*, 51, 341–347.

Ardanuy M, Velasco J I, Antunes M, Rodriguez-Perez M A and de Saja J A (2010), 'Structure and properties of polypropylene/hydrotalcite nanocomposites', *Polym Compos*, 870–878.

Ardanuy M, Velasco J I, Maspoch M L, Haurie L and Fernández A I (2009), 'Influence of EMAA compatibilizer on the structure and properties of HDPE/hydrotalcite nanocomposites prepared by melt mixing', *J Appl Polym Sci*, 113, 950–958.

Ardanuy M, Velasco J I, Realinho V, Arencón D and Martínez A B (2008), 'Non-isothermal crystallization kinetics and activity of filler in polypropylene/Mg–Al layered double hydroxide nanocomposites', *Thermochim Acta*, 479, 45–52.

Bao Y Z, Cong L F, Huang Z M and Weng Z X (2008a), 'Preparation and proton conductivity of poly(vinylidene fluoride)-layered double hydroxide nanocomposite gel electrolytes', *J Mater Sci*, 43, 390–394.

Bao Y Z, Huang Z M, Li S X and Weng Z X (2008b), 'Thermal stability, smoke emission and mechanical properties of poly(vinyl chloride)/hydrotalcite nanocomposites', *Polym Degrad Stab*, 93, 448–455.

Bao Y Z, Huang Z M and Weng Z X (2006), 'Preparation and characterization of poly(vinyl chloride)/ layered double hydroxides nanocomposite via *in situ* suspension polymerization', *J Appl Polym Sci*, 102, 1471–1477.

Bugatti V, Costantino U, Gorrasi G, Nocchetti M, Tammaro L and Vittoria V (2010), 'Nano-hybrids incorporation into poly($\varepsilon$-caprolactone) for multifunctional applications: Mechanical and barrier properties', *Eur Polym J*, 46, 418–427.

Bugatti V, Gorrasi G, Montanari F, Nocchetti M, Tammaro L and Vittoria V (2011), 'Modified layered double hydroxides in polycaprolactone as a tunable delivery system: *In vitro* release of antimicrobial benzoate derivatives', *App. Clay Sci*, 52(1–2), 34–40.

Camino G, Maffezzoli A, Braglia M, de Lazzaro M and Zammarano M (2001), 'Effect of hydroxides and hydroxycarbonate structure on fire retardant effectiveness and mechanical properties in ethylene-vinyl acetate copolymer', *Polym Degrad Stab*, 74, 457–464.

Cao F, Wang Y, Ping Q and Liao Z (2011), 'Zn-Al-NO3-layered double hydroxides with intercalated diclofenac for ocular delivery', *Int J Pharma*, 404(1–2), 250–256.

Chan Y N, Juang T Y, Liao Y L, Dai S A and Lin J J (2008), 'Preparation of clay/epoxy nanocomposites by layered-double-hydroxide initiated self-polymerization', *Polymer*, 49, 4796–4801.

Chen W, Feng L and Qu B (2004a), '*In situ* synthesis of poly(methyl methacrylate)/MgAl layered double hydroxide nanocomposite with high transparency and enhanced thermal properties', *Solid State Communications*, 130, 259–263.

Chen W, Feng L and Qu B (2004b), 'Preparation of nanocomposites by exfoliation of ZnAl layered double hydroxides in nonpolar LLDPE solution', *Chem Mater*, 16, 368–370.

Chen W and Qu B (2003), 'Structural characteristics and thermal properties of PE-g-MA/MgAl–LDH exfoliation nanocomposites synthesized by solution intercalation', *Chem Mater*, 15, 3208–3213.

Chen W and Qu B (2005), 'Enhanced thermal and mechanical properties of poly(methyl acrylate)/ZnAl layered double hydroxide nanocomposites formed by *in situ* polymerization', *Polym Degrad Stab*, 90, 162–166.

Chen X G, Wu D D, Lv S S, Zhang L, Ye Y and Cheng J P (2010), 'Layered double hydroxide/NaSb(OH)6–poly(vinyl chloride) nanocomposites: Preparation, characterization, and thermal stability', *J Appl Polym Sci*, 116, 1977–1984.

Choy J H, Choul K B and Hwan P D (Cnpharm Co Ltd) (2010a), *Nanocomposite comprising eicosapentaenoic acid for preventing hair loss and enhancing hair restoration, composition comprising the nanocomposite*. KR20090005980 20090123.

Choy J H, Choul K B and Hwan P D (Univ EWHA Ind Collaboration (KR); Cnpharm Co Ltd (KR)) (2010b), *Nanocomposite comprising vitamin C for preventing hair loss and enhancing hair restoration, composition comprising the nanocomposite* (KR 20100086560-A). Application number: KR20090005825 20090123, Priority number: KR20090005825 2009012.

Choy J H, Choul K B and Hwan P D (Univ EWHA Ind Collaboration (KR)) (2010c), *Indole-3-acetic acid for preventing hair loss and enhancing hair restoration, nanocomposite comprising the compound, and composition comprising the compound or the nanocomposite* (KR 20100086122-A). Application number: KR20090005317 20090122; Priority number: KR20090005317 20090122.

Costa F R, Abdel-Goad M, Wagenknecht U and Heinrich G (2005), 'Nanocomposites based on polyethylene and Mg–Al layered double hydroxide. I. Synthesis and characterization', *Polymer*, 46, 4447–4453.

Costa F R, Satapathy B K, Wagenknecht U, Weidisch R and Heinrich G (2006b), 'Morphology and fracture behaviour of polyethylene/Mg–Al layered double hydroxide (LDH) nanocomposites', *Eur Polym J*, 42, 2140–2152.

Costa F R, Wagenknecht U and Heinrich G (2007), 'LDPE/MgeAl layered double hydroxide nanocomposite: Thermal and flammability properties', *Polym Degrad Stab*, 92, 1813–1823.

Costa F R, Wagenknecht U, Jehnichen D, Goad M A and Heinrich G (2006a), 'Nanocomposites based on polyethylene and Mg–Al layered double hydroxide. Part II. Rheological characterization', *Polymer*, 47, 1649–1660.

Costantino U, Gallipoli A, Nocchetti M, Camino G, Bellucci F and Frache A (2005), 'New nanocomposites constituted of polyethylene and organically modified ZnAl-hydrotalcites', *Polym Degrad Stab*, 90, 586–590.

Das A, Wang D Y, Leuteritz A, Subramaniam K, Greenwell H C, *et al.* (2011), 'Preparation of zinc oxide free, transparent rubber nanocomposites using a layered double hydroxide filler', *J Mater Chem*, 21(20), 7194–7200.

Dimotakis E D and Pinnavaia T J (1990), 'New route to layered double hydroxides intercalated by organic anions: precursors to polyoxometalate-pillared derivates', *Inorg Chem*, 29(13), 2393–2394.

Ding P and Qu B (2006), 'Synthesis of exfoliated PP/LDH nanocomposites via melt-intercalation: structure, thermal properties, and photo-oxidative behavior in comparison with PP/MMT nanocomposites', *Polym Eng Sci*, 14, 1153–1159.

Ding Y, Gui Z, Zhu J, Hu Y and Wang Z (2008), 'Exfoliated poly(methyl methacrylate)/MgFe-layered double hydroxide nanocomposites with small inorganic loading and enhanced properties', *Mater Res Bulletin*, 43, 3212–3220.

Du L and Qu B (2006), 'Structural characterization and thermal oxidation properties of LLDPE/MgAl–LDH nanocomposites', *J Mater Chem*, 16, 1549–1554.

Du L and Qu B (2007), 'Effects of synthesis conditions on crystal morphological structures and thermal degradation behavior of hydrotalcites and flame retardant and mechanical properties of EVA/hydrotalcite blends', *Polym Compos*, 131–138.

Du L, Qu B and Xu Z (2006), 'Flammability characteristics and synergistic effect of hydrotalcite with microencapsulated red phosphorus in halogen-free flame retardant EVA composite', *Polym Degrad Stab*, 91, 995–1001.

Du L, Qu B and Zhang M (2007), 'Thermal properties and combustion characterization of nylon 6/MgAl–LDH nanocomposites via organic modification and melt intercalation', *Polym Degrad Stab*, 92, 497–502.

Dupin J C, Martinez H, Guimon C, Dumitriu E and Fechete I (2004), 'Intercalation compounds of Mg-Al layered double hydroxides with dichlophenac: different methods of preparation and physico-chemical characterization', *Appl Clay Sci*, 27, 95–106.

Ferrara G, Costantini E and Consalvi M (Basell Poliolefine SRL) (2010), *Polyolefin nanocomposites materials* (US 2010304068-A1). Application number: US20080734858 20081015; priority number: US20080734858 20081015; EP20070121629 20071127; US20070005731P 20071207; WO2008EP63819 20081015.

Giannelis E P (1996), 'Polymer layered silicate nanocomposites', *Adv Mater*, 8 29–35.

Gook S K, Kwan L Y, Suk C M and Hee P S (Univ Kyung Hee Univ Ind Coop (KR); Gyeonggi Do (KR)) (2008), *Polymer electrolyte comprising nanocomposite and preparing method thereof (KR 20080105587-A)*. Application number: KR20070053355 20070531; priority number: KR20070053355 20070531.

Gopakumar T G, Lee J A, Kontopoulou M and Parent J S (2002), 'Influence of clay exfoliation on the physical properties of montmorillonite/polyethylene composites', *Polymer*, 43, 5483–5491.

Guo S, Zhang C, Peng H, Wang W and Liu T (2011), 'Structural characterization, thermal and mechanical properties of polyurethane/CoAl layered double hydroxide nanocomposites prepared via *in situ* polymerization', *Compos Sci Tech*, 71, 791–796.

He F A, Zhang L M, Yang F, Chen L S and Wu Q (2006), 'New nanocomposites based on syndiotactic polystyrene and organo-modified ZnAl layered double hydroxide', *J Polym Res*, 13, 483–493.

Herrero M, Benito P, Labajos F M, Rives V, Zhu Y D, *et al.* (2010), 'Structural characterization and thermal properties of polyamide 6.6/Mg, Al/adipate–LDH nanocomposites obtained by solid state polymerization', *J Solid State Chem*, 183, 1645–1651.

Hibino T and Tsunashima A (1998), 'Characterization of repeatedly reconstructed Mg-Al hydrotalcite-like compounds: gradual segregation of aluminium from the structure', *Chem Mater*, 10, 4055–4061.

Hintze-Bruening H, Troutier A L and Leroux F (2011), 'Layered particle based polymer composites for coatings: Part III – Textured coatings obtained via lyotropic liquid crystals', *Prog Org Coat*, 70(4), 240–244.

Hsueh H B and Chen C Y (2003a), 'Preparation and properties of LDHs/epoxy nanocomposites', *Polymer*, 44, 5275–5283.

Hsueh H B and Chen C Y (2003b), 'Preparation and properties of LDHs/polyimide nanocomposites', *Polymer*, 44, 1151–1161.

Hu L, Yuan Y and Shi W (2011), 'Preparation of polymer/LDH nanocomposite by UV-initiated photopolymerization of acrylate through photoinitiator-modified LDH precursor', *Mater Res Bulletin*, 46(2), 244–251.

Illaik A, Taviot-Gueho C, Lavis J, Commereuc S, Verney V and Leroux F (2008), 'Unusual polystyrene nanocomposite structure using emulsifier-modified layered double hydroxide as nanofiller', *Chem Mater*, 20, 4854–4860.

Jaymand M (2011), 'Synthesis and characterization of an exfoliated modified syndiotactic polystyrene/Mg-Al-layered double-hydroxide nanocomposite', *Polym J*, 43(2), 186–193.

Kakati K, Prakash A and Pugazhenthi G (2011), 'Thermal properties and thermal degradation kinetics of polypropylene/organomodified Ni-Al layered double hydroxide (LDH) nanocomposites prepared by melt intercalation technique', *Key Eng Mater*, 471–472, 209–214.

Kornmann X (2001), 'Synthesis and characterisation of thermoset–clay nanocomposites', PhD dissertation, Division of Polymer Engineering, Lulea University of Technology, Lulea, Sweden.

Kotal M, Kuila T, Srivastava S K and Bhowmick A K (2009), 'Synthesis and characterization of polyurethane/Mg-Al layered double hydroxide nanocomposites', *J Appl Polym Sci*, 114, 2691–2699.

Kotal M, Srivastava S K, Bhowmick A K and Chakraborty S K (2011), 'Morphology and properties of stearate-intercalated layered double hydroxide nanoplatelet-reinforced thermoplastic polyurethane', *Polym Int*, 60(5), 772–780.

Kuila T, Acharya H, Srivastava S K and Bhowmick A K (2007), 'Synthesis and characterization of ethylene vinyl acetate/Mg–Al layered double hydroxide nanocomposites', *J Appl Polym Sci*, 104, 1845–1851.

Kuila T, Srivastava S K, Bhowmick A K and Saxena A K (2008), 'Thermoplastic polyolefin based polymer – blend-layered double hydroxide nanocomposites', *Compos Sci Tech*, 68, 3234–3239.

Lee W D and Im S S (2007), 'Thermomechanical properties and crystallization behavior of layered double hydroxide/poly(ethylene terephthalate) nanocomposites prepared by *in situ* polymerization', *J Polym Sci B: Polym Phys*, 45, 28–40.

Lee W D, Im S S, Lim H M and Kim K J (2006), 'Preparation and properties of layered double hydroxide/poly(ethyleneterephthalate) nanocomposites by direct melt compounding', *Polymer*, 47, 1364–1371.

Leroux F, Illaik A, Stimpfling T, Troutier-Thuilliez A L, Fleutot S, *et al.* (2010), 'Percolation network of organo-modified layered double hydroxide platelets into polystyrene showing enhanced rheological and dielectric behavior', *J Mater Chem*, 20, 9484–9494.

Leroux F, Illaik A and Verney V (2009), 'A comprehensive study of an unusual jammed nanocomposite structure using hybrid layered double hydroxide filler', *J Coll Inter Sci*, 332, 327–335.

Leroux F, Meddar L, Mailhot B, Morlat-Therias S and Gardette J L (2005), 'Characterization and photooxidative behaviour of nanocomposites formed with polystyrene and LDHs organo-modified by monomer surfactant', *Polymer*, 46, 3571–3578.

Liu J, Chen G and Yang J (2008), 'Preparation and characterization of poly(vinyl chloride)/ layered double hydroxide nanocomposites with enhanced thermal stability', *Polymer*, 49, 3923–3927.

Liu J, Chen G, Yang J and Ma Y (2007), 'New facile preparation of a poly(vinyl chloride)/ layered double hydroxide nanocomposite via solution intercalation', *Chem Lett*, 36(12), 1454–1455.

Lonkar S P, Morlat-Therias S, Caperaa N, Leroux F, Gardette J L and Singh RP (2009), 'Preparation and nonisothermal crystallization behavior of polypropylene/layered double hydroxide nanocomposites', *Polymer*, 50, 1505–1515.

Lonkar S P, Therias S, Capera N, Leroux F and Gardette J L (2010), 'Photooxidation of polypropylene/layered double hydroxide nanocomposites: Influence of intralamellar cations', *Eur Polym J*, 46, 1456–1464.

Lv S, Yuan Y and Shi W (2009a), 'Strengthening and toughening effects of layered double hydroxide and hyperbranched polymer on epoxy resin', *Prog Org Coat*, 65, 425–430.

Lv S, Zhou W, Miao H and Shi W (2009b), 'Preparation and properties of polymer/LDH nanocomposite used for UV curing coatings', *Prog Org Coat*, 65, 450–456.

Manzi-Nshuti C, Chen D, Su S and Wilkie C A (2009a), 'Structure–property relationships of new polystyrene nanocomposites prepared from initiator-containing layered double hydroxides of zinc aluminum and magnesium aluminum', *Polym Degrad Stab*, 94, 1290–1297.

Manzi-Nshuti C, Hossenlopp J M and Wilkie C A (2008a), 'Fire retardancy of melamine and zinc aluminum layered double hydroxide in poly(methyl methacrylate)', *Polym Degrad Stab*, 93, 1855–1863.

Manzi-Nshuti C, Hossenlopp J M and Wilkie C A (2009b), 'Comparative study on the flammability of polyethylene modified with commercial fire retardants and a zinc aluminum oleate layered double hydroxide', *Polym Degrad Stab*, 94, 782–788.

Manzi-Nshuti C, Wang D, Hossenlopp J M and Wilkie C A (2008b), 'Aluminum-containing layered double hydroxides: the thermal, mechanical, and fire properties of (nano) composites of poly(methyl methacrylate)', *J Mater Chem*, 18, 3091–3102.

Marangoni R, da Costa-Gardolinski J E F, Mikowski A and Wypych F (2011), 'PVA nanocomposites reinforced with $Zn_2Al$ LDHs, intercalated with orange dyes', *J Solid State Electrochem*, 15(2), 303–311.

Marangoni R, Taviot-Gueho C, Illaik A, Wypych F and Leroux F (2008), 'Organic inorganic dye filler for polymer: Blue-coloured layered double hydroxides into polystyrene', *J Coll Inter Sci*, 326, 366–373.

Martínez A B, Realinho V, Antunes M, Maspoch M L and Velasco J I (2011), 'Microcellular foaming of layered double hydroxide-polymer nanocomposites', *Ind Eng Chem Res*, 50(9), 5239–5247.

Martínez-Gallegos S, Herrero M and Rives V (2008), '*In situ* microwave-assisted polymerization of polyethylene terephtalate in layered double hydroxides', *J Appl Polym Sci*, 109, 1388–1394.

Matusinovic Z, Rogosic M and Sipusic J (2009), 'Synthesis and characterization of poly(styrene-co-methyl methacrylate)/layered double hydroxide nanocomposites via *in situ* polymerization', *Polym Degrad Stab*, 94, 95–101.

Meyn M, Beneke K and Lagaly G (1990), 'Anion-exchange reactions of layered double hydroxides', *Inorg Chem*, 29, 5201–5207.

Miyata S (1980), 'Physico-chemical properties of synthetic hydrotalcites in relation to compositions', *Clay Miner*, 28, 50–56.

Miyata S and Kumura T (1973), 'Synthesis of new hydrotalcite-like compounds and their physico-chemical properties', *Chem Lett*, 843–848.

Morgan A B and Gilman J W (2003), 'Characterization of polymer-layered silicate (clay) nanocomposites by transmission electron microscopy and X-ray diffraction: A comparative study', *J Appl Polym Sci*, 87, 1329–1338.

Newman S P and Jones W (1998), 'Synthesis, characterization and applications of layered double hydroxides containing organic guests', *New J Chem*, 105–115.

Nyambo C and Wilkie C A (2009), 'Layered double hydroxides intercalated with borate anions: Fire and thermal properties in ethylene vinyl acetate copolymer', *Polym Degrad Stab*, 94, 506–512.

Nyambo C, Chen D, Su S and Wilkie C A (2009a), 'Does organic modification of layered double hydroxides improve the fire performance of PMMA?', *Polym Degrad Stab*, 94, 1298–1306.

Nyambo C, Kandare E and Wilkie C A (2009b), 'Thermal stability and flammability characteristics of ethylene vinyl acetate (EVA) composites blended with a phenyl phosphonate-intercalated layered double hydroxide (LDH), melamine polyphosphate and/or boric acid', *Polym Degrad Stab*, 94, 513–520.

Nyambo C, Songtipya P, Manias E, Jimenez-Gasco M M and Wilkie C A (2008), 'Effect of MgAl-layered double hydroxide exchanged with linear alkyl carboxylates on fire-retardancy of PMMA and PS', *J Mater Chem*, 18, 4827–4838.

Okada A, Kawasumi M, Usuki A, Kojima Y, Kurauchi T and Kamigaito O (1990), 'Synthesis and properties of nylon-6/clay hybrids', in Schaefer D W and Mark J E, *Polymer based molecular composites*. MRS Symposium Proceedings, Pittsburgh, 171, 45–50.

Pavlidoua S and Papaspyrides C D (2008), 'A review on polymer-layered silicate nanocomposites', *Progr Polym Sci*, 33, 1119–1198.

Peng H, Han Y, Liu T, Tjiu W C and He C (2010), 'Morphology and thermal degradation behavior of highly exfoliated CoAl-layered double hydroxide/polycaprolactone nanocomposites prepared by simple solution intercalation', *Thermochim Acta*, 502, 1–7.

Peng H D, Tjiu W C, Shun L, Huang S, He C B and Liu T X (2009), 'Preparation and mechanical properties of exfoliated CoAl layered double hydroxide (LDH)/polyamide 6 nanocomposites by *in situ* polymerization', *Compos Sci Tech*, 69, 991–996.

Pinnavaia T J and Beall G W (2001), *Polymer-clay Nanocomposites*, England, John Wiley & Sons.

Pradhan S, Costa F R, Wagenknecht U, Jehnichen D, Bhowmick A K and Heinrich G (2008), 'Elastomer/LDH nanocomposites: Synthesis and studies on nanoparticle dispersion, mechanical properties and interfacial adhesion', *Eur Polym J*, 44, 3122–3132.

Pucciariello R, Tammaro L, Villani V and Vittoria V (2007), 'New nanohybrids of poly(e-caprolactone) and a modified Mg/Al hydrotalcite: mechanical and thermal properties', *J Polym Sci B: Polym Phys*, 45, 945–954.

Qiu L, Chen W and Qu B (2005), 'Structural characterisation and thermal properties of exfoliated polystyrene/ZnAl layered double hydroxide nanocomposites prepared via solution intercalation', *Polym Degrad Stab*, 87, 433–440.

Qiu L, Chen W and Qu B (2006), 'Morphology and thermal stabilization mechanism of LLDPE/MMT and LLDPE/LDH nanocomposites', *Polymer*, 47, 922–930.

Qiu L and Qu B (2006), 'Preparation and characterization of surfactant-free polystyrene/layered double hydroxide exfoliated nanocomposite via soap-free emulsion polymerization', *J Coll Inter Sci*, 301, 347–351.

Ramaraj B, Nayak S K and Yoon K R (2010), 'Poly(vinyl alcohol) and layered double hydroxide composites: Thermal and mechanical properties', *J Appl Polym Sci*, 116, 1671–1677.

Ramaraj B and Yoon K R (2008), 'Thermal and physicomechanical properties of ethylene-vinyl acetate copolymer and layered double hydroxide composites', *J Appl Polym Sci*, 108, 4090–4095.

Realinho V, Antunes M, Arencón D, Fernández A I and Velasco J I (2009), 'Effect of a dodecylsulfate-modified magnesium-aluminum layered double hydroxide on the morphology and fracture of polystyrene and poly(styrene-coacrylonitrile) composites', *J Appl Polym Sci*, 111, 2574–2583.

Rives V (2001), *Layered Double Hydroxides: Present and Future*. New York, Nova Science Publishers.

Rives V (2002), 'Characterisation of layered double hydroxides and their decomposition products', *Mater Chem Phys*, 75, 19–25.

Romeo V, Gorrasi G and Vittoria V (2007), 'Encapsulation and exfoliation of inorganic lamellar fillers into polycaprolactone by electrospinning', *Biomacromol*, 8(10), 3147–3152.

Schonhals A, Goering H, Costa F R, Wagenknecht U and Heinrich G (2009), 'Dielectric properties of nanocomposites based on polyethylene and layered double hydroxide', *Macromol*, 42, 4165–4174.

Shi Y, Chen F, Yang J and Zhong M (2010), 'Crystallinity and thermal stability of LDH/polypropylene nanocomposites', *Appl Clay Sci*, 50, 87–91.

Sorrentino A, Gorasi G, Tortora M, Vittoria V, Constantino U, *et al.* (2005), 'Incorporation of Mg-Al hydrotalcite into biodegradable poly(ε-caprolactone) by high energy ball milling', *Polymer*, 46, 1601–1608.

Tagaya H, Sato S, Morioka H, Kadokawa J, Karasu M and Chiba K (1993), 'Preferential intercalation of isomers of naphthalenecarboxylate ions into the interlayer of layered double hydroxides', *Chem Mater*, 5, 1431–1433.

Taviot-Gueho C, Illaik A, Vuillermoz C, Commereuc S, Verney V and Leroux F (2007), 'LDH–dye hybrid material as coloured filler into polystyrene: Structural characterization and rheological properties', *J Phys Chem Solids*, 68, 1140–1146.

Trujillano R, Holgado M J and Rives V (2002), 'Alternative synthetic routes for NiAl layered double hydroxides with alkyl and alkylbenzene sulfonates', in Aiello R, Giordano G and Testa F, *Studies in Surface Science and Catalysis*, 142, Elsevier Science.

Tsai T Y, Lu S W and Li F S (2008), 'Preparation and characterization of epoxy/layered double hydroxides nanocomposites', *J Phys Chem Solids*, 69, 1386–1390.

Tseng C H, Hsueh H B and Chen C Y (2007), 'Effect of reactive layered double hydroxides on the thermal and mechanical properties of LDHs/epoxy nanocomposites', *Compos Sci Tech*, 67, 2350–2362.

Ulibarri M A, Labajos F M, Rives V, Trujillano R, Kagunya W and Jones W (1994), 'Comparative study of the synthesis and properties of vanadate-exchanged layered double hydroxides', *Inorg Chem*, 33, 2592–2599.

Vaia R A (2001), 'Structural characterization of polymer-layered silicate nanocomposites', in Pinnnavaia T J and Beall G W, *Polymer-Clay Nanocomposites*, Wiley Series in Polymer Science, 229–263.

Wagner R and Reisinger T J G (2003), 'A rheological method to compare the degree of exfoliation of nanocomposites', *Polymer*, 44, 7513–7518.

Wang D Y W, Das A, Costa F R, Leuteritz A, Wang Y Z, *et al.* (2010), 'Synthesis of organo cobalt-aluminum layered double hydroxide via a novel single-step self-assembling method and its use as flame retardant nanofiller in PP', *Langmuir*, 26(17), 14162–14169.

Wang D Y W, Leuteritz A, Kutlu B, Landwehr M A D, Jehnichen D, *et al.* (2011), 'Preparation and investigation of the combustion behavior of polypropylene/organomodified MgAl–LDH micro-nanocomposite', *J Alloys and Compounds*, 509(8), 3497–3501.

Wang D Y, Liu L, Wang Y Z, Stec A A, Hull T R and Price D (2008), 'Effect of metal chelates on the ignition and early flaming behaviour of intumescent fire-retarded polyethylene systems', *Polym Degrad Stab*, 93, 1024–1030.

Wang L, Su S, Chen D and Wilkie C A (2009), 'Variation of anions in layered double hydroxides: Effects on dispersion and fire properties', *Polym Degrad Stab*, 94, 770–781.

Wang G A, Wang C C and Chen C Y (2005), 'The disorderly exfoliated LDHs/PMMA nanocomposite synthesized by *in situ* bulk polymerization', *Polymer*, 46, 5065–5074.

Wang G A, Wang C C and Chen C Y (2006), 'The disorderly exfoliated LDHs/PMMA nanocomposites synthesized by *in situ* bulk polymerization: The effects of LDH-U on thermal and mechanical properties', *Polym Degrad Stab*, 91, 2443–2450.

Xu Z P, Saha S K, Braterman P S and d'Souza N (2006), 'The effect of Zn, Al layered double hydroxide on thermal decomposition of poly(vinyl chloride)', *Polym Degrad Stab*, 91, 3237–3244.

Yongzhong Y B, Zhiming H and Zhixue W (Univ Zhejiang) (2005), *Nano composite resin ofpoly(vinyl chloride)/hydrotalcite* (CN 1563179-AApplicant number: CN20041017944 20040421; priority number: CN20041017944 20040421.

Yuan Y and Shi W (2011), 'Preparation and properties of UV-cured acrylated silane intercalated polymer/LDH nanocomposite', *Mater Res Bul*, 46(1), 124–129.

Zammarano M, Bellayer S, Gilman J W, Franceschi M, Beyer F L, Harris R H and Meriani S (2006), 'Delamination of organo-modified layered double hydroxides in polyamide 6 by melt processing', *Polymer*, 47, 652–662.

Zammarano M, Franceschi M, Bellayer S, Gilman J W and Meriani S (2005), 'Preparation and flame resistance properties of revolutionary self-extinguishing epoxy nanocomposites based on layered double hydroxides', *Polymer*, 46, 9314–9328.

Zhang M, Ding P, Du L and Qu B (2008a), 'Structural characterization and related properties of EVA/ZnAl–LDH nanocomposites prepared by melt and solution intercalation', *Mater Chem Phys*, 109, 206–211.

Zhang M, Ding P, Qu B and Guan A (2008b), 'A new method to prepare flame retardant polymer composites', *J Mater Proces Tech*, 208 (1–3), 342–347.

Zhang G, Ding P, Zhang M and Qu B (2007), 'Synergistic effects of layered double hydroxide with hyperfine magnesium hydroxide in halogen-free flame retardant EVA/HFMH/LDH nanocomposites', *Polym Degrad Stab*, 92, 1715–1720.

Zhao Y, Yang W, Xue Y, Wang X and Lin T (2011), 'Partial exfoliation of layered double hydroxides in DMSO: A route to transparent polymer nanocomposites', *J Mater Chem*, 21(13), 4869–4874.

Zhu Y D, Allen G C, Adams J M, Gittins D, Herrero M, *et al.* (2008), 'Dispersion characterization in layered double hydroxide/nylon 66 nanocomposites using FIB imaging', *J Appl Polym Sci*, 108, 4108–4113.

Zubitur M, Gómez M A and Cortázar M (2009), 'Structural characterization and thermal decomposition of layered double hydroxide/poly(p-dioxanone) nanocomposites', *Polym Degrad Stab*, 94, 804–809.

# 5

# Cellulose nanoparticles as reinforcement in polymer nanocomposites

E. C. RAMIRES, Universidade de São Paulo (USP), Brazil and
A. DUFRESNE, Grenoble Institute of Technology – INP, France

**Abstract:** This chapter describes the preparation, morphological features, and physical properties of cellulose nanocrystals and nanofibrillated cellulose and their incorporation in nanocomposite materials, including processing methods and ensuing properties. Cellulose nanoparticles have remarkable properties such as light weight, low cost, and availability of the raw material, renewability, nanoscale dimension, and unique morphology. Because of these properties, cellulose nanoparticles have been largely applied as reinforcing fillers in nanocomposite materials.

**Key words:** cellulose nanocrystal, nanofibrillated cellulose, polymer nanocomposite, natural fiber.

## 5.1 Introduction

Cellulose nanoparticles have attracted significant interest in the last 15 years because of their impressive mechanical properties and nanoscale dimensions, which result in very high surface area to volume ratios (Eichhorn *et al.*, 2010). These characteristics, along with the remarkable suitability for surface functionalization, make them ideal candidates to improve the mechanical properties of the host material, in the preparation of nanocomposites. The use of cellulose nanoparticles has other advantages, e.g. low cost, low density, renewable nature, biodegradability, wide variety of available filler, low energy consumption, high performance, and modest abrasivity during processing (Siqueira *et al.*, 2011).

Aqueous suspensions of cellulose nanoparticles can be prepared by mechanical treatment (leading to the production of nanofibrillated cellulose or NFC) or acid hydrolysis of the biomass (leading to the production of cellulose nanocrystals or CNCs). The object of acid hydrolysis is to dissolve away regions of low lateral order so that the water-insoluble, highly crystalline residue may be converted into a stable suspension by subsequent vigorous mechanical shearing. The resulting nanocrystals are rod-like particles or whiskers, whose dimensions depend on the nature of the substrate and the hydrolysis conditions, but are in the nanometer scale. Because these nanocrystals contain only a small number of defects, their axial Young's modulus, which has a limit derived from theoretical chemistry of 167.5 GPa, is close to that of steel (200 GPa). These very stiff nanoparticles are therefore suitable for producing green nanocomposite materials.

Cellulose is a high-molecular-weight homopolysaccharide composed of $\beta$–1, 4–anhydro--glucopyranose (Klemm *et al.*, 2005). One of the characteristics of cellulose is that its monomers have three hydroxyl groups. These hydroxyl groups and their hydrogen-bonding ability play a major role in crystalline packing and in governing the important physical properties of these highly cohesive materials. Natural cellulose is referred to as cellulose I, or native cellulose.

In nature, cellulose is slender and rod-like or threadlike, due to a linear association of crystallites. These rods or threads are called microfibrils; they are a collection of cellulose chains and form the basic structural unit of a plant cell wall. Each microfibril is a string of cellulose crystallites, linked along the chain axis by amorphous domains. Their structure consists of a predominantly crystalline cellulosic core, covered by a sheath of paracrystalline polyglucosan material surrounded by hemicelluloses (Whistler and Richards, 1970). These microfibrils are cemented by other polymers such as lignin and hemicelluloses and aggregate further to form lignocellulosic fibers. Depending on their origin, the microfibril diameters range from about 2 to 20 nm (Lu *et al.*, 2008a). According to Sjoström, native cellulose has a degree of polymerization (DP) of approximately 10 000 glucopyranose units in wood cellulose and 15 000 in cotton cellulose (Sjoström, 1981). However, Bledzki and Gassan (1999) showed that purification procedures usually reduce the DP, e.g. a DP of 14 000 in native cellulose can be reduced to about 2500.

In addition to plant origins, cellulose fibers are also secreted extracellularly by certain bacteria belonging to the genera *Acetobacter, Agrobacterium, Alcaligenes, Pseudomonas, Rhizobium*, and *Sarcina* (El-Saied *et al.*, 2004). The most efficient producer of bacterial cellulose is *Acetobacter xylinum* (or *Gluconacetobacter xylinus*), a Gram-negative strain of acetic-acid-producing bacteria (Brown, 2004; Klemm *et al.*, 2006).

The expectation for cellulose-based composites lies in the Young's modulus of the cellulose crystallite. It was first experimentally studied in 1962 in the crystal deformation of cellulose I using highly oriented fibers of bleached ramie (Sakurada *et al.*, 1962) and a value of 137 GPa was found. This value differs from the theoretical estimate of 167.5 GPa reported by Tashiro and Kobayashi (1991). More recently, Raman spectroscopy has been used to measure the elastic modulus of native cellulose crystals (Šturcova *et al.*, 2005) and a value around 143 GPa was found. The elastic modulus of single microfibrils from tunicate was measured by atomic force microscopy (AFM) using a three-point bending test (Iwamoto *et al.*, 2009). Values of 145 and 150 GPa were reported for single microfibrils prepared by 2,2,6,6-tetramethylpiperidine-1-oxyl radical (TEMPO)-oxidation and sulfuric acid hydrolysis, respectively. These values are much higher than the Young's modulus of cellulosic fibers (e.g. 35–45 GPa for flax fibers).

## 5.2 Preparation of cellulose nanoparticles

One of the major drawbacks of natural fibers in composite applications is the large variation in properties inherent to natural products, which are affected by climatic conditions, maturity, and type of soil. One of the basic approaches for avoiding this problem is the elimination of macroscopic flaws by disintegrating the natural fibers and separating out the almost defect-free highly crystalline fibrils. This decreases the size of the natural particles from the microscale to the nanoscale. Conceptually, nanoparticles are particles with at least one dimension smaller than 100 nm (Siqueira *et al.*, 2009; Bras *et al.*, 2010).

This scale shift increases the specific area of the particles, from approximately a few $m^2.g^{-1}$ to a few hundred $m^2.g^{-1}$ (Anglès and Dufresne, 2000; Siqueira *et al.*, 2010c). This, in turn, results in an increase of the interfacial area with the polymeric matrix and a decrease of the average inter-particle distance in the matrix. Thus, particle–particle interactions can be expected. The homogeneous dispersion of nanoparticles in a continuous medium is generally difficult because the surface energy increases with decreasing particle dimensions. Another important feature of nanoparticles is the possibility of improving properties of the material for low filler content without any detrimental effect on impact resistance and plastic deformation. A reduction of gas diffusion (a barrier effect) is also likely to occur. Moreover, cellulosic nanoparticles are distinguished by their liquid-crystal behavior when suspended in water, with birefringence phenomena under polarized light.

The processes for isolating cellulose nanoparticles include simple mechanical shearing disintegration or acid hydrolysis, as described in the following sections.

### 5.2.1 Preparation of nanofibrillated cellulose (NFC)

The process for isolating nanofibrillated cellulose (NFC) consists of the disintegration of cellulose fibers along their long axis. It includes simple mechanical methods or a combination of enzymatic or chemical pretreatment with mechanical processing and high-pressure homogenization. NFC is a cellulosic material, composed of expanded high-volume cellulose, moderately degraded, and greatly expanded in surface area, obtained by homogenization (Nakagaito *et al.*, 2005). In contrast to straight cellulose nanocrystals, nanofibrillated cellulose consists of long and flexible highly elongated nanoparticles. NFC is composed of alternating crystalline and amorphous strings and has a web-like structure (Lu *et al.*, 2008a). Two different pieces of equipment, the Manton Gaulin Homogenizer and the Microfluidizer, are the most commonly used for the mechanical preparation of NFC.

NFC was first obtained by Turbak *et al.* (1983) using, as starting materials, purified cellulose fibers from wood pulp after high-pressure mechanical homogenization. The cellulosic fibers were disintegrated into their sub-structural fibrils and microfibrils having lengths at the micron scale and widths ranging from 10 to a few hundred nanometers depending on the nature of the plant cell walls. The resulting aqueous suspensions exhibited gel-like characteristics in water with pseudo-plastic and thixotropic properties even at low solid content. Initially, this new material was intended to be used as an additive in food, paints, cosmetics, and in medical products.

The major obstacle, for industrial production, is the very high energy consumption involved in processing, because it is necessary to repeat the disintegration and homogenization several times in order to increase the degree of fibrillation (Henriksson *et al.*, 2007; Siró and Plackett, 2010). Thus, chemical and enzymatic pretreatments were introduced in order to facilitate fibrillation and mechanical shearing. NFC can be produced by an environmentally friendly method by combining enzymatic hydrolysis and the mechanical shearing of wood pulp. The enzymatic treatment is considered to be an environmentally friendly process since it does not involve solvents or chemical reactants. The NFC obtained has a more favorable structure, with higher aspect ratio and average molar mass than NFC resulting from acid pretreatment (Henriksson *et al.*, 2007).

A cost-effective chemical pretreatment was also attempted prior to mechanical shearing, by oxidizing cellulose fibers using TEMPO-mediated oxidation, creating carboxyl groups at the surface of fibers. The TEMPO-oxidized cellulose fibers can be converted, by mechanical shearing, to transparent and highly viscous dispersions in water, consisting of highly crystalline individual nanofibers. At pH 10 optimal conditions were reached, giving cellulose nanofibers 3–4 nm in width and a few microns in length (Saito *et al.*, 2006, 2007). Carboxymethylation was also successfully used to chemically pretreat cellulose fibers before mechanical processing to obtain NFC. Clear dispersions with a concentration of 1–2 $g.l^{-1}$ can be prepared directly by this procedure (Wågberg *et al.*, 2008).

Alemdar and Sain (2008) extracted NFC from wheat straw and soy hulls via mechanical treatment involving cryocrushing followed by disintegration and fibrillation. In the cryocrushing method the fibers are first frozen using liquid nitrogen and then high shear forces are applied (Chakraborty *et al.*, 2005). In this method, the ice crystals exert pressure on the cell walls, causing them to disrupt, thereby liberating microfibrils, which can be dispersed uniformly into water suspension using a disintegrator before high-pressure fibrillation.

The processing of cellulosic materials extracted from primary cells such as parenchymal cells from sugar-beet pulp (Dinand *et al.*, 1996, 1999) or cactus cladodes (Malainine *et al.*, 2005) and fruits (Habibi *et al.*, 2009) has been shown to be much easier to mechanically process without any enzymatic or chemical pretreatment. This is because, while in secondary cell walls the cellulose microfibrils are packed in tight arrays, in primary cell walls the cellulose microfibrils are organized in a loose network embedded in an abundant matrix consisting of hemicelluloses and pectins. Since in a primary wall the cellulose microfibrils are either isolated or organized in bundles that contain a limited number of parallel microfibrils, they can easily be separated by mechanical treatment (Dinand *et al.*, 1999).

The origin of the cellulose fibers, which is mainly due to the nature of the plant cell wall, e.g. whether primary or secondary, determines the morphology of the NFC. They are elongated nanoparticles with widths ranging from 3 to 20 nm and lengths of a few microns. NFC from primary cell walls, such sugar-beet pulp or *Opuntia ficus-indica*, are generally thinner and longer, and much easier to produce compared with those extracted from secondary cell walls such as wood, as already explained. Figure 5.1 shows examples of NFC extracted from primary and secondary cell walls.

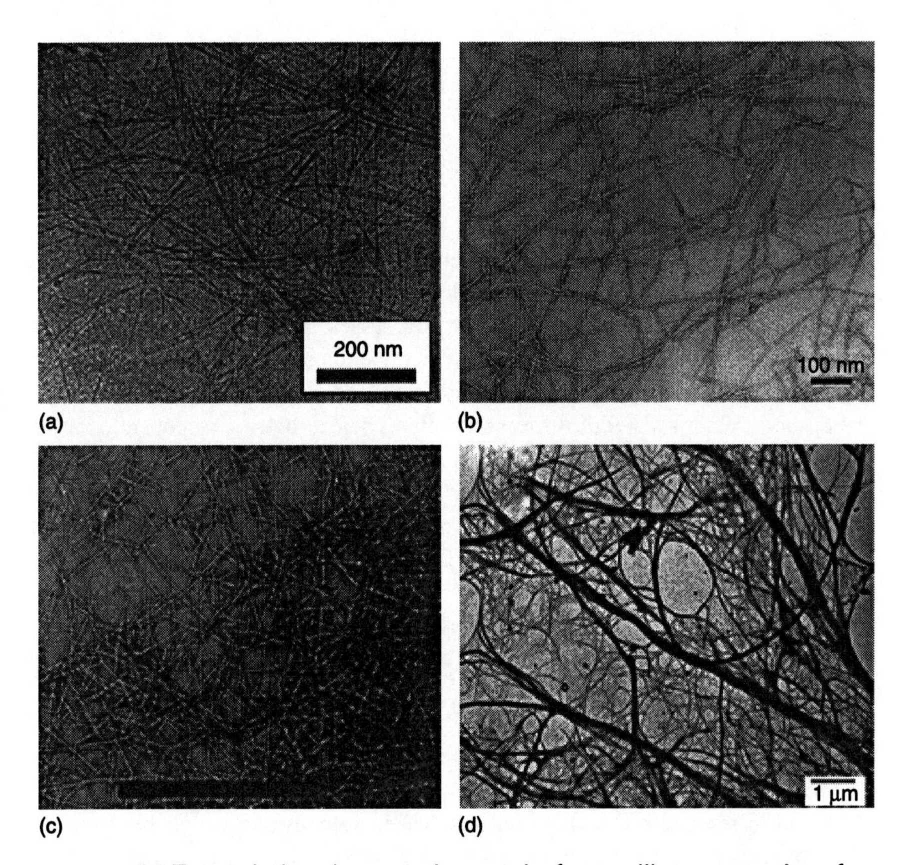


*5.1* Transmission electron micrographs from a dilute suspension of NFC obtained from wood fibers by mechanical processing combined with (a) enzymatic (Pääkkö *et al.*, 2007), (b) TEMPO-mediated oxidation (Saito *et al.*, 2007), (c) carboxymethylation pretreatment (Wågberg *et al.*, 2008), and (d) extracted from *Opuntia ficus-indica* (Malainine *et al.*, 2005). (a), (b), (c) Reprinted with permission from the American Chemical Society, copyright 2007, 2008. (d) Reproduced by permission of Elsevier.

The presence of non-cellulosic polysaccharides, such as hemicellulose and pectin, can also influence the microstructure of the NFC. Using AFM, Ahola *et al.* (2008) compared the microstructure of two types of NFC prepared at Innventia AB (formerly STFI-Packforsk, Stockholm, Sweden). NFC generation 1 was manufactured from a sulfite pulp of high hemicellulose content while NFC generation 2 was produced from a dissolving pulp with lower hemicellulose content. NFC generation 2 has charged groups on the nanofiber surfaces due to the carboxymethylation process. The authors observed that NFC generation 1 formed a more apparent network structure while the charged nanofibers (NFC generation 2) formed smoother and denser films. These observations were confirmed recently in a study by Aulin *et al.* (2009) in which different nanoscale cellulose model films were characterized.

### 5.2.2 Preparation of cellulose nanocrystals (CNC)

Rånby (1951) and Battista *et al.* (1956) were the first to produce stable suspensions of colloidal-sized cellulose crystals by sulfuric acid hydrolysis of cellulose. This process was later optimized (Bondeson *et al.*, 2006; Hamad and Hu, 2010). The extraction or isolation of crystalline cellulosic regions, in the form of rod-like nanocrystals, is a simple process based on acid hydrolysis. The biomass is first submitted to purification and bleaching processes and, after removal of non-cellulosic constituents, such as lignin and hemicelluloses, the bleached material is submitted to a hydrolysis treatment with acid under controlled conditions. The amorphous regions of cellulose act as structural defects and are responsible for the transverse cleavage of the cellulose fibers into short nanocrystals under acid hydrolysis. This transformation consists of the disruption of amorphous regions surrounding and embedding cellulose microfibrils while leaving the crystalline segments intact. It is ascribed to the faster hydrolysis kinetics of amorphous domains compared with crystalline ones. Hydronium ions penetrate the cellulosic material in the amorphous domains, promoting the hydrolytic cleavage of the glycosidic bonds releasing individual crystallites. The resulting suspension is subsequently diluted with water and washed by successive centrifugations. Dialysis against distilled water is then performed to remove free acid in the dispersion. Disintegration of aggregates and complete dispersion of the whiskers is obtained by a sonication step. The dispersions are stored in a refrigerator after filtration to remove residual aggregates. These suspensions are generally much diluted because of the formation of a gel with low nanoparticle contents. The exact determination of the whisker content can be found by gravimetry before and after drying the suspended aliquots. This general procedure has to be adapted depending on the nature of the substrate.

Dong *et al.* (1998) were among the first researchers to study the effect of hydrolysis conditions on the properties of the resulting cellulose nanocrystals. They proved that a longer hydrolysis time leads to shorter monocrystals and also to an increase in their surface charge. The acid concentration was also found to affect the morphology of whiskers prepared from sugar-beet pulp as reported by Azizi Samir *et al.* (2004a). Beck-Candanedo *et al.* (2005) reported the properties of cellulose nanocrystals obtained by the hydrolysis of softwood and hardwood pulp and investigated the influence of hydrolysis time and acid-to-pulp ratio. They explained that the reaction time is one of the most important parameters in the acid hydrolysis of wood pulp. Moreover, they considered that, if the reaction time is too long, the cellulose is completely digested to yield its component sugar molecules. In contrast, lower reaction times yield only large undispersible fibers and aggregates. A higher acid-to-pulp ratio for the long reaction time decreases the nanocrystal dimensions to some extent. At the shorter reaction time, the effect of the acid-to-pulp ratio on the critical concentration and rod dimensions may be more apparent. The effect of the reaction conditions on cellulose nanocrystal surface charge and sulfur content was not significant and they were assumed to be controlled by factors other than the hydrolysis conditions.

It has also been shown that the hydrolysis of amorphous cellulosic chains can be performed simultaneously with the esterification of accessible hydroxyl groups to produce surface-functionalized cellulose nanocrystals in a single step (Braun and Dorgan, 2009). The reaction was carried out in an acid mixture composed of hydrochloric acid and an organic acid (acetic and butyric). The resulting nanocrystals were of similar dimensions compared with those obtained by hydrochloric acid hydrolysis alone. The resulting surface-modified CNCs are dispersible in ethyl acetate and toluene, indicating increased hydrophobicity and presumably higher compatibility with hydrophobic polymers.

The use of sulfuric acid to prepare cellulose nanocrystals leads to a more stable aqueous suspension than that prepared using hydrochloric acid (Araki *et al.*, 1998; Corrêa *et al.*, 2010). It was shown that the $H_2SO_4$-prepared nanoparticles present a negatively charged surface while the HCl-prepared nanoparticles are not charged. During acid hydrolysis via sulfuric acid, acidic sulfate ester groups probably formed on the nanoparticle surfaces. This creates electric double-layer repulsion between the nanoparticles in suspension, which plays an important role in their interaction with a polymer matrix and with each other (Boluk *et al.*, 2011).

If the cellulose nanocrystals are prepared by hydrochloric acid hydrolysis their aqueous suspensions tend to flocculate. Habibi *et al.* performed TEMPO-mediated oxidation of cellulose nanocrystals that were obtained from HCl hydrolysis of cellulose nanoparticles from tunicin to introduce negative charges on their surfaces (Habibi *et al.*, 2006). They showed that, after hydrolysis and TEMPO-mediated oxidation, the nanoparticles kept their initial morphological integrity and native crystallinity, but at their surface the hydroxymethyl groups were selectively converted to carboxylic groups, thus imparting a negative surface charge to the nanocrystals. When dispersed in water these oxidized cellulose nanocrystals did not flocculate, and their suspensions appeared birefringent.

Even though the process of acid hydrolysis of cellulosic material is considered to be a well-known process, Bondeson *et al.* (2006) thought it necessary to optimize it to find a rapid, high-yield process to obtain an aqueous stable colloid suspension of cellulose nanocrystals. They stipulated that large quantities of nanocrystal suspensions are required when they are to be used as nanoreinforcement in polymers. The factors that varied during the hydrolysis process were the concentration of the microcrystalline cellulose (the starting material) and sulfuric acid, the hydrolysis time and temperature, and the ultrasonic treatment time. In this study, cellulose nanocrystals with a length between 200 and 400 nm and a width less than 10 nm were obtained by using 63.5 wt% of sulfuric acid solution for approximately 2 h, with a yield of 30% (of the initial weight).

Cellulose nanocrystals can be prepared from any botanical source containing cellulose and, regardless of the source, cellulose nanocrystals are produced as elongated nanoparticles (Fig. 5.2). The persistence of the spot diffractogram when the electron probe is scanned along the rod during transmission electron microscopy evidences the monocrystalline nature of the cellulosic fragments (Favier *et al.*, 1995a). Therefore, each fragment can be considered as a cellulosic crystal with no apparent defects. Their dimensions depend on several factors, including the source of the cellulose, the exact hydrolysis conditions, and ionic strength. The length and width of hydrolyzed cellulose nanocrystals is generally of the order of a few hundred nanometers and a few nanometers, respectively (Table 5.1).

An important parameter for cellulosic nanocrystals is the aspect ratio, which is defined as the ratio of the length to the width. It determines the anisotropic phase formation and reinforcing properties. The aspect ratio varies between 10 for cotton and 67 for tunicin (cellulose extracted from tunicate) nanocrystals. Relatively large and highly regular tunicin nanocrystals are ideal for modeling rheological and reinforcement behavior and are extensively mentioned in the literature. The shape and dimensions of cellulose nanocrystals can be found from microscopic observations or scattering techniques. The cross sections of nanocrystals observed in transmission electron micrographs are square, whereas their AFM topography shows a rounded profile due to convolution with the shape of the AFM tip (Hanley *et al.*, 1992). Scattering techniques include small-angle light (de Souza Lima *et al.*, 2003) and neutron (Orts *et al.*, 1998) scattering.

## 5.3 Preparation of cellulose nanocomposites from different processes and polymeric matrices

The main challenge with nanoparticles is related to their homogeneous dispersion within a polymeric matrix. The presence of sulfate groups from the acid hydrolysis treatment, when using sulfuric acid to prepare cellulose nanocrystals, induces the stability of the ensuing aqueous suspension. Nanofibrillated cellulose is also usually obtained in aqueous suspension. Water is therefore the initially preferred

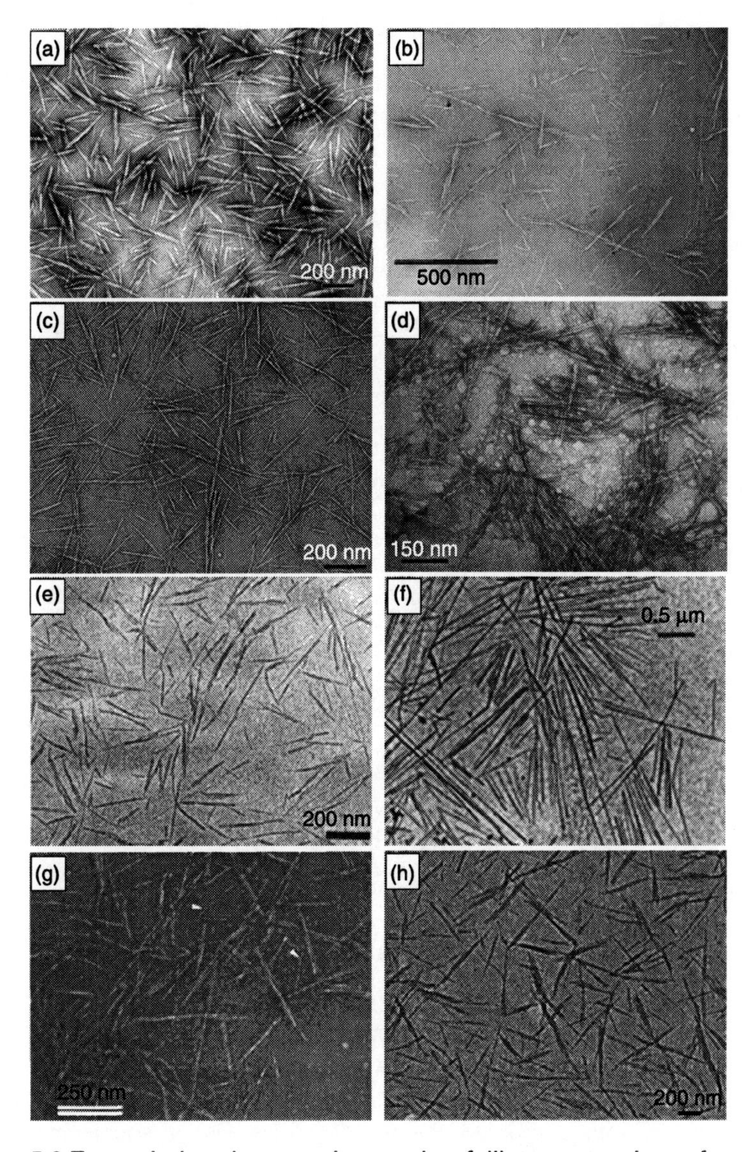


*5.2* Transmission electron micrographs of dilute suspensions of cellulose nanocrystals from: (a) ramie (Habibi *et al.*, 2008), (b) *Luffa cylindrica* (Siqueira *et al.*, 2010a), (c) sisal (Garcia de Rodriguez *et al.*, 2006), (d) microcrystalline cellulose (MCC) (Kvien *et al.*, 2005), (e) sugar-beet pulp (Azizi Samir *et al.*, 2004a), (f) tunicin (Anglès and Dufresne, 2000), (g) wheat straw (Helbert *et al.*, 1996), and (h) cotton (Fleming *et al.*, 2000). (a) Reproduced by permission of the Royal Society of Chemistry. (c) Reproduced by permission of Elsevier. (d), (e), (f), (h) Reprinted with permission from the American Chemical Society, copyright 2008. (g) Reproduced by permission of John Wiley & Sons Inc.

*Table 5.1* Geometrical characteristics of cellulose nanocrystals from various sources: length (*L*) and cross section (*D*)

| Source | *L* (nm) | *D* (nm) | Aspect ratio *L*/*D* | Reference |
|---|---|---|---|---|
| Sisal | 215 | 5 | 43 | Siqueira *et al.*, 2010c |
| Cotton | 172 | 15 | 11–12 | Roohani *et al.*, 2008 |
| Cottonseed linter | 350 | 40 | 9 | Lu *et al.*, 2005 |
| Flax | 327 | 21 | 15.5 | Cao *et al.*, 2007 |
| *Luffa cylindrica* | 242 | 5.2 | 46.8 | Siqueira *et al.*, 2010a |
| MCC | 210 | 5 | 42 | Kvien *et al.*, 2005 |
| Ramie | 150–250 | 6–8 | 18–40 | Habibi *et al.*, 2008; Habibi and Dufresne, 2008 |
| Bacterial | 200–several 1000 | 30–50 | | Grunert and Winter, 2002; Roman and Winter, 2004 |
| Sugar-beet pulp | 210 | 5 | 42 | Azizi Samir *et al.*, 2004a |
| Tunicin | 100–several 1000 | 10–20 | | Favier *et al.*, 1995b |
| Wheat straw | 150–300 | 5 | 30–60 | Helbert *et al.*, 1996 |
| Wood | 105–150 | 4.5–5 | 23–30 | Beck-Candanedo *et al.*, 2005 |

processing medium. A high level of dispersion of the filler within the host matrix in the resulting composite film is expected when processing nanocomposites in aqueous medium.

### 5.3.1 Polymer latex matrices

The first publication reporting the preparation of cellulose-nanocrystal-reinforced polymer nanocomposites described a method using a latex of (poly(S-co-BuA)) and tunicin whiskers (Favier *et al.*, 1995a). The nanocomposite films were obtained by water evaporation and particle coalescence at room temperature, that is, at a temperature higher than $T_g$ of the polymeric matrix (around 0 °C). The mechanical properties (shear modulus) of the obtained nanocomposites increased by more than two orders of magnitude in the rubbery state of the polymeric matrix, when the nanocrystal content was 6% (w/w). This effect was ascribed to the formation of a rigid nanocrystal network, probably linked by hydrogen bonds. The formation of this network was assumed to be governed by the percolation mechanism (Bras *et al.*, 2011). The same copolymer has been used with wheat straw (Helbert *et al.*, 1996) and sugar-beet (Azizi Samir *et al.*, 2004a) cellulose nanocrystals.

Other latexes, such as poly($\beta$-hydroxyoctanoate) (PHO) (Dubief *et al.*, 1999; Dufresne *et al.*, 1999; Dufresne, 2000), polyvinylchloride (PVC) (Chazeau *et al.*, 1999a, 1999b, 1999c, 2000), waterborne epoxy (Matos Ruiz *et al.*, 2001), natural rubber (NR) (Bendahou *et al.*, 2009, 2010; Siqueira *et al.*, 2010b; Bras *et al.*, 2010), and polyvinyl acetate (PVAc) (Garcia de Rodriguez *et al.*, 2006), have also been used as matrices. Recently, stable aqueous nanocomposite dispersions containing cellulose nanocrystals and a poly(styrene-co-hexyl-acrylate) matrix were prepared via miniemulsion polymerization (Ben Elmabrouk *et al.*, 2009). The addition of a reactive silane was used to stabilize the dispersion. Solid nanocomposite films have been obtained by mixing and casting two aqueous suspensions followed by water evaporation. Alternative methods consist of freeze-drying and hot-pressing or freeze-drying, extruding, and hot-pressing the mixture.

### 5.3.2 Hydrosoluble or hydrodispersible polymer matrices

The preparation of cellulosic-particle-reinforced starch (Anglès and Dufresne, 2000; Orts *et al.*, 1998; Mathew and Dufresne, 2002; Mathew *et al.*, 2008; Kvien *et al.*, 2007; Svagan *et al.*, 2009), silk fibroin (Noishiki *et al.*, 2002), poly(oxyethylene) (POE) (Azizi Samir *et al.*, 2004a, 2004b, 2004c, 2004d, 2006), polyvinyl alcohol (PVA) (Zimmerman *et al.*, 2004, 2005; Lu *et al.*, 2008b; Paralikar *et al.*, 2008; Roohani *et al.*, 2008), hydroxypropyl cellulose (HPC) (Zimmerman *et al.*, 2004, 2005), carboxymethyl cellulose (CMC) (Choi and Simonsen, 2006), or soy protein isolate (SPI) (Wang *et al.*, 2006) has been reported in the literature. The hydrosoluble or hydrodispersible polymer is first dissolved in water and this solution mixed with an aqueous suspension of cellulose nanocrystals. The ensuing mixture is generally evaporated to obtain a solid nanocomposite film. It can also be freeze-dried and hot-pressed.

### 5.3.3 Non-aqueous systems: alternative processing methods

An alternative way to prepare non-polar polymer nanocomposites reinforced with cellulose nanocrystals consists of their dispersion in an adequate (with regard to the matrix) organic medium. Coating with a surfactant or surface chemical modification of the nanoparticles can be considered. The overall objective is to reduce their surface energy in order to improve their dispersibility and compatibility with the non-polar media.

The coating of cotton and tunicin nanocrystals by a surfactant such as a phosphoric ester of polyoxyethylene nonyl phenyl ether was found to lead to stable suspensions in toluene and cyclohexane (Heux *et al.*, 2000) or chloroform (Kvien *et al.*, 2005). Coated-tunicin-nanocrystal-reinforced atactic polypropylene (aPP) (Ljungberg *et al.*, 2005), isotactic polypropylene (iPP) (Ljungberg *et al.*,

2006), and poly(ethylene-co-vinyl acetate) (EVA) (Chauve *et al.*, 2005) were obtained by solvent casting using toluene. The same procedure was used to disperse cellulosic nanoparticles in chloroform and process composites with polylactic acid (PLA) (Kvien *et al.*, 2005; Petersson and Oksman, 2006).

Surface chemical modification of cellulosic nanoparticles is another way to decrease their surface energy and disperse them in organic liquids of low polarity. It generally involves reactive hydroxyl groups at the surface. The chemical grafting has to be mild in order to preserve the integrity of the nanoparticles. Goussé *et al.* (2002) stabilized tunicin nanocrystals in tetrahydrofuran (THF) by a partial silylation of their surface. Araki *et al.* (2001) prepared original sterically stabilized aqueous rod-like cellulose microcrystal suspensions by the combination of HCl hydrolysis, oxidative carboxylation, and the grafting of poly(ethylene glycol) (PEG), having a terminal amino group at one end, using water-soluble carbodiimide. The PEG-grafted microcrystals displayed drastically enhanced dispersion stability evidenced through resistance to the addition of 2 M sodium chloride. They also showed an ability to re-disperse into either water or chloroform from the freeze-dried state.

The preparation of stable cellulose nanocrystal suspensions in dimethylformamide (DMF) (Azizi Samir *et al.*, 2004e; Marcovich *et al.*, 2006) and dimethyl sulfoxide (DMSO) or *N*-methyl pyrrolidine (NMP) (van den Berg *et al.*, 2007) without either the addition of a surfactant or any chemical modification has also been reported.

## 5.3.4 Long-chain grafting of cellulose nanoparticles

Long-chain surface chemical modification of cellulosic nanoparticles consisting of grafting agents bearing a reactive end group and a long 'compatibilizing' tail has also been reported in the literature. The general objective is, of course, to increase the apolar character of the nanoparticles. In addition, it has extraordinary possibilities. The surface modifications can act as binding sites for active agents in drug delivery systems or for toxins in purification and treatment systems. These surface modifications may also be able to interdiffuse, on heating, to form a polymer matrix phase. The covalent linkage between the reinforcement and the matrix results in near-perfect stress transfer at the interface, giving exceptional mechanical properties of the composite.

Nanocomposite materials have been processed from polycaprolactone (PCL)-grafted cellulose nanocrystals using the grafting 'onto' (Habibi and Dufresne, 2008) and grafting 'from' (Habibi *et al.*, 2008) approaches. The grafting 'onto' approach consists of mixing the nanoparticles with a polymer and a coupling agent to graft the existing polymer onto the surface of the filler. The grafting 'from' approach consists of mixing the nanoparticles with a monomer and an initiator, and the polymer directly grow on the surface of the filler. The ensuing nanoparticles were used to prepare nanocomposites with PCL as the matrix through a casting/evaporation technique with dichloromethane. A co-continuous crystalline phase around the nanoparticles was observed. Cellulose nanocrystals were also surface-grafted with PCL via microwave-assisted ring-opening polymerization yielding filaceous cellulose nanocrystal-graft-PCL, which was incorporated into PLA as the matrix (Lin *et al.*, 2009). The surface of cellulose nanocrystals has also been chemically modified by grafting organic acid chlorides, having various lengths of their aliphatic chains, by an esterification reaction (de Menezes *et al.*, 2009). These functionalized nanoparticles were extruded with low density polyethylene (LDPE) to prepare nanocomposite materials. Cellulose-nanocrystal-reinforced waterborne polyurethane nanocomposites have been synthesized via *in situ* polymerization using the casting/evaporation technique (Cao *et al.*, 2009). The grafted chains formed a crystalline structure on the surface of the nanoparticles and induced the crystallization of the matrix. Cellulose nanoparticles have been modified with n-octadecyl isocyanate ($C_{18}H_{37}NCO$) using two different methods, one consisting of an *in situ* solvent exchange procedure (Siqueira *et al.*, 2010c). Phenol was also enzymatically polymerized in the presence of TEMPO-oxidized cellulosic nanoparticles to prepare nanocomposites under ambient conditions (Li *et al.*, 2010).

### 5.3.5 Extrusion methods and impregnation

Very few studies have reported the processing of cellulose-nanocrystal-reinforced nanocomposites by extrusion methods. The hydrophilic nature of cellulose causes irreversible agglomeration during drying and aggregation in non-polar matrices because of the formation of additional hydrogen bonds between amorphous parts of the cellulose nanoparticles. Therefore, the preparation of cellulose-whisker-reinforced PLA nanocomposites by melt extrusion has been carried out by pumping the suspension of nanocrystals into the polymer melt during the extrusion process (Oksman *et al.*, 2006). An attempt to use PVA as a compatibilizer to promote the dispersion of cellulose nanocrystals within the PLA matrix has been reported (Bondeson and Oksman, 2007). Organic-acid-chloride-grafted cellulose whiskers were extruded with LDPE (de Menezes *et al.*, 2009). The homogeneity of the ensuing nanocomposite was found to increase with the length of the grafted chains (Fig. 5.3).

Another possible processing technique to prepare nanocomposites using cellulosic nanoparticles in the dry state consists of the filtration of the aqueous suspension to obtain a film or dried mat of particles followed by immersion in a polymer solution. The impregnation of the dried mat is performed under vacuum. Composites have been processed by filling the cavities with transparent thermosetting resins such as phenol formaldehyde (Nakagaito *et al.*, 2005; Nakagaito and Yano, 2004, 2008), epoxy (Shimazaki *et al.*, 2007), acrylic (Yano *et al.*, 2005; Nogi *et al.*, 2005; Iwamoto *et al.*, 2008), and melamine formaldehyde (Henriksson and Berglund, 2007). Non-woven mats of cellulose microfibrils have

*5.3* Photographs of neat LDPE film and extruded nanocomposite films reinforced with 10 wt% of unmodified and C18-acid-chloride-grafted cellulose whiskers (de Menezes *et al.*, 2009). Reproduced by permission of Elsevier.

also been used to prepare polyurethane composite materials using the film-stacking method (Seydibeyoğlu and Oksman, 2008).

Water-re-dispersible nanofibrillated cellulose in powder form was recently prepared from refined bleached beech pulp by carboxymethylation and mechanical disintegration (Eyholzer *et al.*, 2010). However, the carboxymethylated sample displayed a loss of crystallinity and a strong decrease in thermal stability, limiting its use for nanocomposite processing.

### 5.3.6 Electrospinning

Electrostatic fiber spinning or 'electrospinning' is a versatile method for preparing fibers with diameters ranging from several microns down to 100 nm, through the action of electrostatic forces (Huang *et al.*, 2003; Greiner and Wendorff, 2007). It uses an electrical charge to draw a positively charged polymer solution from an orifice to a collector. Electrospinning shares characteristics of both electrospraying and conventional solution dry spinning of fibers. The process is non-invasive and does not require the use of coagulation chemistry or high temperatures to produce solid threads from solution. This makes the process particularly suitable for the production of fibers using large and complex molecules.

Bacterial cellulose whiskers were incorporated into POE nanofibers with a diameter of less than 1 μm by electrospinning to enhance the mechanical properties of the electrospun fibers (Park *et al.*, 2007). The whiskers were found to be globally well embedded and aligned inside the fibers, even though they were partially aggregated. Likewise, electrospun PVA fiber mats loaded with cellulose nanocrystals (Fig. 5.4), with diameters in the nanoscale range and enhanced

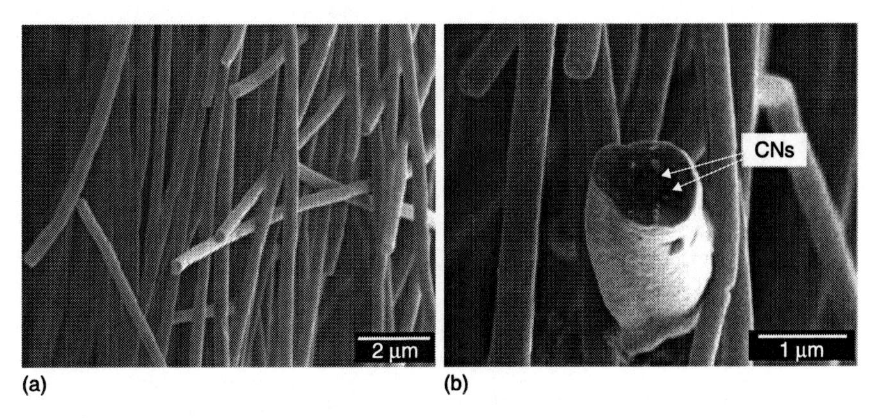


*5.4* Cryo-scanning electron micrographs of electrospun polyvinyl alcohol loaded with 15% of cellulose nanocrystals (CNs) (Peresin *et al.*, 2010). Copyright (2000) the American Chemical Society.

mechanical properties, were successfully produced (Medeiros *et al.*, 2008; Peresin *et al.*, 2010).

Electrospun polystyrene (PS) (Rojas *et al.*, 2009) and PCL (Zoppe *et al.*, 2009) microfibers or nanofibers reinforced with cellulose nanocrystals have been obtained by electrospinning. A non-ionic surfactant, sorbitan monostearate, was used to improve the dispersion of the particles in the hydrophobic PS matrix, while surface grafting of the long chains was used with PCL.

### 5.3.7 Multilayer films

The layer-by-layer (LBL) technique is expected to maximize the interaction between cellulose nanocrystals and the polar polymeric matrix, such as chitosan (de Mesquita *et al.*, 2010). It also allows the incorporation of high numbers of cellulose whiskers, giving a dense and homogeneous distribution in each layer.

Podsiadlo *et al.* (2005) prepared cellulose nanocrystal multilayer composites with a polycation, poly-(dimethyldiallylammonium chloride) (PDDA), using LBL. The authors concluded that the multilayer films had high uniformity and dense packing of the nanocrystals. The preparation of thin films composed of alternating layers of orientated rigid cellulose nanocrystals and flexible polycation chains has been reported (Jean *et al.*, 2008). Alignment of the rod-like nanocrystals was achieved using anisotropic suspensions of cellulose nanocrystals. Green composites based on cellulose nanocrystal/xyloglucan multilayers have been prepared using the non-electrostatic cellulose–hemicellulose interaction (Jean *et al.*, 2009). More recently, biodegradable nanocomposites were obtained using LBL with highly deacetylated chitosan and cellulose nanocrystals (de Mesquita *et al.*, 2010). Hydrogen bonds and electrostatic interactions between the negatively charged sulfate groups on the nanoparticle surfaces and the ammonium groups of the chitosan were the driving forces for the growth of the multilayered films. Self-organized films have also been obtained using only charge-stabilized dispersions of cellulose nanoparticles with opposite charges to the LBL technique (Aulin *et al.*, 2008).

## 5.4 Properties and applications of nanocomposites reinforced with cellulose nanoparticles

### 5.4.1 Mechanical properties

Cellulose nanocrystals and nanofibrillated cellulose have a huge specific surface area and impressive mechanical properties, which make them ideal candidates to improve the mechanical properties of a neat matrix. Dynamic mechanical analysis (DMA) is a powerful tool to investigate the linear mechanical behavior of nanocomposites in different temperature and frequency ranges. Classical tensile or compressive tests can be used to evaluate the non-linear mechanical properties.

The first report of the effect of using cellulose nanocrystals in nanocomposite materials was by Favier *et al.* (1995a, 1995b). Using dynamic mechanical analysis, the authors observed a spectacular increase in the storage modulus of the styrene/butyl acrylate copolymer (poly(S-co-BuA)) matrix with the addition of tunicin nanocrystals even at low content. This increase was especially significant above the glass-rubber transition temperature ($T_g$) of the thermoplastic matrix because of its poor mechanical properties in this temperature range. This outstanding reinforcing effect was ascribed to a mechanical percolation phenomenon (Favier *et al.*, 1995a, 1995b). Above the percolation threshold, the cellulosic nanoparticles can connect and form a three-dimensional continuous pathway through the nanocomposite film. The formation of this cellulose network was assumed to result from strong interactions between the nanoparticles, such as hydrogen bonds (Favier *et al.*, 1997). This phenomenon is similar to the high mechanical properties observed for a paper sheet, which result from the hydrogen-bonding forces that hold the percolating network of fibers. This mechanical percolation effect explains both the high reinforcing effect and the thermal stabilization of the composite modulus for evaporated films.

Any factor that affects the formation of the percolating nano-network or interferes with it changes the mechanical performances of the composite (Dufresne, 2006). Three main parameters have been reported as affecting the mechanical properties of such materials: (i) the morphology and dimensions of the nanoparticles, (ii) the processing method, and (iii) the microstructure of the matrix and matrix–filler interactions.

With regard to the morphology and dimensions, for rod-like particles the geometrical aspect ratio is an important factor since it determines the percolation threshold value. This factor is linked to the source of the cellulose and the preparation conditions for the nanocrystals. A higher reinforcing effect is obtained for nanocrystals with a high aspect ratio. For instance, the rubbery storage tensile modulus was systematically lower for wheat straw whisker/poly(S-co-BuA) composites than for tunicin-whisker-based materials (Dufresne, 2006). The following relation was found between the percolation threshold ($\nu_{Rc}$) and the aspect ratio ($L/d$) of rod-like particles (Dufresne, 2008):

$$\nu_{Rc} = 0.7/(L/d) \qquad [5.1]$$

Also, flexibility and the possibility of tangling the nanofibers plays an important role (Azizi Samir *et al.*, 2004a; Bendahou *et al.*, 2009; Siqueira *et al.*, 2009). Azizi Samir *et al.* (2004a) used poly(S-co-BuA) as a matrix in nanocomposites reinforced with cellulose rod-like nanoparticles extracted from sugar beet. Linear mechanical measurements performed on the nanocomposites did not show significant differences when varying the strength of the hydrolysis step or the length and flexibility of the nanoparticles. However, for non-linear mechanical tensile tests, it was observed that as the hydrolysis strength increased both the modulus and the strength of the nanocomposite decreased, whereas the elongation at break increased. This result shows the strong influence of entanglement on the mechanical behavior of nanocomposites.

The use of higher quantities of NFC in filled nanocomposites was presented by Nakagaito *et al.* (2009). The PLA matrix was filled with up to 90 wt% of NFC using a similar strategy as that for impregnated paper. A linear increase of the Young's modulus when increasing the NFC content was reported. The modulus increased from 5 GPa for a NFC content of 10 wt% to almost 9 GPa for 70 wt% NFC. This phenomenon was observed even at high temperatures. The storage modulus above the glass transition temperature increased as the NFC content in the composites increased. Moreover, for high NFC content such as 70 and 90 wt%, the storage modulus was sustained up to 250 °C. It has been suggested that the preparation of nanocomposites by compression molding of sheets could easily be used by industry.

Regarding processing methods, slow processes such as casting/evaporation have been reported as producing materials with the highest mechanical performance compared with freeze-drying/molding and freeze-drying/extruding/molding. This effect was ascribed to the probable orientation of the rod-like nanoparticles during film processing resulting from shear stress induced by freeze-drying/molding and freeze-drying/extrusion/molding (Dufresne, 2008). During slow water evaporation, because of Brownian motion in the suspension or solution (whose viscosity remains low, up to the end of the process when the latex particle or polymer concentration becomes very high), a rearrangement of the nanoparticles is possible. They have time to interact and connect to form a continuous network, which is the basis of their reinforcing effect. The resulting structure is completely relaxed and direct contact is created between the nanocrystals or nanofibrils. On the other hand, during the freeze-drying/

hot-pressing process, the nanoparticle arrangement in the suspension is first frozen, and then during the hot-pressing stage, because of the polymer melt viscosity, the particle rearrangements are strongly limited.

The microstructure of the matrix and the resulting competition between matrix–filler and filler–filler interactions also affects the mechanical behavior of the cellulose-nanocrystal-reinforced nanocomposites. Classical composite science tends to favor matrix–filler interactions as a fundamental condition for optimal performance. For cellulose-nanocrystal-based composite materials the opposite trend is generally observed when the material is processed via casting/evaporation. Increasing the affinity between the polysaccharide filler and the host matrix leads to a decrease in mechanical performance. This unusual behavior is ascribed to the uniqueness of the reinforcing phenomenon of the cellulosic nanoparticles, resulting from the formation of a hydrogen-bonded percolating network. Grunert and Winter (2002) prepared nanocomposites with a cellulose acetate butyrate matrix and reported a higher reinforcing effect when using unmodified bacterial cellulose nanocrystals compared with trimethylsilylated nanocrystals. They attributed this difference to the restricted filler–filler interaction in the trimethylsilylated-nanocrystal nanocomposites.

The transcrystallization phenomenon reported for poly(hydroxyalkanoate) (PHA) (Dufresne *et al.*, 1999) and plasticized starch (Anglès and Dufresne, 2000) with cellulose nanocrystals resulted in a decrease of the mechanical properties (Anglès and Dufresne, 2001) because the nanoparticles were coated with crystalline domains. When using unhydrolyzed cellulose microfibrils extracted from potato pulp rather than cellulose nanocrystals to reinforce glycerol-plasticized thermoplastic starch, a completely different mechanical behavior was reported (Dufresne and Vignon, 1998; Dufresne *et al.*, 2000) and a significant reinforcing effect was observed. It was suspected that the tangling effect contributed to this high reinforcing effect (Anglès and Dufresne, 2001).

When using a processing route other than casting/evaporation in a water medium, the dispersion of the hydrophilic filler in the polymeric matrix is also involved (de Menezes *et al.*, 2009) and improved filler–matrix interactions generally lead to higher mechanical properties. In non-percolating systems, for instance for materials processed from freeze-dried cellulose nanocrystals, strong matrix–filler interactions enhance the reinforcing effect of the filler (Ljungberg *et al.*, 2005).

### 5.4.2 Thermal properties

Analyses of thermal properties of materials are important to determine their temperature range for processing and use. Important characteristics of polymeric systems, such as the glass-rubber transition temperature ($T_g$), melting point ($T_m$), and thermal stability, can be determined using, for instance, differential scanning calorimetry (DSC) experiments.

In different studies, no change in $T_g$ was reported when increasing the amount of whiskers, regardless of the nature of the polymeric matrix. Studies of POE reinforced with tunicin whiskers have shown that there is no significant influence of whisker content on the $T_g$ of the materials (Azizi Samir *et al.*, 2004c). A similar observation was reported by Grunert and Winter (2002) in their work with cellulose acetate butyrate (CAB) reinforced with bacterial cellulose nanocrystals. Lu *et al.* (2008b) evaluated the thermal properties of polyvinyl alcohol (PVA) reinforced with microfibrillated cellulose and also did not observe changes in $T_g$.

In glycerol-plasticized starch-based nanocomposites, peculiar effects of tunicin whiskers on the $T_g$ of the starch-rich fraction were reported (Anglès and Dufresne, 2000). It was determined that $T_g$ of the nanocomposites depended on the moisture conditions. For low loading level (up to 3.2 wt%), a classical plasticization effect of water was reported. However, an antiplasticization phenomenon was observed for higher whisker content (6.2 wt% and over). These observations were discussed according to the possible interactions between hydroxyl groups on the cellulosic surface and starch, the selective partitioning of glycerol and water in the bulk starch matrix or at a whisker surface, and the restriction of mobility of amorphous starch chains in the vicinity of the starch-crystallite-coated filler surface.

The plasticizing effect of water molecules was responsible for the decrease of $T_g$ in PVA-based nanocomposites reinforced with sisal (Garcia de Rodriguez *et al.*, 2006) and cotton (Roohani *et al.*, 2008) nanocrystals. Regardless of the composition, a decrease of $T_g$ was observed as the humidity increased. However, in a moist atmosphere, the $T_g$ of PVA-based nanocomposites significantly increased when cotton nanocrystals were added (Roohani *et al.*, 2008). For glycerol-plasticized starch reinforced with cellulose crystallites prepared from cottonseed linter (Lu *et al.*, 2005), an increase of $T_g$ with filler content was reported and attributed to cellulose–starch interactions. For tunicin nanocrystals/sorbitol-plasticized starch (Mathew and Dufresne, 2002), $T_g$ was found to increase slightly up to about 15 wt% of nanocrystals and to decrease for higher nanocrystal loading. The crystallization of amylopectin chains on addition of nanocrystals and the migration of sorbitol molecules to the amorphous domains were proposed to explain the observed modifications. The $T_g$ of electrospun PS fiber with added cellulose nanocrystals (and surfactant) tends to decrease with nanocrystal load (Rojas *et al.*, 2009). The authors attributed the reduction of $T_g$ to the plasticizing effect of the surfactant.

The melting temperature, $T_m$, was reported to be nearly independent of the filler content in plasticized starch (Anglès and Dufresne, 2000; Mathew and Dufresne, 2002) and in POE-based materials (Azizi Samir *et al.*, 2004b, 2004c, 2004e) filled with tunicin nanocrystals. The same observation was reported for CAB reinforced with native bacterial cellulose nanocrystals (Grunert and Winter, 2002). However, for bacterial nanocrystals, $T_m$ was found to increase when the number of trimethylsilylated nanocrystals increased. The authors ascribed this difference to the stronger filler–matrix interaction for the chemically modified nanocrystals. A decrease of both $T_m$ and the degree of crystallinity of PVA was reported when adding cellulose nanocrystals (Roohani *et al.*, 2008). However, for electrospun cellulose-whisker-reinforced PVA nanofibres, the crystallinity was found to be reduced on filler addition (Peresin *et al.*, 2010).

A significant increase in crystallinity of sorbitol-plasticized starch (Mathew and Dufresne, 2002) was reported when increasing the cellulose nanocrystal content. This phenomenon was ascribed to an anchoring effect of the cellulosic filler, which probably acted as a nucleating agent. For POE-based composites the degree of crystallinity of the matrix was found to be roughly constant up to 10 wt% tunicin nanocrystals (Azizi Samir *et al.*, 2004b, 2004c, 2004e) and to decrease for a higher loading level (Azizi Samir *et al.*, 2004c). It seems that the nucleating effect of cellulosic nanocrystals is mainly governed by surface chemical considerations (Ljungberg *et al.*, 2006). It was shown from both X-ray diffraction and DSC analysis that the crystallization behavior of films containing unmodified and surfactant-modified nanocrystals displayed two crystalline forms ($\alpha$ and $\beta$), whereas the neat matrix and the nanocomposite reinforced with nanocrystals grafted with maleated polypropylene only crystallized in the $\alpha$ form. It was suspected that the more hydrophilic the nanocrystal surface, the more it appeared to favor the appearance of the $\beta$ phase. Grunert and Winter (2002) observed from DSC measurements that native bacterial fillers impede the crystallization of a CAB matrix whereas silylated ones help to nucleate the crystallization. The crystallinity of *N*-octadecyl isocyanate-grafted sisal-nanocrystal-reinforced PCL was found to increase on filler addition, whereas no influence with *N*-octadecyl isocyanate-grafted sisal NFC was reported (Siqueira *et al.*, 2009). This difference was ascribed to the possibility of entanglement of the NFC, which tends to confine the polymeric matrix and restrict its crystallization.

For tunicin-nanocrystal-filled glycerol-plasticized starch, a transcrystallization phenomenon was reported (Anglès and Dufresne, 2000). The formation of a transcrystalline zone around the nanocrystals was assumed to be due to the accumulation of plasticizer in the cellulose/amylopectin interfacial zones improving the ability of amylopectin chains to crystallize. This transcrystalline zone could originate from a glycerol-starch V structure. In addition, the inherent restricted mobility of the amylopectin chains was put forward to explain the lower water uptake of cellulose/starch composites for increasing filler content.

The presence of sulfate groups introduced at the surface of the nanocrystals during hydrolysis with $H_2SO_4$ reduces the temperature of their thermal decomposition (Roman and Winter, 2004; Li *et al.*, 2009). Thermogravimetric analysis (TGA) experiments were performed to investigate the thermal stability of tunicin nanocrystal/POE nanocomposites (Azizi Samir *et al.*, 2004b, 2004c). No significant influence of the cellulosic filler on the degradation temperature of the POE matrix was reported. Cellulose nanocrystal content appeared to have an effect on the thermal behavior of CMC plasticized with glycerin (Choi and Simonsen, 2006), suggesting a close association between the filler and the matrix.

The thermal degradation of unfilled CMC was observed from its melting point (270 °C) and there was a very narrow temperature range of degradation. Cellulose nanocrystals were found to degrade at a lower temperature (230 °C) than CMC, in a broad degradation temperature range. The degradation of the nanocomposite was observed in a unique step at a temperature between that for the degradation of cellulose nanocrystals and CMC.

### 5.4.3 Swelling properties

The swelling or kinetics of solvent absorption can highlight interactions between the filler and the matrix. It consists of, generally, first drying and weighing the sample, and then immersing it in a liquid solvent or exposing it to a vapor medium. The sample is then removed at specific intervals and weighed until equilibrium is reached. The swelling rate of the sample can be calculated by dividing the gain in weight by the initial weight. Generally, the short-time behavior displays a fast absorption phenomenon, whereas at longer times the kinetics of absorption is low and leads to a plateau, corresponding to the solvent uptake at equilibrium. The diffusion coefficient can be determined from the initial slope of the solvent uptake curve as a function of time (Dufresne, 2008). For nanocomposites reinforced with cellulosic nanoparticles, it is generally of interest to investigate water absorption of the material because of the hydrophilic nature of the reinforcing phase. When a non-polar polymeric matrix is used, the absorption of a non-aqueous liquid can be investigated.

A higher resistance of thermoplastic starch to water was reported when increasing the cellulose nanoparticle content (Anglès and Dufresne, 2000; Lu *et al.*, 2005; Svagan *et al.*, 2009). Both water uptake and the diffusion coefficient of water were found to decrease with the addition of nanoparticles. These phenomena were ascribed to the presence of strong hydrogen-bonding interactions between the particles and between the starch matrix and the cellulose nanocrystals. The hydrogen-bonding interactions in the composites tend to stabilize the starch matrix when it is in a highly moist atmosphere. Moreover, the high crystallinity of cellulose also might be responsible for the decreased water uptake at equilibrium and the diffusion coefficient of the material. A lower water uptake and a dependence on cellulose nanocrystal content were reported when using sorbitol rather than glycerol as a plasticizer for the starch matrix (Mathew and Dufresne, 2002). An explanation was proposed based on the chemical structure of both plasticizers, the number of more accessible end hydroxyl groups in glycerol being about twice that of sorbitol. The addition of sisal nanocrystals was found to stabilize PVA-based nanocomposites with no benefit seen when increasing the nanocrystal content beyond the percolation threshold (Garcia de Rodriguez *et al.*, 2006).

A lower water uptake was observed when using NFC instead of cellulose nanocrystals as a reinforcing phase in NR (Bendahou *et al.*, 2010). This observation was explained by the difference in the structure and composition of the two types of nanoparticle, and in particular by the presence of residual lignin, extractive substances, and fatty acids at the surface of NFC, which limits, comparatively, the hydrophilic character of the filler. In addition, assuming that filler–matrix compatibility was consequently lower for nanocrystal-based nanocomposites, one can imagine that water infiltration is easier at the filler–matrix interface. For NFC-based nanocomposites, despite a higher amorphous cellulose content, the higher hydrophobic character of the filler favors compatibility with NR and restricts, therefore, the interfacial diffusion pathway for water.

The swelling behavior in toluene of poly(S-co-BuA) reinforced with cellulose fibrils from *Opuntia ficus-indica* was reported by Malainine *et al.* (2005). They observed strong resistance to the toluene even at very low filler loading. While the unfilled matrix completely dissolved in toluene, only 27 wt% of the polymer was able to dissolve when filled with only 1 wt% of cellulose microfibrils. This phenomenon was ascribed to the presence of a three-dimensional entangled cellulosic network, which strongly restricted the swelling capability and dissolution of the matrix. For higher microfibril content no significant evolution was observed because of both the overlapping of the microfibrils restricting the filler–matrix interfacial area and the decrease of the entrapping matrix fraction due to the densification of the microfibril network. The swelling behavior of cellulose-nanocrystal-reinforced NR in toluene has also been reported (Bendahou *et al.*, 2010). Swelling was found to strongly decrease even with only 1 wt% of cellulose nanoparticles and to be almost independent of filler content and nature (NFC or CNC).

### 5.4.4 Barrier properties

One promising application of cellulosic nanoparticles is in barrier membranes, where the nano-sized fillers impart enhanced mechanical and barrier properties, because molecules penetrate with difficulty into the crystalline domains of cellulose nanoparticles. Moreover, the cellulosic nanoparticles form a dense percolating network held together by strong inter-particle bonds, which reinforce their use as barrier films.

Cellulose nanoparticles prepared by the drop-wise addition of ethanol/HCl aqueous solution into a NaOH/urea/$H_2O$ solution of MCC were found to decrease the water vapor permeability of glycerol-plasticized starch films (Chang *et al.*, 2010a). The water vapor barrier of mango-puree-based edible films (Azeredo *et al.*, 2009) and chitosan films (Azeredo *et al.*, 2010) was successfully improved by adding nanofibrillated cellulose. The water vapor transmission rate of cotton-nanocrystal-reinforced CMC films was reported to have decreased (Choi and Simonsen, 2006).

TEMPO-oxidized cellulose nanocrystals obtained from Whatman paper were investigated as barrier membranes in a poly(vinyl alcohol) (PVA) matrix (Paralikar *et al.*, 2008). It was reported that the incorporation of cellulose nanocrystals reduced the water vapor transmission rate (WVTR) of the membranes. However, the results were more interesting when nanocomposites were prepared by mixing the PVA matrix with low quantities (10 or 20 wt%) of poly(acrylic acid) (PAA) and nanocrystals. This expected result was explained based on crosslinking with PAA, which reduces the number of hydroxyl groups in the composites and thus the hydrophilicity. The cellulose nanocrystals are a physical barrier to moisture flow, creating a tortuous path for the permeating moisture. As a result the addition of 10 wt% of cellulose nanocrystals was more efficient in the reduction of WVTR than 10 wt% of PAA, lowering the matrix WVTR to less than 50%. The combination of 10 wt% whiskers and 10 wt% PAA was reported to display the best performance. Regarding the chemical vapor transmission rate (CVPR), a composition of 10 wt% nanocrystals and 10 wt% PAA has shown the same synergetic effects with the time lag increasing and flux decreasing. It has been demonstrated that the barrier properties of the membranes were improved by the addition of cellulose nanocrystals in combination with PAA.

Syverud and Stenius (2009) studied the barrier properties of pure NFC films. In this study, the gas permeability (oxygen) of the samples was measured and the results showed that the oxygen transmission rate (OTR) was quite low. It was reported that for pure NFC films the OTR values were 17.0 and 17.8 ml $m^{-2}$ $day^{-1}$. These values are in the range for the recommended OTR for modified atmosphere packaging (below 10–20 ml $m^{-2}$ $day^{-1}$). The good barrier properties of pure NFC films were ascribed to the low permeability of cellulose generally enhanced with the crystalline structure of the fibrils.

Belbekhouche *et al.* (2011) compared the barrier properties of cellulose nanocrystal film and nanofibrillated cellulose film. They observed that the nanocrystal film was more permeable to gases than the NFC film. The gas molecules penetrate much more slowly in the NFC film, probably because of a longer diffusion path. In addition to a different surface chemistry, it was supposed that the entanglement of these long flexible nanoparticles and the lower porosity of the films are barrier domains leading to increased tortuosity of the diffusion pathway in NFC films.

## 5.5 Conclusions and future trends

Due to their abundance, high strength and stiffness, low weight, and biodegradability, nanoscale cellulose fiber materials are considered promising candidates in the preparation of bionanocomposites. A broad range of applications of nanocellulose exists even if there is currently a high number of unknowns. Tens of scientific publications and experts have demonstrated its potential, even if most studies have focused on the mechanical properties as a reinforcing phase and the liquid crystal self-ordering properties. The homogeneous dispersion of cellulosic nanoparticles in a polymer matrix is challenging. Packaging is one area in which nanocellulose-reinforced polymeric films are of interest because of the possibility of producing films with high transparency and improved mechanical and barrier properties. A high oxygen barrier is often a requirement for food and pharmaceutical packaging and an improvement may be key for capturing new markets. Besides packaging, the electronic device industry can also profit by using cellulose nanoparticles. The low thermal expansion of nanocellulosics combined with high strength, high modulus, and transparency make them a potential reinforcing material in different applications, such as in flexible displays, solar cells, electronic paper, and panel sensors.

An important program for the production of nanocellulose began in 2008 in Finland: the 'Suomen Nanoselluloosakeskus' or 'Finnish Centre for Nanocellulosic Technologies.' The goal of the center is to develop an industrial-scale production process for nanocellulose and to develop new uses for cellulose as a material. The center is funded by both public and private investment. The forest industry in Finland is going through a major transition, and the utilization of new technologies is expected to provide a means for strengthening competitiveness in the sector.

Another important program, supported by the Canadian government and represented by FPInnovations and NanoQuébec, is the ArboraNano network. The objective of this program for the valorization of nanocellulose (as cellulose nanocrystals) is also to revive the forestry sector in Canada, strongly affected by growing competition from emerging countries in Asia and South America. In June 2011, Domtar Corporation and FPInnovations created the enterprise CelluForce, which will be responsible for the manufacture of nanocrystalline cellulose in the world's first plant of its kind, located in Windsor, Québec. It is expected to start production in the first quarter of 2012.

## 5.6 Sources of further information and advice

Nanocrystals can be prepared from other polysaccharides, such as chitin and starch. Aqueous suspensions of such nanocrystals can also be prepared by acid hydrolysis of the substrate, as already explained for cellulose. Chitin nanocrystals are rod-like particles, with typical dimensions around a few hundred nanometers in length and between 5 and 20 nm in diameter. Starch nanocrystals are platelet-like nanoparticles with length ranging between 20 and 40 nm, a width ranging between 15 and 30 nm, and a thickness ranging between 5 and 7 nm (Dufresne, 2008). In the literature it is possible to find several studies on the preparation and characterization of nanocomposites reinforced with chitin (Nair and Dufresne, 2003; Feng *et al.*, 2009; Chang *et al.*, 2010b; Huang *et al.*, 2011) and starch (Angellier *et al.*, 2006; García *et al.*, 2009, 2011; Mélé *et al.*, 2011). Le Corre *et al.* (2010) and Lin *et al.* (2011) are interesting reviews on the preparation of starch nanocrystals and their application in nanocomposites.

## 5.7 Acknowledgements

The authors gratefully acknowledge the post-doctoral fellowship for ECR (Proc. 5184-09-6) from the Capes Foundation.

## 5.8 References

Ahola S, Salmi J, Johansson L-S, Laine J and Österberg M (2008), 'Model films from native cellulose nanofibrils. Preparation, swelling, and surface interactions', *Biomacromolecules*, 9, 1273–1282.

Alemdar A and Sain M (2008), 'Isolation and characterization of nanofibers from agricultural residues – Wheat straw and soy hulls', *Bioresour Technol*, 99, 1664–1671.

Angellier H, Molina-Boisseau S, Dole P and Dufresne A (2006), 'Thermoplastic starch-waxy maize starch nanocrystals nanocomposites', *Biomacromolecules*, 7, 531–539.

Anglès MN and Dufresne A (2000), 'Plasticized starch/tunicin whiskers nanocomposites. 1. Structural analysis', *Macromolecules*, 33, 8344–8353.

Anglès MN and Dufresne A (2001), 'Plasticized starch/tunicin whiskers nanocomposite materials. 2. Mechanical behavior', *Macromolecules*, 34, 2921–2931.

Araki J, Wada M, Kuga S and Okana T (1998), 'Flow properties of microcrystalline cellulose suspension prepared by acid treatment of native cellulose', *Colloids Surf A*, 142, 75–82.

Araki J, Wada M and Kuga S (2001), 'Steric stabilization of a cellulose microcrystal suspension by poly(ethylene glycol) grafting', *Langmuir*, 17, 21–27.

Aulin C, Varga I, Claesson PM, Wågberg L and Lindström T (2008), 'Buildup of polyelectrolyte multilayers of polyethyleneimine and microfibrillated cellulose studied by *in situ* dual-polarization interferometry and quartz crystal microbalance with dissipation', *Langmuir*, 24, 2509–2518.

Aulin C, Ahola S, Josefsson P, Nishino T, Hirose Y, *et al.* (2009), 'Nanoscale cellulose films with different crystallinities and mesostructures – Their surface properties and interaction with water', *Langmuir*, 25, 7675–7685.

Azeredo HMC, Mattoso HLC, Wood D, Willians TG, Avena-Bustillos RJ and McHug TH (2009), 'Nanocomposite edible films from mango puree reinforced with cellulose nanofibers', *J Food Sci*, 74, N31–N35.

Azeredo HMC, Mattoso HLC, Avena-Bustillos RJ, Ceotto Filho G, Munford ML, *et al.* (2010), 'Nanocellulose reinforced chitosan composite films as affected by nanofiller loading and plasticizer content', *J Food Sci*, 75, N1–N7.

Azizi Samir MAS, Alloin F, Paillet M and Dufresne A (2004a), 'Tangling effect in fibrillated cellulose reinforced nanocomposites', *Macromolecules*, 37, 4313–4316.

Azizi Samir MAS, Alloin F, Gorecki W, Sanchez J-Y and Dufresne A (2004b), 'Nanocomposite polymer electrolytes based on poly(oxyethylene) and cellulose nanocrystals', *J Phys Chem B*, 108, 10845–10852.

Azizi Samir MAS, Alloin F, Sanchez J-Y and Dufresne A (2004c), 'Cellulose nanocrystals reinforced poly(oxyethylene)', *Polymer*, 45, 4149–4157.

Azizi Samir MAS, Mateos AM, Alloin F, Sanchez J-Y and Dufresne A (2004d), 'Plasticized nanocomposite polymer electrolytes based on poly(oxyethylene) and cellulose whiskers', *Electrochim Acta*, 49, 4667–4677.

Azizi Samir MAS, Alloin F, Sanchez J-Y, El Kissi N and Dufresne A (2004e), 'Preparation of cellulose whiskers reinforced nanocomposites from an organic medium suspension', *Macromolecules*, 37, 1386–1393.

Azizi Samir MAS, Alloin F and Dufresne A (2006), 'High performance nanocomposite polymer electrolytes', *Compos Interfaces*, 13, 545–559.

Battista OA, Coppick S, Howsmon JA, Morehead FF and Sisson WA (1956), 'Level-off degree of polymerization', *Ind Eng Chem*, 48, 333–335.

Beck-Candanedo S, Roman M and Gray DG (2005), 'Effect of reaction conditions on the properties and behavior of wood cellulose nanocrystal suspensions', *Biomacromolecules*, 6, 1048–1054.

Belbekhouche S, Bras J, Siqueira G, Chappey C, Lebrun L, *et al.* (2011), 'Water sorption behavior and gas barrier properties of cellulose whiskers and microfibrils films', *Carbohydr Polym*, 83, 1740–1748.

Ben Elmabrouk A, Thielemans W, Dufresne A and Boufi S (2009), 'Preparation of poly(styrene-co-hexylacrylate)/cellulose whiskers nanocomposites via miniemulsion polymerization', *J Appl Polym Sci*, 114, 2946–2955.

Bendahou A, Habibi Y, Kaddami H and Dufresne A (2009), 'Physico-chemical characterization of palm from *Phoenix dactylifera* – L, preparation of cellulose whiskers and natural rubber–based nanocomposites', *J Biobased Mat Bioenergy*, 3, 81–90.

Bendahou A, Kaddami H and Dufresne A (2010), 'Investigation on the effect of cellulosic nanoparticles' morphology on the properties of natural rubber based nanocomposites', *Eur Polym J*, 46, 609–620.

Bledzki AK and Gassan J (1999), 'Composites reinforced with cellulose based fibres', *Prog Polym Sci*, 24, 221–274.

Boluk Y, Lahiji R, Zhao L and McDermott M T (2011), 'Suspension viscosities and shape parameters of cellulose nanocrystals (CNC)', *Colloids Surf, A*, 377, 297–303.

Bondeson D and Oksman K (2007), 'Polylactic acid/cellulose whisker nanocomposites modified by polyvinyl alcohol', *Composites Part A*, 38, 2486–2492.

Bondeson D, Mathew A and Oksman K (2006), 'Optimization of the isolation of nanocrystals from microcrystalline cellulose by acid hydrolysis', *Cellulose*, 13, 171–180.

Bras J, Hassan ML, Bruzesse C, Hassan EA, El-Wakik NA and Dufresne A (2010), 'Mechanical, barrier, and biodegradability properties of bagasse cellulose whiskers reinforced natural rubber nanocomposites', *Ind Crops Prod*, 32, 627–633.

Bras J, Viet D, Bruzzese C and Dufresne A (2011), 'Correlation between stiffness of sheets prepared from cellulose whiskers and nanoparticles dimensions', *Carbohydr Polym*, 84, 211–215.

Braun B and Dorgan J R (2009), 'Single-step method for the isolation and surface functionalization of cellulose nanowhiskers', *Biomacromolecules*, 10, 334–341.

Brown RM (2004), 'Bacterial cellulose: Its potential for new products of commerce', *Abstr Pap Am Chem Soc*, 227, U303.

Cao X, Dong H and Li CM (2007), 'New nanocomposite materials reinforced with flax cellulose nanocrystals in waterborne polyurethane', *Biomacromolecules*, 8, 899–904.

Cao X, Habibi Y and Lucia L A (2009), 'One-pot polymerization, surface grafting, and processing of waterborne polyurethane-cellulose nanocrystal nanocomposites', *J Mater Chem*, 19, 7137–7145.

Chakraborty A, Sain M and Kortschot M (2005), 'Cellulose microfibrils: A novel method of preparation using high shear refining and cryocrushing', *Holzforschung*, 59, 102–107.

Chang PR, Jian R, Zheng P, Yu J and Ma X (2010a), 'Preparation and properties of glycerol plasticized-starch (GPS)/cellulose nanoparticle (CN) composites', *Carbohydr Polym*, 79, 301–305.

Chang PR, Jian R, Yu J and Maa X (2010b), 'Starch-based composites reinforced with novel chitin nanoparticles', *Carbohydr Polym*, 80, 420–425.

Chauve G, Heux L, Arouini R and Mazeau K (2005), 'Cellulose poly(ethylene-co-vinyl acetate) nanocomposites studied by molecular modeling and mechanical spectroscopy', *Biomacromolecules*, 6, 2025–2031.

Chazeau L, Cavaillé JY, Canova GR, Dendievel R and Boutherin B (1999a), 'Viscoelastic properties of plasticized PVC reinforced with cellulose whiskers', *J Appl Polym Sci*, 71, 1797–1808.

Chazeau L, Cavaillé JY and Terech P (1999b), 'Mechanical behaviour above $T_g$ of a plasticized PVC reinforced with cellulose whiskers, a SANS structural study', *Polymer*, 40, 5333–5344.

Chazeau L, Paillet M and Cavaillé J Y (1999c), 'Plasticized PVC reinforced with cellulose whiskers: I. Linear viscoelastic behavior analyzed through the quasi-point defect theory', *J Polym Sci B: Polym Phys*, 37, 2151–2164.

Chazeau L, Cavaillé JY and Perez J (2000), 'Plasticized PVC reinforced with cellulose whiskers. II. Plastic behavior', *J Polym Sci B: Polym Phys*, 38, 383–392.

Choi YJ and Simonsen J (2006), 'Cellulose nanocrystal-filled carboxymethyl cellulose nanocomposites', *J Nanosci Nanotechnol*, 6, 633–639.

Corrêa AC, Teixeira EM, Pessan LA and Mattoso LHC (2010), 'Cellulose nanofibers from curaua fibers', *Cellulose*, 17, 1183–1192.

de Menezes A., Siqueira G, Curvelo AAS and Dufresne A (2009), 'Extrusion and characterization of functionalized cellulose whisker reinforced polyethylene nanocomposites', *Polymer*, 50, 4552–4563.

de Mesquita JP, Donnici CL and Pereira FV (2010), 'Biobased nanocomposites from layer-by-layer assembly of cellulose nanowhiskers with chitosan', *Biomacromolecules*, 11, 473–480.

de Souza Lima MM, Wong JT, Paillet M, Borsali R and Pecora R (2003), 'Translational and rotational dynamics of rodlike cellulose whiskers', *Langmuir*, 19, 24–29.

Dinand E, Chanzy H and Vignon M (1996), 'Parenchymal cell cellulose from sugar beet pulp: Preparation and properties', *Cellulose*, 3, 183–188.

Dinand E, Chanzy H and Vignon R (1999), 'Suspensions of cellulose microfibrils from sugar beet pulp', *Food Hydrocolloids*, 13, 275–283.

Dong XM, Revol J-F and Gray DG (1998), 'Effect of microcrystallite preparation conditions on the formation of colloid crystals of cellulose', *Cellulose*, 5, 19–32.

Dubief D, Samain E and Dufresne A (1999), 'Polysaccharide microcrystals reinforced amorphous poly($\beta$-hydroxyoctanoate) nanocomposite materials', *Macromolecules*, 32, 5765–5771.

Dufresne A (2000), 'Dynamic mechanical analysis of the interphase in bacterial polyester/ cellulose whiskers natural composites', *Compos Interfaces*, 7, 53–67.

Dufresne A (2006), 'Comparing the mechanical properties of high performances polymer nanocomposites from biological sources', *J Nanosci Nanotechnol*, 6, 322–330.

Dufresne A (2008), 'Polysaccharide nanocrystal reinforced nanocomposites', *Can J Chem*, 86, 484–494.

Dufresne A and Vignon MR (1998), 'Improvement of starch films performances using cellulose microfibrils', *Macromolecules*, 31, 2693–2696.

Dufresne A, Kellerhals MB and Witholt B (1999), 'Transcrystallization in Mcl-PHAs/ cellulose whiskers composites', *Macromolecules*, 32, 7396–7401.

Dufresne A, Dupeyre D and Vignon MR (2000), 'Cellulose microfibrils from potato cells: Processing and characterization of starch/cellulose microfibrils composites', *J Appl Polym Sci*, 76, 2080–2092.

Eichhorn SJ, Dufresne A, Aranguren M, Marcovich NE, Capadona JR, *et al.* (2010), 'Review: Current international research into cellulose nanofibres and nanocomposites', *J Mater Sci*, 45, 1–33.

El-Saied H, Basta AH and Gobran RH (2004), 'Research progress in friendly environmental technology for the production of cellulose products (bacterial cellulose and its application)', *Polym Plast Technol Eng*, 43, 797–820.

Eyholzer C, Bordeanu N, Lopez-Suevos F, Rentsch D, Zimmermann T and Oksman K (2010), 'Preparation and characterization of water-redispersible nanofibrillated cellulose in powder form', *Cellulose*, 17, 19–30.

Favier V, Canova GR, Cavaille JY, Chanzy H, Dufresne A and Gauthier C (1995a), 'Nanocomposite materials from latex and cellulose whiskers', *Polym Adv Technol*, 6, 351–355.

Favier F, Chanzy H and Cavaillé JY (1995b), 'Polymer nanocomposites reinforced by cellulose whiskers', *Macromolecules*, 28, 6365–6367.

Favier V, Canova GR, Shrivastava SC and Cavaillé JY (1997), 'Mechanical percolation in cellulose whiskers nanocomposites', *Polym Eng Sci*, 37, 1732–1739.

Feng L, Zhou Z, Dufresne A, Huang J, Wei M and An L (2009), 'Structure and properties of new thermoforming bionanocomposites based on chitin whiskers-graft-polycaprolactone', *J Appl Polym Sci*, 112, 2830–2837.

Fleming K, Gray D, Prasannan S and Matthews (2000), 'Cellulose crystallites: A new and robust liquid crystalline medium for the measurement of residual dipolar couplings', *J Am Chem Soc*, 122, 5224–5225.

García NL, Ribba L, Dufresne A, Aranguren MI and Goyanes S (2009), 'Physico-mechanical properties of biodegradable starch nanocomposites', *Macromol Mater Eng*, 294, 169–177.

García NL, Ribba L, Dufresne A, Aranguren MI and Goyanes S (2011), 'Effect of glycerol on the morphology of nanocomposites made from thermoplastic starch and starch nanocrystals', *Carbohydr Polym*, 84, 203–210.

Garcia de Rodriguez NL, Thielemans W and Dufresne A (2006), 'Sisal cellulose whiskers reinforced polyvinyl acetate nanocomposites', *Cellulose*, 13, 261–270.

Goussé C, Chanzy H, Exoffier G, Soubeyrand L and Fleury E (2002), 'Stable suspensions of partially silylated cellulose whiskers dispersed in organic solvents', *Polymer*, 43, 2645–2651.

Greiner A and Wendorff JH (2007), 'Electrospinning: A fascinating method for the preparation of ultrathin fibers', *Angew Chem Int Ed*, 46, 5670–5703.

Grunert M and Winter WT (2002), 'Nanocomposites of cellulose acetate butyrate reinforced with cellulose nanocrystals', *J Polym Environ*, 10, 27–30.

Habibi Y and Dufresne A (2008), 'Highly filled bionanocomposites from functionalized polysaccharide nanocrystals', *Biomacromolecules*, 9, 1974–1980.

Habibi Y, Chanzy H and Vignon MR (2006), 'TEMPO-mediated surface oxidation of cellulose whiskers', *Cellulose*, 13, 679–687.

Habibi Y, Goffin A-L, Schiltz N, Duquesne E, Dubois P and Dufresne A (2008), 'Bionanocomposites based on poly(ε-caprolactone)-grafted cellulose nanocrystals by ring-opening polymerization', *J Mater Chem*, 18, 5002–5010.

Habibi Y, Mahrouz M and Vignon MR (2009), 'Microfibrillated cellulose from the peel of prickly pear fruits', *Food Chem*, 115, 423–429.

Hamad WY and Hu TQ (2010), 'Structure-process-yield interrelations in nanocrystalline cellulose extraction', *Can J Chem Eng*, 88, 392–402.

Hanley SJ, Giasson J, Revol JF and Gray DG (1992), 'Atomic force microscopy of cellulose microfibrils: Comparison with transmission electron microscopy', *Polymer*, 33, 4639–4642.

Helbert W, Cavaille JY and Dufresne A (1996), 'Thermoplastic nanocomposites filled with wheat straw cellulose whiskers. Part I: Processing and mechanical behavior', *Polym Compos*, 17, 604–611.

Henriksson M and Berglund LA (2007), 'Structure and properties of cellulose nanocomposite films containing melamine formaldehyde', *J Appl Polym Sci*, 106, 2817–2824.

Henriksson M, Henriksson G, Berglund LA and Lindstrom T (2007), 'An environmentally friendly method for enzyme-assisted preparation of microfibrillated cellulose (MFC) nanofibers', *Eur Polym J*, 43, 3434–3441.

Heux L, Chauve G and Bonini C (2000), 'Nonflocculating and chiral-nematic self-ordering of cellulose microcrystals suspensions in nonpolar solvents', *Langmuir*, 16, 8210–8212.

Huang ZM, Zhang YZ, Kotaki M and Ramakrishna S (2003), 'A review on polymer nanofibers by electrospinning and their applications in nanocomposites', *Compos Sci Technol*, 63, 2223–2253.

Huang J, Zou JW, Chang PR, Yu JH and Dufresne A (2011), 'New waterborne polyurethane-based nanocomposites reinforced with low loading levels of chitin whiskers', *eXPRESS Polym Lett*, 5, 362–373.

Iwamoto S, Abe K and Yano H (2008), 'The effect of hemicelluloses on wood pulp nanofibrillation and nanofiber network characteristics', *Biomacromolecules*, 9, 1022–1026.

Iwamoto S, Kai W, Isogai A and Iwata T (2009), 'Elastic modulus of single cellulose microfibrils from tunicate measured by atomic force microscopy', *Biomacromolecules*, 10, 2571–2576.

Jean B, Dubreuil F, Heux L and Cousin F (2008), 'Structural details of cellulose nanocrystals/polyelectrolytes multilayers probed by neutron reflectivity and AFM', *Langmuir*, 24, 3452–3458.

Jean B, Heux L, Dubreuil F, Chambat G and Cousin F (2009), 'Non-electrostatic building of biomimetic cellulose-xyloglucan multilayers', *Langmuir*, 25, 3920–3923.

Klemm D, Heublein B, Fink H-P and Bohn A (2005), 'Cellulose: Fascinating biopolymer and sustainable raw material', *Angew Chem Int Ed*, 44, 3358–3393.

Klemm D, Schumann D, Kramer F, Hessler N, Hornung M, *et al.* (2006), 'Nanocelluloses as innovative polymers in research and application', *Adv Polym Sci*, 205, 49–96.

Kvien I, Tanem BS and Oksman K (2005), 'Characterization of cellulose whiskers and their nanocomposites by atomic force and electron microscopy', *Biomacromolecules*, 6, 3160–3165.

Kvien I, Sugiyama J, Votrubec M and Oksman K (2007), 'Characterization of starch based nanocomposites', *J Mater Sci*, 42, 8163–8171.

Le Corre D, Bras J and Dufresne A (2010), 'Starch nanoparticles: A review', *Biomacromolecules*, 11, 1139–1153.

Li R, Fei J, Cai Y, Li Y, Feng J and Yao J (2009), 'Cellulose whiskers extracted from mulberry: A novel biomass production', *Carbohydr Polym*, 76, 94–99.

Li Z, Renneckar S and Barone JR (2010), 'Nanocomposites prepared by *in situ* enzymatic polymerization of phenol with TEMPO-oxidized nanocellulose', *Cellulose*, 17, 57–68.

Lin N, Chen G, Huang J, Dufresne A and Chang PR (2009), 'Effects of polymer-grafted natural nanocrystals on the structure and mechanical properties of poly(lactic acid): A case of cellulose whisker-graft-polycaprolactone', *J Appl Polym Sci*, 113, 3417–3425.

Lin N, Huang J, Chang PR, Anderson DP and Yu J (2011), 'Effect of glycerol on the morphology of nanocomposites made from thermoplastic starch and starch nanocrystals', *J Nanomater*, 2011, 1–13.

Ljungberg N, Bonini C, Bortolussi F, Boisson C, Heux L and Cavaillé JY (2005), 'New nanocomposite materials reinforced with cellulose whiskers in atactic polypropylene: Effect of surface and dispersion characteristics', *Biomacromolecules*, 6, 2732–2739.

Ljungberg N, Cavaillé JY and Heux L (2006), 'Nanocomposites of isotactic polypropylene reinforced with rod-like cellulose whiskers', *Polymer*, 47, 6285–6292.

Lu Y, Weng L and Cao X (2005), 'Biocomposites of plasticized starch reinforced with cellulose crystallites from cottonseed linter', *Macromol Biosci*, 5, 1101–1107.

Lu J, Askeland P and Drzal LT (2008a), 'Surface modification of microfibrillated cellulose for epoxy composite application', *Polymer*, 49, 1285–1296.

Lu J, Wang T and Drzal LT (2008b), 'Preparation and properties of microfibrillated cellulose polyvinyl alcohol composite materials', *Composites Part A*, 39, 738–746.

Malainine ME, Mahrouz M and Dufresne A (2005), 'Thermoplastic nanocomposites based on cellulose microfibrils from *Opuntia ficus-indica* parenchyma cell', *Compos Sci Technol*, 65, 1520–1526.

Marcovich NE, Auad ML, Belessi NE, Nutt SR and Aranguren MI (2006), 'Cellulose micro/nanocrystals reinforced polyurethane', *J Mater Res*, 21, 870–881.

Mathew AP and Dufresne A (2002), 'Morphological investigation of nanocomposites from sorbitol plasticized starch and tunicin whiskers', *Biomacromolecules*, 3, 609–617.

Mathew AP, Thielemans W and Dufresne A (2008), 'Mechanical properties of nanocomposites from sorbitol plasticized starch and tunicin whiskers', *J Appl Polym Sci*, 109, 4065–4074.

Matos Ruiz M, Cavaillé JY, Dufresne A, Graillat C and Gérard JF (2001), 'New waterborne epoxy coatings based on cellulose nanofillers', *Macromol Symp*, 169, 211–222.

Medeiros ES, Mattoso LHC, Ito E., Gregorski KS, Robertson GH, *et al.* (2008), 'Electrospun nanofibers of poly(vinyl alcohol) reinforced with cellulose nanofibrils', *J Biobased Mater Bioenergy*, 2, 1–12.

Mélé P, Angellier-Coussy H, Molina-Boisseau S and Dufresne A (2011), 'Reinforcing mechanism of starch nanocrystals in a nonvulcanized natural rubber matrix', *Biomacromolecules*, 12, 1487–1493.

Nair KG and Dufresne A (2003), 'Crab shell chitin whiskers reinforced natural rubber nanocomposites. 3. Effect of chemical modification of chitin whiskers', *Biomacromolecules*, 4, 1835–1842.

Nakagaito AN and Yano H (2004), 'The effect of morphological changes from pulp fiber towards nano-scale fibrillated cellulose on the mechanical properties of high-strength plant fiber based composites', *Appl Phys A*, 78, 547–552.

Nakagaito AN and Yano H (2008), 'Toughness enhancement of cellulose nanocomposites by alkali treatment of the reinforcing cellulose nanofibers', *Cellulose*, 15, 323–331.

Nakagaito AN, Iwamoto S and Yano H (2005), 'Bacterial cellulose: The ultimate nano-scalar cellulose morphology for the production of high-strength composites', *Appl Phys A*, 80, 93–97.

Nakagaito AN, Fujimura A, Sakai T, Hama Y and Yano H (2009), 'Production of microfibrillated cellulose (MFC)-reinforced polylactic acid (PLA) nanocomposites from sheets obtained by a papermaking-like process', *Compos Sci Technol*, 69, 1293–1297.

Nogi M, Handa K, Nakagaito AN and Yano H (2005), 'Optically transparent bionanofiber composites with low sensitivity to refractive index of the polymer matrix', *Appl Phys Lett*, 87, 243110–243112.

Noishiki Y, Nishiyama Y, Wada M, Kuga S and Magoshi J (2002), 'Mechanical properties of silk fibroin-microcrystalline cellulose composite films', *J Appl Polym Sci*, 86, 3425–3429.

Oksman K, Mathew AP, Bondeson D and Kvien I (2006), 'Manufacturing process of cellulose whiskers/polylactic acid nanocomposites', *Compos Sci Technol*, 66, 2776–2784.

Orts WJ, Godbout L, Marchessault RH and Revol JF (1998), 'Enhanced ordering of liquid crystalline suspensions of cellulose microfibrils: A small angle neutron scattering study', *Macromolecules*, 31, 5717–5725.

Pääkkö M, Ankerfors M, Kosonen H, Nykänen A, Ahola S, *et al.* (2007), 'Enzymatic hydrolysis combined with mechanical shearing and high-pressure homogenization for nanoscale cellulose fibrils and strong gels', *Biomacromolecules*, 8, 1934–1941.

Paralikar SA, Simonsen J and Lombardi J (2008), 'Poly(vinyl alcohol)/cellulose nanocrystals barrier membranes', *J Membrane Sci*, 320, 248–258.

Park I, Kang M, Kim HS and Jin HJ (2007), 'Electrospinning of poly(ethylene oxide) with bacterial cellulose whiskers', *Macromol Symp*, 249–250, 289–294.

Peresin MS, Habibi Y, Zoppe JO, Pawlak JJ and Rojas OJ (2010), 'Nanofiber composites of polyvinyl alcohol and cellulose nanocrystals: Manufacture and characterization', *Biomacromolecules*, 11, 674–681.

Petersson L and Oksman K (2006), 'Biopolymer based nanocomposites: Comparing layered silicates and microcrystalline cellulose as nanoreinforcement', *Compos Sci Technol*, 66, 2187–2196.

Podsiadlo P, Choi SY, Shim B, Lee J, Cuddihy M and Kotov NA (2005), 'Molecularly engineered nanocomposites: Layer-by-layer assembly of cellulose nanocrystals', *Biomacromolecules*, 6, 2914–2918.

Rånby BG (1951), 'Fibrous macromolecular systems. Cellulose and muscle. The colloidal properties of cellulose micelles', *Discuss Faraday Soc*, 11, 158–164.

Rojas OJ, Montero GA and Habibi Y (2009), 'Electrospun nanocomposites from polystyrene loaded with cellulose nanowhiskers', *J Appl Polym Sci*, 113, 927–935.

Roman M and Winter WT (2004), 'Effect of sulfate groups from sulfuric acid hydrolysis on the thermal degradation behavior of bacterial cellulose', *Biomacromolecules*, 5, 1671–1677.

Roohani M, Habibi Y, Belgacem NM, Ebrahim G, Karimi AN and Dufresne A (2008), 'Cellulose whiskers reinforced polyvinyl alcohol copolymers nanocomposites', *Eur Polym J*, 44, 2489–2498.

Saito T, Nishiyama Y, Putaux J-L, Vignon M and Isogai A (2006), 'Homogeneous suspensions of individualized microfibrils from TEMPO-catalyzed oxidation of native cellulose', *Biomacromolecules*, 7, 1687–1691.

Saito T, Kimura S, Nishiyama Y and Isogai A (2007), 'Cellulose nanofibers prepared by TEMPO-mediated oxidation of native cellulose', *Biomacromolecules*, 8, 2485–2491.

Sakurada I, Nukushina Y and Ito T (1962), 'Experimental determination of the elastic modulus of crystalline regions in oriented polymers', *J Polym Sci*, 57, 651–660.

Seydibeyoğlu ÖM and Oksman K (2008), 'Novel nanocomposites based on polyurethane and micro fibrillated cellulose', *Compos Sci Technol*, 68, 908–914.

Shimazaki Y, Miyazaki Y, Takezawa Y, Nogi M, Abe K, *et al.* (2007), 'Excellent thermal conductivity of transparent cellulose nanofiber/epoxy resin nanocomposites', *Biomacromolecules*, 8, 2976–2978.

Siqueira G, Bras J and Dufresne A (2009), 'Cellulose whiskers versus microfibrils: Influence of the nature of the nanoparticle and its surface functionalization on the thermal and mechanical properties of nanocomposites', *Biomacromolecules*, 10, 425–432.

Siqueira G, Bras J and Dufresne A (2010a), '*Luffa cylindrica* as a lignocellulosic source of fiber, microfibrillated cellulose, and cellulose nanocrystals', *BioResources*, 5, 727–740.

Siqueira G, Abdillahi H, Bras J and Dufresne A (2010b), 'High reinforcing capability cellulose nanocrystals extracted from *Syngonanthus nitens* (capim dourado)', *Cellulose*, 17, 289–298.

Siqueira G, Bras J and Dufresne A (2010c), 'New process of chemical grafting of cellulose nanoparticles with a long chain isocyanate', *Langmuir*, 26, 402–411.

Siqueira G, Tapin-Lingua S, Bras J, Perez DS and Dufresne A (2011), 'Mechanical properties of natural rubber nanocomposites reinforced with cellulosic nanoparticles obtained from combined mechanical shearing, and enzymatic and acid hydrolysis of sisal fibers', *Cellulose*, 18, 57–65.

Siró I and Plackett D (2010), 'Microfibrillated cellulose and new nanocomposite material: A review', *Cellulose*, 17, 459–494.

Sjoström E (1981), *Wood Chemistry: Fundamentals and Applications*, New York, Academic.

Šturcova A, Davies GR and Eichhorn S J (2005), 'Elastic modulus and stress-transfer properties of tunicate cellulose whiskers', *Biomacromolecules*, 6, 1055–1061.

Svagan AJ, Hedenqvist MS and Berglund L (2009), 'Reduced water vapour sorption in cellulose nanocomposites with starch matrix', *Compos Sci Technol*, 69, 500–506.

Syverud K and Stenius P (2009), 'Strength and barrier properties of MFC films', *Cellulose*, 16, 75–85.

Tashiro K and Kobayashi M (1991), 'Theoretical evaluation of three-dimensional elastic constants of native and regenerated celluloses: Role of hydrogen bonds', *Polymer*, 32, 1516–1526.

Turbak A, Snyder F and Sandberg K (1983), 'Microfibrillated cellulose: A new cellulose product: Properties, uses, and commercial potential', *J App Polym Sci*, 37, 815–827.

van den Berg O, Capadona JR and Weder C (2007), 'Preparation of homogeneous dispersions of tunicate cellulose whiskers in organic solvents', *Biomacromolecules*, 8, 1353–1357.

Wågberg L, Decher G, Norgren M, Lindström T, Ankerfors M and Axnäs K (2008), 'The build-up of polyelectrolyte multilayers of microfibrillated cellulose and cationic polyelectrolytes', *Langmuir*, 24, 784–795.

Wang Y, Cao X and Zhang L (2006), 'Effects of cellulose whiskers on properties of soy protein thermoplastics', *Macromol Biosci*, 6, 524–531.

Whistler RL and Richards EL (1970), *The Carbohydrates*, New York, Academic Press.

Yano H, Sugiyama J, Nakagaito AN, Nogi M, Matsuura T, *et al.* (2005), 'Optically transparent composites reinforced with networks of bacterial nanofibers', *Adv Mat*, 17, 153–155.

Zimmermann T, Pöhler E and Geiger T (2004), 'Cellulose fibrils for polymer reinforcement', *Adv Eng Mat*, 6, 754–761.

Zimmermann T, Pöhler E and Schwaller P (2005), 'Mechanical and morphological properties of cellulose fibril reinforced nanocomposites', *Adv Eng Mat*, 7, 1156–1161.

Zoppe JO, Peresin MS, Habibi Y, Venditti RA and Rojas OJ (2009), 'Reinforcing poly(epsilon-caprolactone) nanofibers with cellulose nanocrystals', *ACS Appl Mater Interfaces*, 1, 1996–2004.

# 6
# Metal–polymer nanocomposites

G. HENESS, University of Technology, Sydney. Australia

**Abstract:** Polymers have low strength and stiffness, high dielectric strength and poor thermal properties. They are poor conductors of electricity and heat and are non-magnetic. Attempts have been made to change and improve some of these to increase the range of applications. Recently, researchers have shown interest in using nanoparticles. This chapter looks at metal nanoparticles, which exhibit uniquely different electronic and optical properties from their bulk counterparts. It is these interesting new properties that are one of the driving forces for their incorporation in polymer matrices. Methods of incorporation of metal nanoparticles into a polymer, the resultant properties, applications and future trends will be discussed.

**Key words:** metal–polymer nanocomposite, incorporating metal nanoparticles into polymer, application of metal polymer nanocomposites.

## 6.1 Introduction: the role of nanoparticles as reinforcement

Polymers generally have low strength and stiffness, poor thermal properties and good dielectric properties. They are good insulators of electricity and heat, and are non-magnetic. Attempts over the years have been made to change some of these properties to increase the range of applications. There have been two approaches: by changing the chemistry of the polymers or by the incorporation of non-polymeric materials into the polymer. The latter, adding a wide range of particles and fibres, has been an ongoing preoccupation with materials scientists and materials engineers for decades.

Metal nanoparticles exhibit uniquely different electronic and optical properties from their bulk counterparts. These properties are inherently related to their size and shape (Hicks *et al.*, 2001; Shevchenko *et al.*, 2006; Collier *et al.*, 1998; You *et al.*, 2005). It is these interesting new properties that are one of the major driving forces for incorporation into polymer matrices (Carotenuto and Nicolais, 2003).

Nanoparticles can show quite different properties from their bulk counterparts. In general, this may be attributed to the nanoparticles having a much larger proportion of their atoms on the surface compared with their bulk counterparts. For example, 3-nm nanoparticles, about 1000 atoms, would have around 40% of their atoms on the surface. In addition to these surface effects, there are changes to the behaviour of the metal due to quantum confinement effects. The electronic structure of these nanoparticles has been found to be quite different from that of the bulk metal. For example, nanoparticles made from ferromagnetic materials can become paramagnetic, nanoparticles made from conducting bulk material may become poor conductors. These changes in behaviour are quite interesting for utilizing their unique properties in nanocomposites. In particular, these property changes are a function of particle size and, thus, the properties are tunable, creating a range of properties that can be 'dialled up'. The so-called quantum effect gives these metal particles unique properties such as plasmon absorption, superparamagnetism, and IR photoluminescence, to name a few.

Only a somewhat abbreviated account is provided here; greater technical detail may be found in the references provided at the end of the chapter.

## 6.2 Techniques for reinforcement

The literature shows that metal nanoparticle–polymer composites are generally in the form of thin polymer films as these are often the easiest to produce and they make use of the flexibility of the films.

A number of methods have been used to successfully place metal nanoparticles into a wide variety of polymers. These can be divided into two approaches. The first is an extrinsic or *ex situ* approach, in which metal nanoparticles are produced, often by some chemical approach or vapour deposition, then introduced into the polymer, which may be a polymer solution, liquid monomer, dissolved polymer solution or polymer powder, for example, by adding nanoparticles during polymerization (Oliveira *et al.*, 2006).

The second is an intrinsic or *in situ* approach, in which the metal nanoparticles are grown within the polymer. This can be through the chemical reduction of a metallic precursor that has been already dissolved in the polymer (Selvan *et al.*, 1998), thermolysis decomposition (Pomogailo *et al.*, 2003), photolysis decomposition (Peng *et al.*, 2009), photochemical preparation (Breimer *et al.*, 2001; Lu *et al.*, 2006), incorporation during polymer electrosynthesis (Zhou *et al.*, 2006) or nanoparticle formation during polymerization (Sarma *et al.*, 2002).

In each case, there is a need to develop methods by which particle size, particle morphology and surface energy are controlled, as these properties determine the ultimate properties of the composite.

### 6.2.1 Extrinsic methods: physical methods

For *ex situ* placing of metal nanoparticles into polymer matrices the following should be considered:

- the metal nanoparticle size and morphology
- the metal nanoparticle chemistry
- the polymer type
- the dispersion of the metal nanoparticles in the polymer.

Whilst not within the scope of this chapter, the preparation of the metal nanoparticles is important since it determines the first two points on the previous page. There are many methods of preparing metal nanoparticles but most are based on the controlled precipitation and stabilisation of colloids. Two common routes are the controlled micelle or reverse micelle reactions and the reduction of a metal salt in a solvent (Tröger *et al.*, 1997; Nicolais and Carotenuto, 2005).

The points mentioned earlier will determine what techniques can be used to incorporate the metal nanoparticles into the polymer. Their physical size, high surface area and their, usually, very different surface energies from the polymer tend to cause agglomeration of the metal and difficulties in achieving uniform dispersion. Depending on the viscosity of the polymer solution and the relative densities of the nanoparticles and polymer, rising or settling of the nanoparticles may occur. Of course, depending on the application, these apparent problems may be used to advantage to control the internal architecture of the composite, for example, through a gradation in particle dispersion.

The surface chemistry of the particles is important. Generally, the metal surfaces are passivated to produce a hydrophobic surface to aid dispersion in the polymer, thus avoiding aggregation. The metal nanoparticles can be functionalized with organic tails to make them hydrophobic. One example of passivation is the use of n-alkanethiol molecules.

A graphene coating has also been shown to make the nanoparticles compatible with the polymer and leads to good dispersion (Luechinger *et al.*, 2008). Passivation surface treatments may also be employed to reduce oxidation of the metal.

Of course, there are some cases, as mentioned in this chapter, that do not require uniform dispersion.

Once the surfaces have been treated, the nanoparticles can be introduced into the polymer, which is most commonly in a powder or liquid form. The simplest form of mixing metal nanoparticles into a polymer matrix is to take the metal nanoparticles and mix them with a polymer powder, followed by standard polymer processing techniques such as hot pressing, extrusion or injection moulding. Mixing techniques include rotary blade mixers and ball mills.

For thermosetting resins such as epoxy and polyester that start off life as two (or more) liquids, simple mixing of the powder into the liquid can be performed. Common processing techniques to produce thin films, such as spin casting, can then be applied. However, the problems mentioned above often create agglomeration. High-speed blender-type mixers can be employed, as can sonication, to try and break up these agglomerates before setting commences. Surface modifications of the metal particles may also be used to assist dispersion. Complete dispersion is, nevertheless, hard to achieve.

These techniques can also be used for some thermoplastics. For example, polymethylmethacrylate can be dissolved in chloroform to form a liquid to which the nanoparticles are added (Luechinger *et al.*, 2008). With this approach some degree of control can be attained by varying the polymer to solvent ratio. This may allow easier dispersion. Care needs to be taken that the solvent does not adversely affect the metal nanoparticles through reaction at the nanoparticle surface. For conducting polymers the dissolution of the polymer in a solvent is not viable as they are insoluble in common solvents (Gangopadhyay and De, 2002). Oliveira *et al.* (2004) used a two-phase polymerisation technique to produce dodecanethiol-capped silver nanoparticle–polyaniline nanocomposites. The polyaniline was dissolved in a silver nanoparticle and toluene solution mixed with an ammonium persulphate aqueous solution. In 2006, Oliveira *et al.* (2006) further developed this route where, by controlling some of the synthesis variables, the structure of the nanocomposites was varied from a thin sheet of polyaniline around the silver nanoparticles to silver nanoparticles dispersed in a polyaniline matrix.

A number of physical deposition processes lend themselves to the production of metal nanoparticle composites (Faupel *et al.*, 2010). These include vapour, plasma, evaporation, sputter, cluster sources and laser ablation. Often the polymer and the metal are co-deposited and control of this co-deposition is critical for controlling the reinforcement architecture. The reader is recommended to read reviews by Biederman (2004), Heilmann (2003) and Faupel *et al.* (2010).

Vapour deposition processes used to prepare composites are often based on co-deposition of the metal and polymer components. Both chemical vapour deposition (CVD) and physical vapour deposition (PVD) can be used. An advantage of this system is that the feedstock flow rates, for both the nanoparticles and the polymer, can be used to control the particle concentration and thus the architecture of the nanocomposites, so that continuous, layered or even gradated distributions of the particles can be made. For CVD, monomers and organometallic complexes can be used as precursors, which react to form the nanocomposite as they deposit. Hydrocarbon, fluorocarbon and organosilicon precursors are used for the polymer, with almost any metal being able to be used for the metal nanoparticles. Physical vapour deposition is a far simpler process in which pure polymers and the metal are thermally evaporated and deposited.

The versatility shown by the plasma process in producing designer films was also shown by Choukourov *et al.*, who used angled deposition to lay down nanostructured composite coatings (Choukourov *et al.*, 2010). Generally a homogeneous distribution of metal nanoparticles can be obtained for a large number of metals within polymer matrices. The technique involves sputtering a metal target with the simultaneous polymerization of polymer precursors within the plasma. The flexibility of this method allows the production of a wide range of nanocomposites. The processing parameters control polymerisation and therefore the matrix properties to some extent (d'Agostino *et al.*, 2005).

It has been found that particular polymers may be deposited by sputtering (Roy *et al.*, 1983) or evaporation (Kubono and Okui, 1994; Gill *et al.*, 2003). Within the broad area of this technique the following are examples of the versatility:

- Monomer evaporation and polycondensation on a substrate (Behnke *et al.*, 2000).
- Thermal cracking and polymerisation onto a substrate (Biswas *et al.*, 2003, 2006).
- Co-deposition of various polymers and noble metals where the metal–polymer deposition rate ratio controls the volume fraction and microstructure of the composite (Takele *et al.*, 2006).

Co-sputtering is used to produce metal nanoparticle polymer composites. The polymer undergoes radio-frequency (RF) magnetron sputtering whilst the metal or alloy is deposited via DC sputtering (Biederman, 2000; Schurmann *et al.*, 2005; Greve *et al.*, 2006; Hillenkamp *et al.*, 2006). If the metal is magnetic then RF sputtering can be used (Greve *et al.*, 2006). Crosslinking of the matrix material can occur through this technique. In a variation on evaporation and sputtering, metal clusters can be formed by evaporating or sputtering into a flow of cold inert gas at relatively light pressure (Haberland *et al.*, 1994), producing clusters in the range 3–15 nm.

Laser ablation has been used for some time to produce nanoparticles from a wide range of ceramics, metals and semiconductors. One method is to mechanically reduce the material to micron size then place it in a liquid, often water, which is circulated through a cell. The solution is subjected to a pulsed laser that ablates the particles to form nano-sized particles. If the metal and the polymer are ablated together then a composite structure can be deposited onto a surface.

Another novel technique is the layer-by-layer (LbL) adsorption technique. This involves alternating the adsorption of anion-stabilized gold nanoparticles and positively charged conducting polymer chains. This produces a laminated composite (Tian *et al.*, 2004; Stoyanova *et al.*, 2011).

### 6.2.2 Intrinsic methods

Often the deposition techniques are limited to thin film production. Whilst wet chemical methods can also produce these thin films, they also lend themselves to the manufacture of bulk samples (Nicolais and Carotenuto, 2005; Murray *et al.*, 2000; Gradess *et al.*, 2009).

These methods generally produce a more uniform distribution of nanoparticles but are more complex than the extrinsic methods mentioned above. These are basically chemical methods, often using metal–organic precursors to develop monodispersed alloy particles in polymers (Sun, 2006). In general the polymer is dissolved along with the metal complex or salt in a co-solvent. Bockstaller *et al.* (2005) used block polymers to control the arrangement of the nanoparticles in the matrix. This was done by treating the surface of the nanoparticles so that they have affinity to one of the copolymer blocks. This surface treatment is often the attachment of a polymer that will interact with the desired part of the copolymer.

There are essentially two stages to these processes. The first places the metal nanoparticle precursor, often a metal salt, in the polymer precursor and the second is the reduction and growth of the metal nanoparticles. This may involve the chemical reduction of a metallic precursor, which has already been dissolved in the polymer (Selvan *et al.*, 1998), by thermolysis or photolysis decomposition, photochemical preparation (Breimer *et al.*, 2001; Lu *et al.*, 2006), incorporation during polymer electrosynthesis (Zhou *et al.*, 2006) or nanoparticle formation during polymerisation (Sarma *et al.*, 2002). The choice of precursor is an important aspect of this technique.

Supercritical fluid methods, commonly $scCO_2$ impregnation (Hasell, 2008), produce metal nanoparticles within the polymer by impregnating metal complex precursors, which are then reduced. An important advantage is that the impregnation can be performed on already formed polymer parts, which may alleviate nanoparticle aggregation during part processing. In addition, the method has several advantages over other techniques in that no organic solvents are used, the nanoparticles produced do not have to have surface coatings and the number of possible polymer matrices available is greater than with other methods. It is believed to be easily up-scalable.

As mentioned earlier, the thermal decomposition of metal precursors is another intrinsic method with a large number of precursors (Mayer, 1998). Even though they can be produced by large-scale production, many of these precursors perform less than ideally when decomposing. In general the metal salts should be added to the polymer before casting, as further processing can degrade the composite's properties. Transition metal mercaptides exhibit affinity with many hydrophobic polymers (Carotenuto *et al.*, 2003). Decomposition of mercaptides occurs between 110 and 180 °C, which is a suitable temperature range for processing many polymers. This results in the reduction to the metal or a metal sulphide. Similar work has been reported for metal salts (Koizumi *et al.*, 2002) and nitrates (Matsuda and Ando, 2005; Porel *et al.*, 2005; Matsuda *et al.*, 2001). The use of mercaptides overcomes the problem of residual organic byproducts and polymer matrix damage by undecomposed metal salts.

If an ion beam is directed at a polymer surface, the incident ions displace atoms. Thus, the polymer atoms lose electrons and de-ionise the metal salts, producing the metal (Stepanov *et al.*, 1998). Particle size can be controlled fairly well but implantation is limited to a few micrometres and care must be taken not to form carbon spheres around the metal particles.

## 6.3 Properties of polymer composites reinforced with metal nanoparticles

The properties of metal nanoparticle composites vary widely depending on the nanoparticle–polymer combination and a detailed description is beyond the scope of this chapter. As with most composites, a reinforcement is added to improve some property of the matrix. Most nanocomposites seek to impose some degree of metallic behaviour on the polymer. Thermal, plasmonic, electrical, optical, catalytic and magnetic are all well-studied properties (Gao *et al.*, 2008; Xu *et al.*, 2009).

From a simple rule of mixtures approach, one would expect these properties to increase linearly but there is often a percolation threshold at which, for example, electrical conductivity may jump several orders of magnitude. Luechinger *et al.* found that for a cobalt-nanoparticle-reinforced polyethylene oxide composite this percolation level could be as low as 1–2%. At this level the authors found that conductivity jumped several orders of magnitude, but this depended on the processing technique since it controlled the distribution of the nanoparticles (Luechinger *et al.*, 2008). This result has been found with other conductive filler–metal composites using aluminium nanoparticles (Xu and Wong, 2005).

Another important factor is that, in general, only a relatively small number of nanoparticles needs to be added, which is important from a manufacturing point of view. The melt flow index is not severely affected and as a result standard manufacturing methods can be used for thermoplastics.

Often the addition of nanoparticles can provide improvements in a number of properties. Mbhele *et al.* (2003) investigated the influence of silver nanoparticles on a number of properties for poly(vinyl alcohol) (PVA). Colloidal silver was mixed with an aqueous solution of PVA. Thermal stability was found to improve and the glass transition temperature increased by 20 °C depending on nanoparticle content. Stiffness (modulus of elasticity) and strength were also found to increase; however, stress relaxation measurements showed a decrease in long-term stability of the PVA matrix.

This research showed the importance of determining the other properties that are affected by the addition of the reinforcement. For example, little research has been performed on the effect of metal nanoparticles on the mechanical properties of composites. This is of some concern, as in the haste to develop or control the physical properties of these composites the mechanical properties are often neglected. The end result could be good physical properties but poor mechanical properties, such as strength, flexibility and impact strength.

## 6.4 Alternative applications of metal–polymer nanocomposites

A variety of metal nanoparticles have been incorporated for a wide range of applications, for example, gold (Kishore *et al.*, 2007; Chen *et al.*, 2011), copper (Delgado *et al.*, 2011), silver (Kishore *et al.*, 2007), palladium (Islam *et al.*, 2011) and cobalt (Luechinger *et al.*, 2008). Metal nanoparticles have been added to polymers for uses such as chemical sensors, photocatalysis (O'Mullane *et al.*, 2004), microelectronic devices (Li *et al.*, 1995; Sargent *et al.*, 1999; Croissant *et al.*, 1998; Murphy *et al.*, 1996), biosensors (Tian *et al.*, 2004; Ngece *et al.*, 2011), memory devices (Tseng *et al.*, 2005), photovoltaic cells (Kishore *et al.*, 2007), corrosion protection (Malik *et al.*, 1999) and sensors (Sharma *et al.*, 2002;

Majumdar *et al.*, 2005). In general, nanocomposites provide superior performance, whether detection levels, faster switching or longer life. Some examples follow.

Malik *et al.* (1999) showed that incorporation of platinum nanoparticles into polyaniline (PANI) film improved the longevity of the corrosion protection provided to stainless steel in strong acid solutions compared with the polyaniline film on its own.

Xian *et al.* (2006) is a novel approach for a biosensor in which Au nanoparticles and conducting PANI nanofibres were used to immobilise glucose oxidase (GOx). The design had low detection limits and a rapid response with high stability and excellent reproducibility.

There are a number of optical and photonic applications for metal nanoparticles embedded into suitable polymers: plasmon waveguides (Grady *et al.*, 2004), polarisers (Dirix *et al.*, 1999), optical information processing and communications (Koch *et al.*, 1988; Yu *et al.*, 2004), stable light filters (Carotenuto and Nicolais, 2004), and eye and sensor protection (Shi *et al.*, 1994).

Sophisticated optical detection systems are becoming increasingly popular for a range of applications such as night-vision devices, laser eye protection, etc. Optical limiting devices, whereby the transmittance of light diminishes, have, thus, been of considerable interest. Protection devices need to react or switch back and forth in quick response to an external signal. Most materials become more transparent under strong light but silver and gold have been found to exhibit good optical limitation (Francois *et al.*, 2000). Porel *et al.* showed that silver nanoparticles can be effectively incorporated into transparent polymers for use as non-linear optical limitation (Porel *et al.*, 2007).

In a quite different application, Liu *et al.* (2004) studied the effect of nano-sized Ni, Cu, NiCu, NiB, NiCuB and Al powders on the decomposition of ammonium perchlorate (AP) and ammonium perchlorate/hydroxy terminated polybutadiene (AP/HTPB) composite solid propellants. They compared their results with micron-sized Ni, Cu and Al powders. Particle sizes ranged from 26 to 90 nm. They found that the nanopowders increased the thermal decomposition of AP and AP/HTPB more than micron-sized metallic powders.

The medicinal effects of silver have been known for centuries, dating back to Hippocrates (Grammaticos and Diamantis, 2008) and even further to the Phoenicians. Before the use of antibiotics, colloidal silver (nano-silver particles) was used as a disinfectant and as a germicide (Searle, 1920). In recent years, silver has been incorporated into polymers for its antimicrobial properties. In most cases these composites are thin films or surface layers in which the silver nanoparticles release silver ions when in contact with water. Many refrigerators, clothes dryers, even air conditioners contain this sort of coating these days.

## 6.5 Future trends

Future research will work to exploit the unique properties of these nanoparticles, and will develop new types of metal nanoparticle. The future of metal-nanoparticle-reinforced composites depends on two things: control of the formation of the nanoparticles and improvements in techniques for incorporating them into matrices, i.e. control of the nano-architecture. Of course new metal nanoparticles, perhaps interesting alloys of metals, will also be developed.

Magnetic nanoparticles are a very useful tool, particularly in biomedicine. Ferrous oxide is the most widely used, though cobalt and nickel can be used as well. Drug targeting and techniques such as cell sorting make use of the attraction of magnetic particles to high magnetic flux densities. The heat capacity and magnetic properties of metal nanoparticles means they can be used as contrast agents in magnetic resonance imaging and for heat mediators in cancer treatment. These metal nanopowders can be encapsulated in some interesting biopolymers to extend their use in diagnostics and cancer treatment. Silver and gold nanoparticles, for example, could be treated the same way to make use of their plasmonic response to trigger drug release, to attack tumours or to detect biomarkers.

Control of the morphology of the nano-reinforcement is also a very interesting challenge. It is already possible to grow a range of geometric shapes at the nanoscale. Taking these shapes and dispersing them in a polymer matrix is an interesting area of research, especially with regard to the mechanical properties (Biswas *et al.*, 2006). Indeed, the exploitation of the unique properties of these metal nanoparticles to control their distribution will be another trend.

I also believe that there will need to be significant investigation of the nanotoxicology of these nanoparticles. For example, silver is an antimicrobial agent because it denatures or unfolds cell membrane proteins by cleaving hydrogen bonds. If silver can get inside a cell it can bind to phosphate groups on DNA and to many other intra-cellular chemicals, blocking their function. It is important to remember that the reason we use nanoparticles is because they behave differently from their larger scale counterparts.

## 6.6 Sources of further information and advice

For more detailed information on these composites the reader is encouraged to look at the following articles and books:

- Conducting polymer nanocomposites (Gangopadhyay and De, 2000).
- Metal nanoparticle–polymer composites (Pomogailo and Kestelman, 2005).
- Metal nanoparticles (Perepichka and Rosei, 2007).
- Metal–polymer nanocomposites (Nicolais and Carotenuto, 2005).
- Metal–polymer nanocomposites for functional applications (Faupel *et al.*, 2010).
- Physical properties and applications of polymer nanocomposites (Tjong and Mai, 2010).
- Polymer film composites (Heilmann, 2003).

## 6.7 References


Behnke, K., Strunskus, T., Zaporojtchenko, V. and Faupel, F. 2000. In: *Proc. 3rd Int. Conf. MicroMat 2000*. Verlag ddp goldenbogen.

Biederman, H. 2000. RF sputtering of polymers and its potential application. *Vacuum*, 59, 594–599.

Biederman, H. 2004. *Plasma Polymer Films*, London, Imperial College Press.

Biswas, A., Marton, Z., Kanzow, J. J. K., Zaporojtchenko, V. and Faupel, F. 2003. Controlled generation of Ni nanoparticles in the capping layers of Teflon AF by vapor-phase tandem evaporation. *Nano Letters*, 3, 69–73.

Biswas, A., Faupel, F. and Zaporojtchenko, V. 2006. Self-organization of ultrahigh-density Fe-Ni-Co nanocolumns in Teflon AF. *App. Phys. Lett.*, 88, 123103/1–123103/2.

Bockstaller, M. R., Mickiewicz, R. A. and Thomas, E. L. 2005. Block copolymer nanocomposites: Perspectives for tailored functional materials. *Advanced Materials*, 17, 1331–1349.

Breimer, M., Yevgeny, G., Sy, S. and Sadik, O. 2001. Incorporation of metal nanoparticles in photopolymerized organic conducting polymers: A mechanistic insight. *Nano Lett.*, 1, 305–308.

Carotenuto, G. and Nicolais, L. 2003. Nanocomposites, metal-filled. In: Mark, H. (ed.) *Encyclopedia of Polymer Science and Technology*, New York, Wiley.

Carotenuto, G. and Nicolais, L. 2004. Synthesis and characterization of gold-based nanoscopic additives for polymers. *J. Compos. B-Eng.*, 35, 385–391.

Carotenuto, G., Martorana, B., Perlo, P. and Nicolais, L. 2003. A universal method for the synthesis of metal and metal sulfide clusters embedded in polymer matrices. *J. Mater. Chem.*, 13, 2927–2930.

Chen, X. J., Zhang, J., Ma, D. F., Hui, S. C., Liu, Y. L. *et al.* 2011. Preparation of gold–poly(vinyl pyrrolidone) core–shell nanocomposites and their humidity-sensing properties. *Journal of Applied Polymer Science*, 121, 1685–1690.

Choukourov, A., Solar, P., Polonskyi, O., Hanus, J., Drabik, M. *et al.* 2010. Structured Ti/hydrocarbon plasma polymer nanocomposites produced by magnetron sputtering with glancing angle deposition. *Plasma Processes and Polymers*, 7, 25–32.

Collier, C. P., Vossmeyer, T. and Heath, J. R. 1998. Nanocrystal superlattices *Annu. Rev. Phys. Chem.*, 49, 371–404.

Croissant, M., Napporn, T., Leger, J. and Lamy, C. 1998. *Electrochim. Acta*, 43, 2447–2557.

d'Agostino, R., Favia, P., Oehr, C. and Wertheimer, M. (eds.) 2005. *Plasma Processes and Polymers*, Berlin: Wiley-VCH.

Delgado, K., Quijada, R., Palma, R. and Palza, H. 2011. Polypropylene with embedded copper metal or copper oxide nanoparticles as a novel plastic antimicrobial agent. *Letters in Applied Microbiology*, 53, 50–54.

Dirix, Y., Darribere, C., Heffels, W., Bastiaansen, C., Caseri, W. *et al.* 1999. Optically anisotropic polyethylene gold nanocomposites. *J. Appl. Optics*, 38, 6581–6586.

Faupel, F., Zaporojtchenko, V., Strunskus, T. and Elbahri, M. 2010. Metal-polymer nanocomposites for functional applications. *Advanced Engineering Materials*, 12, 1177–1190.

Francois, L., Mostafavi, M., Belloni, J., Delouis, J. and Feneyrou, P. 2000. Optical limitation induced by gold clusters. 1. Size effect. *J. Phys. Chem. B*, 104, 6133–6137.

Gangopadhyay, R. and De, A. 2000. Conducting polymer nanocomposites: A brief overview. *Chemistry of Materials*, 12, 608–622.

Gangopadhyay, R. and De, A. 2002. An electrochemically synthesized conducting semi-IPN from polypyrrole and poly (vinyl alcohol). *J. Chem. Mater.*, 12, 608–622.

Gao, Y., Chen, C.-A., Gau, H.-M., Bailey, J., Akhadov, E. A. *et al.* 2008. Facile synthesis of polyaniline-supported Pd nanoparticles and their catalytic properties toward selective hydrogenation of alkynes and cinnamaldehyde. *J. Chem. Mater.*, 20, 2839–2844.

Gill, W., Rogojevic, S. and Lu, T. 2003. Vapor deposition of low-k polymeric dielectrics. In: *Advanced Microelectronics*, Springer.

Gradess, R., Abargues, R., Habbou, A., Canet-Ferrer, J., Pedreuza, E. *et al.* 2009. Localised surface plasman resonance sensor based on Ag-PVA nanocomposite thin films. *J. Mater. Chem.*, 51, 9233–9240.

Grady, N., Halas, N. and Nordlander, P. 2004. Influence of dielectric function properties on the optical response of plasmon resonant metallic nanoparticles. *Chemical Physics Letters*, 399, 167–171.

Grammaticos, P. C. and Diamantis, A. 2008. Useful known and unknown views of the father of modern medicine, Hippocrates and his teacher Democritus. *Hellenic Journal of Nuclear Medicine*, 11, 2–4.

Greve, H., Pochstein, C., Takele, H., Zaporojtchenko, V., Gerber, A. *et al.* 2006. Nanostructured magnetic Fe–Ni–Co/Teflon multilayers for high-frequency applications in the gigahertz range. *J. Appl. Phys. Lett.*, 89, 42501–42503.

Haberland, H., Mall, M., Moseler, M., Qiang, Y., Reiners, T. *et al.* 1994. Filling of micron-sized contact holes with copper by energetic cluster impact. *J. Vac. Sci. Technol.* A, 12, 2925–2930.

Hasell, T. 2008. *Synthesis of Metal-Polymer Nanocomposites*. PhD, University of Nottingham.

Heilmann, A. 2003. *Polymer Films with Embedded Metal Nanoparticles*, Berlin, Springer-Verlag.

Hicks, J. F., Zamborini, F. P., Osisek, A. and Murray, R. W. 2001. The dynamics of electron self-exchange between nanoparticles. *J. Am. Chem. Soc.*, 123, 7048–7053.

Hillenkamp, M., Di Domenicantonio, G. and Felix, C. 2006. Monodispersed metal clusters in solid matrices: A new experimental setup. *Rev. Sci. Instr.*, 77, 025104.

Islam, R. U., Witcomb, M. J., Van Der Lingen, E., Scurrell, M. S., Van Otterlo, W. *et al.* 2011. *In situ* synthesis of a palladium-polyaniline hybrid catalyst for a Suzuki coupling reaction. *Journal of Organometallic Chemistry*, 696, 2206–2210.

Kishore, P., Viswanathan, B. and Varadarajan, T. 2007. Synthesis and characterization of metal nanoparticle embedded conducting polymer–polyoxometalate composites. *Nanoscale Research Letters*, 3, 14–20.

Koch, S., Peyghambarian, N. and Gibbs, H. 1988. Band-edge nonlinearities in direct-gap semiconductors and their application to optical bistability and optical computing. *J. Appl. Phys.*, 63, R1–R12.

Koizumi, S., Matsuda, S. and Ando, S. 2002. Synthesis, characterization, and optical properties of uniaxially drawn and gold nanoparticle dispersed fluorinated polyimide films. *J. Photopolymer Sci. and Technol.*, 15, 231–236.

Kubono, A. and Okui, N. 1994. Polymer thin films prepared by vapor deposition. *Prog. Polym. Sci.*, 19, 389–438.

Li, H.-S., Josowicz, M., Baer, D. R., Engelhard, M. H. and Janata, J. 1995. Preparation and characterization of polyaniline-palladium composite films. *J. Electrochem. Soc.*, 142, 798–805.

Liu, L., Li, F., Tan, L., Li, M. and Yang Y. 2004. Effect of metal and composite metal nanopowders on the thermal decomposition of ammonium perchlorate (AP) and the

ammonium perchlorate/hydroxyterminated polybutadiene (AP/HTPB) composite solid propellant. *Chinese J. Chem. Eng.*, 12, 595–598.
Lu, J., Moon, K.-S. and Wong, C. 2006. Development of novel silver nanoparticles/ polymer composites as high K polymer matrix by *in situ* photochemical method. In: *Proc. 2006 Electronic Components Technology Conference*, 1841–1846.
Luechinger, N. A., Booth, N., Heness, G., Bandyopadhyay, S., Grass, R. N. *et al.* 2008. Surfactant-free, melt-processable metal–polymer hybrid materials: Use of graphene as a dispersing agent. *Advanced Materials*, 20, 3044–3049.
Majumdar, G., Goswami, M., Sarma, T., Paul, A. and Chattopadhyay, A. 2005. Au nanoparticles and polyaniline coated resin beads for simultaneous catalytic oxidation of glucose and colorimetric detection of the product. *Langmuir*, 21, 1663–1667.
Malik, M. A., Galkowski, M. T., Bala, H., Grzybowska, B. and Kulesza, P. J. 1999. Evaluation of polyaniline films containing traces of dispersed platinum for protection of stainless steel against corrosion. *Electrochimica Acta*, 44, 2157–2163.
Matsuda, S. and Ando, S. 2005. Generation behaviors of optical anisotropy caused by silver nanoparticles precipitated in uniaxially drawn polyimide films. *Japanese J. Appl. Phys. Part 1– Regular Papers Short Notes and Review Papers*, 44, 187–192.
Matsuda, S., Ando, S. and Sawada, T. 2001. Thin flexible polariser of Ag-nanoparticle-dispersed fluorinated polyimide. *Electronics Letters*, 37, 706–707.
Mayer, A. 1998. Formation of noble metal nanoparticles within a polymeric matrix: nanoparticle features and overall morphologies. *Mater. Sci. and Eng. C: Biomimetic and Supramolecular Systems*, 6, 155–166.
Mbhele, Z. H., Salemane, M. G., Van Sittert, C. G. C. E., Nedeljković, J. M., Djoković, V. *et al.* 2003. Fabrication and characterization of silver-polyvinyl alcohol nanocomposites. *Chemistry of Materials*, 15, 5019–5024.
Murphy, O., Hitchens, G., Hodko, D., Clarke, E., Miller, D. *et al.* 1996. Method of using conductive polymers to manufacture printed circuit boards. US patent application 5545308.
Murray, C., Kagan, C. and Bawendi, M. 2000. Synthesis and characterization of monodisperse nanocrystals and close-packed nanocrystal assemblies. *Annu. Rev. Mater. Sci.*, 30, 545–610.
Ngece, R., West, N., Ndangili, M., Olowu, R., Williams, A. *et al.* 2011. Impedimetric response of a label-free genosensor prepared on a 3-mercaptopropionic acid capped gallium selenide nanocrystal modified gold electrode. *Int. J. Electrochem. Sci.*, 6, 1820–1834.
Nicolais, L. and Carotenuto, G. 2005. *Metal-Polymer Nanocomposites*, Wiley UK.
Oliveira, M., Zanchet, D., Ugarte, D. and Zarbin, A. 2004. Synthesis and characterization of silver nanoparticle/polyaniline nanocomposites. *Prog. Colloid Polym. Sci.*, 128, 126–130.
Oliveira, M. M., Castro, E. G., Canestraro, C. D., Zanchet, D., Ugarte, D. *et al.* 2006. A simple two-phase route to silver nanoparticles/polyaniline structures. *Journal of Physical Chemistry B*, 110, 17063–17069.
O'Mullane, A., Dale, S., Macpherson, J. and Unwin, P. R. 2004. Fabrication and electrocatalytic properties of polyaniline/Pt nanoparticles composites. *Chem. Commun.*, 21, 1606–1607.
Peng, Z., Zhang, J., Sun, X., Yang, J. and Diao, J. 2009. The thermolysis behavior of Ag/ PAMAMs nanocomposites. *J. Coll. Poly. Sci.*, 287, 604–614.
Perepichka, D. F. and Rosei, F. 2007. Metal nanoparticles: from 'artificial atoms' to 'artificial molecules'. *Angew. Chem. Int. Ed.*, 46, 6006–6008.
Pomogailo, A. and Kestelman, V. 2005. *Metallopolymer Nanocomposites*, Berlin, Springer.

Pomogailo, A., Rosenberg, A. and Dzhardimalieva, G. 2003. Self-organized metal-polymer nanocomposites. *Solid State Phenomena*, 94, 313–318 .

Porel, S., Singh, S., Harsha, S., Rao, D. and Radhakrishnan, T. 2005. Nanoparticle-embedded polymer: *In situ* synthesis, free-standing films with highly monodisperse silver nanoparticles and optical limiting. *Chem. Mater. (Communication)*, 17.

Porel, S., Venkatrarn, N., Rao, D. and Radhakrishnan, T. 2007. Optical power limiting in the femtasecond regime by silver nanoparticulate-embedded polymer film. *J. Appl. Phys.*, 102, 33107–33112.

Roy, R., Messier, R. and Krishnaswamy, V. 1983. Preparation and properties of r.f.-sputtered polymer-metal thin films. *Thin Solid Films*, 109, 27–35.

Sargent, A., Loi, T., Gal, S. and Sadik, O. 1999. The electrochemistry of antibody-modified conducting polymer electrodes. *J. Electroanal. Chem.*, 470, 144–156.

Sarma, T. K., Chowdhury, D., Paul, A. and Chattopadhyay, A. 2002. Synthesis of Au nanoparticle-conductive polyaniline composite using $H_2O_2$ as oxidising as well as reducing agent. *Chemical Communications*, 1048–1049.

Schurmann, W., Hartung, A., Takele, H., Zaporojtchenko, V. and Fauple, F. 2005. Controlled syntheses of Ag–polytetrafluoroethylene nanocomposite thin films by co-sputtering from two magnetron sources. *Nanotechnology*, 16, 1078–1082

Searle, A. B. 1920. Chapter VIII: Germicides and Disinfectants. In: *Disease, the Use of Colloids in Health and Disease* (ed.). Gerstein-University of Toronto, Toronto Collection, London Constable & Co.

Selvan, T., Spatz, J. P., Klok, H.-A. and Möller, M. 1998. Gold–polypyrrole core–shell particles in diblock copolymer micelles. *Advanced Materials*, 10, 132–134.

Sharma, S., Nirkhe, C., Pethkar, S. and Athawale, A. 2002. Chloroform vapor sensor based on copper/polyaniline nanocomposites. *Sens. Actuators*, B 85, 131–136.

Shevchenko, E. V., Talapin, D. V., Kotov, N. A., O'Brien, S. and Murray, C. B. 2006. Structural diversity in binary nanoparticle superlattices. *Nature*, 439, 55–59.

Shi, S., Ji, W., Tang, S., Lang, J. and Xin, X. 1994. Synthesis and optical limiting capability of cubane-like mixed metal clusters (n-Bu4N)3[MoAg3BrX3S4] (X = Cl and I). *J. Amer. Chem. Soc.*, 116, 3615–3616.

Stepanov, A., Abdullin, S. and Khaibullin, I. 1998. Optical properties of polymer layers with silver particles. *Journal of Non-Crystalline Solids*, 223, 250–253.

Stoyanova, A., Ivanov, S., Tsakova, V. and Bund, A. 2011. Au nanoparticle–polyaniline nanocomposite layers obtained through layer-by-layer adsorption for the simultaneous determination of dopamine and uric acid. *Electrochimica Acta*, 56, 3693–3699.

Sun, S. H. 2006. Recent advances in chemical synthesis, self-assembly, and applications of FePt nanoparticles. *Adv. Mater.*, 18, 393–403.

Takele, H., Greve, H., Pochstein, C., Zaporojtchenko, V. and Faupel, F. 2006. Plasmonic properties of Ag nanoclusters in various polymer matrices. *Nanotechnology*, 17, 3499–3505.

Tian, S., Liu, J., Zhu, T. and Knoll, W. 2004. Polyaniline/gold nanoparticle multilayer films: assembly, properties, and biological applications. *Chem. Mater.*, 16, 4103–4108.

Tjong, S. and Mai, Y.-W. (eds.) 2010. *Physical Properties and Applications of Polymer Nanocomposites*, Cambridge, UK, Woodhead Publishing Ltd.

Tröger, L., Hünnefeld, H., Nunes, S., Oehring, M. and Fritsch, D. 1997. Structural characterization of catalytically active metal nanoclusters in poly(amide imide) films with high metal loading. *Journal of Physical Chemistry B*, 101, 1279–1291.

Tseng, R., Huang, J., Ouyang, J., Kaner, R. and Yang, Y. 2005. Polyaniline nanofiber/gold nanoparticle nonvolatile memory. *Nano Lett.*, 5, 1077–1080.

Xian, Y. Z., Hu, Y., Liu, F., Xian, Y., Wang, H. T. and Jin, L. T. 2006. Glucose biosensor based on Au nanoparticles-conductive polyaniline nanocomposite. *Biosensors and Bioelectronics*, 21, 1996–2000.

Xu, J. and Wong, C. 2005. Low-loss percolative dielectric composite. *App. Phys. Lett.*, 87, 82907–82909.

Xu, P., Han, X., Zhang, B., Mack, N., Jeon, S.-H. *et al.* 2009. Synthesis and characterization of nanostructured polypyrroles: Morphology-dependent electrochemical responses and chemical deposition of Au nanoparticles. *Polymer*, 50, 2624–2629.

You, C. C., De, M., Han, G. and Rotello, V. M. 2005. Tunable inhibition and denaturation of a chymotrypsin with amino acid-functionalized gold nanoparticles. *J. Am. Chem. Soc.*, 127, 12873–12881.

Yu, D., Sun, X., Bian, J., Tong, Z. and Qian, Y. 2004. Gamma-radiation synthesis, characterization and nonlinear optical properties of highly stable colloidal silver nanoparticles in suspensions. *Physica E: Low-Dimensional Systems and Nanostructures*, 23, 50–55.

Zhou, H. H., Ning, X. H., Li, S. L., Chen, J. H. and Kuang, Y. F. 2006. Synthesis of polyaniline-silver nanocomposite film by unsymmetrical square wave current method. *Thin Solid Films*, 510, 164–168.

# Part II

## Types of polymer nanocomposites according to base

# 7
# Polyolefin-based polymer nanocomposites

T. KUILA, Chonbuk National University, Republic of Korea,
T. TRIPATHY, Midnapore College, West Bengal, India and
J. HEE LEE, Chonbuk National University, Republic of Korea

**Abstract:** This chapter describes the preparation, characterization and properties of polyolefin-based nanocomposites, including ethylene vinyl acetate, polyethylene, polypropylene, and ethylene propylene diene monomer, which are widely used. To prepare nanocomposites, layered clay minerals (sodium montmorillonite, bentonite, Cloisite, Laponite, layered double hydroxide) and carbon-based nanomaterials (graphite, carbon nanofibers, exfoliated graphite, carbon nanotubes, graphene) are used as fillers. The good mechanical and thermal properties of nanocomposites are due to interfacial interactions and homogeneous dispersion of the nanofiller in the polymer matrix. Carbon-based nanofillers have excellent electrical conductivity and electromagnetic interference shielding compared with the pure polymer. Conductive composites are used to fabricate memory devices, shielding materials, nanoelectronics, etc.

**Key words:** polyolefin, nanocomposite, mechanical properties, thermal properties, electrical conductivity.

## 7.1 Introduction

Polymer nanocomposites are a multiphase solid material in which one of the phases has dispersed in the other phase at a nanometer level. Work on these materials was initiated by Toyota in the late 1980s, when they manufactured a nylon/clay nanocomposite for the timing belt cover for the Toyota Camry automobile.[1,2] The nanocomposite showed significantly improved physicochemical properties normally absent in pure polymers, or traditional microcomposites and macrocomposites based on conventional fillers like talc, glass fiber, calcium carbonate, aluminum (and magnesium) hydroxide, etc.[3–6] Ongoing research shows that nanocomposites exhibit improved mechanical, thermal, gas-barrier, flame retardant, and swelling properties.[3–7] Based on these improved physicochemical properties they have been widely used in the automobile industry, aerospace and construction engineering, electronics, biomedical applications, etc.[3–13] The properties of polymer nanocomposites depend on the local chemistry, degree of curing, polymer chain mobility, and degree of crystallinity.[14] Till now, extensive research has been carried out with different organic and inorganic polymers to explore these materials.[15–22] Sodium montmorillonite (Na-MMT), sodium bentonite, hectorite, Laponite, etc. have been widely used as fillers in a polymer matrix.[3,4,14] In most cases, organo-modified clay (OMC) has been used as a nanofiller to give homogeneous dispersion and a better filler/polymer interaction. With the progress in nanoscience and technology, different types of nanofillers, such as layered double hydroxide (LDH), exfoliated graphite (EG), exfoliated graphite nanoplatelets (xGnPs), metal nanoparticles, carbon black (CB), carbon nanotubes (CNTs), carbon nanofibers (CNFs), and graphene, have been used in the preparation of polymer nanocomposites.[6,23–30] The successful application of clay minerals and LDH as a nanofiller in the preparation of polymer nanocomposites is well established. In contrast, EG, xGnP, CNT, and CNF as nanofillers have been studied less often. However, the high production cost of CNTs limits their large-scale application as nanofillers.[31] The recent development of graphene has added a new dimension to the field of multifunctional polymer nanocomposites.

This chapter reviews the preparation, characterization, and properties of polyolefins and polyolefin/inorganic nanocomposites. The polyolefins considered are ethylene vinyl acetate copolymer (EVA), ethylene propylene diene rubber (EPDM), polyethylene (PE), and polypropylene (PP). EVA is available in different grades, such as rubber, thermoplastic elastomer, and plastic. This copolymer has countless uses in different fields, such as for electrical cable insulation, cable jacketing and repair, waterproofing, packaging, etc.[32] EPDM, a terpolymer of ethylene, propylene, and an unconjugated diene, is one of the most versatile elastomers in use today and it is an important material used in a wide variety of engineering applications.[33,34] These polymers derive their usefulness and versatility from their inherent toughness, chemical resistance, and good mechanical and electrical properties. PE and PP have the ability to elongate under stress, allowing them to maintain their integrity under differential settlement conditions without puncturing, tearing, or cracking.[34,35] Both of these polymers have a wide variety of applications including packaging, textiles (e.g., ropes, thermal underwear, and carpets), stationery, plastic parts, reusable containers of various types, laboratory equipment, loudspeakers, automotive components, and polymer banknotes.[34] In order to extend these applications, the development of mechanically durable and thermally stable nanocomposite materials is required.

## 7.2 Preparation of polymer nanocomposites

Polymer/inorganic nanocomposites can be prepared by *in situ* polymerization and by blending processes, of which there are basically two types: solution blending and melt blending.[3,5,6] With *in situ* polymerization, the nanofillers are dispersed in the monomers and polymerization occurs in the presence of initiators. In solution blending, the nanofillers are first dispersed in the solvent and then the polymer is dissolved in another batch of the same solvent. Finally, the dispersed nanofiller solution is added to the polymer solution, stirred for the desired time and the solvent is removed at reduced pressure to produce the polymer composites.

In contrast, with melt blending the nanofillers are added to a molten polymer. The distribution of the nanofiller is better with solution blending and *in situ* polymerization compared with melt blending. However, melt blending is regarded as a cost-effective, environmentally friendly, and industrially accepted method in comparison with the others.

## 7.3 Different kinds of nanofillers used in the preparation of polymer nanocomposites

Different types of inorganic fillers, e.g. clay minerals, LDH, CNT, EG, graphene, etc., have often been used in the preparation of polyolefin nanocomposites. Of these, clay and LDH are hydrophilic in nature while CNT, EG, and graphene are purely carbonaceous.[15–30] In contrast, the polymeric materials are hydrophobic in nature. Therefore, all of these nanofillers will form phase-separated composites with a polymer. So, surface modification is required.

### 7.3.1 Surface modification of montmorillonite

The surface modification of clay minerals is a well-known method and has been used frequently to obtain organically modified clay (OMC). MMT is made organophilic by exchanging the inorganic cations (usually $Na^+$ or $Ca^{2+}$) in the interlayer space with organic cations such as alkyl or aryl ammonium (or phosphonium) ions.[36] In a typical process, inorganic clay minerals are dispersed in water by stirring at 80 °C. The aliphatic amine/ammonium or phosphonium ions are dissolved in water. Then the organic amine/ammonium or phosphonium solution is added to the clay dispersion and stirred for about 2–3 hours. The incorporation of ammonium ions not only changes the hydrophilic nature of the clay but also expands the interlayer spacing, which, in turn, depends on the number of carbon atoms in the organic ammonium ions.[37,38]

### 7.3.2 Surface modification of layered double hydroxide (LDH)

Pure LDH materials are unsuitable for intercalation by large species like polymers due to the small interlayer gallery spacing of pure LDH (approximately 0.76 nm). The small gallery spacing will not allow polymer chains to penetrate and the charged metal-hydroxide layer makes them incompatible with the non-polar species. In order to overcome these issues, different anionic surfactants like sodium dodecylbenzene sulfonate (SDBS), sodium dodecyl sulfate (SDS), aromatic carboxylate anions, aliphatic carboxylate anions, phosphonates, and polymeric anions, etc., have been used for organo modification.[39,40]

### 7.3.3 Surface modification of carbon nanotubes (CNTs)

The surface modification of CNTs can be performed by covalent or non-covalent techniques. In non-covalent modification, small molecular weight polymers or aromatic molecules are attached to the surface of a CNT by the $\pi/\pi$ interaction. The advantage of non-covalent attachment is that the perfect structure of the nanotube is not altered. Therefore, its physicochemical properties are not expected to change significantly. However, the disadvantage of non-covalent attachment is the weak force of attraction between the wrapping molecule and the nanotube. This may lead to poor load transfer when CNTs are used for composite applications. In contrast, a strong chemical bond is formed between the carboxyl functionality of CNT and an organic modifier. In both cases, the functionalized graphene disperses well in different organic solvents, facilitating its homogeneous dispersion in a polymer matrix.[41–44]

### 7.3.4 Surface modification of graphene

Graphene can be made water dispersible or organic-solvent dispersible by surface modification with graphene oxide. The reduction of functionalized graphene produces organo-modified graphene. Different kinds of surface-modifying agents such as organic amine, amino acids, amine-terminated ionic liquids, biomolecules, low molecular weight polymers, etc., have been used for the surface modification of graphene.[6,45–48] Graphene can be prepared, with an organophilic surface, from graphite using electrochemical methods in the presence of ionic liquids.[49] Very recently, organo-modified graphene has been synthesized by the sonication of flake graphite in the presence of 9-anthracene carboxylic acid.[50]

## 7.4 Characterization of polymer nanocomposites

The properties of polymer nanocomposites are highly dependent on the nature of the polymer (polar or non-polar), the nanofiller, the types of interaction between the filler and polymer, the nature of surface modification, and, certainly, on the distribution of the nanofiller in the polymer matrix. Depending on the distribution of the nanofiller, polymer inorganic nanocomposites are classified into three categories: intercalated nanocomposites, exfoliated nanocomposites, and immiscible nanocomposites. It has been reported that the improvement in physicochemical properties of exfoliated nanocomposites is much better in comparison to intercalated nanocomposites or conventional composites. The characterization and determination of the nanostructure of polymer/inorganic composites can be achieved using X-ray diffraction (XRD), transmission electron microscopy (TEM), and Fourier transform infrared spectroscopy (FT-IR).

### 7.4.1 X-ray diffraction (XRD) and transmission electron microscopy (TEM) analysis

The structure of a nanocomposite can be established by XRD. Due to the ease of use and its availability, XRD has been used widely to probe the structure of nanocomposites. When the layers are exfoliated and far apart, XRD will not show a peak in the overall diffraction diagram. In contrast, a small and broad peak is obtained at a lower angle compared with the pure nanofiller for intercalated or partially exfoliated nanocomposites. For immiscible composites, the original peaks of the clay layers will be retained completely in the composite. Figure 7.1(a) shows the XRD pattern of intercalated, exfoliated, and partially exfoliated nanocomposites.[3] However, XRD results cannot always be taken as absolute evidence for nanolayer exfoliation, e.g. if clay/LDH layers are partially exfoliated in the polymer matrix and do not have sufficiently well-defined layered structures for XRD, then the characteristic diffractions for the basal spacing cannot be detected by bulk XRD apparatus. Moreover, if the basal spacing between layers is enlarged due to the intercalation of polymer molecules, the characteristic peak will shift to a lower angle, beyond the XRD scan range ($<2°$). Therefore, TEM studies have to be made before drawing any definite conclusion. TEM provides a qualitative understanding of the internal structure of a nanocomposite material. In the case of intercalated nanocomposites, the clay layers will remain parallel to each other. In contrast, for exfoliated nanocomposites, the clay layers are homogeneously and randomly dispersed in the matrix polymer. The morphology of different kinds of nanocomposite is shown in Fig. 7.1(b).[3]

### 7.4.2 Fourier transform infra red spectroscopy (FT-IR) analysis

The presence of a nanofiller in a polymer matrix and their interaction can be evidenced by FT-IR data analysis. Acharya *et al.* showed that hexadecylamine-modified clay (16Me-MMT) can be successfully doped onto an EPDM matrix. FT-IR spectra of $Na^+$-MMT, 16Me-MMT, pure EPDM and its hybrid with 4 wt.% 16Me-MMT are shown in Fig. 7.2.[51] Kuila *et al.* showed that dodecyl-sulfate-modified LDH (DS-LDH) can be successfully grafted onto the surface of an EVA-28 matrix.[52,53] The appearance of new peaks for the EVA-28/DS-LDH nanocomposites at about 3500 $cm^{-1}$ and lattice vibration bands in the region 500–800 $cm^{-1}$, which are absent for EVA-28, indicated that DS-LDH was grafted onto the EVA surfaces.

## 7.5 Properties of polymer nanocomposites

The important aspects of nanocomposite materials are their improved mechanical, thermal, gas-barrier, swelling, and flame-retardant properties. These improved properties make them the best type of composite and they find application in

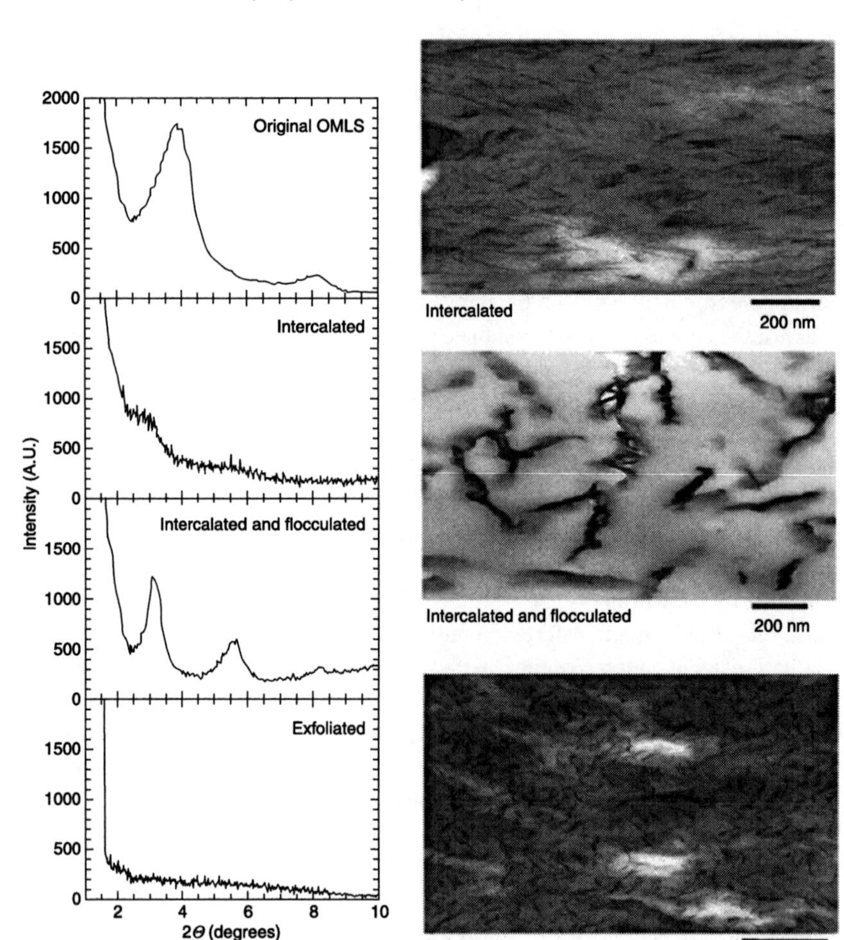


*7.1* (a) XRD patterns and (b) TEM images of three different types of nanocomposite. Reproduced from Ray and Okamoto[3] by permission of Elsevier Science Ltd, UK. OMLS: Organomodified layered silicate.

many different technological fields. However, performance during use is a key feature of any composite material and determines the real fate of the products, especially for outdoor applications. Therefore, in this section, we will discuss the significant properties of polyolefin-based nanocomposites.

### 7.5.1 Mechanical properties

Polymeric materials play an important role in the research and development of modern science and technology. For specific applications, such as in medical

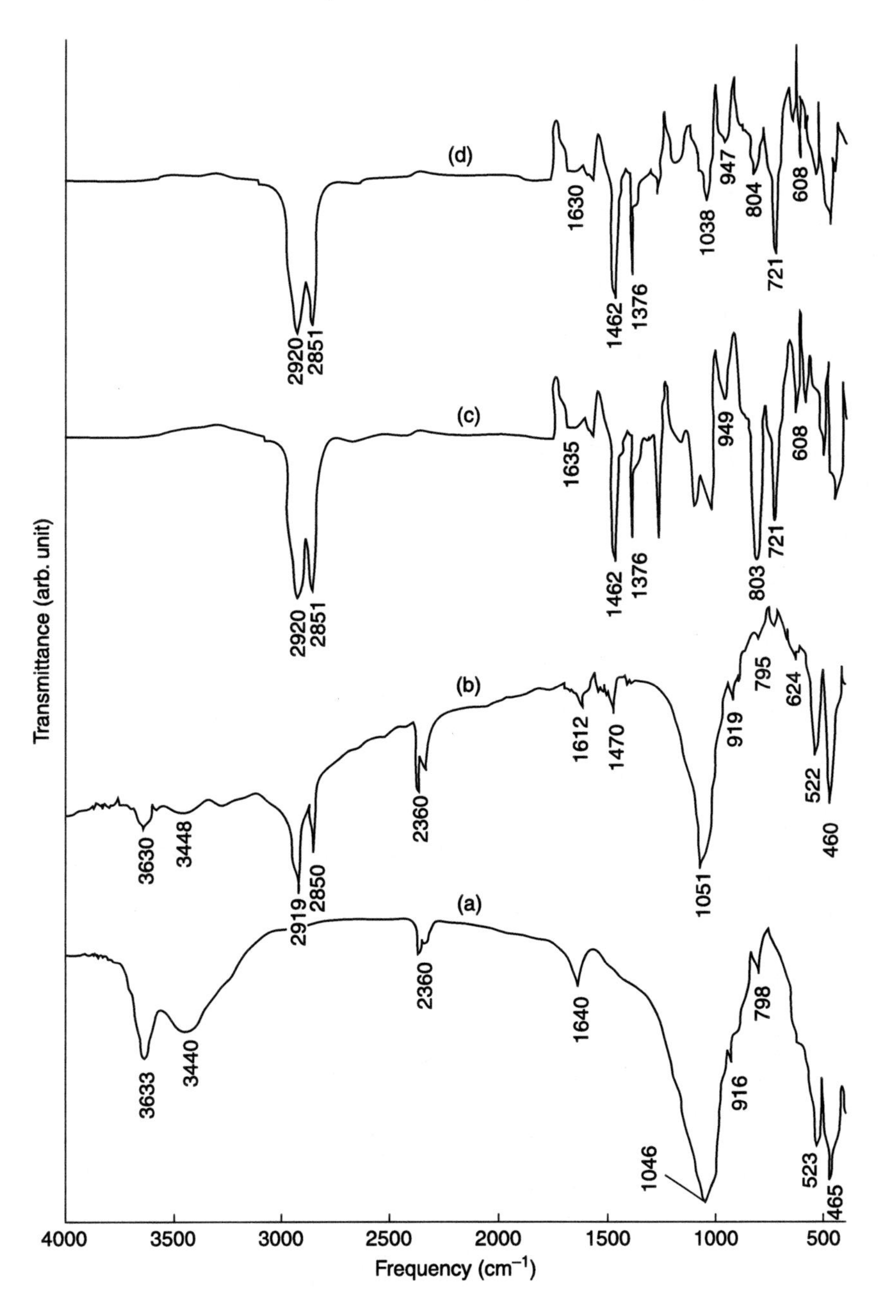


7.2 FT-IR spectra of (a) $Na^+$-MMT, (b) 16Me-MMT, (c) EPDM, and (d) EPDM 4 wt.% 16Me-MMT. Reproduced from Acharya *et al.*[51] by permission of Wiley-VCH, Germany.

science, industrial conveyor belts, hosepipes, car bodies, component packaging, etc., special types of polymer with specific mechanical properties are required. It is necessary to know mechanical properties such as tensile strength (TS), elongation at break (EB), and tensile modulus before their application in these specific areas. Dynamic mechanical analysis (DMA) is used to study the mechanical properties of materials as they deform under periodic forces. DMA measures changes in the elastic modulus ($E'$), loss modulus ($E''$), as well as tan $\delta(E''/E')$, with temperature. A sinusoidal mode of deformation is applied to a sample, as the temperature is scanned from well below the glass transition temperature ($T_g$) to the point when the sample becomes too soft to test. DMA provides the first- and second-order transitions ($T_m$ and $T_g$, respectively), the degree of phase separation, crystallinity, cross-linking of the polymer, and mechanical properties such as the glassy state and rubbery plateau moduli.

*Tensile properties*

Pramanik *et al.* showed that TS increases when increasing the matrix polarity. Figure 7.3 shows that the TS of EVA/dodecyl-amine-modified montmorillonite (12Me-MMT) nanocomposites are higher in comparison to neat EVA-45.[19,54] But the EB of the nanocomposites decreases compared with pure EVA. However, for EVA-28/12Me-MMT nanocomposites, both TS and EB increase with up to 8 wt.% of filler loadings. According to Cui *et al.*, the stress at any given strain level increases as the content of OMC increases.[55] As the vinyl acetate (VA) content in EVA increases, crystallinity decreases, causing the modulus to decrease and the EB to increase. Cardenas *et al.* used bentonite and aluminum trihydrate (ATH) as a filler and studied the effect of filler loading on mechanical properties.[56] It was noted that the EB values are higher for uncoated and fatty-acid-coated ATH fillers, while TS did not change for all the nanocomposites. In contrast, a silane-coated ATH-filler-based nanocomposite showed an important loss of deformability. This is attributed to lower compatibility between the silane modifier and the other components of the nanocomposite. Zhang *et al.* proved that Mg-Al LDH is an efficient filler, which can increase the EB of EVA-28 nanocomposites. But the TS of nanocomposites decreases with the addition of Mg-Al LDH.[57] Kuila *et al.* observed that the TS and EB of EVA/organo-modified LDH (OLDH) nanocomposites increases when increasing the polarity of the EVA matrix. At lower loading (1, 3, 5 wt.%), both the TS and EB are higher than for pure EVA. However, at higher filler loading the mechanical properties were found to deteriorate due to the aggregation of OLDH in the EVA.[58,59] According to George and Bhowmick, the TS and tensile modulus at 100% elongation are always higher for EVA (EVA-40, 50, 60, and 70)/multi-walled carbon nanotube (MWCNT) (4 wt.%) nanocomposites. But the EB for all the nanocomposites was less than for pure EVA. George and Bhowmick also studied the effect of EG and CNF on the mechanical properties of EVA nanocomposites.[60] They found that the reinforcing

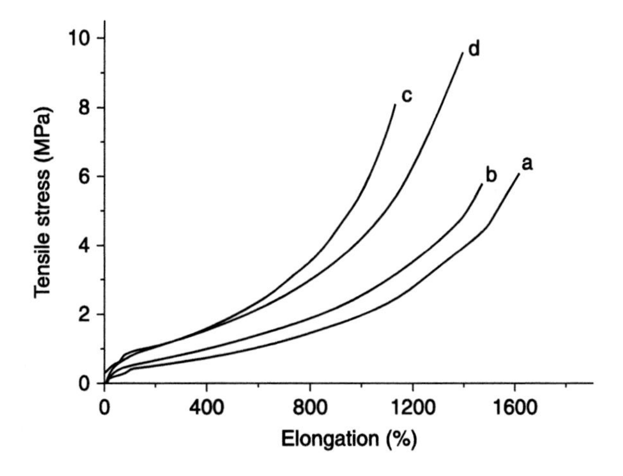


*7.3* Tensile stress versus elongation curves of pure EVA (a), 12Me-MMT (2 wt.%) and EVA (b), 12Me-MMT (4 wt.%) and EVA (c), and 12Me-MMT (6 wt.%) and EVA (d). Reproduced from Pramanik *et al.*[19] by permission of Wiley-VCH, Germany.

efficiency of EG and CNF is much higher than with MWCNTs. This may be attributed to the relatively poor dispersion of MWCNTs compared with the other two fillers. It was also noted that the increment is more prominent in the partially amorphous to fully amorphous transition region, which occurs beyond 50% VA content.

Acharya *et al.* found that the TS and EB of EPDM/clay nanocomposites increases when increasing the 16Me-MMT content up to 4 wt.%. Further addition of OMC does not influence either the TS or the EB significantly.[51] Acharya *et al.* also studied the effect of DS-LDH on the mechanical properties of EPDM/DS-LDH nanocomposites.[18] The TS and EB increase gradually when increasing DS-LDH content in the nanocomposites. Ismail *et al.* observed that the TS of EPDM/halloysite nanotube (HNT) composites increases gradually on increasing the HNT content.[61] The TS and EB of maleic-anhydride-grafted EPDM (MAH-g-EPDM)/EPDM/HNT nanocomposites was much higher than that of EPDM/HNT nanocomposites.[62] Kang *et al.* showed the effect of CNTs and OMC on the mechanical properties of EPDM nanocomposites. The TS increased gradually up to 50 wt.% of CNT loading in the nanocomposites. In contrast, the TS started to decrease with only 20 wt.% of CNT and OMC (mixed) loading.[63]

The addition of OMC can efficiently improve TS, flexural strength, and flexural modulus of PE nanocomposites.[64] However, the degree of improvement is highly dependent on the nature of the organic modifier. The reinforcing efficiency of short alkyl chain modifiers is much better compared with long chain modifiers. This may be due to the plasticization effect of the long chain modifier. Organic-copolymer-modified clay is not regarded as a suitable filler because of its mechanical properties relevant for reinforcement.[65,66] A deterioration of TS and EB was noted after increasing the content of copolymer-modified clay in PE nanocomposites. However, the moduli of the nanocomposites increased significantly with OMC loading. Similar observations were made when pristine LDH was used as a filler in the preparation of PE/LDH nanocomposites.[67] The addition of OLDH was also unable to enhance the TS and EB of linear low-density polyethylene (LLDPE)/OLDH nanocomposites. However, Young's modulus of the nanocomposites increased significantly on increasing the concentration of OLDH.[68] Kanagaraj *et al.* studied the effect of CNT loading on the mechanical behavior of high-density polyethylene (HDPE)/CNT composites, as shown in Table 7.1.[69] Note that CNT can act as a reinforcing filler to improve the mechanical properties of HDPE/CNT composites. The effects of exfoliated graphite on the mechanical properties of LLDPE and HDPE have been studied extensively.[26,70–72] For HDPE, the presence of EG slashes the ductility of the HDPE, which showed necking without breaking in a tensile test. Kim *et al.* observed that the addition of xGnPs can efficiently increase the TS and Young's modulus of LLDPE/xGnP nanocomposites, but with a deterioration of EB.[72] Figure 7.4 shows stress–strain plots of LLDPE/xGnP nanocomposites with different loadings of xGnP. Kim *et al.* also studied the effect of paraffin-coated

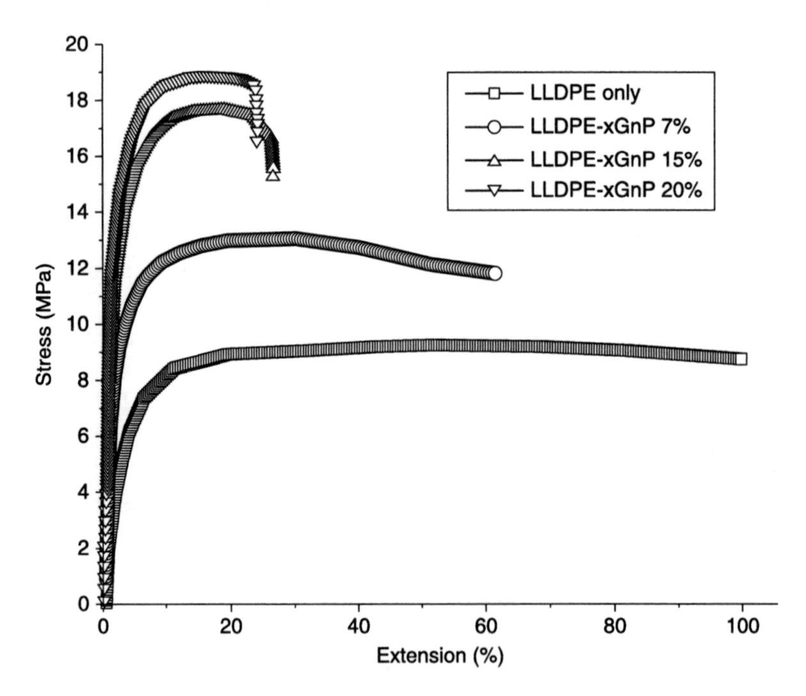


*7.4* Stress–strain curves of LLDPE/xGnP nanocomposites produced by solution mixing against xGnP loading. Reproduced from Kim *et al.*[71] by permission of Wiley-VCH, Germany.

*Table 7.1* Mechanical properties of HDPE/CNT nanocomposites

| Volume fraction CNT concentration (%) | Young's modulus (GPa) | % Increment of $E$ | Ultimate stress (MPa) | Strain ($\varepsilon$) at fracture (%) | % increment of $\varepsilon$ | Toughness (J) | % Increment of toughness |
|---|---|---|---|---|---|---|---|
| 0.00 | 1.095 | 0.00 | 105.80 | 863.4 | 0.00 | 634.53 | 0.00 |
| 0.11 | 1.169 | 6.74 | 105.51 | 948.5 | 9.86 | 743.35 | 17.15 |
| 0.22 | 1.228 | 12.13 | 106.67 | 978.5 | 13.34 | 756.24 | 19.18 |
| 0.33 | 1.287 | 17.56 | 106.38 | 1020.4 | 18.19 | 776.27 | 22.34 |
| 0.44 | 1.338 | 22.23 | 109.86 | 1069 | 23.82 | 842.47 | 32.77 |

Source: Reproduced from Kanagaraj *et al.*[69] by permission of Elsevier Science Ltd, UK.

xGnP on the mechanical behavior of LLDPE/paraffin-coated xGnP nanocomposites.[26] It was found that both the TS and Young's modulus of the LLDPE/paraffin-coated xGnP nanocomposites increased gradually with up to 30 wt.% of filler loading. The mechanical behavior of the nanocomposites was correlated in terms of tensile fracture of the tested specimens. Figure 7.5 shows scanning electron microscopy (SEM) images of LLDPE/paraffin-coated xGnP nanocomposites for different filler loadings. Graphene has also been suggested as a reinforcing filler to improve the mechanical properties of PE-based nanocomposites. Kuila *et al.* showed that dodecyl amine functionalized graphene (DA-G) can enhance the TS and Young's modulus of LLDPE/DA-G nanocomposites with up to 5 wt.% of DA-G content.[35] Wang *et al.* found that reinforcement by graphene resulted in an increase of up to 27.0 and 92.8% in the TS and Young's modulus of the nanocomposites, respectively.[73] Young's modulus of thermally reduced graphene oxide/LLDPE composites was also higher in comparison to neat LLDPE.

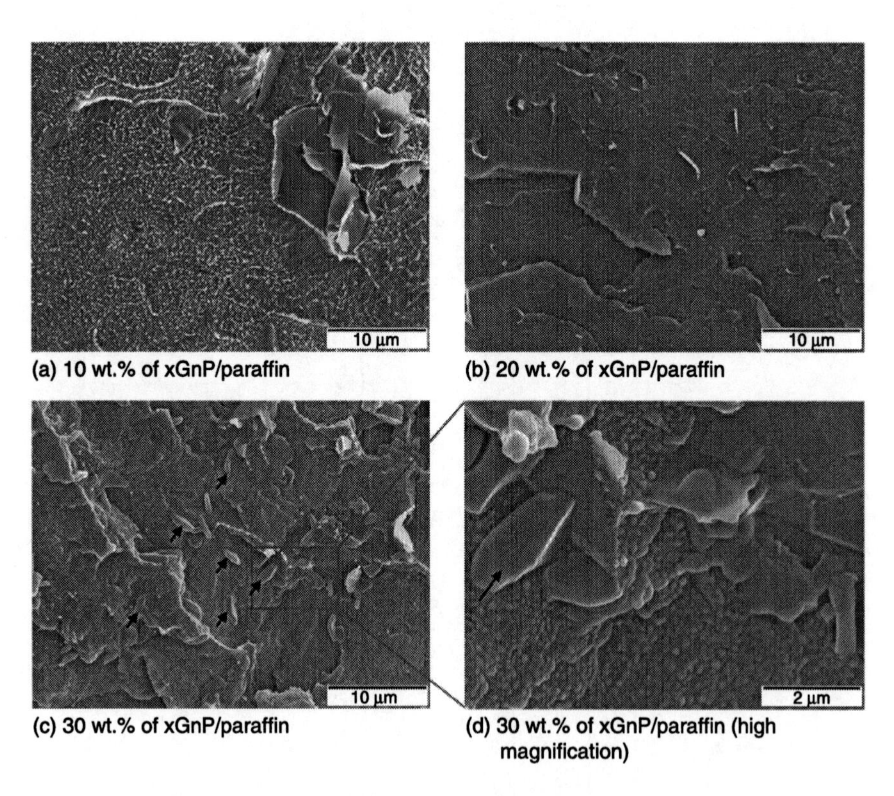


*7.5* SEM images of the fracture surface of LLDPE/paraffin/xGnP nanocomposites with (a) 10 wt.%, (b) 20 wt.%, and (c, d) 30 wt.% of xGnP (5 wt.%)/paraffin loading. Reproduced from Kim and Drzal[26] by permission of Wiley-VCH, Germany.

Cui *et al.* and Kim *et al.* noticed that the TS of PP/OMC and maleic-anhydride-grafted polypropylene (PP-g-MA)/OMC nanocomposites decreased compared with pure PP.[74,75] However, in both cases Young's modulus of the nanocomposites was significantly higher than that of the pure PP. Sharma and Nayak found that the TS of PE/clay composites was higher than that of pure PP. At higher loading (7 wt.%) of nanoclay, the TS of the nanocomposites started to decrease but remained higher than that of the pure PP.[76] Kotsilkova *et al.* showed that the TS of PP/CNT nanocomposites reached a maximum with only 0.5 wt.% of CNT. Beyond this concentration, the TS started to decrease and fell to the level of pure PP.[77] In contrast, the EB of the nanocomposites decreased gradually with the addition of CNTs.

*Dynamic mechanical analysis*

Pramanik *et al.* reported DMA results for EVA-45/12Me-MMT nanocomposites.[19] They showed that, with the addition of clay, the storage modulus ($E'$) of the nanocomposites increased but the glass transition temperature ($T_g$) decreased. The clay restricted the chain mobility of the polymer matrix, which resulted in the enhancement of $E'$ of the nanocomposites. The plasticization effect of modified clay is mainly responsible for the decrease in $T_g$. It was also noted that the height of the tan $\delta$ peak ($\beta$ relaxation) for the nanocomposites decreased compared with neat EVA-45. This was attributed to the restricted mobility of the polymer chains between the clay layers. Zhang *et al.* measured the dynamic mechanical properties of different EVA and EVA/OMC nanocomposites.[78] They showed that $E'$ for EVA-28/SIOM nanocomposites was higher but for EVA-28/unmodified clay composites it was lower than that of pure EVA-28. (SIOM is $CH_3(CH_2)_{17}N^+(CH_3)_2Br^-$.) $E'$ for EVA-28/SIOM (5 wt.%) is 1381 MPa, which is approximately 50% higher than that of pure EVA-28. This can be interpreted as due to the interaction of the clay layers with the EVA-28 chains, strengthening $E'$ for EVA-28. However, $T_g$ for all the nanocomposites and pure EVA did not shift to a higher temperature. $E'$ for the EVA-80/SIOM (5 wt.%) was 3053 MPa at −50 °C, which is about 62% higher than that of pure EVA-80. Note that the degree of enhancement in $E'$ and $T_g$ for EVA-80/SIOM nanocomposites is a little more than that of EVA-28/SIOM. This was attributed to the better dispersion of the filler in the highly polar EVA-80 matrix.[78]

The storage modulus ($E'$) of LDPE/Cloisite nanocomposites is much higher than that of pure LDPE.[79] Exfoliated graphite is regarded as a very effective filler for the reinforcement of the dynamic mechanical properties of HDPE/EG nanocomposites. Almost a twofold increase in $E'$ was noted with 10 wt.% of EG. At 5 wt.% loading, the increase in $E'$ was estimated to be 50%.[70] Kim *et al.* studied the effect of xGnP on the dynamic mechanical behavior of LLDPE nanocomposites.[72] Figure 7.6 shows that, with the addition of only 7 wt.% of xGnP, $E'$ for the composite was 2.5 times higher than the control LLDPE.[72]

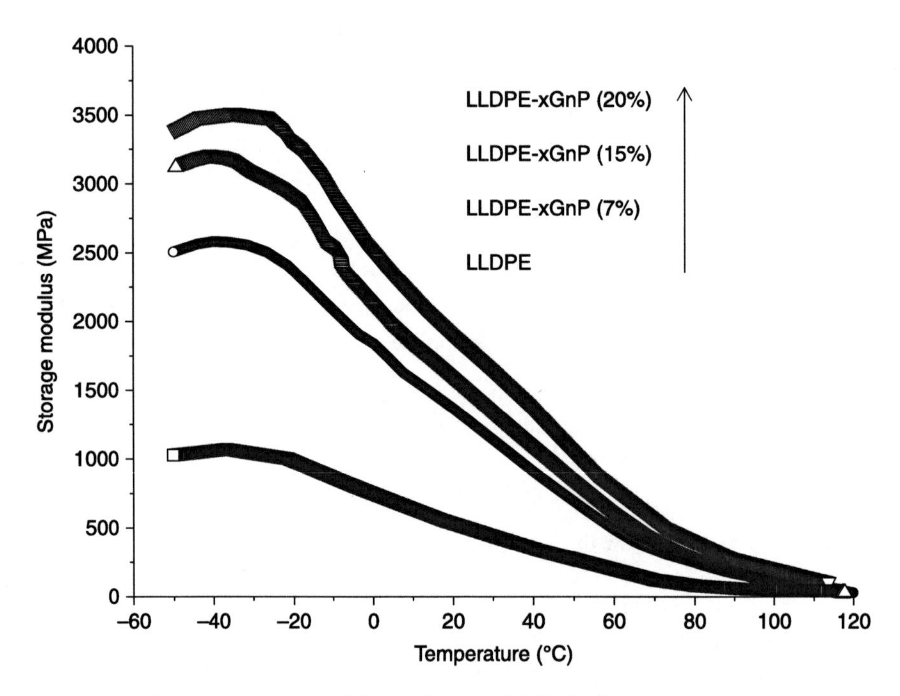


*7.6* The storage modulus ($E''$) of LLDPE/xGnP nanocomposites prepared by solution mixing and injection molded against xGnP loading. Reproduced from Kim *et al.*[72] by permission of Wiley-VCH, Germany.

$E'$ for neat PP at 20 °C is 1450 MPa. $E'$ for specimens infused with 3 and 5 wt.% was 1750 and 1500 MPa, respectively.[80] $E'$ for the nanocomposites gradually increased with the addition of OMC into PP-g-MA. $E'$ for nanocomposites containing 1 wt.% of MWCNT was similar to that of the pure matrix PP. However, an improvement of about 30% of the $E'$ in the glassy region was observed for 3 wt.% MWCNT composites compared with pure PP.[77]

## 7.5.2 Thermal properties

Thermal analysis is a well-established technique for obtaining qualitative and quantitative information about the effects of various heat treatments on materials. The heating is performed under strictly controlled conditions and can reveal changes in the structure and other important properties of the materials under investigation, thereby determining the practical importance of the material. Polymer nanocomposites constitute an important class of material due to their unusual improved properties, which are normally unavailable in the pure forms or in conventional composites. It is therefore necessary to evaluate their thermal stability, the effect of heat and any phase changes. In this regard, thermogravimetric analysis (TGA) and differential scanning calorimetry (DSC) have proved to be very useful.

*Thermogravimetric analysis*

Zhang *et al.* studied the effect of different types of OMC on the thermal degradation of EVA-28 in a nitrogen and air atmosphere.[78] It was found that the thermal degradation of EVA-28/SIOM, EVA-28/DIOM, and EVA-28/TRIOM (in $N_2$) nanocomposites is much faster than pure EVA-28. (DIOM is $(CH_3(CH_2)_{17})_2N^+(CH_3)_2Br^-$ and TRIOM is $(CH_3(CH_2)_{17})_3N^+(CH_3)_2Br^-$.) Above 460 °C, the nanocomposites have a higher thermal decomposition temperature. In contrast, the EVA-28/Na-MMT microcomposite has better thermal stability than pure EVA-28. The thermal degradation of all the nanocomposites was obviously different when in nitrogen. At 30% weight loss, the thermal degradation temperature of EVA-28/SIOM nanocomposites is about 60 °C higher compared with pure EVA-28. TGA thermograms of EVA and EVA/rectorite nanocomposites under nitrogen and air were produced by Fang and co-workers.[81] The mass loss of EVA nanocomposites at the beginning of combustion was higher than that of pure EVA in air, which indicates that the deacylation reaction of EVA nanocomposites was accelerated. This is probably due to catalysis by strongly acidic sites of the layered silicates deriving from thermal decomposition of the organic alkylammonium cations. The maximum weight loss temperature of the nanocomposites with 7.5 wt.% of rectorite was 468 °C, whereas for pure EVA it was 429 °C. This increase in thermal stability is mainly due to the strong barrier effect of layered silicates in the EVA matrix.

Acharya *et al.* studied the effect of dodecyl-sulfate-modified MgAl LDH on the thermal stability of EPDM nanocomposites. It was found that the thermal stability of the EPDM nanocomposites containing 3 wt.% of LDH was about 40 °C higher than pure EPDM (at 10% weight loss). The enhanced thermal stability of the nanocomposites was attributed to the lower permeability of oxygen and the diffusibility of the degradation products from the bulk of the polymer caused by the partially exfoliated LDH in the composites.[18]

Polyethylene/clay nanocomposites have better thermal properties compared with the original PE due to the special structure of MMT.[82] During the first stage of degradation (below 400 °C) (the Hofmann elimination reaction and clay catalyzed degradation), PE/clay nanocomposites degrade faster than pure PE.[64] At higher temperatures (>400 °C), PE/clay nanocomposites are more stable than pure PE. This is attributed to the formation of a charred layer on the surface of the nanocomposites, which disrupts the oxygen supply and prevents the emission of thermally degraded small gaseous molecules. Figure 7.7 shows that the thermal stability of LDPE/LDH nanocomposites is significantly higher than that of pure LDPE.[83] It is evident that both the onset degradation and final degradation

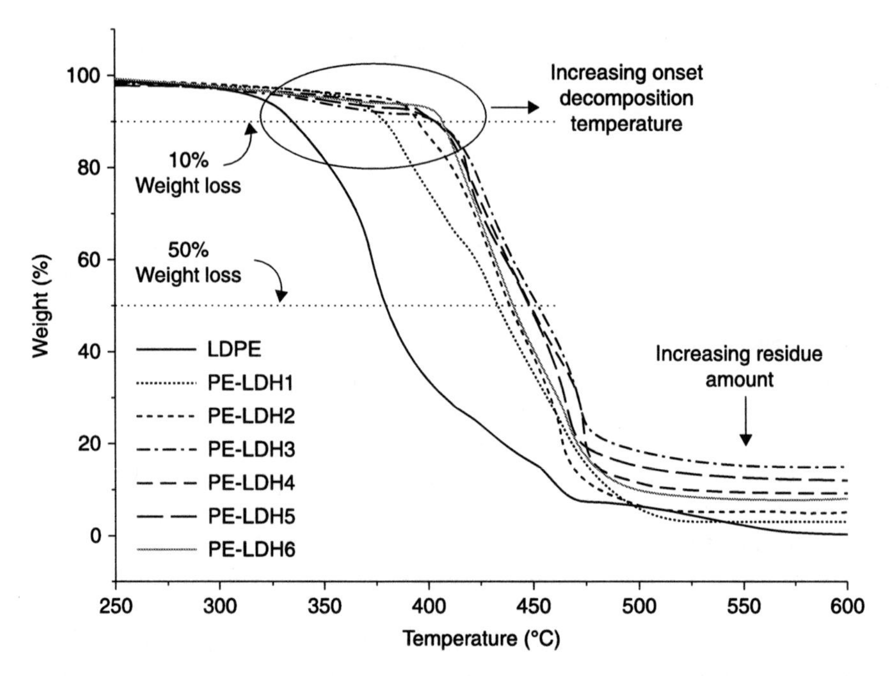


*7.7* TGA plots of LDPE/LDH nanocomposites prepared by melt mixing in an extruder. Reproduced from Costa *et al.*[83] by permission of Elsevier Science Ltd, UK.

temperature have increased with the addition of LDH into the nanocomposites. The enhanced thermal degradation temperature of the nanocomposites was attributed to the endothermic degradation of LDH, which produces smoke and water vapor and disrupts the oxygen supply from the surface to the bulk of the specimen. A charred layer of metal oxide forms a protective coating on the surface of the nanocomposite and hinders the emission of thermally degraded small gaseous molecules. By increasing the content of LDH in the nanocomposites, the size of the charred layer increased, as is evident in Fig. 7.8.[83] Chen *et al.* also found a similar effect while using dodecyl-sulfate-modified ZnAl LDH as a filler in an LLDPE matrix.[68] The thermal stability of CNT-filled PE nanocomposites was found to be higher than that of pure PE. The onset and final degradation temperatures of PE nanocomposites containing 0.5 wt.% of CNT were ~ 50 °C higher than pure PE.[84] Kuila *et al.* found that the onset degradation temperature for a nanocomposite with 3 wt.% DA-G was ~ 40 °C higher than that of neat LLDPE.[35] Thermal stability decreased at higher loadings of DA-G (5 and 8 wt.%). This was due to the presence of a larger number of DA-G layers in the nanocomposite, which act as a heat source domain inducing thermal degradation of the composite. In addition, the larger amount of DA-G produces a less stable charred layer during thermal decomposition.

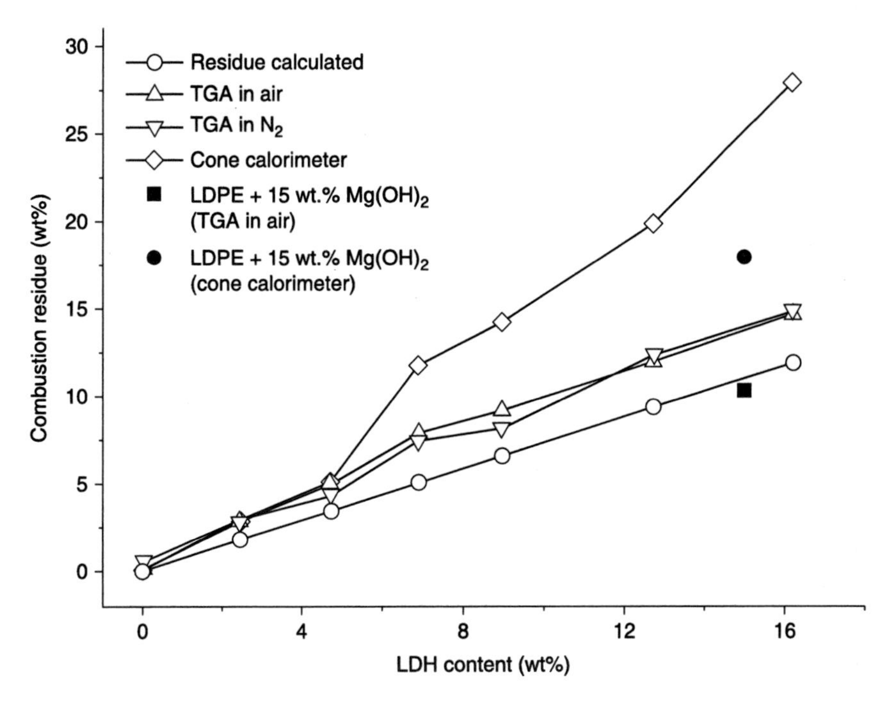


*7.8* Influence of LDH concentration on the amount of residue obtained from TGA (heated to 750 °C), from cone calorimeter experiments and theoretically calculated. Reproduced from Costa *et al.*[83] by permission of Elsevier Science Ltd, UK.

The thermal degradation behavior of pure PP and its nanocomposites with OMC is shown in Fig. 7.9 and the data derived from this is shown in Table 7.2.[76] There was only 10 °C of improvement in thermal stability for the PP/unmodified clay nanocomposites compared with unfilled PP. In contrast, the thermal stabilities of PP/OMC nanocomposites are 35–50 °C higher compared with pure PP. In comparison to unmodified clay, the OMC disperses homogeneously in the nanocomposites and can act as a barrier, preventing the emission of volatile products from the bulk matrix. Fereidoon *et al.* showed that the thermal stability of PP/single-walled carbon nanotube (SWCNT) nanocomposites increased due to: (1) the higher thermal stability of the SWCNTs than that of the PP matrix, (2) the thermal conductivity of the SWCNTs, and (3) the interaction between the matrix and the SWCNTs, which, if it were better, would cause a more homogeneous spread and conduction of heat.[85]

*Differential scanning calorimetric analysis*

Figure 7.10 shows the DSC cooling scan curve of PE/OMC nanocomposites. As seen from this figure, the crystallization temperatures ($T_c$) of pure PE and its

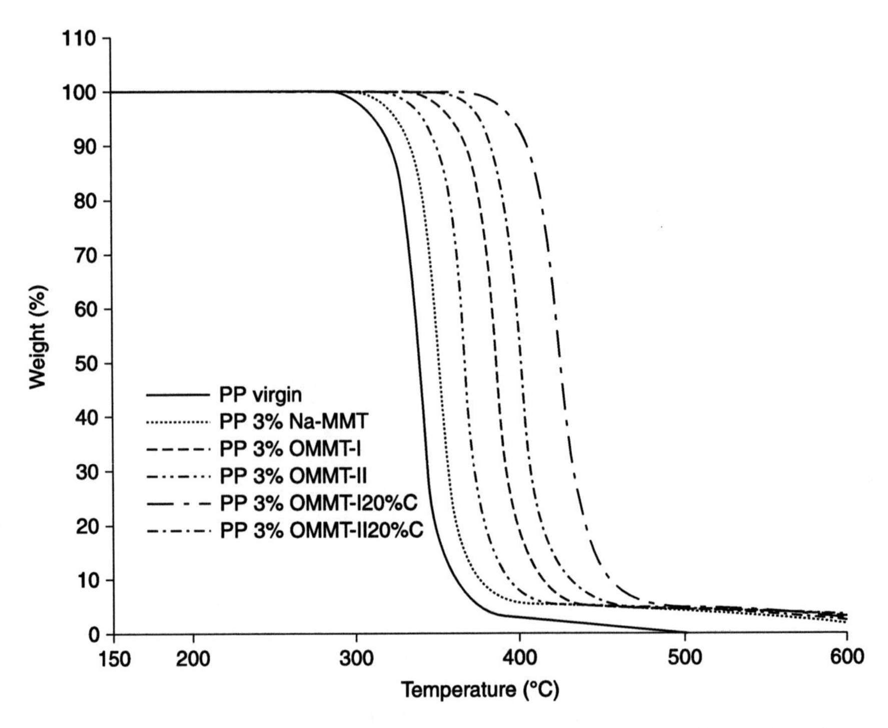


*7.9* TGA thermograms of PP and PP nanocomposites. Reproduced from Sharma and Nayak[76] by permission of Elsevier Science Ltd, UK.

*Table 7.2* TGA values of PP and PP nanocomposites

| Serial No. | Composition (wt.%) | Initial degradation temperature (°C) | Temperature at maximum degradation (°C) |
|---|---|---|---|
| 1 | Virgin PP | 302 | 385 |
| 2 | PP 3% Na-MMT | 313 | 398 |
| 3 | PP 3% OMMT-I | 350 | 436 |
| 4 | PP 3% OMMT-II | 335 | 416 |
| 5 | PP 3% OMMT-I 20%PP-g-MA | 389 | 475 |
| 6 | PP 3% OMMT-II 20%PP-g-MA | 366 | 447 |

Source: Reproduced from Sharma and Nayak[76] by permission of Elsevier Science Ltd, UK.

nanocomposites with 1 wt.% OMC are 117.3 and 119.3 °C, respectively. Further addition of clay has a slight influence on the $T_c$ of the composites. This can be explained by the assumption that the silicate layers act as efficient nucleating agents for crystallization.[82] Yang *et al.* showed that the addition of CNT has a similar effect on the $T_c$ and $T_m$ of the HDPE matrix.[86]

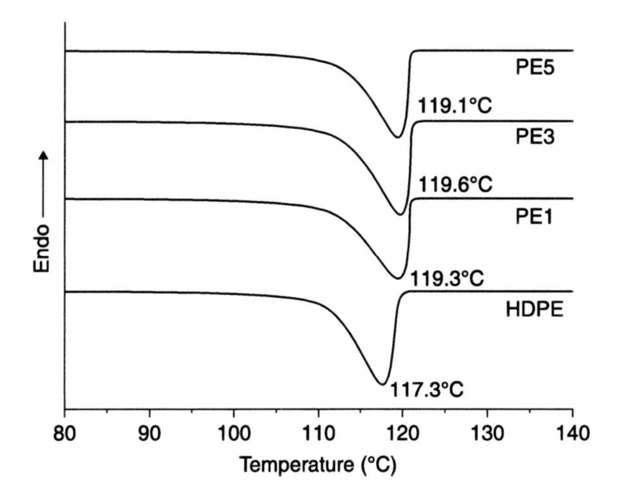


*7.10* DSC curve of PE/OMC nanocomposites. Reproduced from Zhai *et al.*[82] by permission of Elsevier Science Ltd, UK.

The melting point of pure PP has been determined by DSC analysis to be 155 °C.[87] As the clay content increases, the melting point has been observed to increase. Ko *et al.* showed that the glass transition temperatures ($T_g$) of PP hybrids are virtually constant for CNT loadings in the range 0.5–3.0 wt.%.[88] Thus, the effects of small amounts of dispersed CNTs on the free volume of PP are not significant, and the $T_g$ of PP hybrids is not affected by increasing the CNT loading. The melting transition temperatures ($T_m$) of hybrids increase with an increase in CNT content up to 0.5 wt.%, and then remain constant with further increases in CNT loading.

## 7.5.3 Flame-retardant properties

The flame retardancy of polyolefinic nanocomposites has been measured by limiting oxygen index (LOI), cone calorimetry, and Underwriters Laboratories (UL94) testing. Shi *et al.* prepared EVA/layered silicate nanocomposites using melt mixing and observed that the heat release capacity (HRC) and total heat release (THR) are reduced by 21–24% and 16%, respectively.[89] The heat release rate of EVA nanocomposites containing 5 wt.% clay was reduced by 80% and the mass loss rate (MLR) plot was spread over a much longer period of time. Beyer synthesized flame-retardant nanocomposites by melt blending ethylene vinyl acetate copolymers (EVA) with modified layered silicates.[90] A roughly 47% decrease in peak heat release rate (PHRR) as well as a shift towards longer times for burning were detected for a nanocomposite containing 5 wt.% of nanofiller. No significant decrease in PHRR was observed for further addition of nanofiller up to 10 wt.%. It was found that the limiting oxygen index values of the EVA/LDH composites gradually increased on increasing the hydrotalcite content in the EVA. Cone calorimeter characterization showed that pure EVA has a very sharp peak with a PHRR of 1813 $kW/m^2$ and an ignition time of 92 s. However, the peaks of all the samples were relatively smooth with PHRR below 597 $kW/m^2$ and the ignition times were longer. PHRR decreased when the hydrotalcite content was increased.[91] Gao *et al.* found that CNT plays an important role in the reduction of PHRR by forming low permeability char containing graphitic carbon.[92] Peeterbroeck *et al.* showed that the time to ignition (TTI) for EVA/MWCNT nanocomposites was delayed in comparison to unfilled EVA.[93] The decrease of the TTI for composites based on MWCNTs in comparison with the EVA matrix is probably due to the presence of acidic functions that have formed on the surface of the MWCNTs during purification. Ye *et al.* showed that MWCNTs can considerably decrease the heat release rate (HRR) and MLR by about 50–60%; they can almost double the combustion time and increase LOI values by 5% when 2 wt.% MWCNTs substitute for magnesium hydroxide (MH) in EVA/MH/MWCNT samples.[94]

Barbosa *et al.* prepared polyethylene/Brazilian clay nanocomposites and PE/commercial flame-retardant systems by direct melt compounding.[95] The PE/flame-retardant systems had a lower rate of combustion in comparison to the pure polymer and also a lower tendency for dripping. The burning rate of the nanocomposites was significantly reduced by the addition of 3 wt.% of clay compared with a neat PE matrix. The result demonstrates that the flammability resistance of PE/clay nanocomposites is much better compared with PE/commercial flame-retardant composite systems. This is due to the significant contribution of the OMC in decreasing the flammability of the systems, suggesting the barrier properties of the clay minerals. Zhao and co-workers prepared polyethylene nanocomposites with different kinds of OMC. In comparison to pure PE, the PHRR of PE/OMC nanocomposites was significantly reduced and the value decreased on increasing the clay content (Fig. 7.11).[64] The presence of 5% of ZnAl hydrotalcite (HTlc) caused a 55% reduction in PHRR and a delay of about 50 s.[22] The fire performance index of the nanocomposites decreased to 37.6 $kW/m^2s$ from 70.5 $kW/m^2s$ for the pure polymeric matrix. The formation of a protective layer on the specimen surface acted as a good barrier during combustion in the cone calorimeter test. Lu *et al.* prepared[96] flame-retardant maleated PE/magnesium hydroxide sulfate hydrate (MHSH) whisker composites containing OMC by direct melt intercalation. Cone calorimetry results indicated that a synergistic flame-retardant effect on reducing HRR occurred when MHSH and OMC were present in the nanocomposites.

Wagenknecht *et al.* showed that the burning behavior of PP/clay nanocomposites can be decreased by increasing the nanofiller content in the polymer matrix.[97] Tang and co-workers prepared PP/clay nanocomposites starting from pure clay and the reactive compatibilizer hexadecyltrimethylammonium bromide (C16).[98]

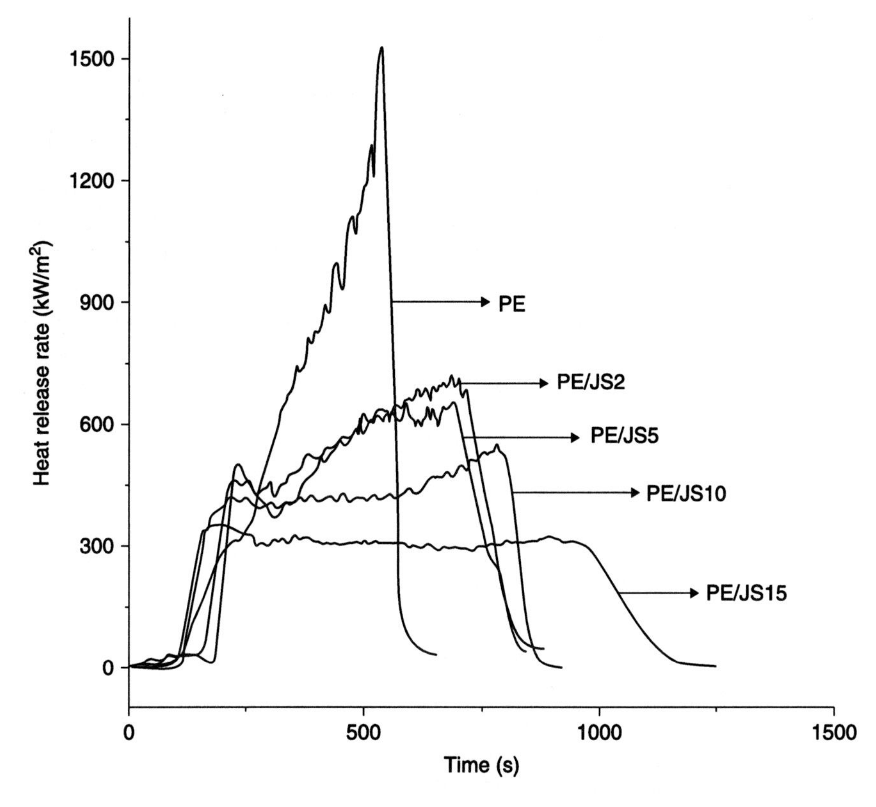


*7.11* Heat release rate for PE and PE/clay nanocomposites. Reproduced from Zhao *et al.*[64] by permission of Elsevier Science Ltd, UK.

The PHRR of PP/MMT hybrids containing 4 wt.% MMT and a mixture of MMT (4 wt.%) and C16 (2 wt.%) was reduced by about 37% and 27%, respectively. Song *et al.* investigated the effect of OMC on the fire retardancy of PP/OMC nanocomposites.[99] It was proposed that the better dispersion of OMC in the matrix is useful at reducing the flame retardancy of PP/OMC nanocomposites. Zhang *et al.* studied the effect of OLDH on the flame-retardant behavior of PP/intumescent-flame-retardant (IFR) nanocomposites. Cone calorimeter tests showed that pure PP burns very quickly (about 300 s) after ignition.[100] A very sharp HRR curve appears in the range 50–300 s with a PHRR of 1275 kW/m$^2$. In contrast, the samples with IFR show a dramatic decline in the HRR curves and the prolongation of the combustion time. Of all the flame-retardant samples, the PP/IFR/ZnAl-LDH sample had the lowest PHRR of 318 kW/m$^2$. Pure PP has a PHRR of 1071 kW/m$^2$ whilst in the nanocomposites with PP-g-MA (6 phr) and OLDH (6 phr) this value is only 754 kW/m$^2$.[101] This is more than a 25% reduction, and, even after taking into account the dilution effect of the filler, it can be considered a significant improvement in flame resistance. Total heat release, which is calculated as the area under a HRR curve, is another important parameter for fire-hazard evaluation. The nanocomposites had THR values of 37.7, 37.4, and 32.2 kJ/g, respectively, corresponding to the increasing OLDH content. In contrast, 38.8 kJ/g was found for neat PP. Pristine CNTs and hydroxylated CNTs (CNT-OHs) have been employed to enhance the thermal stability and flame retardancy of PP/wood flour composites compatibilized by PP-g-MA.[102] Kashiwagi and co-workers measured the flammability properties of PP/MWCNT nanocomposites as the MWCNT content varied from 0.5 to 4% by weight.[103,104] The lowest HRR was observed with a PP/MWCNT (1%) sample due to the balance between the effect of thermal conductivity and the shielding performance of the external radiant flux (and heat feedback from the flame), depending on the concentration of MWCNT in the sample.

### 7.5.4 Thermomechanical analysis

Kim *et al.* measured the thermomechanical properties of LLDPE/xGnP nanocomposites to determine the coefficient of thermal expansion (CTE). An obvious change in the slope in the range 80–85 °C for the xGnP-loaded composites was detected. The CTE was calculated for two temperature ranges, 45–80 °C and 85–100 °C. The CTE of the control LLDPE was $299.8\times10^{-6}$/°C below 80 °C and $365.3\times10^{-6}$/°C above 85 °C. Figure 7.12 shows that the CTE for LLDPE/xGnP composites in the range 45–80 °C is much lower than that of pure LLDPE.[72] However, there was no obvious change in CTE with the addition of xGnP. Kalaitzidou *et al.* measured the CTE of PP/xGnP composites along each direction for two temperature regimes.[105] A decrease in CTE along the longitudinal direction was observed for all the fillers (xGnP-1, xGnP-2, xGnP-15, CB, carbon fiber, and clay) at both below and above $T_g$. In particular, for a temperature below $T_g$, xGnP-1 exhibited a similar reduction in CTE as the carbon fibers, i.e., a reduction of CTE by ~25%. At higher $T_g$, the effect of carbon fibers on the CTE was more than that of xGnP. This is attributed to the high degree of alignment of the carbon fibers along the flow direction and the fact that these fibers are stiffer compared with the other fillers.

### 7.5.5 Gas-barrier properties

Although extensive research has been carried out on polymer/inorganic nanocomposites, the gas-barrier properties of polyolefin nanocomposites have been less well studied. Mittal studied the effect of organically modified MMT (OMMT) on the oxygen permeability of PP/clay nanocomposites.[106] The relative oxygen permeability of the nanocomposites decreased gradually with the addition of OMC. Figure 7.13 shows that the relative oxygen permeability of the nanocomposites with 4 vol.% of filler was half that of pure PP. The much better oxygen permeation behavior for OMC can be attributed to its better thermal stability. Permeation

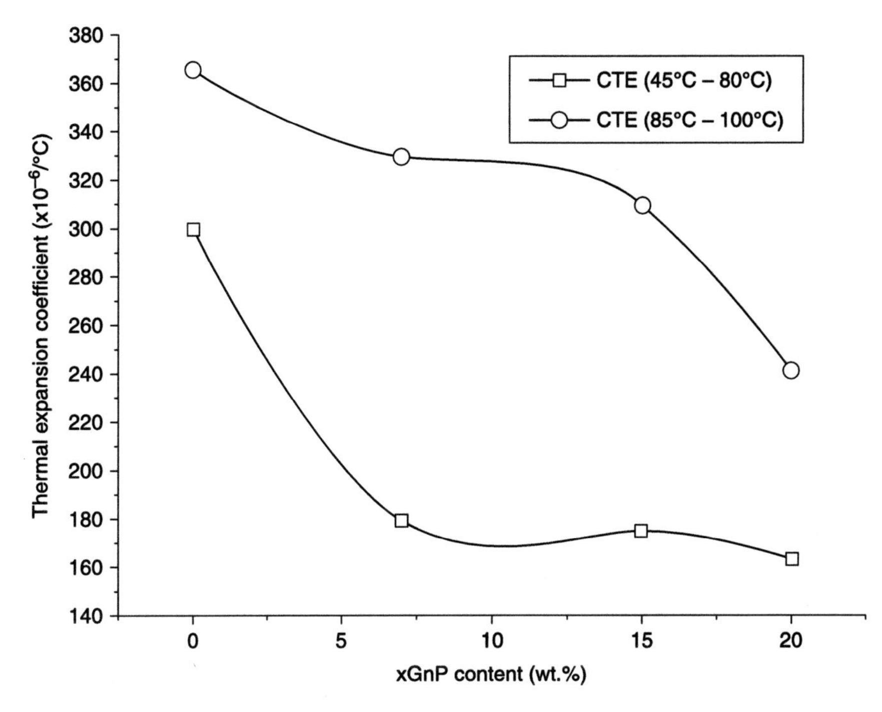


*7.12* Coefficient of thermal expansion of LLDPE/xGnP nanocomposites prepared by solution mixing and injection molded against xGnP content in the ranges 45–80 °C and 85–100 °C. Reproduced from Kim *et al.*[72] by permission of Wiley-VCH, Germany.

properties are sensitive to interfacial interactions; therefore, the improved permeation behavior can be attributed to the thermal stability of the nanocomposites. Kalaitzidou *et al.* studied the effect of xGnP and other fillers (CB, carbon fiber, and clay) on the $O_2$ permeability of PP composites with 3 vol.% of nanofiller.[105] They found that CB did not improve the $O_2$ barrier of PP due to the irregular shape of the CB agglomerates. In contrast, both xGnP-1 and carbon fiber caused a similar decrease in $O_2$ permeability, i.e., 10% at a reinforcement loading of 3 vol.%. For xGnP-15, a 20% improvement in $O_2$ permeability was observed with 3 vol.% loading.

## 7.5.6 Electrical conductivity

EVA, PE, EPDM, and PP are insulating in nature. The electrical conductivity of these polymers lies in the range $10^{-14}$ to $10^{-16}$ S/m. Therefore, clay minerals and LDH are not suitable fillers for conductive polymer composites, due to their insulating nature. Carbon-based nanofillers such as EG, CNT, CNF, and graphene are very effective in this respect. Das *et al.* showed that the percolation in electrical

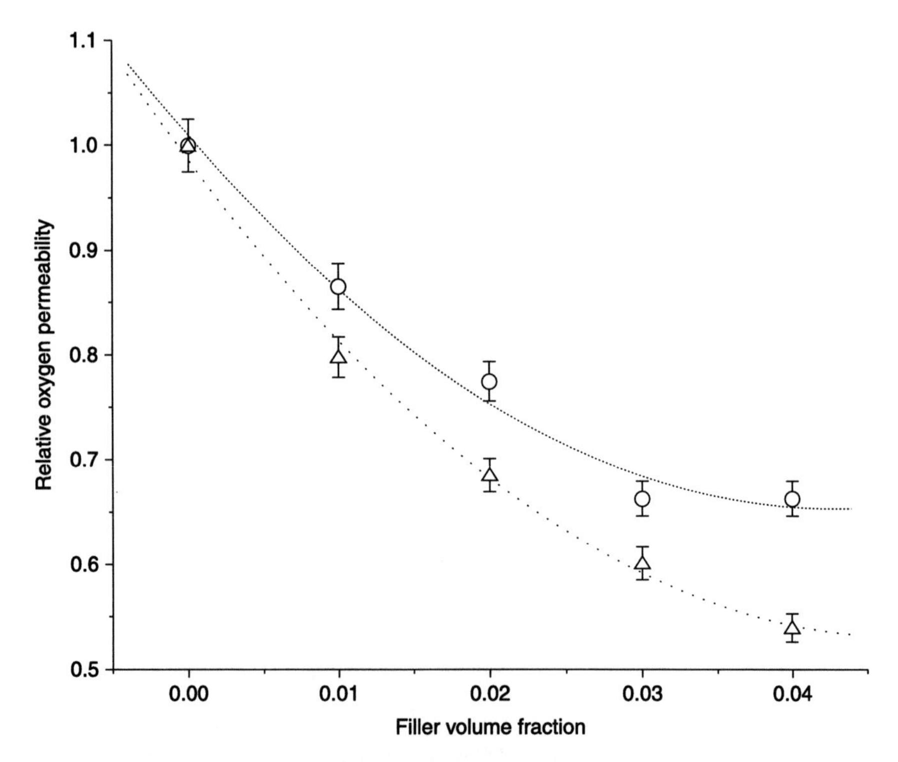


*7.13* Relative oxygen permeability of imidazolium OMMT/PP composites as a function of filler volume fraction compared with 2C18 ammonium OMMT/PP composites. O: 2C18 ammonium OMMT composites; Δ : imidazolium OMMT composites. The dotted lines serve only as guides. Reproduced from Mittal[106] by permission of Elsevier Science Ltd, UK.

conductivity of CNT-filled EVA composites was achieved with 5 wt.% of CNTs.[107] The electrical conductivity of nanocomposites with 30 wt.% of CNTs was 1 S/m at room temperature. Ciselli *et al.* studied the DC conductivity of ultra-high molecular weight PE/MWCNT composite films as a function of MWCNT weight fraction.[108] They observed that the percolation in electrical conductivity was achieved with only 3 wt.% of MWCNTs. The electrical conductivity of nanocomposites with 5 wt.% of MWCNTs increased to 0.1 S/m, indicating the formation of conductive composites. The percolation in electrical conductivity occurred with 3 wt.% for EG-filled PE nanocomposites.[70] The electrical conductivity of PE/EG composites was significantly higher than that of PE/graphite composites. This was attributed to the large surface area of EG compared with graphite. Kim and co-workers studied the effect of paraffin-oil-coated xGnP on the electrical conductivity of LLDPE nanocomposites.[63] Conductivity increased gradually on increasing the concentration of xGnP in the nanocomposites. The electrical conductivity of the

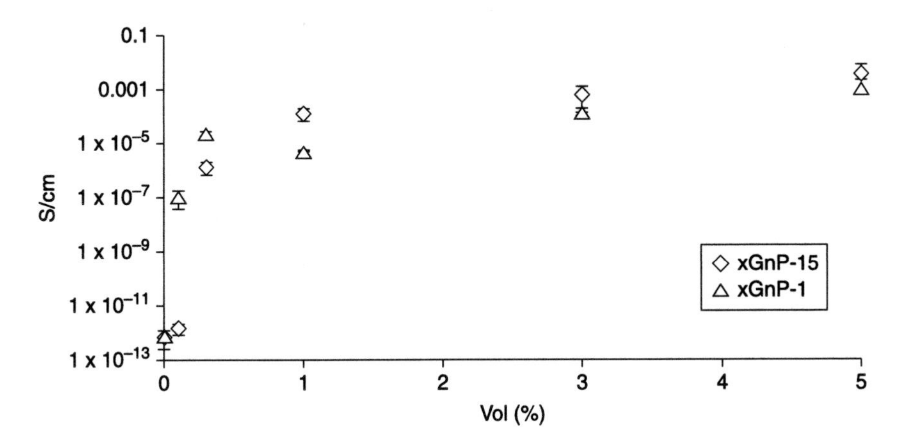


*7.14* Effect of xGnP aspect ratio on the percolation threshold and electrical conductivity of PP/xGnP made by coating and compression molding. Reproduced from Kalaitzidou *et al.*[110] by permission of Elsevier Science Ltd, UK.

nanocomposites reached a maximum when the weight ratio of xGnP, paraffin, and LLDPE was 1:1:8. Kuila *et al.* showed that the percolation in electrical conductivity occurred with 3 wt.% of DA-G in LLDPE/DA-G nanocomposites.[35] The conductivity of the sample reached $10^{-4}$ S/m with 8 wt.% DA-G. Kim *et al.* studied the effect of thermally reduced graphene oxide (TRG) on the electrical conductivity of LLDPE/TRG nanocomposites prepared using melt mixing.[109] Surprisingly, the percolation in electrical conductivity was achieved with only 0.4 vol.% TRG in the LLDPE matrix. The conductivity of LLDPE changed sharply even below 0.4 vol.% TRG. Kalaitzidou *et al.* studied the effect of compounding on the electrical conductivity of PP/xGnP nanocomposites.[110] It was evident that the combination of coating and compression molding yielded composites with a lower percolation threshold and higher conductivity. The coated and compression molded PP composites had a percolation threshold at ~0.3 vol.% xGnP-15 (Fig. 7.14), whereas the coated and injection molded composites had a percolation threshold of ~5 vol.% (Fig. 7.15). Chen *et al.* showed that the percolation threshold for PP/EG composites is 0.67 vol.%.[111] The electrical conductivity reached 0.1 S/m when the EG concentration was 3.90 vol.%.

## 7.5.7 Electromagnetic shielding

Electromagnetic interference (EMI) shielding refers to the reflection or absorption of electromagnetic radiation by a material which acts as a shield against the penetration of the radiation.[112] When plastic materials are used as shielding, noise signals are produced due to their very poor electrical conductivity. Carbon-based

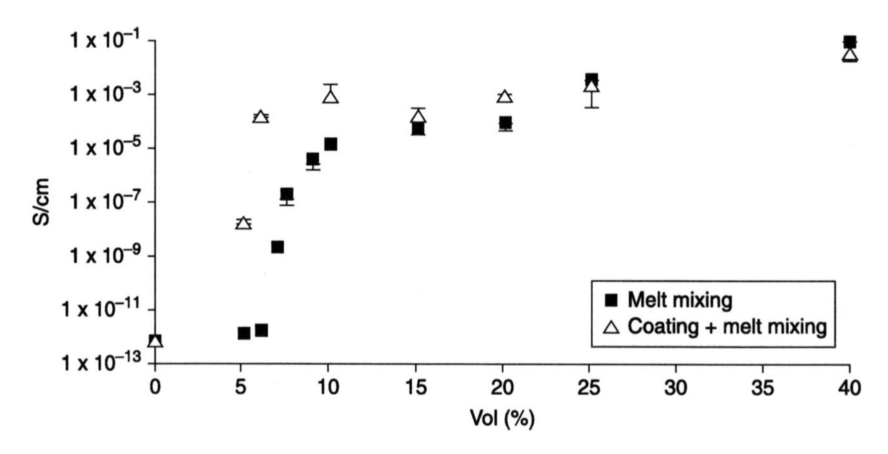


*7.15* Effect of compounding on the percolation threshold and conductivity of PP/xGnP-15 made by injection molding. Reproduced from Kalaitzidou *et al.*[110] by permission of Elsevier Science Ltd, UK.

conductive fillers such as CB, graphite, EG, CNT, CNF, and graphene are very good at improving the EMI shielding efficiency (SE) of plastics. Das *et al.* showed that CB and short carbon fibers (SCFs) are very effective in improving the EMI SE of EVA composites. The target value of the EMI SE needed for commercial applications is around 20 dB, which is only obtained with 20 wt.% of CB/SCF loading. The EMI SE of the composites reached 50 dB with the addition of 50 wt.% filler.[113] Das *et al.* showed that conductive polyaniline is a very effective filler in achieving the target value for EVA-based composites.[114] Das *et al.* also studied the effect of CNT loading on the EMI SE of EVA/CNT composites.[107] The EMI SE of a composite with 15 wt.% SWCNTs was ~22–23 dB in the gigahertz frequency range, showing promise for its commercial use as an EMI shielding material for different electronic devices. Due to its very high electrical conductivity, CNF has been used in the preparation of PE/CNF composites for EMI shielding.[115] Panwar and Mehra studied the effect of graphite loading on the EMI SE of PE/graphite composites.[116] All the composites showed good reproducibility and EMI SE up to 90 °C. Ghosh and Chakrabarti developed EPDM/CB composites for EMI shielding applications.[117] Mahapatra *et al.* showed that the EMI SE of EPDM/Vulcan XC 72 composites increased on increasing the content of Vulcan XC 72. It was also found that there was an initial increase in shielding with an increase in frequency, with a peak at 9.5 GHz, before showing a decreasing trend.[118]

## 7.5.8 Other properties

Polymer/inorganic nanocomposites have improved thermal conductivity, rheological properties, solvent resistance properties, etc. The thermal conductivity of EVA, PE, and PP can be improved by using CNTs, CNFs, EG, or graphene as a nanofiller due to their inherent high thermal conductivity values. The thermal conductivity of CNTs, CNFs, EG, and graphene has been found to be 3500, 2000, 380, $(4.84 \pm 0.44)$ to $(5.30 \pm 0.48) \times 10^3$ W/mK, respectively.[6,119] The thermal conductivity of EVA-50 can be improved by 70 and 188% with the incorporation of 1 and 4 wt.% CNF, respectively.[120] Ghose *et al.* showed that CNTs can enhance the thermal conductivity of an EVA sample.[121] The thermal conductivity of PP/xGnP composites has been measured as a function of xGnP aspect ratio and concentration.[105] The high aspect ratio of xGnPs enhances the thermal conductivity of a pure PP matrix. Melt rheological analysis shows that the incorporation of a nanofiller in a polymer matrix effectively increases the storage modulus ($G'$) and loss modulus ($G''$) of the nanocomposites.[122] It was also found that, generally, dynamic complex viscosity increased substantially on increasing the nanoclay content. This was attributed to the interaction and dispersion of the nanoclay in the polymer matrix. However, in some cases the viscosity may decrease due to the plasticization effect of the nanoclay.

## 7.6 Application of polyolefin nanocomposites

Polyolefin/inorganic hybrids or nanocomposites always exhibit improved properties over their pure form. Therefore, these nanocomposites should exhibit better performance compared with pure polymer when used for the same applications. This makes nanocomposite materials potentially useful in numerous automotive, general and industrial applications, including mirror housings for various vehicle types, door handles, engine covers, intake manifolds, timing belt covers, and many more. More general applications currently being considered include use in impellers and blades for vacuum cleaners, power-tool housings, aerospace, mower hoods, and covers for portable electronic equipment such as mobile phones, pagers, etc.[5,9–13,123]

The ability of clay, LDH, hydrotalcites, and CNTs to reduce the flammability of polymeric materials is a major point of interest to many researchers. Research has demonstrated the extent to which flammability can be restricted in polymers such as polyolefins with as little as 2% filler loading. In particular, heat release rates, as obtained from cone calorimetry experiments, have been found to diminish substantially with the incorporation of a nanofiller. Although the incorporation of conventional microparticles, together with flame-retardant and intumescent agents, would also minimize flammability, it is usually accompanied by reductions in various other important properties. With LDH, a reduction in flammability is usually achieved whilst maintaining or enhancing other properties and characteristics.

Carbon-based conductive fillers can convert insulating plastic materials to electrically conductive materials. Near the percolation threshold, high values of conductivity and the dielectric constant have been achieved with these composites, which means they can be used in charge-storage devices. The observed high dissipation factor makes them suitable for decoupling capacitor applications. This type of composite could be utilized in self-controlled heaters, current-limiters, etc.

## 7.7 Conclusions and future trends

This chapter has reviewed in detail nanocomposites made with polyolefins (EVA, EPDM, PE, and PP). Different kinds of nanofillers including layered clay (Na-MMT, bentonite, Cloisite, Laponite), ATH, LDH, nano-$Mg(OH)_2$, graphite, CNFs, EG, CNTs, graphene, etc., have been used to prepare polyolefin-based nanocomposites. The nanostructure of the composites has been established by XRD, FT-IR, and TEM. Finally, the properties of the nanocomposites have been discussed in detail in terms of their mechanical behavior, thermal stability, thermomechanical stability, flammability, gas-barrier capability, and electrical conductivity. Unmodified fillers are not suitable for enhancing mechanical properties. In this context, OMC, LDH, CNT, and graphene are very effective fillers. The thermal stability of the nanocomposites has also been found to be better compared with their pure counterparts. However, the thermal stability of EVA-based nanocomposites is almost comparable with pure EVA. This is due to the catalytic effect of the nanofiller and the presence of labile acetate groups of EVA. The flame retardancy of clay minerals, LDH, and ATH is significantly higher than for carbon-based nanofillers. The synergistic effect of CNTs with LDH or ATH on the flame retardancy of the nanocomposites has attracted extensive interest from researchers. The thermomechanical properties of nanocomposites are improved with the addition of EG as a nanofiller. Carbon-based nanofillers can be used to convert insulating polyolefins into conductive composites, which are very useful in applications.

Graphene-based composite materials have been studied less well in comparison to polymeric composites based on clay minerals, LDH, EG, and CNTs. There is a possibility of employing graphene as a conductive filler in the preparation of polymer composites for electronic and electrochemical devices. The addition of a very small amount of graphene (0.5 vol.%) can increase the electrical conductivity of polystyrene from $10^{-16}$ S/m to 0.01 S/m.[124] Graphene also has the ability to enhance the EMI SE of polyolefins at much lower loading. The very high aspect ratio and presence of different polar groups on the surface of graphene also help to improve the mechanical properties of composite materials. Flammability characteristics can also be increased with the addition of graphene. The synergistic effect of graphene and other flame-retardant fillers (ATH, LDH, etc.) on the flame-retardant properties of composites has not been studied. Therefore, much work is needed on the preparation, characterization, and application of graphene/polyolefin composite materials.

## 7.8 References

1. Kojima, Y.; Usuki, A.; Kawasumi, M.; Okada, A.; Fukushima, Y.; *et al.* Mechanical properties of nylon 6-clay hybrid. *J. Mater. Res.* 1993, *8*, 1185–1189.
2. Kojima, Y.; Usuki, A.; Kawasumi, M.; Okada, A.; Fukushima, Y.; *et al.* Sorption of water in nylon 6-clay hybrid. *J. Appl. Polym. Sci.* 1993, *49*, 1259–1264.
3. Ray, S. S.; Okamoto, M. Polymer/layered silicate nanocomposites: A review from preparation to processing. *Prog. Polym. Sci.* 2003, *28*, 1539–1641.
4. Pavlidou, S.; Papaspyrides, C. D. A review on polymer–layered silicate nanocomposites. *Prog. Polym. Sci.* 2008, *33*, 1119–1198.
5. Hussain, F.; Hojjati, M.; Okamoto, M.; Gorga, R. E. J. Polymer-matrix nanocomposites, processing, manufacturing, and application: An overview. *Compos. Mater.* 2006, *40*, 1511–1575.
6. Kuila, T.; Bhadra, S.; Yao, D.; Kim, N. H.; Bose, S.; *et al.* Recent advances in graphene based polymer composites. *Prog. Polym. Sci.* 2010, *35*, 1350–1375.
7. Gao, F. Clay/polymer composites: The story. *Mater. Today* 2004, *7*, 50–55.
8. Majumdar, D.; Blanton, T. N.; Schwark, D. Clay–polymer nanocomposite coatings for imaging application. *Appl. Clay Sci.* 2003, *23*, 265–273.
9. Cox, H.; Dearlove, T.; Rodgers, W.; Verbrugge, M.; Wang, C.S., Nanocomposite systems for automotive applications. In: *4th World Congress in Nanocomposites*, EMC, San Francisco (1–3 September 2004).
10. Transparent nanocomposites for aerospace applications. *Advanced Composites Bulletin*, Feb 2004.
11. Hule, R. A.; Pochan, D. J. Polymer nanocomposites for biomedical application. *MRS Bulletin* 2007, *32*, 354–358.
12. Boccaccini, A. R.; Erol, M.; Stark, W. J.; Mohn, D.; Hong, Z.; *et al.* Polymer/bioactive glass nanocomposites for biomedical applications: A review. *Compos. Sci. Technol.* 2010, *70*, 1764–1776.
13. Godovsky, D. Y. Device applications of polymer-nanocomposites. *Adv. Polym. Sci.* 2000, *153*, 163–205.
14. Kornmann, X. Synthesis and characterization of thermoset-layered silicate nanocomposites. PhD thesis 2001, Lulea Tekniska Universitet, Sweden.
15. Mu, O.; Feng, S. Thermal conductivity of graphite/silicone rubber prepared by solution intercalation. *Thermochim. Acta* 2007, *462*, 70–75.
16. Dong, W.; Zhang, X.; Liu, Y.; Wang, Q.; Guia, H.; *et al.* Flame retardant nanocomposites of polyamide 6/clay/silicone rubber with high toughness and good flowability. *Polymer* 2006, *47*, 6874–6879.
17. Okutan, E.; Aydin, G. O.; Hacrivelioglu, F.; Lilica, A. Synthesis and characterization of soluble multi-walled carbon nanotube/poly(organophosphazene) composites. *Polymer* 2011, *52*, 1241–1248.
18. Acharya, H.; Srivastava, S. K.; Bhowmick, A. K. Synthesis of partially exfoliated EPDM/LDH nanocomposites by solution intercalation: Structural characterization and properties. *Compos. Sci. Technol.* 2007, *67*, 2807–2816.
19. Pramanik, M.; Srivastava, S. K.; Samantaray, B. K.; Bhowmick, A. K. Rubber-clay nanocomposites by solution blending. *J. Appl. Polym. Sci.* 2003, *87*, 2216–2220.
20. Wang, G. A.; Wang, C. C.; Chen, C. Y. The disorderly exfoliated LDHs/PMMA nanocomposite synthesized by *in situ* bulk polymerization. *Polymer* 2005, *46*, 5065–5074.

21. Lee, W. D.; Im, S. S. Thermomechanical properties and crystallization behavior of layered double hydroxide/poly(ethylene terephthalate) nanocomposites prepared by *in situ* polymerization. *J. Polym. Sci. Part B: Polym. Phys.* 2007, *45*, 28–40.
22. Costantino, U.; Gallipoli, A.; Noccheti, M.; Camino, G.; Bellucci, F.; Frache, A. New nanocomposites constituted of polyethylene and organically modified ZnAl-hydrotalcite. *Polym. Degrad. Stabil.* 2005, *90*, 586–590.
23. Costa, F. R.; Saphiannikova, M.; Wagenknecht, U.; Heinrich, G. Layered double hydroxide based polymer nanocomposites. *Adv. Polym. Sci.* 2008, *210*, 101–168.
24. Brindley, G. W.; Kikkawa, S. Thermal behavior of hydrotalcite and of anion-exchanged forms of hydrotalcite. *Clays Clay Miner.* 1980, *28*, 87–91.
25. Sengupta, R.; Bhattacharya, M.; Bandyopadhyay, S.; Bhowmick, A. K. A review on the mechanical and electrical properties of graphite and modified graphite reinforced polymer composites. *Prog. Polym. Sci.* 2011, *36*, 638–670.
26. Kim, S.; Drzal, L. T. Improvement of electric conductivity of LLDPE based nanocomposites by paraffin coating on exfoliated graphite nanoplatelets. *Compos Part A: Appl. S.* 2010, *41*, 581–587.
27. Spitalsky, Z.; Tasis, D.; Papagelis, K.; Galiotis, C. Carbon nanotube–polymer composites: Chemistry, processing, mechanical and electrical properties. *Prog. Polym. Sci.* 2010, *35*, 357–401.
28. Li, Q.; Park, O. K.; Lee, J. H. Positive temperature coefficient behavior of HDPE/EVA blends filled with carbon black. *Adv. Mater. Res.* 2009, *79*, 2267–2270.
29. Jeevananda, T.; Kim, N. H.; Lee, J. H.; Basavarajaiah, S.; Urs, M. V. D.; Ranganathaiah, C. Investigation of multi-walled carbon nanotube reinforced high-density polyethylene/carbon black nanocomposites using electrical, DSC and positron lifetime spectroscopy techniques. *Polym. Int.* 2009, *58*, 755–780.
30. Tibbetts, G. G.; Lake, M. L.; Strong, K. L.; Rice, B. P. A review of the fabrication and properties of vapor-grown carbon nanofiber/polymer composites. *Compos. Sci. Technol.* 2007, *67*, 1709–1718.
31. Kotov, N. A. Carbon sheet solutions. *Nature* 2006, *442*, 254–255.
32. Kuila, T. Preparation, characterization and properties of ethylene vinyl acetate copolymer/Mg-Al layered double hydroxide nanocomposites. PhD thesis 2009, IIT Kharagpur, India.
33. Acharya, H. Synthesis, characterization and properties of polyolefinic elastomer nanocomposites. PhD thesis 2007, IIT Kharagpur, India.
34. Bhowmick, A. K.; Stephens, H. *Handbook of elastomers*. Second edition, CRC Press, USA.
35. Kuila, T.; Bose, S.; Hong, C. E.; Uddin, M. E.; Khanra, P.; *et al.* Preparation of functionalized graphene/linear low density polyethylene composites by a solution mixing method. *Carbon* 2011, *49*, 1033–1037.
36. Pramanik, M. Studies on layered materials and polymer nanocomposites. PhD thesis 2004, IIT Kharagpur, India.
37. Huang, J. C.; Zhu, Z. K.; Ma, X. D.; Qian, X. F.; Yin, J. Preparation and properties of montmorillonite/organo-soluble polyimide hybrid materials prepared by a one-step approach. *J. Mater. Sci.* 2001, *36*, 871–877.
38. Bala, P.; Samantaray, B. K.; Srivastava, S. K. Synthesis and characterization of Na-montmorillonite-alkylammonium intercalation compounds. *Mater. Res. Bull.* 2000, *35*, 1717–1724.
39. Newman, S. P.; Jones, W. Synthesis, characterization and applications of layered double hydroxides containing organic guests. *New J. Chem.* 1998, 105–115.

40. Vucelic, M.; Moggridge, G. D.; Jones, W. Thermal properties of terephthalate- and benzoate-intercalated LDH. *J. Phys. Chem.* 1995, *99*, 8328–8337.
41. Banerjee, S.; Benny, T. H.; Wong, S. S. Covalent surface chemistry of single walled carbon nanotubes. *Adv. Mater.* 2005, *17*, 17–29.
42. Chen, J.; Hamon, M. A.; Hu, H.; Chen, Y. S.; Rao, A. M.; *et al.* Solution properties of single-walled carbon nanotubes. *Science* 1998, *282*, 95–98.
43. Mickelson, E. T.; Huffman, C. B.; Rinzler, A. G.; Smalley, R. E.; Hauge, R. H.; *et al.* Fluorination of single-wall carbon nanotubes. *Chem. Phys. Lett.* 1998, *296*, 188–194.
44. Dyke, C. A.; Tour, J. M. Solvent-free functionalization of carbon nanotubes. *J. Am. Chem. Soc.* 2003, *125*, 1156–1157.
45. Bekyarova, E.; Itkis, M. E.; Ramesh, P.; Berger, C.; Sprinkle, M.; *et al.* Chemical modification of epitaxial graphene: Spontaneous grafting of aryl groups. *J. Am. Chem. Soc.* 2009, *131*, 1336–1337.
46. Shan, C.; Yang, H.; Han, D.; Zhang, Q.; Ivaska, A.; *et al.* Water soluble graphene covalently functionalized by biocompatible poly-L-lysine. *Langmuir* 2009, *25*, 12030–12033.
47. Bai, H.; Xu, Y.; Zhao, L.; Li, C.; Shi, G. Non-covalent functionalization of graphene sheets by sulfonated polyaniline. *Chem. Commun.* 2009, 1667–1669.
48. Salavagione, H. J.; Gomez, M. A.; Martinez, G. Polymeric modification of graphene through esterification of graphite oxide and poly(vinyl alcohol). *Macromolecules* 2009, *42*, 6331–6334.
49. Liu, N.; Luo, F.; Wu, H.; Liu, Y.; Zhang, C. *et al.* One step ionic-liquid-assisted electrochemical synthesis of ionic-liquid-functionalized graphene sheets directly from graphene. *Adv. Func. Mater.* 2008, *18*, 1518–1525.
50. An, X.; Simmons, T.; Shah, R.; Wolfe, C.; Lewis, K. M.; *et al.* Stable aqueous dispersions of noncovalently functionalized graphene from graphite and their multifunctional high-performance applications. *Nano Lett.* 2010, *10*, 4295–4301.
51. Acharya, H.; Pramanik, M.; Srivastava, S. K.; Bhowmick, A. K. Synthesis and evaluation of high-performance ethylene-propylene-dene terpolymer/organoclay nanoscale composites. *J. Appl. Polym. Sci.* 2004, *93*, 2429–2436.
52. Kuila, T.; Acharya, H.; Srivastava, S. K.; Bhowmick, A. K. Synthesis and characterization of ethylene vinyl acetate/Mg-Al layered double hydroxide nanocomposites. *J. Appl. Polym. Sci.* 2007, *104*, 1845–1851.
53. Kuila, T.; Acharya, H.; Srivastava, S. K.; Bhowmick, A. K. Effect of vinyl acetate content on the mechanical and thermal properties of ethylene vinyl acetate/MgAl layered double hydroxide nanocomposites. *J. Appl. Polym. Sci.* 2008, *108*, 1329–1335.
54. Pramanik, M.; Srivastava, S. K.; Samantaray, B. K.; Bhowmick, A. K. Synthesis and characterization of organosoluble, thermoplastic elastomer/clay nanocomposite. *J. Polym. Sci. Part B: Polym. Phys.* 2002, *40*, 2065–2072.
55. Cui, L.; Ma, X.; Paul, D. R. Morphology and properties of nanocomposites formed from ethylene-vinyl acetate copolymers and organoclays. *Polymer* 2007, *48*, 6325–6339.
56. Cardenas, M. A.; Lopez, D. G.; Mitre, G.; Merino, J. C.; Pastor, J. M.; *et al.* Mechanical and fire retardant properties of EVA/clay/ATH nanocomposites – effect of particle size and surface treatment of ATH filler. *Polym. Degrad. Stab.* 2008, *93*, 2032–2037.
57. Zhang, G.; Ding, P.; Zhang, M.; Qu, B. Synergistic effects of layered double hydroxide with hyperfine magnesium hydroxide in halogen-free flame retardant EVA/HFMH/LDH nanocomposites. *Polym. Degrad. Stab.* 2007, *92*, 1715–1720.
58. Kuila, T.; Srivastava, S. K.; Bhowmick, A. K. Rubber/LDH nanocomposites by solution blending. *J. Appl. Polym. Sci.* 2009, *111*, 635–641.

59. Kuila, T.; Acharya, H.; Srivastava, S. K.; Bhowmick, A. K. Ethylene vinyl acetate/Mg-Al LDH nanocomposites by solution blending. *Polym. Compos.* 2009, *30*, 497–502.
60. George, J. J.; Bhowmick, A. K. Ethylene vinyl acetate/expanded graphite nanocomposites by solution intercalation: Preparation, characterization and properties. *J. Mater. Sci.* 2008, *43*, 702–708.
61. Ismail, H.; Pasbakhsh, P.; Fauzi, M. N. A.; Bakar, A. A. Morphological, thermal and tensile properties of halloysite nanotubes filled ethylene propylene diene monomer (EPDM) nanocomposites. *Polym. Test.* 2008, *27*, 841–850.
62. Pasbakhsh, P.; Ismail, H.; Fauzi, M. N. A.; Bakar, A. A. Influence of maleic anhydride grafted ethylene propylene diene monomer (MAH-g-EPDM) on the properties of EPDM nanocomposites reinforced by halloysite nanotubes. *Polym. Test.* 2009, *28*, 548–559.
63. Kang, I.; Khaleque, M. A.; Yoo, Y.; Yoon, P. J.; Kim, S. Y.; Lim, K. T. Preparation and properties of ethylene propylene diene rubber/multi walled carbon nanotube composites for strain sensitive materials. *Compos. Part A: Appl. S.* 2011, *42*, 623–630.
64. Zhao, C.; Qin, H.; Gong, F.; Feng, M.; Zhang, S.; *et al.* Mechanical, thermal and flammability properties of polyethylene/clay nanocomposites. *Polym. Degrad. Stab.* 2005, *87*, 183–189.
65. Zhang, J.; Jiang, D. D.; Wilkie, C. A. Thermal and flame properties of polyethylene and polypropylene nanocomposites based on an oligomerically-modified clay. *Polym. Degrad. Stab.* 2006, *91*, 298–304.
66. Zhang, J.; Jiang, D. D.; Wilkie, C. A. Polyethylene and polypropylene nanocomposites based upon an oligomerically modified clay. *Thermochim. Acta* 2005, *430*, 107–113.
67. Costa, F. R.; Satapathy, B. K.; Wagenknecht, U.; Weidisch, R.; Heinrich, G. Morphology and fracture behaviour of polyethylene/Mg-Al layered double hydroxide (LDH) nanocomposites. *Eur. Polym. J.* 2006, *42*, 2140–2152.
68. Chen, W.; Qu, B. LLDPE/ZnAl LDH-exfoliated nanocomposites: Effects of nanolayers on thermal and mechanical properties. *J. Mater. Chem.* 2004, *14*, 1705–1710.
69. Kanagaraj, S.; Varanda, F. R.; Zhil'tsova, T. V.; Oliveira, M. S. A.; Simoes, J. A. O. Mechanical properties of high density polyethylene/carbon nanotube composites. *Compos. Sci. Technol.* 2007, *67*, 3071–3077.
70. Zheng, W.; Lu, X.; Wong, S. C. Electrical and mechanical properties of expanded graphite-reinforced high-density polyethylene. *J. Appl. Polym. Sci.* 2004, *91*, 2781–2788.
71. Kim, S.; Do, I.; Drzal, L. T. Multifunctional xGnP/LLDPE nanocomposites prepared by solution compounding using various screw rotating systems. *Macromol. Mater. Eng.* 2009, *294*, 196–205.
72. Kim, S.; Do, I.; Drzal, L. T. Thermal stability and dynamic mechanical behavior of exfoliated graphite nanoplatelets-LLDPE nanocomposites. *Polym. Compos.* 2010, *31*, 755–761.
73. Wang, J.; Xu, C.; Hu, H.; Wan, L.; Chen, R.; Zheng, H.; *et al.* Synthesis, mechanical, and barrier properties of LDPE/graphene nanocomposites using vinyl triethoxysilane as a coupling agent. *J. Nanopart. Res.* 2011, *13*, 869–878.
74. Cui, L.; Hunter, D. L.; Yoon, P. J.; Paul, D. R. Effect of organoclay purity and degradation on nanocomposite performance, Part 2: Morphology and properties of nanocomposites. *Polymer* 2008, *49*, 3762–3769.
75. Kim, D. H.; Fasulo, P. D.; Rodgers, W. R.; Paul, D. R. Structure and properties of polypropylene-based nanocomposites: Effect of PP-g-MA to organoclay ratio. *Polymer* 2007, *48*, 5308–5323.

76. Sharma, S. K.; Nayak, S. K. Surface modified clay/polypropylene (PP) nanocomposites: Effect on physico-mechanical, thermal and morphological properties. *Polym. Degrad. Stab.* 2009, *94*, 132–138.
77. Kotsilkova, R.; Ivanov, E.; Krusteva, E.; Silvestre, C.; Cimmino, S.; *et al.* Isotactic polypropylene composites reinforced with multiwall carbon nanotubes, Part 2: Thermal and mechanical properties related to the structure. *J. Appl. Polym. Sci.* 2010, *115*, 3576–3585.
78. Zhang, W.; Chen, D.; Zhao, Q.; Fang, Y. Effects of different kinds of clay and different vinyl acetate content on the morphology and properties of EVA/clay nanocomposites. *Polymer* 2003, *44*, 7953–7961.
79. Malucelli, G.; Ronchetti, S.; Lak, N.; Priola, A.; Dintcheva, N. D.; *et al.* Intercalation effects in LDPE/o-montmorillonites nanocomposites. *Eur. Polym. J.* 2007, *43*, 328–335.
80. Kanny, K.; Jawahar, P.; Moodley, V. K. Mechanical and tribological behavior of clay-polypropylene nanocomposites. *J. Mater. Sci.* 2008, *43*, 7230–7238.
81. Fang, P.; Chen, Z.; Zhang, S.; Wang, S.; Wang, L.; *et al.* Microstructure and thermal properties of ethylene-(vinyl acetate) copolymer/rectorite nanocomposites. *Polym. Int.* 2006, *55*, 312–318.
82. Zhai, H.; Xu, W.; Guo, H.; Zhou, Z.; Shen, S.; *et al.* Preparation and characterization of PE and PE-g-MAH/montmorillonite nanocomposites. *Eur. Polym. J.* 2004, *40*, 2539–2545.
83. Costa, F. R.; Wagenknecht, U.; Heinrich, G. LDPE/MgAl layered double hydroxide nanocomposite: Thermal and flammability properties. *Polym. Degrad. Stab.* 2007, *92*, 1813–1823.
84. Barus, S.; Zanetti, M.; Bracco, P.; Musso, S.; Chiodoni, A.; *et al.* Influence of MWCNT morphology on dispersion and thermal properties of polyethylene nanocomposites. *Polym. Degrad. Stab.* 2010, *95*, 756–762.
85. Fereidoon, A.; Ahangari, M. G.; Saedodin, S. Thermal and structural behaviors of polypropylene nanocomposites reinforced with single-walled carbon nanotubes by melt processing method. *J. Macromol. Sci. Part: B* 2008, *48*, 196–211.
86. Yang, J.; Wang, K.; Deng, H.; Chen, F.; Fu, Q. Hierarchical structure of injection-molded bars of HDPE/MWCNTs composites with novel nanohybrid shish–kebab. *Polymer* 2010, *51*, 774–782.
87. Isci, S.; Unlu, C. H.; Atici, O. Polypropylene nanocomposites prepared with natural and ODTABr-modified montmorillonite. *J. Appl. Polym. Sci.* 2009, *113*, 367–374.
88. Ko, J. H.; Yoon, C. S.; Chang, J. H. Polypropylene nanocomposites with various functionalized-multiwalled nanotubes: Thermomechanical properties, morphology, gas permeation, and optical transparency. *J. Polym. Sci. Part: B* 2011, *49*, 244–254.
89. Shi, Y.; Kashiwagi, Y.; Walters, R. N.; Gilman, J. W.; Lyon, R. E.; *et al.* Ethylene vinyl acetate/layered silicate nanocomposites prepared by a surfactant free method: Enhanced flame retardant and mechanical properties. *Polymer* 2009, *50*, 3478–3487.
90. Beyer, G. Flame retardant properties of EVA-nanocomposites and improvements by combination of nanofillers with aluminium trihydrate. *Fire Mater.* 2001, *25*, 193–197.
91. Jiao, C. M.; Wang, Z. Z.; Ye, Z.; Hu, Y.; Fan, W. C. Flame retardation of ethylene vinyl acetate copolymer using nano magnesium hydroxide and nano hydrotalcite. *J. Fire Sci.* 2006, *24*, 47–63.
92. Gao, F.; Beyer, B.; Yuana, O. A mechanistic study of fire retardancy of carbon nanotube/ethylene vinyl acetate copolymers and their clay composites. *Polym. Degrad. Stab.* 2005, *89*, 559–564.

93. Peeterbroeck, S.; Laoutid, F.; Swoboda, B.; Lopez-Cuesta, J. M.; Moreau, N.; *et al.* How carbon nanotube crushing can improve flame retardant behaviour in polymer nanocomposites? *Macromol. Rapid Commun.* 2007, *28*, 260–264.
94. Ye, L.; Wu, Q.; Qu, B. Synergistic effects and mechanism of multiwalled carbon nanotubes with magnesium hydroxide in halogen-free flame retardant EVA/MH/MWNT nanocomposites. *Polym. Degrad. Stab.* 2009, *94*, 751–756.
95. Barbosa, R.; Araújo, E. M.; Melo, T. J. A.; Ito, E. N. Comparison of flammability behavior of polyethylene/Brazilian clay nanocomposites and polyethylene/flame retardants. *Mater. Lett.* 2007, *61*, 2575–2578.
96. Lu, H.; Hu, Y.; Xiao, J.; Wang, Z. Magnesium hydroxide sulfate hydrate whisker flame retardant polyethylene/montmorillonite nanocomposites. *J. Mater. Sci.* 2006, *41*, 363–367.
97. Wagenknecht, U.; Kretzschmar, B.; Reinhardt, G. Investigation of fire retardant properties of polypropylene-clay-nanocomposite. *Macromol. Symp.* 2003, *194*, 207–212.
98. Tang, Y.; Hu, Y.; Li, B.; Liu, L.; Wang, Z.; *et al.* Polypropylene/montmorillonite nanocomposites and intumescent, flame retardant montmorillonite synergism in polypropylene nanocomposites. *J. Polym. Sci: Part A* 2004, *42*, 6163–6173.
99. Song, R.; Wang, Z.; Meng, X.; Zhang, B.; Tang, T. Influences of catalysis and dispersion of organically modified montmorillonite on flame retardancy of polypropylene nanocomposites. *J. Appl. Polym. Sci.* 2007, *106*, 3488–3494.
100. Zhang, M.; Ding, P.; Qu, B. Flammable, thermal, and mechanical properties of intumescent flame retardant PP/LDH nanocomposites with different divalent cations. *Polym. Compos.* 2009, 1000–1006.
101. Wang, D. Y.; Das, A.; Costa, F. R.; Leuteritz, A.; Wang, Y. Z.; *et al.* Synthesis of organo cobalt-aluminum layered double hydroxide via a novel single-step self-assembling method and its use as flame retardant nanofiller in PP. *Langmuir* 2010, *26*, 14162–14169.
102. Fu, S.; Song, P.; Yang, H.; Jin, Y.; Lu, F.; *et al.* Effects of carbon nanotubes and its functionalization on the thermal and flammability properties of polypropylene/wood flour composites. *J. Mater. Sci.* 2010, *45*, 3520–3528.
103. Kashiwagi, T.; Grulke, E.; Hilding, J.; Harris, R.; Awad, W.; *et al.* Thermal degradation and flammability properties of poly(propylene)/carbon nanotube composites. *Macromol. Rapid Commun.* 2002, *23*, 761–765.
104. Kashiwagi, T.; Grulke, E.; Hilding, J.; Groth, K.; Harris, R.; *et al.* Thermal and flammability properties of polypropylene/carbon nanotube nanocomposites. *Polymer* 2004, *45*, 4227–4239.
105. Kalaitzidou, K.; Fukushima, H.; Drzal, L. T. Multifunctional polypropylene composites produced by incorporation of exfoliated graphite nanoplatelets. *Carbon* 2007, *45*, 1446–1452.
106. Mittal, V. Gas permeation and mechanical properties of polypropylene nanocomposites with thermally-stable imidazolium modified clay. *Eur. Polym. J.* 2007, *43*, 3727–3736.
107. Das, N. C.; Maiti, S. Electromagnetic interference shielding of carbon nanotube/ethylene vinyl acetate composites. *J. Mater. Sci.* 2008, *43*, 1920–1925.
108. Ciselli, P.; Zhang, R.; Wang, Z.; Reynolds, C. T.; Baxendale, M.; *et al.* Oriented UHMW-PE/CNT composite tapes by a solution casting-drawing process using mixed-solvents. *Eur. Polym. J.* 2009, *45*, 2741–2748.

109. Kim, H.; Kobayashi, S.; AbdurRahim, M. A.; Zhang M. J.; Khusainova, A.; *et al.* Graphene/polyethylene nanocomposites: Effect of polyethylene functionalization and blending methods. *Polymer* 2011, *52*, 1837–1846.
110. Kalaitzidou, K.; Fukushima, H.; Drzal, L. T. A new compounding method for exfoliated graphite-polypropylene nanocomposites with enhanced flexural properties and lower percolation threshold. *Compos. Sci. Technol.* 2007, *67*, 2045–2051.
111. Chen, X. M.; Shen, J. W.; Huang, W. Y. Novel electrically conductive polypropylene/graphite nanocomposites. *J. Mater. Sci. Lett.* 2002, *21*, 213–214.
112. Chung, D. D. L. Electromagnetic interference shielding effectiveness of carbon materials. *Carbon* 2001, *39*, 279–285.
113. Das, N. C.; Chaki, T. K.; Khastgir, D.; Chakraborty, A. Electromagnetic interference shielding effectiveness of ethylene vinyl acetate based conductive composites containing carbon fillers. *J. Appl. Polym. Sci.* 2001, *80*, 1601–1608.
114. Das, N. C.; Yamazaki, S.; Hikosaka, M.; Chaki, T. K.; Khastgir, D.; *et al.* Electrical conductivity and electromagnetic interference shielding effectiveness of polyaniline-ethylene vinyl acetate composites. *Polym. Int.* 2005, *54*, 256–259.
115. Al-Saleh, M. H.; Sundararaj, U. Electrically conductive carbon nanofiber/polyethylene composite: Effect of melt mixing conditions. *Polym. Adv. Technol.* 2011, *22*, 246–253.
116. Panwar, V.; Mehra, R. M. Analysis of electrical, dielectric, and electromagnetic interference shielding behavior of graphite filled high density polyethylene composites. *Polym. Eng. Sci.* 2008, *48*, 2178–2187.
117. Ghosh, P.; Chakrabarti, A. Conducting carbon black filled EPDM vulcanizates: Assessment of dependence of physical and mechanical properties and conducting character on variation of filler loading. *Eur. Polym. J.* 2000, *36*, 1043–1054.
118. Mahapatra, S. P.; Sridhar, V.; Tripathy, D. K. Impedance analysis and electromagnetic interference shielding effectiveness of conductive carbon black reinforced microcellular EPDM rubber vulcanizates. *Polym. Compos.* 2008, *29*, 465–472.
119. Xiang, J.; Drzal, L. T. Thermal conductivity of a monolayer of exfoliated graphite nanoplatelets prepared by liquid-liquid interfacial self-assembly. *J. Nanomater.* 2010, Article ID 481753.
120. George, J. J.; Bhowmick, A. K. Influence of matrix polarity on the properties of ethylene vinyl acetate-carbon nanofiller nanocomposites. *Nanoscale Res. Lett.* 2009, *4*, 655–664.
121. Ghose, S.; Watson, K. A.; Working, D. C.; Connell, J. W.; Smith Jr J. G.; *et al.* Thermal conductivity of ethylene vinyl acetate copolymer/nanofiller blends. *Compos. Sci. Technol.* 2008, *68*, 1843–1853.
122. Chafidz, A.; Ali, M. A.; Elleithy, R. Morphological, thermal, rheological, and mechanical properties of polypropylene-nanoclay composites prepared from masterbatch in a twin screw extruder. *J. Mater. Sci.* 2011, *46*, 6075–6086.
123. Njuguna, J.; Pielichowski, K. Polymer nanocomposites for aerospace applications: Properties. *Adv. Eng. Mater.* 2003, *5*, 769–778.
124. Stankovich, S.; Dikin, D. A.; Dommett, G. H. B.; Kohlhaas, K. V.; Zimney, E. J.; *et al.* Graphene-based composite materials. *Nature* 2006, *442*, 282–286.

# 8

# Poly(vinyl chloride)(PVC)-based nanocomposites

M. GILBERT, Loughborough University, UK

**Abstract:** This chapter first discusses the structure, processing and formulation of poly(vinyl chloride) (PVC), which are rather different from those of other thermoplastic materials. In particular, a PVC formulation typically contains a number of different additives, and exists in rigid (unplasticised PVC or uPVC) and flexible (plasticised PVC or pPVC) forms. The different nanofillers, which have been used in PVC, are then introduced individually. Next the most important properties that can be modified in PVC are considered. Finally, brief statements about the opportunities for PVC nanocomposites and future trends are provided. An extensive list of references is included.

**Key words:** poly(vinyl chloride), uPVC, pPVC, nanofiller.

## 8.1 Introduction

Poly(vinyl chloride) (PVC) exists as two distinctly different thermoplastics, rigid PVC and flexible PVC, which are used in a wide variety of applications as discussed later. Approximately two-thirds of the PVC used is the rigid form. Research has been carried out into the addition of nanofillers into both of these materials, with the earliest papers published by Wang and co-workers in 2001 and 2002.[1,2]

As discussed below, there are a number of specific problems associated with introducing nanofillers into PVC, so there are no significant commercial nanocomposites.

This chapter will be divided into the following sections. Section 8.2 is concerned with the polymerisation and performance of PVC as relevant to the formation of nanocomposites. PVC is primarily made by suspension polymerisation, but also by emulsion polymerisation. The morphology and applications of these two polymerisation methods will be considered, as will their fusion to produce a final product. A key feature of PVC is poor thermal stability, which leads to degradation at relatively low temperatures. It is therefore necessary to add stabilisers in order to process PVC, so that a mixing process is required. In practice, PVC is mixed with a wide variety of additives, making it an extremely versatile polymer, with different formulations used for different applications. Typical additives will be discussed.

Section 8.3 considers the methods which have been used to produce PVC nanocomposites. Section 8.4 details the types of nanofillers that have been used, while the property changes that can be achieved are considered in Section 8.5. The remaining sections will consider opportunities, problems and future trends and provide sources of information. A comprehensive reference list is provided.

## 8.2 Poly(vinyl chloride) (PVC)

### 8.2.1 PVC polymerisation, structure and morphology

The basic repeat unit for PVC is:

$$-CH_2CHCl-$$

All types of PVC are formed via free-radical polymerisation of vinyl chloride monomers using peroxide catalysts. Use of this type of catalyst normally produces an atactic polymer (e.g. polystyrene). However, due to the relatively large size and electronegativity of the chlorine atom, commercial PVC contains about 55% syndiotacticity. This is located in sequences around 5–12 repeat units long; these sequences are long enough to crystallise, producing ~10% crystallinity. Crystallites are small (~10 nm), and do not form spherulites, so PVC can be transparent. They vary considerably in size and perfection, melting over a temperature range from 110 to 240 °C.

Most PVC is produced by suspension polymerisation. Because PVC is insoluble in its own monomer, the morphology of PVC grains is complex. The grains themselves are 100–150 μm in diameter, and irregular in shape, as shown in the scanning electron micrographs in Fig. 8.1. When a grain is cut open (Fig. 8.1(b)) primary particles (~1 μm in diameter) are revealed together with a considerable amount of porosity. Further examination reveals a pericellular membrane, produced from the suspending agent, which coats each grain, while the primary particles consist of smaller particles known as domains, around 10 nm in size. The crystallinity exists within these domains.

In emulsion polymerisation a latex containing spherical primary particles (usually with a size range of around 0.3–1.0 μm) is produced. This is then dried to produce aggregates of primary particles, described as secondary particles or grains, which are 30–60 μm in diameter.

Clearly, when nanofillers are added to PVC, their location with respect to the structures described is of considerable importance, as discussed later.

### 8.2.2 PVC fusion and processing

All rigid PVC and some flexible PVC is melt processed, mainly by extrusion, but compression moulding, injection moulding and calendering are also used. During processing, it is necessary to convert individual grains into a continuous solid structure. This process is known as fusion or gelation. The fusion, or gelation, of PVC can be defined simply as the conversion of PVC grains, which will be mixed with the required additives, to produce a product with the desired properties. In practice, the process is significantly more complex. Important work in the area was carried out by Allsopp,[3] who studied morphology changes during the processing of PVC by a variety of techniques, and proposed two possible

(a)

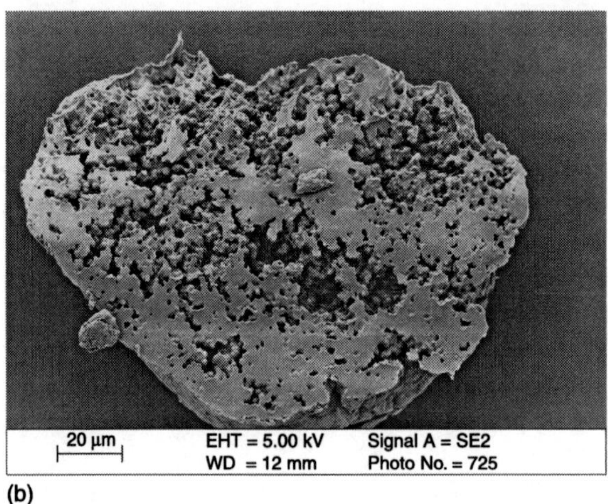

(b)

*8.1* PVC grain structure: (a) Whole grains. (b) Grain cut to show internal structure.

mechanisms for PVC fusion. Fusion requires both heat and shear, and the level of shear determines which mechanism occurs. In the comminution mechanism, the PVC grains are broken down into primary particles, additives are distributed, and the primary particles are then fused together to form a melt. This occurs in more aggressive high-shear equipment such as Brabender rheometers and Banbury mixers. In compression moulding, which is a very low-shear process, primary particles can be fused together within grains, but it is difficult to fuse grains together adequately, except at very high temperatures at which degradation can become a problem. For extrusion and two-roll milling, shear is higher than in compression moulding, and fusion occurs through the mechanism described by Allsopp as the compaction, elongation, densification and fusion (CDFE) mechanism.

The CDFE mechanism, as illustrated in Fig. 8.2, has the following stages:

- compaction – individual grains are packed together and some porosity is removed. Additives remain distributed around the grains
- densification – further porosity is removed
- elongation – grains become elongated in the direction of shear
- fusion – grains fuse together to form a melt, and additives become distributed.

An unusual feature of PVC is the fact that it is processed below its highest melting temperature. Thermal analysis shows that melting and reformation of crystallinity have an important role in fusion. This led Covas *et al.*[4] to develop the CDFE mechanism further to incorporate crystallisation. The results are summarised in Fig. 8.3. During processing, typically at 170 °C to 210 °C, some of the crystals will melt, whilst the remaining more perfect crystals, referred to as primary crystals, will be annealed. When a sufficient number of crystals have melted, the material will recrystallise during cooling to produce secondary crystallinity, which links the original structures to produce a strong material. Summers *et al.*[5] also used crystallisation to explain the results obtained by a rheological method used to measure fusion.

Although initial studies in this area related to rigid PVC, it has subsequently been shown that similar considerations apply to the fusion of flexible PVC.[6] It should be noted that some flexible PVC is processed via a plastisol process using

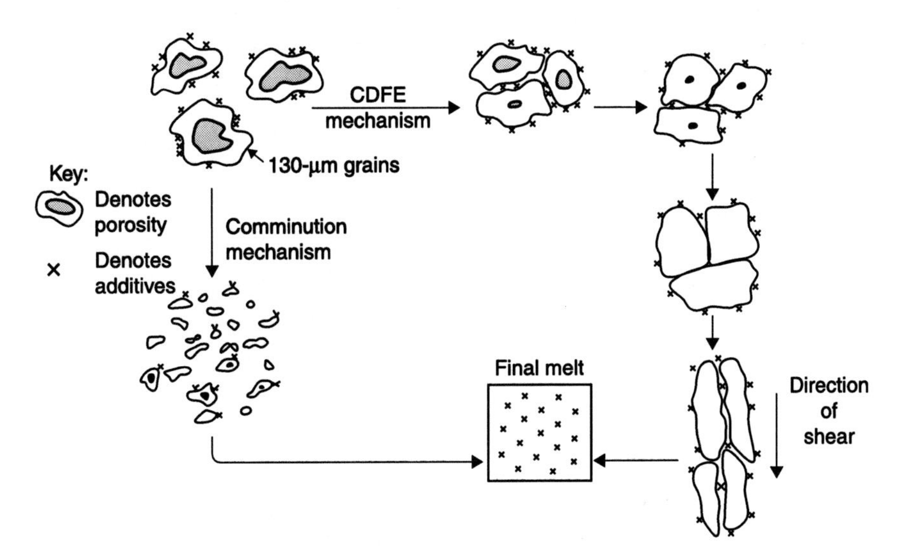


*8.2* Mechanisms of PVC fusion (Fig. 8.21 in Allsopp[3]).

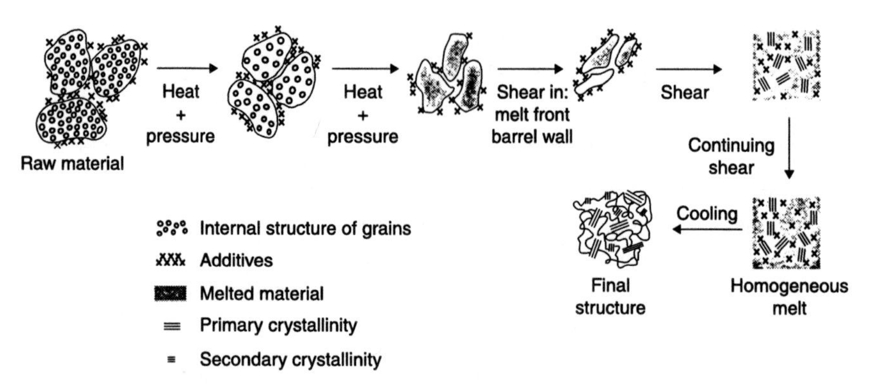


*8.3* Modified CDFE mechanism.[4]

emulsion polymer. Although the process is different, the melting and crystallisation processes discussed previously still occur. When nanofillers are added, it is important that they are well dispersed in the homogeneous melt.

### 8.2.3 PVC degradation

If PVC were made up solely of (-$CH_2CHCl$-) repeat units, its stability would be reasonably good. However, in practice, chemical defect structures are present – labile chlorine, unsaturation, head-to-head units, and peroxide and hydroperoxide groups. The latter in particular act as sites for the initiation of degradation. The first stage of degradation is the formation of macroradicals, but these undergo a rapid 'unzipping' reaction leading to the production of hydrogen chloride, which is both corrosive and acts as a catalyst for degradation, and results in the formation of conjugated unsaturation along the polymer chain, i.e.

$$\sim\!\!\sim CH_2CHCl\ CH_2CHCl\ CH_2CHCl\sim\!\!\sim \quad \rightarrow$$

$$\sim\!\!\sim CH_2CHCl\ CH_2CHClCH_2^{\cdot} \quad \rightarrow$$

$$\sim\!\!\sim CH_2CHCl\ CH_2C^{\cdot}H = CH_2 + Cl^{\cdot} \quad \rightarrow$$

$$\sim\!\!\sim CH_2CHCl\ CH_2CH = CH_2 + HCl \quad \rightarrow$$

$$\sim\!\!\sim CH{=}CH{-}CH{=}CH{-}CH{=}CH{-}\ \text{etc.}$$

Conjugated chains absorb UV radiation, but as the number of conjugated units increases, the conjugation starts to absorb in the visible region, and the PVC changes in appearance, producing the well-known sequence of colours white/yellow/orange/brown/black. Without a stabiliser, this degradation can start at

temperatures as low as 100 °C. It will be seen later that certain nanofillers can accelerate this process, while others can act as stabilisers.

### 8.2.4 PVC formulation and additives

As a plastic material, PVC is unique in the extent to which it is formulated. Additives are normally introduced by high-speed mixing in which the powder is stirred at speeds of the order of 3000 rpm. Solid additives are usually added first, followed by liquid ones. Solid additives coat the PVC grains, becoming located in hollows on the grain surfaces, while liquid additives are absorbed and located in porous regions. The high speed results in shear heating to above the glass transition temperature of the PVC (~80 °C), and mixing is normally continued until the temperature reaches 120 °C, when the mix is discharged into a cooling chamber. The resulting powder blend may be processed directly to make a product or extruded to make a pelletised compound for subsequent processing.

When plastisol technology is used the PVC is suspended in a liquid plasticiser, and additives are introduced by mixing with the liquid state.

The main additives used in PVC formulations are summarised below.

*Heat stabilisers*

A variety of different compounds have been used for this purpose. Lead compounds are inexpensive and very common, and were popular in the past, but they are now being phased out in Europe, due to concerns about toxicity.[7] Mixed metal soaps are widely used. At one time these contained various combinations of calcium, zinc, barium and cadmium, but cadmium has now been phased out in Europe,[7] and stabilisers are based on Ca and Zn. These stabilisers are less effective, particularly during the later stages of degradation, and nanofillers that could enhance performance would provide significant benefits. The so-called organo-based stabilisers include calcium. Another group of stabilisers is based on organotin compounds. These are expensive but can be used at lower concentrations. In addition to the above there are a number of secondary stabilisers, such as epoxy compounds and hydrotalcites.

*Lubricants*

Both external and internal lubricants are required in rigid PVC formulations, while internal or multifunctional lubricants are used when flexible formulations are melt compounded. Although the distinction between the two types is not precise, in general external lubricants migrate to the melt surface and reduce the friction between the PVC and the surface of the processing equipment; an example of this type of lubricant would be paraffin wax. Internal lubricants reduce the friction between the PVC grains and improve particle flow. The most common internal lubricant is calcium stearate. Stabiliser/lubricant packages are frequently used in PVC formulations.

*Processing aids*

Processing aids are used in rigid formulations to increase melt strength, particularly for extrusion processes. The most common type of processing aid is a high molecular weight poly(methyl methacrylate), which entangles with the PVC molecules in the melt and prevents tearing when the material exits from the extruder die.

*Impact modifiers*

Impact modifiers are added to improve toughness for demanding applications such as pressure pipes. They are either rubber based (acrylic polymers or chlorinated polyethylene) or finely divided minerals, in particular calcium carbonate. There is considerable scope for using nanofillers for this purpose.

*Fillers*

Fillers have been incorporated in PVC to improve mechanical strength, hardness and stiffness, flame retardancy and electrical properties, to reduce thermal expansion and sometimes just to reduce costs. The most common filler for PVC is calcium carbonate, although various silicates (talc, kaolin and mica) have also been used. Nanofillers could replace conventional fillers for some of these purposes.

*Plasticisers*

Plasticisers are compatible and involatile organic liquids that reduce the glass transition temperature of the PVC to below room temperature, thus producing a flexible polymer. The more plasticiser added, the more flexible the plastic. Apart from specific applications containing fillers, the additives described above are used at levels $<10$ parts per hundred of PVC resin (phr). However, typical plasticiser levels are 30–100 phr. Of the plasticisers used $>90\%$ are phthalates, which are good general purpose plasticisers with low cost. Because of perceived health problems associated with phthalate plasticisers, risk assessments on the commonly used phthalates have been carried out as part of the Vinyl 2010 voluntary commitment,[7] and some of the lower molecular weight phthalates are no longer permitted for use in children's toys and childcare articles, but higher molecular weight phthalates ($\geq C_9$) have a clean bill of health for all applications. Specialist plasticisers include phosphates for fire retardancy, polymeric plasticisers (low migration) and some newer 'non-phthalate' plasticisers such as citrates, benzoate esters, alkyl sulphonic phenyl esters (Mesamoll®), di-isononyl cyclohexane-1,2-dicarboxylate (Hexamoll® DINCH) and biobased plasticisers (e.g. Grindsted® Soft-n-Safe).

*Others*

A variety of other additives are used for specific purposes. These include colourants, foaming agents, biocides, flame retardants and UV absorbers.

## 8.3 Manufacturing techniques

The standard addition techniques for nanofillers have all been used for PVC, namely during polymerisation, solution blending and melt blending. An additional technique exists for PVC because nanofillers can be added to plastisols. There are also a number of examples cited later in which a nanofiller is produced as a suspension in a liquid, which is then added to PVC. Specific applications of these techniques are discussed below.

## 8.4 Nanofillers

### 8.4.1 Montmorillonite (MMT)

The most widely used nanofiller for PVC is the layered clay, montmorillonite (MMT), both in its natural form as sodium montmorillonite (Na-MMT), and as organomodified (OMMT) grades. PVC/MMT nanocomposites were reviewed by Pagacz and Pielichowski in 2009.[8] Organomodification of MMT was discussed in detail in this paper, which also gives a useful summary of salts used for MMT modification. Most researchers have used melt compounding to disperse nanoclays, with varying degrees of success. However, solution blending,[1,9,10] polymerisation[11–17] and plastisols have also been used.[18] One problem associated with these nanofillers arises because the most common organomodifiers for MMT are quaternary ammonium compounds, and the ammonium cations can actually contribute to PVC degradation.[8] A variety of other modifiers have therefore been investigated.

Another of the problems associated with the formation of PVC nanocomposites is related to the morphology of PVC. This is illustrated in Fig. 8.4, which shows transmission electronic microscopy (TEM) micrographs of (a) melt compounded, (b) solution blended and (c) plastisol based nanocomposites containing Cloisite 30B OMMT.

Figure 8.4(a) shows intercalation and some exfoliation of nanoclay. Light areas without filler are PVC primary particles, which have not been destroyed. These areas could have been reduced by increasing the processing temperature, but full intercalation was not achieved because the PVC did not fully melt. When a melt-processed sample is dissolved in tetrahydrofuran, all morphology and crystallinity are destroyed, and a high level of exfoliation can be achieved (Fig. 8.4(b)). This

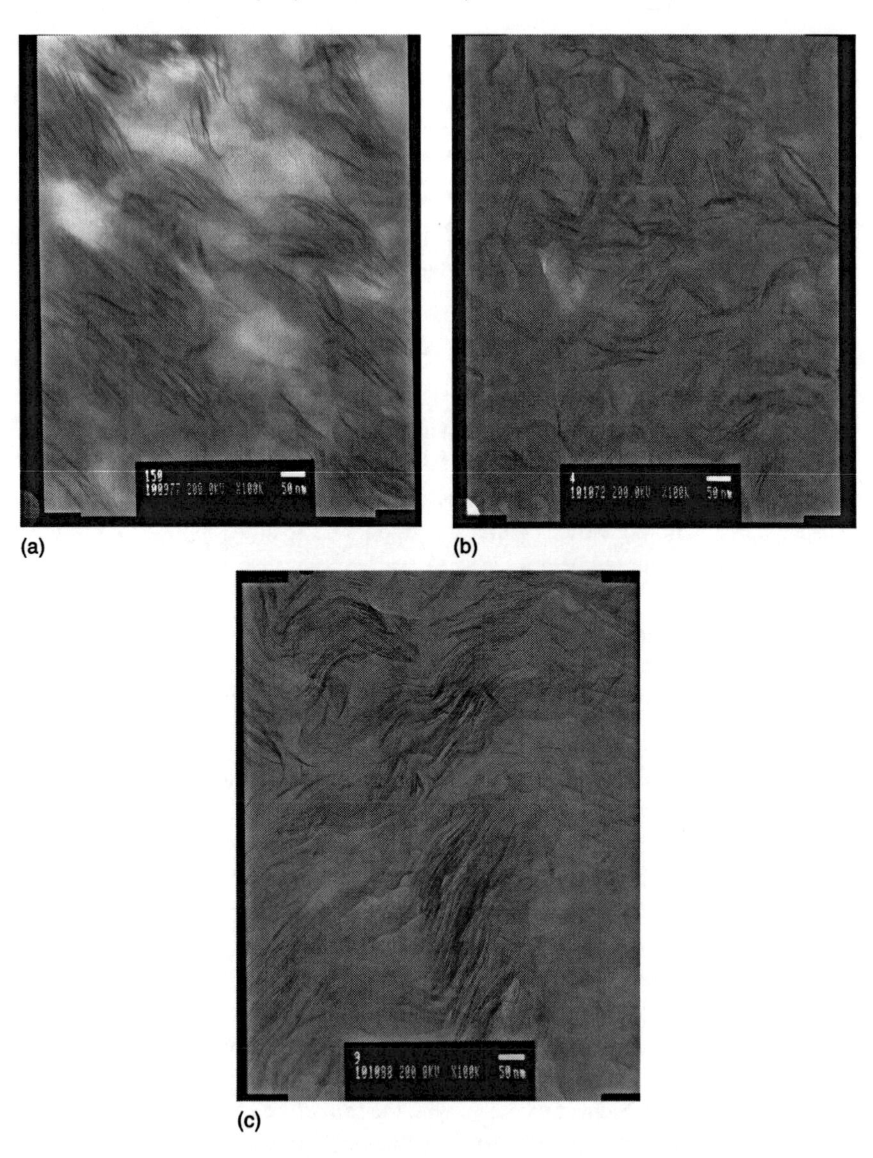


*8.4* TEM micrographs of nanocomposites. (a) Melt blended in a Haake rheometer at 150 °C. (b) Melt blend dissolved in tetrahydrofuran, then cast. (c) Produced from plastisol containing 70 phr di-isodecyl phthalate.[19] (Equal magnifications, scale bar equals 50 nm.)

is confirmed by the disappearance of the low-angle peaks observed by X-ray diffraction. When a nanocomposite is produced from a plastisol containing emulsion PVC, again there is some exfoliation (Fig. 8.4(c)), but less than in the solution-blended sample. In other words, the formation of PVC nanocomposites with high levels of exfoliation is difficult via melt compounding. Matuana[20] addressed the problem of achieving the appropriate PVC morphology for good nanoclay dispersion, using a torque rheometer, and showed that when nanoclay is introduced at the onset of fusion, when the PVC particles are reduced in size, better dispersion and enhanced mechanical properties are obtained. Industrial applications of this approach in, for example, extrusion need to be addressed.

When nanofillers are added during polymerisation, exfoliation can be achieved much more readily.[12,13,16,17]

### 8.4.2 Other clays

Other clays which have been used as nanofillers in PVC include Laponite,[21] bentonite,[22–25] hectorite[23,25] and kaolinite.[26] Vandevyver and Eicholz[24] recognised the problems associated with dispersing nanofillers in suspended PVC, so they focussed on mixing sodium bentonite with an emulsion PVC latex, taking advantage of the ability of water to cause exfoliation. The mixture produced was spray dried, and its behaviour after milling and pressing, and in plastisols, was investigated.

### 8.4.3 Calcium carbonate

Various types of calcium carbonate (ground calcite, ground whiting and ground limestone) with mean particle sizes ranging from 1 to 30 μm are the most common fillers for PVC. More specialist (and expensive) precipitated calcium carbonates (mean particle size <1 μm) are also used when good mechanical properties are required, particularly for rigid formulations. It is therefore unsurprising that nano-sized calcium carbonates have received a considerable amount of attention. Nanocomposites have been produced by polymerisation,[27,28] solution blending[29] and melt processing.[30–39] A novel polymerisation method involved synthesising nano-porous PVC particles then reacting calcium hydroxide and carbon dioxide inside these to produce PVC/$CaCO_3$ nanocomposites, which were finally melt blended.[40]

### 8.4.4 Silica

Silica is another type of nanofiller which has been used in PVC. Incorporation techniques include polymerisation.[17,28] There has been a specific focus on the surface coating of silica particles to improve compatibility with the PVC matrix[41–46] in melt-compounded nanocomposites.

### 8.4.5 Layered double hydroxides

Layered double hydroxides (LDHs), specifically hydrotalcite, although not necessarily in the form of nanoscale particles have, since 2000, found a significant application as secondary stabilisers in PVC.[47,48] They are usually found in stabiliser/lubricant packages, which are added to PVC formulations. Hydrotalcites are layered structures, but, unlike MMTs, have anions between the layers, which can be replaced by $Cl^-$ ions produced during PVC degradation, thus retarding PVC degradation. Because of the benefits of hydrotalcites, hydrotalcite nanofillers are obviously of interest. In one of the earliest papers in this area, Wang *et al.*[49] compared nanocomposites prepared by *in situ* polymerisation and melt blending, and demonstrated that the former performed better in all respects. The same group also investigated nanocomposites containing various LDHs, and carried out further work on PVC/hydrotalcite nanocomposites prepared by *in situ* polymerisation.[50,51] LDH nanocomposites have also been prepared by melt processing,[52] solution intercalation[53] and using a novel method in which the filler was delaminated by lauryl ether phosphate in tetrahydrofuran and then the suspension produced was added to PVC.[54] LDH nanofillers have also been modified by grafting with toluene diisocyanate to improve dispersion.[55]

### 8.4.6 Carbon nanotubes

There are some reports of the use of carbon nanotubes (CNTs) in PVC, although the addition of a costly filler to an inexpensive commodity thermoplastic polymer can only be justified for very specialist applications, as discussed later. Solution blending has been used for the preparation of nanocomposites.[56]

### 8.4.7 Other nanofillers

The use of a variety of other nanofillers in PVC has been reported. These include antimony trioxide,[57] polyhedral oligomeric silsesquioxane (POSS),[58,59] calcium sulphate,[60,61] calcium phosphate[62] and zinc oxide.[63] Finally, combinations of two different fillers have been used on a number of occasions. These include CNTs with wood-flour, to improve the performance of commercially available wood-filled PVC,[64] and nano-calcium carbonate and wood-flour.[65] Nanoclays have been used with metallic oxides, which affect combustion and smoke suppressant properties,[66] and with zeolite to improve thermal stability.[67]

## 8.5 Effects of nanofillers

### 8.5.1 Thermal stability

The thermal stability of PVC compounds may be improved or reduced by nanofillers. Frequently thermal stability has been assessed by thermogravimetric analysis,[8] but in practice tests routinely used to assess the thermal stability of PVC compounds (e.g. the Congo red test, oven ageing and torque rheometry) are more important. Unsurprisingly, different nanofillers have different effects. Generally, MMTs reduce thermal stability, calcium carbonate and other clays can improve performance, but the effect of hydrotalcite is the most positive.

Commercially available organically modified MMTs tend to have adverse effects on PVC thermal stability,[8] although effects can be reduced to some extent by changing the stabilisers present in the PVC formulation.[68] Wan *et al.*[69] suggested that discolouration rather than degradation occurred when OMMT was used in PVC, but this is not a widely held view. Attempts have been made to avoid the problem of poor stability by developing alternative modifiers. It has been shown that the incorporation of silica into a modifier improves thermal stability.[67,70] Sterky *et al.*[71] showed that, unlike normal cationic modifiers, non-ionic intercalants did not affect thermal stability. Another approach is the use of chelating agents to modify surface-treated MMT.[72] Added metallic oxides also produced favourable results.[66] Other modifications have been discussed in the literature.[8]

Nano-calcium carbonate,[35,73] kaolinite[26] and sodium bentonite[24] also appear to improve the thermal stability of PVC compounds to some extent.

Consistently with the use of layered double hydroxides, such as hydrotalcite, as secondary stabilisers, these compounds appear to be the most effective nanofillers for enhancing PVC stability.[49,51,53,55,72–76]

### 8.5.2 Mechanical properties

One of the most common benefits sought by the incorporation of nanofillers in a polymer is an improvement in mechanical properties. It should be remembered that rigid PVC itself has good strength and stiffness compared with other commodity thermoplastics (tensile strength 60 MPa and Young's modulus 2.8 GPa), providing that it is processed correctly. Toughness is typically improved by the addition of impact modifiers, as discussed in Section 8.2.4. The mechanical properties of PVC nanocomposites have been reported extensively. As with any filler, the modulus normally increases. Improvements in tensile properties can be achieved, but, unless dispersion is good, toughness and elongation can often decrease.

The mechanical properties of MMT/PVC nanocomposites have been discussed previously.[8] It is generally concluded that improvements can be achieved at low levels of MMT, but they are relatively modest compared with improvements seen for other polymers. In recent work the use of OMMT with a silane coupling agent has produced impact strengths higher than that of PVC alone when the clay content was $<7$ phr.[77]

Nano-calcium carbonate was found to be effective in increasing toughness in nanocomposites.[37] In order to enhance toughness, specific filler treatments have been used on a number of occasions. Examples include grafted nanotubes[78] and nano-calcium carbonate coated with silicone rubber.[79]

### 8.5.3 Flammability and smoke suppression

Due to the presence of chlorine, rigid PVC does not burn readily but does produce smoke on burning. Flexible PVC burns more readily due to the presence of relatively large quantities of plasticiser (Section 8.2.4). Nanofillers that reduce the amount of smoke produced or make flexible PVC, particularly in applications like cables, more difficult to burn are of significant interest. A number of workers have investigated the use of MMTs to improve the fire performance of PVC.[8] A considerable amount of research has been carried out by Beyer and co-workers[25,80–82] but results remain inconclusive, and, as observed previously, significant discolouration (dehydrochlorination) of the PVC occurs. However, modified flexible PVC nanocomposites also containing halloysite, an aluminosilicate (nanotube-based) filler, were found to exhibit enhanced flame retardancy.[83] It has also been shown using cone calorimetry that flame retardancy and smoke suppression were significantly improved by the addition of OMMT to PVC/wood-flour composites.[84]

Modified OMMTs in which sodium ions were replaced by $Cu^{2+}$ have been shown to have improved flame retardancy.[85,86] The $Cu^{2+}$ ions were also found to be effective in smoke suppression. Fe-modified OMMTs have also shown improved flame retardant properties.[87] A detailed study of OMMT with added zinc, copper and molybdenum oxides has been carried out, and has shown that the metal oxides have a more significant effect than OMMT in improving fire and smoke suppression.[88]

The beneficial effects of hydrotalcite on the thermal stability of PVC have already been discussed. It also appears to improve, very often in a modified form, fire performance. Nanofillers that appear promising include LDH/ZnO combinations, which reduced smoke density and increased the limiting oxygen index.[89,90] Phosphate-modified hydrotalcite has been shown to reduce smoke emissions.[91]

Zinc hydroxystannate ($ZnSn(OH)_6$) is well known as a flame-retardant/smoke suppressant for PVC, and nano-($ZnSn(OH)_6$) has been shown to provide even better performance.[92]

### 8.5.4 Permeability

Nanofillers have been used to reduce gas and liquid permeation through a polymer by increasing the tortuosity of the diffusion path. Hectorite and bentonite clays have been shown to decrease oxygen permeation in flexible PVC by up to 77%. A fivefold improvement in oxygen barrier properties by the addition of OMMT has been reported.[93] PMMA-grafted silica has also been used to reduce oxygen and water permeation.[43] Recent work has investigated the use of OMMTs to reduce plasticiser migration in flexible PVC.[94] It was found that the nanoclay reduced plasticiser migration slightly when compounds were melt compounded, but the effect was considerably greater after dissolving the nanocomposite in tetrahydrofuran, and producing a cast film, because clay dispersion was much better, as discussed in Section 8.4. A novel application of nanofillers is the coextrusion of a carbon nanotube coating to reduce the ingress of moisture into wood-filled composites.[95]

### 8.5.5 Transition temperatures

PVC has three transition temperatures. The broad melting temperature discussed in Section 8.2.1 is generally unaffected by additives, being controlled by the syndiotacticity of the PVC and the processing conditions. The glass transition temperature of rigid PVC is approximately 80 °C, depending on the precise formulation; this reduces to temperatures from below ambient down to temperatures as low as −50 °C as plasticiser concentration increases. PVC also has a $\beta$ transition, detected by dynamic mechanical analysis. This is split into two, with maxima at –50 °C and 0 °C (using a frequency of 110 Hz), attributed to the movement of small chain segments in the amorphous and crystalline regions, respectively.

The effect of nanofillers on PVC transition temperatures has been reported extensively in the literature, with research mainly focussing on the glass transition temperature in rigid PVC. One paper[96] commented on the effects of MMT on the $\beta$ transition in a lightly plasticised PVC, which was found to become broader with a similar peak value as Na-MMT but moved to a lower temperature; this transition also broadened in the presence of OMMT. Reasons for this are unclear because the glass transition temperature was virtually unchanged in these compounds. Mkhabela *et al.*[97] suggested that the PVC melting temperature was increased in carbon nanotube/PVC composites, but the differential scanning calorimetry traces reported do not illustrate PVC melting.

Nanofillers have been reported as increasing,[13,14,22,27,30,33,36,50,77,97–99] decreasing[16,54,100] and having no significant effect[99–101] on the glass transition temperature of rigid PVC, but where changes are reported they generally seem to be no more than 2–3 °C, so probably the latter is nearer the truth in most cases.

Benderly *et al.*[23] observed no change in the glass transition temperature of pPVC with the addition of hectorite and bentonite clays.

The Vicat softening point is a penetrometer technique, and measures the temperature at which a loaded probe penetrates a sample by a specific amount. It is a technological measure of transition temperature and, in an unfilled largely amorphous polymer such as PVC, it would be expected to approximate to the glass transition temperature. In filled polymers, the fillers themselves restrict penetration of the probe, such that Vicat softening points of filled polymers are often much higher than conventional glass transition temperatures. The $T_g$ of rigid PVC is relatively low at 80 °C, so additives that increase this are beneficial. There are significant increases in the Vicat softening point with the addition of Na-MMT[13,102] and nanoscale calcium carbonate.[30]

### 8.5.6 Electrical properties

Multi- and single-walled carbon nanotubes have been used to reduce electrical resistance.[56] Increased electrical conductivity with a very low percolation threshold has been achieved by the use of multi-walled carbon nanotubes.[103,104] PVC/graphite nanosheet/nickel nanocomposites have been developed for electrostatic charge dissipation.[105]

### 8.5.7 Other properties

PVC nanocomposites have been developed to enhance various other properties. Magnetite[106–108] has been used to impart magnetic properties. UV resistance was achieved by using phthalocyanine-modified Laponite clay.[21]

Various commercially available nanofillers have been used to improve specific compressive strength, specific flexural modulus and the density of PVC foams.[109] It has been shown that nanoparticles can increase the orientation of crystalline and amorphous phases in oriented PVC films[110] and modify plastisol rheology.[24] Nanofillers can also be effective during recycling. The properties of recycled PVC were shown to be improved by the addition of nanoclay,[111] and surface-modified nano-calcium carbonate can improve compatibility with polypropylene in mixed plastic waste so that material containing 10–20% polypropylene can be produced.

## 8.6 Opportunities and problems

Mixing PVC with additives, together with an industry very familiar with the issue of formulation, provides an excellent opportunity for introducing nanofillers. However, their dispersion is hampered by the complex morphology of PVC. The areas in which nanofillers are most likely to offer benefits are for impact modification, stabilisation (thermal and UV), smoke suppression and electrical and thermal applications.

## 8.7 Future trends

At present only one commercial PVC compound containing nanofillers has been made available. This product, a paste PVC grade called NanoVin, contained bentonite, and was launched some years ago by Solvin. Unfortunately, despite potential benefits,[24] the market for this product did not develop, but it still remains on standby.

The areas in which PVC nanocomposites are likely to be of most interest in the future are in foams,[112] as smoke suppressants and flame retardants, to improve thermal stability (particularly nanohydrotalcites) and to aid recycling.

## 8.8 Sources of further information and advice

This article provides a thorough coverage of the literature available on PVC nanocomposites. Due to the lack of commercial materials, it is difficult to identify further sources of information at the present time.

## 8.9 References

1. Wang D. Y., Parlow D., Yao Q. and Wilkie C. A. (2001), 'PVC-clay nanocomposites: Preparation, thermal and mechanical properties', *J. Vinyl and Additive Tech.*, **7**, 203–213.
2. Wang D. Y., Parlow D., Yao Q. and Wilkie C. A. (2002), 'PVC-clay nanocomposites: Preparation, thermal and mechanical properties', *J. Vinyl and Additive Tech.*, **8**, 139–150.
3. Allsopp M. W. (1982), 'Mechanism of gelation of rigid PVC' in *Manufacture and Processing of PVC*, R. H. Burgess, Applied Science Publishers, London.
4. Covas J. A., Gilbert M. and Marshall D. E. (1988), 'Twin screw extrusion of a rigid PVC compound – effect on fusion and properties', *Plast. & Rubb. Process. Proc. and Appl.*, **9**, 107.
5. Summers J. W., Rabinovitch E. B. and Booth P. C. (1986), 'Measurement of PVC fusion (gelation)', *J. Vinyl Tech.*, **8**, 6.
6. Patel S. V. and Gilbert M. (1985), 'Effect of processing on the fusion of plasticised PVC', *Plast. and Rubb. Proc. and Appl.*, **5**, 85.
7. The European PVC Industry's Sustainable Development Programme. Vinyl 2010. Progress report 2010.
8. Pagacz J. and Pielichowski K. (2009), 'Preparation and characterization of PVC/montmorillonite nanocomposites – A review', *J. Vinyl and Additive Tech.*, **15**, 61–76.
9. Wang D. Y. and Wilkie C. A. (2002), 'Preparation of PVC-clay nanocomposites by solution blending', *J. Vinyl and Additive Tech.*, **8**, 238–245.
10. Magdaleno L., Schjodt-Thomsen J. and Pinto J. C. (2010), 'Morphology, thermal and mechanical properties of PVC/MMT nanocomposites prepared by solution blending and solution blending plus melt compounding', *Composites Sci. and Tech.*, **70**, 804–814.
11. Aguilar-Solis C., Xu Y. and Brittain W. (2002), 'Polymer-layered silicate nanocomposites by suspension and emulsion polymerizations: PVC-MMT nanocomposites', *Polym. Prepr. (ACMS. Div. Polym. Chem)*, **43**, 1019.
12. Gong F. I., Feng M., Zhao C. G., Zhang S. M. and Yang M. S. (2004), 'Particle configuration and mechanical properties of poly(vinyl chloride)/montmorillonite nanocomposites via *in situ* suspension polymerization', *Polymer Testing*, **23**, 847–853.
13. Pan M. W., Shi X. D., Li M. C., Hu H. Y. and Zhang L. C. (2004), 'Morphology and properties of PVC/clay nanocomposites via *in situ* emulsion polymerisation', *J. Appl. Polym. Sci.*, **94**, 277–286.
14. Gong F. I., Feng M., Zhao C. G., Zhang S. M. and Yang M. S. (2004), 'Thermal properties of poly(vinyl chloride)/montmorillonite nanocomposites', *Polym. Deg. and Stab.*, **84**, 289–294.
15. Hu H. Y., Pan M. W., Li X. C., Shi X. D. and Zhang L. C. (2004), 'Preparation and characterization of poly(vinyl chloride)/organoclay by *in situ* intercalation', *Polymer International*, **53**, 225–231.
16. Yang D. Y., Liu Q. X., Xie X. L. and Zeng F. D. (2006), 'Structure and thermal properties of exfoliated PVC/layered silicate nanocomposites via *in situ* polymerisation', *J. Thermal Analysis and Calorimetry*, **84**, 355–359.
17. Obloj-Muzaj M., Zielecka M., Kozakiewicz J., Abramowicz A., Szulc A. *et al.* (2006), 'Polymerization of vinyl chloride in the presence of nanofillers – effects on the shape and morphology of PVC grains', *Polimery*, **51**, 133–137.
18. Peprnicek T., Kalendova A., Pavlova E., Simonik J., Duchet J. *et al.* (2006), 'Poly(vinyl chloride-paste/clay nanocomposites: Investigation of thermal and morphological characteristics', *Polym. Deg. and Stab.*, **91**, 3322–3329.

19. Zheng X. (2009), 'Development of plasticised PVC/clay nanocomposites', PhD thesis, Loughborough University.
20. Matuana L. M. (2009), 'Rigid PVC/(layered silicate) nanocomposites produced through a novel melt-blending approach', *J. Vinyl and Additive Tech.*, **15**, 77–86.
21. Essawy H. A., El-Wahab N. A. A. and El-Ghaffar M. A. A. (2008), 'PVC-Laponite nanocomposites: Enhanced resistance to UV radiation', *Polym. Deg. and Stab.*, **93**, 1472–1478.
22. Romero-Guzman M. E., Romo-Uribe A., Ovalle-Garcia E., Olayo R. and Cruz-Ramos C. A. (2008), 'Microstructure and dynamic mechanical analysis of extruded layered silicate PVC nanocomposites', *Polymers for Advanced Technologies*, **19**, 1168–1176.
23. Benderly D., Osorio F. and Ijdo W. L. (2008), 'PVC nanocomposites-nanoclay chemistry and performance', *J. Vinyl and Additive Tech.*, **14**, 155–162.
24. Vandevyver E. and Eicholz E. (2008), 'Latest advancements in PVC/clay composites: Potential applications in plastisols', 10th International PVC Conference, Brighton, England.
25. Awad W. H., Beyer G., Benderly D., Ijdo W. L., Songtipya P. *et al.* (2009), 'Material properties of nanoclay PVC composites', *Polymer*, **50**, 1857–1867.
26. Turhan Y., Dogan M. and Alkan M. (2010), 'Poly(vinyl chloride)/kaolinite nanocomposites: Characterization and thermal and optical properties', *Ind. and Eng. Chem. Res.*, **49**, 1503–1513.
27. Xie X. L., Liu Q. X., Li R. X. Y., Zhou X. P., Zhang Q. X. *et al.* (2004), 'Rheological and mechanical properties of PVC/$CaCO_3$ nanocomposites prepared by *in situ* polymerization', *Polymer*, **45**, 6665–6673.
28. Georgiadou S., Thomas N. L., Gilbert M. and Brooks B. W. (2008), 'Suspension polymerisation of vinyl chloride in presence of ultra fine filler particles', *Plastics, Rubber and Composites*, **37**, 431–435.
29. Liu P., Zhao M. and Guo J. (2006), 'Thermal stabilities of poly(vinyl chloride)/calcium carbonate (PVC/$CaCO_3$) composites', *J. Macromol. Sci. Part B: Phys.*, **45**, 1135–1140.
30. Chen N., Wan C. Y., Chang Y. and Zhang Y. X. (2004), 'Effect of nano-$CaCO_3$ on mechanical properties of PVC and PVC/blendex blend', *Polym. Test.*, **23**, 169–174.
31. Wu D. Z., Wang X. D., Song Y. Z. and Jin R. G. (2004), 'Nanocomposites of poly(vinyl chloride) and nanometric calcium carbonate particles: Effects of chlorinated polyethylene on mechanical properties, morphology, and rheology', *J. Appl. Polym. Sci.*, **92**, 2714–2723.
32. Chen N., Wan C. Y., Zhang Y., Zang Y. X. and Zhang C. M. (2005), 'Fracture behavior of PVC/blendex/nano-$CaCO_3$ composites', *J. Appl. Polym. Sci.*, **95**, 953–961.
33. Chen C. H., Teng C. C., Su S. F., Wu W. C. and Yang C. H. (2006), 'Effects of microscale calcium carbonate and nanoscale calcium carbonate on the fusion, thermal, and mechanical characterizations of rigid poly(vinyl chloride)/calcium carbonate composites', *J. Polym. Sci. Part B: Polym. Phys.*, **44**, 451–460.
34. Sun S. S., Li C. Z., Zhang L., Du H. L. and Burnell-Gray J. S. (2006), 'Interfacial structures and mechanical properties of PVC composites reinforced by $CaCO_3$ with different particle sizes and surface treatments', *Polym. Int.*, **55**, 158–164.
35. Zheng X. F., Wang W. Y., Wang G. Q. and Chen J. F. (2007), 'Influence of the diameter of $CaCO_3$ particles on the mechanical and rheological properties of PVC composites', *J. Mat. Sci.*, **43**, 3505–3509.
36. Patil C. B., Kapadi U. R., Hundiwale D. G. and Mahulikar P. P. (2009), 'Preparation and characterization of poly(vinyl chloride) calcium carbonate nanocomposites via melt intercalation', *J. Mat. Sci.*, **44**, 3118–3124.

37. Kemal I., Whittle A., Burford R. and Vodenitcharova T. (2009), 'Toughening of unmodified polyvinylchloride through the addition of nanoparticulate calcium carbonate', *Polymer*, **50**, 4066–4079.
38. Shimpi N. G., Verma J. and Mishra S. (2010), 'Dispersion of nano $CaCO_3$ on PVC and its influence on mechanical and thermal properties', *J. Comp. Mat.*, **44**, 211–219.
39. Zhang L., Luo M. F., Sun S. S., Ma J. and Li C. (2010), 'Effect of surface structure of Nano-$CaCO_3$ particles on mechanical and rheological properties of PVC composites', *J. Macromol. Sci. Part B: Phys.*, **49**, 970–982.
40. Xiong C. X., Liu S. J., Wang D. Y., Dong L. J., Jiang D. D. *et al.* (2005), 'Microporous polyvinyl chloride: Novel reactor for PVC/$CaCO_3$ nanocomposites', *Nanotechnology*, **16**, 1787–1792.
41. Sun S. S., Li C., Zhang L., Du H. L. and Burnell-Gray J. S. (2006), 'Effects of surface modification of fumed silica on interfacial structures and mechanical properties of poly(vinyl chloride) composites', *Europ. Polym. J.*, **42**, 1643–1652.
42. Guo Y. K., Wang M. Y., Zhang H. Q. and Qu H. (2008), 'The surface modification of nanosilica, preparation of nanosilica/acrylic core-shell composite latex, and its application in toughening PVC matrix', *Polym. Eng. Sci.*, **107**, 2671–2680.
43. Zhu A. P., Cai A. Y., Zhang J., Jia H. W. and Wang J. Q. (2008), 'PMMA-grafted-silica/PVC nanocomposites: Mechanical performance and barrier properties', *J. Appl. Polym. Sci.*, **108**, 2189–2196.
44. Zhu A. P., Cai A. Y., Zhou W. D. and Shi Z. H. (2008), 'Effect of flexibility of grafted polymer on the morphology and property of nanosilica/PVC composites', *Appl. Surf. Sci.*, **254**, 3745–3752.
45. Ziong Y., Chen G. S. and Guo S. Y. (2008), 'Solid mechanochemical preparation of core-shell $SiO_2$ particles and their improvement on the mechanical properties of PVC composites', *J. Polym. Sci. Part B: Polym. Phys.*, **46**, 938–948.
46. Zhu A. P., Shi Z. H., Cai A. Y., Zhao F. and Liao T. Q. (2008), 'Synthesis of core-shell PMMA-$SiO_2$ nanoparticles with suspension-dispersion-polymerization in an aqueous system and its effect on mechanical properties of PVC composites', *Polym. Test.*, **27**, 540–547.
47. Van der Ven L., Van Gemert M. L. M., Batenburg L. F., Keern J. J., Gielgens L. H. *et al.* (2000), 'On the action of hydrotalcite-like clay minerals as stabilizers in poly(vinyl chloride)', *Appl. Clay Sci.*, **17**, 25–34.
48. Grossman R. F. (2000), 'Acid absorbers as PVC costabilizers', *J Vinyl and Additive Tech.*, **6**, 4–6.
49. Wang H., Bao Y. Z., Huang Z. M. and Weng Z. X. (2006), 'Morphology and mechanical properties of poly(vinylchloride)/nano-hydrotalcite composites', *Acta Polymerica Sinica*, **44**, 451–461.
50. Bao Y. Z., Huang Z. M. and Weng Z. X. (2006), 'Preparation and characterization of poly(vinyl chloride)/layered double hydroxides nanocomposite via *in situ* suspension polymerization', *J. Appl. Polym. Sci.*, **102**, 1471–1477.
51. Bao Y. Z., Huang Z. M. and Weng Z. X. (2008), 'Thermal stability, smoke emission and mechanical properties of poly(vinyl chloride)/hydrotalcite nanocomposites', *Polym. Degrad. and Stab.*, **93**, 448–455.
52. Chen G. M. and Chen G. (2007), 'Preparation of a poly(vinyl chloride)/layered double hydroxide nanocomposite with a reduced heavy-metal thermal stabilizer', *J. Appl. Polym. Sci.*, **106**, 817–820.

53. Liu J., Chen G. and Yang J. (2008), 'Preparation and characterization of poly(vinyl chloride)/layered double hydroxide nanocomposites with enhanced thermal stability', *Polymer*, **49**, 3923–3927.
54. Huang N. H. and Wang J. Q. (2009), 'A new route to prepare nanocomposites based on polyvinyl chloride and MgAl layered double hydroxide intercalated with laurylether phosphate', *Express Polym. Lett.*, **3**, 595–604.
55. Liu J., Chen G. M., Yang J. P. and Ding L. P. (2009), 'Improved thermal stability of poly(vinyl chloride) by nanoscale layered double hydroxide particles grafted with toluene-2,4-di-isocyanate', *Mat. Chem. Phys.*, **118**, 405–409.
56. Broza G., Piszczek K. and Styerzynski T. (2005), 'Nanocomposites of poly(vinyl chloride) with carbon nanotubes (CNT)', *Compos. Sci. Tech.*, **67**, 890–894.
57. Xie X. L., Li R. K. Y., Liu Q. X. and Mai Y. W. (2004), 'Structure-property relationships of *in situ* PMMA modified nano-sized antimony trioxide filled poly(vinyl chloride) nanocomposites', *Polymer*, **45**, 2793–2802.
58. Soong S. Y., Cohen R. E., Boyce M. C. and Mulliken A. D. (2006), 'Rate-dependent deformation behavior of POSS-filled and plasticized poly(vinyl chloride)', *Macromolecules*, **39**, 2900–2908.
59. Gao J., Du Y. and Dong C. (2010), 'Rheological behavior and mechanical properties of blends of poly(vinylchloride) with CP-POSS', *Int. J. Polymeric Mat.*, **39**, 15–24.
60. Shimpi N. G., Verma J. and Mishra S. (2009), 'Preparation, characterization and properties of poly(vinylchloride)/CaSO4 nanocomposites', *Polym.-Plast. Tech. Eng.*, **48**, 997–1001.
61. Patil C. B., Shisode P. S., Kapadi U. R., Hundiwale D. G. and Mahulikar P. P. (2010), 'Effect of calcium sulphate nanoparticles on fusion, mechanical and thermal behaviour of polyvinyl chloride (PVC)', *Int. J. Modern Phys.*, **24**, 64–75.
62. Patil C. B., Shisode P. S., Kapadi U. R., Hundiwale D. G. and Mahulikar P. P. (2010), 'Preparation and characterization of poly(vinyl chloride) calcium phosphate nanocomposites', *Mat. Sci. Eng. B – Advanced Functional Solid State Mat.*, **168**, 231–236.
63. Li X., Chen W., Xing Y. and Zhang P. (2010), 'Effect of ZnO nanoparticles on the UV light fastness and climate resistance of PVC film', *Manufacturing Sci. Eng.*, **97–101**, 2197–2200.
64. Faruk O. and Matuana L. M. (2008), 'Reinforcement of rigid PVC/wood-flour composites with multi-walled carbon nanotubes', *J. Vinyl Additive Tech.*, **19**, 60–64.
65. Jia M. Y., Xue P., Zhao Y. S. and Wang K. J. (2009), 'Creep behaviour of wood flour/poly(vinyl chloride) composites', *J. Wuhan University of Tech.-Materials*, **24**, 440–447.
66. Rodolfo A. and Innocentini-Mei L. H. (2010), 'Poly(vinyl chloride)/metallic oxides/organically modified montmorillonite nanocomposites: Preparation, morphological characterization, and modeling of the mechanical properties', *J. Appl. Polym. Sci.*, **116**, 422–432.
67. Thongpin C., Juntum J., Sa-Nguan-Moo R., Siksa-Ard A. and Sombatsompop N. (2010), 'Thermal stability of PVC with gamma-APS-g-MMT and zeolite stabilizers by TGA technique', *J. Thermoplast. Comp. Mat.*, **23**, 435–445.
68. Zheng X. and Gilbert M. (2011), 'An investigation into the thermal stability of PVC/MMT composites', *J. Vinyl Additive Tech.* **17**, 77–84.
69. Wan C. Y., Tian G. H., Cui N., Zhang Y. X. and Zhang Y. (2004), 'Processing thermal stability and degradation kinetics of poly(vinylchloride)/montmorillonite composites', *J. Appl. Polym. Sci.*, **92**, 1521–1526.

70. Wu B., Qi S. H. and Wang X. (2010), 'Thermal behaviour of poly(vinyl chloride) treated montmorillonite-silica-3-triethoxysilyl-1-propanamine (K-Si-MMT) nanocomposites', *Polym. Test.*, **29**, 717–722.
71. Sterky K., Hjertberg T. and Jacobsen H. (2009), 'Effect of montmorillonite treatment on the thermal stability of poly(vinyl chloride) nanocomposites', *Polym. Deg. Stab.*, **94**, 1564–1570.
72. Yarahmadi N., Jakubowicz I. and Hjertberg T. (2010), 'Development of poly(vinyl chloride)/montmorillonite nanocomposites using chelating agents', *Polym. Deg. Stab.*, **95**, 132–137.
73. Chen T. Y., Li W. F., Jun J. D., Pen J. H. and Chao J. X. (2010), 'Modification of nanometre calcium carbonate and its application on PVC composites *in situ* suspension polymerisation', *Mat. Sci. Tech.*, **26**, 871–874.
74. Gu Z., Liu W., Dou W. and Tang F. (2010), 'Preparation of a novel heat stabilizer for poly(vinyl chloride)-Zn, Mg, Al-layered double hydroxide', *Polym. Comp.*, **31**, 928–932.
75. Liu J., Chen G. M., Yang J. and Ding L. (2010), 'Thermal stability of poly(vinyl chloride)/ layered double hydroxide nanocomposites', *J. Appl. Polym. Sci.*, **116**, 2058–2064.
76. Yarahmadi N., Jakubowicz I. and Hjertberg T. (2010), 'Development of poly(vinyl chloride)/montmorillonite nanocomposites using chelating agents', *Polym. Deg. Stab.*, **95**, 132–137.
77. Ge M. L. and Jia D. M. (2008), 'Influence of organoclay prepared by solid state method on the morphology and properties of polyvinyl chloride/organoclay nanocomposites', *J. Elast. Plast.*, **40**, 223–235.
78. Shi J. H., Yang B. X., Pramoda K. P. and Goh S. H. (2007), 'Enhancement of the mechanical performance of poly(vinyl chloride) using poly(n-butyl methacrylate)-grafted multi-walled carbon nanotubes', *Nanotechnology*, **18**, 19.
79. Yang L., Hu Y., Guo H., Song L., Chen Z. Y. *et al.* (2006), 'Toughening and reinforcement of rigid PVC with silicone rubber/nano-$CaCO_3$ shell-core structured fillers', *J. Appl. Polym. Sci.*, **102**, 2560–2567.
80. Beyer G. (2007), 'Flame retardancy of thermoplastic polyurethane and polyvinyl chloride by organoclays', *J. Fire Sciences*, **25**, 65–78.
81. Beyer G. (2008), 'Organoclays as flame retardants for PVC', *Polymers for Adv. Tech.*, **19**, 485–488.
82. Beyer G. (2008), 'Organoclays as flame retardants for PVC and new application of nanocomposites for simplifying flame retardant cable designs', *PMSE Preprints*, **98**, 822–823.
83. Beyer G. (2007), 'PVC nanocomposites and new nanostructured flame retardants.' *Proc. 13th Int. Plastics and Additives and Compounding Conf.*, Addcon World, Frankfurt, Germany, Paper 19.
84. Zhao Y., Wang K., Zhu F., Xue P. and Jia M. (2006), 'Properties of poly(vinyl chloride)/ wood flour/montmorillonite composites: Effects of coupling agents and layered silicate', *Polym. Deg. Stab.*, **91**, 2874–2883.
85. Yang Z., Li B. and Tang F. (2007), 'Influence of $Cu^{2+}$-organic montmorillonites on thermal decomposition and smoke emission of poly(vinyl chloride) by cone calorimetric study', *J. Vinyl Addit. Tech.*, **13**, 31–39.
86. Li B. and Yang Z. (2009), 'An Investigation of the flammability, morphology and torque rheology of poly(vinyl chloride) with silanes and $Cu^{2+}$ modified montmorillonites', *Polym. and Polym. Comp.*, **17**, 291–301.

87. Kong Q., Zhang J., Ma J., Li F., Liu H. *et al.* (2008), 'Flame retardant and smoke suppressant of Fe-organophilic montmorillonite in polyvinyl chloride nanocomposites', *Chinese J. Chem.*, **26**, 2278–2284.
88. Rodolfo A. and Innocenti-Mei L. H. (2010), 'Poly(vinyl chloride)/metallic oxides/ organically modified montmorillonite nanocomposites: Fire and smoke behaviour', *J. Appl. Polym. Sci.*, **116**, 946–958.
89. Zhang Z., Zhu M., Sun B., Zhang Q., Yan C. *et al.* (2006), 'The effect of hydrotalcite and zinc oxide on smoke suppression of commercial rigid PVC', *J. Macromol. Sci., Pt. A: Pure and Appl. Chem.*, **43**, 1807–1814.
90. Yan C., Zhang Z., He L., He Z. and Zhang Z. (2007), 'The effect of nano-hydrotalcite composites on smoke suppression and flame retardant of flexible PVC', *Suliao*, **36**, 8–11 and 93.
91. Bao Y., Huang Z., Li S. and Weng Z. (2008), 'Thermal stability, smoke emission and mechanical properties of poly(vinyl chloride)/hydrotalcite nanocomposites', *Polym. Deg. Stab.*, **93**, 448–455.
92. Zhang Y., Li B., Xu X., Li Y., Wu Z. *et al.* (2007), 'Influences of ($ZnSn(OH)_6$) on flame retardancy and smoke suppression of flexible poly(vinyl chloride)', *YingyongHuaxue*, **24**, 286–290.
93. Francis N. and Schmidt D. F. (2007), 'PVC/layered silicate nanocomposites: Preparation, characterization, and properties', *SPE ANTEC 2007, Proc. 65th SPE Annual Conference*, Cincinnati.
94. Zheng X. and Gilbert M. (2011), 'The effect of processing on the structure of PVC/ montmorillonite composites. Melt and solution process', *J. Vinyl Addit. Tech.*, **17**, 231–238.
95. Jin S. and Matuana L. M. (2010), 'Wood/plastic composites co-extruded with multi-walled carbon nanotube-filled rigid poly(vinyl chloride) cap layer', *Polym. Int.*, **59**, 648–657.
96. Wan C. Y., Qiao X. Y., Zhang, Y. and Zhang Y. X. (2003), 'Effect of different clay treatment on morphology and mechanical properties of PVC-clay nanocomposites', *Polymer Testing*, **22**, 453–461.
97. Mkhabela V. J., Mishra A. K., Mbianda X. Y., Mkhabela, V. J., Mishra A. K. *et al.* (2011), 'Thermal and mechanical properties of phosphorylated multiwalled carbon nanotube/polyvinyl chloride composites', *Carbon*, **49**, 610–617.
98. Elashmawi I. S., Hakeem N. A., Marei L. K. and Hanna, F. F. (2010), 'Structure and performance of ZnO/PVC nanocomposites', *Physica B: Condensed Matter*, **405**, 4163–4169.
99. Sterzynskia T., Tomaszewska J., Piszczek K. and Skorczewska K. (2010), 'The influence of carbon nanotubes on the PVC glass transition temperature', *Composite Science and Technology*, **70**, 966–969.
100. Xu W. B., Zhou Z. F., Ge M. L. and Pan W. P. (2004), 'Polyvinyl chloride/ montmorillonite nanocomposites – Glass transition temperature and mechanical properties', *J. Thermal Anal. & Calorimetry*, **78**, 91–99.
101. Ren T. B., Yang J., Huang Y. X., Ren J. and Liu Y. (2006), 'Preparation, characterization, and properties of poly(vinylchloride)/organophilic-montmorillonite nanocomposites', *Polymer Composites*, **27**, 55–64.
102. Shi X. D., Pan M. W., Li X. C., Zhang, L. C. and Ding, H. L. (2004), 'Studies on the morphology and properties of PVC/Na+-MMT nanocomposites prepared by *in situ* emulsion polymerization', *Acta Polymerica Sinica*, **1**, 149–1520.

103. Mamunya Y., Bodenne A., Lebovka N., Ibos L., Candeau Y. *et al.* (2008), 'Electrical and thermophysical behaviour of PVC-MWCNT nanocomposites', *Comp. Sci. Tech.*, **68**, 1981–1988.
104. Mamunya Y. P., Levchenko V. V., Rybak A., Boiteux G., Lebedev E. V. *et al.* (2010), 'Electrical and thermomechanical properties of segregated nanocomposites based on PVC and multiwalled carbon nanotubes', *J. Non-crystalline Sol.*, **356**, 635–641.
105. Al-Ghamdi A. A., El-Tantawy F., Aal N. A., Mossalamy E. H. and Mahmoud W. E. (2009), 'Stability of new electrostatic discharge protection and electromagnetic wave shielding effectiveness from poly(vinyl chloride)/graphite/nickel nanoconducting composites', *Polym. Degrad. Stab.*, **94**, 980–986.
106. Yanez-Flores I. G., Betancort-Galindo R., Aquino J. A. M. and Rodriguez-Fernandez O. S. (2007), 'Preparation and characterization of magnetic PVC nanocomposites', *J. Non-Cryst. Sol.*, **353**, 799–801.
107. Rodriguez-Fernandez O. S., Rodriguez-Calzadiaz C. A., Yanez-Flores I. G. and Montemayor S. M. (2008), 'Preparation and characterization of a magneto-polymeric nanocomposite: $Fe_3O_4$ nanoparticles in a grafted, cross-linked and plasticized poly(vinyl chloride) matrix', *J. Mag. Mag. Mat.*, **320**, E81–E84.
108. Servin-Hernandez E., Rodriguex-Fernandez O. S. and Garcia-Cerda L. A. (2010), 'Synthesis of plasticizer-based ferrofluid and its use in the preparation of magnetic PVC nanocomposites', *Adv. Electron Microscopy and Nanomaterials*, **644**, 13–16.
109. Alian A. M. and Abu-Zahra N. H. (2009), 'Mechanical Properties of Rigid Foam PVC-Clay Nanocomposites', *Polym. Plast. Tech. Eng.*, **48**, 1014–1019.
110. Yalcin B. and Cakmak M. (2005), 'Molecular orientation behavior of poly(vinyl chloride) as influenced by the nanoparticles and plasticizer during uniaxial film stretching in the rubbery stage', *J. Polym. Sci. Part B: Polym. Phys.*, **43**, 724–742.
111. Yoo Y., Kim S. S., Won J. C., Choi K. Y. and Lee J. H. (2004), 'Enhancement of the thermal stability, mechanical properties and morphologies of recycled PVC/clay nanocomposites', *Polym. Bull.*, **52**, 373–380.
112. Thomas N. L. (2007), 'Cellular PVC-U: Current technology and future challenges', *J. Cell. Plast.*, **43**, 237–255.

# 9
# Nylon-based polymer nanocomposites

S.-H. LIM, Singapore Institute of Manufacturing Technology (Simtech), Singapore and A. DASARI, Nanyang Technological University (NTU), Singapore and Madrid Institute for Advanced Studies of Materials (IMDEA Materials Institute), Spain

**Abstract:** This chapter highlights fundamental and recent major developments in the application of various nanoparticles in the design and performance of nylon-based materials. Specifically, it deals with techniques employed to prepare these nanocomposites, their physico-chemical characterization, mechanical and functional properties as well as applications and future prospects.

**Key words:** polymer nanocomposites, nanoparticles, nylon, barrier, toughness.

## 9.1 Introduction

Nylon, a family of synthetic polymers known generically as polyamide (or aliphatic polyamide), is one of the most commonly used engineering plastics. It is a semi-crystalline polymer, which was used to replace silk in military applications such as parachutes, flak vests, and many types of vehicle tire during World War II. It also replaced many mechanical and structural parts previously cast in metal, such as machine screws, gears, and other low- to medium-stress components. To further improve the mechanical properties of these materials, microscale particles like glass/carbon fibers, talc, and wollastonite are generally incorporated. But, due to the increasing emphasis on 'multi-functionality,' nanostructured materials have taken the limelight. Polymer nanocomposites are a good example of this class of nanostructured materials with the potential to exhibit unique combinations of mechanical, physical, optical, and thermal properties at relatively lower particle loadings than those in typical traditional composites. Moreover, many inherent properties of polymers, such as crystal structure, melt viscosity, and glass transition, are affected by the incorporation of nanoparticles. Hence, understanding the processing/structure/property relations in polymer nanocomposites is crucial for fully exploiting the fundamental characteristics of nanoparticles (large surface area and aspect ratio) in achieving combinatorial properties as well as improving the magnitude of property enhancements.

Over the last two decades, the study of these nanocomposite materials has been a mushrooming field of research yielding innovative advanced materials with high added value. This chapter highlights the major developments and applications of various nanoparticles in the design and performance of nylon-based materials.

The various techniques employed to prepare these nanocomposites, their physico-chemical characterization, mechanical and functional properties as well as future prospects will be outlined.

## 9.2 Types of nanoparticle and their modification

All nanoparticles have at least one dimension smaller than 100 nm and can be classified as one-, two-, or three-dimensional depending on their size and the number of dimensions at the nanoscale;[1] that is, nanoplatelets are treated as one-dimensional, nanotubes two-dimensional, and equiaxed nanoparticles three-dimensional, where all three dimensions are less than 10 nm. Schematic representations along with representative examples are given in Table 9.1.

In the development of polymer nanocomposites, the main challenge is to disperse the nanoparticles homogeneously and individually within the matrix because of their high tendency to agglomerate. Equally challenging is to control the interactions between these nanoparticles and the polymer matrix via physical interaction or chemical bonding. The extent of dispersion and interfacial interactions with the matrix will affect the inherent properties of the polymer and determine the magnitude of ultimate property improvement. So, the first step is to improve the dispersion of the nanoparticles in the matrix. Creating high shear forces during processing of polymer nanocomposites was considered as a way of breaking up the strongly bound aggregates of nanoparticles. But it was found that, if the matrix polymer lacks sufficient

*Table 9.1* Different classes of nanoparticles along with representative examples

| Size and dimension | Schematic representation | Examples |
|---|---|---|
| a, b, c: 1–100 nm<br>3-dimensional | a, b, c | Silica, carbon black, fullerenes |
| a, b: 1–100 nm<br>c > 100 nm<br>2-dimensional | a, b | Carbon nanotubes, metallic nanowires, sepiolite nanofibers |
| c: 1–100 nm<br>a, b > 100 nm<br>1-dimensional | c | Smectites, kaolinites, layered double hydroxides |

Source: Adapted from Ruiz-Hitzky and Van Meerbeek (2006).[1]

compatibility with the nanoparticles, stress alone cannot achieve fine dispersion of the nanoparticles.[2] Therefore, depending on the matrix–filler combination, their compatibility has to be enhanced by functionalizing the fillers with organic surfactants (and other chemical agents). For this purpose, both physical and chemical methods have been used, such as the surface treatment of nanoparticles and the addition of coupling or grafting agents and compatibilizers. The following paragraphs will describe the various prevalent and potential approaches adopted in the surface treatment of nanoparticles, with representative examples.

### 9.2.1 Three-dimensional (3-D) nanoparticles

During the last decade, many three-dimensional (3-D) nanoparticles including $Al_2O_3$,[3–5] zinc oxide (ZnO),[6,7] calcium carbonate ($CaCO_3$),[8–12] and $SiO_2$[13–18] have been incorporated into nylon matrices either as preformed nanoparticles or synthesized *in situ* by the sol-gel process. The former is common for nylon/$CaCO_3$ nanocomposites, in which the nanoparticles are often surface coated with saturated or unsaturated fatty acids like stearic acid, aminocaproic acid, or amino acid.[8–11] The basic idea is to reduce the surface energy of the $CaCO_3$ nanoparticles and the particle–particle interaction, thus facilitating dispersion of the nanoparticles. But it should be noted that, in general, this surface treatment simultaneously weakens the particle–matrix interaction. However, by varying the number of reactive or functional groups in the fatty acid, the extent of the particle–matrix interfacial interaction can be tuned. For instance, amino acid and aminocaproic acid, which are more reactive compared with stearic acid, have a stronger interaction with the nylon matrix.[11,12] In stearic acid, the carboxylic (–COOH) group at one end of the molecule forms a strong ionic bond with the calcium in the $CaCO_3$, while the $-CH_3$ non-polar group at the other end of the molecule interacts far less strongly with the nylon matrix, resulting in a much weaker particle–matrix interaction. In amino acid and aminocaproic acid, there are more functional groups to interact with $CaCO_3$ and the nylon matrix (Fig. 9.1). In fact, even in the absence of any surface treatment, the interaction between $CaCO_3$ nanoparticles and the nylon matrix is sufficient to provide a slight improvement in the mechanical properties. However, the agglomeration of untreated $CaCO_3$ nanoparticles in the nylon matrix inevitably affects many other properties required of the composite.

Surface treatment with organosilane (also termed silanization) is another popular strategy in the modification of nanoparticles. Organosilane is a silicon-based chemical compound with both organic- and inorganic-compatible functionalities within the same molecule. One end of the molecule bears a hydrolyzable group such as alkoxy or acetoxy while the other end of the molecule has an organo-functional group like epoxy or amine. The inorganic compatibility comes from the alkoxy groups and the organic compatibility stems from the

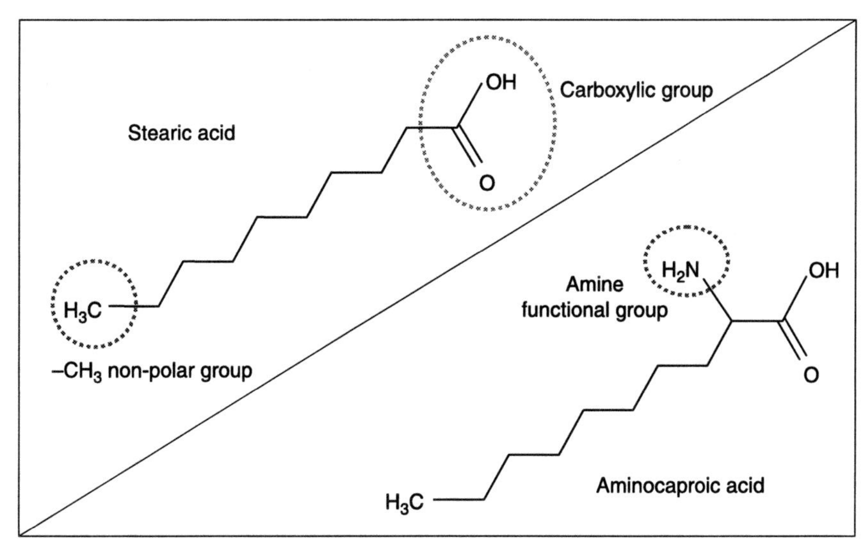


*9.1* Chemical structure of stearic acid and aminocaproic acid with an additional amine functional group.

organo-functional group. Silanization is commonly used in the surface modification of $SiO_2$ nanoparticles[13–16] and, during silanization, the alkoxy group on a silane molecule reacts with an hydroxyl group on the $SiO_2$ nanoparticle surface, thus forming stable covalent –Si–O–Si– bonds. Typical silane-coupling agents that have been used in the preparation of nylon/$SiO_2$ nanocomposites are γ-aminopropyl-triethoxysilane (APS),[15] hexamethyl-disilazane (HMDS),[15] and γ-glycidoxypropyl-trimethoxysilane (GPS).[16] Silanization of silica nanoparticles can be carried out *in situ*[13,15] or pretreated.[14,16]

*In situ* surface modification of $SiO_2$ involves condensation polymerization, in which the hydrolysis product, sodium metasilicate, is used as a monomer and the organosilane is used as a chain terminator. First, sodium metasilicate (the monomer) is hydrolyzed to form silicic acid in the presence of hydrochloric acid. Condensation polymerization takes place by anhydration during hydrolysis of silicic acid. The Si and O atoms bond to each other to form three-dimensional tetrahedral network structures. Large numbers of hydroxyls remain on the surface of the silica nanoparticles. At this point, an organosilane molecule like APS is introduced and the silanol groups of APS generated by hydrolysis can react rapidly with the hydroxyl groups of $SiO_2$ to form a modified layer on the silica surface. The organic chains substitute for a portion of the active groups on the $SiO_2$ surface, thus resulting in steric hindrance. This prevents the $SiO_2$ from continuously growing or agglomerating. By controlling the reaction conditions, nano-$SiO_2$ particles capped with different organic compound(s) can then be obtained. For the surface treatment for preformed nano-$SiO_2$ particles, the process is much easier. An organosilane solution, in ethanol[14] or toluene,[16] is first prepared, followed by the addition of $SiO_2$ nanoparticles under rapid stirring. The treated nanoparticles are collected and dried for the subsequent preparation of nanocomposites.

### 9.2.2 Two-dimensional (2-D) nanoparticles

Examples of two-dimensional (2-D) nanoparticles are carbon nanotubes (CNTs), halloysite nanotubules, ZnO nanofibres, etc. In the following discussion, we will only consider CNTs as they are an ideal reinforcing agent for high-strength multi-functional polymer composites because of their exceptionally high Young's modulus ~1 TPa,[19] which is comparable to that of diamond ~1.2 TPa. Moreover, CNTs have unique structural and physical properties: electronically they can be either metallic or semi-conductive depending on their geometrical structure; good ballistic transport properties; extremely high thermal conductivity; and high optical polarizability. Depending on their structural configuration, CNTs are either single-walled carbon nanotubes (SWCNTs) or multi-walled carbon nanotubes (MWCNTs). SWCNTs are formed from only a single layer of a graphene sheet while MWNTs consist of several layers of coaxial graphene sheets.

As with other types of nanoparticle, a critical issue in taking advantage of the superior properties of CNTs is the ability to disaggregate and control their dispersion in the polymer as well as the molecular interaction with the matrix. As such, various modification strategies for CNTs have been established. One strategy involves the physical adsorption of surfactants, biomacromolecules, or polymers on the surface of the CNTs so that these functionalized CNTs can be dispersed easily in a wide variety of polar/non-polar solvents and polymer matrices. Anionic surfactants like sodium dodecyl sulfate (SDS)[20] and sodium dodecylbenzene sulfonate (NaDDBS),[21,22] cationic surfactants such as hexadecyltrimethylammonium bromide (CTAB),[23] and non-ionic surfactants like triton-X100[24] have been used to decrease the aggregation of CNTs and improve solubility. In general, the interaction between the surfactants and the CNTs depends on the nature of the surfactants, which includes alkyl chain length, head-group size, and charge. This is clearly exemplified when comparing the surfactants SDS and NaDDBS, which has a benzene ring. The $\pi$-stacking interaction of the benzene rings on the surface of the CNTs increases the binding and surface coverage of NaDDBS surfactant molecules significantly.[22] Besides using surfactants, non-covalent functionalization of CNTs has also been accomplished with the use of polymers, which physically wrap around the CNTs, forming supramolecular complexes.[25–30] In these cases, the graphitic sidewalls of the CNTs provide the possibility for $\pi$-stacking interactions with organic polymers. CNTs can be wrapped or encapsulated with poly(p-phenylenevinylene-co-2,5-dioctyloxy-m-phenylenevinylene)[26–28] and amphiphilic poly(styrene)-block-poly(acrylic acid)[29] copolymers. Encapsulation significantly enhances the dispersion of modified CNTs in different solvents and polymer matrices because of the permanently fixed copolymer shell. However, with this type of non-covalent surface modification, the forces between the wrapping molecule and the CNTs might be weak and, when acting as a filler in a polymer composite, the efficiency of load transfer is low.

Another strategy for CNT modification is the covalent grafting of polymers to surface-bound carboxylic acid groups (–COOH) or hydroxyl groups (–OH) on CNTs. Typically, these functional groups are created on CNT surfaces during oxidation by oxygen, air, concentrated sulfuric acid, nitric acid, aqueous hydrogen peroxide, or an acid mixture. Depending on the acid treatment temperature and time, oxidation procedures and oxidizing agents, the extent of the induced –COOH and –OH functionality differs.[30,31] Their presence on a nanotube surface helps the attachment of organic or inorganic materials, which enhance the dispersion and solubility of functionalized CNTs in solvents and in polymer matrices. This is because these functional groups are very reactive and a number of chemical reactions can occur with these groups. Covalent functionalization takes place either through the attachment of as-prepared or commercially available polymer molecules onto the CNT surface via amidation, esterification, or radical coupling ('grafting to') or through the *in situ* polymerization of monomers onto the reactive surfaces of a CNT with the aid of active compounds (initiators or comonomers) via radical, cationic, anionic, ring-opening, and condensation polymerization ('grafting from').[32] This approach has been used successfully to graft many polymers, including nylon, onto CNT surfaces.[33,34] Grafting of CNTs with nylon-6 can be achieved in a two-step process.[34] The first step is the formation of the acyl caprolactam initiator by the addition of ε-caprolactam to isocyanate-functionalized CNTs. The second step is the formation of polymer covalently functionalized CNTs in which the nanotube-bound acyl caprolactam initiates anionic ring-opening polymerization of ε-caprolactam in the presence of sodium caprolactamate as a catalyst. This approach is highly efficient and the resulting functionalized CNT surfaces are uniformly coated with nylon-6.

### 9.2.3 One-dimensional (1-D) nanoparticles

In contrast to the 3-D and 2-D nanofillers discussed earlier, plate-like or layer structured nanofillers are one-dimensional (1-D) with a thickness of, typically, a few nanometers while the other two dimensions are in the sub-micron range. But, as with other nanofillers, layered fillers are difficult to disperse individually in polymers due to the large contact area between the particles and the enhanced particle–particle interaction. As a result, most of the layered fillers (and in particular silicates) exist naturally as stacks and hundreds or thousands of these layers are stacked together by van der Waals forces (and, in some cases, electrostatic forces). Nevertheless, this configuration provides an excellent opportunity for fine-tuning their surface chemistry through ion exchange reactions with organic and inorganic cations. This in turn leads to many possibilities and prospects for tailoring various properties required for specific end applications, other than just merely dispersing the layers in a polymer matrix.

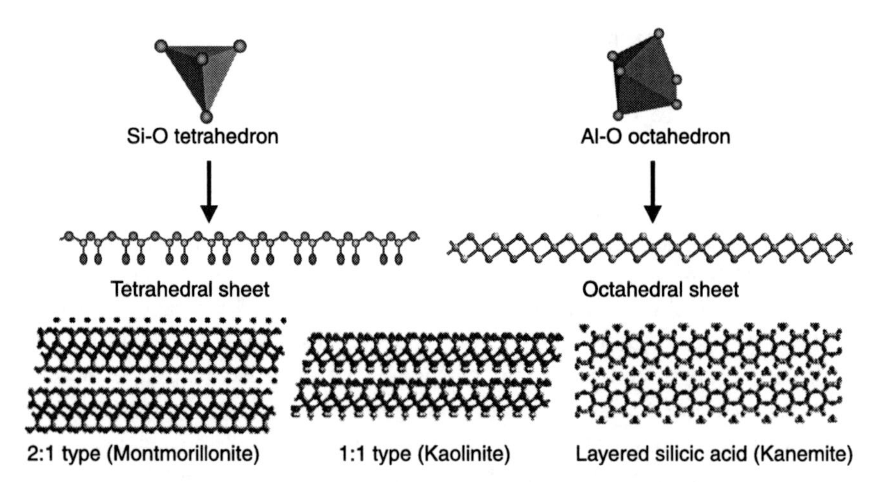


*9.2* Classification of clay minerals and their structures. Reprinted with permission from Zeng *et al.*[35] Copyright (2005) American Scientific Publishers.

The clay minerals used for polymer nanocomposites can be classified into three groups, as schematically shown in Fig. 9.2:[35] the 2:1 type, the 1:1 type, and layered silicic acids. Of these, the most commonly used for the preparation of polymer nanocomposites belongs to the family of 2:1 phyllosilicates and, in particular, smectite clays because of their swelling properties and high cation exchange capacities. For a 2:1 type clay, the crystal structure is made up of layers of two tetrahedrally coordinated silicon atoms fused to an edge-shared octahedral sheet of either aluminum or magnesium hydroxide. The layer has a thickness ~0.94 nm and lateral dimensions varying from 30 nm to several microns, depending on the particular layered silicate (for example, saponite ~50–60 nm, montmorillonite ~100–150 nm, and hectorite ~200–300 nm), thus providing a large surface area ~750 $m^2/g$.[36] These layers organize themselves into stacks interspaced with interlayer or intra-gallery space leading to a regular van der Waals gap between the layers. Isomorphic substitution within the layers (for instance, $Al^{3+}$ replaced by $Mg^{2+}$ or by $Fe^{2+}$, or $Mg^{2+}$ replaced by $Li^{+}$) generates a net negative charge. If the layers are negatively charged, this charge is generally counterbalanced by alkali or alkaline earth cations located in the interlayer, such as $Na^{+}$ and $K^{+}$. Therefore, this type of layered silicate is characterized by a moderate negative surface charge, termed the cation exchange capacity (CEC).

The CEC of layered silicates is an important factor during the synthesis of polymer nanocomposites as it determines the amount of cationic surfactants that can be intercalated into the layered silicate intra-gallery through ion-exchange reactions to render the hydrophilic phyllosilicates more organophilic and compatible with organic polymers. Typical cationic surfactants for layered silicates include, but are not limited to, primary, secondary, tertiary, and quarternary alkylammonium or alkylphosphonium cations.[37] These organic cations not only diminish the surface energy of the layered silicates and improve the wetting characteristics of the polymer matrix, but also increase the interlayer spacing of the layered silicates. The basal spacing of pristine clay layers depends on the type of clay; for example, pristine montmorillonite has a basal spacing of ~1.7 nm. Upon modification with organic cationic surfactants, the spacing (interlayer or intra-gallery) could be in the range ~1.9–4.0 nm, resulting in an organically modified clay (Fig. 9.3).[38] However, the space between the silicate layers depends greatly on the length of the alkyl chain and the ratio of cross-sectional area to available area per cation.[36,39,40]

Apart from the above-mentioned alkylammonium-based surfactants, silanization was also adopted to improve the compatibility and the interfacial interaction between nylon-6 and layered silicates.[41] Organosilanes with various functional groups were used by Kim *et al.*,[41] i.e., APS, GPS, and 3-isocyanate-propyltriethoxysilane (IPS). The experimental results indicated that epoxy functional groups on GPS were more effective in terms of enhancing the nanoparticle–matrix interaction than amino groups on APS.

Further, with non-polar polymers, proper functionalization of the polymer matrix or the addition of compatibilizers is necessary to enhance compatibility

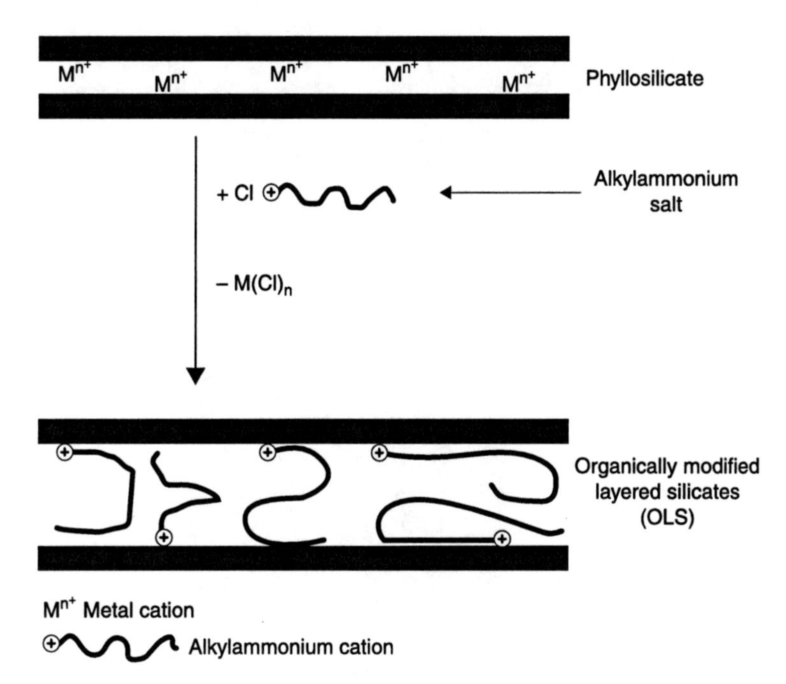


*9.3* Cation-exchange reaction between silicate and an alkylammonium salt. Reprinted with permission from Zanetti *et al.*[38] Copyright (2000) John Wiley & Sons Inc.

with the organoclay. For example, with polypropylene (PP) or polyethylene (PE), maleic anhydride grafted polypropylene or polyethylene (PP-g-MA or PE-g-MA) is added, respectively, to improve the dispersion and swelling of the clay layers. The underlying reason is attributed to the possible interaction via hydrogen bonding between the oxygen group of the silicates and the grafted maleic anhydride group, which would assist the desired nanoscale dispersion of the organoclay in the polypropylene or polyethylene matrices.[36,42–44] Figure 9.4 shows

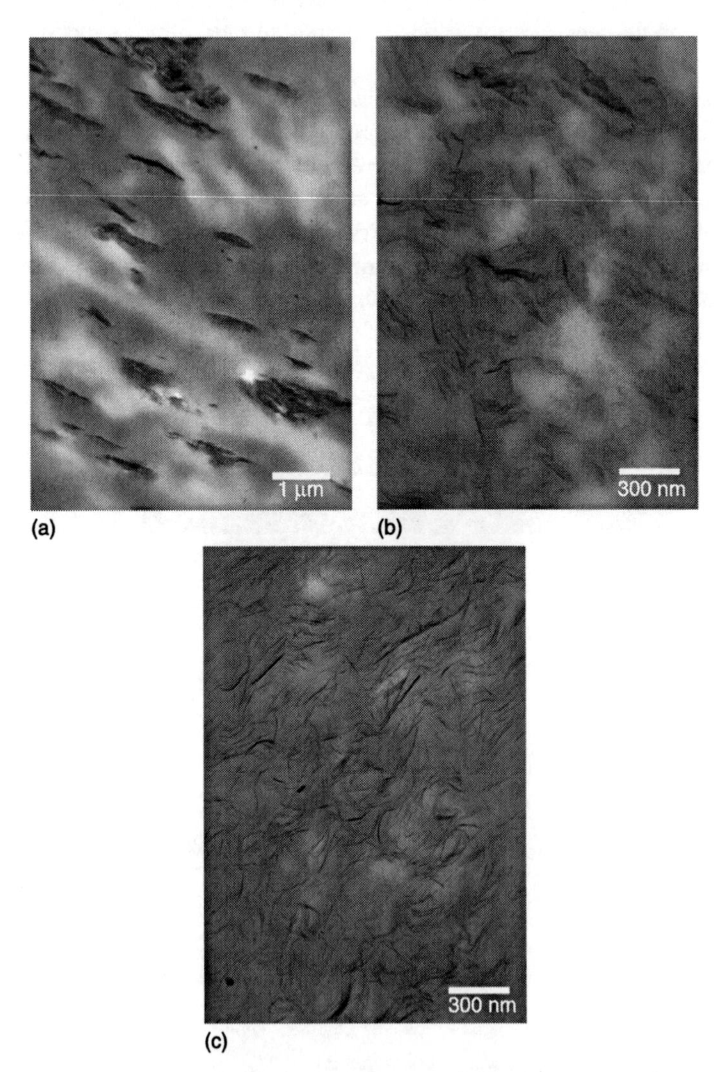


*9.4* Transmission electron micrographs (TEMs) showing (a) poor compatibility of the organoclay layers with the polypropylene matrix resulting in agglomerates and good dispersion of the organoclay layers in a mixture of polypropylene and PP-g-MA with (b) 50 wt% and (c) 100 wt% PP-g-MA.

the differences in dispersion and exfoliation of organoclay in polypropylene without and with various contents of PP-g-MA.[45]

Several other approaches for the surface modification of nanoparticles have been established but are not discussed here. Although inherently different nanoparticle surface-modified technologies have been developed, they all address important issues that affect nanocomposite properties. The key to success for any nanoparticle surface-modification technology is to disperse the nanoparticles individually and homogeneously within the polymer matrix and to achieve an optimized chemical affinity with the surrounding polymer matrix so as to facilitate stress transfer from the 'softer' matrix to the 'harder' nanoparticles. An optimized interfacial interaction between the two components is crucial to ensure superior mechanical properties, particularly fracture toughness. If the interfacial interaction between the two components is too strong, the resulting nanocomposites may fail and undergo plastic deformation, while if it is too weak the interfacial interaction may lead to lower tensile strength and premature brittle fracture.

## 9.3 Current manufacturing techniques

A fundamental knowledge of the nature of the different manufacturing techniques is important in determining the ultimate performance of polymer nanocomposites. In this section, typical processing techniques for polymer nanocomposites and, in particular, nylon-based nanocomposites will be described. The three most common techniques to fabricate nanocomposites are solution processing, *in situ* polymerization, and melt processing.

### 9.3.1 Solution processing

The general protocol for all solution-processing methods consists of the dispersion of nanoparticles in a liquid medium by vigorous stirring or ultra-sonication, followed by mixing of the nanoparticle suspension with a polymer solution and, subsequently, controlled evaporation of the solvent with or without vacuum. For example, Vaia and coworkers[46] adopted this approach and compared the dispersion of clay layers with and without melt-processed nylon-6/quarternary alkylammonium-modified montmorillonite (organoclay) nanocomposite pellets. That is, they used film casting of solution-processed nylon-6 and organoclay using 1,1,1,3,3,3-hexa-fluoro-2-propanol (HFIP) as a solvent and film casting of reconstituted, melt-processed nylon-6/organoclay nanocomposite pellets in HFIP. The former yielded an aggregated structure since the nylon-6 chains were not intercalated into the layered silicates. There was a small decrease in gallery height (from 1.85 nm to 1.7 nm), which might be due to the de-intercalation of excess ammonium modifier and dissolution within the nylon-6 matrix. In contrast, film casting from reconstituted, melt-processed nylon-6/organoclay nanocomposite pellets in HFIP resulted in a well-intercalated morphology. During melt processing, there is strong physico-absorption and hydrogen bonding between the nylon-6 chains and the organoclay layers. Specifically, amide and carbonyl groups on the polymer probably interacted with the oxygen plane of the layer surface as well as the weakly acidic SiOH and the strongly acidic bridging hydroxyl groups present at the layer edge. Due to these strong interfacial interactions in melt processing, the physico-absorbed nylon chains were not displaced on addition of HFIP during solution processing, and the HFIP effectively dispersed the individual layers by dissolution of the surface-absorbed nylon chains. This result underscores the importance of process history and the crucial balance of the interfacial interactions between the layered silicate and the surrounding medium.

Nylon-based nanocomposites are more commonly prepared by melt processing and *in situ* polymerization approaches rather than by solution processing. More often than not, these two approaches yield better quality nylon-based nanocomposites. Other drawbacks of solution processing are the need for suitable solvent or polymer solvent pairs and the high costs associated with the solvents, their disposal, and their negative impact on the environment.[40]

## 9.3.2 *In situ* polymerization

*In situ* polymerization has been demonstrated to be a successful approach for the preparation of nylon-based nanocomposites. Typically, in this process, the nanoparticle is firstly dispersed in the monomer or precursor and the mixture is subsequently polymerized by adding an appropriate catalyst or an organic initiator. A higher percentage of nanoparticles may be easily dispersed by this method and the nanoparticles can form strong interactions with the polymer matrices. Unlike solution or melt processing, *in situ* polymerization involves mixing the nanoparticles with short chain molecules instead of macromolecules. Thus, there is less possibility for nanoparticle agglomeration in the composites as there is less entanglement and steric hindrance between the short chain molecules. Indeed, *in situ* polymerization was first employed at the Toyota Central Research Laboratories to synthesize a nylon-6/clay nanocomposite.[47–52] The researchers showed that the $\varepsilon$-caprolactam monomer was able to swell the $\omega$-amino-acid-modified layered silicates and subsequently initiated ring-opening polymerization to obtain the nylon-6 layered silicate nanocomposite. Also, they showed that the carbon number of the amino acid should be greater than 8 for significant swelling of the modified layered silicates.[50] A more detailed description of the process can be found in Usuki *et al.*[52]

The preparation of nylon-based nanocomposites with 3-D nanoparticles using *in situ* polymerization has also been reported.[4,16,18,53] Some researchers even treated the nanoparticles with organosilane agents before *in situ* polymerization, so as to achieve a better dispersion of the nanoparticles in the matrix.[4,16,53] By employing this approach, Ou and coworkers[53] prepared nylon-6/silica nanocomposites in which the silica nanoparticles, pretreated with aminobutyric acid, were dispersed in $\varepsilon$-caproamide and aminocaproic acid at 90 °C, which were

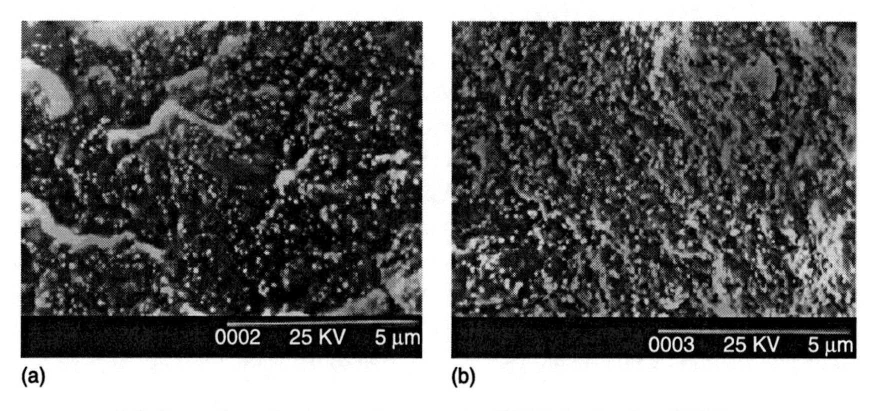


*9.5* Scanning electron micrographs (SEMs) of nylon 6/$SiO_2$ nanocomposites with (a) 10 wt% unmodified and (b) modified $SiO_2$ nanoparticles prepared by *in situ* polymerization. Reprinted with permission from Yang *et al.*[53] Copyright (1998) John Wiley & Sons Inc.

the monomer and initiator, respectively, followed by *in situ* polymerization at a high temperature under a nitrogen atmosphere. With this method, the modified silica nanoparticles, of diameter ~50 nm, were found to be well dispersed within the matrix (Fig. 9.5).[53]

Apart from directly mixing treated nanoparticles with a caprolactam monomer in a conventional *in situ* polymerization process, another route for achieving homogeneous dispersion of nanoparticles within a polymer matrix is through the integration of the well-established sol-gel process and *in situ* polymerization. The sol-gel process, as described earlier, is a wet-chemical technique primarily used for the fabrication of metal-oxide nanoparticles, starting from organic precursors such as metal alkoxides and metal chlorides, which eventually form a three-dimensional cross-linked inorganic network structure via various forms of hydrolysis-condensation reactions. Sengupta *et al.*[17] added tetraethoxysilane (TEOS) to polyamide-66 (PA66) dissolved in formic acid under vigorous stirring with a magnetic stirrer bar at room temperature for 1 h. The solution was then kept at room temperature for 12 h to allow the hydrolysis and condensation reactions of TEOS. Finally, the solution was poured onto glass plates for the solvent to evaporate under a fume hood. This method permitted the synthesis of PA66/$SiO_2$ nanocomposite with no aggregation of nanoparticles.

The preparation of nylon/CNT composites by *in situ* polymerization has also been very successful.[54–58] Gao and coworkers[54] functionalized MWCNTs using a solution of sulfuric acid and nitric acid to add –COOH groups to the nanotube surfaces. The MWCNTs/nylon-1010 composites were then prepared by polymerization of a nylon-1010 monomer salt in the presence of oxidized MWCNTs. A similar approach was used by Zhao *et al.*[55] to prepare nylon-6/MWCNT composites using ε-caprolactam monomers. Xu *et al.*[56]

prepared nylon-6/functionalized-CNT composites from hexanolactam and 6-aminoaldehexose acid. Surprisingly, they found that their synthesized CNTs hindered the polymerization of nylon-6 and they later modified the sequence of the process by polymerizing hexanolactam and 6-aminoaldehexose acid for 2.5 h before adding CNTs. The underlying causes of the inability of the monomer to polymerize in their initial processing approach were investigated using Fourier transform infrared spectroscopy. The analysis revealed that the reason might be because the –COOH groups on the oxidized CNTs created a strong nylon-6/CNT interface linked by C–O–N chemical bonds. These chemical bonds consisted of more stable nylon-6/CNT molecule chains and further growth of nylon-6 molecules could be obstructed by the formation of these bonds.

## 9.3.3 Melt processing

Due to the fact that thermoplastic semi-crystalline polymers like nylon soften when heated above their melting point, the fabrication of nanocomposites based on these polymers by melt compounding has proved to be a very valuable technique. This processing technique is compatible with conventional industrial processes; additionally the process is solvent-free, thereby reducing the associated costs and effect on the environment. In general, this process involves the mixing of a polymer melt with nanoparticles above the softening point of the polymer, under shear (or even in the absence of shear, in some cases). The melt-blended materials can then be processed by several methodologies such as film extrusion, injection molding, or compression molding, depending on the final morphology and shape of the products required. Nonetheless, there are some limitations that should be taken into consideration when using melt processing to prepare nylon-based nanocomposites. First, the polymer matrix degrades during processing, in general, at high temperatures and shear stresses. Second, the success of this technique requires the polymer to be highly compatible with the nanoparticles.

The preparation of nylon-6 layered silicate nanocomposites by melt processing was first demonstrated by Liu *et al.*[59] following work reported by Vaia *et al.*[60] in the preparation of polystyrene (PS) layered silicate nanocomposites. Subsequently, melt processing has become the mainstream method in the manufacture of numerous polymer nanocomposites incorporating a wide range of nanoparticles, from spherical[3,9–11] to plate-like,[39,61] and fibrous nanoparticles.[62–64] For nylon-based nanocomposites, much work has been spent on understanding the dispersion behavior of the nanoparticles in the resulting nanocomposites with respect to parameters such as processing conditions, the chemical nature of the matrix, the effect of nanoparticle surface modifications, the type of surface modifier, and melt rheology. Obviously, it is impossible to make an exhaustive list of works that have covered these topics in this chapter. Therefore, characteristic and important examples will be provided during the discussion of the structure–property relation in the next section.

## 9.4 Structures and properties

As mentioned previously, nanoparticles have the potential to exhibit a combination of different properties (physical, chemical, thermal, and mechanical). In addition, the exceptionally large surface-area/volume ratio of the nanoparticles, which is available for interaction with the polymer matrix, offers a chance to fine-tune the specific properties of nanocomposites required for end applications. An overview of some of these properties is given below, with particular emphasis on nylon-based matrices.

### 9.4.1 Mechanical properties

In general, the stiffness of a nanocomposite, even with agglomerated particles within the matrix, has a certain degree of improvement, though not to a large extent. This is because of its strong dependence on the orientation of anisotropic particles.[65] Further, the effect of aspect ratio is evident by comparing the magnitude of the improvement in stiffness for various polymer nanocomposite systems. With high-aspect-ratio nanoparticles like CNTs and clays, when well dispersed in the polymers, significantly improved stiffness has been achieved compared with the neat polymers. For instance, in the presence of only 2 wt% of MWCNTs, the elastic modulus of the resulting melt-processed nylon-6 nanocomposite was substantially improved by about 214% from 396 MPa to 1242 MPa (Fig. 9.6).[64]

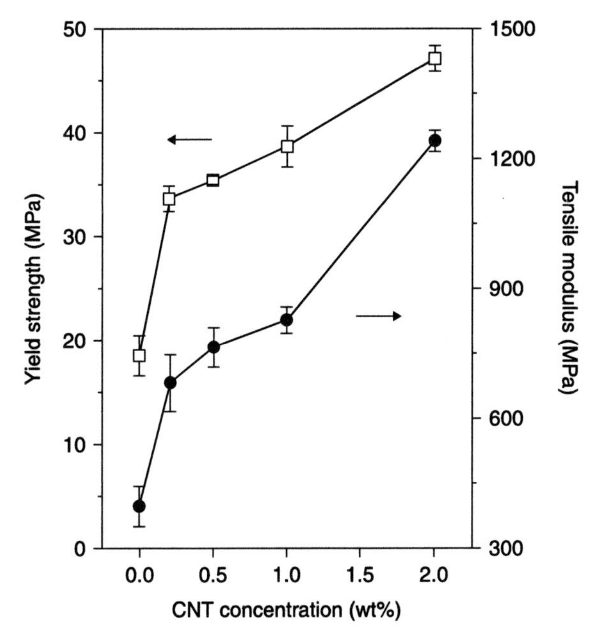


*9.6* Variation of tensile properties for nylon-6 and its nanocomposites against MWCNT loading. Reprinted with permission from Liu *et al.*[64] Copyright (2004) American Chemical Society.

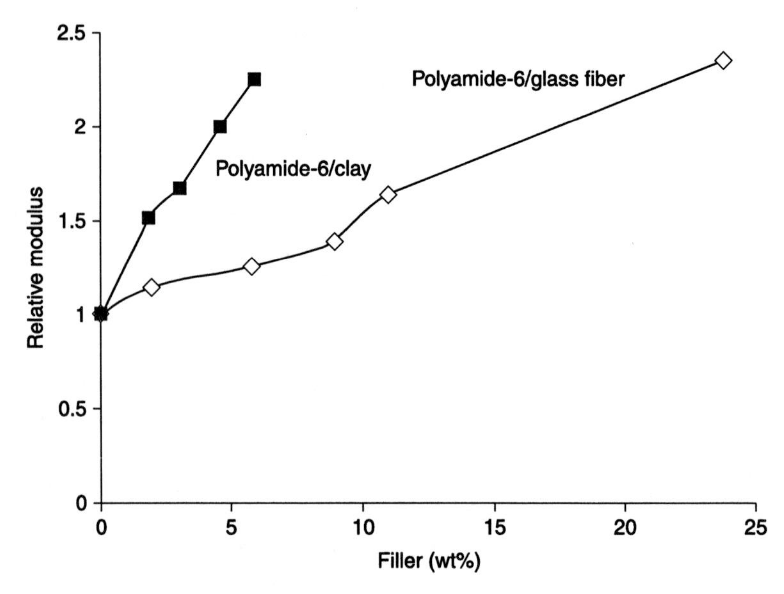


*9.7* Comparison of elastic modulus (relative to a polyamide-6 matrix polymer) for nanocomposites based on clay (montmorillonite) and glass fibers. Reprinted with permission from Fornes and Paul.[66] Copyright (2003) Elsevier.

Moreover, the advantage of high-aspect-ratio nanoparticles over conventional reinforcements is evident in Fig. 9.7, which compares the tensile modulus of nylon-6 materials reinforced with organoclay and glass fibers.[66–68] Even at very low loadings of filler, the nanoclay-reinforced nylon-6 shows a tremendous positive difference in the relative Young's modulus ($E/E_m$, where $E$ and $E_m$ are the Young's modulus of the composite and neat matrix, respectively). As explained earlier, the substantial improvement originates from the exceptionally high interfacial surface area, and it was also shown that, if the magnitude of intercalation is higher, greater improvements in elastic modulus and tensile strength can be achieved, as intercalation provides higher aspect ratios and optimizes the number of reinforcing elements.[69]

In some cases, however, enhancements in stiffness and strength were not realized even with the incorporation of high-aspect-ratio fillers like CNTs in the nylon matrix. This was seen in styrene maleic anhydride (SMA) copolymer encapsulated SWCNTs where a ~10% drop in tensile modulus was noted for nylon-6 composites with 1.5 wt% SMA-modified SWCNTs.[70] No improvements in yield strength were noticed in the nanocomposites. This was attributed to the loose network-like structure of the SWCNTs or the existence of SWCNT agglomerates acting as defects. These unusual experimental results for the mechanical properties of nylon composites imply that further research in terms of processing methods and CNT functionalization, as well as a better understanding and control of the polymer–CNT interfacial interaction, is necessary for the design and tailoring of nylon–CNT composites with desirable and predictable performance.

Specifically with regard to the mechanical properties of polymer layered silicate nanocomposites, comparative results have been reported for intercalated and exfoliated structures. Some mathematical models have also been proposed to explain the reinforcement efficiency of layered silicates in polymers by considering factors such as aspect ratio and orientation.[66,71] Much work has also been spent on understanding the influence of several other parameters like processing conditions,[72] the effects of the origin of the clay, the surfactants used,[39] the cation exchange capacity,[73] and the properties of the polymer matrices.[74] These aspects have been comprehensively reviewed in several publications[36,37,40,75–77] and thus will not be reiterated here.

On the other hand, properties measured at larger deformation, notably tensile yield strength, as well as the corresponding deformation mechanisms depend strongly on the microstructure and interfacial interaction. This is shown in Table 9.2, which compares three types of MWCNTs with different surface functionalization, used in the preparation of nylon-6/MWCNT nanocomposites by melt processing.[78] The first type of MWCNTs were synthesized in an ethanol flame, designated as F-MWCNTs; the second type are F-MWCNTs grafted with long-chain molecules of *n*-hexadecylamine (H-MWCNTs) and the third type are commercially available MWCNTs used as a benchmark, denoted as C-MWCNTs. The inherent defects in F-MWCNTs make them chemically active, and abundant functional groups such as carboxyl and hydroxyl groups are readily found on the surfaces of the F-MWCNTs. This facilitated further functionalization of the F-MWCNTs without oxidation or detriment to the primary structures of the nanotubes. Hence, the surface grafting of F-MWCNTs with long-chain molecules of *n*-hexadecylamine (H-MWCNTs) resulted in better nanoscopic dispersion and interfacial adhesion in the polyamide-6 (PA6) matrix, so that it had the highest tensile modulus of the three nanocomposites.

Based on the majority of studies, there is general agreement that the presence of layered silicates or CNTs results in either no improvement or even deterioration

*Table 9.2* Tensile properties of PA6 and its composites with various MWCNTS

| Sample | Young's modulus (MPa) | Tensile strength (MPa) | Elongation at break (%) |
|---|---|---|---|
| Neat nylon-6 | 601.4 ± 22.6 | 64.8 ± 1.6 | 352 ± 13 |
| Nylon-6/C-MWCNTs | 746.3 ± 13.3 | 60.9 ± 0.3 | 187 ± 28 |
| Nylon-6/F-MWCNTs | 747.3 ± 9.6 | 70.0 ± 1.2 | 287 ± 19 |
| Nylon-6/H-MWCNTs | 776.8 ± 10.9 | 68.8 ± 2.2 | 249 ± 11 |

Source: Adapted from Liu *et al.*[78]

in fracture toughness and ductility of nylon-based nanocomposites.[74,79–82] Chen *et al.*[80] investigated the embrittlement mechanisms of nylon-66 layered silicate nanocomposites and found that micro-sized and sub-micron voids developed around the layered silicates, which coalesced and formed premature cracks without any significant matrix plastic deformation, thus slashing the fracture toughness of the nanocomposites. He *et al.*[81] also showed that, in nylon-6 with 2.5 wt% organoclay, crazes were initiated in the matrix adjacent to the interface between the clay platelets and polymer, which developed into micro-cracks on continued deformation to failure. This resulted in decreased $K_{IC}$. At higher clay loadings, small-angle X-ray scattering results revealed a lower density of crazes; it was suggested that the large number of micro-cracks (due to the higher clay loading) had prevented the crazing mechanism from becoming fully operative before failure, thus leading to even lower toughness. We also reported similar observations for a nylon-6/organoclay layer nanocomposite in which only a limited number of micro-voids were found along the crack paths in the vicinity of an arrested crack tip, as shown in Fig. 9.8.[82]

It has also been shown that the presence of stiff layered silicates restricts the mobility of the surrounding chains, thus limiting their ability to undergo plastic deformation.[83] Due to the strong tethering junctions between individual clay layers and the matrix (due to the ionic interaction between the ends of the polymer chains and the surface of the negatively charged clay layers), full-scale debonding at the polymer–clay interface was rarely observed under stress conditions, indicating that the constraint on the polymer adjacent to the clay was not relieved, limiting the ability of the polymer to undergo plastic deformation. In many studies

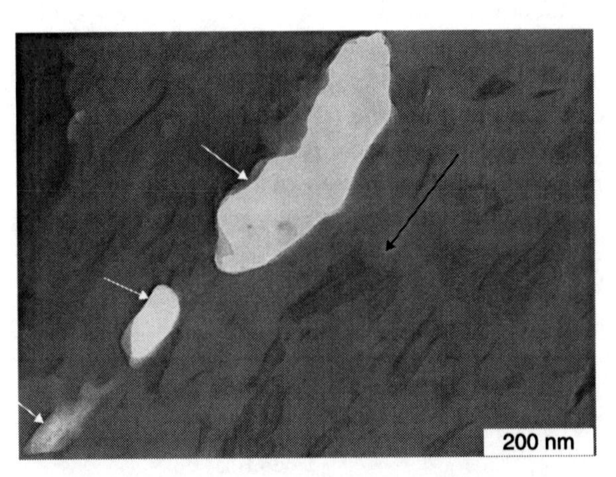


*9.8* TEM micrograph of polyamide-6/organoclay (95/5) nanocomposite showing voids along the path of an arrested crack tip. Black arrows indicate the direction of crack growth and the white arrow shows the delamination of the clay layers. Reprinted with permission from Lim *et al.*[82] Copyright (2007) Elsevier.

with CNTs, the toughness of the composites fell. For instance, the presence of even 2 wt% CNTs reduced the toughness from 12.4 kJ/m$^2$ to 10.0 kJ/m$^2$.[56] This reduction was attributed to the nucleation of cracks at the interface between encapsulated nylon-6 layers on CNT surfaces and the nylon-6 matrix.

But the high flexibility of CNTs, in a few cases, was shown to deflect cracks via matrix crack bridging under loading and to the pull-out of the nanotubes. Bridging by the nanotubes prevents a crack from opening and thus restrains catastrophic crack propagation in the polymer matrix. An example of bridging of a matrix crack is shown in Fig. 9.9. This mechanism is believed to be the main energy absorbing mechanism in nylon-6/MWCNT composites.[64] The pull-out mechanism, inspired by conventional polymer/fiber composites (including work done against sliding friction in pulling out the fiber), governs the extent of energy absorption. Hence, the very high interfacial areas in polymer/nanotube composites are expected to result in drastic improvements in the work of fracture due to nanotube pull-out. But recently, based on a scaling argument[84] correlating the radius ($r$), fiber strength ($\sigma$), and interface strength ($\tau$) with the energy absorbed per unit cross-sectional area by fiber pull-out (i.e., $G_{pull-out} \sim \frac{r\sigma^2}{\tau}$) ), it has been shown that improvements in toughness in polymer/CNT nanocomposites cannot be attributed to the nanotube pull-out mechanism, as the pull-out energy significantly decreases when the fiber radius is scaled down to the nanoscale. Wichmann *et al.*[85] argued that this conventional correlation is not valid for nanotubes by simply

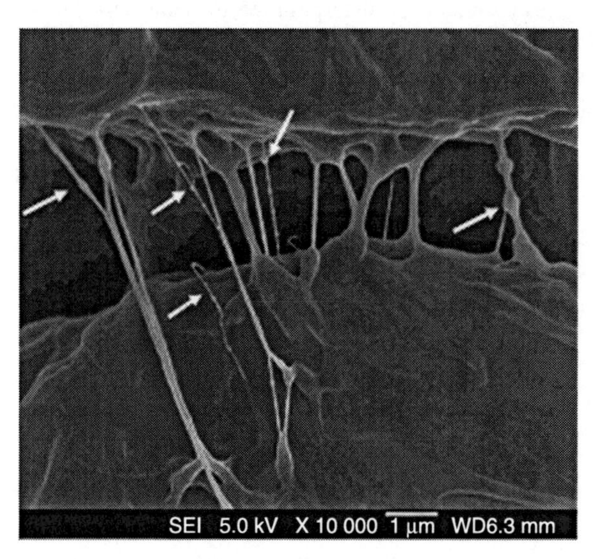


*9.9* SEM image of a micro-crack linked by MWCNTs sheathed with polymer beads as indicated by the white arrows in a nylon-6 nanocomposite. Reprinted with permission from Liu *et al.*[64] Copyright (2004) American Chemical Society.

considering the Kelly–Tyson expression (critical length, $l_c = \frac{r\sigma}{\tau}$), that is, it is impossible to vary the fiber radius independently without changing other parameters. They further suggested that, if spatial or only local bonding existed between the nanotubes and the matrix, there would be partial debonding of the interface allowing crack bridging similar to conventional polymer/fiber composites, as shown and analyzed by Gao *et al.*[86] two decades ago.

Aside from these few cases where improvements in toughness and ductility have been reported, the majority of studies show reduced ductility and toughness when high-aspect-ratio nanoparticles are incorporated in nylon. Even in those few cases that report improvements, the properties of the nanocomposites were not consistent and the failure mechanisms were not well understood. Nevertheless, to achieve a balance between stiffness and toughness in nanocomposites, ternary nanocomposites (prepared by adding a soft elastomeric dispersed phase to binary polymer nanocomposite systems) have been developed.[2,82,87–92] Though this is the best-known approach to date for counteracting the embrittlement of polymer nanocomposites,[2] its associated disadvantages must also be considered. The final microstructures are generally complex and control of the location of the rigid fillers (either in the matrix or as elastomeric particles) is necessary for achieving enhanced properties.

With 3-D nanoparticles, the enhancement in stiffness and strength of the nanocomposites is not as significant. But, generally, a positive ductility and toughness response is noted with these materials. Li *et al.*[16] observed a ~25% improvement in impact strength on incorporation of nano-$SiO_2$ (10 nm), treated with amine and epoxy-functionalized silanes, into a nylon-6 matrix compared with neat nylon-6. They also noted that with unmodified nano-$SiO_2$ the impact strength was reduced by 50%. To increase the fracture toughness of a material, it is necessary to initiate several effective energy dissipation processes like crazing and shear yielding around the crack tip so that the total amount of plastic energy absorbed during fracture is high. Zhang and coworkers[93] studied the toughness of nylon-66/$TiO_2$ nanocomposites using the essential work of fracture (EWF). They found that the essential work of fracture (initiation) increased with increasing $TiO_2$ content up to 3 vol% while the non-essential work of fracture (propagation) decreased. Fractographic analysis revealed that the nanoparticles acted as stress concentrators, triggering debonding and then inducing a large amount of local plastic deformation of the matrix. With an increasing number of $TiO_2$ nanoparticles, the plastic zone became more and more constrained and thus the non-essential work of fracture decreased. The reduction of resistance to crack propagation was due to the inevitably agglomerated nanoparticles at higher nanoparticle loadings, leading to a high level of stress concentration, which favored catastrophic crack propagation. In another work, similar results for the toughness property and deformation mechanism were obtained for polyamide-66 with $SiO_2$ and $Al_2O_3$ nanoparticles.[94]

## 9.4.2 Wear and scratch resistance

Wear generally originates from damage induced by rubbing bodies due to repeated applications of mechanical, impact, and other kinds of forces. Therefore, the surface loses mechanical cohesion and debris is formed from material dislodged from the contact zone.[95–98] For reliable and effective functioning of materials, it is necessary to reduce material damage and material removal, or, in some cases, to control the extent of the material removed. Generally, the specific wear rate $K$ ($mm^3\ N^{-1}\ m^{-1}$) is determined by

$$K = \frac{V}{Lvt} \qquad [9.1]$$

where $V$ ($mm^3$) is the wear volume loss calculated from

$$V = 2\pi RA \qquad [9.2]$$

Note that $t$ (s) is the test duration, $A$ ($mm^2$) cross-sectional area of the wear track, $L$ (N) normal load, $R$ (mm) the wear track radius, and $v$ (m/s) the sliding speed.

However, as polymers are highly sensitive to scratch and wear damage, they exhibit various modes of deformation, e.g., abrasive, adhesive, fatigue, corrosive, erosive, and delamination, even within a narrow range of contact variables.[99–103] Moreover, there is always an overlap of different mechanisms in any particular contact process and any combination of the different mechanisms may represent the actual situation. This indicates the complexity of wear and scratch damage in polymers, and the difficulty in accurately quantifying and predicting the extent of damage and the damage modes, and thus limits their application. Nonetheless, for polymers, different wear mechanisms can be grouped into two main categories: cohesive and interfacial wear processes (schematically shown in Fig. 9.10).[104] In

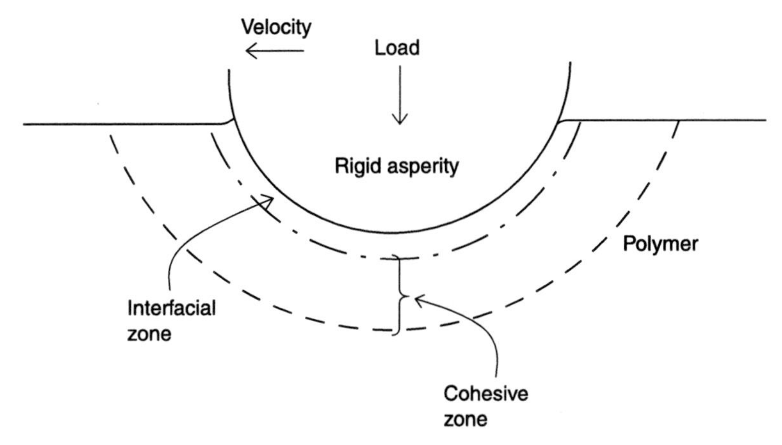


*9.10* Different wear mechanisms under cohesive and interfacial wear processes. Adapted from Briscoe and Sinha.[104]

cohesive wear processes, the frictional work is dissipated in relatively large volumes adjacent to the interface, either through the interaction of surface forces and the resultant traction stresses or simply via geometric interlocking of interpenetrating contacts. The extent of this surface zone is defined by the contact geometry and contact stresses generated in the surface. Cohesive wear processes are mainly controlled by the mechanical properties of the interacting bodies. Most mechanical wear processes can be grouped under this category, such as abrasive, fatigue, and fretting. In contrast, interfacial wear processes involve dissipation of frictional work in much thinner regions and at greater energy densities. This creates a large increase in local temperature. The chemistry of the surfaces and the forces emanating from them should be considered, instead of the mechanical properties of the interacting materials, to determine the extent of wear damage. Transfer wear and chemical or corrosive wear are interfacial wear processes.

Fortunately, from the tribological viewpoint, the major benefits of polymer nanocomposites relative to micro-sized particle composites are: (i) material removal is expected to be less as the nanoparticles have a similar size to the segments of the surrounding polymer chains; and (ii) the bonding between the nanoparticles and the matrix is expected to be better due to their large specific surface areas.[105–107] However, there are many important factors affecting the tribological behavior of polymer nanocomposites, including particle size, aspect ratio, hardness, concentration, orientation, nature of the interface between the polymer matrix and the particles, tribochemical reactions, and transfer films that may arise due to the interaction of particles and counterfaces. Additionally, the indirect influence of nano-additives on polymeric materials should be considered when investigating their tribological response.

Many studies have described wear and scratch damage in polymers filled with different nanoparticles and the majority of them reported improved tribological properties. This was attributed to the presence of the nanoparticles themselves or to improved mechanical properties like the modulus and hardness or to the formation of transfer films on the slider contact surface. However, when dealing with polymer nanocomposites, it is not valid to assume that nanoparticles irrevocably improve wear and scratch (and friction) properties. Material properties like the modulus, hardness, fracture toughness, or the wear rate or scratch penetration depth, are not the sole indicators for comparing and ranking candidate materials.[98,108,109] It was also found that nanoparticles by themselves, even if uniformly dispersed with good interfacial interaction with the matrix, do not irrevocably improve the wear (and friction) properties. Although it is important to consider these factors, it is necessary to thoroughly understand all nanostructural parameters and the relations between penetration depth and material deformation and sub-surface damage in the wear and scratch track. This knowledge is critical for understanding the surface integrity of these materials. The correct physical and spatial characterization of nanocomposites is required. For example, while the orientation and dispersion of clay layers are independent structural features, they must be simultaneously considered in determining the effective structural reinforcement in polymer/clay nanocomposites; otherwise, the results can be misleading if only dispersion or orientation of the structures is considered.[109]

Moreover, considering the complexities involved in evaluating the tribological response of polymer nanocomposites, only a limited effort has been made to model the stress fields induced by different slider geometries. Hence, to date, no explicit correlations between material parameters and wear and scratch damage, particularly for polymer nanocomposites at the nanoscale, are available. For a more in-depth review of the wear and scratch response of polymer nanocomposites, refer to Dasari *et al.*[98]

### 9.4.3 Barrier properties and permeability

Barrier properties are of prime importance in bottling, food packaging, packaging hydrocarbon solvents, and the protective coating industries. In comparison with other physical and mechanical properties of polymer/clay nanocomposites, which have mixed results, barrier properties are mostly positive, that is, dramatic decreases in permeability to different media can be achieved at relatively low loadings of nanoparticles (particularly with high-aspect-ratio fillers like layered silicates) in contrast to the neat polymers and polymer micro-composites.[110–116] This is due to the high aspect ratios of impermeable silicate platelets forcing the gas or liquid molecules to traverse a tortuous path in the polymer matrix surrounding these silicate particles. This obviously increases the effective path length for diffusion (Fig. 9.11).[37,117] As is evident, aspect ratio, loading, orientation, and degree of dispersion have a significant effect on barrier properties. In addition, many factors, like the surface treatment of the nanoparticles, processing techniques, and the degree of crystallinity and cross-linking of the polymer, can have an indirect bearing on the barrier properties of these materials.

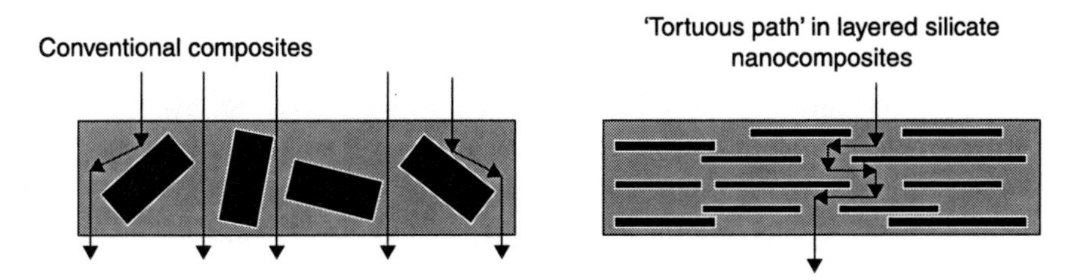


*9.11* Comparison of tortuosity for the diffusion path in a polymer filled with micro-sized particles and polymer/clay nanocomposites. Reprinted with permission from Ray and Okamoto.[37] Copyright (2003) Elsevier.

According to the tortuous-path model,[117] the oxygen gas permeability of a film decreases to 1/5 and 1/2400, compared with the pure film, for 5 and 95 wt% of inorganic platelets, respectively. The tortuosity factor $\tau$ is calculated as[117]

$$\tau = 1 + (L/2W)V_f \quad [9.3]$$

where $L/W$ is the aspect ratio of the layered filler and $V_f$ is the volume fraction. This is for an ideal case where clay particles are completely exfoliated, uniformly dispersed, and oriented such that the direction of diffusion is normal to the direction of the sheets (at a preferred angle of 0°). This arrangement will obviously result in higher tortuosity and the model also suggests that platelet-shaped particles are more efficient at maximizing the path length than particles with other shapes. Recently, Bharadwaj,[118] based on geometric arguments, considered a range of orientations of the clay layers relative to the flow direction of the gas or liquid across the film and studied their effect on the relative permeability. As expected, the permeability is highest when the clay layers are aligned in the flow direction. He also concluded that dispersing longer sheets ($L$ >500 nm) than the normal length of clay platelets in a polymer matrix not only increased the tortuosity, but also reduced the dependence of the relative permeability on the orientation of the layers.

Further, the effect of tortuosity on the permeability can be expressed as

$$\frac{P_{\text{PCN}}}{P_p} = \frac{1 - V_f}{\tau} \quad [9.4]$$

where $P_{\text{PCN}}$ and $P_{\text{P}}$ are the permeability coefficients of the nanocomposite and the pure polymer, respectively, and $V_f$ is the volume fraction of clay.[37,117]

Yano *et al.*[119] studied the effect of clay length on the relative permeability coefficients of polyimide with 2 wt% of organoclay. They compared experimental data and theoretical values predicted using Eq. 9.4 (Fig. 9.12) and found a good fit, that is, as the length of the clay increased, the relative permeability decreased drastically.

Nevertheless, huge permeability resistance to $O_2$, He, $CO_2$, water vapor, and many other solvents and gases has been reported with polymer/clay (montmorillonite) nanocomposites (even with intercalated structures). For example, in polyethylene terephthalate/clay nanocomposites, a twofold reduction in $O_2$ permeability with only 1 wt% clay was noted.[120] Even for hydrogen and water vapor, the permeability coefficients of nylon-6 with only 0.74 vol% of montmorillonite were less than 70% of the equivalent coefficients for nylon-6. In polyimide nanocomposites, the permeability coefficient of water vapor showed a tenfold reduction with 2 wt% synthetic mica relative to pristine polyimide.[121] Moreover, studies have indicated that a percolation threshold exists in polymer layered silicate nanocomposites, after which permeability does not decrease or has a negative effect.[122,123] Similar behavior has also been observed for other physical properties of polymer nanocomposites.[124,125] In this regard, Lu and Mai[126]

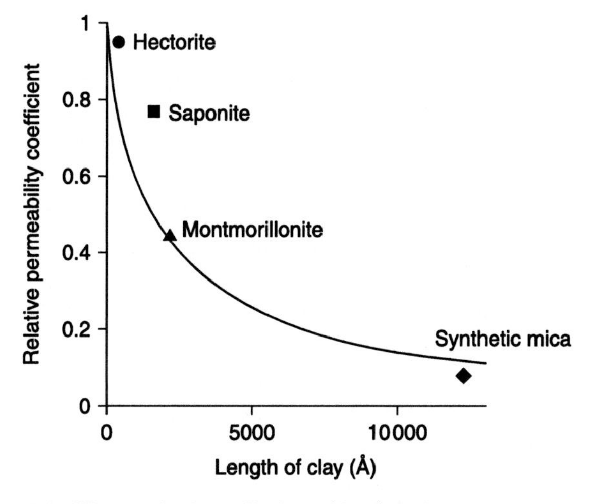


*9.12* Theoretical predictions (line) and experimental data (symbols) showing the effect of clay-layer size on relative permeability for polyimide/clay nanocomposites. Reprinted with permission from Yano *et al.*[119] Copyright (1993) John Wiley & Sons Inc.

proposed a renormalization group (RG) model and identified the influence of aspect ratio, orientation, and extent of exfoliation of clay layers on the barrier properties of polymer/clay nanocomposites. The basic hypothesis of the RG approach is that the probability $p$ that a cell acts as a barrier is the same at all orders. It was also concluded that the aspect ratio of clay layers is the controlling and determining factor of barrier properties.

But it is important to note that the tortuous path theory, and other proposed relations (e.g. the Cussler[127] formula or the power law equation[128]), are based on the assumption that the presence of nanoparticles does not affect the diffusivity of the polymer matrix, which in reality is not generally true. Some liquids and vapors often have noticeable solubility in polymers, plasticizing them. Moreover, organophilic clay gives rise to superficial adsorption and to specific interactions with some of the solvents. Further, the presence of clay layers, as is well known, induces polymorphism, alters crystallization kinetics (temperature, rate, crystal fraction, and percentage crystallinity), and promotes the formation of small, irregular crystallites of polymers due to their heterogeneous nucleation effect. These increases of the surface area of crystals and the fraction changes in the crystalline and amorphous parts affect the permeability properties, as the crystalline regions are generally more impermeable to penetrant molecules than amorphous regions. Furthermore, as polymers usually have a wide range of relaxation times associated with the motion of the polymer segments, the barrier properties are affected by changes in temperature and the concentration of the permeant.

### 9.4.4 Thermal properties

Apart from the mechanical properties, thermal properties like dimensional stability, heat distortion temperature (HDT), and thermal stability dictate the widespread applicability of polymeric materials. For instance, the low thermal deformation tolerance of polylactic acid (PLA), a widely marketed biodegradable polymer, has restricted its application in electronics, microwavable packaging, etc. Therefore, it is important to understand and resolve these issues to widen the applicability of polymeric materials.

First, the dimensional stability of polymeric materials during molding is a key parameter, particularly in automobile applications, where polymers with essentially high coefficients of thermal expansion (CTEs) are integrated with metals, which have much lower CTEs.[67] Moreover in batteries, a polymer electrolyte is employed as both electrolyte and separator. The heat generated during charging and discharging may cause deformation of the polymer electrolyte (temporary or even permanent dimensional changes), which may result in reduced efficiency or even a short circuit: good dimensional stability is essential. Layered silicate platelets, due to their shape and size, have proved to be useful in reducing the CTEs of polymers in two directions.[129–131] That is, with changes in temperature, expansion or contraction of the matrix occurs but the fillers resist this process by inducing opposing stresses in the two phases. Yoon *et al.*[130] measured the linear thermal expansion behavior in the flow direction (FD), the transverse direction (TD), and the normal direction (ND) in ~$6.3 \times 12.7 \times 3.2\,mm^3$ rectangular bars of nylon-6/clay nanocomposites. As shown in Fig. 9.13, the addition of clay reduced the thermal expansion coefficient in FD and TD but increased it in ND. They attributed this to the high constraint effect of nylon-6 in FD and TD, which resulted in a 'squeezing' effect in ND, thus increasing expansion in this direction. Hwang and Liu[132] showed that the dimensional stability of polyacrylonitrile (PAN) electrolyte improves even with just 3 wt% clay. Within the temperature range 30–75°C, the thermal expansion coefficient of the PAN/clay electrolyte was $778 \times 10^{-6}$ mm/°C whereas for the neat PAN electrolyte it was $3400 \times 10^{-6}$ mm/°C.

HDT is another thermal property and, as the name indicates, it quantifies heat resistance on external loading. An increase in HDT has been observed in clay-based nanocomposites for many polymer systems. The HDT of nylon-6 nanocomposites has been shown to increase from 65 °C (pristine nylon-6) to 150 °C with 5 wt% of clay. Such an increase in HDT is very difficult to achieve in conventional polymer composites reinforced by micro-particles. Ray *et al.*[133] also concluded that the addition of clay layers to PLA increased its HDT; with a load of 0.98 MPa, HDT increased from 76 °C to 93 °C with 4 wt% of organoclay and it further increased to 115 °C with 10 wt% of organoclay. Even the incorporation of $CaCO_3$ nanoparticles in a PP copolymer considerably increased its HDT along with the flexural modulus and impact strength.[134]

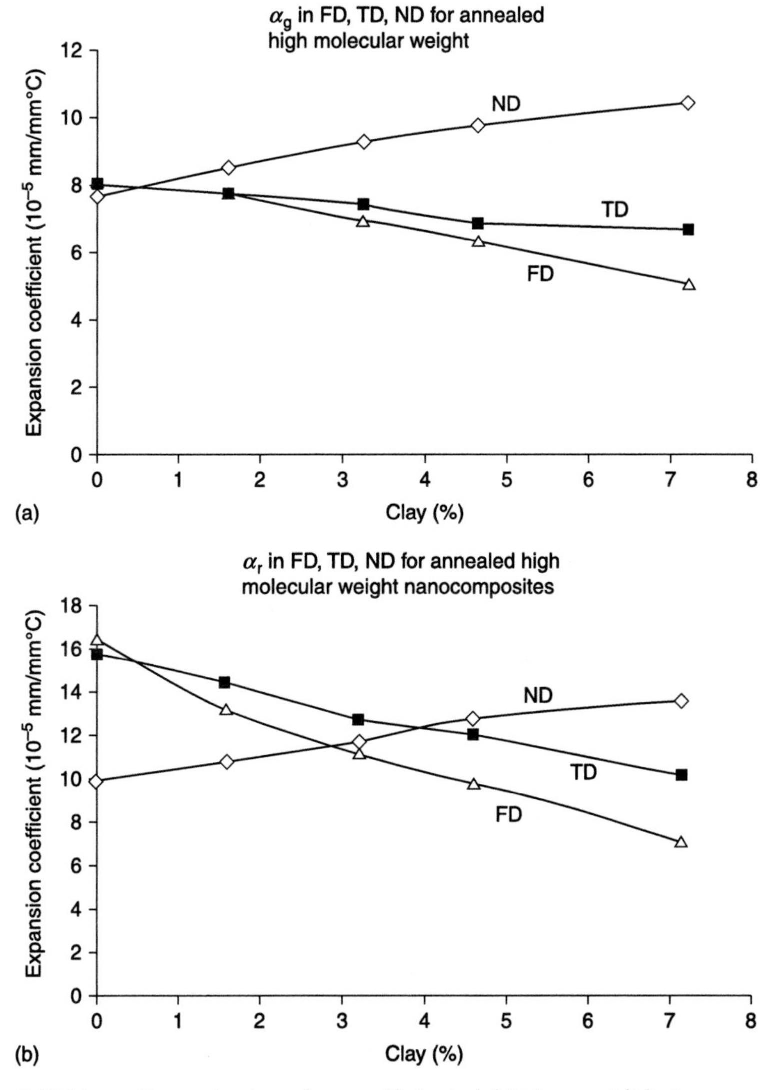


*9.13* Linear thermal expansion coefficients (a) below and (b) above the nylon-6 glass transition temperature for annealed nylon-6/clay nanocomposites in three directions. Reprinted with permission from Yoon *et al.*[130] Copyright (2002) Elsevier.

Further, it is important to understand the effect of additional functional additives and organic surfactants for modifying nanoparticles on the thermal stability of the polymeric matrix. For instance, despite the advantages of the organic modification of nanoparticles, it considerably degrades the polymer's thermal stability. In particular, the commonly used long-chain alkylammonium surfactants have poor thermal (and photochemical) stability[135] and, normally, surfactants account for

(~20–30% of the inorganic filler loading. Amine groups start to decompose usually from ~180 °C,[136] which is below the processing temperatures of most engineering polymers, like nylon-6. Thus, the presence of such a large amount of low molecular weight bound or unbound surfactant may adversely affect thermal properties in addition to the mechanical and physical properties of the resultant nanocomposites.

Much effort has been directed towards developing alternative routes for producing nanocomposites without using the conventional alkylammonium surfactants. These include modifying the clay layers with thermally stable surfactants (usually aromatic compounds), water-assisted approaches, partially exchanged systems that decrease the amount of surfactant required,[137] etc. Thermally stable surfactants (like phosphonium, pyridinium, iminium, and imidazolium salts) decompose at higher temperatures than conventional alkylammonium surfactants and are stable, particularly during processing. In particular, the treatment of clay with imidazolium salts (Fig. 9.14) via standard ion-exchange methods showed a tremendous improvement in thermal stability of the modified clays and ultimately polymer nanocomposites.[61] Gilman *et al.* obtained an improvement of ~100 °C (in terms of peak decomposition temperature) by using 1-alkyl-2,3-dimethylimidazolium-treated clay in nitrogen in comparison to that of common alkylammonium clay.[61] Further polyhedral oligomeric silsesquioxane (POSS)-treated clays were also used in polymers to obtain a positive effect on thermal stability, as POSS molecules are stable up to 300 °C.[138–140] POSS molecules have hybrid (organic–inorganic) architecture and their structure contains a stable inorganic Si–O core, which is intermediate between silica and silicones. This core is covered externally by organic substituents which can be modified to yield a wide range of polarities and functionalities. This great variety gives a diversity of silsesquioxanes.

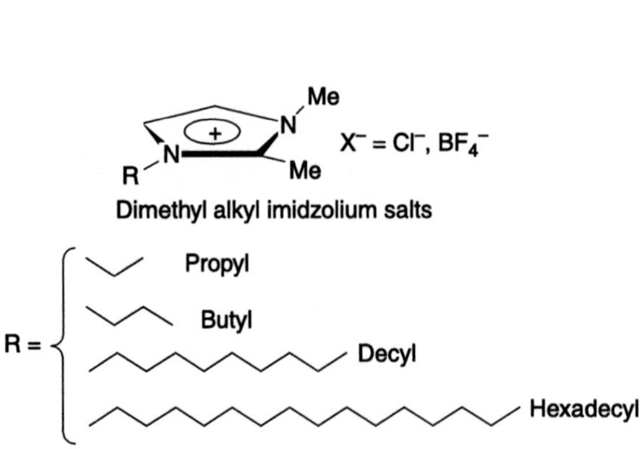

*9.14* Structures of different imidazolium salts used to treat sodium montmorillonite (from Gilman *et al.*[61]).

### 9.4.5 Other properties

There are many other properties of polymer nanocomposites, which have not been discussed, but are important in many commercial applications. These include biodegradability (the ability of functionalized nanoparticles to speed the process), optical properties, corrosion, photo-degradation resistance, etc. The flame-retardant properties of polymer nanocomposites are also worth mentioning. A significant amount of recent research has been on this topic. See, for example, various publications[141–143] and references therein. This is because, even at low loadings of nanoparticles and with no additional flame retardants in the system, the heat-release and mass-loss rates were greatly reduced, in addition to a huge delay in burning compared with the neat polymers. The reduction in the peak heat-release rate is an important parameter for fire safety, as it represents the point in a fire where heat is likely to propagate further (flame spread) or ignite adjacent objects. The improved flame retardancy was mainly attributed to the structural collapse of the nanocomposite during combustion and the formation of a multi-layered inorganic barrier on the polymer surface.

## 9.5 Applications of nylon-based polymers

Before discussing the various applications of polymer nanocomposites, it is important to realize that the scientific literature on these materials is immense and many research and development centers have studied these materials only in the past couple of decades. As with other major discoveries and inventions, it generally takes a few decades for them to have a large commercial effect, as economics is beyond the realm of scientific investigation and additional advances may be required to make the materials commercially viable. Moreover, biomedical applications require an in-depth understanding of the side effects of nanoparticles on humans, which may take even longer to establish. Nevertheless, many polymer-based nanocomposites are increasingly being used in applications including structural, transportation, biomedical, and sports.

There is wide speculation that the enhancements in tensile strength, stiffness, and the heat distortion temperature through the application of nanoparticle-filled engineering thermoplastics like nylon will challenge metal and glass applications in many new areas. The automotive and aerospace industries are leading the way in the development of nanocomposite applications. This is because of improved reinforcement, corrosion resistance, noise dampening, dimensional stability, scratch and wear resistance, and superior barrier properties.[144,145] An additional advantage of polymer nanocomposites is the ease of processing (extrusion and molding into different shapes as required). Some applications are shown in Fig. 9.15, with particular emphasis on layered silicates. The first application for polymer layered silicate nanocomposites was an automotive application by Toyota for a timing belt cover.[37] It had good rigidity, excellent thermal stability, and no

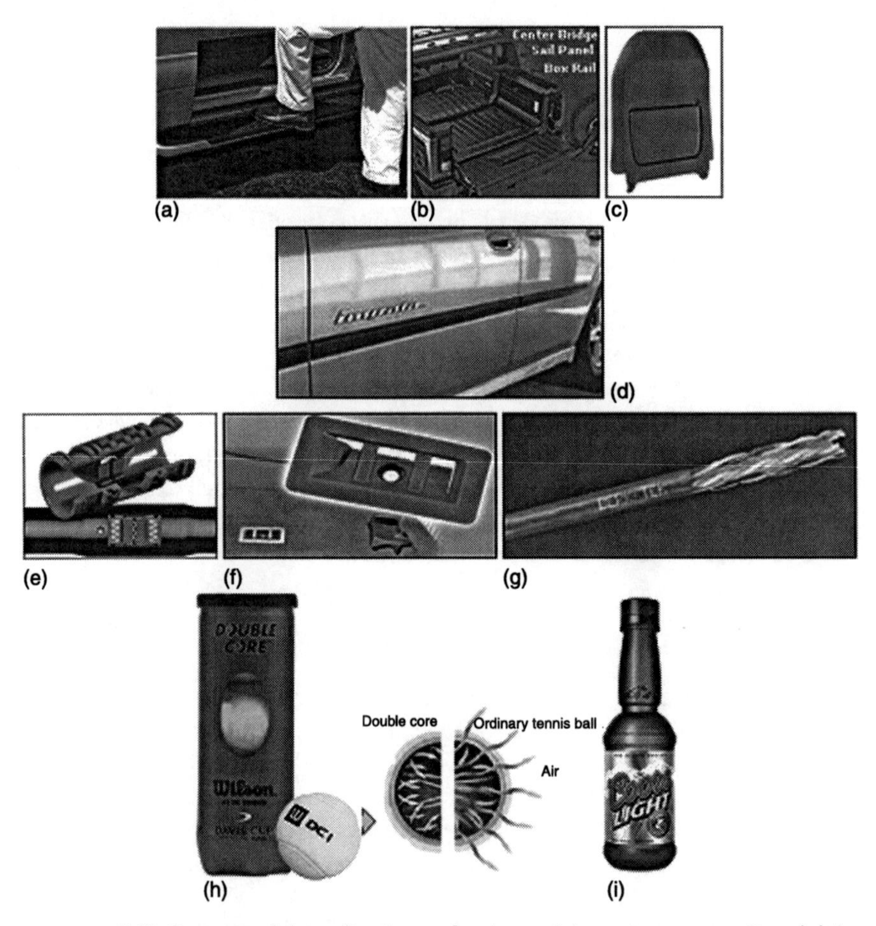


*9.15* Commercial applications of polymer/clay nanocomposites. (a) A step-assist on a GMC Safari van that incorporates a thermoplastic olefin/clay nanocomposite. (b) GM's 2005 Hummer H2 cargo bed uses ~7 lb of a molded-in-color thermoplastic olefin nanocomposite. (c) Noble Polymers' Forte polypropylene/clay nanocomposite seat backs for the 2004 Acura TL. (d) Thermoplastic olefin nanocomposite in the body side molding of GM's 2004 Chevrolet Impala. (e) Geoflow's linear low-density polyethylene/clay nanocomposite drip emitter for irrigation tubing ensures the timed release of herbicide. (f) Putsch and Süd-Chemie jointly prepared Elan XP, a compound of polypropylene and polystyrene compatibilized by clay, which is used as an interior air vent for the Audi A4 and a Volkswagen van. (g) Nexans' cable jacketing nanocomposite, the first such product for plenum cable used in office buildings due to the good flame retardancy of the polymer/clay materials. (h) Wilson's double-core tennis balls have an inner core coated with a butyl-based rubber filled with nanoclay particles, which acts as a barrier and stops air escaping from the core. (i) Coors' beer bottle, which has a layer of a polymer/clay nanocomposite because of the high barrier properties. Pictures are courtesy of the relevant company and technology websites.

warpage. Soon after, Mitsubishi introduced nylon-6/clay nanocomposite-based engine covers on their gasoline direct injection (GDI) models and General Motors added an external thermoplastic polyolefin/clay nanocomposite step assist to its Safari/Chevrolet Astro vans.[141]

Moreover, since nanocomposites are light, there is an average weight saving of 25% compared with highly filled-conventional polymers and 80% over steel,[144] which means large reductions in fuel costs. The savings are predicted to be 1.5 billion liters of gasoline over the life of one year's production of vehicles with the added advantage of reduced $CO_2$ emissions of more than 5 billion kilograms.[35] Apart from the automotive industry, polymer/clay nanocomposites in molded parts are also used in construction (building sections and structural panels) and aerospace (flame-retardant panels and high-performance components) applications.

Other promising applications of polymer nanocomposites are due to their outstanding barrier, permeability (particularly with high-aspect-ratio nanoparticles), and optical transparency properties. They are currently used in packaging for processed meats, cheese, confectionery, cereals, boil-in-the-bag foods, and extrusion-coating applications with paperboard for fruit juice and dairy products, and with co-extrusion processes in the manufacture of bottles for beer and carbonated drinks. Commercial grades of packaging films based on nylon-6 nanocomposites are available from Ube Industries and from Bayer AG.

Nanofibers are another type of nylon-based material that are used in applications. The outstanding properties of nanofibers (large surface area to volume ratio, flexibility in surface functionalities, and superior mechanical performance) make them suitable candidates for applications such as filtration, protective clothing, biomedical (including wound dressing), and as supports for enzymes and catalysts. A number of processing techniques[146] such as drawing, template synthesis, phase separation, self-assembly, and electrospinning have been used to prepare polymer nanofibers. Of these, electrospinning and electrostatic spinning have many advantages. For an extensive comparison of these techniques, the fundamentals behind them, their advantages and disadvantages, refer to the literature.[147–150]

While nano-layer silicates add muscle to nylon, CNTs impart electrical and thermal conductivity. Electro-conductive polymer nanocomposites are the preferred replacement for traditional steel in fuel delivery lines because they prevent the build-up of static electricity. They have also been developed for electro-paintable exterior body panel applications. In summary, though the main applications for nanocomposites are currently in packaging and transportation, their application will increase as R&D progresses.

## 9.6 Future trends

The critical challenge in the development of polymer nanocomposites with remarkable improvements in thermo-mechanical, electrical, and optical properties lies in the dispersion of the nanoparticles in the polymer matrix. In addition to the various surface-modified technologies in the preparation of polymer nanocomposites, there are still opportunities and challenges for improving dispersion and the interfacial properties of these materials. Thus, further research and development from these perspectives is required. Frequently, the poor characterization of these materials and inadequate quantitative descriptions of observed phenomena have led to contradictions and misleading impressions for these nanocomposites. Thus, much more work is needed to develop analytical and modeling tools to predict and verify the behavior of these types of nanocomposite.

## 9.7 Acknowledgements

The permission from various publishers to reproduce the figures in this chapter is much appreciated.

## 9.8 References

1. Ruiz-Hitzky E and Van Meerbeek A, 'Clay mineral and organoclay-polymer nanocomposite', In: Bergaya F, Theng BKG and Lagaly G, editors, *Handbook of Clay Science – Developments in Clay Science*, Vol. 1, Elsevier Ltd, The Netherlands, 2006, p. 583.
2. Dasari A, Lim SH, Yu ZZ and Mai YW, 'Fracture properties and mechanisms of polyamide/clay nanocomposites', In: Karger-Koscis J and Fakirov S, editors, *Nano- and micromechanics of polymer blends and composites*, Munich: Hanser, 2009, p. 377.
3. Tabuani D, Ceccia S and Camino G, 'Nylon-6 nanocomposites, study of the influence of the nanofiller nature on morphology and material properties', *J Polym Sci B Polym Phys*, 2009 **47** (19) 1935–1948.
4. Zheng LY, Lau KT, Zhao LX, Zhang YQ and Hui D, 'Mechanical and thermal properties of nano-$Al_2O_3$/nylon 6 composites', *Chem Eng Comm*, 2010 **197** (3) 343–351.
5. Chen CH, Li HY, Chien CY, Yen FS, Chen HY and Lin JM, 'Preparation and characterization of alpha-$Al_2O_3$/nylon 6 nanocomposite masterbatches', *J Appl Polym Sci*, 2009 **112** (2) 1063–1069.
6. Wang SB, Ge SR and Zhang DK, 'Comparison of tribological behavior of nylon composites filled with zinc oxide particles and whiskers', *Wear*, 2009 **266** (1–2) 248–254.
7. Zheng JR, Siegel RW and Gregory Toney C, 'Polymer crystalline structure and morphology changes in nylon 6/ZnO nanocomposites', *J Polym Sci Part B-Polym Phys*, 2003 **41** (10) 1033–1050.
8. Moussa MA, Ghoneim AM, Hakim AAA and Turhy GM, 'Electrical and thermal properties of nylon 6/calcium carbonate composites', *Adv Polym Techn*, 2009 **28** (4) 257–266.
9. Avella M, Carfagna C, Cerruti P, Errico ME and Gentile G, 'Nylon based nanocomposites: Influence of calcium carbonate nanoparticles on the thermal stability', *Macromol Symposia*, 2006 **234** 163–169.

10. Avella M, Errico ME and Gentile G, 'Nylon 6/calcium carbonate nanocomposites: Characterization and properties', *Macromol Symposia*, 2006 **234** 170–175.
11. Cayer-Barrioz J, Ferry L, Frihi D, Cavalier K, Suguela R and Vigier G, 'Microstructure and mechanical behavior of polyamide 66-precipitated calcium carbonate composites: Influence of the particle surface treatment', *J Appl Polym Sci*, 2006 **100** (2) 989–999.
12. Tomlinson WJ and Coulson CN, 'Strength of surface-modified calcite nylon-6 composites', *J Mater Sci Lett*, 1992 **11** (9) 531–534.
13. Fang XW, Li XH, Yu LG and Zhang ZJ, 'Effect of *in situ* surface-modified nano-$SiO_2$ on the thermal and mechanical properties and crystallization behavior of nylon 1010', *J Appl Polym Sci*, 2010 **115** (6) 3339–3347.
14. Mahfuz H, Hasan M, Dhanak V, Beamson G, Stewart J, *et al.*, 'Reinforcement of nylon 6 with functionalized silica nanoparticles for enhanced tensile strength and modulus', *Nanotechnol*, 2008 **19** (44) 445702.
15. Xu XM, Li BJ, Lu HM, Zhang ZJ and Wang HG, 'The effect of the interface structure of different surface-modified nano-$SiO_2$ on the mechanical properties of nylon 66 composites', *J Appl Polym Sci*, 2008 **107** (3) 2007–2014.
16. Li Y, Yu J and Guo ZX, 'The influence of silane treatment on nylon 6/nano-$SiO_2$ *in situ* polymerization', *J Appl Polym Sci*, 2002 **84** (4) 827–834.
17. Sengupta R, Bandyopadhyay A, Sabharwal S, Chaki TK and Bhowmick AK, 'Polyamide-6,6/*in situ* silica hybrid nanocomposites by sol-gel technique: synthesis, characterization and properties', *Polym*, 2005 **46** (10) 3343–3354.
18. Reynaud E, Jouen T, Gauthier C, Vigier G and Varlet J, 'Nanofillers in polymeric matrix: a study on silica reinforced PA6', *Polym*, 2001 **42** (21) 8759–8768.
19. Treacy MMJ, Ebbesen TW and Gibson JM, 'Exceptionally high Young's modulus observed for individual carbon nanotubes', *Nature*, 1996 **381** (6584) 678–680.
20. Duesberg GS, Burghard M, Muster J, Philipp G and Roth S, 'Separation of carbon nanotubes by size exclusion chromatography', *Chem Comm*, 1998 **3** 425–436.
21. Moore VC, Strano MS, Haroz EH, Hauge RH, Smalley RE, *et al.*, 'Individually suspended single-walled carbon nanotubes in various surfactants', *Nano Lett*, 2003 **3** (10) 1379–1382.
22. Islam MF, Rojas E, Bergey DM, Johnson AT and Yodh AG, 'High weight fraction surfactant solubilization of single-wall carbon nanotubes in water', *Nano Lett*, 2003 **3** (2) 269–273.
23. Das D and Das P K, 'Superior activity of structurally deprived enzyme-carbon nanotube hybrids in cationic reverse micelles', *Langmuir*, 2009 **25** (8) 4421–4428.
24. Wang H, Zhou W, Ho DL, Winey KI, Fischer JE, *et al.*, 'Dispersing single-walled carbon nanotubes with surfactants: A small angle neutron scattering study', *Nano Lett*, 2004 **4** (9) 1789–1793.
25. Cheng F, Imin P, Maunders C, Botton G and Adronov A, 'Soluble, discrete supramolecular complexes of single-walled carbon nanotubes with fluorene-based conjugated polymers', *Macromol*, 2008 **41** (7) 2304–2308.
26. Curran SA, Ajayan PM, Blau WJ, Carroll DL, Coleman JN, *et al.*, 'A composite from poly(m-phenylenevinylene-co-2,5-dioctoxy-p-phenylenevinylene) and carbon nanotubes: A novel material for molecular optoelectronics', *Adv Mater*, 1998 **10** (14) 1091–1093.
27. Coleman JN, Dalton AB, Curran S, Rubio A, Davey AP, *et al.*, 'Phase separation of carbon nanotubes and turbostratic graphite using a functional organic polymer', *Adv Mater*, 2000 **12** (3) 213–216.

28. Dalton AB, Stephan C, Coleman JN, McCarthy B, Ajayan PM, *et al.*, 'Selective interaction of a semiconjugated organic polymer with single-wall nanotubes', *J Phys Chem B*, 2000 **104** (43) 10012–10016.
29. Kang YJ and Taton TA, 'Micelle-encapsulated carbon nanotubes: A route to nanotube composites', *J Am Chem Soc*, 2003 **125** (19) 5650–5651.
30. Georgakilas V, Kordatos K, Prato M, Guldi DM, Holzingger M and Hirsch A, 'Organic functionalization of carbon nanotubes', *J Am Chem Soc*, 2002 **124** (5) 760–761.
31. Datsyuk V, Kalyva M, Papagelis K, Parthenios J, Tasis D, *et al.*, 'Chemical oxidation of multiwalled carbon nanotubes', *Carbon*, 2008 **46** (6) 833–840.
32. Sahoo NG, Rana S, Cho JW, Li L and Chan SH, 'Polymer nanocomposites based on functionalized carbon nanotubes', *Prog Polym Sci*, 2010 **35** (7) 837–867.
33. Qu LW, Veca LM, Lin Y, Kitaygorodskiy A, Chen BL, *et al.*, 'Soluble nylon-functionalized carbon nanotubes from anionic ring-opening polymerization from nanotube surface', *Macromol*, 2005 **38** (24) 10328–10331.
34. Yang M, Gao Y, Li HM and Adronov A, 'Functionalization of multiwalled carbon nanotubes with polyamide 6 by anionic ring-opening polymerization', *Carbon*, 2007 **45** (12) 2327–2333.
35. Zeng QH, Yu AB, Lu GQ and Paul DR, 'Clay-based polymer nanocomposites: research and commercial development', *J Nanosci Nanotechnol*, 2005 **5** (10) 1574–1592.
36. Tjong SC, 'Structural and mechanical properties of polymer nanocomposites', *Mater Sci Eng R: Reports*, 2006 **53** (3–4) 73–197.
37. Ray SS and Okamoto M, 'Polymer/layered silicate nanocomposites: A review from preparation to processing', *Prog Polym Sci*, 2003 **28** (11) 1539–1641.
38. Zanetti M, Lomakin S and Camino G, 'Polymer layered silicate nanocomposites', *Macromol Mater Eng*, 2000 **279** (6) 1–9.
39. Fornes TD, Yoon PJ, Hunter DL, Keskkula H and Paul DR, 'Effect of organoclay structure on nylon 6 nanocomposite morphology and properties', *Polym*, 2002 **43** (22) 5915–5933.
40. Nguyen QT and Baird DG, 'Preparation of polymer-clay nanocomposites and their properties', *Advances in Polym Technol*, 2006 **25** (4) 270–285.
41. Kim ES, Shim JH, Woo JY, Yoo KS and Yoon JS, 'Effect of the silane modification of clay on the tensile properties of nylon 6/clay nanocomposites', *J Appl Polym Sci*, 2010 **117** (2) 809–816.
42. Wang KH, Choi MH, Koo CM, Choi YS and Chung IJ, 'Synthesis and characterization of maleated polyethylene/clay nanocomposites', *Polym*, 2001 **42** (24) 9819–9826.
43. Kato M, Okamoto H, Hasegawa N, Tsukigase A and Usuki A, 'Preparation and properties of polyethylene-clay hybrids', *Polym Eng Sci*, 2003 **43** (6) 1312–1316.
44. Ristolainen N, Vainio U, Paavola S, Torkkeli M, Serimaa R and Seppala J, 'Polypropylene/organoclay nanocomposites compatibilized with hydroxyl-functional polypropylenes', *J Polym Sci B-Polym Phys*, 2005 **43** (14) 1892–1903.
45. Lim SH, Dasari A, Yu ZZ and Mai YW, unpublished work.
46. Fong H, Liu WD, Wang CS and Vaia RA, 'Generation of electrospun fibers of nylon 6 and nylon 6-montmorillonite nanocomposite', *Polym*, 2002 **43** (3) 775–780.
47. Kojima Y, Usuki A, Kawasumi M, Okada A, Kurauchi T and Kamigaito O, 'Synthesis of nylon-6-clay hybrid by montmorillonite intercalated with ε-caprolactam', *J Polym Sci Part A: Polym Chem*, 1993 **31** (4) 983–986.

48. Kojima Y, Usuki A, Kawasumi M, Okada A, Fukushima Y, *et al.*, 'Mechanical properties of nylon-6-clay hybrid', *J Mater Res*, 1993 **8** (5) 1185–1189.
49. Kojima Y, Usuki A, Kawasumi M, Okada A, Kurauchi T and Kamigaito O, 'One-pot synthesis of nylon-6 clay hybrid', *J Polym Sci Part A: Polym Chem*, 1993 **31** (7) 1755–1758.
50. Usuki A, Kawasumi M, Kojima Y, Okada A, Kurauchi T and Kamigaito O, 'Swelling behavior of montmorillonite cation exchanged for omega-amino acids by ε-caprolactam', *J Mater Res*, 1993 **8** (5) 1174–1178.
51. Fukushima Y and Inagaki S, 'Synthesis of an intercalated compound of montmorillonite and 6-polyamide', *J Inclusion Phenom*, 1987 **5** (4) 473–482.
52. Usuki A, Kojima Y, Kawasumi M, Okada A, Fukushima Y, *et al.*, 'Synthesis of nylon 6-clay hybrid', *J Mater Res*, 1993 **8** (5) 1179–1184.
53. Yang F, Ou YC and Yu ZZ, 'Polyamide 6 silica nanocomposites prepared by *in situ* polymerization', *J Appl Polym Sci*, 1998 **69** (2) 355–361.
54. Zeng HL, Gao C, Wang YP, Watts PCP, Kong H, *et al.*, '*In situ* polymerization approach to multiwalled carbon nanotubes-reinforced nylon 1010 composites: Mechanical properties and crystallization behavior', *Polym*, 2006 **47** (1) 113–122.
55. Zhao CG, Hu GJ, Justice R, Schaefer DW, Zhang SM, *et al.*, 'Synthesis and characterization of multi-walled carbon nanotubes reinforced polyamide 6 via *in situ* polymerization', *Polym*, 2005 **46** (14) 5125–5132.
56. Xu C, Jia Z, Wu D, Han Q and Meek T, 'Fabrication of nylon-6/carbon nanotube composites', *J Electronic Mater*, 2006 **35** (5) 954–957.
57. Kang M, Myung SJ and Jin HJ, 'Nylon 610 and carbon nanotube composite by *in situ* interfacial polymerization', *Polym*, 2006 **47** (11) 3961–3966.
58. Moniruzzaman M, Chattopadhyay J, Billups WE and Winey KI, 'Tuning the mechanical properties of SWNT/Nylon 6, 10 composites with flexible spacers at the interface', *Nano Lett*, 2007 **7** (5) 1178–1185.
59. Liu LM, Qi ZN and Zhu XG, 'Studies on nylon 6 clay nanocomposites by melt-intercalation process', *J Appl Polym Sci*, 1999 **71** (7) 1133–1138.
60. Vaia RA, Ishii H and Giannelis EP, 'Synthesis and properties of 2-dimensional nanostructures by direct intercalation of polymer melts in layered silicate', *Chem Mater*, 1993 **5** (12) 1694–1696.
61. Gilman JW, Awad WH, Davis RD, Shields J, Harris RH, *et al.*, 'Polymer/layered silicate nanocomposites from thermally stable triakylimidazolium-treated montmorillonite', *Chem Mater*, 2002 **14** (9) 3776–3785.
62. Radhakrishnan VK, Davis EW and Davis VA, 'Influence of initial mixing methods on melt-extruded single-walled carbon nanotube-polypropylene nanocomposites', *Polym Eng Sci*, 2010 **50** (9) 1831–1842.
63. Shofner ML, Khabashesku VN and Barrera EV, 'Processing and mechanical properties of fluorinated single-walled carbon nanotube-polyethylene composites', *Chem Mater*, 2006 **18** (4) 906–913.
64. Liu TX, Phang IY, Shen L, Chow SY and Zhang WD, 'Morphology and mechanical properties of multiwalled carbon nanotubes reinforced nylon-6 composites', *Macromol*, 2004 **37** (19) 7214–7222.
65. Kiss A, Fekete E and Pukánszky B, 'Aggregation of $CaCO_3$ particles in PP composites: Effect of surface coating', *Composites Sci Tech*, 2007 **67** (7–8) 1574–1583.
66. Fornes TD and Paul DR, 'Modeling properties of nylon 6/clay nanocomposites using composite theories', *Polym*, 2003 **44** (17) 4993–5013.

67. Paul DR and Robeson LM, 'Polymer nanotechnology: Nanocomposites', *Polym*, 2008 **49** (15) 3187–3204.
68. Lee HS, Fasulo PD, Rodgers WR and Paul DR, 'TPO based nanocomposites. Part 1. Morphology and mechanical properties', *Polym*, 2005 **46** (25) 11673–11689.
69. Fornes TD, Hunter DL and Paul DR, 'Nylon-6 nanocomposites from alkylammonium-modified clay: The role of alkyl tails on exfoliation', *Macromol*, 2004 **37** (5) 1793–1798.
70. Bhattacharyya AR, Potschke P, Haussler L and Fischer D, 'Reactive compatibilization of melt mixed PA6/SWNT composites: Mechanical properties and morphology', *Macromol Chem Phys*, 2005 **206** (20) 2084–2095.
71. Weon JI and Sue HJ, 'Effects of clay orientation and aspect ratio on mechanical behavior of nylon-6 nanocomposites', *Polym*, 2005 **46** (17) 6325–6334.
72. Dennis HR, Hunter DL, Chang D, Kim S, White JL, *et al.*, 'Effect of melt processing conditions on the extent of exfoliation in organoclay-based nanocomposites', *Polym*, 2001 **42** (23) 9513–9522.
73. Chavarria F, Nairn K, White P, Hill AJ, Hunter DL and Paul DR, 'Morphology and properties of nanocomposites from organoclays with reduced cation exchange capacity', *J Appl Polym Sci*, 2007 **105** (5) 2910–2924.
74. Fornes TD, Yoon PJ, Keskkula H and Paul DR, 'Nylon 6 nanocomposites: The effect of matrix molecular weight', *Polym*, 2001 **42** (25) 9929–9940.
75. Alexandre M and Dubois P, 'Polymer-layered silicate nanocomposites: Preparation, properties and uses of a new class of materials', *Mater Sci Eng R: Reports*, 2000 **28** (1–2) 1–63.
76. LeBaron PC, Wang Z and Pinnavaia TJ, 'Polymer-layered silicate nanocomposites: An overview', *Appl Clay Sci*, 1999 **15** (1–2) 11–29.
77. Chen B, Evans JRG, Greenwell HC, Boulet P, Coveney PV, *et al.*, 'A critical appraisal of polymer-clay nanocomposites', *Chem Soc Reviews*, 2008 **37** (3) 568–594.
78. Liu HM, Wang X, Fang PF, Wang SJ, Qi X, *et al.*, 'Functionalization of multi-walled carbon nanotubes grafted with self-generated functional groups and their polyamide 6 composites', *Carbon*, 2010 **48** (3) 721–729.
79. Yu ZZ, Yan C, Yang MS and Mai YW, 'Mechanical and dynamic mechanical properties of nylon 66/montmorillonite nanocomposites fabricated by melt compounding', *Polym International*, 2004 **53** (8) 1093–1098.
80. Chen L, Phang IY, Wong SC, Lv PF and Liu TX, 'Embrittlement mechanisms of nylon 66/organoclay nanocomposites prepared by melt-compounding process', *Mater Manufacturing Processes*, 2006 **21** (2) 153–158.
81. He CB, Liu TX, Tjiu WC, Sue HJ and Yee AF, 'Microdeformation and fracture mechanisms in polyamide-6/organoclay nanocomposites', *Macromol*, 2008 **41** (1) 193–202.
82. Lim SH, Dasari A, Yu ZZ, Mai YW, Liu SL and Yong MS, 'Fracture toughness of nylon 6/organoclay/elastomer nanocomposites', *Compos Sci Technol*, 2007 **67** (14) 2914–2923.
83. Dasari A, Yu ZZ and Mai YW, 'Transcrystalline regions in the vicinity of nanofillers in polyamide-6', *Macromol*, 2007 **40** (1) 123–130.
84. Windle AH, 'Two defining moments: A personal view by Prof Alan H Windle', *Compos Sci Technol*, 2007 **67** (5) 929–930.
85. Wichmann MHG, Schulte K and Wagner HD, 'On nanocomposite toughness', *Compos Sci Technol*, 2008 **68** (1) 329–331.

86. Gao YC, Mai YW and Cotterell B, 'Fracture of fiber-reinforced materials', *Zeitschrift Fur Angewandte Mathematik Und Physik*, 1988 **39** (4) 550–572.
87. Dasari A, Yu ZZ and Mai YW, 'Effect of blending sequence on microstructure of ternary nanocomposites', *Polym*, 2005 **46** (16) 5986–5991.
88. Sue HJ, Gam KT, Bestaoui N, Clearfield A, Miyamoto M and Miyatake N, 'Fracture behavior of $\alpha$-zirconium phosphate-based epoxy nanocomposites', *Acta Materialia*, 2004 **52** (8) 2239–2250.
89. Liu WP, Hoa SV and Pugh M, 'Morphology and performance of epoxy nanocomposites modified with organoclay and rubber', *Polym Eng Sci*, 2003 **44** (6) 1178–1186.
90. González I, Eguiazábal JI and Nazábal J, 'Rubber-toughened polyamide 6/clay nanocomposites', *Compos Sci Technol*, 2006 **66** (11–12) 1833–1843.
91. Tjong SC and Meng YZ, 'Impact-modified polypropylene/vermiculite nanocomposites', *J Polym Sci B-Polym Phys*, 2003 **41** (19) 2332–2341.
92. Khatua BB, Lee DJ, Kim HY and Kim JK, 'Effect of organoclay platelets on morphology of nylon-6 and poly(ethylene-ran-propylene) rubber blends', *Macromol*, 2004 **37** (7) 2454–2459.
93. Yang JL, Zhang Z and Zhang H, 'The essential work of fracture of polyamide 66 filled with $TiO_2$ nanoparticles', *Compos Sci Technol*, 2005 **65** (15–16) 2374–2379.
94. Zhang H, Zhang Z, Yang JL and Friedrich K, 'Temperature dependence of crack initiation fracture toughness of various nanoparticles filled polyamide 66', *Polym*, 2006 **47** (2) 679–689.
95. Briscoe BJ and Sinha SK, 'Scratch resistance and localised damage characteristics of polymer surfaces – a review', *Materialwissenschaft Und Werkstofftechnik*, 2003 **34** (10–11) 989–1002.
96. Lim SC and Ashby MF, 'Wear-mechanism maps', *Acta Metallurgica*, 1987 **35** (1) 1–24.
97. Sayles RS, 'Basic principles of rough surface contact analysis using numerical methods', *Tribology International*, 1996 **29** (8) 639–650.
98. Dasari A, Yu ZZ and Mai YW, 'Fundamental aspects and recent progress on wear/scratch damage in polymer nanocomposites', *Mater Sci Eng R: Reports*, 2009 **63** (2) 31–80.
99. Bahadur S, 'The development of transfer layers and their role in polymer tribology', *Wear*, 2000 **245** (1–2) 92–99.
100. Fischer TE, 'Tribochemistry', *Annual Review of Mater Sci*, 1988 **18** 303–323.
101. Zhang MQ, Rong MZ, Yu SL, Wetzel B and Friedrich K, 'Effect of particle surface treatment on the tribological performance of epoxy based nanocomposites', *Wear*, 2002 **253** (9–10) 1086–1093.
102. Friedrich K, 'Wear models for multi-phase materials and synergistic effects in polymeric hybrid composites', In: Friedrich K, editor, *Advances in Composite Tribology*, Amsterdam: Elsevier Science Publishers B V, 1993, p 209.
103. Dasari A, Yu ZZ, Mai and YW, 'Wear and scratch damage in polymer nanocomposites', In: Friedrich K and Alois KS, editors, *Tribology of Polymeric Nanocomposites, Tribology and Interface Engineering Series*, Vol 55, Elsevier, 2008, p 374.
104. Briscoe BJ and Sinha SK, 'Wear of polymers', *Proceedings of the Institution of Mechanical Engineers J: J Eng Tribology*, 2002 **216** (J6) 401–413.
105. Chang L, Zhang Z, Breidt C and Friedrich K, 'Tribological properties of epoxy nanocomposites – I. Enhancement of the wear resistance by nano-$TiO_2$ particles', *Wear*, 2005 **258** (1–4) 141–148.

106. Kietzke T, Neher D, Landfester K, Montenegro R, Guntner R and Scherf U, 'Novel approaches to polymer blends based on polymer nanoparticles', *Nature Mater*, 2003 **2** (6) 408-U7.
107. Xing XS and Li RKY, 'Wear behavior of epoxy matrix composites filled with uniform sized sub-micron spherical silica particles', *Wear*, 2004 **256** (1–2) 21–26.
108. Dasari A, Yu ZZ and Mai YW, 'Nanoscratching of nylon 66-based ternary nanocomposites', *Acta Materialia*, 2007 **55** (2) 635–646.
109. Dasari A, Yu ZZ, Mai YW and Kim JK, 'Orientation and the extent of exfoliation of clay on scratch damage in polyamide 6 nanocomposites', *Nanotechnol*, 2008 **19** (5) 055708.
110. Messersmith PB and Giannelis EP, 'Synthesis and barrier properties of poly(ε-caprolactone)-layered silicate nanocomposites', *J Polym Sci A: Polym Chem*, 1995 **33** (7) 1047–1057.
111. Osman MA, Mittal V, Morbidelli M and Suter UW, 'Polyurethane adhesive nanocomposites as gas permeation barrier', *Macromol*, 2003 **36** (26) 9851–9858.
112. Gorrasi G, Tortora M, Vittoria V, Galli G and Chiellini E, 'Transport and mechanical properties of blends of poly(ε-caprolactone) and a modified montmorillonite-poly(ε-caprolactone) nanocomposite', *J Polym Sci B: Polym Phys*, 2002 **40** (11) 1118–1124.
113. Doppers LM, Breen C and Sammon C, 'Diffusion of water and acetone into poly(vinyl alcohol)-clay nanocomposites using ATR-FTIR', *Vibrational Spectroscopy*, 2004 **35** (1–2) 27–32.
114. Xu B, Zheng Q, Song YH and Shangguan Y, 'Calculating barrier properties of polymer/clay nanocomposites: Effects of clay layers', *Polym*, 2006 **47** (8) 2904–2910.
115. Frounchi M, Dadbin S, Salehpour Z and Noferesti M, 'Gas barrier properties of PP/EPDM blend nanocomposites', *J Membrane Sci*, 2006 **282** (1–2) 142–148.
116. Nazarenko S, Meneghetti P, Julmon P, Olson BG and Qutubuddin S, 'Gas barrier of polystyrene montmorillonite clay nanocomposites: Effect of mineral layer aggregation', *J Polym Sci B: Polym Phys*, 2007 **45** (13) 1733–1753.
117. Nielsen LE, 'Platelet particles enhance barrier of polymers by forming tortuous path', *J Macromol Sci: Chem*, 1967 **A1** (5) 929–942.
118. Bharadwaj RK, 'Modeling the barrier properties of polymer-layered silicate nanocomposites', *Macromol*, 2001 **34** (26) 9189–9192.
119. Yano K, Usuki A, Okada A, Kurauchi T and Kamigaito O, 'Synthesis and properties of polyimide clay hybrid', *J Polym Sci A: Polym Chem*, 1993 **31** (10) 2493–2498.
120. Kim SH and Kim SC, 'Synthesis and properties of poly(ethylene terephthalate)/clay nanocomposites by *in situ* polymerization', *J Appl Polym Sci*, 2007 **103** (2) 1262–1271.
121. Yano K, Usuki A and Okada A, 'Synthesis and properties of polyimide-clay hybrid films', *J Polym Sci A: Polym Chem*, 1997 **35** (11) 2289–2294.
122. Choi WJ, Kim HJ, Yoon KH, Kwon OH and Hwang CI, 'Preparation and barrier property of poly(ethylene terephthalate)/clay nanocomposite using clay-supported catalyst', *J Appl Polym Sci*, 2006 **100** (6) 4875–4879.
123. Ray SS, Okamoto K and Okamoto M, 'Structure-property relationship in biodegradable poly(butylene succinate)/layered silicate nanocomposites', *Macromol*, 2003 **36** (7) 2355–2367.
124. Koo CM, Kim SO and Chung IJ, 'Study on morphology evolution, orientational behavior, and anisotropic phase formation of highly filled polymer-layered silicate nanocomposites', *Macromol*, 2003 **36** (8) 2748–2757.

125. Bharadwaj RK, Mehrabi AR, Hamilton C, Trujillo C, Murga M, *et al.*, 'Structure-property relationships in cross-linked polyester-clay nanocomposites', *Polym*, 2002 **43** (13) 3699–3705.
126. Lu CS and Mai YW, 'Influence of aspect ratio on barrier properties of polymer-clay nanocomposites', *Phys Rev Lett*, 2005 **95** (8) 088303.
127. Cussler EL, Hughes SE, Ward WJ and Aris R, 'Barrier membranes', *J Membrane Sci*, 1988 **38** (2) 161–174.
128. Fukuda M and Kuwajima S, 'Molecular-dynamics simulation of moisture diffusion in polyethylene beyond 10 ns duration', *J Chem Phys*, 1997 **107** (6) 2149–2159.
129. Kim DH, Fasulo PD, Rodgers WR and Paul DR, 'Structure and properties of polypropylene-based nanocomposites: Effect of PP-g-MA to organoclay ratio', *Polym*, 2007 **48** (18) 5308–53323.
130. Yoon PJ, Fornes TD and Paul DR, 'Thermal expansion behavior of nylon 6 nanocomposites', *Polym*, 2002 **43** (25) 6727–6741.
131. Chow TS, 'Effect of particle shape at finite concentration on thermal expansion of filled polymers', *J Polym Sci B: Polym Phys*, 1978 **16** (6) 967–970.
132. Hwang JJ and Liu HJ, 'Influence of organophilic clay on the morphology, plasticizer-maintaining ability, dimensional stability, and electrochemical properties of gel polyacrylonitrile (PAN) nanocomposite electrolytes', *Macromol*, 2002 **35** (19) 7314–7319.
133. Ray SS, Yamada K, Okamoto M and Ueda K, 'New polylactide-layered silicate nanocomposites. 2. Concurrent improvements of material properties, biodegradability and melt rheology', *Polym*, 2003 **44** (3) 857–866.
134. Zhu WP, Zhang GP, Yu JY and Dai G, 'Crystallization behavior and mechanical properties of polypropylene copolymer by *in situ* copolymerization with a nucleating agent and/or nano-calcium carbonate', *J Appl Polym Sci*, 2004 **91** (1) 431–438.
135. Yu ZZ, Hu GH, Varlet J, Dasari A and Mai YW, 'Water-assisted melt compounding of nylon-6/pristine montmorillonite nanocomposites', *J Polym Sci B: Polym Phys*, 2005 **43** (9) 1100–1112.
136. Xie W, Gao ZM, Pan WP, Hunter D, Singh A and Vaia R, 'Thermal degradation chemistry of alkyl quaternary ammonium montmorillonite', *Chem Mater*, 2001 **13** (9) 2979–2990.
137. Lan T, Kaviratna D and Pinnavaia TJ, 'Epoxy self-polymerization in smectite clays', *J Phys Chem Solids*, 1996 **57** (6–8) 1005–1010.
138. Zhang Y, Lee S, Yoonessi M, Liang K and Pittman CU, 'Phenolic resin-trisilanolphenyl polyhedral oligomeric silsesquioxane (POSS) hybrid nanocomposites: Structure and properties', *Polym*, 2006 **47** (9) 2984–2996.
139. Jash P and Wilkie CA, 'Effects of surfactants on the thermal and fire properties of poly(methyl methacrylate)/clay nanocomposites', *Polym Degradation and Stability*, 2005 **88** (3) 401–416.
140. Fina A, Tabuani D, Carniato F, Frache A, Boccaleri E and Camino G, 'Polyhedral oligomeric silsesquioxanes (POSS) thermal degradation', *Thermochimica Acta*, 2006 **440** (1) 36–42.
141. Laoutid F, Bonnaud L, Alexandre M, Lopez-Cuesta JM and Dubois P, 'New prospects in flame retardant polymer materials: From fundamentals to nanocomposites', *Mater Sci Eng R*, 2009 **63** (3) 100–125.
142. Kiliaris P and Papaspyrides CD, 'Polymer/layered silicate (clay) nanocomposites: An overview of flame retardancy', *Prog Polym Sci*, 2010 **35** (7) 902–958.

143. Dasari A, Cai G, Yu ZZ and Mai Y-W, 'Flame retardancy of polymer-clay nanocomposites', In: Tjong SC and Mai Y-W, editors, *Physical properties of polymer nanocomposites*, Cambridge, Woodhead Publishing Ltd, 2010, p. 347.
144. Garces JM, Moll DJ, Bicerano J, Fibiger R and McLeod DG, 'Polymeric nanocomposites for automotive applications', *Adv Mater*, 2000 **12** (23) 1835–1839.
145. Njuguna B and Pielichowski K, 'Polymer nanocomposites for aerospace applications: Properties', *Adv Eng Mater*, 2003 **5** (11) 769–778.
146. Huang ZM, Zhang YZ, Kotaki M and Ramakrishna S, 'A review on polymer nanofibers by electrospinning and their applications in nanocomposites', *Compos Sci Technol*, 2003 **63** (15) 2223–2253.
147. Reneker DH and Chun I, 'Nanometre diameter fibres of polymer, produced by electrospinning', *Nanotechnol*, 1996 **7** (3) 216–223.
148. Doshi J and Reneker DH, 'Electrospinning process and applications of electrospun fibers', *Special Technical Session on Electrostatics in Polymer Processing and Charge Monitoring*, IEEE Industry Applications Society 28th Annual Meeting, 2–8 October, 1993, Toronto, Canada, 1698–1703.
149. Li D and Xia YN, 'Electrospinning of nanofibers: Reinventing the wheel?', *Adv Mater*, 2004 **16** (14) 1151–1170.
150. Reneker DH, Yarin AL, Fong H and Koombhongse S, 'Bending instability of electrically charged liquid jets of polymer solutions in electrospinning', *J Appl Phys*, 2000 **87** (9) 4531–4537.

# 10

# Clay-containing poly(ethylene terephthalate) (PET)-based polymer nanocomposites

J. BANDYOPADHYAY, National Centre for Nanostructured Materials, Republic of South Africa and S. SINHA RAY, National Centre for Nanostructured Materials, Republic of South Africa and University of Johannesburg, Republic of South Africa

**Abstract:** This chapter is an extensive overview of clay-containing nanocomposites of poly(ethylene terephthalate) (PET). The various techniques used to prepare clay-based PET nanocomposites, their structural and morphological characterization, their improved mechanical and material properties, their melt-state rheological and crystallization behavior, and, finally, current applications and future prospects of PET-based nanocomposite materials are discussed.

**Key words:** poly(ethylene terephthalate), clay, nanocomposites, preparation, characterization, properties, review.

## 10.1 Introduction: the importance of poly(ethylene terephthalate) (PET)-based nanocomposites

Over the past few years, the mixing of nanoparticles into a polymer matrix has been an area of great research interest since such composite materials can exhibit a concurrent improvement of properties compared with the neat polymer. Such improvements include higher thermal stability and flame retardancy, mechanical properties, gas barrier properties, chemical resistance, and so on.[1,2] The nanoscale dispersion of fillers or controlling the nanostructures in the composite can introduce new physical properties and novel behavior that are absent in the unfilled matrices. In general, the properties of nanocomposites depend on the following parameters: the properties of the matrix polymer and filler, the nature and the strength of the interfacial interaction, and the area of interfacial bonds. For a nanoparticle-filled composite, the area of interfacial bonds is determined by the aspect ratio (i.e., the length to thickness ratio). Therefore, the nano-level dispersion of the filler in a polymer matrix plays a key role in achieving a concurrent improvement of properties in the nanocomposites. Hence, significant research efforts are underway to control the nanostructures via innovative synthetic approaches.

Because of these improved properties, commercial applications of nanocomposites have been growing at a rapid rate. Improvements in mechanical properties have resulted in major interest in nanocomposite materials for automotive and industrial applications.[3–5] Improved barrier properties make nanocomposites suitable for packaging materials.[1,2,5] Coating with nanocomposite films can enhance the toughness and hardness of a material without interfering with light transmission characteristics.[1] In some cases the optical transparency of a nanocomposite film is better than the neat polymer, which in turn makes nanocomposites suitable for producing new scratch- and abrasion-resistant materials.[1] Furthermore, the use of nanocomposites in the next few years will include key areas such as drug delivery systems, anticorrosion barrier coatings, UV protection gels, scratch-free paints, superior strength fiber and films, etc.

Along with many other polymers, the global demand for PET, including its use as a matrix polymer in the preparation of nanocomposites, has grown rapidly over the last decade. The demand for PET is highest in Asia. Europe is the second largest consumer of PET. The demand for PET in South and Central America is also growing strongly. The majority of the world's PET production is for synthetic fibers (in excess of 60%) with bottle production accounting for around 30% of global demand.[6] PET is generally referred to as 'polyester' in textile applications, while 'PET' is used most often to refer to packaging. About 18% of the world's polymer production comprises polyester, which ranks third after polyethylene (PE) and polypropylene (PP). The question is why are PET and PET-based nanocomposites so important from an industrial point of view?

PET is used in a variety of applications, such as fibers, bottles, films, and engineering plastics for automobiles and electronics, due to its relatively low cost and high performance.[7] PET has high impact resistance and is strong and naturally colorless with high transparency. It has good gas barrier properties. For this reason, PET is the most popular material for soft drinks bottles. Sometimes polyvinyl alcohol (PVA) is sandwiched between PET films to give a further improvement in the gas barrier properties to meet the most challenging applications for sensitive beverages. A biaxially oriented PET film (known as Mylar) can be aluminized by evaporating a thin film of metal onto it to reduce its permeability and to make it reflective and opaque. This property is useful for flexible food packaging and thermal insulation, such as space blankets.

Because of its high mechanical strength (its tensile strength is 55–75 MPa), PET film is also used in tape applications, such as the carrier for magnetic tape or the backing for pressure-sensitive additive tapes.

While most thermoplastics, in principle, can be recycled, the recycling of PET bottles is more practical than recycling other thermoplastics, such as PP, polystyrene, etc. For example, recycled PET bottles can be used to prepare fleece material. Fleece is a soft, warm, comfortable fabric usually used in sweaters, jackets, gym clothes, hats, etc.

PET is also an excellent candidate for thermal disposal (incineration) since it is composed of carbon, hydrogen, and oxygen, with only trace amounts of catalyst elements (but no sulfur). PET has the energy content of soft coal.[8]

It is clear that PET is a very useful and important plastic from an industrial point of view. For this reason, worldwide research is being conducted on PET to make it more useful. One approach is to prepare nanocomposites of PET by using different types of nanofiller. Among the different nanofillers, such as clay, carbon nanotubes, and metal-oxide nanoparticles, the clay-based nanocomposites are very important since they often exhibit enhanced thermal and mechanical properties, heat resistance, gas permeability, and flammability characteristics.[1] Hence, this chapter will focus mainly on the preparation, characterization, and properties of clay-based PET nanocomposites. The primary objective in the development of a PET/clay nanocomposite is to improve the gas barrier property of the matrix for beverage and food packaging.[9] Another expectation for PET/clay nanocomposites is as an alternative to glass-fiber-reinforced PET. It is well known that the properties of a nanocomposite are directly related to the nanoscale dispersion of the clay platelets in the PET matrix. Therefore, it is a challenge for researchers to achieve good dispersion. It is necessary to understand how to control the agglomeration of dispersed nanoparticles (through the choice of nanoparticle), how processing conditions affect the properties, and, finally, the effect of nano-confinement on the properties.

## 10.2 Types of PET-based nanocomposite

Depending on the nanofiller, PET-based composites can be classified into three categories: clay-based PET nanocomposites, PET composites containing carbon nanotubes as a filler material, and PET composites containing metal-oxide nanoparticles. Usually most clays are environmentally friendly and possess a very high modulus and aspect ratio, which are the basic requirements for physical property improvement after nanocomposite preparation. Besides clay, the other popular nanofillers are carbon nanotubes or nanofibers. They are widely studied because of their high aspect ratio and outstanding Young's modulus, which are favorable properties for making multifunctional polymer nanocomposites. However, significant agglomeration, difficult surface modification, and the high cost of carbon-based nanofillers restrict their widespread use in applications as advanced nanofillers.[10] That is why clay is the most popular nanofiller.

## 10.3 Preparative methods

The methods used to prepare nanocomposite can be divided into three main groups according to the starting materials and processing techniques:[1]

1. *Intercalation of the polymer or pre-polymer from solution.* This method uses a solvent system in which the polymer or pre-polymer is soluble and the silicate layers are swellable. The clay is first swollen in a suitable solvent, such as water, chloroform, or toluene. When the polymer and clay solutions are mixed, the polymer chains intercalate and displace the solvent within the

interlayers of the silicate. On solvent removal, the intercalated structure remains, resulting in a nanocomposite.

2. In situ *intercalative polymerization*. In this method, the clay is swollen within the liquid monomer or a monomer solution so polymer formation can occur between the intercalated sheets. Polymerization can be initiated either by heat or radiation, by the diffusion of a suitable initiator, or by an organic initiator or catalyst fixed through cation exchange inside the interlayer before the swelling step.
3. *Melt intercalation*. This method involves annealing, statically or under shear, a mixture of the polymer and organically modified clay at the softening point of the polymer. This method has great advantages over either *in situ* intercalative polymerization or polymer solution intercalation. First, this method is environmentally benign due to the absence of organic solvents. Second, it is compatible with current industrial processes, such as extrusion and injection molding. Melt intercalation can use polymers that are not suitable for *in situ* polymerization or solution intercalation.

Generally, layered silicate minerals are divided into three major groups: (a) the kaolinite group, (b) the smectite group, and (c) the illite or the mica group. Of these three major groups, smectite types, or more precisely montmorillonite (MMT), saponite, and hectorite, are the most commonly used layered silicates in polymer nanocomposite technology. Of these MMT is the most commonly used layered silicate for the preparation of polymer nanocomposites because it is highly abundant and inexpensive.[1,2]

In the primary structure of MMT, neighboring hydrophilic platelets attract each other through multiple anionic charges and exchangeable metal counter ions. As a result, there is an enormous force of ionic attraction in the layered structure of MMT, which makes it difficult to break and disperse homogeneously in hydrophobic polymer matrices.[5] This problem is usually solved by modifying MMT with certain organic compounds, such as quaternary ammonium or phosphonium salts, which has proven to be an effective way of improving the compatibility between the polymer matrix and the clay. The processing temperature of PET, as well as the temperature at which this polymer is synthesized by a polycondensation reaction, is about 280°C. This is well above the decomposition temperature of the ammonium surfactants commonly used for organic modification of pristine MMT.[11] Therefore, it is a challenging job to disperse MMT particles nicely into a PET matrix.

So far researchers have used several innovative techniques to modify pristine MMT in order to enhance the compatibility between the MMT surface and matrix PET. A simple way to prepare organically modified MMT (OMMT) is as follows: stir pristine MMT in 200 ml deionized water at 70 °C for 12 h. Then slowly add a solution of 7.69 g (19.6 mequiv) *N,N,N*-trimethyloctadecylammonium bromide in 50 ml deionized water to the MMT solution with vigorous stirring at 70 °C. After mixing, stir the mixture at 70°C for 12 h. After recovering the OMMT by filtration, it must be washed with 70°C deionized water several times to remove freely existing ionic intercalants and impurities. Finally, centrifugation and vacuum drying at 70°C for 24 h are required to obtain OMMT ready for nanocomposite preparation.[7]

### 10.3.1 Intercalation of PET from solution

Although a solvent casting method has been used to prepare a PET/clay nanocomposite, the most popular ways to prepare PET/clay nanocomposites are *in situ* intercalation and melt blending. Ou *et al.*[12] modified MMT with cetylpyridinium chloride (CPC) (CPC/clay = 2/1 by equivalent) and then prepared a PET/clay nanocomposite using a phenol/chloroform mixture as the solvent. The paper contains more details of the method used to produce the nanocomposites from PET.

### 10.3.2 *In situ* intercalative polymerization

*In situ* polymerization to prepare PET/clay nanocomposite can be carried out in several ways. The first method is the intercalation and ring-opening polymerization of ethylene terephthalate cyclic oligomers (ETC) in pre-swelled OMMT galleries. Usually, OMMT is swelled in dichloromethane (DCM). The ETC is dissolved in DCM separately. Then the OMMT solution is added to the ETC solution under vigorous stirring followed by solvent extraction and drying resulting in an intercalated nanocomposite.[7] Several sessions of extensive nitrogen purging and high vacuum treatment are necessary during polymerization. A PET/clay nanocomposite has been synthesized via the ester exchange reaction of ethylene glycol (EG) and dimethyl terephthalate (DMT) in the clay gallery in the presence of a zinc acetate catalyst followed by, finally, polycondensation in the presence of an antimony (III) oxide catalyst.[2] Here, the clay was ultrasonicated with ethylene glycol and zinc acetate before the ester interchange reaction was initiated due to the addition of DMT to this mixture.

The clay can also be added in the polycondensation step after transesterification of EG and DMT in the presence of a manganese acetate catalyst, but the reaction temperature has to be increased.[3] Instead of DMT, the combination of purified terephthalic acid (PTA), antimony (III) oxide as a catalyst, and additives, such as triethyl phosphate (TEP) as a thermal stabilizer and cobalt (II) acetate tetrahydrate (CoAc) as a color inhibitor, can also be used.[13]

Jung *et al.*[14] prepared a PET/organically modified mica hybrid by transesterification of EG and DMT in the presence of isopropyl titanate. They cooled the sample to room temperature as soon as the ester interchange reaction was finished, repeatedly washed it with water and, finally, vacuum dried it in order to obtain the PET hybrid. After compression molding of the sample they extruded it through the die of a capillary rheometer.[13]

Since polyamide 6 (PA6) has better compatibility with clay, Li *et al.*[15] expected that, if PA6 chains were introduced into the PET, the compatibility between PET and the clay could be improved. So they synthesized a PET/PA6 copolymer/MMT nanocomposite by transesterification of DMT and EG in the presence of a zinc acetate catalyst at 180°C, then dissolved the PA6/MMT master batch sample with an antimony trioxide catalyst in the reaction system and increased the reaction temperature to 280°C. The results of $^1H$ nuclear magnetic resonance (NMR) proved that the ester amide exchange reaction had taken place and an average 3.75 repeat units of PA6 were dispersed randomly in the PET molecular chains.[15]

With the aim of exfoliating the clay platelets in a polyester/clay nanocomposite, Tsai *et al.*[16] proposed a new preparative technique. They prepared three agents: Agent A contained antimony acetate dissolved in EG, Agent B contained purified clay swollen in EG, and Agent C contained sodium cocoampho-hydroxypropylsulfonate dissolved in deionized water. After mixing Agents A and B, calcination resulted in modified clay intercalated with antimony oxide ($Sb_2O_3$). The $Sb_2O_3$-modified clay was added to Agent C to give $Sb_2O_3$-SB modified clay. Then the modified clays were mixed with terephthalic acid bis(2-hydroxyethylester) (BHET) at its melting point for 2 h to prepare the nanocomposite named PET/PK805/Sb-SB.[16]

In order to understand the effect of different *in situ* intercalation techniques on the dispersion of clay platelets in a PET matrix, transmission electron micrographs (TEM images) were used. Figure 10.1 shows TEM micrographs of (a) PET nanocomposite containing 2 wt% organically modified mica,[14] (b) PET/PA6 copolymers/organically modified MMT nanocomposite containing 2 wt% clay loading,[15] and (c) a PET nanocomposite based on organically modified clay (PET/PK805/Sb-SB composite) with antimony acetate as a catalyst.[16] From these images, we can see that the dispersion of clay platelets in the PET/PA6 copolymers/OMMT nanocomposite (Fig. 10.1(b)) is better than the composite prepared via conventional *in situ* intercalation (Fig. 10.1(a)). The antimony-acetate-treated clay-based composite has better delaminated clay layers (Fig. 10.1(c)) compared with the PET/PA6 copolymers/OMMT nanocomposite. Note that different authors used different types of OMMT. However, the improvement in the dispersion of clay layers in the PET/PK805/Sb-SB composite may be attributed to the catalyst-treated clay, since antimony acetate catalyzes the polymerization reaction inside the clay galleries.

### 10.3.3 Melt intercalation

Since melt processing is compatible with current industrial processes, such as extrusion and injection molding, Costache *et al.*[11] modified natural clays, such as MMT, hectorite, and magadiite, with the highly thermally stable surfactants hexadecyl-quinolinium (Q16) and vinylbenzyl-ammonium chloride-lauryl-acrylate copolymer (L-surfactant). They prepared PET/clay nanocomposites by melt blending in a Brabender plasticorder for 7 min at 280°C. The authors found

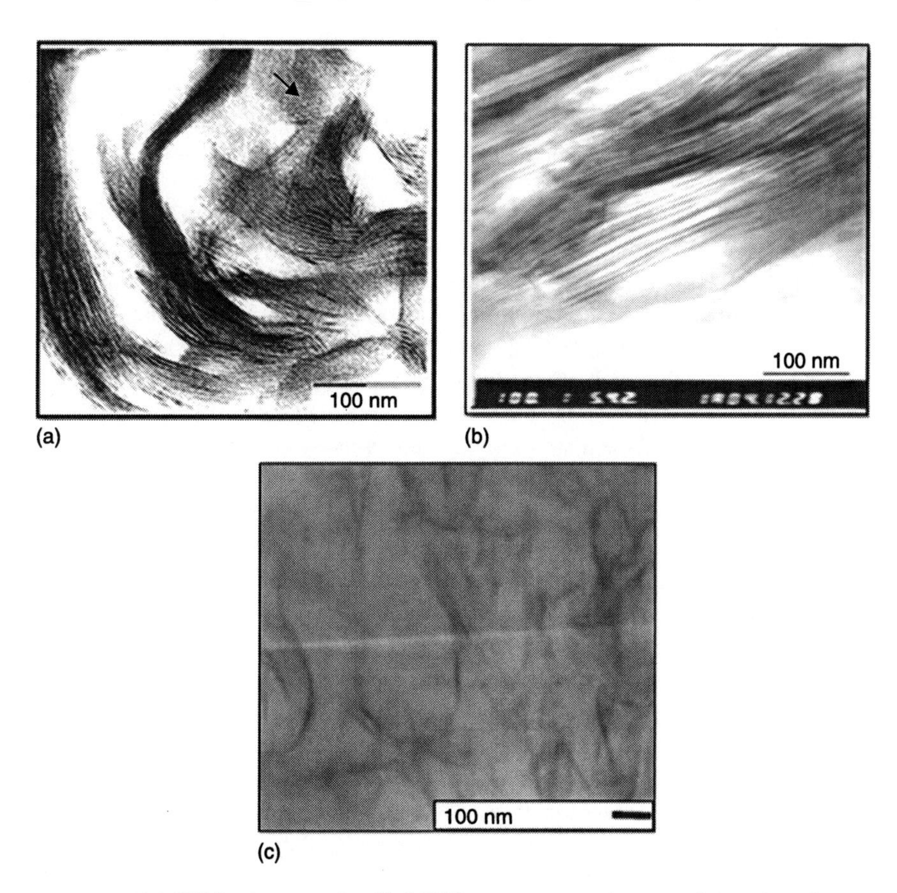


*10.1* TEM micrograph of (a) PET nanocomposite containing 2 wt% organically modified mica. Reproduced with permission from Jung *et al.*[14] (b) PET/PA6 copolymers/organically modified MMT nanocomposite containing 2 wt% clay loading. Reproduced with permission from Li *et al.*[15] (c) PET nanocomposite based on organically modified and antimony-acetate-treated clay (PET/PK805/Sb-SB composite). Reproduced with permission from Tsai *et al.*[16]

that, although the modified clay has better thermal stability, the nanocomposites did not show any improvement in stability, though in some cases it decreased. This can be explained from the thermal stability results of clays at the processing temperature for the processing time. The current authors believe that during processing at high temperature the clay starts to degrade and hence the thermal stability decreases.[11]

Processing time can be reduced significantly by preparing the nanocomposite sample in an extruder. The PET and the clay must be sufficiently dry before extrusion in order to reduce decomposition due to the moisture in the sample. So far various types of commercially available clay, such as Cloisite®Na$^+$ (CNa),[17] Cloisite®10A (C10A),[17] Cloisite®15A (C15A),[17,18] Cloisite®20A (C20A),[19] and Cloisite®25A (C25A),[20] have been used to prepare PET/clay nanocomposites via melt extrusion. The processing temperature (the temperature of different zones of the extruder to the die) of a PET/clay nanocomposite can vary between 240 and 285°C. Giraldi *et al.*[20] reported that a low screw speed during processing results in composites with a high Young's modulus, low elongation at break, strength, and impact. This is because the polymer matrix is degraded in the long residence time due to the slow speed.[20] Further, a nitrogen environment must be maintained in the hopper and the feeder to keep the materials dry[17] and, of course, to avoid possible degradation at such a high processing temperature. Another important parameter in extrusion through a die is the draw ratio. A large draw ratio results in debonding of the organoclay and the matrix polymer and creates many nano-sized voids due to excess stretching of the fibers. These voids (bubbles) hinder energy dissipation and stress transfer under a certain strain during the measurement of mechanical properties. As a result, the modulus and strength of the material decrease. It can also affect the flexibility of the material, measured by elongation at break. *N,N*-dimethyl-*N,N*-dioctadecyl ammonium modified clay is not suitable for the preparation of PET/clay nanocomposites since degradation of this modifier makes the nanocomposite brittle. On the other hand, 1,2-dimethyl-3-*N*-hexadecyl imidazolium tetrafluoroborate exhibits good dispersion in the PET matrix.[20]

Barber *et al.*[17] melted an extruded PET ionomer and organically modified MMT directly to prepare a nanocomposite. The preparation scheme is shown in Fig. 10.2(a). The concentration of the C10A clay in the composite was 5 wt%. The ionomer used was sulfonated poly(ethylene terephthalate).[17,18]

Ammala *et al.*[21] tried to improve the compatibility between the clay and the PET matrix by incorporating the PET ionomer (AQ55S) in a different way. They compared the effect of the PET ionomer on the dispersion of C10A, a synthetic fluoromica clay (Somasif ME100) before modification, and quaternary ammonium modified synthetic fluoromica clay (Somasif MEE). In order to prepare the nanocomposite, they dispersed the clay in water with the PET ionomer, coated the suspension onto the solid PET, followed by removal of water, and then extruded. The preparation scheme is shown in Fig. 10.2(b). The concentration of clay in the composites was 5 wt%. The composite with the modified clay showed some discoloration compared with the composite with Somasif ME100. This may be due to degradation of the quaternary ammonium modifier during extrusion. Although the optical transparency improved in the PET/Somasif MEE nanocomposite compared with the PET/Somasif ME100 nanocomposite, the optical transparency attained its maximum value with the PET/C10A nanocomposite. This is due to the smaller particle size of C10A in comparison to the mica clays. Due to its small size, C10A will scatter less light and, hence, the optical transparency increased.[21]

Barber *et al.*[17] and Ammala *et al.*[21] both used 5 wt% C10A clay and a PET ionomer to prepare a nanocomposite with PET. The Barber group used

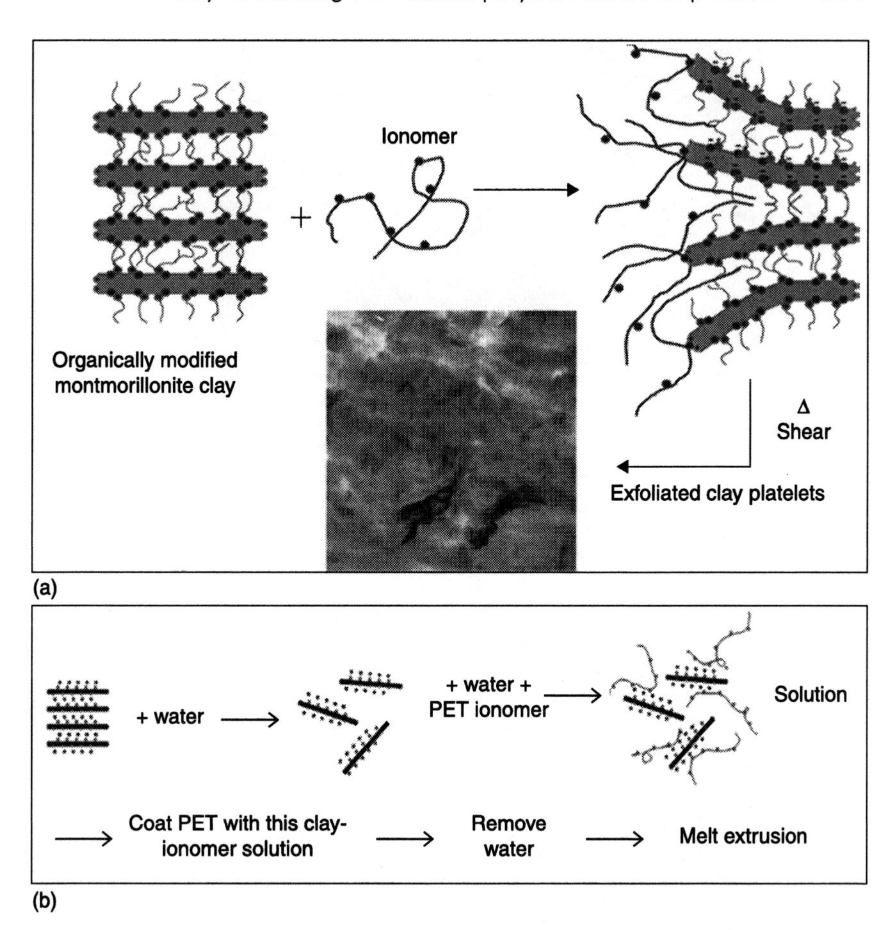


*10.2* (a) Conventional melt blending. The ionomer is sulfonated poly(ethylene terephthalate). Clay concentration in the nanocomposite is 5 wt%. Reproduced with permission from Barber *et al.*[17] (b) Modified melt blending. Clay concentration in the nanocomposite is 5 wt%. Reproduced with permission from Ammala *et al.*[21]

conventional melt blending (Fig. 10.2(a)), whereas the Ammala group used melt blending in a different way (Fig. 10.2(b)). Now the question is which method shows better dispersion of the clay layers in the PET matrix, since the properties of the nanocomposites are directly related to the dispersion characteristics in the PET matrix. To compare the dispersion of the clay layers in the PET matrix, TEM images of the nanocomposites prepared via two different melt blending techniques are shown in Fig. 10.3. According to this figure, the dispersion of clay platelets in the PET matrix prepared by modified melt blending is more homogeneous compared with that by the conventional melt blending method.

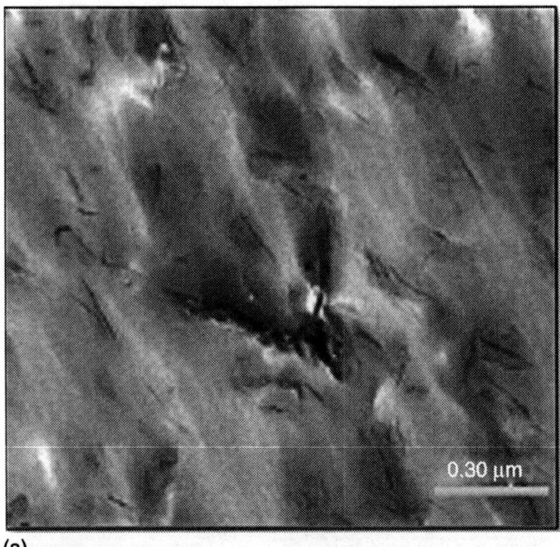


(a)

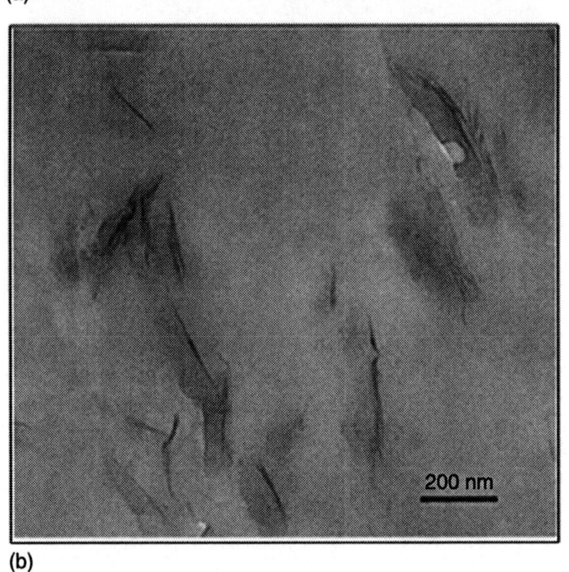


(b)

*10.3* PET nanocomposite prepared via (a) conventional and (b) modified melt blending. Reproduced with permission from Barber *et al.*[17] and Ammala *et al.*,[21] respectively.

The dispersion of clay in a PET matrix can also be improved by equibiaxial stretching when drawing the composite through the die of the extruder. The stretching may increase the aspect ratio due to slippage of the clay platelets during stretching and preferential orientation of the clay platelets. Hence, the mechanical and barrier properties of the nanocomposites improve, which in turn is advantageous for packaging.[22,23] For example, for the same stretch ratio, the oxygen permeability coefficient of neat PET (~2.35 $cm^3$-mm/($m^2$-day-bar)) reduces after nanocomposite preparation (~1.82 $cm^3$-mm/($m^2$-day-bar)).[22]

Therefore, when preparing a PET/clay nanocomposite one should consider the higher thermal stability, large *d*-spacing, and compatibility of the clay surface with the PET matrix. Compatibility can be determined by the polar solubility parameter using group contribution methods, such as the Fedors approach.[24] The incorporation of a PET ionomer during melt blending is also effective in improving the dispersion of clay platelets in the nanocomposite.[21] With *in situ* intercalation, it is better to treat the clay with a catalyst and then allow polymerization.[16]

## 10.4 Structural characterization

It is well known that the most efficient techniques for analyzing the structure of a clay-containing composite are wide-angle X-ray diffraction (WXRD) and transmission electron microscopy (TEM). Usually WXRD patterns start from a diffraction angle ($2\theta$) ~2°. Therefore, it is hard to believe that the clay layers are exfoliated in the matrix polymer if the clay peak is absent above the 2° diffraction angle in the nanocomposite sample. For this reason, sometimes WXRD results contradict TEM results.[9,14,16] The dispersion of nanoparticles can be directly visualized from TEM images. However, it is best not to use only an image of the most well-dispersed area; instead it is good research practice to use the average dispersion characteristics of low-magnification TEM images. To provide good contrast in TEM, osmium tetroxide can be used to stain the clay. To confirm the true extent of exfoliation it is necessary to perform small-angle X-ray scattering (SAXS) measurements. From a SAXS study, the probability of finding neighboring clay stacking within a certain distance as well as the electron density profile can be estimated. This type of analysis provides a much better understanding of the average dispersion characteristics.

In order to produce a clay-based polymer nanocomposite with improved properties, it is necessary to add ~ 3 to 5 wt% of organically modified clays.[25] As the clay loading in a PET nanocomposite increases, the degree of delamination of the clay platelets is affected due to geometric constraints.[25] Therefore, it is quite difficult to achieve full delamination of clay platelets in a PET matrix. The large clay clusters (0.5–1 μm) dispersed in a matrix can also be visualized from the images obtained by scanning electron microscopy (SEM).[26] However, from an image of a fractured surface it is difficult to confirm whether the color contrasts really represent clay particles or only voids (bubbles formed during sample preparation) or both. On the other hand, atomic force microscopy (AFM) images provide a much clearer and better view of the dispersed clay layers in a polymer matrix compared with SEM.[26,27] Figure 10.4 shows an AFM image of the surface of a nanocomposite. In this three-dimensional topography, the clay layers are

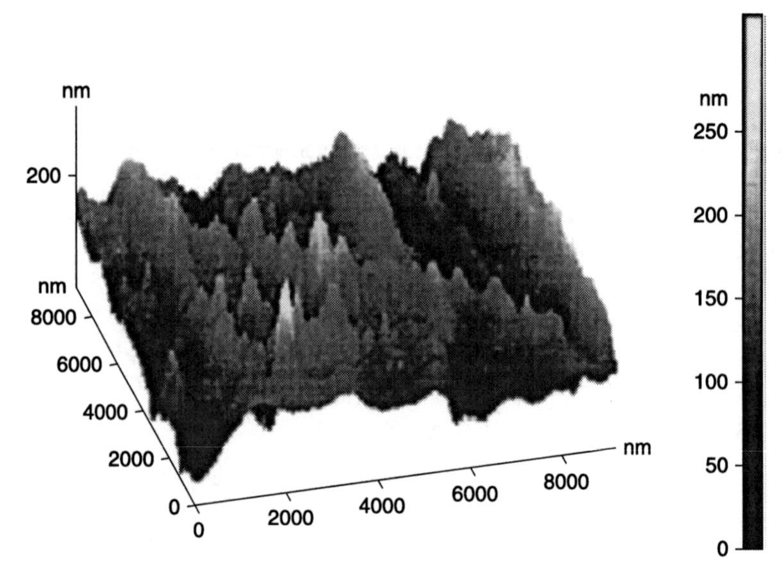


*10.4* AFM image of a clay-based PET nanocomposite. Reproduced with permission from Monemian *et al.*[27]

represented by the hills and the matrix PET by the valleys.[27] Using mainly TEM images, it can be inferred that most PET/clay nanocomposites possess an intercalated structure;[28,29] of course, some of them have very delaminated clay layers.[16] According to TEM images, the average thickness of dispersed clay layers ($D$) can be determined by Eq. 10.1:[30]

$$D = 1.56d_n \qquad [10.1]$$

where $d_n$ is the arithmetic mean of the thicknesses of a few hundreds of particles observed in a TEM image and 1.56 is a coefficient.

## 10.5 Properties of nanocomposites

### 10.5.1 Thermal properties and kinetics of crystallization

The most common technique for evaluating thermal properties, such as phase changes, is differential scanning calorimetry (DSC). DSC determines the amount of heat change (either absorbed or released) when a substance undergoes physical or chemical changes. Such changes alter the internal energy ($U$) of the substance. At a constant pressure, the internal energy is known as enthalpy ($H$). Any thermal transition, e.g., a phase change, can be described in terms of the change of enthalpy between the two phases. When the internal energy of a solid is increased by the application of an external energy source, the molecular vibrations in the substance increase. As a result, the substance becomes less and less ordered. Hence the enthalpy and entropy ($S$) of the material increase. Melting occurs when the Gibbs free energy ($G = H-TS$) of the liquid is lower than that of the solid since every system seeks to attain a minimum free energy state. Crystallization is the opposite of melting. During crystallization, entropy decreases because the ordering of molecules within the system is overcompensated by the thermal randomization of the surroundings. Due to the release of heat of fusion, the entropy of the universe increases. Hence, crystallization is an exothermic process. Usually crystallization occurs at a lower temperature than melting, known as supercooling. This means a crystal can be destroyed more easily than it is formed. Crystallization consists of two phenomena: nucleation and growth. Nucleation is the onset of a phase transition in a small region. The presence of foreign materials (e.g., dispersed nanoparticles in nanocomposites) can facilitate nucleation. Crystal growth is the subsequent growth of nuclei that have reached a critical cluster size.

Hamzehlou and Katbab[29] described the properties of clay-based nanocomposites with bottle grade PET, and the flash PET that is left at the gate of an injection mold during preform manufacturing of PET bottles from neat PET by injection molding (which is known as recycled PET). The clay used for this purpose was MMT organically modified by methyl tallow bis(2-hydroxyethyl) ammonium salt. According to these authors, the recycled-PET-based nanocomposites showed an improvement in thermal properties. Although there was no change in melting temperature, the glass transition temperature ($T_g$) and crystallization temperature ($T_c$) shifted towards higher temperatures because they are quite thermally stable systems.[29] Even for this particular case, the intrinsic viscosity remained unaltered in the nanocomposites when compared with the recycled PET. However, a change in the organic modifier of the clay to dimethyl dehydrogenated tallow ammonium salt resulted in a decrease in $T_g$ due to clay agglomeration and may be due to degradation of the polymer matrix during processing.[31] $T_g$ for the nanocomposite may remain the same as that of the neat polymer if the composite has a homogeneous intercalated structure. In such cases, the degree of crystallinity decreases in the nanocomposite relative to the neat polymer.[18] For any industrial application, it is essential to achieve high nucleation efficiency and a low degree of crystallinity in the polymer nanocomposite compared with the matrix polymer because higher nucleation efficiency allows preform injection molding, blow molding, etc., at higher temperatures. On the other hand, a reduction in the degree of crystallinity makes the material tougher.

The melting and crystallization behaviors of neat PET and PET/C20A nanocomposite samples were investigated by both conventional and temperature modulated DSC.[13] The results indicate that a clay particle acts as a nucleating agent in the crystallization of the PET. To find the effect of clay on the cold crystallization of neat PET, conventional DSC and temperature-modulated DSC (TMDSC) of melt-quenched samples were also carried out. With nanocomposites, cold crystallization was accompanied by a small degree of fusion and subsequent crystallization in the reversal component during TMDSC. This may be due to the presence of short polymer chains formed by the degradation of the matrix at a high temperature in the presence of clay. These small chains undergo a small amount of fusion during crystallization of the bulk. The shift of the cold crystallization peak temperature ($T_{cc}$) of PET toward lower temperatures in nanocomposites suggested the intercalated silicate layers act as nucleating agents, so that crystallization of the matrix starts sooner. However, the decrease in enthalpy of cold crystallization ($\Delta H_{cc}$) of the nanocomposites confirmed that, although the clay particles act as nucleating agents, the nanocomposites lose some crystallizable moiety due to the intercalation of the polymer chains in the clay galleries. A further increase in temperature resulted in melting with re-crystallization. In the quenched state, the initial percentages of crystallinity in the nanocomposites were higher than in the neat PET.

Crystallization can simultaneously involve varying rates of nucleation, growth, aggregation breakage, re-dissolution of unstable crystal clusters, and filler concentration. The result is a complex process that is irreversible and often difficult to predict accurately – especially during scale-up or in technology transfer to alternative equipment. For polymers, however, kinetic control is the dominant factor. Polymer crystallization only occurs at a reasonable rate at temperatures well below the equilibrium melting temperature. The reasons for this behavior will emerge from the discussion of crystallization kinetics. There are several different kinds of experimental method that are commonly used to observe the time development of crystallinity (crystallization kinetics) in polymers. One method assesses the cooling rate at which the total amount of crystallinity develops from the supercooled liquid, known as non-isothermal crystallization. Another method assigns a time for crystal growth at a certain temperature near or above the supercooled phase, known as isothermal crystallization. The study of non-isothermal crystallization is more important because most current processing techniques, such as injection molding, blow molding, etc., with polymeric materials follow non-isothermal crystallization. The Avrami model[32] was developed for isothermal crystallization kinetics; Ozawa[33] extended the Avrami equation to non-isothermal crystallization.[34] The Liu model, which is applicable for non-isothermal crystallization kinetics, is a combination of the Avrami and Ozawa models.[35] Prior to the study of crystallization kinetics using these models, it is necessary to determine the relative degree of crystallinity as a function of temperature and time.

The relative degree of crystallinity, $X_T$, can be defined as:

$$X_T = \frac{\int_{T_o}^{T} \left(dH_c / dT\right) dT}{\int_{T_o}^{T_\alpha} \left(dH / dT\right) dT} \quad [10.2]$$

where $T_o$ ($T_{c,on}$) and $T_\alpha$ ($T_{c,fin}$) are, respectively, the initial and final temperatures of crystallization. For example, $X_T$ as a function of temperature for neat PET and nanocomposites with different clay loadings at different cooling rates during

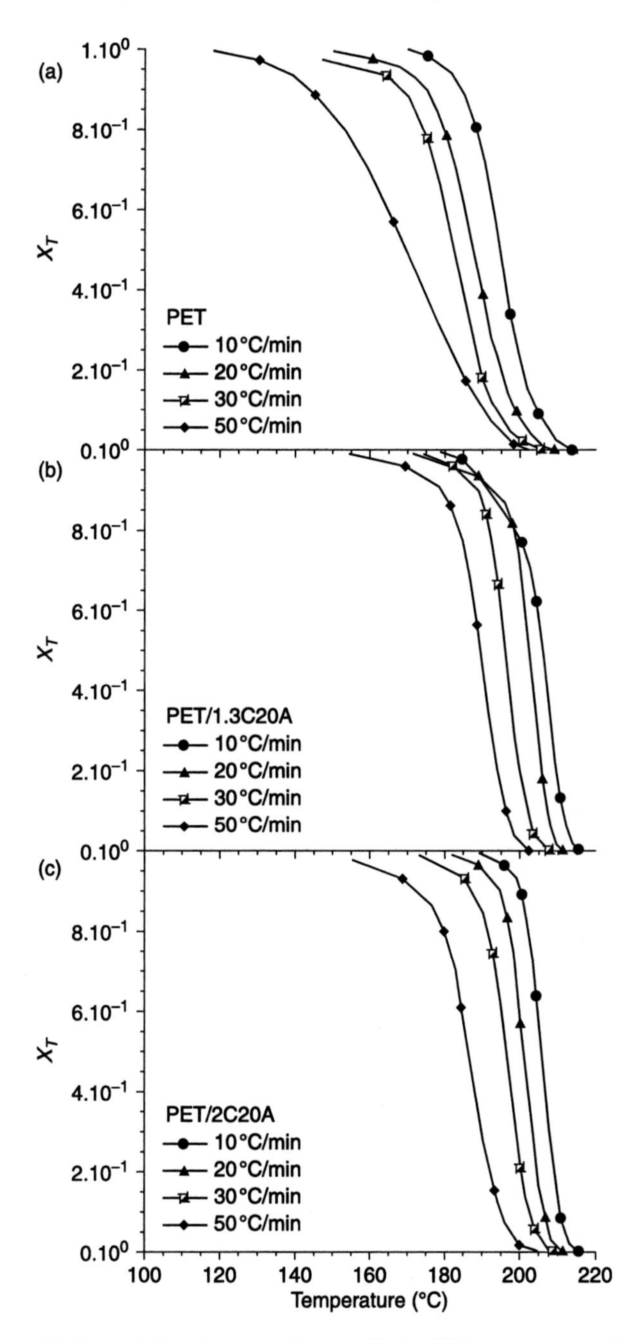


*10.5* (a–c) The degree of crystallinity ($X_T$) plotted as a function of temperature for neat PET and nanocomposites with different clay content. The cooling rates were varied during non-isothermal crystallization.

non-isothermal crystallization are plotted in Fig. 10.5. To see the change of the relative degree of crystallinity as a function of time ($X_t$), the temperature scale was transformed into a time scale using Eq. 10.3 and the equivalent time dependences of the fractional crystallinity are plotted in Fig. 10.6.

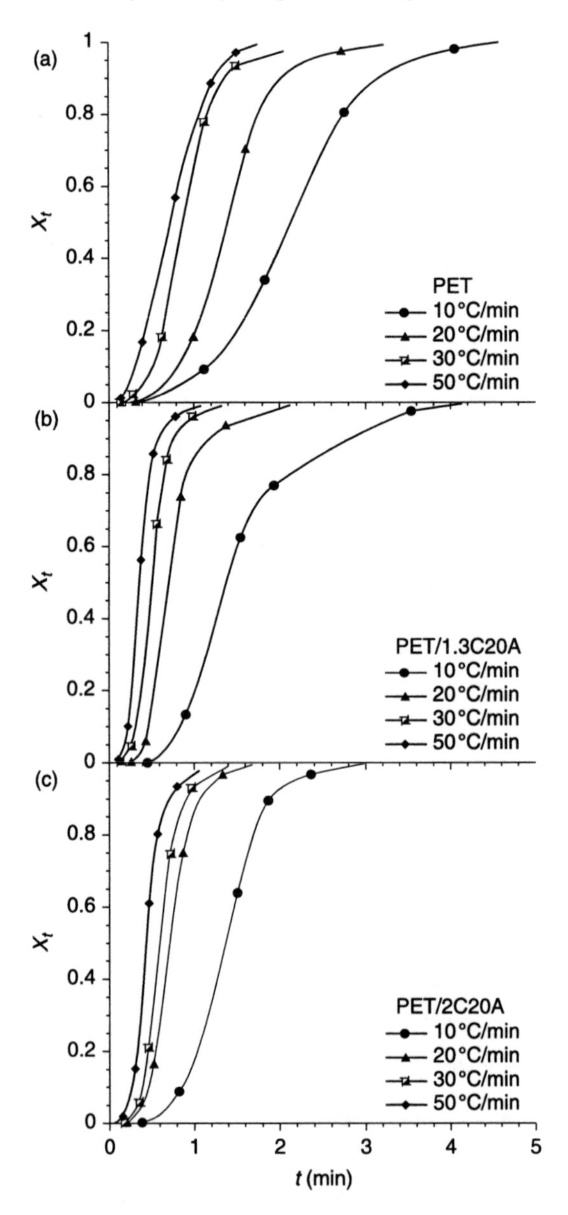


*10.6* (a–c) The degree of crystallinity ($X_t$) plotted as a function of time for neat PET and nanocomposites with different clay content. The cooling rates were varied during non-isothermal crystallization.

$$t = (T_o - T)/\phi \quad [10.3]$$

where $T_o$ is the onset temperature of crystallization and $T$ is the same temperature used to determine $X_T$. As a function of time or temperature, the relative degree of crystallinity remains the same; the suffix only denotes the abscissa.

By truncating non-isothermal crystallization into an infinitesimally small isothermal process, Ozawa extended the Avrami model for isothermal crystallization to analyze non-isothermal crystallization kinetics. According to this model, $X_T$ can be written as a function of the cooling rate:

$$1 - X_T = \exp[-K(T)/\phi^m] \quad [10.4]$$

where $K(T)$ is the Ozawa crystallization rate constant and $m$ is the Ozawa exponent depending on the dimension of crystal growth. Taking a double logarithm on both sides of Eq. 10.4:

$$\ln[-\ln(1-X_T)] = \ln K(T) - m \ln \phi \quad [10.5]$$

Therefore, a plot of $\ln[-\ln(1-X_T)]$ versus $\ln \phi$ should be a straight line if this model is valid. $K(T)$ and $m$ can be estimated from the antilogarithms of the $y$-intercept and slope, respectively.

Although the Ozawa model is designed to analyze non-isothermal crystallization kinetics, this model is not valid for PET/C20A nanocomposites since this model does not take into account the secondary crystallization process (impingement of crystals). Therefore, the Ozawa model fails to describe non-isothermal crystallization kinetics with secondary crystallization. As an alternative approach, the Avrami model can be used to explain crystallization kinetics for such systems.[34,36]

According to the Avrami model, the equivalent time-dependent crystallinity can be expressed as:

$$X_t = 1 - \exp(-Z_t t^n) \quad [10.6]$$

where $Z_t$ is a composite rate constant involving both nucleation and growth rate parameters and the Avrami exponent, $n$, is a constant depending on the type of nucleation and the growth process. Taking a double logarithm on both sides of Eq. 10.6:

$$\ln[-\ln(1-X_t)] = \ln Z_t + n \ln t \quad [10.7]$$

Equation 10.7 should be a straight line if this model is valid, and $Z_t$ and $n$ can be determined from the antilogarithms of the $y$-intercept and slope, respectively. Note that, during analysis of non-isothermal crystallization by this model, $Z_t$ and $n$ do not possess the same physical meaning as in the original Avrami analysis for isothermal crystallization because the temperature changes instantaneously in the non-isothermal process. Here they are adjustable parameters to fit the experimental results and they help to analyze the crystallization kinetics. However, the variation of the Avrami exponent with cooling rate implies that crystallization of the sample has occurred in various growth forms.[37] According to this model, the primary and secondary crystallization of PET produce almost the same size of crystallites. Hence, there is almost no deviation from straight lines in the Avrami plot as shown in Fig. 10.7(a). In nanocomposites, although crystallization starts by nucleation in

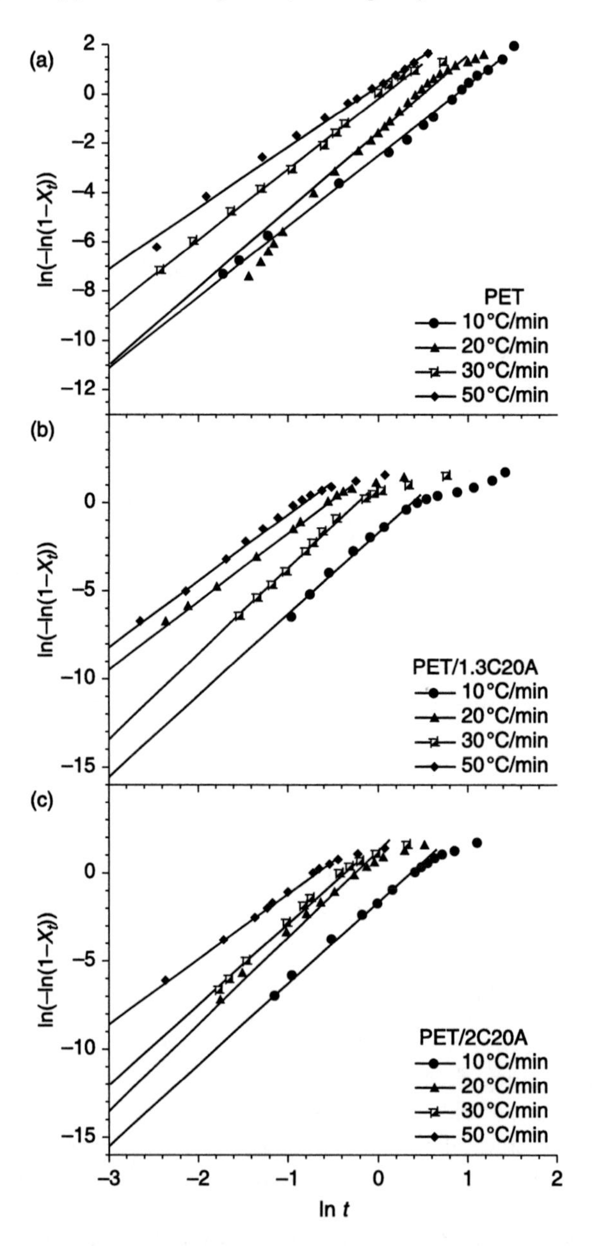


*10.7* (a–c) Validity of the Avrami model for neat PET and clay-based PET nanocomposites.

the presence of foreign materials, the dispersed clay layers hinder impingement and, as a result, the growth of small crystallites occurs due to secondary crystallization. Actually, the different size distribution of crystallites is responsible for the deviation of the linear portion of the Avrami plot as shown in Fig. 10.7(b) and Fig. 10.7(c).

Jeziorny suggested the parameter $Z_t$ should be modified when Avrami analysis is applied to non-isothermal crystallization kinetics. Assuming a constant or almost constant cooling rate, the final form of this parameter as suggested by Jeziorny is:[38]

$$\ln Z_c = \ln Z_t/\phi \tag{10.8}$$

The combined Avrami and Ozawa model, i.e., the Liu model, is given by Eqs 10.9 and 10.10, and is also valid for the current system as shown in Fig. 10.8.

$$\ln Zt + n \ln t = \ln K_T - m \ln \varphi \tag{10.9}$$

By rearranging Eq. 10.9, the final form of the Liu model is:[35,39]

$$\ln \varphi = \ln F(t) - a \ln t \tag{10.10}$$

where $F(T) = [K(T)/Zt]^{1/m}$ is for the cooling rate to reach a defined degree of crystallinity and $a$ is the ratio of the Avrami exponent to the Ozawa exponent, i.e., $a = n/m$. For a given degree of crystallinity, $F(T)$ and $a$ can be determined from the $y$-intercept and slope of the straight lines defined by Eq. 10.10.

The different kinetic parameters determined from these models proved that, in nanocomposites, organoclay is efficient at initiating crystallization earlier by nucleation, but crystal growth will decrease in nanocomposites due to intercalation of the polymer chains in the silicate galleries. Wang *et al.*[40] also observed similar non-isothermal crystallization kinetics as for PET/C20A nanocomposites in their systems. If there is a strong interfacial interaction between the clay and the matrix PET, the polymer segmental motion reduces and hence the crystal growth rate decreases either because the crystals form more slowly or because there is less overall crystallization, i.e., the $n$ value reduces.[27]

Similar crystal growth with secondary crystallization has also been observed during isothermal crystallization.[30] For neat PET, the average value of the Avrami exponent, $n$, is 3. This implies that crystal growth is three dimensional (spherulitic). This value changes in nanocomposites depending on secondary crystallization.[41,42] Fourier transform infrared spectroscopy (FTIR) can be used to determine crystalline structure and perfection. In some nanocomposites, the crystal lamellar thickness and perfection of the crystal is reduced, since the ethylene glycol segments are more likely to sit in the polymer/clay interface.[41]

From a technical point of view, the study of non-isothermal cold crystallization kinetics is important since it is frequently encountered in processing methods, such as the reheat stretch blow molding of bottles, heat setting, production of films and fibers, etc. The physical and mechanical properties of such products are directly or indirectly controlled by the crystallization processes.

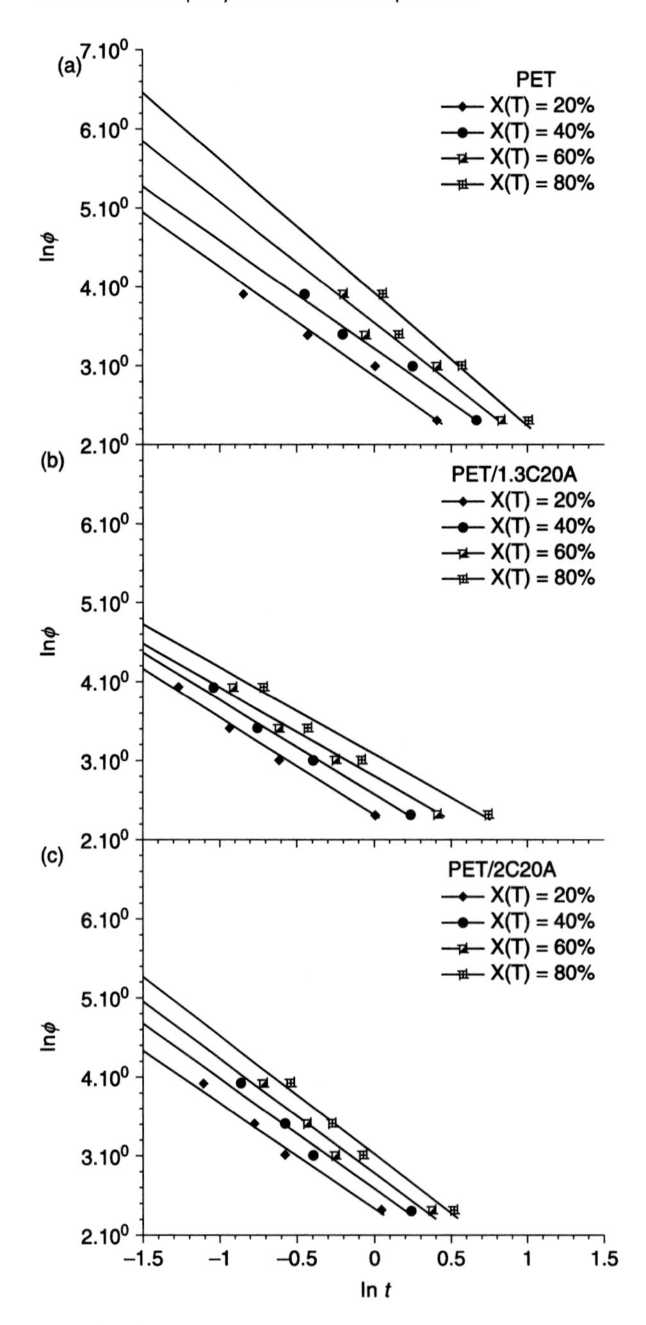


*10.8* (a–c) Validity of the Liu model for neat PET and clay-based PET nanocomposites.

In order to study non-isothermal cold crystallization kinetics, it is necessary to quench samples from the molten state to a temperature below $T_g$, and then heat them at different rates.[43]

Crystal growth can be directly visualized using polarized optical microscopy (POM) images taken at the same isothermal or non-isothermal crystallization conditions as used in DSC.[34,41,42,44,45]

The activation energy ($\Delta E$) for non-isothermal crystal growth can be determined using the Augis–Bennett, Kissinger, and Takhor methods represented by the following equations.

Augis–Bennett method:[46]

$$\frac{d\left[\ln\left(\phi/\left(T_o - T_c\right)\right)\right]}{d\left(1/T_c\right)} = -\frac{\Delta E}{R} \qquad [10.11]$$

Kissinger method:[47]

$$\frac{d\left[\ln\left(\phi/T_c^2\right)\right]}{d\left(1/T_c\right)} = -\frac{\Delta E}{R} \qquad [10.12]$$

Takhor method:[48]

$$\frac{d\left[\ln\left(\phi\right)\right]}{d\left(1/T_c\right)} = -\frac{\Delta E}{R} \qquad [10.13]$$

The reduction of the activation energy in nanocomposites compared with the neat polymer leads to the conclusion that the nanoparticles in the nanocomposite facilitate crystal growth.[34]

The determination of the activation energy by the Kissinger method is more reliable than that determined by the Augis–Bennett method since the Kissinger method is model free. On the other hand, there are two assumptions in the Augis–Bennett method: first, $\Delta E >> RT$ and second, $(T_{1/2})_1.(T_{1/2})_2 \approx T_c^2$;[49] where $(T_{1/2})_1$ and $(T_{1/2})_2$ are the temperatures at the half maximum before and after the crystallization peak, respectively. Determination of the activation energy by the Augis–Bennett method is really fruitful if the reaction obeys the Avrami law.[49]

## 10.5.2 Thermal stability

Although some surfactants used to modify pristine clays are very thermally stable, there is a chance of degradation of the surfactant during nanocomposite preparation due to the high processing temperature of PET.[19,28] Therefore, to reduce the possibility of organoclay degradation, it is better to dry the clay overnight under vacuum before processing and to minimize the processing time. The processing time can be reduced significantly by preparing the sample in an extruder. The processing time can be estimated from a study of isothermal degradation of the organoclay at the processing temperature.[19] Further, under oxidative conditions a PET/C20A nanocomposite sample exhibits a two-step decomposition. In the first half of the degradation process, the nanocomposite exhibits less onset thermal stability than neat PET. This is due to the degradation of the surfactant used for the modification of MMT,[50,51] as alkyl ammonium modifiers are known to undergo Hoffman degradation at around 200 °C.[52] As for the oxidative condition, under an inert atmosphere the nanocomposite samples also exhibit less onset thermal stability (at 10% weight loss) than neat PET. However, the main degradation temperature for the nanocomposite samples increased in air compared with a nitrogen atmosphere. It is possible that the different types of char formation mechanism under an oxidative environment actually slow down oxygen diffusion, thus hindering oxidation under thermo-oxidative conditions. This observation suggests that there is improved flame retardancy for nanocomposites.[13] The same behavior was also obtained for PET/C30B nanocomposites.[28] On the other hand, phosphonium-modified MMT with low phosphonium content exhibits an improvement in thermal stability of the PET/clay nanocomposite when compared with the neat PET.[28,53,54] Different phosphonium salts, such as (4-carboxybutyl) triphenylphosphonium bromide[53] and dodecyltriphenyl phosphonium chloride,[54,55] can be used to modify the natural MMT. The thermal stability of the PET/ammonium-salt-modified MMT nanocomposites can be improved by using recycled PET instead of neat PET.[29] An imidazolium-surfactant-modified MMT also enhances the thermal stability of PET/clay nanocomposites.[56,57] Aminosilane- and imidosilane-modified palygorskite clay-based PET nanocomposites also show an improvement in thermal stability compared with the neat PET resin.[58] However, it has already been proved, at the same level of loading, that clay-based PET composites exhibit much more thermal stability compared with composites based on silica nanoparticles.[59] Because of its high aspect ratio, the clay has a much higher surface area for polymer–filler interactions, which in turn allows better dispersion of the clay layers in the polymer matrix than spherical nanoparticles. Due to this better dispersion, the tortuous path decelerates the permeation of oxygen gas and enhances the thermal stability of the PET/clay nanocomposite.

The main factors for polymer degradation in clay-based nanocomposites are the number of hydroxyl groups on the edge of the clay platelets and the ammonium linkage on the clay.[60] For example, acid-treated sodium MMT (H-MMT) has a reduced thermal stability of the PET matrix due to the larger number of Brønsted acid sites generated by the acid treatment (Fig. 10.9). On the other hand, silane-modified MMT (S-MMT) shows less degradation of the PET matrix during preparation of the nanocomposite. After silane modification, the gallery spacing of the clay remains unaltered since the silane coupling agent was grafted onto the sides of the clay layers as shown in Fig. 10.9. Therefore, ammonium modification is necessary. However, the preparation of ammonium-modified clay (A-MMT) further accelerates the degradation of the PET since the ammonium modifiers take part in the Hoffman elimination reaction, which produces more Brønsted acid sites. The effect of ammonium modification on the degradation of PET can be lowered by washing the modified clay with ethanol (W-A-MMT, Fig. 10.9) or adding a silane grafting agent (S-A-MMT in Fig. 10.9).

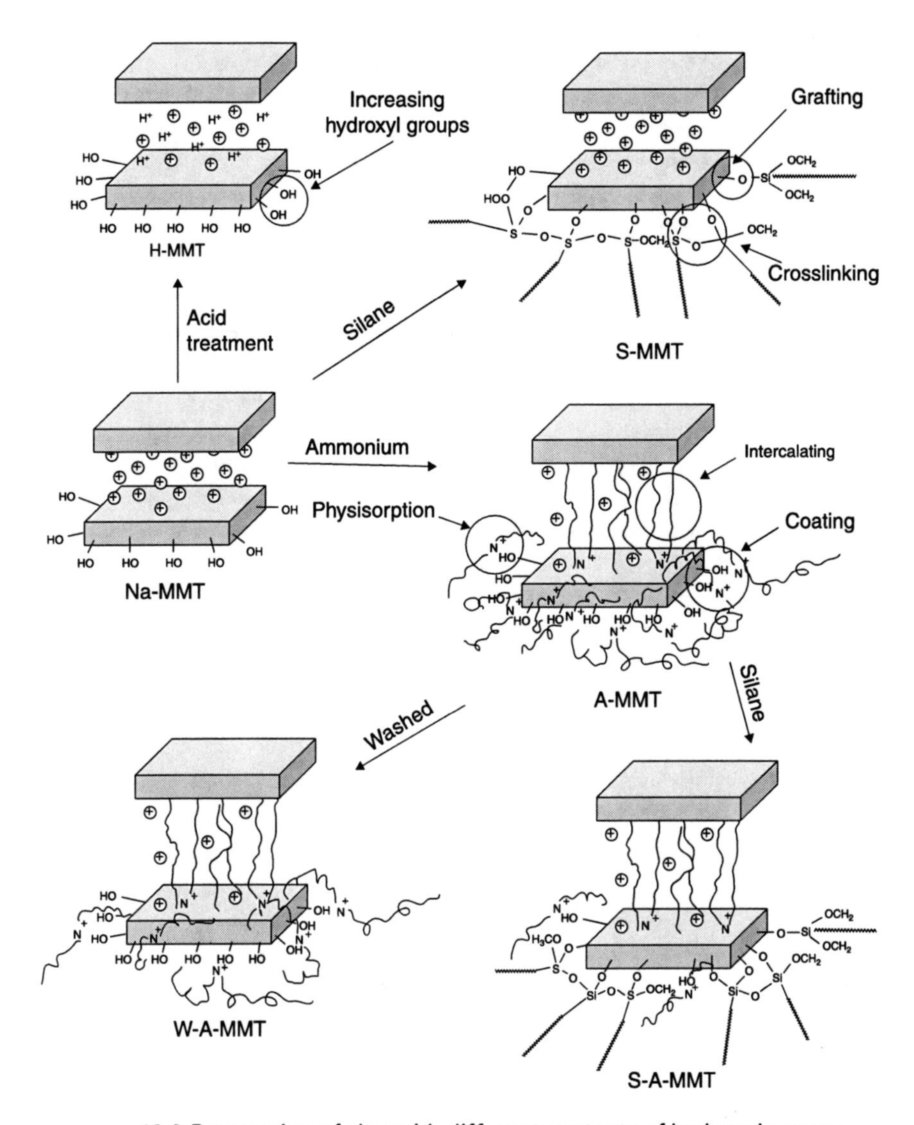


*10.9* Preparation of clay with different contents of hydroxyl groups on the edge of the clay platelets and ammonium linkage on clay. Reproduced with permission from Xu *et al.*[60]

We have discussed the thermal stability of clay-based PET nanocomposites. What happens if the clay has fibrous rod-like structures instead of layered ones? Yuan *et al.*[61] described the thermal stability of organically modified fibrous silicate (palygorskite) (modified by water-soluble polyvinylpyrrolidone)/PET nanocomposites. According to these researchers, the thermal stability of a nanocomposite containing 3 wt% of clay exhibited an improvement in the onset degradation temperature for all heating rates under air, as shown in Table 10.1. On the other hand, under a nitrogen atmosphere the nanocomposite exhibited a reduction of the onset degradation temperature for slower heating rates; however, stability increased at higher heating rates. Due to the coexistence of the barrier effect and catalytic decomposition, such nanocomposites exhibit a complex degradation mechanism. For both atmospheres, the catalytic decomposition reaction relied on the metal cations dissociating from the chemical constitution of the clay crystal structure, which mainly exists in the inner part of the matrix. Hence, the thermal degradation reaction should principally originate from the inner part of the matrix. Regardless of the atmospheres examined, the well-dispersed silicate layers in the PET matrix hindered thermal transport from the surface to the inner matrix and weakened the thermal transduction rate. Furthermore, the barrier effect of the nanocomposite was also ascribed as being due to blocking by the thermal degradation products volatized from the inner matrix to the external environment. However, compared with nitrogen, in air there was an additional path for oxygen from the outside to the interior of the nanocomposite, as shown in Fig. 10.10 by the black solid arrows. The delaminated clay also acted as an obstacle in the transfer of oxygen gas, which enhanced the barrier effect of the nanocomposite compared with the nitrogen environment. The additive barrier effect of the nanocomposite is the primary reason for the increasing thermal oxidative stability of the nanocomposite at the first stage for all heating rates in air compared with nitrogen.

*Table 10.1* Heating rate dependency of the onset degradation temperatures of neat PET and a palygorskite-based PET nanocomposite under air and a nitrogen atmosphere

| Sample | Heating rate | Onset degradation temperature under air | Onset degradation temperature under nitrogen |
|---|---|---|---|
| PET | 5 | 381.5 | 409.8 |
| PET | 10 | 408.2 | 425.2 |
| PET | 20 | 418.2 | 436.2 |
| PET | 40 | 424.8 | 449.5 |
| PET nanocomposite | 5 | 402.9 | 401.8 |
| PET nanocomposite | 10 | 420 | 421.5 |
| PET nanocomposite | 20 | 423 | 441.6 |
| PET nanocomposite | 40 | 439.5 | 473.4 |

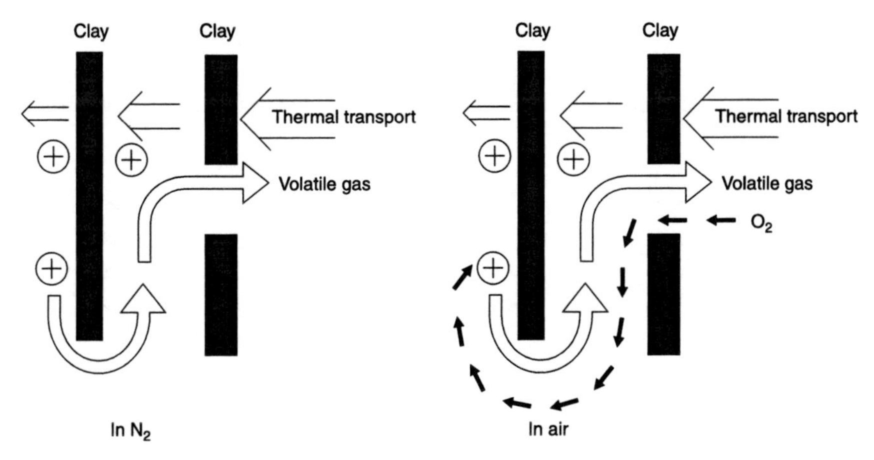


*10.10* Degradation of nanocomposites in nitrogen and air. Reproduced with permission from Yuan *et al.*[61]

In order to avoid decomposition of the organic modifier of organoclay at the processing temperature of PET, Wang *et al.*[62] first prepared a master batch of PET and sodium MMT (PET:MMT = 50:50) by solid-state shear milling. Then they prepared the nanocomposites by extruding the neat PET with a varying proportion of the master batch. Although the nanocomposites had intercalated structures, the thermal stability of the nanocomposites improved compared with the neat PET resin.[62] They also compared the thermal stability of PET/sodium-MMT nanocomposites prepared by simple extrusion (PETCNx where x is the clay content) with nanocomposites prepared by extrusion of the PET/MMT master batch (PETSNx). PETSNx showed better thermal stability compared with PETCN with the same amount of clay loading.

Therefore, to prepare a PET/organically modified MMT nanocomposite one must choose a surfactant (used for organic modification) with a high thermal stability. Otherwise, degradation of the surfactant will facilitate the collapse of the clay layers. As a result, intercalation of the polymer chains in the clay gallery will be difficult. This will result in an intercalated nanocomposite, possibly even a phase-separated composite material. The collapse of the clay structures results in much better heat transfer between the polymer and the clay. This accelerates the degradation of the matrix polymer during processing.

## 10.5.3 Mechanical properties

Generally, the mechanical properties of nanocomposites depend on the extent of delamination of the nanoparticles (particularly nanoclay) in the polymer matrix, which actually depends on the processing conditions and the chemical treatment and organic modification of the clay surface.[63] Better dispersion and orientation of the nanoclay in the direction of application of the mechanical force promotes better stress transfer from one end to the other of the sample.[64] As a result, the nanocomposites exhibit enhanced mechanical properties when compared with the neat polymer.

The tensile property of a phosphonium-modified MMT-containing PET nanocomposite exhibited a significant improvement in modulus, but on the other hand the strength and elongation at break decreased a lot. Usually, clay has a very high modulus compared with polymers. Therefore, it is expected that the incorporation of clay in a polymer matrix will result in an improved modulus for the nanocomposite. If the clay layers are not delaminated homogeneously, the energy-dissipation mechanism will be hindered, which in turn will be responsible for a reduction of the elongation at break.

The tensile strength can be improved by preparing the nanocomposite with recycled PET instead of neat PET (Fig. 10.11(a)).[29] In this figure neat PET is labeled as FPET and recycled PET as RPET. The tensile strength is the ultimate

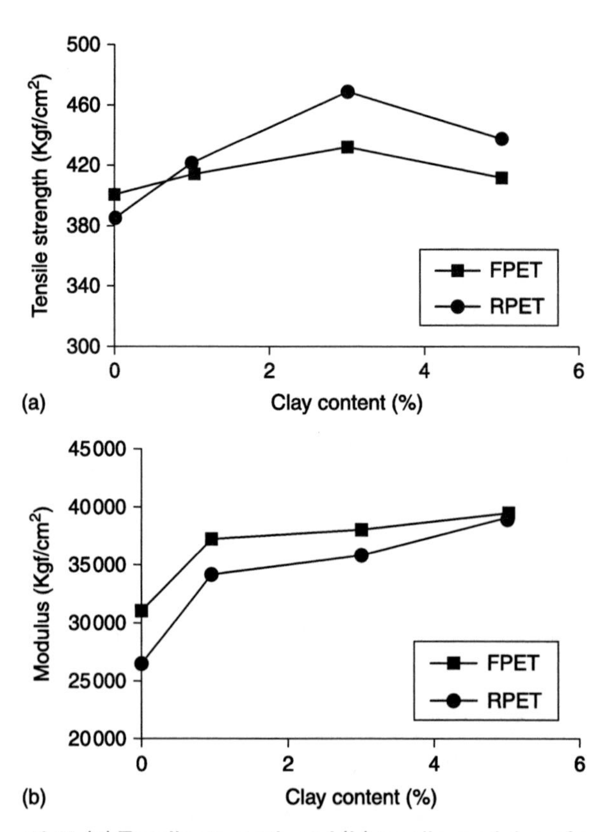


*10.11* (a) Tensile strength and (b) tensile modulus of neat PET (FPET) and RPET-based nanocomposites as a function of clay loading. Reproduced with permission from Hamzehlou and Katbab.[29]

capacity of the material to resist a tensile load regardless of deflection. Figure 10.11(a) shows that the highest value of the tensile strength can be achieved with 3 wt% clay loading. The RPET-based nanocomposite containing 3 wt% clay loading has the maximum tensile strength. According to the current authors, the probable reason for this improvement can be explained as follows. During injection molding or melt stretch molding, the polymer chains become disentangled or partially oriented. This disentangled structure is responsible for a better dispersion of the clay platelets in the recycled PET matrix. However, according to XRD results, the nanocomposites still possess intercalated structures. As a result, the energy dissipation mechanism improves slightly and hence the tensile strength. However, as shown in Fig. 10.11(b), the tensile modulus is lower for a RPET-based composite compared with a FPET-based composite when the clay loading is below 5 wt%. The tensile modulus is a measure of the stiffness of an isotropic elastic material. It is defined as the ratio of the uniaxial stress over the uniaxial strain. It is determined from the slope of the stress-strain curve traced during tensile tests conducted on a sample of the material. Now it is interesting to note that the tensile strength decreases above 3 wt% clay loading but the modulus of the composites still increases above this value. Although the authors only reported the TEM image of the composite containing the 3 wt% clay loading, the dispersion of the clay in the composite containing a higher clay loading can be inferred from the results of tensile property measurements. The improvement of tensile strength up to 3 wt% OMMT loading indicates that the intercalation of PET chains in the clay gallery reaches a maximum for this specific composite. After that the stacking of clay layers increases. This hinders the energy dissipation mechanism during tensile testing. As a result, the tensile strength decreases with higher clay loadings beyond 3 wt%. On the other hand, the modulus increases as the clay loading increases above 3 wt% due to the high modulus of clay itself.

PET/silane-modified organoclay composites exhibit improved tensile properties, which can be useful for fiber and film applications.[65] Nanocomposites prepared via solid-state shear milling (PETSNx, as mentioned above) can exhibit increased strength compared with nanocomposites prepared via simple extrusion (PETCNx) (Fig. 10.12(a)). However, tensile strength increases systematically with an increase in clay loading in both samples. Since, according to the TEM images, the dispersion of PETSNx is better than PETCNx, it is expected that the modulus and strength of the PETSNx will be higher in this case due to the high modulus of the dispersed silicate layers. However, there is no significant change in elongation at break (Fig. 10.12(b)) between the composites prepared via the two different techniques.[54] For the PET/C20A nanocomposite such improvements can be achieved with 1 wt% clay loading. The reason is that the enhanced amorphous orientation of the intercalated clay layers leads to improvements in modulus and strength.[66]

As well as the tensile properties, the thermomechanical properties of clay-based PET nanocomposites were studied by dynamic mechanical analysis (DMA).

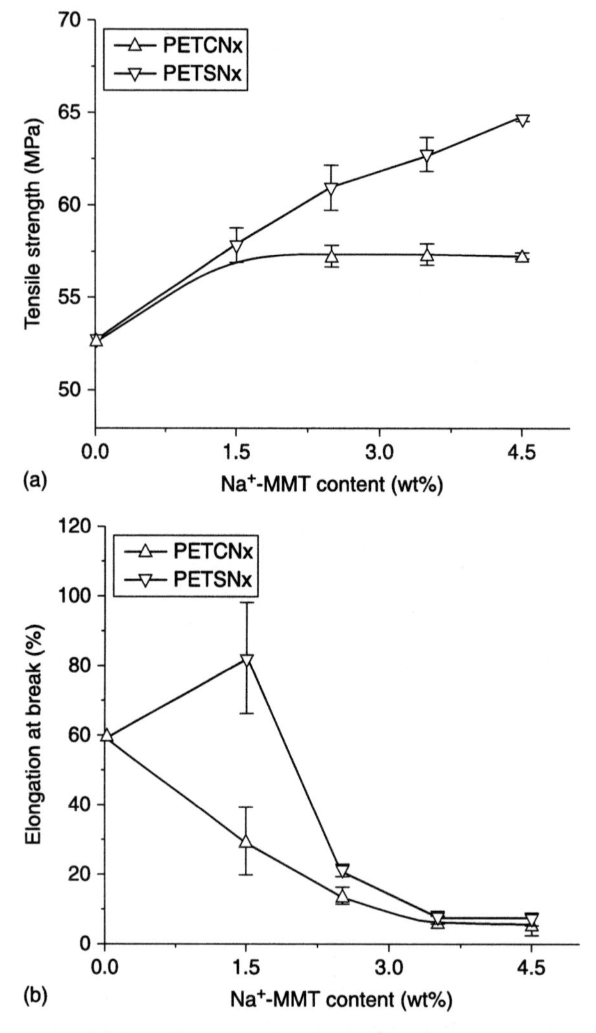


*10.12* (a) Tensile strength and (b) elongation at break of PET nanocomposite samples prepared via solid-state shear milling and via extrusion as a function of clay content. Reproduced with permission from Kráčalík *et al.*[65]

The variation of storage and loss moduli as a function of temperature explains how the rigidity of the nanocomposite material varies with clay content and how clay concentration affects the glass transition temperature of the matrix.[67]

## 10.5.4 Rheological properties

Rheological studies on PET nanocomposites are scarce, but exhibit very interesting features.[55] A rheological analysis of a polymer/clay nanocomposite is a powerful and complementary technique for characterizing the state of clay dispersion due to the sensitivity of the rheological response to the structure and surface characteristics of the dispersed phase.[60] The advantages of rheological characterization over electron microscopy and X-ray scattering reside in the fact that measurements are performed of the molten state and that rheological methods can be used under both linear and non-linear deformation. A disadvantage of rheological methods is that they only provide indirect information on the structures of nanocomposites.[68]

Vidotti *et al.*[68] studied the rheological properties of C20A (o-MMT)-based PET nanocomposites in the presence of a polyester ionomer (PETi). The melt-state rheological properties were measured by a stress-controlled rheometer with parallel plate geometry (25 mm diameter) at a temperature of 280 °C. Figure 10.13 show the complex viscosity ($\eta^*$) as a function of angular frequency ($\omega$) for the neat and extruded PET and PET/PETi/o-MMT nanocomposites containing 1, 3, and 5 wt% of o-MMT. Figure 10.13(a) shows that both the neat and extruded PET exhibited a Newtonian behavior ($\eta^*$ being nearly constant) over the whole frequency range studied. The complex viscosity of all the PET/PETi/o-MMT nanocomposites containing 1 wt% of clay was smaller than that of the neat PET, but similar to or higher than that of the extruded PET. Moreover, the complex viscosity increased when increasing the amount of PETi in the nanocomposite. This observation indicates that an increase in the amount of PETi in a nanocomposite improves the state of dispersion of the clay particles in the PET matrix. A decrease in viscosity of the extruded PET compared with that of the neat PET indicates that the PET had undergone degradation during extrusion.[63,68] Parts (b) and (c) of Fig. 10.13 show that at very low frequencies the complex viscosity of the PET nanocomposites was higher than that of the neat PET and much higher than that of the extruded PET. As shown in Fig. 10.14, the elastic modulus ($G'$) at low frequencies for the nanocomposites is almost independent of frequency. This is typical of solid-like behavior and indicates the formation of a percolated network, which in the case of nanocomposites implies good dispersion of the organoclay.

Giraldi *et al.*[20] described how the intrinsic viscosity of recycled PET changes with the screw speed in the extruder and also in the presence or absence of an antioxidant (Table 10.2). Higher screw speeds lead to higher viscosity. Moreover, the viscosity increases in the presence of an antioxidant for both screw speeds. However, they did not report the same result for nanocomposites.[20]

Usually a homopolymer exhibits Newtonian behavior in a complex viscosity versus frequency plot. In comparison, the complex viscosity of nanocomposites increases significantly in the low frequency region. This shear thinning phenomenon observed for nanocomposites in the low frequency region is initially due to the disruption of the network structure and then in a later stage by the orientation of filler particles in the flow direction.[65] For a silane-modified commercially available organoclay, the complex viscosity either increases or

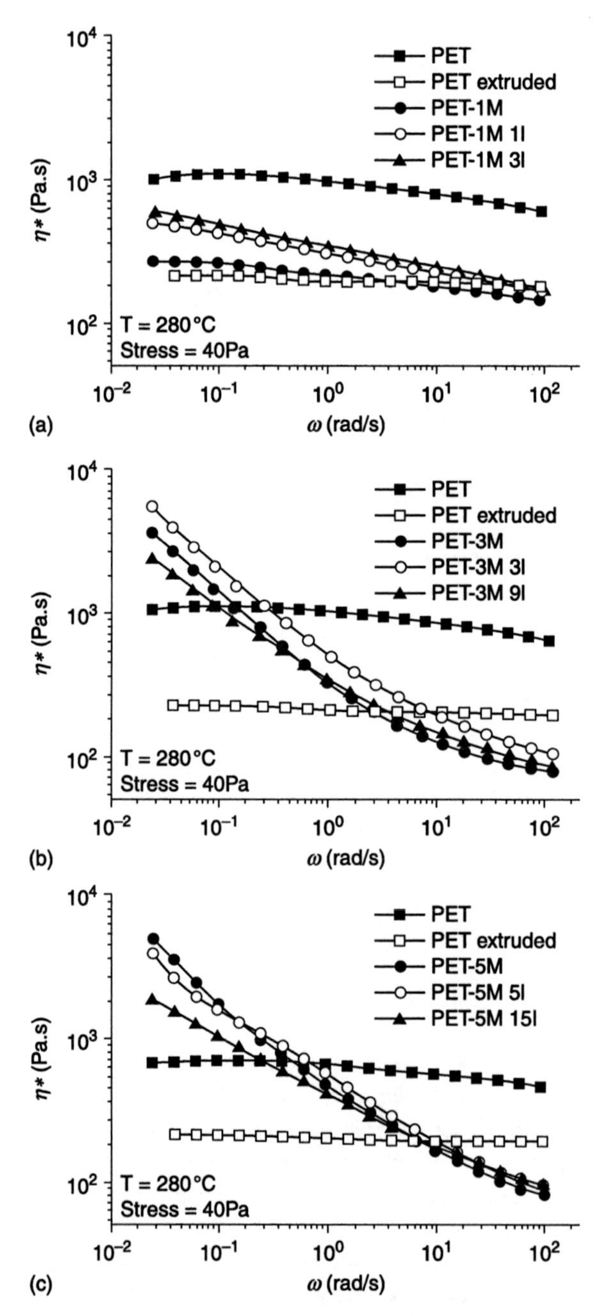


*10.13* Complex viscosity of the PET/PETi/o-MMT nanocomposites containing (a) 1 wt%, (b) 3 wt%, and (c) 5 wt% of o-MMT and different PETi concentrations. Reproduced with permission from Vidotti *et al.*[68]

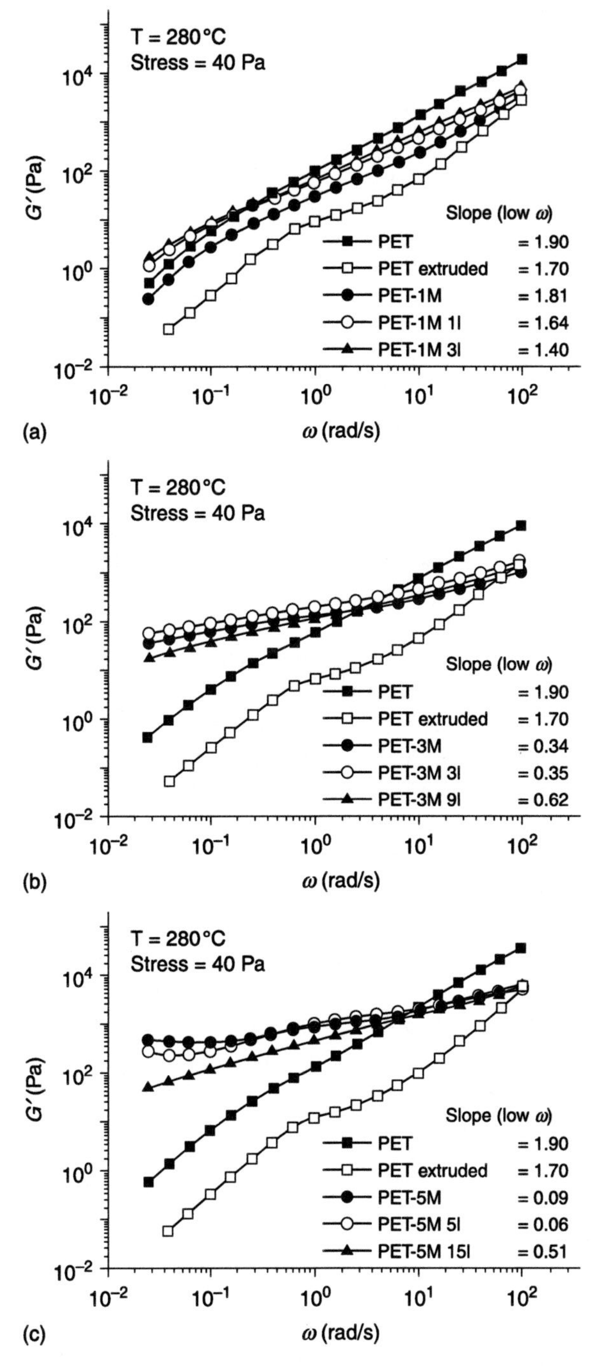


*10.14* Storage modulus (G′) of the PET/PETi/o-MMT nanocomposites containing (a) 1 wt%, (b) 3 wt%, and (c) 5 wt% of o-MMT and different PETi concentrations. Reproduced with permission from Vidotti *et al.*[68]

*Table 10.2* Change of intrinsic viscosity of recycled PET for two screw speeds and with or without an antioxidant

| Screw speed (rpm) | Antioxidant (1 wt%) | Intrinsic viscosity ($dL.g^{-1}$) |
|---|---|---|
| 250 | Without | 0.53 |
| 250 | With | 0.57 |
| 150 | Without | 0.46 |
| 150 | With | 0.50 |

decreases depending on the silane coupling agent. However, if the complex viscosity of a nanocomposite exceeds the complex viscosity of the neat polymer, then it is expected that this is because the degradation of the matrix was significantly reduced during processing and rheological property measurements.[65] An inert (e.g., nitrogen) atmosphere also plays an important role in achieving this. However, during silanization, particularly in the case of C10A and C30B, the retention of water within silicate layers via a chemical reaction between the organic groups of the organoclay and the silane modifier can reduce the melt strength of the composite material.[65]

It has been reported that clay nanoparticles suppress crystallization and orientation of the PET in high-speed melt spinning. On the other hand, polyhedral oligomeric silsesquioxane (POSS) nanoparticles slightly promote the crystallization of PET even in high-speed melt spinning. Hence, Lee *et al.*[69] studied the shear-induced crystallization behavior of PET/clay nanocomposites and PET/POSS nanocomposites. They melted samples at 280°C for 5 min in the 1 mm gap of parallel plates (diameter of the plates was 25 mm) and then performed time sweep experiments at 220, 210, and 200 °C over an angular frequency range 0.5 to 1 rad/s. A comparison of the shear-induced crystallization behavior of PET/clay nanocomposites and PET/POSS nanocomposites is shown in Fig. 10.15. According to this figure, the increase of $G'$ at the early stage of crystallization is almost negligible. An abrupt increase of $G'$ follows in a few minutes because the homogeneous melt becomes a heterogeneous system with the formation and growth of crystallites. Hence $G'$ increases with time. In addition, the clay nanoparticles are more effective in the crystallization process than POSS nanoparticles.[69]

### 10.5.5 Flame retardancy

The wide application of PET is sometimes hindered due to its combustibility, and the increasing uses of PET also place emphasis on improving flame retardancy. Incorporating a chemically reactive phosphorus-containing monomer into the polymer chain is one of the most efficient methods of improving the flame retardancy of polyesters.[70]

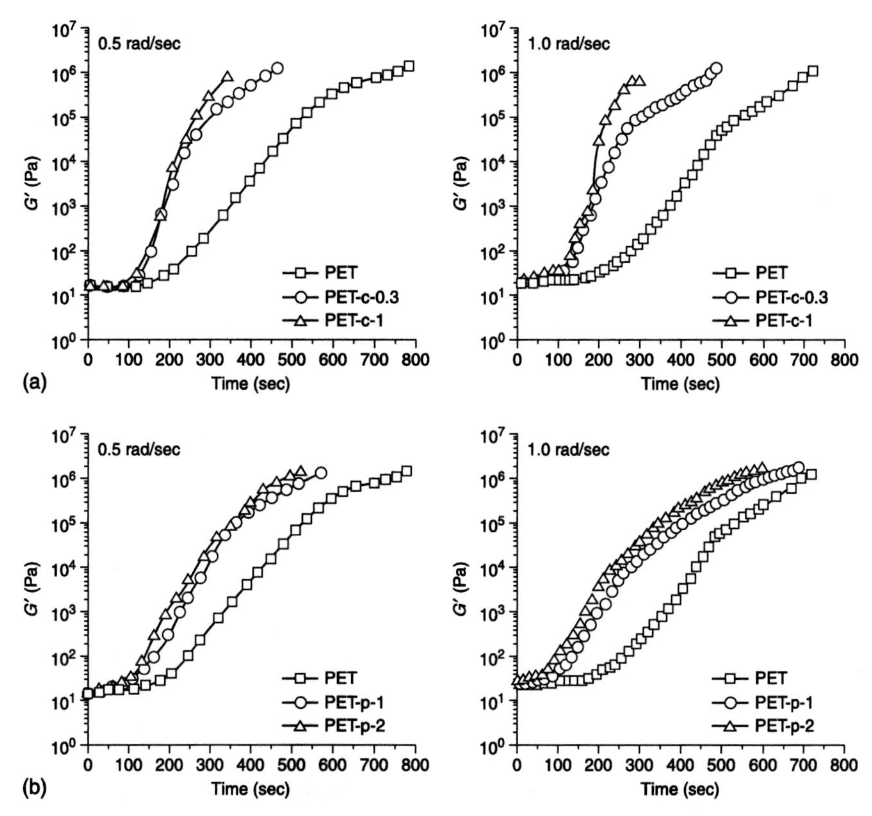


*10.15* Variation of storage modulus ($G'$) with time at 0.5 rad/s (left) and 1.0 rad/s (right) at 220 °C for PET and PET nanocomposites; (a) PET/clay nanocomposites, (b) PET/POSS nanocomposites. Reproduced with permission from Lee *et al.*[69]

On the other hand, the study of polymer/clay nanocomposites has become even more attractive due to the demonstrations of their flame-retardant properties, which mainly demonstrate a significant decrease in the peak heat release rate (PHRR), a change in char structure, and a decrease in the rate of mass loss during combustion in a cone calorimeter.[70] Not only was reduced flammability obtained at very low organoclay contents (2–5 wt%) without increasing carbon monoxide or smoke yields, but also the physical properties of the polymers improved simultaneously. Moreover, the incorporation of nanoclay particles with other flame retardants allows a significant portion of a conventional flame retardant to be removed from the formulated polymer products while maintaining or improving flammability performance and enhancing the physical properties. So, the preparation of nanocomposites is another tool for chemists who are engaged in the research and development of flame-retardant materials.[70]

A novel phosphorus-containing flame-retardant copolyester/MMT nanocomposite (PET-co-HPPPA/O-MMT) was synthesized by *in situ* intercalation polycondensation of terephthalic acid, ethylene glycol, and 2-carboxyethyl(phenyl phosphinic) acid (HPPPA) with OMMT.[70] The flame retardancy of the samples was characterized by the limiting oxygen index test (LOI) and the UL-94 vertical test. The LOI is defined as the minimum fraction of oxygen ($O_2$) in a mixture of $O_2$ and nitrogen ($N_2$) that will just support flaming combustion. It is an important and representative parameter for describing the flame-retardant properties of a material. UL-94 ratings are used to describe the ease with which a polymer can be burned or extinguished. LOI and UL-94 test results are shown in Fig. 10.16 and Table 10.3, respectively. It can be seen that PET-co-HPPPA/O-MMT nanocomposites exhibit much better flame retardancy than PET-co-HPPPA. It is interesting that there is a considerable increase in LOI (from 31.4 to 34.0) at very low OMMT content (1 wt%), while for a further increase in OMMT content (from 1 wt% to 3 wt%) no further increase in LOI was observed. Furthermore, the UL-94 vertical test results show that the after-flame time of each sample was 0 s. When the OMMT content is lower than 2 wt%, the samples cannot reach a rating of V-0 because the surgical cotton ignites. However, when the amount of OMMT increases to 2 wt%, the sample (PET5-2) can achieve the rating of V-0 in the UL-94 test, which also indicates that the incorporation of nanoclay particles with HPPPA improves flame retardancy. The higher loadings of MMT are not necessary for improving the LOI, but, by increasing the viscosity of the burning polymer, they reduce its tendency to drip with fire. This phenomenon can be explained by the following flame-retardant mechanism: the carbonaceous–silicate char builds up on the surface during burning, and insulates the underlying material, reducing the mass loss rate of the decomposition products.[70]

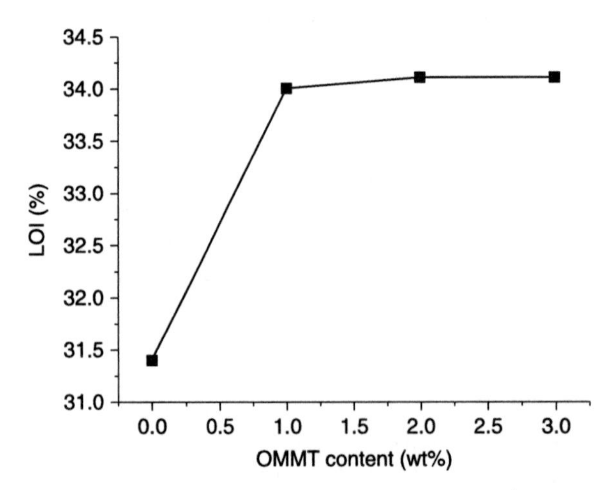


*10.16* LOI versus OMMT content for PET-co-HPPPA/O-MMT nanocomposites. Reproduced with permission from Ge *et al.*[70]

*Table 10.3* Flame retardancy of PET-co-HPPPA/O-MMT nanocomposites determined by UL-94 tests

| Sample | UL-94 standard | After-flame time (s) | Cotton ignited |
|---|---|---|---|
| PET5-0 | V-2 | 0 | Yes |
| PET5-1 | V-2 | 0 | Yes |
| PET5-2 | V-0 | 0 | No |
| PET5-3 | V-0 | 0 | No |

## 10.5.6 Gas barrier properties

PET is a widely used packaging material for beverages and drugs. However, its application could be broadened if packaging materials could be made more sensitive to oxygen and carbon dioxide by improving the barrier property. One approach for achieving this is in the preparation of PET/clay nanocomposites. If the clay layers are dispersed nicely in the polymer matrix, a tortuous path will be created by the homogeneously dispersed clay layers in the nanocomposite, as shown in Fig. 10.17,[1] which actually reduces the permeability of oxygen and carbon dioxide. For example, the Nanolin $DK_2$ clay-based PET nanocomposite exhibits an improved gas barrier property compared with a C15A-based nanocomposite.[18] Tsai *et al.*[16] showed that the barrier property against carbon dioxide gas can be improved by *in situ* polymerization of a monomer/oligomer in the clay gallery in the presence of an antimony acetate catalyst. According to Fig. 10.18, it is clear that the addition of a few wt% OMMT in the PET matrix can effectively reduce the oxygen ($O_2$) gas permeability of a PET film. When the content of the OMMT reached 3 wt%, the permeation of $O_2$ was reduced to half of the neat PET film.[71] Equibiaxial stretching sometimes promotes the gas barrier property since the effective length of the clay tactoids increases due to stretching.[22]

On the other hand, recycled PET-based nanocomposites show some property improvement, but the oxygen gas permeability decreases significantly for such composites.[29]

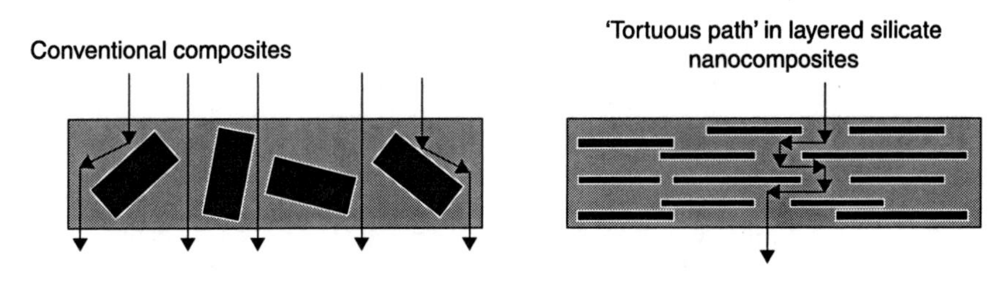


*10.17* Tortuous path in polymer/clay nanocomposites.

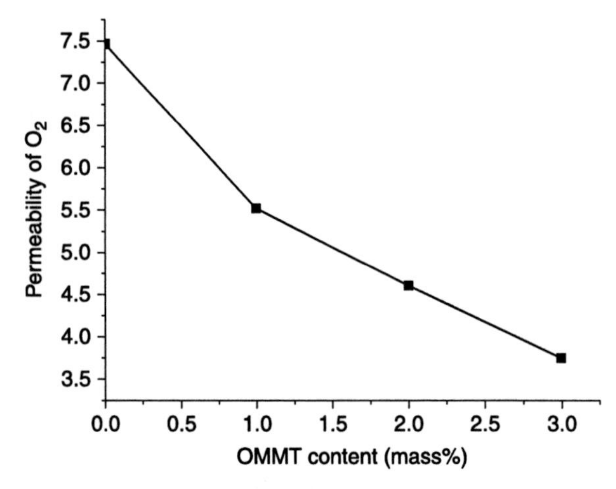


*10.18* Oxygen permeability of a PET nanocomposite. Reproduced with permission from Tsai *et al.*[16]

## 10.6 Applications of PET-based nanocomposites

PET is commonly used for textile applications and has the largest percentage of synthetic fibers in the market. It is used as a neat material or mixed with cellulose. PET fiber has the most compact and crystalline structure, and is markedly hydrophobic. Its hydrophobic nature, which accounts for its inability to wick moisture away from the body and poor launderability, can be a disadvantage in the underlying applications.[7,9,10] To change this passive and lazy role of PET, surface modification of PET fiber without loss of bulk properties has been an oft-sought goal in the fiber industry. Alkaline hydrolysis of PET fibers is often used to improve the soil release property and moisture regain. Recently, PET/$SiO_2$ nanocomposites have been prepared via *in situ* polymerization and melt spun to fibers. Such nanocomposites exhibit tougher superfine structures (e.g., cracks, craters, and cavities), which facilitate certain applications like deep dyeing.[72] In order to improve the mechanical properties of a PET/cotton fabric, different mixtures of resin/clay were deposited with different clay percentages on a PET/cotton fabric sheet. Finally, different mechanical tests, such as tensile, breakout, abrasion, and tear tests, were performed. Furthermore, it appears that a fabric's mechanical performance globally increases versus the amount of clay for all resins except polyvinyl acetate, which showed a decrease in fabric tear resistance when using more than 20 wt% clay. It was also noticed that the mechanical characteristics of the coated fabric without clay decreased significantly in comparison with the reference, but became better when clay was added to the resin. Consequently, it is obvious that the type of resin is important, because the results differ from one resin to another.[73]

In addition to textiles, the other important application of PET is packaging. The shelf life of any product depends on the gas barrier properties of the packaging material. This property can be improved by incorporation of clay in the PET matrix as discussed in the previous section. A thin layer of hybrid organic/inorganic (creamer) dramatically improves the oxygen barrier properties of PET film. The apparent gas permeability of a multilayer film can be reduced by three orders of magnitude compared with neat PET. The pure coating properties are calculated taking into account the effect of the different diffusion resistances in the various layers. The effect of aging has also been investigated by comparing oxygen permeability before and after immersion in water for films that are suspected of being degraded by water.[74]

Smart materials are defined as materials that sense and react to environmental conditions or stimuli (e.g., mechanical, chemical, electrical, or magnetic signals). In the last decade, a wide range of novel smart materials has been produced for aerospace, transportation, telecommunications, and domestic applications. Based on their advantages over other forms of material, such as their high specific area and superior mechanical and handling properties, fibers are being used more and more in applications in different areas. Preliminary results for PET and poly(methacrylic acid) or poly(2-acrylamido-2-methyl-1-propanesulfonic acid) showed that the selective transport of water vapor over dimethyl methylphosphonate vapor increased 12 times for both nanocomposites in comparison to a PET dense membrane.[75] It is well known that liquid crystal polymers (LCPs) possess some excellent properties, such as high-strength and high-stiffness fibers, lower melt viscosity, and highly resistant chemical and thermal properties. For this reason, it is expected that, depending on the interaction between the LCP, the thermoplastic polymer, and the organoclay, LCPs can facilitate the dispersion of clay in a thermoplastic matrix or induce a particular orientation in the dispersed clay layers of polymer/clay nanocomposites. Further, it has already been proved that LCP molecules align themselves according to the direction of an applied mechanical, electric or magnetic field. So, when polymer/clay/LCP nanocomposites are exposed to different fields, the LCP molecules will tend to align themselves in accordance with the field direction and will induce further orientation of the clay particles. In a recent study, small-angle X-ray scattering showed that the degree of anisotropy and mean orientation angles of clay platelets in a PET/LCP blend matrix were altered significantly after solid-state rheological property measurements (frequency and temperature sweep tests) were carried out in bending mode.[76]

In order to make a polymer surface resistant to calcification, multiblock copolymers containing PET (30 wt%) as a hard segment and dilinoleic acid (DLA) as a soft segment were prepared in the presence of two concentrations of $TiO_2$ particles (23 nm diameter) as a nanofiller (0.2 and 0.4 wt%) via *in situ* polycondensation. Changes in the thermal properties as well as investigations of the topography of these nanocomposites suggest that the addition of a small number of $TiO_2$ nanoparticles resulted in higher crystallinity and roughness of the copolymer surface. Incubation in a simulated body fluid (SBF) increased the roughness of all copolymers, as characterized by root mean square (rms) values. However, no hydroxyapatite layer formed on the sample surface. The highest difference of rms values was found for the neat PET/DLA material. The addition of nanocrystalline $TiO_2$ seems to protect the material surface from calcification, which represents a positive effect when a nanocomposite is intended for medical applications, especially when it will come into contact with soft tissue.[77]

An inorganic electroluminescent (EL) device on a flexible PET substrate was fabricated by Kim *et al.*[78] It was found that the brightness of the inorganic EL device was strongly dependent on the quality of the carbon nanotube (CNT) composite films. After treatment of the PET substrate with 3-aminopropyltriethoxysilane, the CNTs were uniformly dispersed and showed good adhesion to the substrate, and the resulting EL device showed better performance. The flexible EL device had a brightness of 96.8 $cd/m^2$ at 28 kHz and 50 V.

## 10.7 Future trends

The main downstream industries based on PET are the production of polyester fibers, accounting for around 65% of global consumption, and PET bottle resins consuming around 30%. Other applications are for polyester film and polyester engineering resins. Although much research has been carried out to improve the mechanical, barrier, and thermal properties of PET resin, it is still necessary to improve these properties to meet the high demands on this resin. The enhanced barrier properties of PET nanocomposites are not only useful in the bottle industry; they are also important for sports shoes, which have gas-filled bladders for shock damping, and tennis balls, which are claimed to maintain their internal pressure for a longer time. Other commercial products dependent on the barrier properties of nanocomposites include fuel hoses, where environmental regulations require low levels of leakage by diffusion through the hose.

A current problem in the commercialization of polymer nanocomposites is the homogeneous dispersion of clay layers in the polymer matrix. Particle aggregation reduces the effective particle aspect ratio and limits the surface area. This prevents the improvement of properties of the resultant nanocomposites.[79] The commercial technique most preferred for preparing nanocomposites is melt blending. For PET nanocomposites, *in situ* intercalation is also very important. During the preparation of a PET nanocomposite, via either melt blending or *in situ* intercalation, a high processing temperature is required. Hence it is necessary to choose clay with a surfactant with high thermal stability. Among such thermally stable surfactants, the most interesting are alkyl chain imidazolium and phosphonium halides. However, the dispersion characteristics obtained are not sufficient for commercialization. So far, most PET/clay nanocomposites exhibit higher strength and stiffness, but not higher elongation at break (or tenacity or toughness). Therefore, there is still a need to improve the toughness of these materials for defense applications.

It should always be kept in mind that the improvement of properties of a polymer nanocomposite is directly related to the dispersion characteristics of the nanofillers in the polymer matrix. So far the best dispersion of clay platelets in a polymer matrix has been achieved by *in situ* intercalation[16] and melt blending.[21] Those procedures are mentioned in the preparative methods section of this chapter. In both references the authors described the preparation and structural characterization of the clay-based PET nanocomposites. But in our opinion, on the basis of one TEM result for a specific region, and XRD, it is difficult to get an impression of the overall dispersion of clay layers in a PET matrix. It is therefore important to study the dispersion of silicate particles in a PET matrix by SAXS in detail. As a complementary technique, the three-dimensional reconstruction of images obtained from focused ion beam SEM is also very useful. It is necessary to explore the properties of such composites in detail to move PET nanocomposites one step closer to commercialization.

## 10.8 Acknowledgements

The authors would like to thank the Department of Science and Technology and the Council for Scientific and Industrial Research in South Africa for financial support.

## 10.9 References

1. Sinha Ray S and Okamoto M (2003) ‘Polymer/clay nanocomposites: A review from preparation to processing’, *Prog Polym Sci*, 28, 1539–1641.
2. Sinha Ray S and Bousmina M (2003) ‘Biodegradable polymers/layered silicate nanocomposites: in greening the 21st century materials science’, *Prog Polym Sci*, 50, 962–1035.
3. Bogue R (2011) ‘Nanocomposites: A review of technology and application’, *Assembly Automation*, 31, 106–112.
4. Camargo PHC, Satyanarayana KG and Wypych F (2009) ‘Nanocomposites: Synthesis, structure, properties and new application opportunities’, *Mat Res*, 12, Print version ISSN 1516–1439.
5. Yin M, Li C, Guan G, Yuan X, Zhang D and Xiao Y (2009) ‘In-situ synthesis of poly(ethylene terephthalate)/clay nanocomposites using $TiO_2/SiO_2$ sol-intercalated montmorillonite as polycondensation catalyst’, *Polym Eng Sci*, 1562–1572.
6. www.worldofplastic.net/PolyethyleneTerephthalate.htm, Types of Plastic – Worldofplastic.net, downloaded on 24 October 2011.
7. Lee S-S, Ma YT, Rhee H-W and Kim J (2005) ‘Exfoliation of clay facilitated by ring-opening reaction of cyclic oligomers in PET-clay nanocomposites’, *Polymer*, 46, 2201–2210.

8. www.petmachine.in/advantages_of_pet.htm, Advantages of PET – Plastic PET, downloaded on 24 October 2011.
9. Hwang SY, Lee WD, Lim JS, Park KH and Im SS (2008) 'Dispersibility of clay and crystallization kinetics for *in situ* polymerized PET/pristine and modified montmorillonite nanocomposites', *J Polym Sci: Part B: Polym Phys*, 46, 1022–1035.
10. Yuan X, Li C, Guan G, Liu X, Xiao Y and Zhang D (2007) 'Synthesis and characterization of poly(ethylene terephthalate)/attapulgite nanocomposites', *J Appl Polym Sci*, 103, 1279–1286.
11. Costache MC, Heidecker MJ, Manias E and Wilkie CA (2006) 'Preparation and characterization of poly(ethylene terephthalate)/clay nanocomposites by melt blending using thermally stable surfactants', *Polym Adv Techno*, 17, 764–771.
12. Ou CF, Ho MT and Lin JR (2004) 'Synthesis and characterization of poly(ethylene terephthalate) nanocomposites with organoclay', *J Appl Polym Sci*, 91, 140–145.
13. Kim S-H, Park S-H and Kim S-C (2005) 'Novel clay treatment and preparation of poly(ethylene terephthalate)/clay nanocomposite by *in situ* polymerization', *Polym Bull*, 53, 285–292.
14. Jung M-H, Chang J-H and Kim J-C (2007) 'Poly(ethylene terephthalate) nanocomposite fibers with organomica via *in situ* intercalation', *Polym Eng Sci*, 1820–1826.
15. Li C, Xiao Y, Guan G, Liu X and Zhang D (2006) 'Preparation and properties of PET/PA6 copolymer/montmorillonite hybrid nanocomposite', *J Appl Polym Sci*, 101, 2512–2517.
16. Tsai T-Y, Li C-H, Chang C-H, Cheng W-H, Hwang C-L and Wu R-J (2005) 'Preparation of exfoliated polyester/clay nanocomposites', *Adv Mater*, 17, 1769–1773.
17. Barber GD, Calhoun BH and Moore RB (2005) 'Poly(ethylene terephthalate) ionomer based clay nanocomposites produced via melt extrusion', *Polymer*, 46, 6706–6714.
18. Frounchi M and Dourbash A (2009) 'Oxygen barrier properties of poly(ethylene terephthalate) nanocomposite films', *Macromol Mater Eng*, 294, 68–74.
19. Bandypadhyay J, Sinha Ray S and Bousmina M (2007) 'Thermal and thermo-mechanical properties of poly(ethylene terephthalate) nanocomposites', *J Ind Eng Chem*, 13, 614–623.
20. Giraldi ALF de M, Bizarria MTM, Silva AA, Velasco JI, d'Ávila MA and Mei LHI (2008) 'Effects of extrusion conditions on the properties of recycled poly(ethylene terephthalate)/nanoclay nanocomposites prepared by a twin-screw extruder', *J Appl Polym Sci*, 108, 2252–2259.
21. Ammala A, Bell C and Dean K (2008) 'Poly(ethylene terephthalate) clay nanocomposites: improved dispersion based on an aqueous ionomer', *Compo Sci Tech*, 68, 1328–1337.
22. Rajeev RS, Soon K, McNally T, Menary G, Armstrong CG and Martin PJ (2009) 'Studies on the effect of equi-biaxial stretching on the exfoliation of nanoclays in poly(ethylene terephthalate)', *Euro Polym J*, 45, 332–340.
23. Rajeev RS, Harkin-Jones E, Soon K, McNally T and Menary G, *et al.* (2008) 'A method to study the dispersion and orientation of nanoclay tactoids in PET matrix-focused ion beam milling combined with electron microscopy', *Mater Lett*, 62, 4118–4120.
24. Krevelen DWV (1990) *Properties of Polymers*. Amsterdam, the Netherlands: Elsevier.
25. Bandypadhyay J, Maity A, Khatua BB and Sinha Ray S (2010) 'Thermal and rheological properties of biodegradable poly[(butylene succinate)-co-adipate] nanocomposites', *J Nanosci Nanotechnol*, 10, 4184–4195.

26. Kim SH and Kim SC (2007) 'Synthesis and properties of poly(ethylene terephthalate)/clay nanocomposites by *in situ* polymerization', *J Appl Polym Sci*, 103, 1262–1271.
27. Monemian SA, Goodarzi V, Zahedi P and Angaji MT (2007) 'PET/imidazolium based nanocomposites via *in situ* polymerization: Morphological, thermal, and non-isothermal crystallization studies', *Adv Polym Technol*, 26, 247–257.
28. Patro TU, Khakhar DV and Misra A (2009) 'Phosphonium-based clay–poly(ethylene terephthalate) nanocomposites: Stability, thermal and mechanical properties', *J Appl Polym Sci*, 113, 1720–1732.
29. Hamzehlou Sh and Katbab AA (2007) 'Bottle-to-bottle recycling of PET via nanostructure formation by melt intercalation in twin screw compounder: Improved thermal, barrier, and microbiological properties', *J Appl Polym Sci*, 106, 1375–1382.
30. Ke Y-C, Yang Z-B and Zhu C-F (2002) 'Investigation of properties, nanostructure, and distribution in controlled polyester polymerization with clay', *J Appl Polym Sci*, 85, 2677–2691.
31. Bizarria MTM, Giraldi ALF de M, de Carvalho CM and Velasco JI (2007) 'Morphology and thermomechanical properties of recycled PET-organoclay nanocomposites', *J Appl Polym Sci*, 104, 1839–1844.
32. Avrami MJ (1940) 'Kinetics of phase change. II Transformation-time relations for random distribution of nuclei', *Chem Phys*, 8, 212–225.
33. Ozawa T (1971) 'Kinetics of non-isothermal crystallization', *Polym*, 12, 150–158.
34. Bandypadhyay J, Sinha Ray S and Bousmina M (2007) 'Nonisothermal crystallization kinetics of polyethylene terephthalate nanocomposites', *J Nanosci Nanotechnol*, 8, 1–11.
35. Liu TX, Mo ZS, Wang SE and Zhang HF (1997) 'Nonisothermal melt and cold crystallization kinetics of poly(Aryl Ether Ether Ketone Ketone)', *Poly Eng Sci*, 37, 568–575.
36. Chen G-X and Yoon J-S (2005) 'Nonisothermal crystallization kinetics of poly(butylene succinate) composites with a twice functionalized organoclay', *J Polym Sci: Part B: Polym Phys*, 43, 817–826.
37. Durmus A, Ercan N, Soyubol G, Deligöz H and Kaşgöz A (2010) 'Nonisothermal crystallization kinetics of poly(ethylene terephthalate)/clay nanocomposites prepared by melt processing', *Polym compos*, DOI 10.1002/pc.20892, 1056–1066.
38. Jeziorny A (1978) 'Parameters characterizing the kinetics of the nonisothermal crystallization of poly(ethylene terephthalate) determined by d.s.c.', *Polymer*, 19, 1142–1144.
39. Yao X, Tian X, Zheng K, Zhang X, Zheng J, Wang R, Liu C, Li Y and Cui P (2009) 'Non-isothermal crystallization kinetics of poly(butylene terephthalate)/silica nanocomposites', *J Macromol Sci*, 48, 537–549.
40. Wang Y, Shen C, Li H and Chen J (2004) 'Nonisothermal melt crystallization kinetics of poly(ethylene terephthalate)/clay nanocomposites', *J Appl Polym Sci*, 91, 308–314.
41. Wan T, Chen L, Chua Y C and Lu X (2004) 'Crystalline morphology and isothermal crystallization kinetics of poly(ethylene terephthalate)/clay nanocomposites', *J Appl Polym Sci*, 94, 1381–1388.
42. Guan G, Li C, Yuan X, Xiao Y, Liu X and Zhang D (2008) 'New insight into the crystallization behavior of poly(ethylene terephthalate)/clay nanocomposites', *J Polym Sci: Part B: Polym Phys*, 46, 2380–2394.
43. Wang Y, Shen C and Chen J (2003) 'Nonisothermal cold crystallization kinetics of poly(ethylene terephthalate)/clay nanocomposites', *Polym J*, 35, 884–889.

44. Calcagno CIW, Mariani CM, Teixeira SR and Mauler RS (2007) 'The effect of organic modifier of the clay on morphology and crystallization properties of PET nanocomposites', *Polymer*, 48, 966–974.
45. Stoeffler K, Lafleur PG and Denault J (2008) 'Thermal decomposition of various alkyl onium organoclays: effect on poly(ethylene terephthalate) nanocomposites' properties', *Polym Degrad Stabil*, 93, 132–1350.
46. Augis JA and Bennett JE (1978) 'Calculation of the Avrami parameters for heterogeneous solid state reactions using a modification of the Kissinger method', *J Therm Anal Cal*, 13, 283–292.
47. Kissinger HE (1957) 'Reaction kinetics in differential thermal analysis', *Anal Chem*, 29, 1702–1706.
48. Takhor RL (1971) *Advances in Nucleation and Crystallization of Glasses,* American Chemical Society: Columbus, OH, 166–172.
49. Bandypadhyay J, Sinha Ray S and Bousmina M (2010) 'Effect of nanoclay incorporation on the thermal properties of poly(ethylene terephthalate)/liquid crystal polymer blends', *Macromol Mater Eng*, 295, 822–837.
50. Xie W, Gao Z, Pan W-P, Hunter D, Singh A and Vaia R (2001) 'Thermal degradation chemistry of alkyl quaternary ammonium montmorillonite', *Chem Mater*, 13, 2979–2990.
51. Xie W, Gao Z, Liu K-L, Pan W-P and Vaia R, *et al.* (2001) 'Thermal characterization of organically modified montmorillonite', *Thermochim Acta*, 367, 339–350.
52. Sinha Ray S, Bousmina M and Okamoto K (2005) 'Structure and properties of nanocomposites based on poly(butylene succinate-co-adipate) and organically modified montmorillonite', *Macromol Mater Eng*, 290, 759–768.
53. Lai M-C, Chang K-C, Huang W-C, Hsu S-C and Yeh J-M (2008) 'Effect of swelling agent on the physical properties of PET-clay nanocomposite materials prepared from melt intercalation approach', *J Phys Chem Solids*, 69, 1371–1374.
54. Chang J-H and Mun MK (2007) 'Nanocomposite fibers of poly(ethylene terephthalate) with montmorillonite and mica: thermomechanical properties and morphology', *Polym Int*, 56, 57–66.
55. Chang J-H, Mun MK and Lee IC (2005) 'Poly(ethylene terephthalate) nanocomposite fibers by *in situ* polymerization: The thermomechanical properties and morphology', *J Appl Polym Sci*, 98, 2009–2016.
56. Guan G, Li C, Zhang D and Jin Y (2006) 'The effects of metallic derivatives released from montmorillonite on the thermal stability of poly(ethylene terephthalate)/montmorillonite nanocomposites', *J Appl Polym Sci*, 101, 1692–1699.
57. Kráčalík M, Studenovský M, Mikešová J, Sikora A, Thomann R and Friedrich C (2007) 'Recycled PET-organoclay nanocomposites with enhanced processing properties and thermal stability', *J Appl Polym Sci*, 106, 2092–2100.
58. Yuan X, Li C, Guan G, Xiao Y and Zhang D (2008) 'Thermal stability of surfactants with amino and imido groups in poly(ethylene terephthalate)/clay nanocomposites', *J Appl Polym Sci*, 109, 4112–4120.
59. Vassiliou AA, Chrissafis K and Bikiaris DN (2010) '*In situ* prepared PET nanocomposites: Effect of organically modified montmorillonite and fumed silica nanoparticles on PET physical properties and thermal degradation kinetics', *Thermochimica Acta*, 500, 21–29.
60. Xu X, Ding Y, Qian Z, Wang F, Wen B and Zhou H (2009) 'Degradation of poly(ethylene terephthalate)/clay nanocomposites during melt extrusion: effect of clay catalysis and chain extension', *Polym Degrad Stabil*, 94, 113–123.

61. Yuan X, Li C, Guan G, Xiao Y and Zhang D (2008) 'Thermal degradation investigation of poly(ethylene terephthalate)/fibrous silicate nanocomposites', *Polym Degrad Stabil*, 93, 466–475.
62. Wang G, Chen Y and Wang Q (2008) 'Structure and properties of poly(ethylene terephthalate)/$Na^+$-montmorillonite nanocomposites prepared by solid state shear milling ($S^3M$) method', *J Polym Sci: Part B: Polym Phys*, 46, 807–817.
63. Sánchez-Solís A, Romero-Ibarra I, Estrada MR, Calderas F and Manero O (2004) 'Mechanical and rheological studies on poly(ethylene terephthalate)/montmorillonite nanocomposites', *Polym Eng Sci*, 44, 1094–1102.
64. Bandypadhyay J and Sinha Ray S (2010) 'Mechanism of enhanced tenacity in a polymer nanocomposite studied by small-angle X-ray scattering and electron microscopy', *Polymer*, 51, 4860–4866.
65. Kráčalík M, Studenovský M, Mikešová J, Sikora A and Thomann R, *et al.* (2007) 'Recycled PET nanocomposites improved by silanization of organoclay', *J Appl Polym Sci*, 106, 926–937.
66. Litchfield DW and Baird DG (2008) 'The role of nanoclay in the generation of poly(ethylene terephthalate) fibers with improved modulus and tenacity', *Polymer*, 49, 5027–5036.
67. Hao J, Lu X, Liu S, Lau SK and Chua Y C (2006) 'Synthesis of poly(ethylene terephthalate)/clay nanocomposites using aminododecanoic acid-modified clay and a bifunctional compatibilizer', *J Appl Polym Sci*, 101, 1057–1064.
68. Vidotti SE, Chinellato AC, Hu G-H and Pessan LA (2007) 'Preparation of poly(ethylene terephthalate)/organoclay nanocomposites using a polyester ionomer as a compatibilizer', *J Polym Sci: Part B: Polym Phys*, 45, 3084–3091.
69. Lee SJ, Hahm WG, Kikutani T and Kim BC (2009) 'Effects of clay and POSS nanoparticles on the quiescent and shear-induced crystallization behavior of high molecular weight poly(ethylene terephthalate)', *Polym Eng Sci*, DOI 10.1002/pen.21262, 317–323.
70. Ge X-G, Wang D-Y, Wang C, Qu M-H and Wang J-S, *et al.* (2007) 'A novel phosphorous containing copolyester/montmorillonite nanocomposites with improved flame retardancy', *Euro Polym J*, 43, 2882–2890.
71. Ke Z and Yongping B (2005) 'Improve the gas barrier property of the PET film with montmorillonite by *in situ* interlayer polymerization', *Mater Lett*, 59, 3348–3351.
72. Yang Y and Gu H (2006) 'Superfine structure, physical properties, and dyeability of alkaline hydrolyzed poly(ethylene terephthalate)/silica nanocomposite fibers prepared by *in situ* polymerization', *J Appl Polym Sci*, 102, 3691–3697.
73. Abid K, Dhouib S and Sakli F (2010) 'Addition effect of nanoparticles on the mechanical properties of coated fabric', *J Textile Institute*, 101, 443–451.
74. Minelli M, Rocchetti M, De Angelis MG, Montenero A and Doghieri F (2006) 'Gas barrier properties of polymeric films with hybrid organic/inorganic coatings for food packing applications', AIChE Annual Meeting, San Francisco.
75. Chen H (2005) 'Smart nanotechnology in biomaterials, sensors, actuators and textiles', AIChE Annual Meeting and Fall Showcase, Cincinnati.
76. Bandyopadhyay J and Sinha Ray S (2011) 'Determination of structural changes of dispersed clay platelets in a polymer blend during solid-state rheological property measurement by small-angle X-ray scattering', *Polymer*, submitted.
77. Piegat A, El Fray M, Jawad H, Chen Qz and Boccaccini AR (2008) 'Inhibition of calcification of polymer-ceramic composites incorporating nanocrystalline $TiO_2$', *Adv Appl Ceram*, 107, 287–292.

78. Kim MJ, Shin DW, Kim J-Y, Park SH, Han I and Yoo JB (2009) 'The production of a flexible electroluminescent device on polyethylene terephthalate films using transparent conducting carbon nanotube electrode', *Carbon*, 47, 3461–3465.
79. Litchfield DW, Baird DG, Rim PB and Chen C (2010) 'Improved mechanical properties of poly(ethylene terephthalate) nanocomposite fibers', *Polym Eng Sci*, DOI 10.1002/pen.21758.

# 11
# Thermoplastic polyurethane (TPU)-based polymer nanocomposites

D. J. MARTIN, A.F. OSMAN, Y. ANDRIANI and G. A. EDWARDS, University of Queensland, Australia

**Abstract:** Thermoplastic polyurethanes (TPU) are a commercially important class of thermoplastic elastomers, which have an inherent nanostructured morphology. This chapter introduces thermoplastic polyurethanes before summarizing the benefits of introducing various types of nanofillers using various processing methods and highlighting the engendered properties and performance. A detailed discussion of the influence of nanofillers on TPU morphology is provided, as well as a section on biomedical applications and nanofiller toxicity.

**Key words:** thermoplastic polyurethane, nanocomposite, morphology, mechanical performance, biomaterials, toxicity.

## 11.1 Introduction: the potential of thermoplastic polyurethane (TPU) nanocomposites

### 11.1.1 Polyurethanes and thermoplastic polyurethanes

Polyurethanes, first discovered by Otto Bayer in 1937, encompass the series of polymers whose molecular backbone contains a significant number of urethane linkages, regardless of the content of the rest of the macromolecule.[1] The urethane linkage is formed via the reaction between an isocyanate group and a hydroxyl group. Initial studies into these polymers focused on forming linear polyurethanes by reacting diisocyanates with diols, but it was very quickly realized that a multitude of polymers with wide-ranging properties could be produced.[2] Early work focused on polyester-based polyols; however, the enhanced hydrolytic stability and immense versatility afforded by polyether-based polyols saw them become a preferred precursor in polyurethane synthesis.[2] These days, the vast selection of polyols, isocyanates, and chain extenders allows polyurethanes to be varied from soft thermoplastic elastomers to adhesives, coatings, flexible foams, and rigid thermosets. This chapter will focus solely on thermoplastic polyurethane (TPU) elastomers as host polymers for polymer nanocomposite systems. These TPUs are often called 'segmented polyurethane block copolymers,' and form a subset of a large and commercially important family of thermoplastic elastomers (TPEs) comprising styrene block copolymers (SBCs), thermoplastic olefins (TPOs), thermoplastic vulcanizates (TPVs), and copolyester elastomers (COPEs). TPEs combine the flexibility and resilience of rubbers with the processability of plastics. Global demand for TPEs is forecast to increase 6.3% pa through 2011 to 3.7 million tonnes. TPU materials currently comprise approximately 15% of the volume of TPE polymers sold annually.[3] They are at the premium end of this class of materials and are known for their high strength, durability, and tremendous versatility in terms of tailored formulations.

Some common applications for TPUs include automotive interiors, footwear, flexible hose and tubing, cellphone buttons, closures, seals and o-rings, adhesives, cable jacketing, sport and leisure items, textiles and textile coatings, implantable medical device components, mining and mineral processing equipment, laminates for impact glazing, photovoltaic cell encapsulation, and wastewater treatment equipment, just to name a few.

The clever incorporation of functional nanofillers of various types into TPU host polymers can extend typical TPU property profiles, improving mechanical and tribological performance,[4,5] dimensional stability at higher operating temperatures,[6] gas barrier and flame retardancy performance,[7,8] electrical properties,[6,9] and biological properties.[10–12] This chapter reviews some of the exciting progress being made in these areas.

### 11.1.2 TPU chemistry, morphology, and properties

TPUs are linear, block copolymers of alternating hard and soft segments. The soft segments (SSs) are composed of long-chain diols (most typically polyester, polyether, or polycaprolactone-based) with a molecular weight of 1000–4000 g/mol, whilst the hard segments (HSs) commonly consist of alternating diisocyanates and short-chain extender sequences.[1] Thermodynamic incompatibility between the segments results in phase separation, and subsequent organization into hard and soft domains with a nanoscale texture, which gives TPUs their distinct mechanical properties and thermoplastic utility.[13–15]

The flexible, amorphous soft domains primarily influence the elastic nature of the TPU, but still have some influence on the hardness, tear strength, and elastic modulus of the polymer. They are typically above their glass transition temperature ($T_g$) at room temperature and thus control TPU performance at low temperatures. In contrast, the rigid and typically semi-crystalline hard domains primarily influence elastic modulus, hardness, tear strength, and melt processability.[1] However, the hard domains also act as physical cross-links, imparting elastomeric properties to the soft domain phase.[16] Thus, the mechanical properties of TPUs are strongly influenced by the complex microphase morphology of the hard and soft domains. These are, in turn, dependent on such factors as the hard/soft segment composition ratios, the segmental solubility parameters and associated compatibility, molecular weight and polydispersity of the hard and soft segments, and the thermal and processing history of the TPU.[17–23] The unique microphase morphology confers on TPUs a higher tensile strength and toughness when compared with most other elastomers,[22] and the absence of covalent cross-linking allows TPUs to be both melt and solution processed. However, under continuous or cyclic loading the absence of chemical cross-linking gives way to a degree of plastic flow whereby the hard domains can restructure, which results in large hysteretic losses and poor creep resistance.[24,25] The clever introduction of nanofillers into this complex nanoscale morphology, and the characterization of the level of interplay between the various system components, represents a fascinating challenge for current researchers. Similarly, the typically moderate hard domain $T_g$ in these systems seriously limits the upper use temperatures and high temperature performance and chemical or dimensional stability in aggressive environments. Again this is believed to be an area where innovation in TPU nanocomposite systems can effectively address these limitations.

### 11.1.3 TPU nanocomposites versus traditional microcomposites

The polymer industry is continually researching to find new materials that offer increased performance at lower costs, often necessitating the introduction of fillers. There have been various attempts to improve the mechanical performance, creep resistance, and permanent set of TPUs either by varying the composition of the hard and soft segments[19,26] or by introducing macro- or micro-sized particulate fillers,[27,28] with the majority of these studies yielding suboptimal results. Generally microfillers significantly stiffen TPUs and also substantially reduce elongation to break. They can also be associated with major wear of polymer processing equipment and dies due to the high loading levels required (up to 40% by weight) and the highly abrasive nature of these fillers.

Nanofillers offer significant advantages over macro-sized or micro-sized fillers, including vastly superior reinforcement efficiency, a greater surface area to mass ratio, low percolation threshold, and often very high aspect ratios, and as such there has been significant research and investment into the production of polymeric nanocomposites. Recently, a range of nanoparticles has been used in TPU-based composites, with varying levels of success. The bulk of this work has involved the use of carbon nanotube or layered silicate-based nanofillers.

## 11.2 TPU nanocomposites: structure, processing, properties, and performance

### 11.2.1 Mechanical influence of nanofillers on TPU stiffness and ultimate tensile strength

Figure 11.1 and 11.2 compare the mechanical properties of many of the best literature examples of TPU nanocomposites reinforced with layered silicate nanofillers or carbon nanotubes or nanofibres. We have chosen only the most impressive results from the academic and commercial literature. This is because

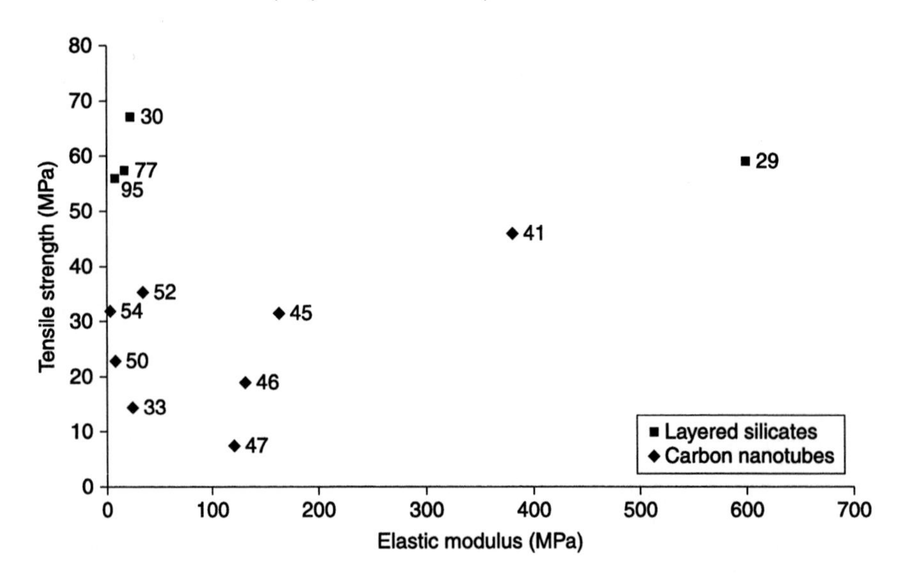


*11.1* Tensile strength versus elastic modulus for various TPU nanocomposites reinforced with carbon nanotubes (◆) or layered silicates (■). Source references are given beside each point.

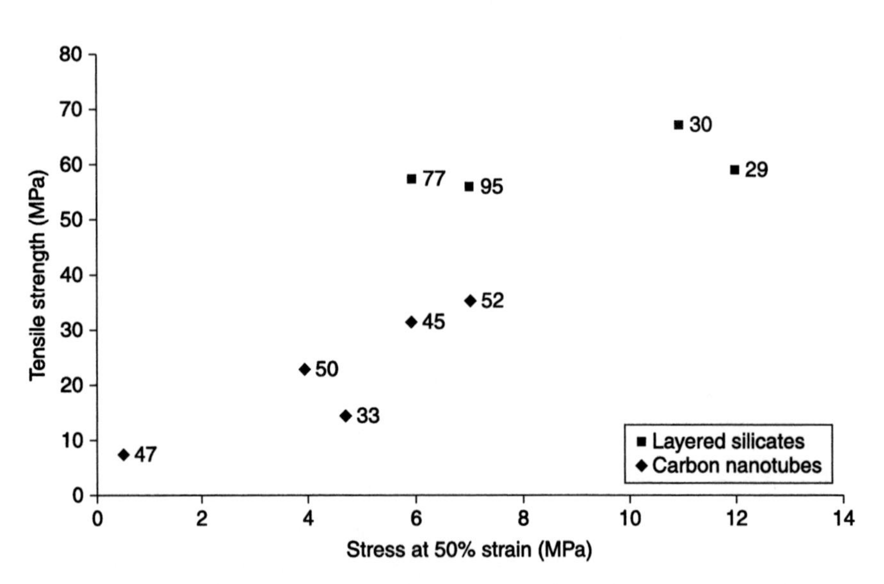


*11.2* Tensile strength versus stress at 50% strain for various TPU nanocomposites reinforced with carbon nanotubes (◆) or layered silicates (■). Source references are given beside each point.

we found that many studies, while presenting large percentage changes as a result of nanofiller reinforcement, did not benchmark particularly well to typical commercial TPU property ranges. While clearly there are many other material properties of great interest for TPU nanocomposites which will be discussed elsewhere in this chapter, these figures do present some very noteworthy trends.

First, Fig. 11.1 plots ultimate tensile strength versus elastic modulus. Even taking into account obvious differences in the particular TPU matrix materials, nanofillers and nanofiller loadings used in each study, the data clearly shows a more pronounced stiffening effect for carbon nanotubes and a more pronounced strengthening effect for layered silicates. The one exception is the very stiff nanocomposite prepared by solvent exchange at an unusually high loading (20 wt%) of unmodified synthetic hectorite.[291] The highest tensile strengths achieved (anything above 45 MPa is considered very good in the industry) were all layered silicate reinforced systems, with the strongest being prepared by reactive extrusion.[30]

Figure 11.2 plots tensile strength against the more industrially acceptable parameter for TPUs of stress at 50% strain. The grouping of the layered silicate data in the top right quadrant tells us that nanofiller-TPU molecular interactions are more substantial and perhaps 'cooperative' during deformation in layered-silicate reinforced systems, and that carbon nanotube (CNT)-based nanofillers need further optimization for utility in TPU systems.

A TPU property profile which engenders strength and toughness while maintaining flexibility and resilience has been traditionally very difficult to achieve via the composite approach due to substantial stiffening, and moving to softer TPU matrix polymers is not possible because they tend to suffer from a lack of strength due to the limits of segmental demixing evident at low hard segment composition ratios. In addition, typically if a stiffer, harder TPU is required, a higher hard-segment composition ratio can readily be formulated. Therefore stiffening of this class of polymers without the accompaniment of other improved functional properties or performance benefits is not generally of commercial interest.

## 11.2.2 TPU nanocomposites employing nanofillers other than layered silicates

There are reports in the literature of TPU nanocomposites incorporating nanosilica in the form of either fumed silica,[31,32] to improve the mechanical or adhesive properties of TPUs, or in the form of polyhedral oligomeric silsesquioxane POSS components to improve either biostability[10] or resistance to thrombosis.[10,11] Magnetite nanofillers have recently been dispersed and solution cast to produce TPU nanocomposite films with superparamagnetic properties,[9] and zinc oxide-TPU nanocomposites have recently been prepared and fabricated into electrospun web constructs, which demonstrated enhanced UV absorption and antimicrobial performance for potential use in protective clothing.[12]

There is also a growing body of work on TPU nanocomposites incorporating single-walled carbon nanotubes (SWNTs),[33] double-walled carbon nanotubes,[34] and multi-walled carbon nanotubes (MWNTs).[8] The manipulation and incorporation of large numbers of individual CNTs into polymeric systems is a difficult task because the high molecular weights and strong intertube forces (both van der Waals and electrostatic) promote the formation of micron-sized bundles and ropes,[35] leading to phase separation and poor mechanical properties. As a result both surfactant and covalent functionalizations have been employed in an attempt to fragment and disperse these micron-sized inclusions.[33,36–40] The majority of these TPU-CNT nanocomposites have been formulated via solution processing techniques,[8,33,41–49] although *in situ* polymerization[46,50–53] and melt processing[8] have been reported as well. Despite the superior mechanical performance of SWNTs, MWNTs are the most common variety of CNT studied and a wide variety of surface functionalizations has been investigated. Koerner and co-workers reported the first study into TPU-CNT nanocomposites, in which the researchers attempted to improve the stress-recovery characteristics of Morthane (a commercially available, shape memory TPU) by adding 1–5 vol.% MWNT via solution processing.[45] The researchers demonstrated that the nanocomposite could store and subsequently release up to 50% more recovery stress than the unmodified TPU. The nanocomposite also demonstrated the ability to recover stress, above that of the host TPU, when exposed to increased temperatures, infrared radiation, and electrical current. The authors theorized that the absorption of energy by MWNTs caused localized heating, which in turn melted the strain-induced TPU soft segment crystals, increasing stress recovery.

Sen and co-workers were the first to investigate the fabrication of TPU nanocomposites incorporating functionalized SWNTs.[33] The solution-processed nanocomposites demonstrated significantly increased mechanical properties over the host TPU. Whilst the final material was mechanically inferior to several commercially available TPUs without nanoparticle reinforcement, the study was the first to demonstrate both the possibility of increased dispersion in a TPU-CNT nanocomposite and that TPU-CNT interactions can be enhanced by CNT functionalization.

As a result, the majority of research into TPU-CNT nanocomposites has focused on using functionalized CNTs; however, there have been several studies published describing the use of pristine CNTs.[8,43,48] Importantly, these investigations have demonstrated poor CNT dispersion and phase separation between the SWNTs and the host TPU and suggest that the future of TPU-CNT nanocomposites lies with modification of the CNT surface. A wide variety of functional groups have been attached to CNTs in an attempt to overcome these problems. CNT surface modifiers are typically either short hydrophilic groups like COOH,[47,49] $NH_2$,[46] and OH[41] capable of significant hydrogen bonding with the host polymer, or long-chain hydrophobic groups designed to break apart CNT aggregates and increase dispersion.[33,54] Recently, significant results have been achieved by Deng and co-workers,[42,55] who demonstrated dramatic mechanical enhancements with very small loadings of MWNTs modified with novel functional groups. Several research groups have also attempted to use CNTs as cross-linking agents, covalently bonding individual CNTs to the TPU backbone. However, results have been mixed due to the fine stoichiometric control required to produce nanocomposites with the same molecular weights and hard/soft segment ratios as the control TPU.[50,52]

Despite the abundance of work into TPU-CNT nanocomposites, the only published investigations of TPU-functionalized CNT nanocomposites formulated by melt processing techniques have been conducted by Chen and co-workers[56] and more recently Barick and Tripathy.[57] Chen *et al.* demonstrated that acid-treated MWNTs had dramatically increased mechanical properties with an eightfold increase in Young's modulus and a 2.4-fold increase in tensile strength due to the strong interfacial interactions and good MWNT dispersion within the melt-drawn fibers. Barick and Tripathy reported substantial stiffening and toughening of a soft aliphatic grade of TPU by melt compounding vapor-grown carbon nanofibres (VGCNF), which had been purified, treated with acid and heat, into the host polymer. They also observed significant shifts in TPU thermal transitions and thermal stability, suggesting strong interfacial effects. Significantly, these results corroborate computational studies that suggest melt processing may induce a residual thermal radial stress (through differences in the coefficients of thermal expansion between the CNT and the host polymer), forcing the polymer into closer contact with the tube surface and thereby enhancing molecular interactions.[58]

### 11.2.3 TPU layered silicate nanocomposites

The vast majority of work on TPU nanocomposites has involved the use of layered-silicate-based nanofillers. Therefore, the remainder of this chapter will focus upon this class of nanocomposites and their chemistry, processing, structure, performance and safety or nanotoxicology.

### 11.2.4 Layered silicates as nanofillers

The incorporation of layered silicates into polymer matrices has been performed for over 50 years. During recent years, polymer nanocomposites containing natural clays and synthetic layered silicates have been most widely investigated, since the starting clay materials have been readily available in various sizes and shapes, and because their intercalation chemistry and interactions with surfactants and polymers have been studied for several decades.[59] In particular, the development of instrumentation to characterize polymer nanocomposites at small length scales, such as scanning force, laser scanning fluorescence, and electron microscopes, have spurred researchers into probing the influence of particle size, shape, and surface chemistry on the properties of these new materials.

### 11.2.5 Fabrication techniques for TPU-clay nanocomposites

Typically smectite-type clays are usually used as fillers. Both natural clays (montmorillonite and hectorite) and synthetic clays (synthetic hectorites and fluoromicas) have been investigated, and TPU nanocomposite processing has seen many different approaches. The most common dispersion methods[60] are solvent casting,[61] melt processing,[57] *in situ* polymerization,[62] and reactive extrusion.[30,38]

In solvent casting, a polymer is solubilized in an organic solvent, then the nanoclay is dispersed in the resulting solution, often assisted with high shear homogenization or ultrasonic energy. The solvent is then allowed to evaporate, leaving the nanocomposite behind, typically as a thin film. The solvent imparts the enhanced mobility the polymer needs in order to intercalate between the silicate layers, while thermodynamic compatibility combined with physical mixing gives rise to a dispersed system. There are limitations to solvent casting. The selected solvent must be able to completely dissolve the polymer and disperse the nanoclay. This approach can also lead to poor clay dispersion, as well as other problems such as high costs due to the large amounts of solvent required to achieve appreciable filler dispersion, technical phase separation problems, and health and safety problems.[60,63]

*In situ* polymerization involves the dispersion of clay layers into the matrix by polymerization, mixing the silicate layers with the monomer, in conjunction with the polymerization initiator or the catalyst. This method, which was used by Usuki *et al.*[64] in their pioneering research, takes advantage of the relative ease with which a small monomer molecule can intercalate silicate layers compared with a much larger polymer molecule.

Melt processing refers to the process of single or twin screw extrusion, whereby polymer chains are melted to increase their mobility and then the nanoclays are dispersed into the melt to allow intercalation of the nanoclays. Twin screw extruders or Brabender mixers require much higher shear energy, so care is needed when processing thermally sensitive TPU materials, as aggressive melt processing inevitably leads to a loss in molecular weight and a reduction in properties and performance. The selection of thermally stable organoclay modifiers is paramount, and careful washing away of excess exchanged alkylammonium cations is also important to avoid unwanted TPU degradation in the extruder.[65] Melt processing of TPU nanocomposites can be easily scaled to manufacturing quantities, but care must be taken to avoid TPU degradation.[66]

Reactive extrusion of TPU involves the *in situ* polymerization of TPU precursors (polyol, diisocyanate, and a chain extender) in a twin screw extruder. Nanofillers can be either side fed as dry powders or pre-dispersed into the polyol liquid precursor.[30,38] This method is used for chemical modification of existing polymers and the function of the extruder is as a continuous chemical reactor for polymerization.[67] This technique requires chemical reaction control besides conventional screw extruder parameter control (of materials conveying, melting, and mixing). Compared with conventional solution casting, reactive extrusion has the following advantages: no solvent is required, high-viscosity polymers can be used, and processing conditions, such as mixing time and temperature, are flexible.[67] Importantly, it avoids TPU reprocessing and thermal degradation, potentially avoids the handling of dry, inhalable nanopowder feeds, and effectively shifts the TPU nanocomposite processing route further up the TPU supply chain.[30] The majority of commercial TPU materials available today are manufactured using reactive extrusion.

### 11.2.6 TPU clay nanocomposite structure-property relations

The dispersion of clay nanolayers and the morphology of the nanocomposite depend on various factors, such as the nature of the components used (layered silicate, surface modifier, and polymer matrix) and method of preparation (mixing method, time, and temperature).[63,68,69] Generally, four main nanocomposite structures can be obtained, as illustrated in Fig. 11.3.[60]

In an intercalated structure, the polymer chains are sandwiched in between the silicate layers, which display limited dispersion. On the other hand, exfoliated nanocomposites demonstrate separated, individual silicate layers, which are more or less uniformly dispersed in the polymer matrix. The ordered structure is lost and the uniform dispersion of the anisotropic nano-sized particles can lead to a large interfacial area between the constituents at extremely low loadings of the

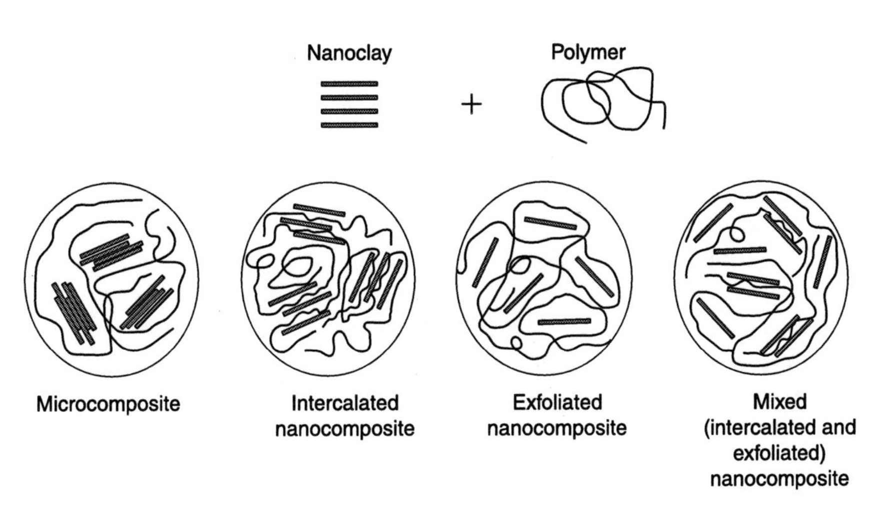


*11.3* Principal polymer layered silicate nanocomposite structures (adapted from Alexandre and Dubois[60]).

nanoparticles. This large interfacial area and the nanoscopic dimensions between constituents differentiate polymer nanocomposites from conventional composites and filled plastics. Hence, new mechanical, optical, and electrical properties can be developed, which may not occur in the macroscopic counterparts.[60] In fact, exfoliation of the platelets is the desirable objective of dispersion, because it achieves optimum enhancement of the properties. As well as these two well-defined structures, other intermediate organizations can exist presenting both intercalation and exfoliation.

Compared with the large body of research into polymer-clay nanocomposites, the number of studies conducted on TPU nanocomposites is relatively small. The use of TPU as the nanocomposite matrix presents some interesting challenges in understanding the nanoscale and microscale morphology, due to the pre-existing nanoscale morphology of the segmented TPU domain (the so-called soft and hard segments). TPU nanocomposites have been prepared predominantly through *in situ* polymerization,[62,70–74] solvent casting,[6,75–77] melt processing,[78–79] and to a lesser extent via reactive extrusion.[80,81]

Previous research shows that TPUs can be tailored to meet the specific property requirements by incorporating organically modified smectic clays. The first examples of TPU nanocomposites with greatly improved performance were reported by Wang and Pinnavaia.[62] The nanocomposites were prepared via *in situ* polymerization and alkylammonium-exchanged montmorillonites were used as nanofillers. They showed that the tensile strength, modulus, and strain at break all increased by more than 100% at a loading of only 10 wt% organoclay.

Meng *et al.*[82] employed different molecular weight polyether-based TPUs and organically modified montmorillonite (OMMT) in their investigation of the exfoliation of OMMT in a TPU matrix by melt processing. They concluded that optimized exfoliated nanocomposites can be obtained if there is adequate shear stress and appropriate molecular diffusion to accommodate interaction between the OMMT and the TPU.

Low permeability TPU nanocomposites were produced by Runt and co-workers[61] using polyether-based TPU and octadecylammonium-modified montmorillonite (MMT) by solvent casting. The water vapor permeabilities of the nanocomposites were reduced by as much as a factor of five at the highest OMMT content (20 wt%) and the enhancement of modulus was achieved without loss of ductility.

Cai *et al.*[81] produced TPU-MMT nanocomposites by reactive extrusion, which involved a one step direct polymerization-intercalation technique with a twin screw extruder. Morphological characterization of the nanocomposite indicated that there were delaminated nanocomposite structures with well-dispersed nanoclays in the TPU matrix. Improvements in thermal stability, flame retardancy, and tensile strength were achieved with the addition of up to 4 wt% MMT.

Other successful TPU nanocomposites have been reported in the following publications. Depending on the system, increases in modulus,[62,75,77,83,84] tensile strength,[70,74,75,78,83] elongation,[6,74] flame resistance and thermal stability,[70,72,83] barrier resistance,[70,75,83] tear strength,[85] abrasion resistance,[72] fracture toughness,[72] fatigue resistance,[86] and corrosion resistance[87] have been reported.

## 11.2.7 Hard and soft segment interactions with nanofillers

Knowledge in the TPU nanocomposite area has progressed recently with several research papers discussing the effects of nanofillers on the underlying complex TPU morphology, highlighting the importance of understanding the specific hard and soft segment interactions with the nanofillers in correlation with the resulting nanocomposite properties. Some of the literature in this field is summarized in Table 11.1.

The nanofiller in a nanocomposite should be distributed on a nanoscale level and therefore undergo substantial interactions with polymer chains and sequences. There is therefore tremendous potential 'reinforcement efficiency' due to the very high nanofiller aspect ratio and surface area, the local confinement of molecular motion,[88,89] and the tailoring of specific hard or soft segment interactions with the nanofiller surface.[34,90,91] However, due to the layered structure of silicate nanofillers, the organization of the resultant nanocomposite depends on the thermodynamic compatibility of the polymer and the layered silicate, the intercalation kinetics, and the processing conditions.[65,92] The types of structures encountered are shown in Fig. 11.3. Virgin layered silicate particles are naturally hydrophilic, and as such are largely immiscible with TPUs, which contain a significant hydrophobic component in the form of the soft segments. This natural repulsion between the untreated silicate and the TPU discourages the formation of an exfoliated system, which is required for an effective nanocomposite. Instead a phase-separated composite or microcomposite is formed, whose properties do not exceed those of traditional composites.

In order to prepare effective nanocomposites, the hydrated cations in the interlayer spacing are replaced with organic cations to render the surface more organophilic and improve wetting by the polymer matrix.[60,93] Furthermore, this substitution allows swelling of the gallery between the silicate layers, which in turn allows the polymer to enter the galleries during the formation of the nanocomposite. An unswollen silicate has an interlayer spacing of the order of 1 nm, which is smaller than the host polymer's radius of gyration.[94,95] The control and enhancement of both the polymer-silicate interaction and the swelling of the gallery layers are thus critical factors in the production of any polymeric nanocomposite, particularly so for TPUs, which generally possess both hydrophilic (hard segments) and hydrophobic (soft segment) components, both of which are of a length scale much smaller than even the smallest synthetic clay-based nanofillers (~25 nm).

In 1998, Wang and Pinnavaia[62] reported the first examples of TPU-layered silicate nanocomposites. The nanocomposites were prepared via *in situ*

*Table 11.1* Summary of studies of TPU/organofiller nanocomposites with specific hard-segment or soft-segment interactions

| Reference | Nanocomposite system | | Processing method | Key findings |
|---|---|---|---|---|
| | TPU | Nanofiller | | |
| 88 | Polyether-based TPU | 1. Laponite® RD modified with dodecyl amine hydrochloride<br>2. Cloisite® 20A modified with dimethyl dehydrogenated tallow quaternary ammonium ions | Solvent casting | Modified Cloisite (with higher hydrophobicity) was found to have affinity towards the SSs while modified Laponite® RD (with lower hydrophobicity) had a preferential association with the HSs.<br>Modified Laponite® RD (higher hydrophilic nature and less modification) has a greater tendency for aggregation and tends to re-aggregate in the matrix if the content is higher than 3%, while modified Cloisite (more hydrophobic) has less tendency to aggregate; hence the optimum content is 5%. |
| 89 | Polyether-based TPU and polyester-based TPU | 1. MMT/OH (MMT modified with quaternary ammonium bromide with a hydroxyl group (QAB/OH))<br>2. Cloisite® 30B (C30B) (MMT modified with a quaternary ammonium salt having one methyl, two short hydroxyalkyls and one tallow, which is a long alkyl chain) | Solvent casting | MMT/OH (longer hydroxyalkyl chain) exhibits better affinity to polyester-TPU than C30B, since polyester-TPU has an ester-polyol, which has higher hydrophilicity than polyether-TPU.<br>The hydroxyl group was observed at the outer surface of the clay in the MMT/OH system, which may provide more hydrogen-bonding sites compared with a clay having a shorter hydroxyalkyl chain modifier (C30B).<br>Greater dispersion was observed in the polyester-TPU/MMT/OH nanocomposites along with excellent tensile properties. |

| | | | | |
|---|---|---|---|---|
| 34 | Polyether-based TPU | Double-walled carbon nanotubes (DWNT) functionalized by reacting methanol, dodecyl amine, or octadecyl amine with a toluene 2,4-diisocyanate linker and referred to, respectively, as: DWNT-TDI-$CH_3$ DWNT-TDI-$C_{12}$ DWNT-TDI-$C_{18}$ | Solvent casting | The octadecyl amine functionality (DWNT-TDI-$C_{18}$), which had the longest hydrophobic alkyl tails, provided the greatest enhancement in the tensile strength of the nanocomposite. It was suggested that the greater dispersion resulting from steric hindrance of the larger functional groups increased the number of individual CNTs interacting with the TPU's hard and soft segments. DWNT-TDI-$C_{18}$ was suggested to have preferential reinforcement with the HS and allowed some interaction with the SSs in the region along the nanotube length where the phase boundaries were crossed. |
| 90 | Polyether-based TPU | Cloisite® 30B (MMT with methyl bis-2-hydroxyethyltallow ammonium modifier (MT2EtOH) | Melt processing | Significantly good dispersion of Cloisite® 30B with the intercalated/exfoliated silicate layers throughout the TPU matrix was observed from transmission electron microscopy (TEM) and scanning electron energy-dispersive X-ray spectroscopy (SEM-EDX) studies. Fourier transform infrared spectra indicated that Cloisite® 30B has inherent affinity towards the SSs of the TPU matrix, and is capable of forming hydrogen bonds with the HS, thus providing a strong polymer/filler interaction in the system. The ultimate tensile strength was found to have increased by about 152.46% with the incorporation of 5wt% Cloisite® 30B. |
| 91 | Polyether-based TPU | 1. Laponite® RDS (Laponite containing a pyrophosphate-based peptizer) modified with cetyl trimethyl ammonium bromide (cLS) | Solvent casting | The morphology of the two modified clay nanocomposites (TPU-cLS and TPU-dLS) was markedly different from each other. From TEM analysis, dLS exhibited preferential association with the HSs whereas cLS did not show such a preference. This was suggested to be due to the more hydrophobic character of cLS compared with dLS (with alkyl chain lengths of 16 and 12, respectively). |

(*Continued*)

*Table 11.1* Continued

| Reference | Nanocomposite system | | Processing method | Key findings |
|---|---|---|---|---|
| | TPU | Nanofiller | | |
| | | 2. Laponite® RDS modified with 1, 3, 5 and 7 wt% dodecyl amine hydrochloride (dLS) | | cLS-based TPU nanocomposites exhibit partly exfoliated, intercalated, and aggregated structures at lower clay content (1 wt%), but a network type of structure (aggregation) at higher clay content (5 wt%). This resulted in an approximately twofold increase in the storage modulus in both the glassy and rubbery states with 1 wt% cLS addition, but gradually decreased with an increase in clay content.<br>dLS-based TPU nanocomposites exhibited a spherical cluster-type of structure for all clay contents studied and a gradual increase in the storage modulus was observed with an increase in clay content. |

polymerization using montmorillonite (ion exchanged with alkylammonium salts) as the nanofiller. X-ray diffraction (XRD) patterns indicated that intragallery polymerization contributed to the dispersion of the organosilicates, and improvements in tensile strength, elongation, and modulus were observed.

The commercially available Cloisite® series of organosilicates have been used in a large number of TPU nanocomposite studies. Earlier studies employed solvent casting[54,61,96,97] to assist in overcoming some of the abovementioned thermodynamic barriers, and also to avoid complications due to thermal degradation of the alkylammonium modifiers.[98,99] Interestingly, the more hydrophobic Cloisite 15A generally did not disperse well, although increases in tensile and barrier properties[96,97] were reported with this nanofiller. Finnigan *et al.* reported that the more hydrophilic Cloisite 30B produced nanocomposites with relatively better dispersion, and very large associated increases in stiffness, tensile hysteresis, and permanent set.[77] This increase in hysteresis in nanocomposites has not been well documented in the literature, but is an important observation because many dynamic applications of TPUs require that resilience is maintained. This study was also the first to directly compare identical TPU nanocomposite formulations prepared via both solution and melt processing. However, the very small lab scale conical twin screw extruder employed caused extreme thermal degradation and loss of ultimate properties.

Due to the more commercially scalable nature of melt compounding, several workers have investigated TPU-Cloisite® nanocomposites prepared via melt compounding on a larger scale.[7,57,100,101] The relatively hydrophilic Cloisite® 30B and Cloisite® 10A generally disperse and delaminate well in TPUs.[7,54,57] This is due to their respective polar hydroxyl and benzyl functionalities, which promote thermodynamically favorable enthalpic driving forces, such as H bonding, for intercalation.[54,34,103] Cloisite® 25A, in the middle of the hydrophilicity/hydrophobicity range, has also resulted in good dispersion and property enhancement.[75,83] Dan *et al.* achieved the best property improvements with Cloisite® 30B, and also found that higher tensile properties could be achieved in an ester-based TPU with respect to an ether-based TPU, despite better dispersion in the latter.[102] They also reported the influence of some nanofillers on hard/soft segment demixing and morphology, as well as a decrease in observed properties after a second melt processing step, implying thermal degradation.

Chavarria and Paul[100] further probed the effect of organoclay structure by modifying a series of montmorillonite-based nanofillers with various alkylammonium surfactants, and studying the resulting structure-property relations for a high-hardness ether-based versus a medium-hardness ester-based host TPU. They concluded that for both host TPU systems: (a) one long alkyl tail on the ammonium ion rather than two, (b) hydroxyl ethyl groups on the amine rather than methyl groups, and (c) a longer alkyl tail as opposed to a shorter one, all lead to improved clay dispersion and TPU nanocomposite stiffness. However, although the harder ether-based (and presumably higher melt viscosity) TPU host polymer had better nanofiller dispersion, the percentage increase in properties remained higher in the softer TPU. Meng *et al.*[82] also showed that the degree of dispersion of the nanofiller (Cloisite 30B) was improved by processing the TPU host so that it had a higher molecular weight and higher viscosity due to improved shear. They also showed that an optimum Brabender melt processing time was required to achieve the highest combination of nanofiller dispersion and tensile stress at 500% strain, due to excessive thermal degradation at longer processing times. Ma *et al.* organically modified a less well-known class of layered silicate, rectorite, and performed melt compounding studies on a polyester-based TPU and reported substantial increases in both tensile strength and tear strength.[101] Most interestingly, the strongest formulations did not necessarily provide the best tear strength, suggesting that toughening mechanisms in these systems are complex and that non-delaminated clay tactoids may play more of a role in toughening than discrete organoclay platelets.

Whilst much of the research has focused on ionic substitution of bulky organic surfactants, several attempts at covalent modification of the silicate surface have been attempted.[74,104] An initial report from Tien and Wei[74] utilized organosilicates with one to three hydroxyl groups per surfactant molecule, which were capable of participating in the polymerization process and promoting delamination from within the layers. Since then, other researchers have used this approach with Cloisite® 30B.[104] The tethering of TPU chains to the organosilicate surface has also been used to achieve a strong interfacial bond. However, it is difficult to reach the full potential for property improvements via this route, because the isocyanate groups can also react with bound water in the silicate interlayer, and with hydroxyl groups on the silicate edges, which, if not correctly accounted for, can lead to reduced molecular weight and matrix cross-link density.[105] The size of the layered silicate has been shown to influence the resulting nanocomposite morphology. Due to the relative scale of the hard domains compared with the most common silicates employed, few groups have investigated this area. However, research into smaller silicates has shown tensile strength can be significantly enhanced without significantly altering Young's modulus or resilience.[78,106] In particular, the number of polymer-filler interactions and entanglements increases as the platelet size decreases, because the smaller platelets have a larger surface area and a smaller interparticle distance. This can have a significant effect on the mechanical properties, particularly at high strains where smaller platelets can more readily rotate without overcrowding, and therefore can more readily interact with the matrix by communicating in shear. Finnigan *et al.*[78] clearly demonstrated this by performing *in situ* strained synchrotron small-angle X-ray scattering measurements on a series of solution-cast TPU nanocomposites incorporating organically modified fluoromica nanofillers of controlled aspect ratio, which had been prepared via high energy milling. In a follow-up study they showed that superior resistance to stress relaxation could also be achieved by the incorporation of low aspect ratio organoclays.[106] They explained that, as the particle size and aspect

*Table 11.2* Overall performance enhancement for a TPU nanocomposite prepared by reactive extrusion, with respect to the unreinforced host polymer[30, 107]

| Property | Host | Nanocomposite |
|---|---|---|
| Tensile strength (MPa) | 53 | 67 |
| Elongation at break (%) | 665 | 748 |
| Shore hardness (D) | 43 | 43 |
| Stress at 100% strain | 10 | 11 |
| Stress at 300% strain | 16 | 16 |
| Tear strength (MPa) | 87 | 106 |
| Creep modulus (MPa) | 2.27 | 3.49 |
| Creep modulus (MPa) 60C | 1.73 | 1.93 |
| Compression set (%) at room temp | 27 | 15 |

ratio of the filler increase, the TPU chains in the interfacial region are more restrained and experience greater localized stresses, which result in an increase in strain-induced slippage at the polymer-filler interface. This technology has been patented,[108] and a startup company, TenasiTech Pty Ltd, formed to commercialize the technology. TenasiTech has now substantially scaled up TPU nanocomposite production via reactive extrusion.[30] Typical performance gains for a common polyether-based aromatic TPU reinforced with a 2(w/w)% fluoromica-based nanofiller are shown in Table 11.2.

Notably, the property improvements reported are evidence that well-engineered nanofiller systems can not only avoid stiffening and hardening effects; they can also have a positive influence on critical viscoelastic properties such as creep and compressive set, in addition to ultimate tensile strength and tear strength.

Using a Monte Carlo simulation, Pandey *et al.*[109] predicted that smaller clay platelets should be more easily delaminated and dispersed in a compatible solvent-host polymer than high aspect ratio clays. Furthermore, computational models by Balazs and co-workers on block copolymers indicated that a change in microphase texture and properties would be expected since nanofiller size scales with the phase domain size of the block copolymer.[110] McKinley and co-workers[29,91] incorporated unmodified, low aspect ratio hectorite (Laponite) into a number of TPUs at up to 20 wt% using a novel solvent-exchange technique. They reported remarkable mechanical properties, heat distortion temperature, and morphological changes in fully delaminated nanocomposite systems. Some degree of 'tuned' segmental association of the Laponite® was demonstrated, depending on the solubility parameter of the chosen soft segment.[91]

## 11.3 TPU nanocomposites as potential biomaterials

Thermoplastic polyurethane is the material of choice for many biomedical applications due to the relative ease of fabrication into devices, flexibility, biocompatibility, biostability, and electrical insulation properties. Polyether-based TPUs have been the materials of choice for certain types of medical implants for many years. However, there have been some cases in which the TPU degraded and led to surface or deep cracking, stiffening, erosion, or the deterioration of mechanical properties such as tensile and flexural strength.[111–114] These eventually caused implant malfunction. Polydimethylsiloxane (PDMS)-based elastomers were then gradually commercialized and introduced to overcome these problems. A PDMS/ poly(hexamethylene oxide)-based TPU based on an optimized formulation (Elast-Eon™) from AorTech International plc exhibited properties comparable to those of medical grade polyether-based TPUs such as Pellethane 80A.[115] Elast-Eon TPUs are now widely accepted as being the most biostable of all TPUs and as such are imminently suitable for long-term implantation.[116]

Few researchers have recognized the potential of TPU nanocomposites as biomaterials even though recent studies have revealed that TPU nanocomposites have the potential to be developed for biomedical device components due to the enhancement of thermal, mechanical, and barrier properties[4,61,117,118] and their capability in tuning cell-material interactions.[119,120] Furthermore, nanofillers can introduce new functionality, such as antimicrobials.[121,122] One research group directly studying TPU-organosilicate nanocomposites as biomaterials is led by James Runt at Pennsylvania State University. This group has published two papers describing the preparation and properties of solution-cast nanocomposites prepared from the biomedical TPU Biospan™ (chemically similar to poly(ether urethane) (PEU)) and Cloisite® 15A.[61,97] The nanocomposites demonstrated an enhancement in tensile strength, modulus, and elongation at break when the nanoclay content increased from 0 wt% to 20 wt%. Styan[123] also prepared biomedical TPU organosilicate nanocomposites from PEU with organically modified MMT loadings in the range of 1 wt% to 15 wt%, using solution casting, and found that partially exfoliated nanocomposites were produced using 15 wt% Cloisite® 30B. These nanocomposites displayed several advantageous properties, namely significant antibacterial activity, reducing *Staphylococcus epidermidis* adherence after 24 h to ~20% and enhanced biostability. Liff *et al.*[29] employed a novel solvent-exchange approach to efficiently exfoliate synthetic smectic clays (Laponite® RD) in a biomedical poly(ether urethane) (Elasthane™ 80A). Elasthane™ 80 reported an ultimate tensile strength increase of 50% when 10 wt% of unmodified Laponite® RD was incorporated into the Elasthane matrix using a labor intensive solvent exchange technique.

The next generation of cochlear-implant electrode arrays will require easily processable insulating materials with substantially improved mechanical performance. Currently, soft silicone materials are employed as insulation for implantable cochlear electrode arrays. In an attempt to produce new materials with improved properties, we recently generated PDMS-based TPU nanocomposites containing low aspect ratio synthetic hectorite (Lucentite SWN from Kobo Products) and high aspect ratio synthetic fluoromica (Soma sif™ ME100 from

*Table 11.3* Tensile properties of E5-325 TPU host and nanocomposites incorporating 2 and 4% low-aspect-ratio (hectorite) and high-aspect-ratio (fluoromica) synthetic organoclays, and Nusil MED-4860 silicone for comparison

| Elastomer | Nanofiller content | Tensile strength (MPa) | Young's modulus (MPa) | Elongation at break (%) | Tear strength (MPa) |
|---|---|---|---|---|---|
| Nusil MED-4860 | – | 7.5 ± 0.1 | 2.7 ± 0.2 | 595 ± 39 | 41 ± 14 |
| TPU | – | 20 ± 2 | 10 ± 2 | 883 ± 63 | 53 ± 3 |
| | 2 wt% organo-fluoromica | 24.5 ± 0.8 | 12.2 ± 0.5 | 1129 ± 30 | 73 ± 6 |
| | 4 wt% organo-fluoromica | 20.5 ± 0.8 | 11.1 ± 0.3 | 941 ± 70 | 65 ± 3 |
| | 2 wt% organo-hectorite | 24 ± 2 | 11.3 ± 0.6 | 1127 ± 52 | 71 ± 3 |
| | 4 wt% organo-hectorite | 8.6 ± 0.2 | 34 ± 4 | 199 ± 9 | 41 ± 9 |

Kobo Products) with a hydrophobic surface modification.[124] The mechanical properties of the solvent-cast TPU (blank and nanocomposites) are summarized in Table 11.3, compared with Nusil MED 4860, a biomaterial that is currently used for insulation in cochlear implants. In general, this PDMS-based TPU displays greater mechanical properties compared with Nusil MED-4860, with an increase of 167% in tensile strength, 48% in elongation at break, and 29% in tear strength. Adding 2 wt% of modified nanofillers further increased the mechanical properties in this system. Both high and low aspect ratio nanofillers successfully increased the tensile strength, elongation at break, and tear strength by ~20%, 28%, and 34% respectively (Table 11.3 and Fig. 11.4). These preliminary results show that PDMS-based TPU nanocomposites are promising materials for biomedical applications, and may allow thinner and more intricate electrode arrays to be designed, while still maintaining structural integrity. To this end, we are currently investigating the morphology of this PDMS-based TPU and associated nanocomposites in attempt to develop an in-depth understanding of the structure-property relations of these materials, most importantly the interplay between the TPU nanophase domains and the engineered low and high aspect ratio nanofillers.

Organomodifiers are commonly ion exchanged with cations in the interlayer galleries of the clay to improve compatibility between the clay and the polyurethane matrix. Pool-Warren *et al.* showed that antimicrobial functionality can be introduced by incorporating chlorhexidine-diacetate-modified montmorillonite in nanocomposites. When dodecyl-amine-modified montmorillonite was used, the adhesion of both platelets and fibroblasts decreased, indicating the potential use of polyurethane nanocomposites as bioactive materials.[125]

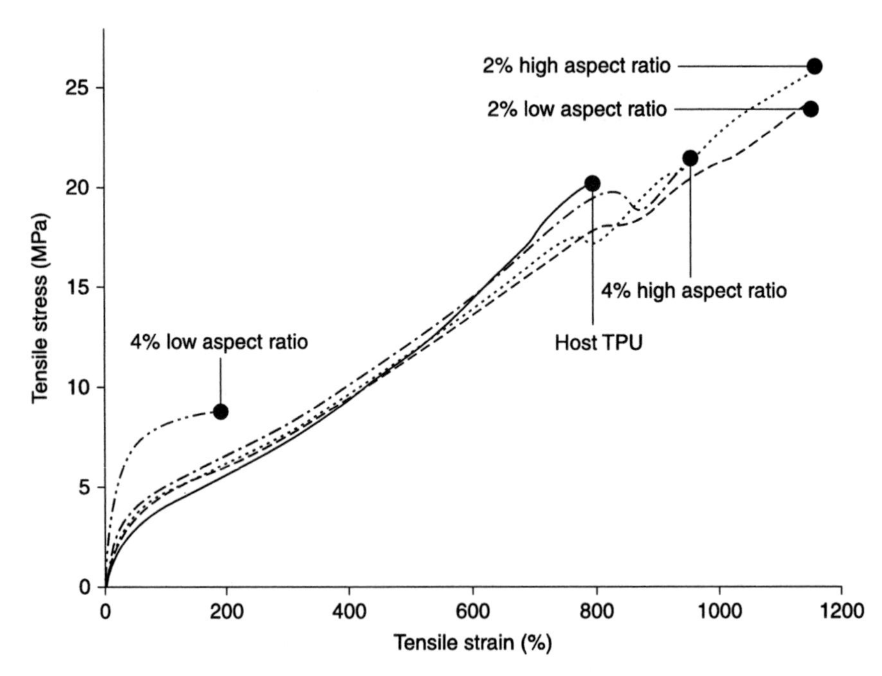


*11.4* Representative stress–strain curves for the TPU host polymer and nanocomposites incorporating 2% and 4% low- and high-aspect-ratio organoclays

Some studies show that the release of quaternary alkyl ammonium from organoclays could induce toxicity.[126,127] Rueda *et al.* reported that the presence of 2 wt% Cloisite® 30B in polyurethane nanocomposites caused a dramatic reduction in the number of proliferating cells in contact with an extract of the nanocomposites.[117] Styan *et al.* mentioned that methyl tallow bis-2-hydroxyethyl ammonium, which was released from Cloisite® 30B, was capable of inhibiting cell growth.[118] In contrast, Mishra *et al.* showed different results. Cells were able to grow on polyurethane nanocomposites containing Cloisite® 30B within 48 h. Furthermore, the authors showed that the nanocomposites were not genotoxic by observing the localization of the enzyme HIPK2. HIPK2 is an apoptosis activator. The presence of HIPK2 in a nucleus indicates genotoxicity stress. In this paper, HIPK2 was found in cytoplasm and thus incapable of transfecting the apoptosis gene.[128] Unfortunately, direct comparison of the previously mentioned studies is not possible because the authors used different cell lines to investigate toxicity.

In order to ensure the biocompatibility of our TPU nanocomposites, we are investigating the cytotoxicity of nanofillers on mouse neuroblasts (neuro N2A) using a CellTiter 96AQ$_{ueous}$ non-radioactive cell proliferation assay (MTS). Preliminary results show that some of organo-fluoromica are is significantly less toxic than

Cloisite 30B. The half maximal inhibitory concentration ($IC_{50}$) of the least toxic organo fluoromica is 90 μg/ml, while that of Cloisite 30B is 20 μg/ml, both concentrations normalized to clay (unpublished work). Styan *et al.* showed that, when montmorillonite modified with amino undecanoic acid was used, cell growth was not severely inhibited because at pH 7.4 this organomodifier was negatively charged. Thus, the possibility of interacting with cell membranes and causing cell death was reduced. These imply that suitable material selection is important when incorporating organoclays in nanocomposites. In addition, the method of processing the materials is another crucial factor. Edwards *et al.* showed that some commercial organoclays contained excess surfactants, which could be released and thus induce toxicity.[99]

The susceptibility of polyurethanes to degrade *in vivo* is useful for the development of bioresorbable implants intended for short term implantation. The hydrophilic nature of clays is utilized in implanted polyurethane nanocomposites to attract water, which induces hydrolysis and consequently leads to polyurethane degradation. Da Silva *et al.* showed that employing unmodified montmorillonite could enhance the rate of biodegradation and subsequently the release of drugs incorporated in the implant.[129,130]

The above findings show that TPU nanocomposites can provide a combination of biostability and advantageous mechanical properties. They are, therefore, promising candidates for use as biomaterials in the fabrication of a wide range of medical devices.

## 11.4 Future trends

While the nanocomposite approach has undoubtedly led to several encouraging examples of new functional and mechanical property profiles, most examples have failed to live up to their theoretical potential. An excellent recent review paper by Schaefer and Justice[131] points out that in the majority of polymer nanocomposites today 'large scale disorder is ubiquitous.' Consistently with some of the reports discussed here, they suggest that the fixation on high aspect ratio nanofillers is unproductive due to: (a) problems with agglomeration and dispersion of the high aspect ratio particles, and (b) the potential for tailored molecular self-assembly involving a higher degree of 'cooperation' between the nanofiller and matrix. The nature of more recent investigations suggests that a more systematic and exacting approach is now being adopted by TPU nanocomposite researchers, guided by a growth in multiscale theoretical modeling. New experimental techniques need to be developed and applied to better probe and understand the subtle structures and conformations at the TPU-nanofiller interface. Synchrotron X-ray and neutron scattering will continue to be important. Highly engineered model nanoparticle and TPU systems may be required to elucidate these questions. There is also a need for the development of new classes of layered inorganic nanofillers that are more readily delaminated and more thermally stable. Metal phosphonates have been proposed as one possible new class[132] and there have also been some efforts to employ covalent organic modification[133] using silane grafting.

Historically, the mechanical testing performed on polymer nanocomposites has been limited to simple tensile experiments, and thereby has largely ignored many of the properties critical to the true service performance of the materials. Therefore, the long-term mechanical, thermal, and other environmental ageing, tear and cut-growth resistance, and fatigue performance of TPU nanocomposite materials are other areas greatly in need of attention.

The effects of processing on nanocomposite structure and properties, and the use of rheology to elucidate the effects of nanocomposite formation, are new and exciting fields of research in nanocomposites. Traditionally much research (highlighted above) has focused on the properties of nanocomposites produced from one type of process or processing condition. Clearly the effects of processing on nanocomposite formation and the understanding of the rheology of nanocomposites are some of the next key scientific steps needed to fully understand and control TPU nanocomposite properties and performance in various applications. An optimistic perspective would see future applications of TPU nanocomposites in biomedical implants (e.g., softer, tougher, implantable electrode insulation for pacemakers),[124] better creep and abrasion resistant mining screens, drive and conveyer belts, more durable TPU seals, o-rings, and hoses for fuels. Further uses are stronger, lighter, melt spun Spandex, thin aliphatic gas barrier coatings for encapsulating the next generation of thin-film photovoltaics, tougher, more scuff-resistant covers on golf balls and protective wear, and many others. Nanocomposite technology could also see TPUs remain at the 'top of their class' with respect to the many other TPEs available.

## 11.5 References


1. Hepburn, C., *Polyurethane Elastomers*. Elsevier, London, New York, 1991: p. 441.
2. Oertel, G. and Abele, L., *Polyurethane Handbook: Chemistry, Raw Materials, Processing, Application, Properties*, 2nd edition. Hanser/Gardner, Munich, New York, Cincinnati, 1994: p. 688.
3. Sullivan, F. A., *U.S. Thermoplastic Elastomers (TPE) Markets*. 2004: p. A750–37.
4. Song, H. J., Qi, H.A., Li, N., Tribological behavior of carbon nanotubes/polyurethane nanocomposite coatings. *Micro & Nano Letters*, 2011. 6(1): pp. 48–51.
5. Yusoh, K., Jin, J. and Song, M., Subsurface mechanical properties of polyurethane/organoclay nanocomposite thin films studied. *Progress in Organic Coatings*, 2010. 67(2): pp. 220–224.
6. Silva G.G., Rodrigues M.F., and Fantini, C., Thermoplastic polyurethane nanocomposites produced via impregnation of long carbon nanotube forests. *Macromolecular Materials and Engineering*, 2011. 296(1): pp. 53–58.
7. Berta, M., Saiani, A., Lindsay, C., and Gunaratne, R., Effect of clay dispersion on the rheological properties and flammability of polyurethane-clay nanocomposite elastomers. *Journal of Applied Polymer Science*, 2008. 112(5): pp. 2847–2853.

8. Heraidan, M., Shishesaz, M. R. and Kassiriha, S. M., Study on the effect of ultrasonication time on transport properties of polyurethane/organoclay nanocomposite coatings. *Journal of Coatings Technology and Research*, 2011. 8(2): pp. 265–274.
9. Ashjari, M., and Mahdavian, A. R., Efficient dispersion of magnetite nanoparticles in the polyurethane matrix through solution mixing and investigation of the nanocomposite properties. *Journal of Inorganic and Organometallic Polymer and Materials*, 2010. 20(2): pp. 213–219.
10. Kannan, R. Y., Salacinski, H. J., De Groot, J., Clatworthy, I., Bozec, L., *et al.*, The antithrombogenic potential of a polyhedral oligomeric silsesquioxane (POSS) nanocomposite. *Biomacromolecules*, 2006. 7(1): pp. 215–223.
11. Kannan, R. Y., Salacinski, H. J., Odlyha, M., Butler, P. E., and Seifalian, A. M., The degradative resistance of polyhedral oligomeric silsesquioxane nanocore integrated polyurethanes: An *in vitro* study. *Biomaterials*, 2006. 27(9): pp. 1971–1979.
12. Lee, S., Multifuctionality of layered fabric systems based on electrospun polyurethane/zinc oxide nanocomposite fibers. *Journal of Applied Polymer Science*, 2009. 114(6): pp. 3652–2658.
13. Chu, B., Gao, T., Li, Y. J., Wang, J., Desper, C. R., *et al.*, Microphase separation kinetics in segmented polyurethanes – effects of soft segment length and structure. *Macromolecules*, 1992. 25(21): pp. 5724–5729.
14. Elwell, M. J., Ryan, A. J., Grunbauer, H. J. M., and VanLieshout, H. C., *In-situ* studies of structure development during the reactive processing of model flexible polyurethane foam systems using FT-IR spectroscopy, synchrotron SAXS, and rheology. *Macromolecules*, 1996. 29(8): pp. 2960–2968.
15. Ryan, A. J., Willkomm, W. R., Bergstrom, T. B., Macosko, C. W., Koberstein, J. T., *et al.*, Dynamics of (micro)phase separation during fast, bulk copolymerization – Some synchrotron SAXS Experiments. *Macromolecules*, 1991. 24(10): pp. 2883–2889.
16. Lilaonitkul, A., and Cooper, S. L., Properties of thermoplastic polyurethane elastomers. *Advances in Urethane Science and Technology*, 1979. 7: pp. 163–183.
17. Abouzahr, S., Wilkes, G. L., and Ophir, Z., Structure property behavior of segmented polyether MDI butanediol based urethanes – Effect of composition ratio. *Polymer*, 1982. 23(7): pp. 1077–1086.
18. Harrell, L. L., Segmented polyurethanes – Properties as a function of segment size and distribution. *Abstracts of Papers of the American Chemical Society*, 1969(SEP): pp. PO57 ff.
19. Martin, D. J., Meijs, G. F., Renwick, G. M., Gunatillake, P. A., and McCarthy, S. J., Effect of soft-segment $CH_2$/O ratio on morphology and properties of a series of polyurethane elastomers. *Journal of Applied Polymer Science*, 1996. 60(4): pp. 557–571.
20. Miller, J. A., Lin, S. B., Hwang, K. K. S., Wu, K. S., Gibson, P. E., *et al.*, Properties of polyether polyurethane block copolymers – Effects of hard segment length distribution. *Macromolecules*, 1985. 18(1): pp. 32–44.
21. Samuels, S. L., and Wilkes, G. L., Rheo-optical and mechanical behavior of a systematic series of hard-soft segmented urethanes. *Journal of Polymer Science Part C: Polymer Symposium*, 1973(43): pp. 149–178.
22. Speckhard, T. A., and Cooper, S. L., Ultimate tensile properties of segmented polyurethane elastomers – Factors leading to reduced properties for polyurethanes based on nonpolar soft segments. *Rubber Chemistry and Technology*, 1986. 59(3): pp. 405–431.
23. Wang, C. B., and Cooper, S. L., Morphology and properties of segmented polyether polyurethaneureas. *Macromolecules*, 1983. 16(5): pp. 775–786.

24. Lee, H. S., and Hsu, S. L., Structural changes and chain orientational behavior during tensile deformation of segmented polyurethanes. *Journal of Polymer Science Part B: Polymer Physics*, 1994. 32(12): pp. 2085–2098.
25. Reynolds, N., Spiess, H. W., Hayen, H., Nefzger, H., and Eisenbach, C. D., Structure and deformation behavior of model poly(ether-urethane) elastomers. 1. Infrared studies. *Macromolecular Chemistry and Physics*, 1994. 195(8): pp. 2855–2873.
26. Martin, D. J., Meijs, G. F., Gunatillake, P. A., McCarthy, S. J., and Renwick, G. M., The effect of average soft segment length on morphology and properties of a series of polyurethane elastomers. 2. SAXS-DSC annealing study. *Journal of Applied Polymer Science*, 1997. 64(4): pp. 803–817.
27. Torro-Palau, A., Fernandez-Garcia, J. C., Orgiles-Barcelo, A. C., Pastor-Blas, M. M., and Martin-Martinez, J. M., Comparison of the properties of polyurethane adhesives containing fumed silica or sepiolite as filler. *Journal of Adhesion*, 1997. 61(1–4): pp. 195–211.
28. Varma, A. J., Deshpande, M. D., and Nadkarni, V. M., Morphology and mechanical properties of silicate filled polyurethane elastomers based on castor oil and polymerix MDI. *Angewandte Makromolekulare Chemie*, 1985. 132(JUN): pp. 203–209.
29. Liff, S., Kumar, N., and McKinley, G., High-performance elastomeric nanocomposites via solvent-exchange processing. *Nature Materials*, 2007. 6(1): pp. 76.
30. Marshall, R., and Martin, D., Nanocomposite polyurethanes for extreme applications in oil and gas. *High Performance Thermoplastic & Composites for Oil and Gas Applications 2011*, Houston, USA, 2011.
31. Petrovic, Z. S., Javni, I., Waddon, A., and Banhegyi, G., Structure and properties of polyurethane-silica nanocomposites. *Journal of Applied Polymer Science*, 2000. 76(2): pp. 133–151.
32. Vega-Baudrit, J., Sibaja-Ballestero, M., Vazquez, P., Torregrosa-Macia, R., and Martin-Martinez, J. M., Properties of thermoplastic polyurethane adhesives containing nanosilicas with different specific surface area and silanol content. *International Journal of Adhesion and Adhesives*, 2007. 27(6): pp. 469–479.
33. Sen, R., Zhao, B., Perea, D., Itkis, M. E., Hu, H., *et al.*, Preparation of single-walled carbon nanotube reinforced polystyrene and polyurethane nanofibers and membranes by electrospinning. *Nano Letters*, 2004. 4(3): pp. 459–464.
34. Balazs, A. C., Singh, C., and Zhulina, E., Modeling the interactions between polymers and clay surfaces through self-consistent field theory. *Macromolecules*, 1998. 31(23): pp. 8370–8381.
35. Tagmatarchis, N., and Prato, M., Functionalization of carbon nanotubes via 1,3-dipolar cycloadditions. *Journal of Materials Chemistry*, 2004. 14(4): pp. 437–439.
36. Barisci, J. N., Tahhan, M., Wallace, G. G., Badaire, S., Vaugien, T., *et al.*, Properties of carbon nanotube fibers spun from DNA-stabilized dispersions. *Advanced Functional Materials*, 2004. 14(2): pp. 133–138.
37. Hilding, J., Grulke, E. A., Zhang, Z. G., and Lockwood, F., Dispersion of carbon nanotubes in liquids. *Journal of Dispersion Science and Technology*, 2003. 24(1): pp. 1–41.
38. Lin, T., Bajpai, V., Ji, T., and Dai, L. M., Chemistry of carbon nanotubes. *Australian Journal of Chemistry*, 2003. 56(7): pp. 635–651.
39. Ortiz-Acevedo, A., Xie, H., Zorbas, V., Sampson, W. M., Dalton, A. B., *et al.*, Diameter-selective solubilization of single-walled carbon nanotubes by reversible cyclic peptides. *Journal of the American Chemical Society*, 2005. 127(26): pp. 9512–9517.

40. Dieckmann, G. R., Dalton, A. B., Johnson, P. A., Razal, J., Chen, J., *et al.*, Controlled assembly of carbon nanotubes by designed amphiphilic peptide helices. *Journal of the American Chemical Society*, 2003. 125(7): pp. 1770–1777.
41. Buffa, F., Abraham, G. A., Grady, B. P., and Resasco, D., Effect of nanotube functionalization on the properties of single-walled carbon nanotube/polyurethane composites. *Journal of Polymer Science Part B: Polymer Physics*, 2007. 45(4): pp. 490–501.
42. Deng, J., Zhang, X., Wang, K., Zou, H., Zhang, Q., *et al.*, Synthesis and properties of poly(ether urethane) membranes filled with isophorone diisocyanate-grafted carbon nanotubes. *Journal of Membrane Science*, 2007. 288(1–2): pp. 261–267.
43. Foster, J., Singamaneni, S., Kattumenu, R., and Bliznyuk, V., Dispersion and phase separation of carbon nanotubes in ultrathin polymer films. *Journal of Colloid and Interface Science*, 2005. 287(1): pp. 167–172.
44. Koerner, H., Liu, W. D., Alexander, M., Mirau, P., Dowty, H., *et al.*, Deformation-morphology correlations in electrically conductive carbon nanotube thermoplastic polyurethane nanocomposites. *Polymer*, 2005. 46(12): pp. 4405–4420.
45. Koerner, H., Price, G., Pearce, N. A., Alexander, M. and Vaia, R. A., Remotely actuated polymer nanocomposites – Stress recovery of carbon nanotube filled thermoplastic elastomers. *Nature Materials*, 2004. 3(2): pp. 115–120.
46. Kuan, H. C., Ma, C. C. M., Chang, W. P., Yuen, S. M., Wu, H. H., *et al.*, Synthesis, thermal, mechanical and rheological properties of multiwall carbon nano tube/ waterborne polyurethane nanocomposite. *Composites Science and Technology*, 2005. 65(11–12): pp. 1703–1710.
47. Kwon, J. Y., and Kim, H. D., Preparation and properties of acid-treated multiwalled carbon nanotube/waterborne polyurethane nanocomposites. *Journal of Applied Polymer Science*, 2005. 96(2): pp. 595–604.
48. Meng, J., Kong, H., Xu, H. Y., Song, L., Wang, C. Y., *et al.*, Improving the blood compatibility of polyurethane using carbon nanotubes as fillers and its implications to cardiovascular surgery. *Journal of Biomedical Materials Research Part A*, 2005. 74A(2): pp. 208–214.
49. Sahoo, N. G., Jung, Y. C., Yoo, H. J., and Cho, J. W., Effect of functionalized carbon nanotubes on molecular interaction and properties of polyurethane composites. *Macromolecular Chemistry and Physics*, 2006. 207(19): pp. 1773–1780.
50. Jung, Y. C., Sahoo, N. G., and Cho, J. W., Polymeric nanocomposites of polyurethane block copolymers and functionalized multi-walled carbon nanotubes as crosslinkers. *Macromolecular Rapid Communications*, 2006. 27(2): pp. 126–131.
51. Xia, H., Song, M., Jin, J., and Chen, L., Poly(propylene glycol)-grafted multi-walled carbon nanotube polyurethane. *Macromolecular Chemistry and Physics*, 2006. 207(21): pp. 1945–1952.
52. Xu, M., Zhang, T., Gu, B., Wu, J. L., and Chen, Q., Synthesis and properties of novel polyurethane-urea/multiwalled carbon nanotube composites. *Macromolecules*, 2006. 39(10): pp. 3540–3545.
53. Xiong, J., Zheng, Z., Qin, X., Li, M., Li, H., *et al.*, The thermal and mechanical properties of a polyurethane/multi-walled carbon nanotube composite. *Carbon*, 2006. 44(13): pp. 2701–2707.
54. Smart, S., Fania, D., Milev, A., Kannangara, G. S. K., Lu, M., *et al.*, The effect of carbon nanotube hydrophobicity on the mechanical properties of carbon nanotube-reinforced thermoplastic polyurethane nanocomposites. *Journal of Applied Polymer Science*, 2010. 117(1): pp. 24–32.

55. Deng, J., Cao, J., Li, J., Tan, H., Zhang, Q., *et al.*, Mechanical and surface properties of polyurethane/fluorinated multi-walled carbon nanotubes composites. *Journal of Applied Polymer Science*, 2008. 108(3): pp. 2023–2028.
56. Chen, W., Tao, X., and Liu, Y., Carbon nanotube-reinforced polyurethane composite fibers. *Composites Science and Technology*, 2006. 66: pp.3029–3034.
57. Barick, A. K., and Tripathy, D. K., Effect of nanofiber on material properties of vapor-grown carbon nanofiber reinforced thermoplastic polyurethane (TPU/CNF) nanocomposites prepared by melt compounding. *Composites Part A: Applied Science and Manufacturing*, 2010. 41(10): pp. 1471–1482.
58. Wong, M., Paramsothy, M., Xu, X. J., Ren, Y., Li, S., *et al.*, Physical interactions at carbon nanotube-polymer interface. *Polymer*, 2003. 44(25): pp. 7757–7764.
59. Akelah, A., and Moet, A., Polymer-clay nanocomposites: Free-radical grafting of polystyrene on to organophilic montmorillonite interlayers. *Journal of Materials Science*, 1996. 31(13): pp. 3589–3596.
60. Alexandre, M., and Dubois, P., Polymer-layered silicate nanocomposites: Preparation, properties and uses of a new class of materials. *Materials Science and Engineering: R: Reports*, 2000. 28(1–2): pp. 1–63.
61. Xu, R. J., Manias, E., Snyder, A. J., and Runt, J., Low permeability biomedical polyurethane nanocomposites. *Journal of Biomedical Materials Research Part A*, 2003. 64A(1): pp. 114–119.
62. Wang, Z., and Pinnavaia, T., Nanolayer reinforcement of elastomeric polyurethane. *Chemistry of Materials*, 1998. 10(12): pp. 3769–3771.
63. Bhattacharya, S., Kamal, M., and Gupta, R., *Polymeric Nanocomposites: Theory and Practice*. Hanser Gardner Pubns, 2007.
64. Usuki, A., Kojima, Y., Kawasumi, M., Okada, A., Fukushima, Y., *et al.*, Synthesis of nylon 6-clay hybrid. *Journal of Materials Research(USA)*, 1993. 8(5): pp. 1179–1184.
65. Dennis, H., Hunter, D., Chang, D., Kim, S., White, J., *et al.*, Effect of melt processing conditions on the extent of exfoliation in organoclay-based nanocomposites. *Polymer*, 2001. 42(23): pp. 9513–9522.
66. Giannelis, E., Polymer-layered silicate nanocomposites: Synthesis, properties and applications. *Applied Organometallic Chemistry*, 1998. 12(10–11): pp. 675–680.
67. Xanthos, M., and Biesenberger, J., *Reactive Extrusion: Principles and Practice*. Hanser Munich, 1992.
68. Mittal, V., *Optimization of Polymer Nanocomposite Properties*. Wiley-VCH, 2010.
69. Takeichi, T., Zeidam, R., and Agag, T., Polybenzoxazine/clay hybrid nanocomposites: Influence of preparation method on the curing behavior and properties of polybenzoxazines. *Polymer*, 2002. 43(1): pp. 45–53.
70. Choi, W., Kim, S., Jin Kim, Y., and Kim, S., Synthesis of chain-extended organifier and properties of polyurethane/clay nanocomposites. *Polymer*, 2004. 45(17): pp. 6045–6057.
71. Ma, J., Zhang, S., and Qi, Z., Synthesis and characterization of elastomeric polyurethane/clay nanocomposites. *J. Appl. Polym. Sci.*, 2001. 82(6): pp. 1444–1448.
72. Pattanayak, A., and Jana, S., Thermoplastic polyurethane nanocomposites of reactive silicate clays: Effects of soft segments on properties. *Polymer*, 2005. 46(14): pp. 5183–5193.
73. Rhoney, I., Brown, S., Hudson, N., and Pethrick, R., Influence of processing method on the exfoliation process for organically modified clay systems. I. Polyurethanes. *J. Appl. Polym. Sci.*, 2003. 91(2): pp. 1335–1343.

74. Tien, Y., and Wei, K., High-tensile-property layered silicates/polyurethane nanocomposites by using reactive silicates as pseudo chain extenders. *Macromolecules*, 2001. 34(26): pp. 9045–9052.
75. Chang, J., and An, Y., Nanocomposites of polyurethane with various organoclays: Thermomechanical properties, morphology, and gas permeability. *Journal of Polymer Science Part B: Polymer Physics*, 2002. 40(7): pp. 670–677.
76. Chen, T. K., Tien, Y. I., and Wei, K. H., Synthesis and characterization of novel segmented polyurethane/clay nanocomposites. *Polymer*, 2000. 41(4): pp. 1345–1353.
77. Finnigan, B., Martin, D., Halley, P., Truss, R., and Campbell, K., Morphology and properties of thermoplastic polyurethane composites incorporating hydrophobic layered silicates. *J. Appl. Polym. Sci.*, 2005. 97(1): pp. 300–309.
78. Finnigan, B., Jack, K., Campbell, K., Halley, P., Truss, R., *et al.*, Segmented polyurethane nanocomposites: Impact of controlled particle size nanofillers on the morphological response to uniaxial deformation. *Macromolecules*, 2005. 38(17): pp. 7386–7396.
79. Mishra, J., Kim, I., and Ha, C., New millable polyurethane/organoclay nanocomposite: Preparation, characterization and properties. *Macromolecular Rapid Communications*, 2003. 24(11): pp. 671–675.
80. Kim, T., Kim, B., Kim, Y., Cho, Y., Lee, S., *et al.*, The properties of reactive hot melt polyurethane adhesives modified with novel thermoplastic polyurethanes. *Journal of Applied Polymer Science*, 2009. 114(2): pp. 1169–1175.
81. Cai, Y., Hu, Y., Song, L., Liu, L., Wang, Z., *et al.*, Synthesis and characterization of thermoplastic polyurethane/montmorillonite nanocomposites produced by reactive extrusion. *Journal of Materials Science*, 2007. 42(14): pp. 5785–5790.
82. Meng, X., Du, X., Wang, Z., Bi, W., and Tang, T., The investigation of exfoliation process of organic modified montmorillonite in thermoplastic polyurethane with different molecular weights. *Composites Science and Technology*, 2008. 68: pp. 1815–1821.
83. Kim, B., Seo, J., and Jeong, H., Morphology and properties of waterborne polyurethane/clay nanocomposites. *European Polymer Journal*, 2003. 39(1): pp. 85–91.
84. Moon, S., Kim, J., Nah, C., and Lee, Y., Polyurethane/montmorillonite nanocomposites prepared from crystalline polyols, using 1, 4-butanediol and organoclay hybrid as chain extenders. *European Polymer Journal*, 2004. 40(8): pp. 1615–1621.
85. Varghese, S., Gatos, K., Apostolov, A., and Karger-Kocsis, J., Morphology and mechanical properties of layered silicate reinforced natural and polyurethane rubber blends produced by latex compounding. *J. Appl. Polym. Sci.*, 2004. 92(1): pp. 543–551.
86. Song, M., Hourston, D., Yao, K., Tay, J., and Ansarifar, M., High performance nanocomposites of polyurethane elastomer and organically modified layered silicate. *J. Appl. Polym. Sci.*, 2003. 90(12): pp. 3239–3243.
87. Chen-Yang, Y., Yang, H., Li, G., and Li, Y., Thermal and anticorrosive properties of polyurethane/clay nanocomposites. *Journal of Polymer Research*, 2005. 11(4): pp. 275–283.
88. LeBaron, P. C., Wang, Z., and Pinnavaia, T. J., Polymer-layered silicate nanocomposites: An overview. *Applied Clay Science*, 1999. 15(1–2): pp. 11–29.
89. Padmanabhan, K. A., Mechanical properties of nanostructured materials. *Materials Science and Engineering A: Structural Materials Properties Microstructure and Processing*, 2001. 304: pp. 200–205.

90. Edwards, G., *Optimisation of organically modified layered silicate based nanofillers for thermoplastic polyurethanes*. 2007.
91. Korley, L. T. J., Liff, S. M., Kumar, N., McKinley, G. H., and Hammond, P. T., Preferential association of segment blocks in polyurethane nanocomposites. *Macromolecules*, 2006. 39(20): pp. 7030–7036.
92. Giannelis, E., Krishnamoorti, R., and Manias, E., Polymer-silicate nanocomposites: Model systems for confined polymers and polymer brushes. *Polymers in Confined Environments*, 1999: pp. 107–147.
93. Giannelis, E. P., Polymer layered silicate nanocomposites. *Advanced Materials*, 1996. 8(1): pp. 29 ff.
94. Bergaya, F., and Lagaly, G., Surface modification of clay minerals. *Applied Clay Science*, 2001. 19(1–6): pp. 1–3.
95. Zilg, C., Thomann, R., Mulhaupt, R., and Finter, J., Polyurethane nanocomposites containing laminated anisotropic nanoparticles derived from organophilic layered silicates. *Advanced Materials*, 1999. 11(1): pp. 49–52.
96. Finnigan, B., Martin, D., Halley, P., Truss, R., and Campbell, K., Morphology and properties of thermoplastic polyurethane composites incorporating hydrophobic layered silicates. *Journal of Applied Polymer Science*, 2005. 97(1): pp. 300–309.
97. Xu, R., Manias, E., Snyder, A., and Runt, J., New biomedical poly (urethane urea) layered silicate nanocomposites. *Macromolecules*, 2001. 34(2): pp. 337–339.
98. He, H. P., Duchet, J., Galy, J., and Gerard, J. F., Influence of cationic surfactant removal on the thermal stability of organoclays. *Journal of Colloid and Interface Science*, 2006. 295(1): pp. 202–208.
99. Edwards, G., Halley, P., Kerven, G., and Martin, D., Thermal stability analysis of organo-silicates, using solid phase microextraction techniques. *Thermochimica Acta*, 2005. 429(1): pp. 13–18.
100. Chavarria, F., and Paul, D. R., Morphology and properties of thermoplastic polyurethane nanocomposites: Effect of organoclay structure. *Polymer*, 2006. 47(22): pp. 7760–7773.
101. Ma, X. Y., Lu, H. J., Liang, G. Z., and Yan, H. X., Rectorite/thermoplastic polyurethane nanocomposites: Preparation, characterization, and properties. *Journal of Applied Polymer Science*, 2004. 93(2): pp. 608–614.
102. Dan, C. H., Lee, M. H., Kim, Y. D., Min, B. H., and Kim, J. H., Effect of clay modifiers on the morphology and physical properties of thermoplastic polyurethane/clay nanocomposites. *Polymer*, 2006. 47(19): pp. 6718–6730.
103. Vaia, R. A., and Giannelis, E. P. Lattice model of polymer melt intercalation in organically-modified layered silicates. *Macromolecules*, 1997. 30(25): pp. 7990–7999.
104. Pattanayak, A., and Jana, S. C., Synthesis of thermoplastic polyurethane nanocomposites of reactive nanoclay by bulk polymerization methods. *Polymer*, 2005. 46(10): pp. 3275–3288.
105. Choi, M. Y., Anandhan, S., Youk, J. H., Baik, D. H., Seo, S. W., *et al.*, Synthesis and characterization of *in situ* polymerized segmented thermoplastic elastomeric polyurethane/layered silicate clay nanocomposites. *Journal of Applied Polymer Science*, 2006. 102(3): pp. 3048–3055.
106. Finnigan, B., Casey, P., Cookson, D., Halley, P., Jack, K., *et al.*, Impact of controlled particle size nanofillers on the mechanical properties of segmented polyurethane nanocomposites. *International Journal of Nanotechnology*, 2007. 4(5): pp. 496–515.

107. McCarthy, S., Meijs, G., Mitchell, N., Gunatillake, P., Heath, G., *et al.*, *In-vivo* degradation of polyurethanes: Transmission-FTIR microscopic characterization of polyurethanes sectioned by cryomicrotomy. *Biomaterials*, 1997. 18(21): pp. 1387–1409.
108. Martin, D., and Edwards, G., *Polymer Composite*. Granted International Patent Application AU 2005279677 2011.
109. Pandey, R. B., Anderson, K. L., and Farmer, B. L., Exfoliation of stacked sheets: Effects of temperature, platelet size, and quality of solvent by a Monte Carlo simulation. *Journal of Polymer Science Part B: Polymer Physics*, 2006. 44(24): pp. 3580–3589.
110. Ginzburg, V. V., Qiu, F., and Balazs, A. C., Three-dimensional simulations of diblock copolymer/particle composites. *Polymer*, 2002. 43(2): pp. 461–466.
111. Lelah, M., and Cooper, S., *Polyurethanes in Medicine*. CRC Press, Inc., 1986: p. 225.
112. Szycher, M., Biostability of polyurethane elastomers: A critical review. *Journal of Biomaterials Applications*, 1988. 3(2): pp. 297.
113. Szycher, M., and McArthur, W. *Surface Fissuring of Polyurethanes Following* in vivo *Exposure*. ASTM International, 1985.
114. Williams, D., Definitions in biomaterials: Proceedings of a consensus conference of the European Society for Biomaterials. Chester, *Progress in Biomedical Engineering*, Elsevier, Amsterdam, 1987.
115. Gunatillake, P., Meijs, G., Mccarthy, S., and Adhikari, R., Poly (dimethylsiloxane)/poly (hexamethylene oxide) mixed macrodiol based polyurethane elastomers. I. Synthesis and properties. *J. Appl. Polym. Sci.*, 2000. 76(14): pp. 2026–2040.
116. http://www.aortech.com/technology/elast-eon.
117. Rueda, L., Garcia, I., Palomares, T., Alonso-Varona, A., Mondragon, I., *et al.*, The role of reactive silicates on the structure/property relationships and cell response evaluation in polyurethane nanocomposites. *Journal of Biomedical Materials Research Part A*, 2011. 97A(4): pp. 480–489.
118. Styan, K. E., Martin, D. J., and Poole-Warren, L. A., *In vitro* fibroblast response to polyurethane organosilicate nanocomposites. *Journal of Biomedical Materials Research Part A*, 2008. 86A(3): pp. 571–582.
119. Kannan, R. Y., Salacinski, H. J., Sales, K. M., Butler, P. E., and Seifalian, A. M., The endothelialization of polyhedral oligomeric silsesquioxane nanocomposites – An *in vitro* study. *Cell Biochemistry and Biophysics*, 2006. 45(2): pp. 129–136.
120. Wang, W., Guo, Y.-l., and Otaigbe, J. U., The synthesis, characterization and biocompatibility of poly(ester urethane)/polyhedral oligomeric silesquioxane nanocomposites. *Polymer*, 2009. 50(24): pp. 5749–5757.
121. Deka, H., Karak, N., Kalita, R. D., and Buragohain, A. K., Bio-based thermostable, biodegradable and biocompatible hyperbranched polyurethane/Ag nanocomposites with antimicrobial activity. *Polymer Degradation and Stability*, 2010. 95(9): pp. 1509–1517.
122. Hsu, S. H., Tseng, H. J., and Lin, Y. C., The biocompatibility and antibacterial properties of waterborne polyurethane-silver nanocomposites. *Biomaterials*, 2010. 31 (26): pp. 6796–6808.
123. Styan, K., *Polyurethane organosilicate nanocomposites for novel use as biomaterials*. Thesis. 2006.
124. Osman, A.F., Andriani, Y., Schiller, T. L., Padsalgikar, A., Svehla, M., *et al.*, *Assessing thermoplastic polyurethane nanocomposites for biomedical devices*. Australasian Polymer Symposium (APS 2011), 2010.

125. Poole-Warren, L. A., Farrugia, B., Fong, N., Hume, E., and Simmons, A., Controlling cell-material interactions with polymer nanocomposites by use of surface modifying additives. *Applied Surface Science*, 2008. 255(2): pp. 519–522.
126. Li, P. R., Wei, J. C., Chiu, Y. F., Su, H. L., Peng, F. C., *et al.*, Evaluation on cytotoxicity and genotoxicity of the exfoliated silicate nanoclay. *ACS Applied Materials & Interfaces*. 2(6): pp. 1608–1613.
127. Lordan, S., Kennedy, J. E., and Higginbotham, C. L., Cytotoxic effects induced by unmodified and organically modified nanoclays in the human hepatic HepG2 cell line. *Journal of Applied Toxicology*, 2011. 31(1): pp. 27–35.
128. Mishra, A., Das Purkayastha, B. P., Roy, J. K., Aswal, V. K., and Maiti, P., Tunable properties of self-assembled polyurethane using two-dimensional nanoparticles: Potential nano-biohybrid. *Macromolecules*, 2010. 43(23): pp. 9928–9936.
129. Da Silva, G. R., Ayres, E., Orefice, R. L., Moura, S. A. L., Cara, D. C., *et al.*, Controlled release of dexamethasone acetate from biodegradable and biocompatible polyurethane and polyurethane nanocomposite. *Journal of Drug Targeting*, 2009. 17(5): pp. 374–383.
130. Da Silva, G. R., Da Silva-Cunha, A., Behar-Cohen, F., Ayres, E., and Orefice, R. L., Biodegradation of polyurethanes and nanocomposites to non-cytotoxic degradation products. *Polymer Degradation and Stability*. 95(4): pp. 491–499.
131. Schaefer, D. W., and Justice, R. S., How nano are nanocomposites? *Macromolecules*, 2007. 40(24): pp. 8501–8517.
132. Rule, M., *Metal Phosphonates and Related Nanocomposites*, Patent no. US 7199172 B2, 2007.
133. He, H., Grafting of swelling clay materials with 3-aminopropyltriethoxysilane. *Spectrochimica Acta Part A: Molecular and Biomolecular Spectroscopy*, 2005. 288(1): pp. 171–176.

# 12
# Soft polymer nanocomposites and gels

K. HARAGUCHI, Kawamura Institute of Chemical Research, Japan

**Abstract:** This chapter describes polymer nanocomposites with organic/inorganic network structures: nanocomposite (NC) gels and soft, polymer nanocomposites (M-NCs). NC gels and M-NCs are synthesized by *in situ* free-radical polymerization in aqueous systems and in the presence of exfoliated clay nanoplatelets. We describe the optical, mechanical, and swelling/deswelling properties, and the characteristics of optical anisotropy, polymer/clay morphology, biocompatibility, stimulus-sensitive surfaces, micro-patterning, etc. M-NCs have improvements in optical and mechanical properties, including ultra-high reversible extensibility and well-defined yielding behavior. The disadvantages (intractability, mechanical fragility, optical turbidity, poor processing ability, low stimulus sensitivity) of conventional polymeric materials have been overcome in NC gels and M-NCs. We discuss possible applications and future trends.

**Key words:** clay, hydrogel, nanocomposite gel, organic/inorganic network, *in situ* free-radical polymerization, mechanical toughness, stimuli-sensitivity.

## 12.1 Introduction

Polymer-based organic/inorganic nanocomposites (P-NCs), composed of an organic polymer and inorganic nanoparticles, have been widely investigated in the last three decades in order to produce new, value-added advanced materials (Saegusa and Chujo, 1992; Giannelis, 1996; Mark, 2003; Okada and Usuki, 2006). To date, many P-NCs consisting of various polymers and inorganic nanoparticles (e.g., silica, silsesquioxane, titania, clay, carbon nanotubes) have been developed by utilizing sol–gel reactions of metal alkoxides or the organic pre-modification of nanoparticles (Saegusa and Chujo, 1992; Giannelis, 1996; Mark, 2003; Okada and Usuki, 2006; Usuki *et al.*, 1993). The resulting P-NCs show significant improvements in some properties such as modulus, heat-distortion temperature, hardness, gas impermeability, etc. However, the P-NCs developed so far have encountered inherent difficulties in preparation and processing as the inorganic content increases. In the case of polymer/clay nanocomposites (NC), in general, only a few weight percent of clay can actually be incorporated into NCs (< 10 wt%, at the most) in the form of organic modified clay (Giannelis, 1996; Okada and Usuki, 2006; Usuki *et al.*, 1993). Further increases in clay content often cause structural inhomogeneities due to inadequate dispersion or irregular aggregation of the clay, and always result in disadvantageous optical and mechanical properties and processability.

In order to overcome these limitations, we extend the concept of 'organic/inorganic nanocomposites' to the field of soft materials, such as 'polymer hydrogels' and 'soft polymers,' under the strategy of fabricating novel organic/inorganic structures. In the present chapter, we give an overview of the development of two types of soft nanocomposites, i.e., nanocomposite gels (Haraguchi and Takehisa, 2002; Haraguchi and Li, 2005; Haraguchi, 2007a, 2007b; Fukasawa *et al.*, 2010) and soft polymer nanocomposites (Haraguchi *et al.*, 2006a, 2010b) with unique organic/inorganic network structures, which break through the previous limitations and exhibit outstanding new characteristics as well as excellent optical and mechanical properties.

## 12.2 Nanocomposite (NC) gels

Polymer hydrogels consisting of three-dimensional polymer networks and large amounts of water filling their interstitial spaces have received extensive attention as transparent, soft, and wet materials (Osada and Khokhlov, 2002; Kopeček, 2007), because their compositions (nearly 90 wt% or more water) and characteristics (including functions related to biomedical, optical, electrical, analysis, etc.) are totally different from those of conventional solid materials. When a polymer with stimulus sensitivity is used as a network constituent, a stimulus-sensitive polymer hydrogel is obtained. In particular, poly(*N*-isopropylacrylamide) (PNIPA), which undergoes a thermoresponsive coil-to-globule transition at a lower-critical-solution temperature (LCST $\approx$ 32 °C) in aqueous media (Heskins and Guillet, 1968), is a typical example of a smart gel. To date, extensive studies have been conducted into the feasibility of using PNIPA hydrogels in various systems such as artificial insulin control systems (Matsumoto *et al.*, 2004), efficient bioseparation devices (Cai *et al.*, 2001), drug-delivery systems (Okano *et al.*, 1990), and biotechnological and tissue engineering devices (Stayton *et al.*, 1995; Yamato and Okano, 2004). In most previous cases, PNIPA hydrogels have been prepared by chemical crosslinking using an organic crosslinker, e.g., *N,N'*-methylenebis(acrylamide) (BIS).

It is known that chemically crosslinked polymer hydrogels (hereinafter abbreviated as OR gels, since they are prepared using organic crosslinkers) have several serious limitations because of their network structure, which consists of a random arrangement of a large number of chemical crosslinks (Fig. 12.1(a)(i)) (Haraguchi and Takehisa, 2002; Haraguchi, 2007a). In these materials, an increase in the crosslink density ($\nu$) is accompanied by a decrease in the inter-crosslink molecular weight ($M_c$), $\nu \propto M_c^{-1}$. Also, since the crosslinking reaction cannot occur at regularly separated positions, $M_c$ is always characterized by a broad distribution of chain lengths between crosslinking points. Furthermore, structural inhomogeneity due to the heterogeneous aggregation of crosslinking points often occurs when the crosslinker concentration is high (Fig. 12.1(b)) (Haraguchi *et al.*, 2002). Therefore, OR gels have some serious disadvantages: (1) optical opacity at

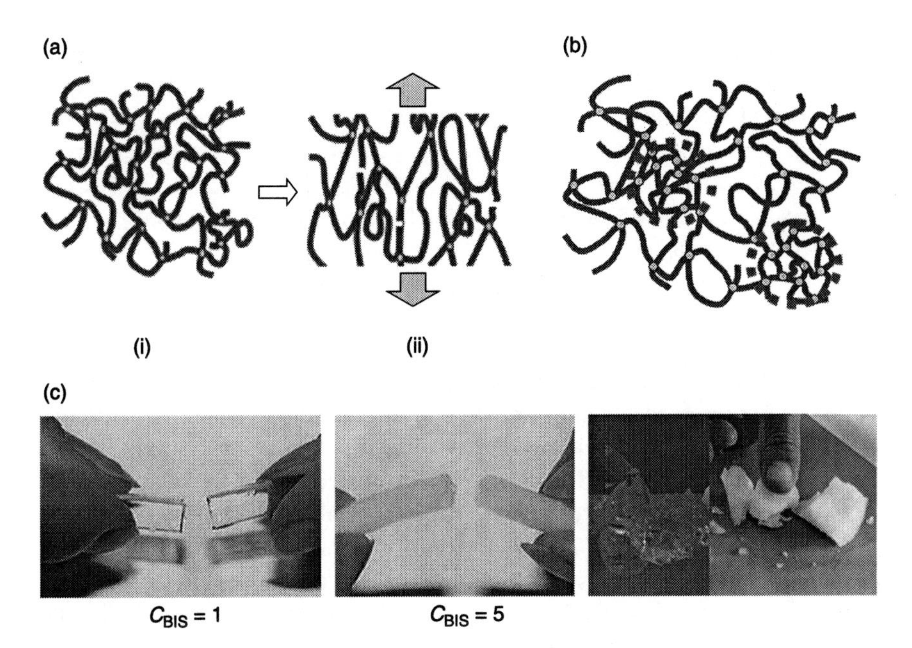


*12.1* (a) (i) Network of a chemically crosslinked OR gel. (ii) Rupture of crosslinked chains on unidirectional extension. (b) Heterogeneous network structure at high concentration of BIS ($C_{BIS}$). (c) Fracture of OR gels with different $C_{BIS}$ (1, 5 mol%) at low stress on elongation and compression.

high $v$; (2) poor mechanical strength (tensile strength $\sigma \leq 10$ kPa) and brittleness (elongation at break $\varepsilon_b < 50\%$); (3) low degree of swelling and slow deswelling rate due to restrictions in the movement of polymer chains by chemical crosslinks, even at the moderate crosslinker concentrations commonly used (Haraguchi and Takehisa, 2002; Haraguchi *et al.*, 2002).

Figure 12.1(c) illustrates brittle fractures in both tensile and compression tests for OR gels prepared with different concentrations of the organic crosslinker (BIS), $C_{BIS}$ in mol%. These weak, brittle properties of OR gels are attributed to their network structure. As shown in Fig. 12.1(a)(ii), on unidirectional extension, polymer chains with a broad distribution of $M_c$ are successively broken due to stress localization in the shorter chains present at any instant. Thus, the chemically crosslinked network structure is a principal factor responsible for the fragility of OR gels. Therefore, to dramatically enhance the mechanical properties of these gels, it is necessary to fabricate new network structures, preferably by employing conditions where $v$ and $M_c$ can be independently controlled.

Recently, the mechanical properties of polymeric hydrogels have been improved remarkably by creating new types of network structures using different strategies (Tanaka *et al.*, 2005; Johnson *et al.*, 2010), such as networks with sliding crosslinks (slide-ring gels) (Okumura and Ito, 2001), organic/inorganic networks (nanocomposite gels) (Haraguchi and Takehisa, 2002), and interpenetrating networks (double-network gels) (Gong *et al.*, 2003). Among these polymeric hydrogels, nanocomposite gels consisting of organic (polymer) and inorganic clay not only overcame all the problems associated with OR gels but also exhibited the best mechanical properties as well as excellent optical, swelling, and stimulus-sensitive properties.

### 12.2.1 Synthesis of NC gels

NC gels have been prepared by *in situ* free-radical polymerization of water-soluble monomers containing amide groups, such as *N*-isopropylacrylamide (NIPA), *N,N*-dimethylacrylamide (DMAA), and acrylamide (AAm), in the presence of inorganic clay, which has been exfoliated and uniformly dispersed in aqueous media (Haraguchi and Takehisa, 2002; Haraguchi and Li, 2005, 2006; Haraguchi *et al.*, 2002, 2003). A variety of clay minerals with layered crystal structures and good water swellability can be used as the inorganic component; examples are the smectite-group clays (hectorite, saponite, montmorillonite, etc.) and mica-group clays (synthetic fluorine mica). Among these, the synthetic hectorite Laponite XLG (Rockwood Ltd, UK; $[Mg_{5.34}Li_{0.66}Si_8O_{20}(OH)_4]Na_{0.66}$; layer size: 30 nm (diameter) × 1 nm (thickness); cation-exchange capacity: 104 mequiv/100 g) (Haraguchi *et al.*, 2003; Rosta and Gunten, 1990), which has a 2:1 layered structure (Fig. 12.2(a)), is the most effective because of its high degree of swellability and exfoliation, high purity, sufficiently small platelet size, and good interaction with PNIPA.

In some cases, commercially modified Laponite XLS, which is mixed with pyrophosphate-Na as an additive to reduce the viscosity of clay aqueous solutions (i.e., increase the dispersion), is used (Haraguchi and Li, 2004; Zhu *et al.*, 2006; Fukasawa *et al.*, 2010). In particular, Laponite XLS has often been used to prepare polyacrylamide (PAAm)-based NC gels (Zhu *et al.*, 2006; Xiong *et al.*, 2008; Li *et al.*, 2009), since the AAm monomer interacts strongly with clay. However, as shown in a later section (Section. 12.2.3), NC gels prepared using Laponite XLS always show lower initial modulus and tensile strength in fixed monomers (e.g., NIPA and DMAA) than those prepared using Laponite XLG (Haraguchi and Li, 2004). This indicates that XLG forms crosslinks with poly(*N*-alkyl acrylamides) more effectively than XLS does. Here, it should be noted that a simple increase in the elongation at break in tensile tests does not indicate improved mechanical properties, since the largest elongations are observed in viscous polymer solutions. In the present review, inorganic clay refers to Laponite XLG, unless otherwise stated. For example, to synthesize an N-NC3 gel (see the nomenclature below), a transparent aqueous solution consisting of water (20 mL), clay (XLG: 0.457 g), monomer (NIPA: 2.26 g), initiator (potassium persulfate (KPS): 0.02 g), and accelerator (*N,N,N′,N′*-tetramethylenediamine (TEMED): 16 $\mu$L) was prepared at

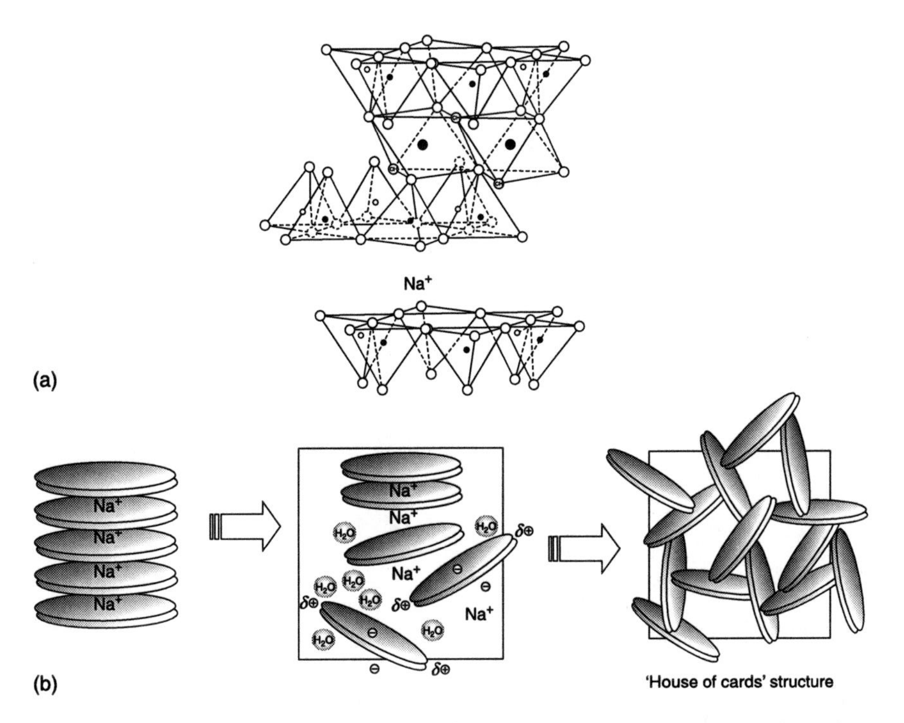


*12.2* (a) Structure of hectorite. (b) Exfoliation in aqueous media and formation of the house of cards structure (Haraguchi *et al.*, 2005a).

ice bath temperature and then the temperature was increased to 20 °C to initiate *in situ* free-radical polymerization.

Consequently, structurally uniform, transparent hydrogels (NC gels) are obtained without syneresis or phase separation. Polymerization yields are almost 100%, regardless of the NC gel composition. The compositions of NC gels can, therefore, be precisely controlled over a wide range by altering those of the initial reaction solutions. It should be noted that, for the synthesis of NC gels, organic crosslinkers, which afford chemically crosslinked networks, are not required. More specifically, such crosslinkers should not be used in conjunction with clay. When two crosslinkers, clay and BIS, are used together ($C_{clay}$ = 5 and $C_{BIS}$ = 1), the resulting hydrogel (NC5-OR1 gel) becomes weak and brittle on elongation, similar to an OR1 gel (Haraguchi and Song, 2007).

The methods used to prepare NC gels are simple and versatile, i.e., injecting reaction solutions into closed vessels followed by polymerization at ambient temperature. Hence, NC gels can be readily formed in various shapes and sizes, such as films, sheets, rods, spheres, and hollow tubes (Haraguchi, 2007a, 2007b). NC gels can also be prepared by photoinitiated free-radical polymerization using very low concentrations (e.g., 0.1 wt% relative to the monomer) of a hydrophobic photoinitiator in aqueous systems (Haraguchi and Takada, 2010). Furthermore, the other type of NC gel, i.e., tetra-TEG-based NC gels, with good biocompatibility can be prepared by incorporating clay nanoparticles into the tetra-PEG network (Fukasawa *et al.*, 2010).

The nomenclature codes used to identify NC and OR gels are based on the monomer used (N- and D- for NIPA and DMAA, respectively) and the concentrations of clay and monomer relative to water; thus, N-NC*n*-M*m* indicates an NC gel prepared using $n \times 10^{-2}$ mol of clay and $m$ mol of NIPA in 1 l-$H_2O$, and D-OR$n'$ indicates an OR gel prepared using $n'$ mol% of BIS relative to the monomer and 1 M of DMAA. When $m$ is 1, the last symbol (−M1) is often omitted for simplicity. Also, the initial symbol (N- or D-) is sometimes omitted in NC and OR gels where there is no possibility of confusion. The clay, polymer, and BIS contents of NC and OR gels ($C_{clay}$, $C_p$, and $C_{BIS}$) are expressed using simplified numerical values of $n$, $m$, and $n'$, respectively.

## 12.2.2 Organic/inorganic network structure of NC gels

*Formation of networks with clay*

Gel formation, i.e., the formation of polymer networks with inorganic clay (in the absence of an organic crosslinker), can be confirmed by their swelling and dynamic mechanical properties. In dynamic mechanical measurements at 22 °C (Haraguchi, 2007a), it has been observed for D-NC5 gel that the storage modulus ($G'$) is always greater than the loss modulus ($G''$) in the frequency range $10^{-1}$–$10^{2}$ rad s$^{-1}$, and that $G'$ and $G''$ change little with frequency (Fig. 12.3(a)). The constantly high $G'$ ($> G''$) indicates that viscoelastic relaxation does not occur on this time scale. These observations are consistent with the dynamic viscoelastic properties that characterize hydrogels with three-dimensional networks. In addition, network formation in NC gels is confirmed by their rubbery nature, such as reversible large deformation and recovery in stretching and compression tests, as shown in a later section.

Network formation in NC gels was further ascertained by swelling measurements. NC gels do not dissolve in water but swell until they reach an equilibrium state, with no free linear polymer chains or free clay particles separating from the network (Haraguchi *et al.*, 2002, 2003). These results indicate that NC gels form three-dimensional networks in which all the polymer chains and clay particles are incorporated. In general, the degree of equilibrium swelling (DES) of a hydrogel is inversely proportional to $v$. In the case of NC gels, the degree of swelling changes with gel composition, i.e., DES tends to decrease with increasing $C_{clay}$ and $C_p$ (Haraguchi *et al.*, 2003). This indicates that the networks in NC gels are formed by crosslinking polymer chains by clay in a specific manner.

*Polymer-clay network structure*

Various analytical studies (transmission electron microscopy (TEM), thermogravimetry, X-ray diffraction (XRD), differential-scanning calorimetry

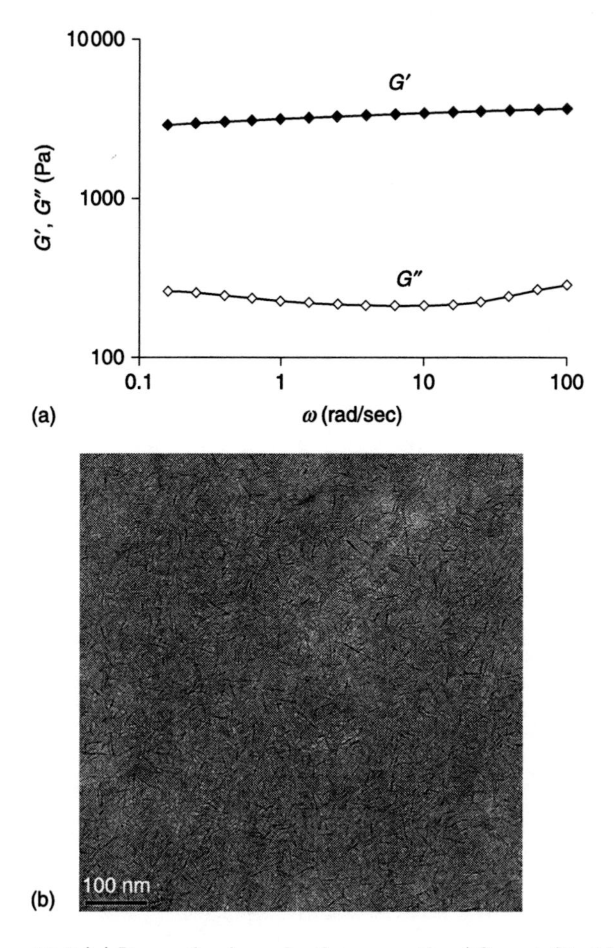


*12.3* (a) Dynamic viscoelastic properties ($G'$ and $G''$ of D-NC5 gel measured at 22 °C in the range 0.2–100 (rad $s^{-1}$) (Haraguchi, 2007a). (b) Transmission electron micrograph of dried N-NC5 gel. The scale bar indicates 100 nm.

(DSC), Fourier transform infrared spectroscopy (FTIR) for dried NC gels; dynamic light scattering (DLS) and small-angle neutron scattering (SANS) for NC gels) have revealed the following structural aspects of NC gels (Haraguchi and Takehisa, 2002; Haraguchi *et al.*, 2002, 2003, 2005a; Shibayama *et al.*, 2004; Miyazaki *et al.*, 2006, 2007; Nie *et al.*, 2006; Haraguchi and Li, 2009).

- Disk-like inorganic clay nanoparticles (30 nm $\phi \times$ 1 nm), resulting from exfoliation of the layered clay mineral (hectorite), are uniformly dispersed in the polymer matrix (XRD, TEM, see Fig. 12.3(b) for dried N-NC5 gel).
- Flexible polymer chains with the same $T_g$ as the linear polymer exist in NC gels (DSC).

- There is no distinct difference between the absorptions of PNIPA and clay in the dried NC gels and those of pure PNIPA or clay (FTIR). This is probably because, in the dried state, PNIPA shows strong hydrogen bonding, which prevents hydrogen bonding with the clay from being clearly identified in dried NC gels.

On the basis of the analytical data obtained and the excellent optical, mechanical, and swelling/deswelling properties of NC gels, it was concluded that NC gels possess a unique organic (polymer)/inorganic (clay) network structure (Fig. 12.4(a)) (Haraguchi and Takehisa, 2002; Haraguchi *et al.*, 2002, 2003, 2005a; Miyazaki *et al.*, 2007; Haraguchi and Li, 2009), in which exfoliated clay nanoparticles (uniformly dispersed in an aqueous medium) are interlinked by a

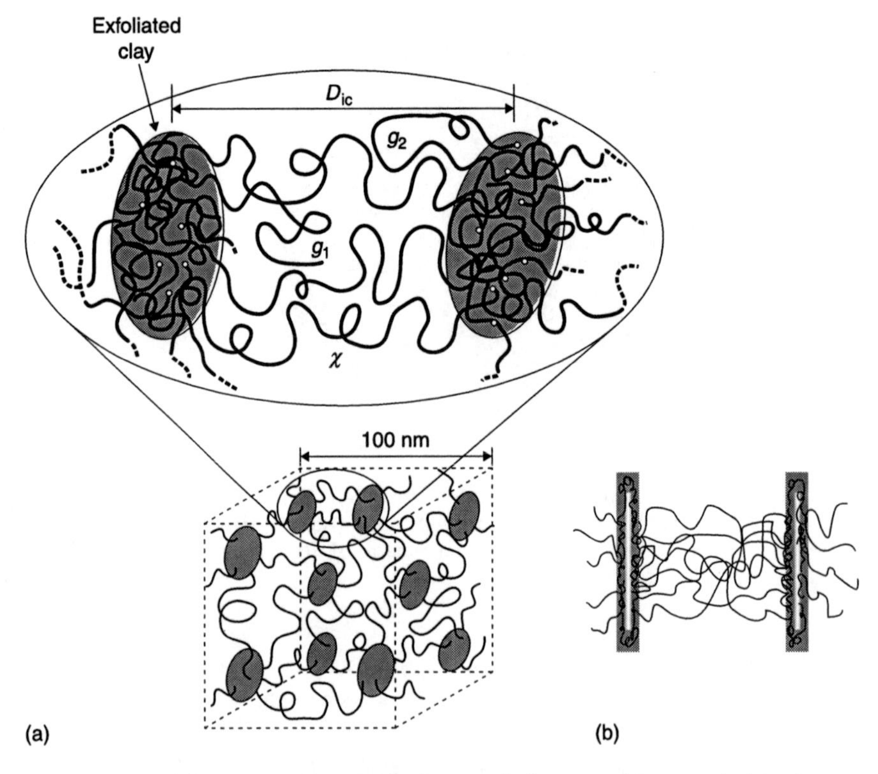


*12.4* (a) Structural model with organic (polymer)/inorganic (clay) networks in an NC gel. $D_{ic}$ is the interparticle distance of exfoliated clay sheets. $\chi$, $g_1$, and $g_2$ are a crosslinked chain, a grafted chain, and a looped chain, respectively. In the model, only a small number of polymer chains are depicted for simplicity (Haraguchi and Takehisa, 2002; Haraguchi *et al.*, 2002; Haraguchi and Li, 2009). (b) Decorated clay platelet model sandwiched by a polymer layer (Miyazaki *et al.*, 2007).

number of flexible polymer chains. Here, $D_{ic}$ is the interparticle distance between exfoliated clay sheets. In the network, various types of polymer chains, such as grafts with free chain ends, looped chains, and topologically crosslinked chains may be present in addition to the crosslinking chains. The interaction between the polymer and the clay nanoparticles is ascribed to non-covalent bonds, probably hydrogen bonds between the amide side groups on the polymer and the surface of the clay (SiOH, Si–O–Si units).

In the polymer/clay network structures, it was considered that a number of polymer chains interact with a single clay platelet and that each polymer chain may interact with the clay surface at multiple points (Haraguchi and Takehisa, 2002; Haraguchi *et al.*, 2002, 2003, 2005a). In other words, the exfoliated clay platelets act as multifunctional crosslinkers, and hence the polymer chains in NC gels are crosslinked by a planar series of crosslinks. These results have been confirmed by molecular characteristics of PNIPA separated from NC gels and contrast-variation SANS measurements. By successful separation of PNIPA from the N-NC gel, without damage, by decomposing the clay in the network using hydrofluoric acid (Fig. 12.5) (Haraguchi *et al.*, 2010c), it became clear that clay

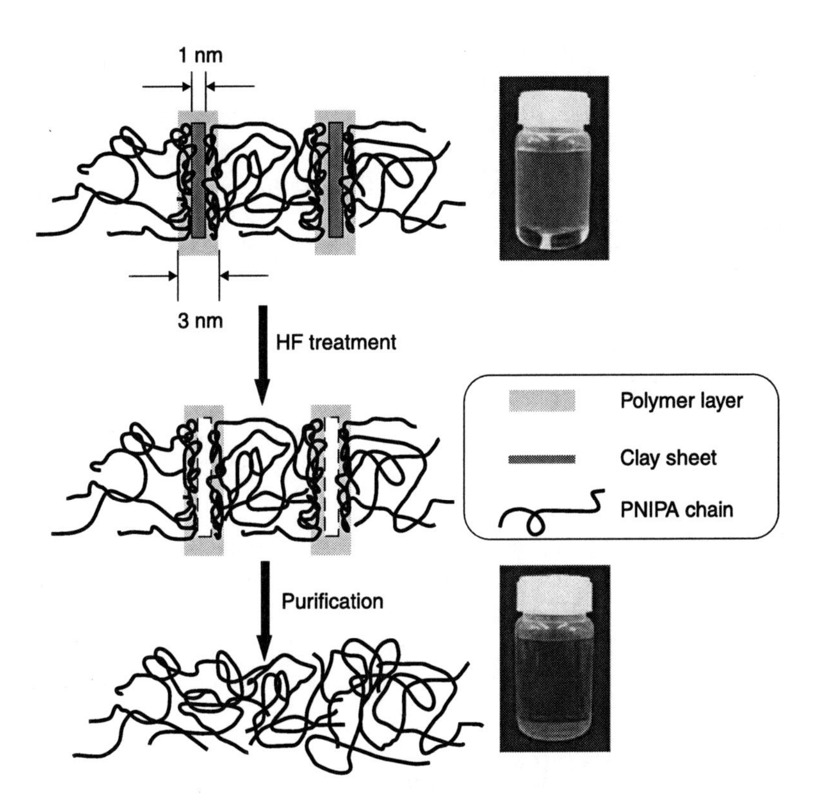


*12.5* Separation of PNIPA from an NC gel through the decomposition of the clay (Haraguchi *et al.*, 2010c). (HF: hydrofluoric acid)

platelets play an important role in preventing the formation of self-crosslinked PNIPA networks (Xu *et al.*, 2010), and that PNIPA in N-NC gels has a high molecular weight ($M_w$ = 5.6 × $10^6$ g/mol) (Haraguchi *et al.*, 2010c), almost regardless of $C_{clay}$ (1–25), indicating that a single polymer chain repeatedly links to neighboring clay sheets. In contrast-variation SANS measurements, it was shown that the polymer chains aggregate to form 1-nm-thick layers on the clay surface (Fig. 12.4(b)) (Miyazaki *et al.*, 2007).

The number of crosslinked polymer chains per unit volume of NC gel, $N^*$, can be estimated by Eq. 12.1, based on the kinetic theory of rubber elasticity (Tobolsky *et al.*, 1961).

$$F = \Phi N^* k_B T\{\alpha-(1/\alpha)^2\} \tag{12.1}$$

where $F$ is the force per unit original (undeformed) cross-sectional area of the swollen network, $\Phi$ is a front factor (= 1), $\alpha$ is the elongation ratio, and $k_B$, $T$ are the Boltzmann constant and the absolute temperature, respectively. From the tensile stress–strain curves and Eq. 12.1, $N^*$ and the number of crosslinked chains per clay sheet were calculated for NC gels with different $C_{clay}$ (Haraguchi *et al.*, 2003). The results revealed that neighboring clay sheets in NC gels are linked by at least tens of flexible polymer chains (a few tens to more than 100).

The effective crosslink densities ($v_e$) of NC gels were calculated using the DES, according to the Flory–Rehner theory (Flory and Rehner, 1943). Although an NC gel is a kind of ionic polymer gel containing ionic clay platelets, the effect of clay on the swelling is totally different from that of the usual organic ionic groups. In NC gels, the swelling decreased with increasing $C_{clay}$ (Haraguchi *et al.*, 2002, 2003), although normal ionic polymer gels exhibit increased swelling with increasing concentrations of ionic groups. This is because the clay platelets act as effective crosslinkers. Therefore, to calculate $v_e$ for a NC gel, Eq. 12.2, which corresponds to an affine network model for non-ionic gel systems, was used to simplify the condition.

$$\phi + \ln(1-\phi) + \chi\phi^2 = -V_s v_e \left[ \left(\frac{\phi}{\phi_0}\right)^{1/3} - \frac{2}{f}\left(\frac{\phi}{\phi_0}\right) \right] \tag{12.2}$$

where, $\phi$ and $\phi_0$ are the network volume fractions at equilibrium swelling and in a reference state, respectively. $(2/f)$ (where $f$ is the functionality) is 0.5 for BIS, and almost 0 for clay in NC gels because of its high functionality. $V_s$ is the molar volume of water. $\chi = \chi_1 + \phi\chi_2$, where $\chi_1 = (\Delta H - T\Delta S)/k_B T$, $\chi_2 = 0.518$, $\Delta H = -12.462 \times 10^{-21}$ J, and $\Delta S = -4.717 \times 10^{-23}$ J/K (Baker *et al.*, 1994). The calculated values of $v_e$ for NC and OR gels are listed in Table 12.1 (Haraguchi *et al.*, 2007b). The effect of $C_{clay}$ on $v_e$ in NC gels is analogous to that of $C_{BIS}$ on $v_e$ in OR gels, i.e., the $v_e$ of NC gels increases with $C_{clay}$. In Table 12.1, the $v_e$ of soft NC1 gel is about one-ninth that of the commonly used OR1 gel; even for the NC10 gel, which has excellent mechanical properties (initial modulus, $E$, and tensile

*Table 12.1* Degree of equilibrium swelling (DES) and effective network density ($\nu_e$) for NC and OR gels, calculated according to the Flory–Rehner theory, Eq. 12.2

| Hydrogel | NC1 | NC5 | NC10 | NC15 | NC20 | OR1 | OR5 |
|---|---|---|---|---|---|---|---|
| DES (Ws/Wdry) | 50.94 | 28.94 | 22.57 | 17.61 | 12.94 | 16.93 | 8.55 |
| $\nu_e$ | 0.0048 | 0.0099 | 0.0127 | 0.0170 | 0.0256 | 0.0431 | 0.1582 |

Source: Haraguchi *et al.*, 2007b.

strength, $\sigma$), $\nu_e$ is only one-third that of the OR1 gel. The fact that values of $\nu_e$ for NC gels are generally smaller than those of OR gels is consistent with the results obtained for the mechanical properties. Thus, it was concluded that, in NC gels, relatively small numbers of crosslinks are effectively formed on the surfaces of the clay platelets.

*Mechanism of polymer-clay network formation*

The mechanism of forming the organic (polymer)/inorganic (clay) network structure of NC gels was elucidated on the basis of changes in viscosity, optical transparency, XRD, and mechanical properties (Haraguchi *et al.*, 2005a). Model structures showing the formation of NC gels are given in Fig. 12.6(a). During preparation of the initial reaction solutions, a specific solution structure is formed, where the monomer (NIPA) largely prevents gel formation of the clay itself (Fig. 12.6(a)(i)), and the initiator and accelerator are located near the clay surface through ionic interactions (Fig. 12.6(a)(ii)). By increasing the solution temperature, free-radical polymerization is initiated by the redox system close to the surface of the clay (Fig. 12.6(a)(iii)). We propose that this is the mechanism for the formation of 'clay-brush particles,' composed of exfoliated clay platelets with numbers of polymer chains grafted to their surfaces, in the early stages of NC gel synthesis, as depicted in Fig. 12.6(a)(iv).

The formation of clay-brush particles was confirmed by the time dependence of optical transmittance during *in situ* polymerization. A distinct drop in transparency was observed at a very early stage of *in situ* polymerization (Fig. 12.6(b): N-NC2 gel) (Haraguchi *et al.*, 2005a), which may correspond to the formation of assemblies of clay-brush particles. Monomer conversion at the point of minimum transparency was approximately 7%, where the polymer/clay weight ratio was about 0.2:1. Transparency was re-established on further polymerization. In contrast, no transparency changes were observed during the polymerization of OR gels or LR, or even for PNIPA containing silica or titania particles (Fig. 12.6(b)). Therefore, it was concluded that the newly observed transparency changes were a characteristic of NC gel synthesis, i.e., the changes are due to the formation of clay-brush particles and the subsequent formation of an organic/inorganic network

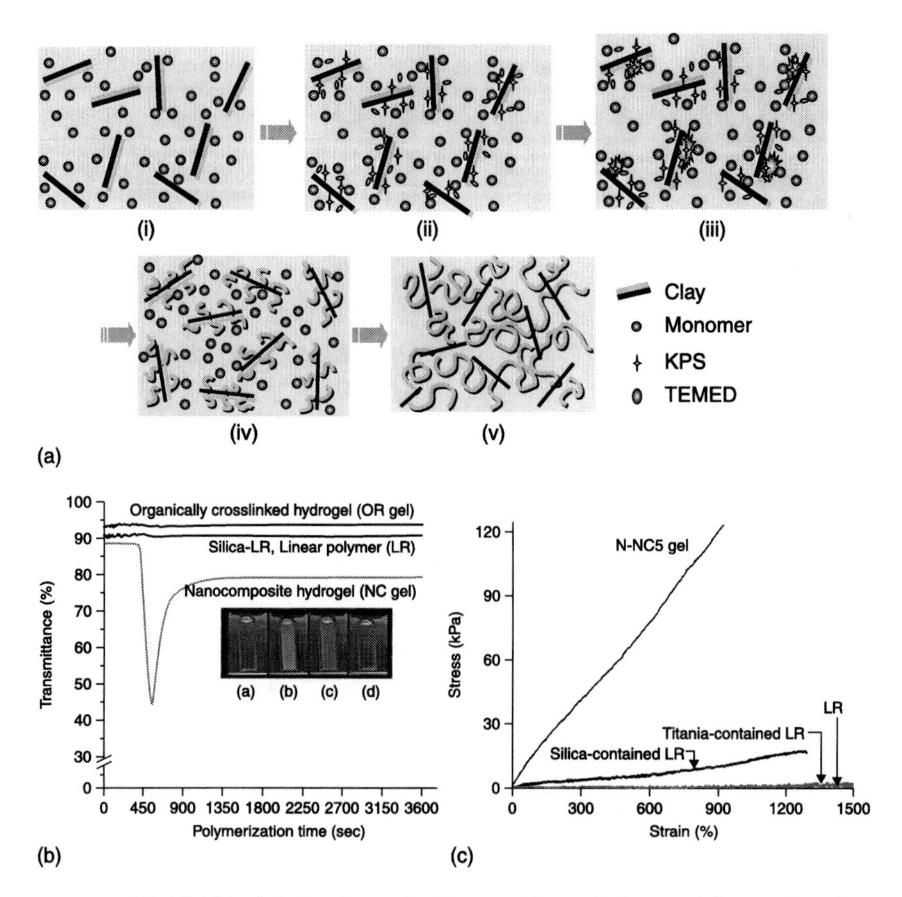


*12.6* (a) Model structures for the reaction solution and the mechanism of forming the organic/inorganic network structure in an NC gel. (i) Aqueous solution consisting of clay and NIPA. (ii) Reaction solution consisting of clay, NIPA, KPS, and TEMED. (iii) Radical formation near the clay surface in the reaction solution. (iv) Formation of clay-brush particles. (v) Formation of organic/inorganic networks. In the models, only a small number of monomer (polymer), KPS, and TEMED are depicted for simplicity. (b) Changes in optical transparency during the polymerization of NC gel (N-NC2-M1 gel), OR gel (N-OR1-M1 gel), linear polymer (LR) and silica-containing LR. Inset photographs (a)–(d) were obtained during the polymerization of the NC gel. (c) Stress–strain curves for N-NC5-M1 gel, LR (N-LR-M1), and silica-contained LR, and titania-contained LR. Silica and titania were contained in the same weight with clay in N-NC5-M1 gels (Haraguchi *et al.*, 2005a).

structure. This was also confirmed by XRD measurements (Haraguchi *et al.*, 2005a). The decrease in transparency was not observed for the AAm-XLS system (Zhu *et al.*, 2006), probably because of the weak interactions between PAAm and XLS clay and the high dispersion of the resulting complex.

The proposed mechanism of forming the organic/inorganic network structure was supported by the fact that the stress–strain curves obtained for NC gels with different polymer contents exhibit characteristic two-step changes for D-NC2.5-M*m* gels (Haraguchi *et al.*, 2005a), which correspond to the formation of a primary network (first step, $m \leq 0.5$) and a subsequent increase in the number of crosslinked polymer chains (second step, $m \geq 1$). Furthermore, DLS and SANS measurements showed that the gelation in NC gels is classified as an ergodic–non-ergodic transition, as that in OR gels; the only difference is that huge clusters of NC microgels (corresponding to clay-brush particles) are formed before the gelation threshold in the former (Miyazaki *et al.*, 2007).

NC gels with excellent mechanical properties and structural homogeneity could not be prepared by other methods, such as those involving mixing of clay and polymer solutions, or by *in situ* polymerization in the presence of the other inorganic nanoparticles such as silica or titania, as shown in Fig. 12.6(c) (Haraguchi *et al.*, 2005a). These results indicate that the formation of organic/inorganic network structures in NC gels is highly specific and is achieved only by *in situ* free-radical polymerization in the presence of clay. The specific role of clay, as distinct from silica, in the preparation of NC gels is also confirmed from investigations of the nature of PNIPA after eliminating clay and silica from the gels (Haraguchi *et al.*, 2010c).

### 12.2.3 Basic properties of NC gels

The optical, mechanical, and swelling/deswelling properties of NC gels are superior to those of conventional OR gels. Furthermore, these basic properties can be controlled as desired by altering the network composition or by modifying the network structures.

#### *Optical transparency*

The optical transparency of hydrogels generally reflects the spatial inhomogeneity in the networks. The main factors affecting the transparency of NC gels are the degree of dispersion of the clay nanoparticles in aqueous media and the clay/monomer or clay/polymer interactions. Translucent or opaque NC gels are obtained when the clay is insufficiently exfoliated in the reaction solution or when microscopic aggregations of the clay and monomers or polymers are formed. The effect of the clay content ($C_{clay}$ = 1–9) in N-NC gels, and that of the BIS content ($C_{BIS}$ = 1–9) in N-OR gels, on transparency measured at 1 °C are shown in Fig. 12.7 (Haraguchi *et al.*, 2002). Transparent OR gels generally become opaque

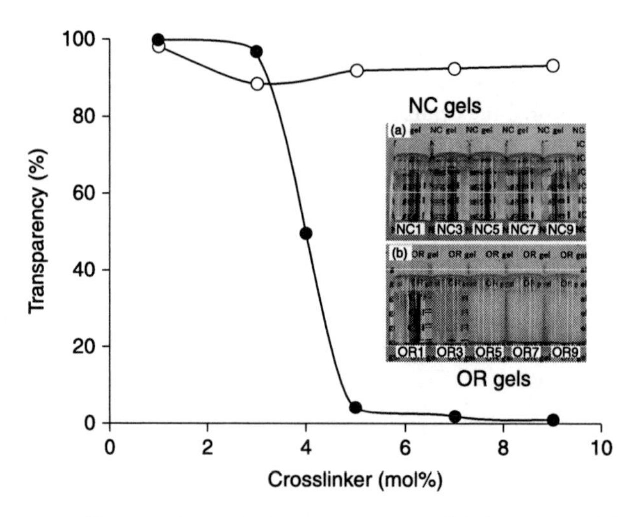


*12.7* Changes in optical transparency of N-NC and N-OR gels produced by altering the crosslinker contents. ○ and picture (a) for N-NC gel and ● and picture (b) for N-OR gel. Optical transmittance was measured at 1 °C for hydrogels (Haraguchi *et al.*, 2002).

with increasing $C_{BIS}$, because of the inhomogeneous distribution of crosslinking points; the critical value above which the transmittance decreases changes with the type of polymer used (e.g., 5 mol% for N-OR and 8 mol% for D-OR). In contrast, NC gels are generally transparent, regardless of $C_{clay}$ and $C_p$ or the type of polymer used.

With regard to the temperature dependence of the gel transparency, D-NC gels are always transparent regardless of the temperature, while N-NC gels exhibit reversible changes in transparency at the LCST as a result of the coil-to-globule transition of PNIPA. Thus, for example, N-NC3 and N-NC5 gels became opaque (white) above the LCST. However, the extent of the change in transparency at the LCST varied dramatically with $C_{clay}$, as shown in Fig. 12.8 (Haraguchi and Li, 2005). The decreases in transparency gradually reduced as $C_{clay}$ increased. Eventually, for $C_{clay}$ greater than $C_{clay}^{crit(opt(1))}$ (= 10), there was no loss in transparency and the gels remained transparent regardless of the temperature (see inset to Fig. 12.8) (Haraguchi and Li, 2005). This indicates that, in NC gels with $C_{clay}$ greater than $C_{clay}^{crit(opt(1))}$, the thermoresponsive dehydration of PNIPA chains, i.e., the conformational change to a globular, hydrophobic form, is hindered in PNIPA chains attached to, or lying close to, hydrophilic clay surfaces. This is the first observation of non-thermosensitive PNIPA hydrogels, where the coil-to-globule transition of PNIPA chains is completely restricted by the presence of inorganic nanoparticles. The transition temperature of NC gels, identified as a decrease in optical transmittance of 50%, increases with increasing $C_{clay}$, although the transmittance starts to decrease at a slightly lower temperature (Haraguchi *et al.*,

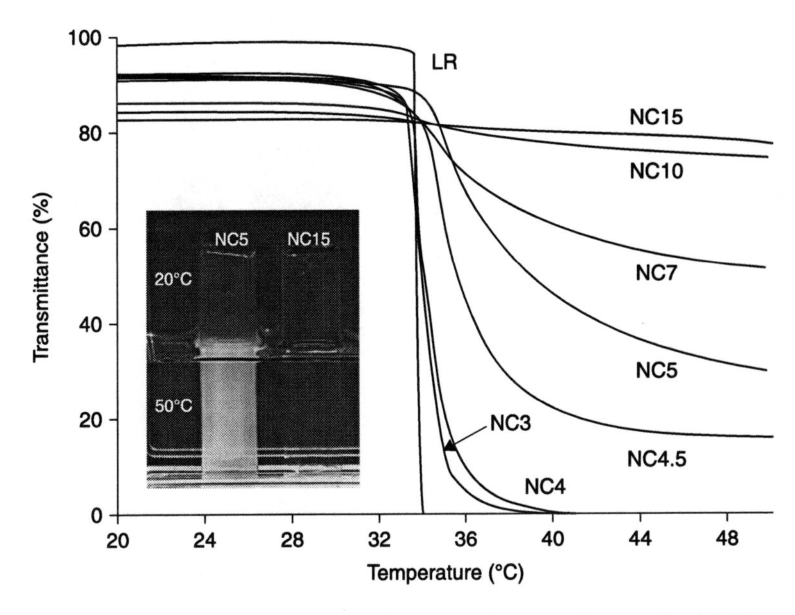


*12.8* Temperature dependency of the optical transmittance for PNIPA solution (LR) and NC gels with different $C_{clay}$. Inset: Transparency of NC5 and NC15 gels below (in air: upper) and above LCST (in water: lower). The transparency change observed for NC5 gel was not observed for NC15 gel (Haraguchi and Li, 2005).

2007b). Furthermore, the transition temperature decreased or increased markedly on adding an inorganic salt (e.g., NaCl, $CaCl_2$, or $AlCl_3$) or a cationic surfactant (e.g., hexadecyltrimethylammonium chloride) to the gel (Haraguchi, 2011).

*Mechanical properties*

For a long time, the tensile mechanical properties of hydrogels such as PNIPA hydrogels were not investigated, because hydrogels with chemically crosslinked networks (OR gels) cannot withstand mechanical stretching or bending as described in the section above. However, the creation of NC gels has made it possible to conduct all kinds of mechanical tests. The most striking characteristics of NC gels are their excellent mechanical properties, such as high $\varepsilon_b$ (>1000%), high tenacity (bending by more than 360°), and high $\sigma$(>100 kPa); these properties were entirely unexpected and distinct from those of OR gels having the same composition, except for the type of crosslinker.

Figure 12.9 shows the changes in the tensile stress–strain curves for NC gels with different $C_{clay}$ ($C_{clay}$ = 0.05–10) (Haraguchi, 2007a). At low $C_{clay}$ (<1), $E$ increased slightly with $C_{clay}$, but $\varepsilon_b$ decreased sharply to about 1000%. However, when $C_{clay}$ was higher than that for NC1, $E$ and $\sigma$ increased greatly, as indicated by the dotted line, while the values of $\varepsilon_b$ were approximately constant

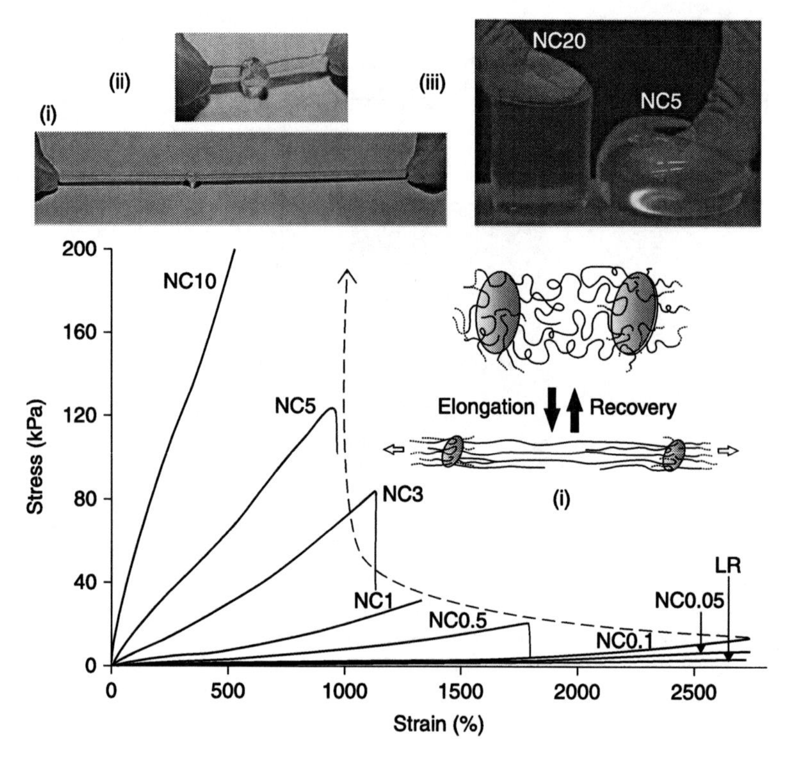


*12.9* Tensile stress–strain curves of N-NC gels with different $C_{clay}$ (0.05–10) and LR ($C_{clay}$ = 0) (Haraguchi, 2007a). (i) Unit structure of an organic/inorganic network (Haraguchi and Takehisa, 2002). (ii) Knotted and stretched NC5 gel (Haraguchi *et al.*, 2002). (iii) Compression of NC5 and NC20 gels (Haraguchi and Li, 2005).

($\varepsilon_b \approx 1000\%$). Thus, the effects of the crosslinker, BIS, and clay on the tensile mechanical properties are totally different from each other. In NC gels, clay platelets play a specific role, acting as effective multifunctional crosslinkers to form polymer/clay networks. The critical concentration $C_{clay}^{crit(1)}$, above which a well-defined organic/inorganic network is formed, is approximately 1, for the present system ($C_{NIPA}$ = 1 M).

Since the number of crosslinking units per unit volume in NC and OR gels are different (e.g., 10 and 5400 units per 100 nm cube for NC3 and OR1, respectively), $D_{ic}$ in NC gels should be very large with a narrower distribution than in OR gels. The reason for the high $\varepsilon_b$ in NC gels, regardless of the high $C_{clay}$, is attributed to the high deformability of the long, flexible polymer chains in the unit structure of the polymer/clay networks (Fig. 12.9(i)). $E$ and $\sigma$ for NC gels also increased with increasing $C_p$ (Haraguchi *et al.*, 2005a), since the number of polymer chains linking neighboring clay platelets increases with $C_p$.

Figure 12.10(a) shows tensile stress–strain curves for N-NC gels with different $C_{clay}$ (NC5–NC25) (Haraguchi and Li, 2005, 2006), where both $E$ and $\sigma$ increase markedly with $C_{clay}$, as shown by the arrows. $\sigma$ exceeds 1000 kPa in the N-NC25 gel. The steep increase in $E$ at high $C_{clay}$ is probably due to the formation of rigid structures involving clay/clay interactions, similar to a house-of-cards structure (Dijkstra *et al.*, 1995) (Fig. 12.2(b)) or a nematic-like clay structure (Haraguchi *et al.*, 2007b). In fact, the optical anisotropy of as-prepared NC gels is pronounced in NC gels with $C_{clay} \geq 10$ (Haraguchi and Li, 2006). Thus, the tensile mechanical properties can be controlled over a wide range by altering $C_{clay}$ and $C_p$. It should be noted that the changes in mechanical properties of NC gels are unique, i.e., totally different from those observed in normal polymeric materials. In the case of the latter, an increase in $E$ is generally accompanied by a decrease in $\varepsilon_b$ because the increase is normally caused by the orientation of polymer chains or modification of the polymer structure to a rigid one. In contrast, in NC gels, $E$ and $\sigma$ can increase without a large sacrifice of $\varepsilon_b$ because of the polymer/clay network structure. From the area under the tensile stress–strain curve, the fracture energies ($\xi_f$) for NC and OR gels and the dependences of $\xi_f$ on the crosslinker content can be determined (Haraguchi and Li, 2005, 2006). $\xi_f$ for NC gels increases in proportion to $C_{clay}$, whereas that for OR gels remains almost constant, regardless of $C_{BIS}$ ($\xi_f = 0.0014$ J). The value of $\xi_f$ for NC25 is nearly 3300 times that of OR gels (Haraguchi and Li, 2006). This is a very striking result, since NC and OR gels differ only in the crosslinker used. NC gels prepared using Laponite XLS also showed similar results for both the tensile mechanical properties, with the exception of large $\varepsilon_b$ ($\equiv$ low crosslink density), and the swelling/deswelling properties (Liu *et al.*, 2006).

NC gels showed characteristic recovery from high elongation. The time dependence of the residual strain after an elongation of 900% followed by immediate stress release is shown in Fig. 12.10(b): region A, instant recovery (within 1 min); region B, time-dependent residual strain; and region C, pseudo-permanent strain (remains for more than 2 weeks). For N-NC gels with a low $C_{clay}$ (<NC10), 90–99% instantaneous recovery was observed; on the other hand, for NC gels with high $C_{clay}$ (NC10–20), the permanent strain ($C$) gradually increased with $C_{clay}$, which is attributed to the irreversible orientation of the clay platelets. These elongation and recovery patterns of NC gels can be explained using a typical four-element mechanical model (inset of Fig. 12.10(b)).

The stress–strain curves of N-NC gels in the second cycle were different from those observed in the first cycle (Fig. 12.10(c)) (Haraguchi and Li, 2006). For NC gels with higher $C_{clay}$, remarkable increases in $E$ and $\sigma$ and a significant decrease in $\varepsilon_b$ were observed in the second cycle. The effects of $C_{clay}$ on the $\sigma$ and $E$ of N-NC gels in the first and second cycles are summarized in Fig. 12.11 (Haraguchi and Li, 2006). The critical value of $C_{clay}^{crit(2)}$, above which the mechanical properties in the second cycle become very different from those in the first cycle, was found to be approximately 10. The change in the relative magnitudes of $E_{10-50}$ and $E_{100-200}$

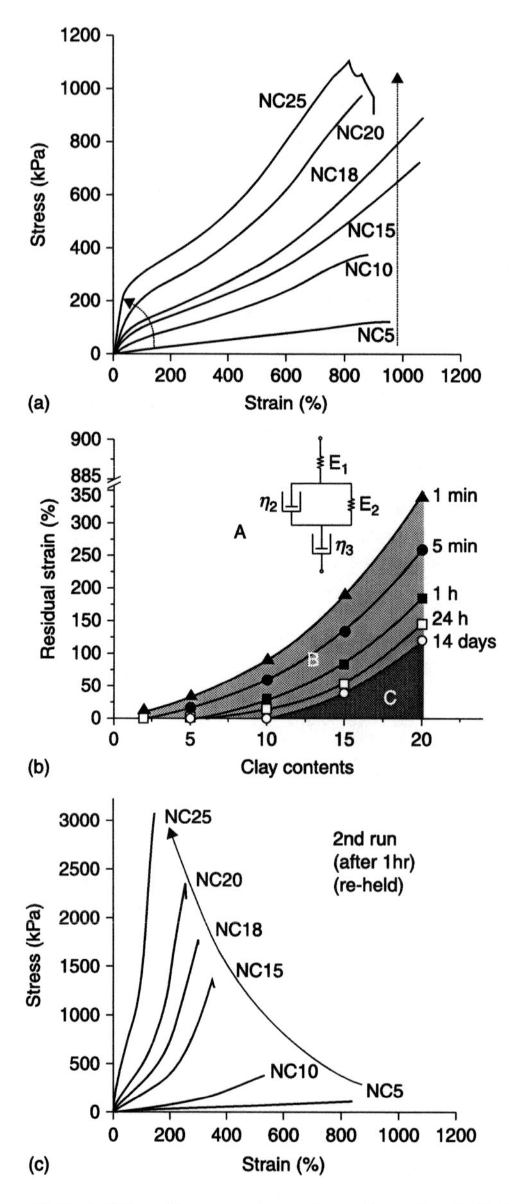


*12.10* (a) Tensile stress–strain curves for N-NC gels with different $C_{clay}$ (NC5–NC25). All samples tested had the same ratio of water:polymer (10:1.13 w/w). (b) Time dependence of residual strain for NC gels with different $C_{clay}$ after release from stress from an initial elongation of 900% (or 800% for NC20 and NC25). A: quick recovery, B: time-dependent recovery, C: pseudo-permanent strain. (Inset) Four-element mechanical model adopted for NC gels. (c) Tensile stress–strain curves for once-elongated NC gels with different $C_{clay}$ which were prepared by elongation to 900% (or 800% for NC20 and NC25) and subsequently relaxed for 1 h (Haraguchi and Li, 2006).

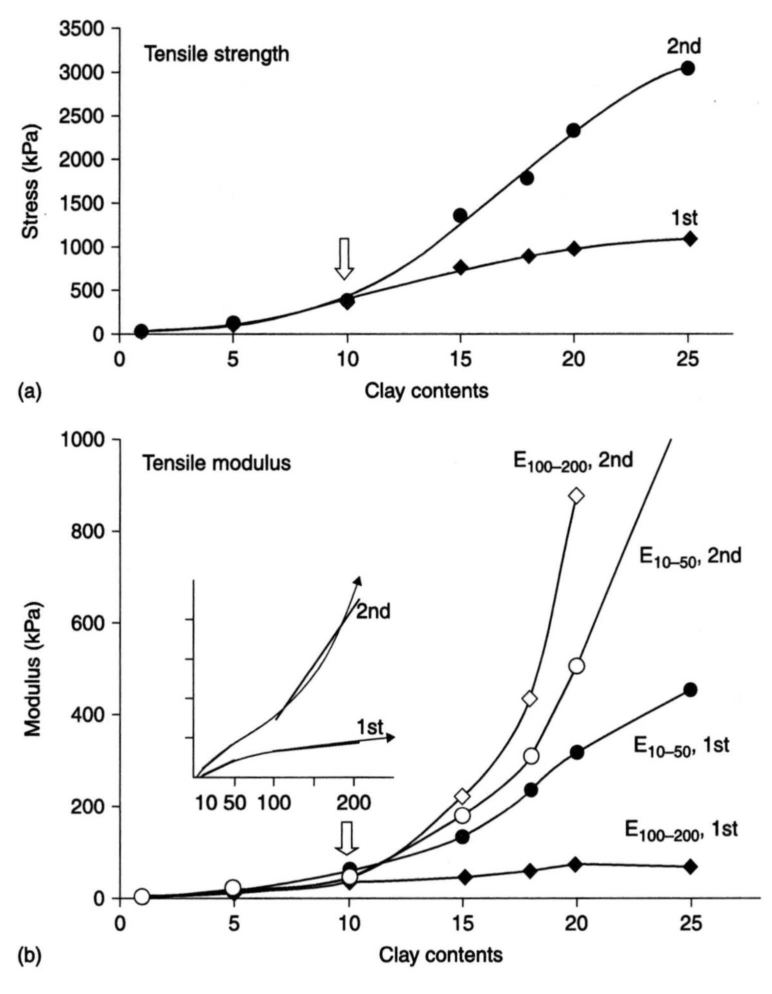


*12.11* Changes in (a) tensile strength and (b) tensile modulus for two series of NC gels. As-prepared NC gels (1st cycle: rhombus) and once-elongated NC gels (2nd cycle: circle). Two tensile moduli, $E_{10-50}$ and $E_{100-200}$, were calculated from the slopes in the ranges 10–50% and 100–200% elongation, respectively. The arrow indicates the critical clay concentration ($C_{clay}^{crit(2)}$), above which the mechanical properties of the 2nd cycle are largely changed (Haraguchi and Li, 2006).

was attributed to the rigidity of the as-prepared gels and the permanent orientation of the clay platelets (and the PNIPA chains attached to them) after the first elongation.

The tensile mechanical properties of NC gels could also be varied by changing the type of clay used (Haraguchi and Li, 2004). The modification of clay (e.g., by fluorine substitution or by the addition of a dispersant such as pyrophosphate-Na) had a marked effect on the mechanical properties, where $E$ and $\sigma$ decreased but $\varepsilon_b$ increased (Fig. 12.12). Similar tendencies were also observed in NC gels prepared using clay with a large particle size (montmorillonite, >300 nm). Further, the tensile mechanical properties can be modified by drying, which is attributed to the additional crosslinks formed by irreversible rearrangement of the polymer/clay network structure in the concentrated state (Haraguchi *et al.*, 2010a).

The use of clay as a multifunctional crosslinking agent is the most crucial point in the fabrication of NC gels with excellent properties. After an NC gel had been prepared successfully for the first time using exfoliated clay platelets (Haraguchi and Takehisa, 2002), different kinds of inorganic or organic materials were adopted as multifunctional crosslinkers for NC gels in systems such as: attapulgite (a fibrillar clay mineral) and copolymer (Xiang *et al.*, 2006), octa (propylglycidyl ether) polyhedral oligomeric silsesquioxane and PNIPA (Mu and Zheng, 2007), layered double hydroxides and agarose (Hibino, 2010), $Fe_3O_4$ (surface treated by silanization) and PNIPA (Chen *et al.*, 2011), organic hydrophobic association and PAAm (Jiang *et al.*, 2010), chitosan nanofiber and PAAm (Zhou and Wu, 2011), and cellulose rods and PAAm (Zhou *et al.*, 2011).

NC gels can withstand high levels of deformation in all modes, including compression, torsion, tearing, and bending, in addition to elongation. For example, NC gels can be tied into a knot without being damaged, and the knotted NC gel can be stretched without breaking at the knot (Fig. 12.9(ii)) (Haraguchi *et al.*, 2002). In compression tests, NC gels generally withstand a compression of up to 90% (Fig. 12.9(iii)) (Haraguchi and Li, 2005, 2006). The compression modulus

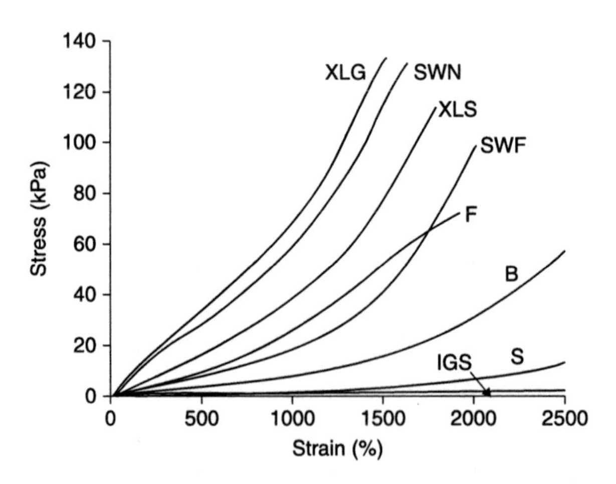


*12.12* Stress–strain curves of D-NC gels prepared with various kinds of clay: hectorite (XLG, SWN), fluorinated hectorite (SWF, B), hectorite modified with a dispersing agent (XLS), montmorillonite (F), B and a dispersing agent (S), and sepiolite (IGS) (Haraguchi and Li, 2004).

and strength of N-NC gels increase almost proportionally with $C_{clay}$. The strength reaches 5 MPa at 80% strain for N-NC20 gels (Haraguchi and Li, 2006).

Furthermore, the ultimate tensile properties of rubbery NC gels changed considerably when the water content ($R_{H_2O} = W_{H_2O}/W_{dry}$) was varied over a wide range (Fig. 12.13) (Haraguchi and Li, 2009). At a high $R_{H_2O}$, where PNIPA chains are fully hydrated, N-NC4 gels retain their rubbery tensile properties. However, when $R_{H_2O}$ is decreased, these gels undergo plastic-like deformation. Consequently, for a series of N-NC4 gels with different $R_{H_2O}$, the rupture points in the stress–strain curves were connected to give a 'failure envelope' (Fig. 12.13). Here, a counterclockwise movement of the rupture point was observed on increasing the strain rate or decreasing $R_{H_2O}$. Thus, a decrease in $R_{H_2O}$ in NC gels produces an effect similar to that produced by a decrease in temperature in an amorphous elastomer (e.g., SBR) (Smith and Stedry, 1960), although the mechanisms are different in the two cases; viz. entropic elasticity in the elastomer and plasticization in the NC gel.

While modifying the network structure by simultaneously using two crosslinkers, i.e., an inorganic clay and very small amounts of organic BIS, it was found that in co-crosslinked PNIPA hydrogels (NC-OR gels) the controllable region in the correlation between the compression strength and the modulus was

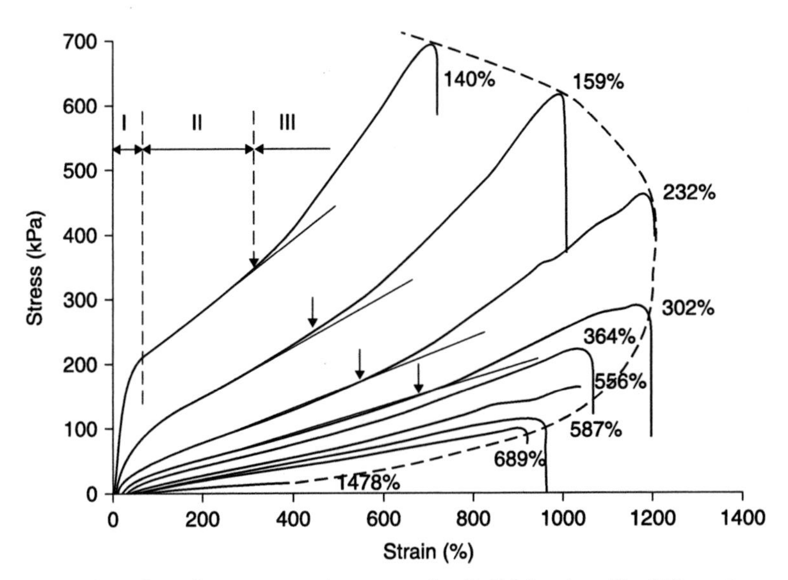


*12.13* A series of stress–strain curves for N-NC4 gels with different $R_{H_2O}$ (wt%). The numerical values are $R_{H_2O}$. $R_{H_2O}$ for the as-prepared NC4 gel is 689 wt%. The dashed line is the failure envelope obtained by connecting the rupture points. The arrows represent the threshold strain points ($\varepsilon_t$) above which strain-hardening occurs (Haraguchi and Li, 2009).

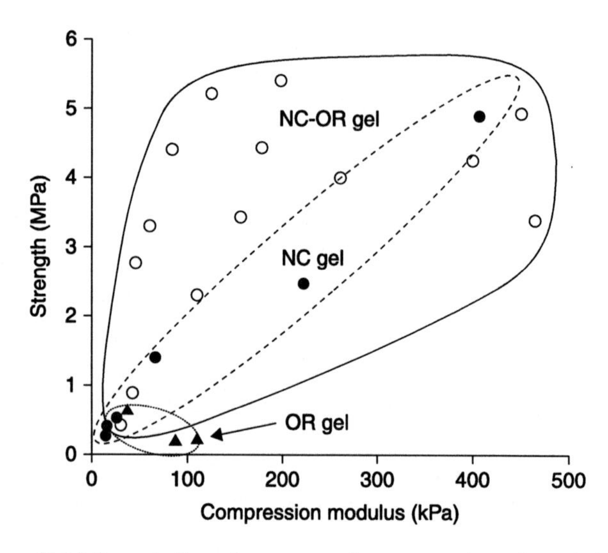


*12.14* Correlation of compression strength and modulus, which is controllable in $NC_n$, OR*m*, and NC*n*-OR*m* gels by altering $n = 0–10$ and $m = 0–5$. The modulus–strength relations can be controlled within each rounded region. Here, the strength for NC and NC-OR gels is at 80% compression (without fracture), whereas the strength for OR gels is the maximum at break (Haraguchi and Song, 2007).

expanded, as shown in Fig. 12.14 (Haraguchi and Song, 2007). This change was attributed to the formation of a micro-complex structure with relatively high chemical crosslink densities in proximity to the clay surface. Here, the combination of high strength and low modulus, which is often required for biomaterials but usually difficult to achieve in materials design, could be obtained.

*Swelling and deswelling properties*

With regard to swelling in water at 20 °C, NC gels normally exhibit greater swelling than OR gels, as shown in Fig. 12.15(a) and Table 12.1 (the dependence of DES on the crosslinker content). These effects are attributed to the lower crosslink density of NC gels compared with OR gels.

In the case of PNIPA hydrogels, i.e., N-NC and N-OR gels, they both exhibited volume changes at the LCST in response to the coil-to-globule transition of PNIPA chains. Thermosensitivity and its control in PNIPA hydrogels have attracted much attention because of the many potential applications (Matsumoto *et al.*, 2004; Cai *et al.*, 2001; Okano *et al.*, 1990; Stayton *et al.*, 1995; Yamato and Okano, 2004). However, the N-OR gels used so far, with chemically crosslinked networks, have several important limitations, such as low volume change and slow deswelling rate, as well as poor mechanical properties. Of these disadvantages,

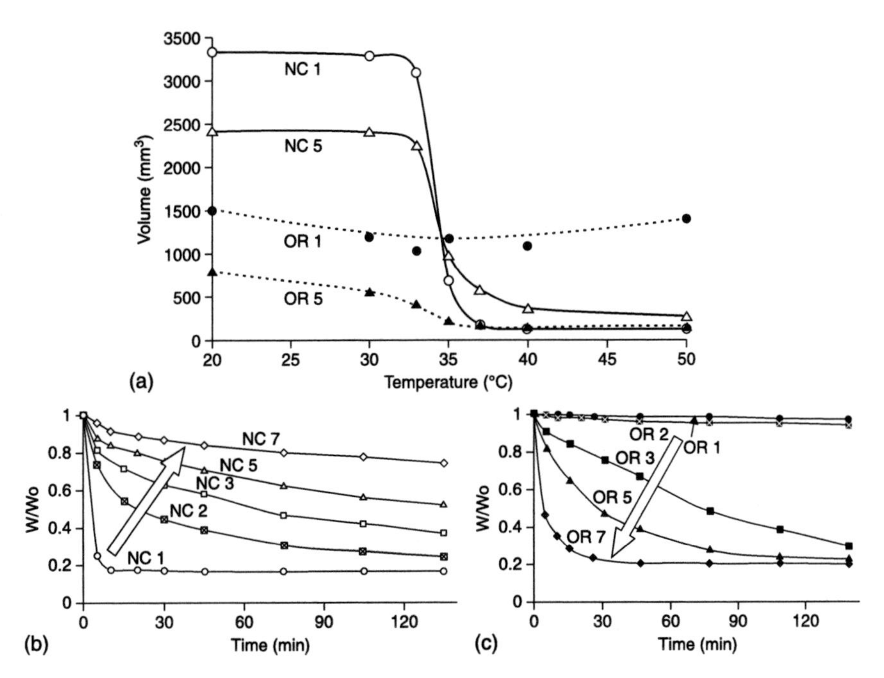


*12.15* (a) Temperature dependence of the gel volume for NC and OR gels with different crosslinker contents. All original gels had the same water/polymer ratio (= 10/1 (w/w)) and the same size (5.5 mm $\phi$ × 30 mm in length). Each gel was first swelled at 20 °C for 48 h and subsequently maintained in a water bath at a specific temperature for 8 h before measurement. (b, c) Deswelling kinetics (time dependency of weight) at 40 °C for (b) NC gels and (c) OR gels with different crosslinker (clay and BIS) contents. All original gels had the same water/polymer ratio (= 10/1 (w/w)) and sample size (5.5 mm $\phi$ × 30 mm in length) (Haraguchi *et al.*, 2002).

the slow rates of deswelling have been studied most extensively. As a result, fast deswelling has been achieved by introducing porosities (Zhang and Zhuo, 2001), structural inhomogeneities (Okajima *et al.*, 2000), or tailored graft structures (Yoshida *et al.*, 1995) into OR gels. However, in most cases, other properties such as mechanical properties, swelling ratio, and optical transparency were not improved and were often made worse.

N-NC gels overcome all these limitations; they show large volume changes and high deswelling rates coupled with excellent mechanical properties (Haraguchi and Takehisa, 2002; Haraguchi *et al.*, 2002). N-NC gels, particularly those with low $C_{clay}$, show outstanding stimulus sensitivity because the PNIPA chains in the network adopt flexible, random conformations, as described in the preceding sections. The temperature dependence of gel volume for NC and OR gels with different crosslinker contents are shown in Fig. 12.15(a). NC gels exhibit larger volume changes than OR gels. For example, NC1–5 gels exhibit about 30- to 50-fold changes in volume, whereas volume changes for OR5 gel are less than 15-fold. Also, OR gels with low BIS contents, e.g., OR1, did not show the apparent swelling/deswelling behavior under the present experimental conditions. This may be due to the very low rate of shrinking of the OR1 gel; OR gels with low BIS contents required far longer to shrink than those used in the current experiments.

The deswelling kinetics for N-NC gels in water at 40 °C (>LCST) and the effect of crosslinker content are shown in Fig. 12.15(b), and those for N-OR gels are shown for comparison in Fig. 12.15(c) (Haraguchi *et al.*, 2002). Since deswelling rates are inversely proportional to the square of typical gel dimensions and the initial samples used have large volumes (712 $mm^3$), the transparent N-OR gels with low $C_{BIS}$, such as N-OR1 and N-OR2, which are most commonly used as PNIPA hydrogels, underwent a low degree of deswelling, i.e., the time taken to shrink was more than a month. In contrast, the N-NC1 gel exhibited a rapid response, and shrinking was almost complete within 10 min. This is because, in the structurally homogeneous N-OR1 gel, the gel initially forms a hydrophobic, collapsed polymer skin on the macroscopic gel surface by eliminating near-surface water, through which subsequent water permeation from the gel interior is very limited. In contrast, in NC gels with low $C_{clay}$, shrinking proceeds rapidly as water is squeezed from the gel interiors through large numbers of water channels formed by micro-phase separation of the flexible PNIPA chains, including grafts. This concept is consistent with the fact that shrinking occurs rapidly when additional grafts are intentionally introduced into the chemically crosslinked networks (Yoshida *et al.*, 1995).

Another interesting point is that N-NC and N-OR gels exhibit entirely opposite tendencies with respect to the effects of crosslinker content ($C_{clay}$ and $C_{BIS}$) on the deswelling rate (Fig. 12.15(b) and Fig. 12.15(c)). This is because, in NC gels, the mobility of PNIPA chains is gradually restricted as $C_{clay}$ increases, while the homogeneous network structure is retained. The deswelling rate, therefore, gradually decreases as $C_{clay}$ increases; this relation differs from that given by Tanaka–Fillmore theory (Tanaka and Fillmore, 1979). In contrast, in N-OR gels, deswelling rates increase with $C_{BIS}$. This is due to the water channels formed in optically turbid (structurally inhomogeneous) N-OR gels with high $C_{BIS}$ as a result of the heterogeneous distribution of crosslinking points. The different characteristics of N-NC and N-OR gels during swelling and deswelling are clearly shown by experiments performed using alternating temperature changes (Haraguchi *et al.*, 2002). While increases in $C_{clay}$ in N-NC gels resulted in decreased volume changes in a fixed time (48 h), increases in $C_{BIS}$ in N-OR gels caused the volume change to increase.

With further increases in $C_{clay}$, the degree of thermosensitive deswelling decreased markedly (Haraguchi and Li, 2005; Haraguchi *et al.*, 2007b). As shown in Fig. 12.16(a), no volume contraction was observed for N-NC gels with $C_{clay}$ greater than 12. Instead, NC gels underwent swelling, even at 50 °C (>LCST), where PNIPA behaved as a hydrophilic polymer with no thermosensitive transition (Haraguchi *et al.*, 2007b). Thus, the volume change of swelling or deswelling N-NC gels is

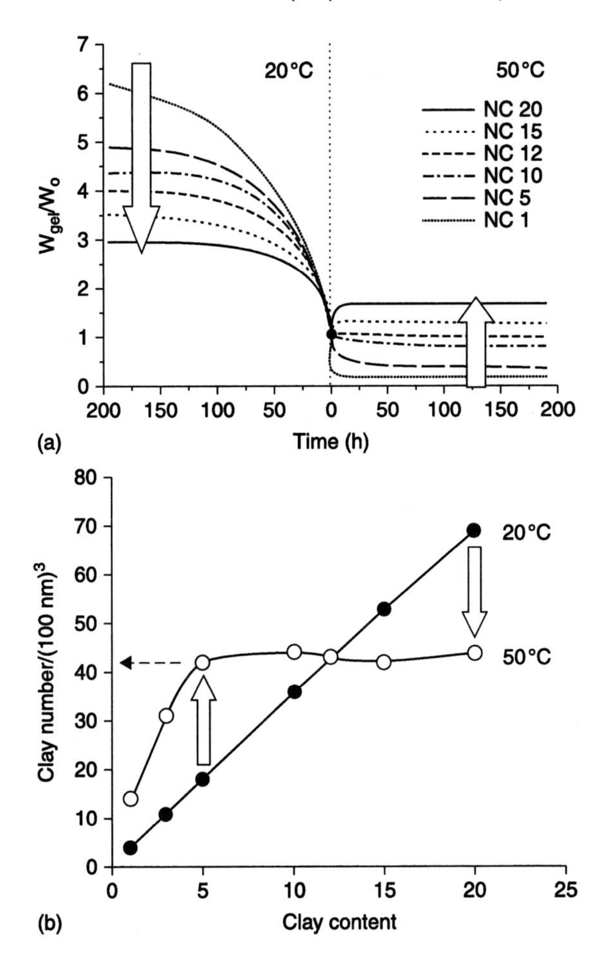


*12.16* (a) Swelling/deswelling behavior of NC gels with different $C_{clay}$, measured in water at 20 and 50 °C. Changes in weight ratio ($W_{gel}/W_o$) with elapsed time are plotted. $W_o$ is the initial weight of the as-prepared gel (initial size: 5.5 mm $\phi$ × 30 mm in length). (b) $C_{clay}$ dependency of the number of clay platelets per unit of gel volume ($10^6$ nm$^3$), calculated for as-prepared NC gels at 20 °C (filled circles) and for corresponding shrunken NC gels at 50 °C (open circles) (Haraguchi *et al.*, 2007b).

controllable through large variations in $C_{clay}$. An analysis of the number of clay platelets per unit gel volume ($n_{clay}$) in as-prepared NC gels and in gels stored at 50 °C showed that $n_{clay}$ at 50 °C was almost the same (≈ 42 per $10^6$ nm$^3$ of gel) for all NC gels with $C_{clay} \geq 5$ (Fig. 12.16(b)); this observation indicates that, during volume contraction, $n_{clay}$ increases to an upper limit of 42 per $10^6$nm$^3$. Also, NC gels with high clay loadings in the as-prepared state, such as NC15 and NC20, tend to expand even at 50 °C, until they reach the same value of $n_{clay}$ (≈ 42) (Haraguchi *et al.*, 2007b).

In order to enhance the swelling capability or introduce pH sensitivity in addition to thermosensitivity, ionic monomers such as acrylic acid (AAc), sodium acrylate (NaAA), methacrylic acid (MAAc), sodium methacrylate (NaMAA), and dimethylaminoethyl methacrylate (DMAEMA) can be introduced into the NC gel network through copolymerization. Several reports have been published on different systems such as NIPA-AAc/Laponite XLG (Song *et al.*, 2008a), NaAA/montmorillonite (Xu *et al.*, 2007), NIPA-MAAc/Laponite XLG+pyrophosphate (Mujumdar and Siegel, 2008), NIPA-AAm-AAc/montmorillonite (Janovak *et al.*, 2009), AAm-NaAA/Laponite XLG (Xiong *et al.*, 2009), and NIPA-NaMAA/Laponite XLS (Hu *et al.*, 2009). Moreover, some studies have been conducted on cationic polymer/clay network systems such as NIPA-DMAEMA/Laponite XLG (Mujumdar and Siegel, 2008), cationic AAm/bentonite (Xu *et al.*, 2008), and AAm-DMAEMA/Laponite XLS (Zhu *et al.*, 2010).

To ensure that NC gels have outstanding swelling/deswelling behavior in response to both temperature and pH, while retaining their remarkable tensile mechanical properties, NC gels were prepared with a semi-interpenetrating organic/inorganic network structure by using linear poly(acrylic acid) (PAAc) (Song *et al.*, 2008a). The PAAc content ($C_{PAAc}$) required to achieve good mechanical properties, as well as good temperature and pH sensitivities, changed with $C_{clay}$, and the upper critical value of $C_{PAAc}/C_{clay}$ was approximately 2.5–3. NC gels with semi-interpenetrating organic/inorganic networks were also prepared using linear sodium carboxymethylcellulose and a PNIPA/clay network (Ma *et al.*, 2007), and temperature and pH sensitivities and good mechanical properties were achieved.

## 12.2.4 New characteristics of NC gels

*Optical anisotropy*

Since polymeric hydrogels consist of amorphous polymer networks swollen with large amounts of water, hydrogels are normally amorphous (optically isotropic). To date, optical anisotropy has only been studied in polymeric hydrogels containing mesogenic groups (Urayama, 2007), or orientated by a flow-gelation process (Yokoyama *et al.*, 1991), or an electric field (Stellwagen and Stellwagen, 1989). In contrast, NC gels exhibit optical anisotropy with increasing $C_{clay}$ and uniaxial deformation, since they contain clay nanoparticles with an anisotropic, disk-like shape and are also capable of large deformations. For example, when $C_{clay}$ exceeds a critical value ($C_{clay}^{crit(opt2)} = 10$) N-NC gels show optical anisotropy in the as-prepared (unstretched) state (Haraguchi and Li, 2006), whereas N-OR gels show no optical anisotropy at any value of $C_{BIS}$. This $C_{clay}^{crit(opt2)}$ is consistent with the critical value calculated for spontaneous clay aggregation (layer stacking) in NC gels (Haraguchi *et al.*, 2007b). It should also be noted that $C_{clay}^{crit(opt2)}$ is consistent with the other critical values of $C_{clay}$, $C_{clay}^{crit(opt(1))}$ (Haraguchi and Li, 2005; Haraguchi *et al.*, 2007b) and $C_{clay}^{crit(2)}$ (Haraguchi and Li, 2006), described in the previous sections.

The optical anisotropy of an NC gel changes in a unique manner on uniaxial deformation, regardless of its optical characteristics in the original (as-prepared) state; that is, when stretched uniaxially, all N-NC gels show remarkable optical anisotropy, as can be seen in Fig. 12.17 (Murata and Haraguchi, 2007). Their

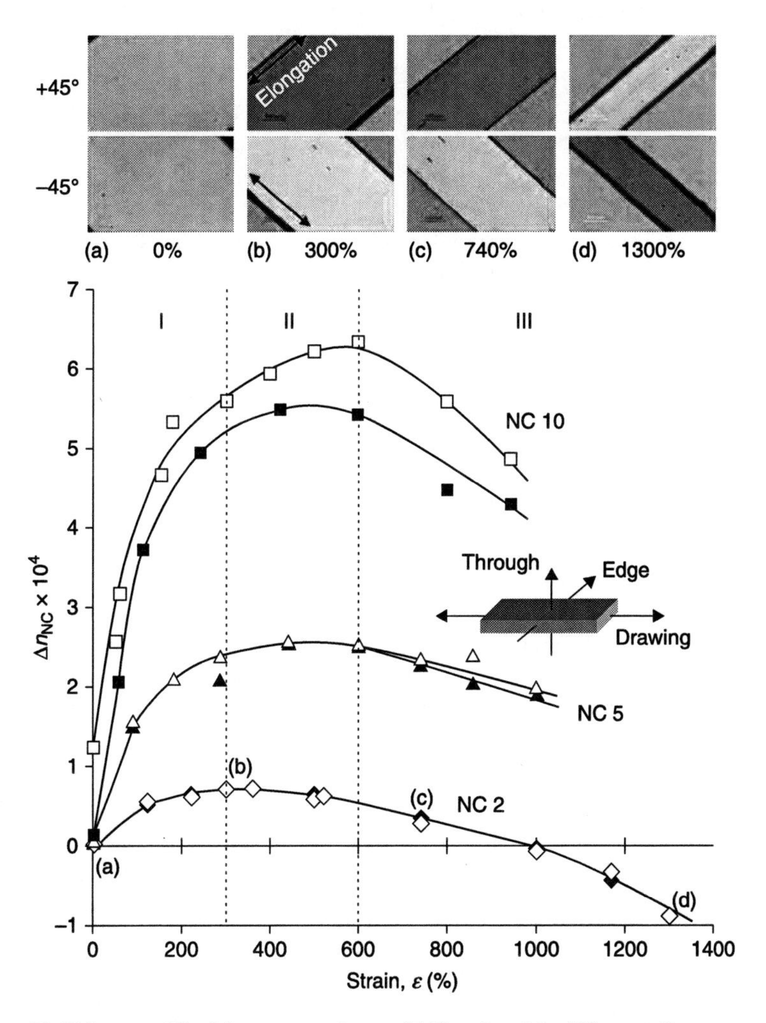


*12.17* Lower: Birefringences, $\Delta n_{NC}$, of NC gels with different $C_{clay}$ (NC2–NC10) as functions of strain. Closed and open symbols are for through and edge directions (see inset), respectively. Upper: Photo images (a–d) show polarized-light micrographs for stretched NC2 gels (through direction) under crossed polarizers in conjunction with a 530-nm retardation plate, where +45° (–45°) orientation is parallel to the slow (fast) axis of the retardation plate. Each photo (a–d) corresponds to the same point on the $\Delta n_{NC}$–strain curve of NC2 gel (Murata and Haraguchi, 2007).

birefringence ($\Delta n_{NC}$) strongly depends on $C_{clay}$. An interesting point is that $\Delta n_{NC}$ shows a distinct maximum at a strain of around 300–600% and a sign inversion on further elongation. An NC2 gel stretched to its intersection point with the strain axis is optically isotropic (990%, $\Delta n_{NC} = 0$), although it has definitely a highly orientated network structure. By evaluating the separate contributions of clay and PNIPA, assuming that $\Delta n_{NC} = \Delta n_{PNIPA} + \Delta n_{clay}$, we concluded that $\Delta n_{clay}$ increases rapidly in the early stages of elongation and saturates at 300–600% strain. On the other hand, $\Delta n_{PNIPA}$ changes monotonically; it becomes more negative when NC gels are stretched. As a result, elongation of the chains decreases the net birefringence and could nullify the contribution from the clay (positive $\Delta n_{clay}$), and eventually reverse the sign of $\Delta n_{NC}$. Also, values of $\Delta n_{NC}$ for NC gels with high $C_{clay}$ ($\geq C_{clay}^{crit(opt2)}$) were different in different directions (Fig. 12.17); this was attributed to the partial plane orientation of the clay platelets (Murata and Haraguchi, 2007). The orientations of clay and PNIPA during stretching were confirmed by contrast-variation SANS measurements (Nishida *et al.*, 2009).

*Sliding friction behavior*

In general, since OR gels are readily damaged by sliding, sliding friction measurements have rarely been conducted at the surfaces of OR gel films, except when the films were wet. In contrast, NC gels are mechanically tough, and hence sliding friction tests have been performed at their surfaces under different environmental conditions and even under high loading. Sliding frictional forces at NC-gel surfaces are sensitive to gel composition, loading, and surrounding environment (wet, in-air, and temperature) (Fig. 12.18) (Haraguchi and Takada, 2005). In air, NC gels show a force profile with a maximum static frictional force (max-SFF) and a subsequent constant dynamic frictional force. In contrast, NC gels show very low frictional forces when wet. The change in sliding friction behavior with the surroundings is most prominent for NC gels with low $C_{clay}$. For example, max-SFF for N-NC1 gel decreases by a factor greater than $10^2$ when its environment is changed from air to wet. Also, under wet conditions, the dynamic frictional coefficient, $\mu_d$, decreases with increasing load and becomes very small at high loads ($\mu_d < 0.01$). Thus, the frictional characteristics of N-NC1 gels can alternate between sticky and slippery depending on their surroundings. The sliding frictional forces for N-NC gels in air also decrease when they are heated to 50 °C (>LCST), because of the coil-to-globule transition of PNIPA chains (Haraguchi and Takada, 2010). The results show the important role played by dangling chains at the gel–air and gel–water interfaces.

*High contact angles with water*

Surface wettability is one of the most important properties of all materials, since it reflects the real structure and chemical composition at the outermost surface. As

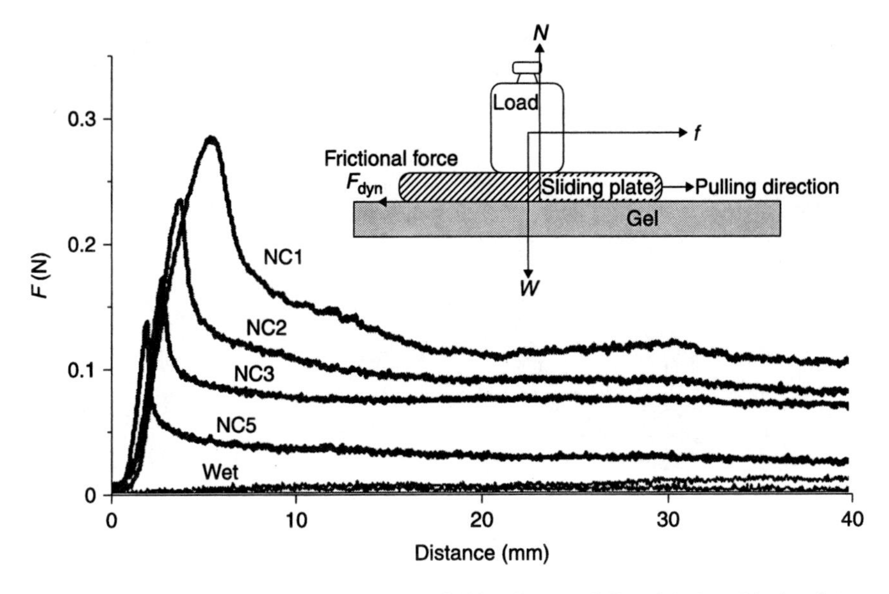


*12.18* Effects of clay content of NC gels on sliding frictional behavior. The force profiles were measured in air and wet for N-NC gels with different clay contents (NC1–NC5) (Haraguchi and Takada, 2005).

readily expected from their compositions, polymeric hydrogels, which consist largely of water and a hydrophilic polymer network, are naturally hydrophilic, and their surfaces generally show very low contact angles for water ($\theta_w$). The relation between the surface wettability for PNIPA hydrogels (N-OR gels) and surface-grafted PNIPA and the hydrophilic (coil)-to-hydrophobic (globule) transition occurring at the LCST has been studied extensively (Zhang *et al.*, 1995; Teare *et al.*, 2005; Sun *et al.*, 2004). It has been reported that $\theta_w$ at the N-OR gel surface is low (e.g., ~60°) below the LCST but relatively high (e.g., ~80°) above the LCST. In contrast, N-NC gel surfaces showed extraordinarily high hydrophobicity (high $\theta_w$) below the LCST (Fig. 12.19) (Haraguchi *et al.*, 2007a), although all individual components of the N-NC gels were hydrophilic under the test conditions. Values of $\theta_w$ for N-NC gels were generally greater than 100° for a broad range of $C_{clay}$ and $R_{H2O}$ and reached a maximum of 151° at a specific composition (Haraguchi *et al.*, 2007a, 2008). These results were astonishing since values of $\theta_w$ for N-NC gels were higher than those of polypropylene and poly(tetrafluoroethylene). The high hydrophobicity of N-NC gels was primarily attributed to the amphiphilicity of PNIPA and, more specifically, to the spontaneous alignment of *N*-isopropyl groups at the gel–air interface (Haraguchi *et al.*, 2008). Hydrophobicity was also enhanced by other factors such as the network structure and water content, whereas the effect of surface roughness was negligible. It was also found that the surfaces of N-NC gels underwent reversible hydrophobic-to-hydrophilic changes when the environment was changed from air to wet, and vice versa (Haraguchi *et al.*, 2008).

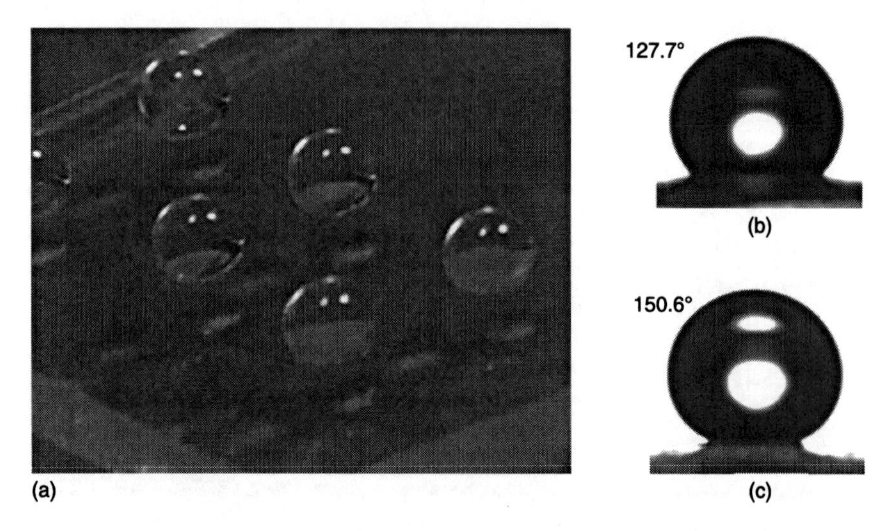


*12.19* (a) Water droplets on N-NC6 gel film. (b), (c) Water contact angles for sessile drops on the surfaces of N-NC6 gels with water contents of (b) 630 wt% and (c) 210 wt% (Haraguchi *et al.*, 2007a).

*Cell cultivation*

Culturing cells on a substrate is one of the most important and indispensable experimental procedures in medical, biological, pharmaceutical, and tissue engineering research. N-OR gels (Takizawa *et al.*, 1990) and PNIPA coatings of thickness greater than 30 nm (Akiyama *et al.*, 2004) cannot be used as substrates. However, it was found that cells can be cultured on the surfaces of normal or dried N-NC gels, regardless of the gel thickness (Fig. 12.20(a) and Fig. 12.20(b)) (Haraguchi *et al.*, 2006c). To the best of our knowledge, this is the first report of the successful culture of cells on a PNIPA hydrogel. Various types of cells, such as human hepatoma (HepG2) cells, normal human dermal fibroblasts (NHDFs), and normal human umbilical vein endothelial (HUVEC) cells, were cultured to be confluent on the surface of an N-NC6 gel. The development of the cell cultures showed little dependence on the water content or thicknesses of the N-NC gel sheets. In contrast, cell cultures did not develop on the surfaces of N-OR gels, regardless of the crosslinker content. Cell culture on the surfaces of D-NC and D-OR gels also failed, probably because of the hydrophilic nature of the gel surfaces. Hence, it was thought that cells adhered to and proliferated only on the surfaces of N-NC gels because of the combined effects of the hydrophobicity of the dehydrated PNIPA chains and the surface charges on the incorporated exfoliated clay. Furthermore, it was found that, when the temperature was decreased to below the LCST (10–20°C), the cell cultures detached from the

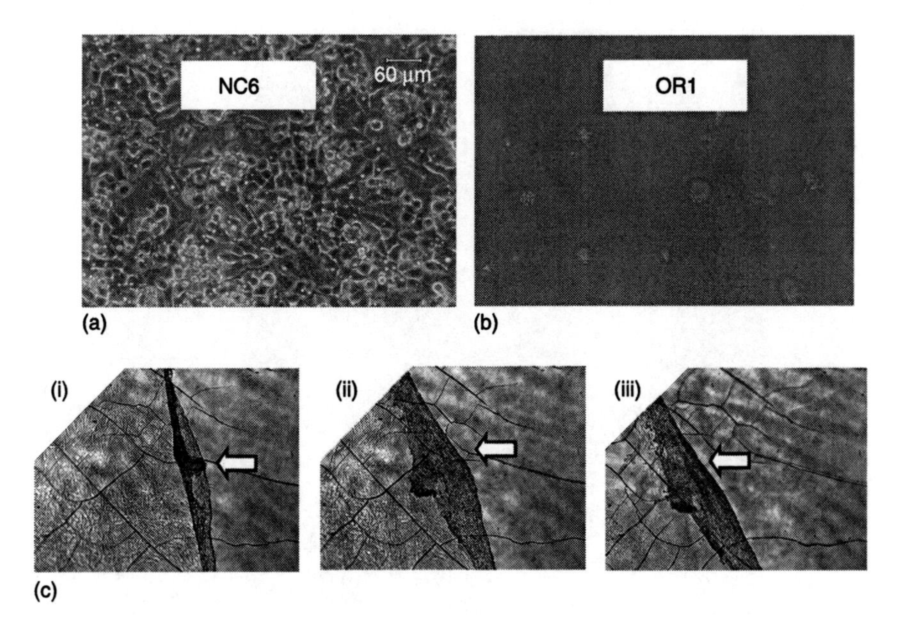


*12.20* (a, b) Phase-contrast photomicrographs of HepG2 cultures on the surfaces of (a) N-NC6 gel and (b) N-OR1 gel after 5 days. (c) Cell sheet detachment of NHDF by decreasing the temperature to 10–20 °C. (i)–(iii) Change in cell sheet detachment from dried N-NC6 gels (Haraguchi *et al.*, 2006c).

surfaces of the N-NC gels without trypsin treatment (Haraguchi *et al.*, 2006c). Confluent layers of HepG2 and NHDF cells could be spontaneously separated as sheets from the gel surface (Fig. 12.20(c)). Recently, an acceleration of cell sheet detachment has been reported for semi-interpenetrating NC gels prepared by incorporating alginate (a water-soluble natural polysaccharide) into PNIPA/clay networks (Wang *et al.*, 2011); this acceleration is possibly due to the rough surface texture and faster water penetration.

*Porous NCs with a layered morphology*

Novel, porous nanocomposites (porous NCs) with characteristic layered morphologies were prepared by freeze-drying NC gels without the use of a porogen (Haraguchi and Matsuda, 2005). The most typical morphology was a concentric three-layer morphology consisting of, successively, a fine-porous layer, a dense layer, and a coarse-porous layer from the exterior to the interior (Fig. 12.21). In the coarse layer, a regular assembly of polyhedral pores had formed spontaneously. A possible mechanism underlying the spontaneous

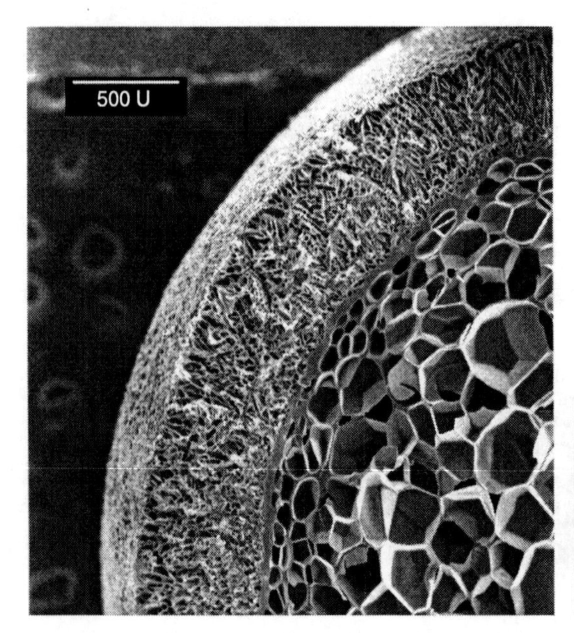


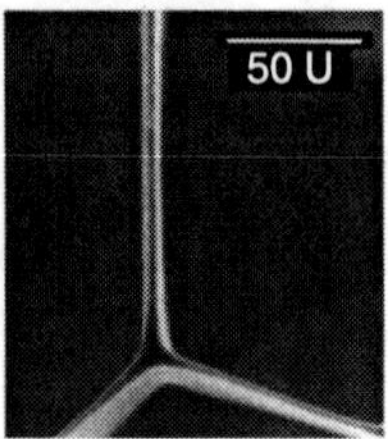


*12.21* Scanning electron micrographs of the three-layer morphology for a cross section of porous NC (freeze-dried NC gel in liquid nitrogen). Fine-porous, dense, and coarse-porous layers were observed from the outer surface to the interior. Right: Junction between adjacent polyhedral pores in the coarse-porous layer. The bars indicates 500 and 50 μm, respectively (Haraguchi and Matsuda, 2005).

formation of the unique three-layer morphology, and other morphologies, during the freeze-drying process was proposed.

*Reversible force generation*

Using temperature-sensitive N-NC gels, we discovered reversible retractive tensile forces in samples constrained to constant length in an aqueous environment in response to alternating temperature changes (Fig. 12.22) (Haraguchi *et al.*, 2005b). This was the first observation of a retractive mechanical force generated as a result of conformational changes (the coil-to-globule transition) in PNIPA chains across the LCST. In contrast, no such force was generated for D-NC and N-OR gels under the same experimental conditions.

*Complicated shapes and surface patterns*

In addition to various shapes, NC-gel films with a wide range of thicknesses ($10^{-3}$–$10^{3}$ mm) and sizes can be prepared (Fig. 12.23). Also, NC gels with uneven

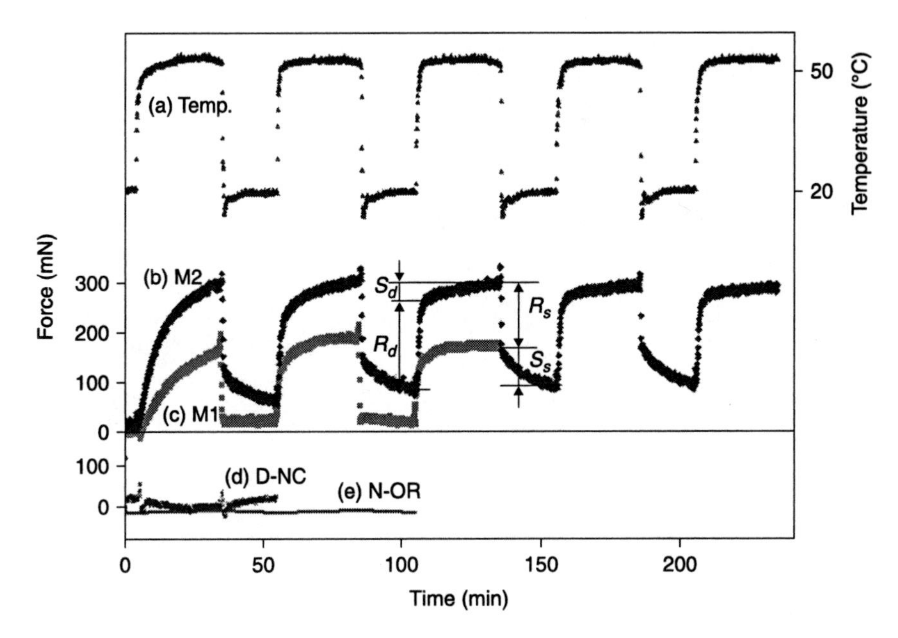


*12.22* Force profiles on alternating the temperature. (a) Temperature change. (b) Force profile for a N-NC4-M2 gel. (c) Force profile for a N-NC4-M1 gel. (d) Force profile for a D-NC4-M2 gel. (e) Force profile for a N-OR2-M2 gel (Haraguchi *et al.*, 2005b).

surfaces, for example, surfaces with a regular array of pillars (Fig. 12.23(d)) or bellows-like rods (Fig. 12.23(e)), can be obtained (Haraguchi, 2007a). Bellows-like rods can be prepared by simply extracting the as-prepared NC gel from the mold without breaking or opening the template; this is because the protuberances on the surface are flexible and can be reversibly deformed during extraction. Further, using direct replica molding, we successfully formed micrometer-scale surface patterns on NC gels (Fig. 12.23(f)) (Song *et al.*, 2008b). The sizes of the patterns could be modified (enlarged or miniaturized) by subsequent swelling, deswelling, and drying. In addition, lightweight NC gels, i.e., porous NC gels, which are very soft (very low modulus) and have a density of 0.05–1.0 g/cm$^3$, were prepared using a porogen or air (Fig. 12.23(g)) (Haraguchi, 2007a). OR gels with the aforementioned shapes and forms are difficult to prepare and handle because of their brittleness.

Photo-NC gels can also be prepared in various forms, including thin films and as nano-coatings on a substrate. The latter exhibit additional characteristics, and have been used in pattern formation using photolithography, cell harvesting, and the fabrication of a microchannel flow system containing a thermosensitive valve made of photo-NC gels (Haraguchi and Takada, 2010).

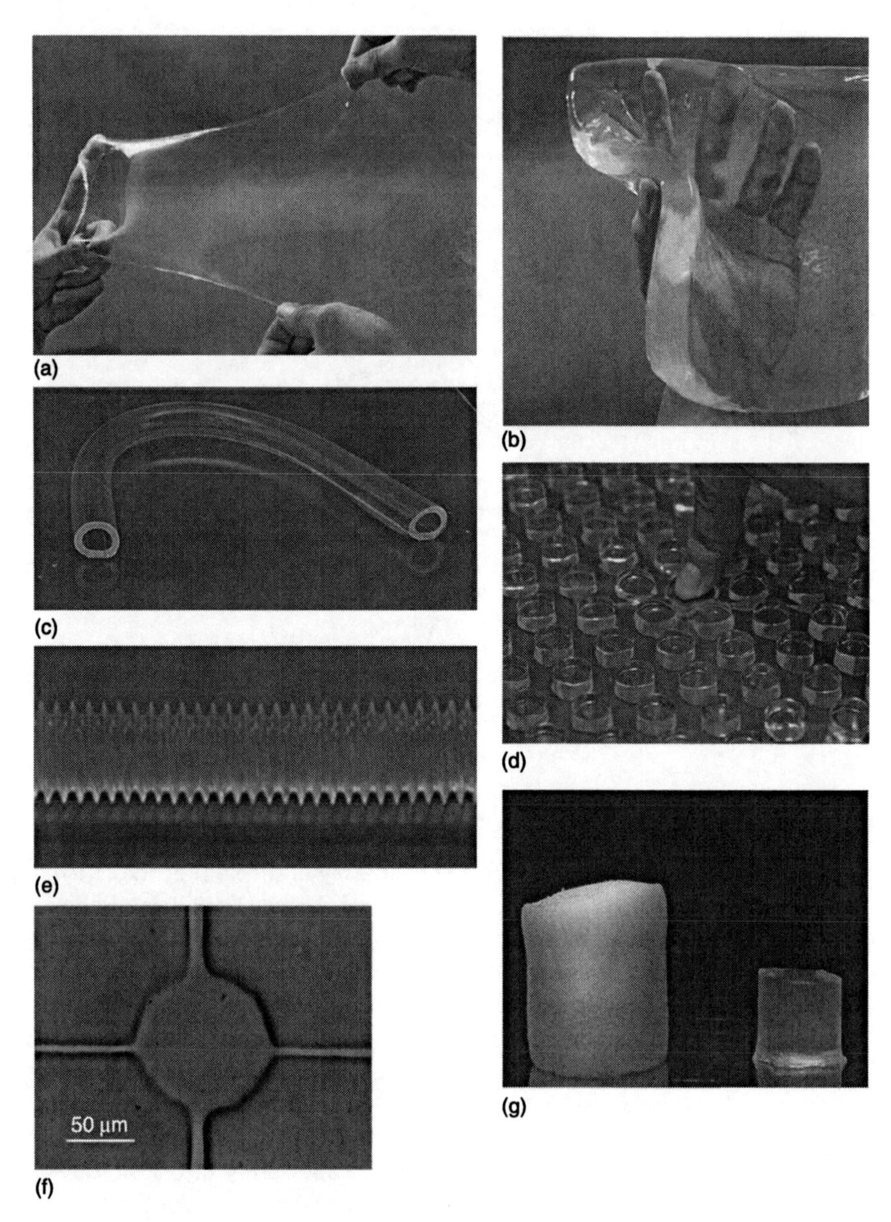


*12.23* NC gels with various shapes and forms. (a) Thin film. (b) Sheet. (c) Hollow tube. (d) Sheet with uneven surface. (e) Bellows. (f) Confocal optical microscopy image of the surface pattern of an NC-gel film prepared by direct replica molding using a quartz template with an array of 3-μm-dot holes at 15-μm intervals. (g) Porous NC gel (Haraguchi, 2007a, 2007b; Song *et al.*, 2008b).

## 12.3 Soft polymer nanocomposites (M-NCs)

As described in the introduction, P-NCs developed so far as advanced composites are mainly obtained by the dispersion of small amounts of inorganic nanoparticles in a polymer matrix (Giannelis, 1996; Okada and Usuki, 2006; Usuki *et al.*, 1993). There are a few reports of P-NCs with high inorganic content, such as nylon66/silica (or clay) P-NCs (Idemura and Haraguchi, 2000), and epoxy resin/silica P-NCs (Goda and Higashino, 2003). However, these P-NCs still have certain limitations, particularly in terms of optical transparency, tractability, and processability.

We developed transparent, soft P-NCs with high contents of inorganic clay by synthesizing solid P-NCs. These novel P-NCs (abbreviated as M-NCs) (Haraguchi *et al.*, 2006a, 2006b), which consist of hydrophobic poly(2-methoxyethyl acrylate) (PMEA) and a hydrophilic inorganic clay (hectorite), have superior optical and mechanical properties, despite their high $C_{clay}$, because of the unique organic/inorganic network structure of aggregated clay and PMEA. PMEA is basically a hydrophobic polymer with a low $T_g$ (−34 °C) (Tanaka *et al.*, 2000), and is considered to have promising applications in medical devices such as in a cardiopulmonary bypass (Saito *et al.*, 2000; Baykut *et al.*, 2001; Tanaka *et al.*, 2002), although in practical applications PMEA has only been used as an ingredient of copolymers or as a thin coating because of its mechanical weakness and intractability.

### 12.3.1 Synthesis of M-NCs

A new type of soft P-NC (M-NC) was synthesized using a modified version of the preparation method for NC gels, i.e., by *in situ* free-radical polymerization of water-soluble 2-methoxyethyl acrylate (MEA) in the presence of exfoliated clay and subsequent drying. The essential point of the synthesis is that hydrophilic (monomer)-to-hydrophobic (polymer) transitions occur during *in situ* polymerization along with micro- and macro-phase separations. The initial transparent solution turned opaque (white) in the early stages of polymerization (within a few minutes) as a result of microscopic-phase separation, which was caused by exclusion of the unaltered hydrophilic clay from the hydrophobic PMEA chains. Subsequently, the system underwent macroscopic-phase separation (after a few tens of minutes), which was accompanied by syneresis and volume contraction. Consequently, a uniform, white gel consisting of nanostructured PMEA/clay with 400 wt% water (relative to the solid component) was obtained. In the subsequent drying process, the white gel first shrank, releasing a large amount of water, and finally turned into a transparent soft solid (M-NC). The resulting M-NCs were always uniform, colorless, and transparent (<90% transmittance), regardless of $C_{clay}$, as shown in Fig. 12.24(a) (M-NC23) (Haraguchi *et al.*, 2006a). The water uptake on immersion in water was only 0.5–15 wt%. This is totally different from NC gels (Haraguchi and Takehisa, 2002; Haraguchi and Li, 2005; Haraguchi *et al.*, 2003), in which dried NC gels can revert to highly

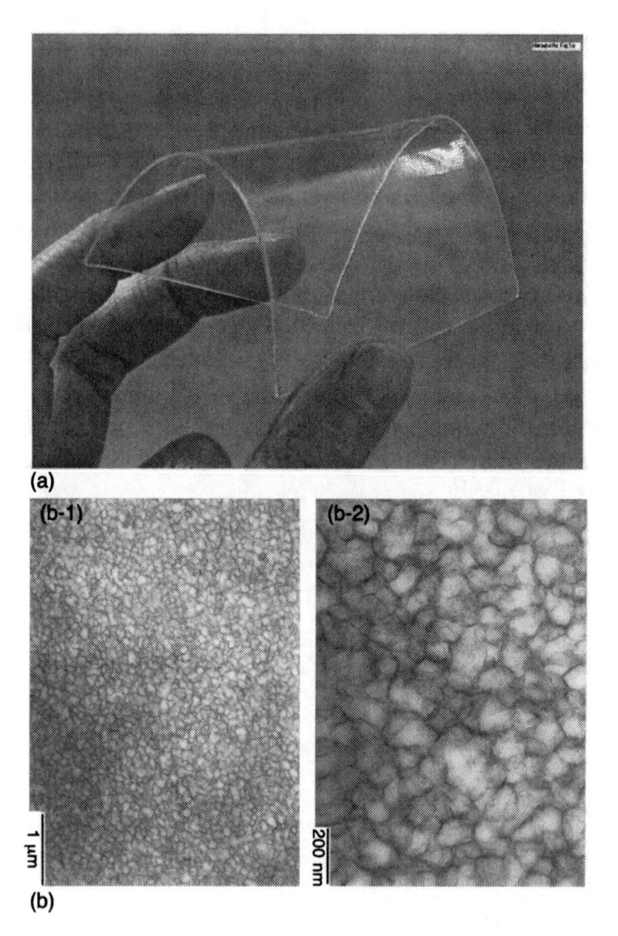


*12.24* (a) Transparent, soft M-NC23 film, which consists of 23 wt% inorganic clay. (b) Transmission electron micrographs for M-NC11. (b-1) and (b-2) are at different magnifications and the scale bars indicate 1 μm and 200 nm, respectively (Haraguchi *et al.*, 2006a).

swollen hydrogels. Also, M-NCs did not dissolve even in good solvents for PMEA, but could swell uniformly and extensively ($W_{solvent}/W_{dry}$ = 2–25). These results strongly indicate that some kind of stable, three-dimensional network is formed in M-NCs, although no organic crosslinking agent was used in their synthesis. The sample code for M-NCs is based on the clay content ($C_{clay}$: *n* wt%), as M-NC*n*. $C_{clay}$ can be varied over a wide range (1–50 wt% or higher) by altering the composition of the reaction solution.

## 12.3.2 Mechanical properties of M-NCs

M-NCs have extraordinary mechanical properties, with two striking aspects: (1) an extremely large $\varepsilon_b$, as high as 1000–3000%, with good recovery on release; and

(2) a well-defined yielding behavior in the early stages of elongation, as shown in Fig. 12.25 (Haraguchi *et al.*, 2006a). The high extensibility of M-NCs is accompanied by a distinctive yielding behavior: a yield point ($I_b$) at maximum stress, a necking region (II) at constant stress, and a strain-hardening region (III) from the end of the necking region ($II_b$). This is evident from the stress–strain curve (Fig. 12.25) and direct observations (the inset to Fig. 12.25). This characteristic yielding behavior is observed for all M-NCs with different $C_{clay}$, although it is more pronounced in M-NCs with $C_{clay}$ greater than 10. On the other hand, M-ORs with the same composition as M-NCs, except for the crosslinker used, were brittle and ruptured at very low elongation (M-OR3 in Fig. 12.25). The fracture energy of M-NC23 could be 200 times greater than that of M-OR3.

In M-NCs, more than 90% of the total elongation is recovered. Once elongated beyond point $II_b$, an M-NC exhibits simple stress–strain behavior, i.e., it undergoes large deformation similar to the original (as-prepared) M-NC, although no well-defined necking phenomenon is observed (Fig. 12.25 (dotted line) for M-NC11). Thus, the large deformation associated with necking becomes reversible on repeated cycling, but the necking phenomenon (i.e., the formation of a neck) itself is irreversible. This is the first report of necking behavior in P-NCs. In extensive studies of polymeric materials such as crystalline polymers and high-impact

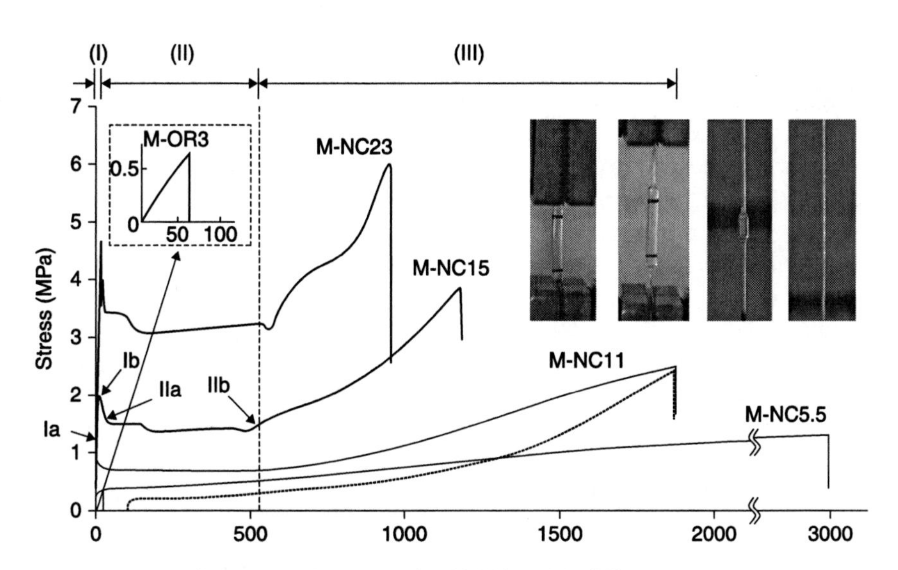


*12.25* Stress–strain curves for M-NCs with different clay contents (M-NC5.5–23) and chemically crosslinked M-OR3. A stress–strain curve for the second cycle of M-NC11 after the first elongation up to 1800% is also shown as a dotted line. Inset: Actual example of yielding behavior observed for M-NC11 showing two necking points, which start at both ends where a high stress concentration initially occurred at the grips (Haraguchi *et al.*, 2006a).

polymer blends, the yielding (necking) behavior and associated large deformation were generally observed as a type of cold drawing, which was irreversible.

### 12.3.3 Network structures in M-NCs

In order to explain the origins of the dramatic changes in the mechanical properties of M-NCs, while retaining high transparency and low water absorption, we proposed the following nanostructure for M-NCs on the basis of analytical data, such as DSC ($T_g$ is the same as for free, uncrosslinked PMEA), XRD (evidence for stacking of clay–polymer–clay (2.7 nm) and clay–clay (1.1 nm) aggregates), FTIR (PMEA/clay interactions), and TEM (Haraguchi *et al.*, 2006a). Since the transmission electron micrographs (Fig. 12.24(b) showing M-NC11) are the same regardless of cutting direction, it was concluded that M-NC11 consists of clay networks, with a large number of connected clay spheres (100–300 nm in diameter), and a 20-nm-thick outer clay shell with the bulk of the flexible PMEA packed inside.

The clay network structure proposed is depicted graphically in Fig. 12.26(a). The following mechanism for its formation is suggested. As the hydrophobic PMEA chains tend to aggregate during the course of *in situ* polymerization, the primary clay platelets are squeezed out of these regions and form clay aggregates in which PMEA chains are attached to their surfaces. These clay aggregates and the bulk of the PMEA form nanometer-sized spherical structures consisting of an outer shell of aggregated clay (Fig. 12.26(a)(iii)) and an inner PMEA core (Fig. 12.26(a)(ii)). Thus, a nanostructured morphology latently forms in the pre-formed gel state; this nanostructure is apparent when water is removed. Since each sphere is connected to neighboring spheres, a three-dimensional clay network forms (Fig. 12.26(a)(i)). The sphere diameter varied with $C_{clay}$ and naturally decreased with an increase in $C_{clay}$. When $C_{clay}$ was very low, a ribbon-like structure was formed instead of a spherical structure, as expected. Despite the formation of such structures, M-NCs are transparent because the clay-shell thickness is very small and the refractive indices of PMEA (1.47) and clay (1.50) are similar.

The characteristic tensile deformation of M-NCs (Fig. 12.25 stages I–III) is also explicable in terms of the proposed model for the clay network, as shown in Fig. 12.26(b). The extraordinarily high elongations (1000–3000%) observed in M-NCs are attributable to two main factors. One is the highly contracted form of the randomly coiled PMEA chains, which is the result of syneresis during polymerization and shrinkage during the drying process. In the case of deswollen polymer networks, it has been reported that extensibility may increase considerably, reaching $\lambda_{max}$ = 18–30 (Urayama and Kohjiya, 1998) because of its supercoiled structure. The results for M-NCs are consistent with those obtained for supercoiled polymer chains. The other factor is the formation of a crosslinked structure between PMEA and the clay. If there are no crosslinks to bear the applied stress, the contracted polymer chains can simply be drawn (at very low stress) by

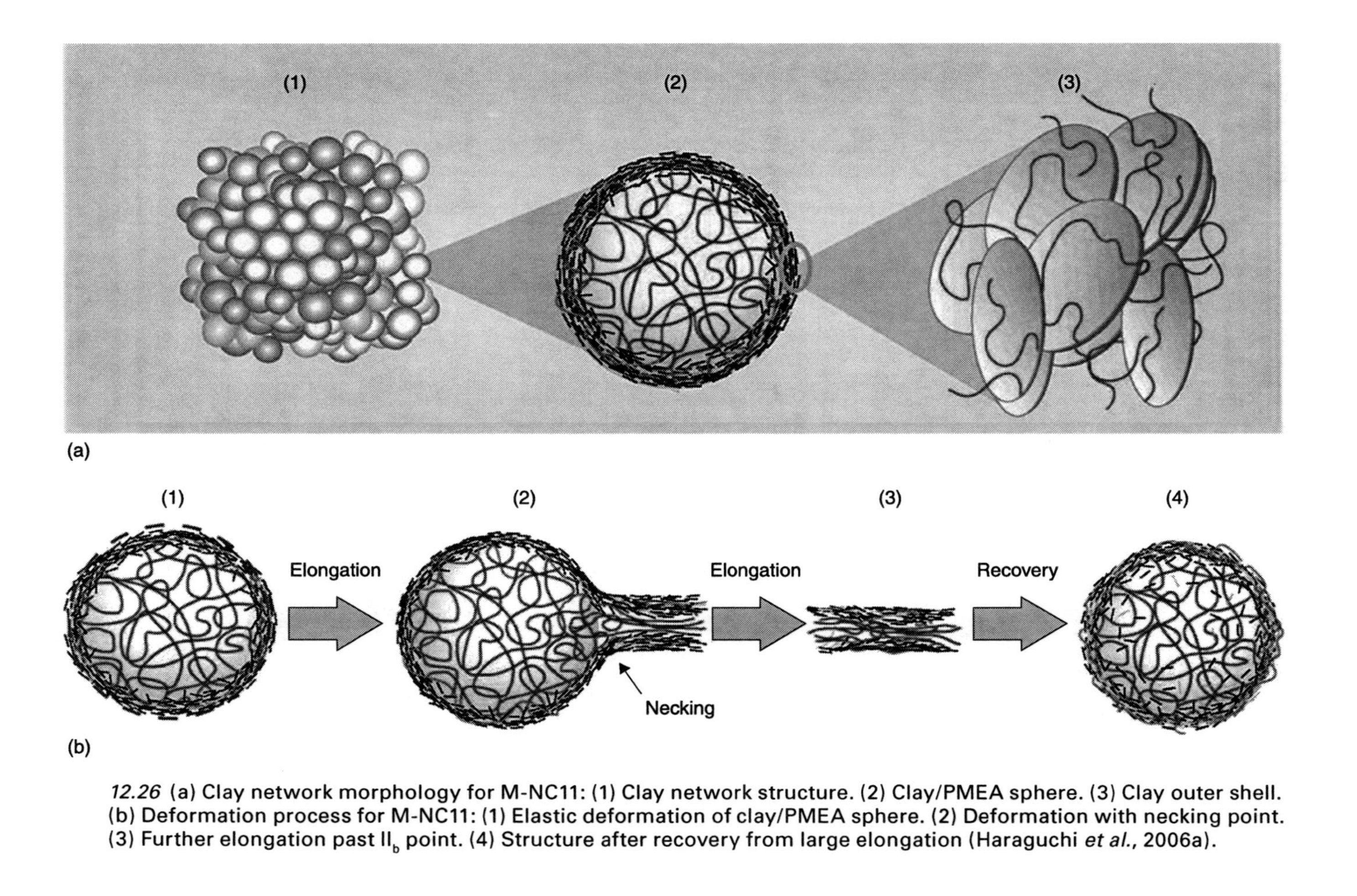


*12.26* (a) Clay network morphology for M-NC11: (1) Clay network structure. (2) Clay/PMEA sphere. (3) Clay outer shell. (b) Deformation process for M-NC11: (1) Elastic deformation of clay/PMEA sphere. (2) Deformation with necking point. (3) Further elongation past $II_b$ point. (4) Structure after recovery from large elongation (Haraguchi *et al.*, 2006a).

slipping. This has been observed for M-LR. In the case of M-NCs, the supercoiled PMEA chains, whose ends may interact with the exfoliated clay, undergo extensive elongation and may retract.

## 12.4 Applications and future trends

### 12.4.1 NC gels

NC gels or chemically modified NC gels can be utilized in various fields:

- Biomedical devices, such as wound dressings for various types of trauma, three-dimensional scaffolds for tissue engineering, injectable or processable hydrogels, blood-compatible materials in various forms such as tubes, sheets, and coatings, mechanically tough, transparent and wet materials (hydrogels) for protection and absorption (water, saline solutions, and exudates), and hydrogels for implantation in living tissue without specific interaction (no inflammation). NC gels can be sterilized in an autoclave (Haraguchi and Takehisa, 2005). Also, the safety of NC gels has been confirmed in a biological test comprising a sensitization test, an irritation test, an intracutaneous test, and an *in vitro* cytotoxicity test.
- Drug-delivery devices, specifically for hydrophilic medicines, and as an immobilizer for bioactive substances.
- Cell-cultivation substrates for various cells, which can be harvested without enzymatic treatment.
- Transparent and wet cosmetic devices, such as sheets, thin films, and colloids, with the capability of large absorption, adhesion to skin with peeling-off for facial masks, drug release, and moisturizing.
- Industrial, building, and agricultural uses: they can be made into soft, transparent, wet, and rubbery materials with various forms and characteristics. Properties include: high heat conductivity and heat capacity; flame resistance; lightweight porosity (e.g., 0.05–1.0 g/cm$^3$, see Fig. 12.23(g)), and high softness (low modulus). Applications include: in actuators, sensors, optical shutters, etc. because of their stimulus-sensitivity; anti-fogging coatings; sealing materials, super absorbents, and for selective absorption of cationic solutes due to their high absorbency; vibration and shock absorbers; electrolyte gels; adhesives; and low friction applications.

### 12.4.2 M-NCs

M-NC gels have various uses:

- Biomedical devices: M-NCs can be utilized in biomedical devices; PMEA has been used to date, such as in components for cardiopulmonary bypass devices.
- Cell-cultivation substrates: Because M-NCs have thermoresponsive cell adhesion and detachment, which is useful for living-cell harvesting systems

(Haraguchi *et al.*, 2010b), they could be used as cell substrates. Also, together with high mechanical properties, such as large, reversible elongation and high strength, M-NCs could be used as transparent, cell-culture substrata (films) with reversible elasticity.
- Cosmetic and industrial materials: They can be made into optically transparent, rubbery films with adhesion to various surfaces including human skin.

### 12.4.3 Future trends

It should be emphasized that NC gels are one of the most environmentally friendly materials. Although NC gels have various characteristics similar to conventional, soft organic materials, such as high mechanical toughness and transparency, in addition to many other functions described above, the primary component of NC gels is water, while 5–65 wt% of the solid is inorganic clay, which can be returned to the soil without harmful effects. Thus, only 5–10 wt% of an NC gel is a polymeric component derived from petroleum. From a resources viewpoint, NC gels are the first rubbery materials in which water is used as a primary raw material. The situation is also the same from the viewpoint of waste, as only 5–10 wt% of NC gels become organic waste. Also, NC gels are environmentally friendly materials, since they are virtually incombustible, even though they have no harmful halogen or phosphate moiety; they show excellent flame retardancy because their primary component is water. Dried NC gels are also expected to show fire retardancy, similarly to polyamide/clay nanocomposites (Gilman *et al.*, 2000), because of their high clay content. M-NCs are also composed of large amounts of inorganic clay (10–60 wt% or more), which is a non-harmful and non-petroleum-based material. Thus, NC gels and M-NCs are promising soft polymeric materials, which will be utilized as advanced (stimulus-responsive) and environmentally friendly materials in various fields. In the future, NC gels will be prepared in different forms, such as microspheres, and fabricated using different kinds of polymer and inorganic particles, which will result in the extension of their characteristics and applications.

## 12.5 Sources of further information and advice

Further information on NC gels is obtainable from http://www.kicr.or.jp/intro/material/index.html.

## 12.6 References

Akiyama Y, Kikuchi A, Yamato M and Okano T (2004), 'Ultrathin poly(*N*-isopropylacrylamide) grafted layer on polystyrene surfaces for cell adhesion/detachment control', *Langmuir*, 20, 5506–5511.

Baker JP, Hong LH, Blanch HW and Prausnitz JM (1994), 'Effect of initial total monomer concentration on the swelling behavior of cationic acrylamide-based hydrogels', *Macromolecules*, 27, 1446–1454.

Baykut D, Bernet F, Wehrle J, Weichelt K, Schwartz P and Zerkowski HR (2001), 'New surface biopolymers for oxygenators: an *in vitro* hemocompatibility test of poly(2-methoxyethylacrylate)', *Eur J Med Res*, 6, 297–305.

Cai W, Anderson EC and Gupta RB (2001), 'Separation of lignin from aqueous mixtures by ionic and nonionic temperature-sensitive hydrogels', *Ind Eng Chem Res*, 40, 2283–2288.

Chen T, Cao Z, Guo X, Nie J, Xu J, *et al.* (2011), 'Preparation and characterization of thermosensitive organic-inorganic hybrid microgels with functional Fe3O4 nanoparticles as crosslinker', *Polymer*, 52, 172–179.

Dijkstra M, Hansen JP and Madden PA (1995), 'Gelation of a clay colloid suspension', *Phys Rev Lett*, 75, 2236–2239.

Flory PJ and Rehner J Jr (1943), 'Statistical mechanics of cross-linked polymer networks II. Swelling', *J Chem Phys*, 11, 521–526.

Fukasawa M, Sakai T, Chung UI and Haraguchi K (2010), 'Synthesis and mechanical properties of a nanocomposite gel consisting of a tetra-peg/clay network', *Macromolecules*, 43, 4370–4378.

Giannelis EP (1996), 'Polymer layered silicate nanocomposites', *Adv Mater*, 8, 29–35.

Gilman JW, Jackson CL, Morgan AB, Harris R, Manias JE, *et al.* (2000), 'Flammability properties of polymer-layered-silicate nanocomposites. Polypropylene and polystyrene nanocomposites', *Chem Mater*, 12, 1866–1873.

Goda H and Higashino T (2003), 'Epoxy resin composition and process for producing silane-modified epoxy resin', USP6525160.

Gong JP, Katsuyama Y, Kurokawa T and Osada Y (2003), 'Double-network hydrogels with extremely high mechanical strength', *Adv Mater*, 15, 1155–1158.

Haraguchi K (2007a), 'Nanocomposite gels: New advanced functional soft materials', *Macromol Symp*, 256, 120–130.

Haraguchi K (2007b), 'Nanocomposite hydrogels', *Curr Opin Solid State Mat Sci*, 11, 47–54.

Haraguchi K (2011), 'Stimuli-responsive nanocomposite gels', *Colloid Polym Sci*, 289, 455–473.

Haraguchi K and Li HJ (2004), 'Mechanical properties of nanocomposite hydrogels consisting of organic/inorganic networks and the effects of clay modification thereto', *J Network Polym Jpn*, 25, 2–12.

Haraguchi K and Li HJ (2005), 'Control of the coil-to-globule transition and ultrahigh mechanical properties of PNIPA in nanocomposite hydrogels', *Angew Chem Int Ed*, 44, 6500–6504.

Haraguchi K and Li HJ (2006), 'Mechanical properties and structure of polymer-clay nanocomposite gels with high clay content', *Macromolecules*, 39, 1898–1905.

Haraguchi K and Li HJ (2009), 'The effect of water content on the ultimate properties of rubbery nanocomposite gels', *J Polym Sci Part B Polym Phys*, 47, 2328–2340.

Haraguchi K and Matsuda M (2005), 'Spontaneous formation of characteristic layered morphologies in porous nanocomposites prepared from nanocomposite hydrogels', *Chem Mater*, 17, 931–934.

Haraguchi K and Song L (2007), 'Microstructures formed in co-cross-linked networks and their relationships to the optical and mechanical properties of PNIPA/clay nanocomposite gels', *Macromolecules*, 40, 5526–5536.

Haraguchi K and Takada T (2005), 'Characteristic sliding frictional behavior on the surface of nanocomposite hydrogels consisting of organic-inorganic network structure', *Macromol Chem Phys*, 206, 1530–1540.

Haraguchi K and Takada T (2010), 'Synthesis and characteristics of nanocomposite gels prepared by *in situ* photopolymerization in an aqueous system', *Macromolecules*, 43, 4294–4299.

Haraguchi K and Takehisa T (2002), 'Nanocomposite hydrogels: A unique organic-inorganic network structure with extraordinary mechanical, optical, and swelling/de-swelling properties', *Adv Mater*, 14, 1120–1124.

Haraguchi K and Takehisa T (2005), 'Novel manufacturing process of nanocomposite hydrogel for bio-applications', *Proc of IMECE (ASME)*, 2005, 80533-1–8.

Haraguchi K, Takehisa T and Fan S (2002), 'Effects of clay content on the properties of nanocomposite hydrogels Composed of Poly (*N*-isopropylacrylamide) and clay', *Macromolecules*, 35, 10162–10171.

Haraguchi K, Farnworth R, Ohbayashi A and Takehisa T (2003), 'Compositional effects on mechanical properties of nanocomposite hydrogels composed of poly (*N,N*-dimethylacrylamide) and clay', *Macromolecules*, 36, 5732–5741.

Haraguchi K, Li HJ, Matsuda K, Takehisa T and Elliot E (2005a), 'Mechanism of forming organic/inorganic network structures during *in situ* free-radical polymerization in PNIPA-clay nanocomposite hydrogels', *Macromolecules*, 38, 3482–3490.

Haraguchi K, Taniguchi S and Takehisa T (2005b), 'Reversible force generation in a temperature-responsive nanocomposite hydrogel consisting of poly (*N*-isopropylacrylamide) and clay', *Chem Phys Chem*, 6, 238–241.

Haraguchi K, Ebato M and Takehisa T (2006a), 'Polymer-clay nanocomposites exhibiting abnormal necking phenomena accompanied by extremely large reversible elongations and excellent transparency', *Adv Mater*, 18, 2250–2254; (2006) 'Stretchy Clay Composites', *Science*, 314, 6 October, 19.

Haraguchi K, Ebato M and Takehisa T (2006b), 'Stretchy clay composites', *Science*, 314, 19.

Haraguchi K, Takehisa T and Ebato M (2006c), 'Control of cell cultivation and cell sheet detachment on the surface of polymer/clay nanocomposite hydrogels', *Biomacromolecules*, 7, 3267–3275.

Haraguchi K, Li HJ and Okumura N (2007a), 'Hydrogels with hydrophobic surfaces: abnormally high contact angles for water on PNIPA nanocomposite hydrogels', *Macromolecules*, 40, 2299–2302; (2007), 'Research Highlights', *Nature*, 446, 350.

Haraguchi K, Li HJ, Song L and Murata K (2007b), 'Tunable optical and swelling/deswelling properties associated with control of the coil-to-globule transition of poly (*N*-isopropylacrylamide) in polymer-clay', *Macromolecules*, 40, 6973–6980.

Haraguchi K, Li HJ and Song L (2008), 'Unusually high hydrophobicity and its changes observed on the newly-created surfaces of PNIPA/clay nanocomposite hydrogels', *J Colloid & Interface Sci*, 326, 41–50.

Haraguchi K, Li HJ, Ren H and Zhu M (2010a), 'Modification of nanocomposite gels by irreversible rearrangement of polymer/clay network Structure through drying', *Macromolecules*, 43, 9848–9853.

Haraguchi K, Masatoshi S, Kotobuki N and Murata K (2010b), 'Thermoresponsible cell adhesion/detachment on transparent nanocomposite films consisting of poly(2-methoxyethyl acrylate) and Clay', *J Biomater Sci Polym Ed*, DOI:10.1163/092050610x540459.

Haraguchi K, Xu Y and Li G (2010c), 'Molecular characteristics of poly(*N*-isopropylacrylamide) separated from nanocomposite gels by removal of clay from the polymer/clay network', *Macromol Rapid Commun*, 31, 718–723.

Heskins M and Guillet JE (1968), 'Solution properties of poly(*N*-isopropylacrylamide)', *J Macromol Sci Part A*, 2, 1441–1455.

Hibino T (2010), 'New nanocomposite hydrogels containing layered double hydroxide', *Appl Clay Sci*, 50, 282–287.
Hu X, Xiong L, Wang T, Lin Z, Liu X and Tong Z (2009), 'Synthesis and dual response of ionic nanocomposite hydrogels with ultrahigh tensibility and transparence', *Polymer*, 50, 1933–1938.
Idemura S and Haraguchi K (2000), 'Glass-polyamide composite and process for producing the same', USP6063862.
Janovak L, Varga J, Kemeny L and Dekany I (2009), 'Swelling properties of copolymer hydrogels in the presence of montmorillonite and alkylammonium montmorillonite', *Appl Clay Sci*, 43, 260–270.
Jiang G, Liu C, Liu X, Chen Q, Zhang G, *et al.* (2010), 'Network structure and compositional effects on tensile mechanical properties of hydrophobic association hydrogels with high mechanical strength', *Polymer*, 51, 1507–1515.
Johnson JA, Turro NJ, Koberstein JT and Mark JE (2010), 'Some hydrogels having novel molecular structures', *Prog Polym Sci*, 35, 332–337.
Kopeček J (2007), 'Hydrogel biomaterials: A smart future?', *Biomaterials*, 28, 5185–5192.
Li P, Kim NH, Siddaramaiah and Lee JH (2009), 'Swelling behavior of polyacrylamide/Laponite clay nanocomposite hydrogels: pH-sensitive property', *Compos Part B Eng*, 40, 275–283.
Liu Y, Zhu M, Liu X, Zhang W, Sun B, *et al.* (2006), 'High clay content nanocomposite hydrogels with surprising mechanical strength and interesting deswelling kinetics', *Polymer*, 47, 1–5.
Ma J, Xu Y, Fan B and Liang B (2007), 'Preparation and characterization of sodium carboxymethylcellulose/poly(*N*-isopropylacrylamide)/clay semi-IPN nanocomposite hydrogels', *Eur Polym J*, 43, 2221–2228.
Mark JE (2003), 'New developments and directions in the area of elastomers and rubberlike elasticity', *Macromolecular Symposia*, 201, 77–83.
Matsumoto A, Yoshida R and Kataoka K (2004), 'Glucose-responsive polymer gel bearing phenylborate derivative as a glucose-sensing moiety operating at the physiological pH', *Biomacromolecules*, 5, 1038–1045.
Miyazaki S, Karino T, Endo H, Haraguchi K and Shibayama M (2006), 'Clay concentration dependence of microstructure in deformed poly(*N*-isopropylacrylamide)-clay nanocomposite gels', *Macromolecules*, 39, 8112–8120.
Miyazaki S, Endo H, Karino T, Haraguchi K and Shibayama M (2007), 'Gelation mechanism of poly(*N*-isopropyl acrylamide)-clay nanocomposite gels', *Macromolecules*, 40, 4287–4295.
Mu J and Zheng S (2007), 'Poly(*N*-isopropylacrylamide) nanocrosslinked by polyhedral oligomeric silsesquioxane: Temperature-responsive behavior of hydrogels', *J Colloid Interface Sci*, 307, 377–385.
Mujumdar SK and Siegel RA (2008), 'Introduction of pH-sensitivity into mechanically strong nanoclay composite hydrogels based on *N*-isopropylacrylamide', *J Polym Sci Part A: Polym Chem*, 46, 6630–6640.
Murata K and Haraguchi K (2007), 'Optical anisotropy in polymer-clay nanocomposite hydrogel and its change on uniaxial deformation', *J Mater Chem*, 17, 3385–3388.
Nie J, Du B and Oppermann W (2006), 'Dynamic fluctuations and spatial inhomogeneities in poly(*N*-isopropylacrylamide)/clay nanocomposite hydrogels studied by dynamic light scattering', *J Phys Chem B*, 110, 11167–11175.

Nishida T, Endo H, Osaka N, Li HJ, Haraguchi K and Shibayama M (2009), 'Deformation mechanism of nanocomposite gels studied by contrast variation small-angle neutron scattering', *Phys Rev E*, 80, 030801(R) 1–4.
Okada K and Usuki A (2006), 'Twenty years of polymer-clay nanocomposites', *Macromol Mater Eng*, 291, 1449–1476.
Okajima T, Harada I, Nishio K and Hirotsu S (2000), 'Discontinuous crossover between fast and slow kinetics at the volume phase transition in poly-*N*-isopropylacrylamide gels', *Jpn J Appl Phys*, 39, L875–L877.
Okano T, Bae YH, Jacobs H and Kim SW (1990), 'Thermally on-off switching polymers for drug permeation and release', *J Control Release*, 11, 255–265.
Okumura Y and Ito K (2001), 'The polyrotaxane gel: A topological gel by figure-of-eight cross-links', *Adv Mater*, 13, 485–487.
Osada Y and Khokhlov AR (2002), *Polymer Gels and Networks*, New York, Marcel Dekker.
Rosta L and Gunten HR von (1990), 'Light scattering characterization of laporite sols', *J Colloid Interface Sci*, 134, 397–406.
Saegusa T and Chujo Y (1992), 'Organic–inorganic polymer hybrids', *Makromol Chem Macromol Symp*, 64, 1–9.
Saito N, Motoyama S and Sawamoto J (2000), 'Effects of new polymer-coated extracorporeal circuits on biocompatibility during cardiopulmonary bypass', *Artificial Organs*, 24, 547–554.
Shibayama M, Suda J, Karino T, Okabe S, Takehisa T and Haraguchi K (2004), 'Structure and dynamics of poly(*N*-isopropylacrylamide)-clay nanocomposite gels', *Macromolecules*, 37, 9606–9612.
Smith TL and Stedry PJ (1960), 'Time and temperature dependence of the ultimate properties of an SBR rubber at constant elongations', *J Appl Phys*, 31, 1892–1898.
Song L, Zhu M, Chen Y and Haraguchi K (2008a), 'Temperature- and pH-sensitive nanocomposite gels with semi-interpenetrating organic/inorganic networks', *Macromol Chem Phys*, 209, 1564–1575.
Song L, Zhu M, Chen Y and Haraguchi K (2008b), 'Surface-patterning of nanocomposite hydrogel film by direct replica molding and subsequent change in pattern size', *Polym J*, 40, 800–805.
Stayton PS, Shimoboji T, Long C, Chilkoti A, Chen G, *et al.* (1995), 'Control of protein-ligand recognition using a stimuli-responsive polymer', *Nature*, 378, 472–474.
Stellwagen J and Stellwagen NC (1989), 'Orientation of the agarose gel matrix in pulsed electric fields', *Nucleic Acids Res*, 17, 1537–1548.
Sun T, Wang G, Feng L, Liu B, Ma Y, *et al.* (2004), 'Reversible switching between superhydrophilicity and superhydrophobicity', *Angew Chem Int Ed*, 43, 357–360.
Takizawa T, Mori Y and Yoshizato K (1990), 'Cell culture on a thermoresponsive polymer surface', *Bio/technology*, 8, 854–856.
Tanaka T and Fillmore DJ (1979), 'Kinetics of swelling of gels', *J Chem Phys*, 70, 1214–1218.
Tanaka M, Motomura T, Ishii N, Shimura K, Onishi M, *et al.* (2000), 'Cold crystallization of water in hydrated poly(2-methoxyethyl acrylate) (PMEA)', *Polym Int*, 49, 1709–1713.
Tanaka M, Mochizuki A, Ishii N, Motomura T and Hatakeyama T (2002), 'Study of blood compatibility with poly(2-methoxyethyl acrylate). Relationship between

water structure and platelet compatibility in poly(2-methoxyethylacrylate-co-2-hydroxyethylmethacrylate)', *Biomacromolecules*, 3, 36–41.

Tanaka Y, Gong JP and Osada Y (2005), 'Novel hydrogels with excellent mechanical performance', *Prog Polym Sci*, 30, 1–9.

Teare DOH, Barwick DC, Schofield WCE, Garrod RP, Beeby A and Badyal JPS (2005), 'Functionalization of solid surfaces with thermoresponsive protein-resistant films', *J Phys Chem B*, 109, 22407–22412.

Tobolsky AV, Carlson DW and Indictor N (1961), 'Rubber elasticity and chain configuration', *J Polym Sci*, 54, 175–192.

Urayama K (2007), 'Selected issues in liquid crystal elastomers and gels', *Macromolecules*, 40, 2277–2288.

Urayama K and Kohjiya S (1998), 'Extensive stretch of polysiloxane network chains with random- and super-coiled conformations', *Eur Phys J*, B-2, 75–78.

Usuki A, Kojima Y, Kawasumi M, Okada A, Fukushima *et al.* (1993), 'Synthesis of nylon 6-clay hybrid', *J Mat Res*, 8, 1179–1184.

Wang T, Liu D, Lian C, Zheng S, Liu X, *et al.* (2011), 'Rapid cell sheet detachment from alginate semi-interpenetrating nanocomposite hydrogels of PNIPAm and hectorite clay', *React Funct Polym*, 71, 447–454.

Xiang Y, Peng Z and Chen D (2006), 'A new polymer/clay nano-composite hydrogel with improved response rate and tensile mechanical properties', *Eur Polym J*, 42, 2125–2132.

Xiong L, Hu X, Liu X and Tong Z (2008), 'Network chain density and relaxation of *in situ* synthesized polyacrylamide/hectorite clay nanocomposite hydrogels with ultrahigh tensibility', *Polymer*, 49, 5064–5071.

Xiong L, Zhu M, Hu X, Liu X and Tong Z (2009), 'Ultrahigh deformability and transparence of hectorite clay nanocomposite hydrogels with nimble pH response', *Macromolecules*, 42, 3811–3817.

Xu K, Wang J, Xiang S, Chen Q, Zhang W and Wang P (2007), 'Study on the synthesis and performance of hydrogels with ionic monomers and montmorillonite', *Appl Clay Sci*, 38, 139–145.

Xu S, Zhang S and Yang J (2008), 'An amphoteric semi-IPN nanocomposite hydrogels based on intercalation of cationic polyacrylamide into bentonite', *Mater Lett*, 62, 3999–4002.

Xu Y, Li G and Haraguchi K (2010), 'Gel formation and molecular characteristics of poly(*N*-isopropylacrylamide)prepared by free-radical redox polymerization in aqueous solution', *Macromol Chem Phys*, 211, 977–987.

Yamato M and Okano T (2004), 'Cell sheet engineering', *Mater Today*, 7, 42–47.

Yokoyama F, Achife EC, Matsuoka M, Shimamura K, Yamashita Y and Monobe K (1991), 'Morphology of oriented calcium alginate gels obtained by the flow-gelation method', *Polymer*, 32, 2911–2916.

Yoshida R, Uchida K, Kaneko Y, Sakai K, Kikuchi A, *et al.* (1995), 'Comb-type grafted hydrogels with rapid de-swelling response to temperature changes', *Nature*, 374, 240–242.

Zhang X and Zhuo R (2001), 'Dynamic properties of temperature-sensitive poly(*N*-isopropylacrylamide) gel cross-linked through siloxane linkage', *Langmuir*, 17, 12–16.

Zhang J, Pelton R and Deng Y (1995), 'Temperature-dependent contact angles of water on poly(*N*-isopropylacrylamide) gels', *Langmuir*, 11, 2301–2302.

Zhou C and Wu Q (2011), 'A novel polyacrylamide nanocomposite hydrogel reinforced with natural chitosan nanofibers', *Colloids Surf B*, 84, 155–162.

Zhou C, Wu Q, Yue Y and Zhang Q (2011), 'Application of rod-shaped cellulose nanocrystals in polyacrylamide hydrogels', *J Colloid Interface Sci*, 353, 116–123.
Zhu M, Liu Y, Sun B, Zhang W, Liu X, *et al.* (2006), 'A novel highly resilient nanocomposite hydrogel with low hysteresis and ultrahigh elongation', *Macromol Rapid Commun*, 27, 1023–1028.
Zhu M, Xiong L, Wang T, Liu X, Wang C and Tong Z (2010), 'High tensibility and pH-responsive swelling of nanocomposite hydrogels containing the positively chargeable 2-(dimethylamino)ethyl methacrylate monomer', *React Funct Polym*, 70, 267–271.

# 13
# Biodegradable polymer nanocomposites

A. R. McLAUCHLIN, University of Exeter, UK
and N. L. THOMAS, Loughborough University, UK

**Abstract:** Fossil carbon reserves are used to produce items whose durability was considered either to be an advantage or irrelevant depending on the application. Economic and environmental concerns are propagating interest in polymers that either are derived from alternative sustainable sources such as biomass or are biodegradable so that the risks of plastic residues in the environment are reduced or removed. Nanocomposite technology can engineer polymers to provide new materials, which is significant because of increasing legislative pressure and consumer demand for biodegradable high-performance materials, particularly in the packaging, biomedical and automotive sectors. This chapter introduces the most significant biodegradable polymers and reviews both methods of production and the physical properties of biodegradable nanocomposites.

**Key words:** biopolymer, poly(lactic acid), poly(hydroxybutyrate), starch, chitin, polycaprolactone, poly(butylene succinate), renewable feedstock, packaging, biomedical, melt processing, solvent casting.

## 13.1 Introduction

Increasing environmental awareness and concern over the disposal of plastic waste has led in recent years to a growing demand from consumers for biodegradable polymers. Many of these polymers are also derived from renewable resources, such as plant-based feedstocks or bacterial fermentation. Hence these materials meet the need to find replacements for fossil-fuel feedstocks as well as addressing growing concerns over end-of-life disposal issues and the necessity of finding alternatives to landfill.

As expected, the vast majority of applications for biodegradable plastics are in short-life, disposable products (Johnson *et al.*, 2003). An important market is in food packaging, such as clear films, trays, clam-shell packaging and blister packs. Other items of packaging are compost sacks and bags. Biodegradable polymers are also used for convenience items such as disposable cutlery, plates and cups. In agriculture there are some obvious uses for compostable materials in plant pots, mulching film and yarn. In the medical sector, biodegradable polymers find use in drug delivery systems, sutures and disposable gloves.

A key barrier to the use of biodegradable polymers is their poor technical performance compared with many conventional polymers, particularly in terms of mechanical properties and barrier properties. For this reason nanocomposite technology has been applied to biodegradable polymers. This chapter reviews biodegradable nanocomposites, describing their methods of production and the effect of nanofillers on mechanical, thermal and barrier properties, as well as the effect on biodegradability.

### 13.1.1 Biopolymers, bio-derived polymers and biodegradable polymers

First it is necessary to define some of the terms used in this chapter and to distinguish between 'biopolymers', 'bio-derived polymers' and 'biodegradable polymers'.

**Biopolymers** are polymers that are produced by the metabolic processes of living cells. This, therefore, includes carbohydrates such as cellulose and starch, and proteins such as keratin and enzymes. Also included in this group are poly(hydroxyalkanoate)s (PHAs), which are polyesters produced *in vivo* by bacteria such as *Ralstonia eutropha*. Another biopolymer of interest is chitin, which is extracted from crustacean shells.

**Bio-derived** polymers are polymers made from renewable, sustainable resources; a definition that excludes fossil carbon sources such as crude oil and coal. Hence this category of polymers includes biopolymers and it also includes two other groups of polymers, both of which are polymerised from monomers that are bio-derived. In the first of these 'synthetic' groups are polymers that retain the source material's property of biodegradability, an important example being poly(lactic acid) or polylactide (PLA). In the production of PLA, lactic acid is first produced by fermentation of a carbohydrate feedstock such as maize sugar and the lactic acid is then extracted and polymerised to give PLA. Another group of bio-derived polymers that has more recently emerged are polymers that are identical to the current fossil-derived polymers, such as polyethylene and polyvinyl chloride, but produced from bio-derived ethylene. These polymers are not biodegradable but have the properties of their fossil-derived equivalents, with which they can be mixed and recycled.

**Biodegradable** polymers have the capability of being broken down into carbon dioxide and water by the action of microorganisms, such as bacteria and fungi. As discussed above, a large number of bio-derived polymers are biodegradable, but so also are a number of fossil-derived polymers. These include polyvinyl alcohol (PVA), polycaprolactone (PCL) and a number of other aliphatic and aromatic polyesters.

### 13.1.2 Degradable, biodegradable and compostable

Some definitions relating to the biodegradation of polymers are given below.

**Degradation** refers to a chemical process in which the long chains of polymer molecules are broken down into shorter lengths, for example by the action of water, oxygen, ultra-violet light or heat.

**Biodegradation** is a biological process that occurs only after the plastics have started to degrade. A biodegradable plastic is defined in EN ISO 472:2001 as a 'degradable plastic in which degradation results in lower molecular weight fragments produced by the action of naturally occurring microorganisms such as bacteria, fungi and algae'. It is stated explicitly that the degradation is brought about by the action of living organisms rather than physical or chemical processes.

A **compostable** plastic is defined by the ASTM D 6400 standard as: 'A plastic that undergoes degradation by biological processes during composting to yield $CO_2$, water, inorganic compounds and biomass at a rate consistent with other known compostable materials and leaves no visible, distinguishable or toxic waste.' The process of composting is defined in the British standard PAS 100:2005 'Specification for Composted Materials' as a 'process of controlled biological decomposition of biodegradable materials under managed conditions that are predominantly aerobic and that allow the development of thermophilic temperatures as a result of biologically produced heat'. Hence compostability is a very specific and commercially important form of biodegradability. To be described as compostable, a plastic must conform to international standards EN13432 and ASTM D6400 and biodegrade within 180 days.

It is often tacitly assumed that because a polymer can be described as 'biodegradable' then no further discussion as to its disposal is necessary and it must have a beneficial impact on the environment. However, this is not the case, and there will clearly be situations where the closed-loop recycling of biodegradable polymers is the preferred option.

Also, it is not helpful to simply describe a product as 'biodegradable' without being more specific. Conversion to carbon dioxide ($CO_2$) is the most direct measurement of biodegradation. Narayan (2009) argues that the measurement of carbon dioxide evolved when a sample is incubated in soil or compost is the true measure of biodegradability, and that it is unacceptable to claim that a plastic will 'eventually biodegrade' without stipulating the disposal environment, time period and extent of biodegradation according to international standards.

### 13.1.3 Sustainability

Biosustainable polymers can be interpreted as those which are derived from renewable resources; a definition that excludes fossil carbon sources such as crude oil and coal. Inherent in this definition is the principle that supply will always meet demand, a situation that is unlikely ever to occur in the case of fossil carbon. In this category, the family of compounds known as poly(hydroxyalkanoate)s (PHA) have received much attention because they are easily produced in quantity through the fermentation of carbon-rich substrates by microorganisms, in particular bacteria. In this family is the polymers poly(*beta*-hydroxybutyrate) (P(3HB)).

The primary carbon source for biopolymers is carbon dioxide, $CO_2$, which is 'fixed' by plants and other autotrophic ('self-feeding') organisms, such as bacteria, into simple sugars, which are then polymerised to give cellulose or starch, or converted into other biomolecules such as proteins (polypeptides) or lipids. For much of the biota, sunlight provides the energy to drive these processes through the mechanism of photosynthesis, although some autotrophs derive energy by facilitating the oxidation of reduced sulphur in minerals such as pyrites: even in the darkest cave or the deepest ocean trench, there can be found organisms that fix carbon dioxide. Notwithstanding current developments in emulating photosynthesis by artificial structures, plant and microbial biosyntheses are likely to remain the most significant source of carbon-based biopolymers for some time.

While plants and microbes are the most efficient source of fixed carbon precursors for polymer production, further energy is required to extract and process the polymers for use. In the production of poly(lactic acid), lactic acid is first produced by fermentation of a carbohydrate feedstock such as maize sugar (Huang, 2005). The lactic acid is then extracted and polymerised to give PLA. By contrast, poly(hydroxyalkanoate)s are produced *in vivo* by bacteria such as *Ralstonia eutropha*. The polymer is then extracted with solvents, which are then recovered. More recently, plants capable of PHA synthesis have been developed by genetic manipulation (Poirier *et al.*, 1992).

These processes were subjected to a life cycle analysis (LCA) by workers at Dartmouth College and Monsanto (Gerngross and Slater, 2000). They found that the production of 1 kg of PHA would require three times the 29 MJ required to produce the same weight of polyethylene (PE) (Table 13.1). In terms of fossil-fuel requirements, 1 kg of PHA requires the power generated by 2.65 kg of a fossil fuel whereas the same amount of PE requires 2.2 kg, of which only 1.3 kg would be used for energy production. In their words, 'What is gained by substituting the renewable resource for the finite one is lost in the additional requirement for energy.' The result of this LCA contributed in part to Monsanto's decision in late

*Table 13.1* Energy required to produce plant-derived and fossil-fuel-derived plastics

| Polymer | Source | Processing energy | Raw material energy | Total energy |
|---|---|---|---|---|
| | | MJ/kg plastic | | |
| PHA | Maize | 90 | | 90 |
| PHA | Bacteria | 81 | | 81 |
| PLA | Bacteria | 56 | | 56 |
| PE | Crude oil | 29 | 52 | 81 |
| PET | Crude oil | 37 | 39 | 76 |
| Polyamide[a] | Crude oil | 93 | 49 | 142 |

Note: [a] The polyamide type was not specified by the authors of this article.
Source: Gerngross and Slater, 2000.

1999 to cease developing PHA production systems. However, realising that energy production was the major hurdle in this exercise, these workers showed that by burning corn 'stover', i.e. waste stem and leaf matter, and using this biomass-derived energy as the power source, the production process became energetically viable. The fossil-fuel requirement of PLA is rather lower (Table 13.1) and NatureWorks are currently marketing it on the basis that it requires 30% to 50% less fossil fuel than the equivalent weight of polymer derived from hydrocarbon feedstock (Leaversuch, 2002).

In 2003 workers at Cargill Dow (Vink *et al*., 2003) published a comprehensive life cycle assessment of NatureWorks™ PLA production, comparing it with a range of thermoplastics on a weight basis, i.e. each parameter was expressed per kilogram of polymer produced. The polymers included: polyamide 6,6 (PA-6,6), polyamide 6 (PA-6), polycarbonate, 'cellophane', low-density polyethylene, poly(propylene) (PP) and poly(ethylene terephthalate) (PET). Two scenarios for PLA production were also considered, one in which the processing energy was derived from fossil carbon sources and one in which it was derived from biomass and wind power. Part of the analysis considered the fossil energy equivalent of the petroleum feedstock used to make the polymers in addition to that used to generate the required power. On this basis, production of all the other polymers consumed more fossil energy than PLA production, although, when the fossil feedstock element was discounted, PET consumed less energy than PLA. If the energy sources were biomass and wind power, however, the energy consumption of PLA production was greatly reduced. This alternative strategy led Cargill Dow to declare an objective to decrease fossil energy consumption from 54 MJ/kg PLA to 7 MJ/kg within 5 to 8 years from 2003, with a similar objective to reduce greenhouse gas emissions from +1.8 to −1.7 kg $CO_2$ equivalents/kg PLA.

An assessment of the contribution of biodegradable polymers to sustainable polymer production (Murphy and Bartle, 2004) listed several major findings: (1) that the LCAs published so far indicated that biopolymers have a particular advantage over petro-chemical-derived polymers where fossil energy consumption and global warming potential was concerned, (2) that the 'eco-profiles' of biopolymers were favourable on account of their plant biomass organic carbon content, (3) that disposal at the end of life was a significant element of the polymer life cycle for maximising the environmental benefit, and (4) that biopolymers offered the potential for adding value to waste and by-products from other industrial processes.

## 13.2 Biodegradable polymers

### 13.2.1 Poly(lactic acid)

Poly(lactic acid), Fig. 13.1, is an aliphatic ester of the poly(2-hydroxy) type, the most significant other example of which is poly(glycolic acid) (PGA). Although

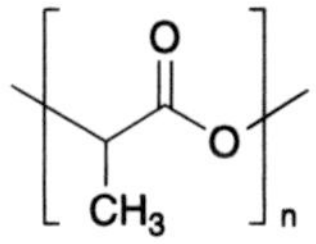

13.1 Structure of poly(lactic acid).

these two polymers were discovered in the mid-1950s, they were initially disregarded on account of their hydrothermal instability, which prevented their use in the injection moulding and extrusion processes. In the 1960s, the first use of PGA to make dissolving artificial sutures was reported (Schmitt and Polistina, 1969). In this application, sensitivity to water was an advantage and thus PGA provided the first practical alternative to denatured collagen. This property of *in vivo* degradability was also demonstrated for poly(L-lactic acid) and poly(DL-lactic acid) (Kulkarni *et al.*, 1966). Consequently PLA/GA copolymers are widely used in biomedical applications because of their biocompatibility and range of properties obtainable by manipulation of the formulation conditions.

Lactic acid, unlike glycolic acid, contains an asymmetric carbon atom and therefore has two optically active enantiomers, D and L (Fig. 13.2). Lactic acid readily forms cyclic esters, or lactides, composed of two lactic acid units or dimers. Three optically isomeric lactides are therefore possible: L-lactide (a dimer of L-lactic acid), D-lactide (a dimer of D-lactic acid) and *meso*-lactide (a dimer of D- and L- lactic acid) (Fig. 13.3). A racemic mixture of D- and L-lactides may also be obtained. It may also be noted that the lactides contain two asymmetric carbon centres.

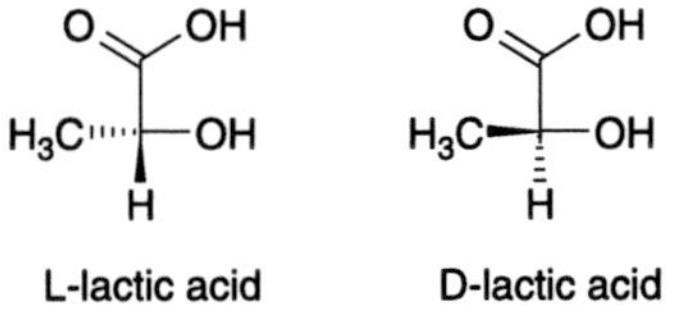

L-lactic acid D-lactic acid

13.2 Chemical structures of L-lactic acid and D-lactic acid.

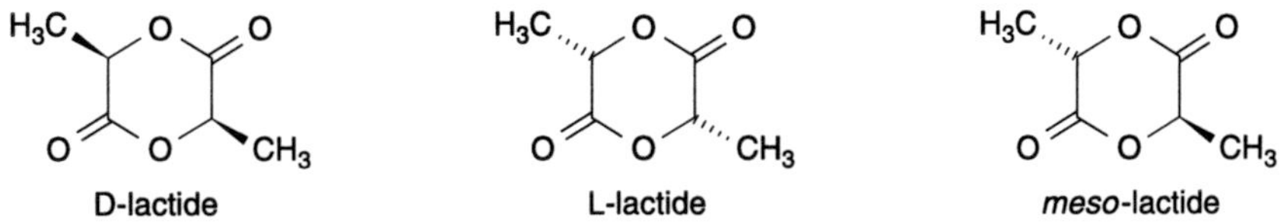

D-lactide L-lactide *meso*-lactide

13.3 Chemical structures of D-lactide, L-lactide and *meso*-lactide.

The standard production method, fermentation, gives L-lactic acid, which can be racemised to give DL-lactic acid if that product is desired. Either material can be converted to the lactide, which is then harvested by recrystallisation. *Meso*-lactide can be obtained from the DL-lactide recrystallisation filtrate (Vert *et al*., 1995).

Poly(lactic acid) may be produced by condensation polymerisation of the free acid or by ring-opening polymerisation of the lactide. In this respect there is an important interplay between the chirality of the monomer and the polymerisation method. Hence, if pure L- or D-lactic acid is polymerised by the condensation method, the product will be poly(D-lactic acid) or poly(L-lactic acid) respectively; condensation polymerisation of a racemic mixture will give a polymer with a random distribution of enantiomers in the chain. The important difference with ring-opening polymerisation is that the chain is built up by a pair addition mechanism, so that a pair of lactate subunits is added at a time. Consequently, polymerisation of L-lactide or D-lactide leads to the corresponding isotactic chain structures whereas polymerisation of *meso*-lactide leads to a non-random distribution of L- and D-enantiomers.

The polymer is produced by ring-opening polymerisation of the lactide in bulk or in organic solution: the former is preferred in industrial production. The process is subject to many factors, such as: initiator type and concentration; temperature; time; the presence of acids and water and the presence of alcohol initiators. In order to achieve high molecular weights, it is necessary to remove water in order to minimise hydrolysis of the polymer. Initiators in common use are generally metals such as tin (in the form of stannous octanoate) or zinc (as a powdered metal). The choice of initiator is partly dictated by the end-use of the polymer; for biomedical applications, biocompatible substances such as zinc are preferred (Vert *et al*., 1995). Other polymerisation catalysts such as lanthanide alkoxides have been reviewed.

While the crude polymer can be processed, it must be further purified before extrusion or injection moulding. Although PLA films can be formed by solvent-casting methods, the high temperatures required to remove all traces of solvent can cause thermal degradation of the product. The optimal processing methods for PLA, therefore, are those that involve low temperatures, no solvents and the absence, or controlled presence, of water.

Free hydroxy acids can also be polymerised by vacuum distillation of water, in the presence or absence of a catalyst, to give oligomers up to 5000 daltons (Fukuzaki *et al*., 1989) with a polydispersity of 2. If the distillation temperature is raised above 180 °C in an attempt to obtain higher molecular weights, the product starts to discolour. However, by post condensation in an organic solution, molecular weights of 50 000 daltons are achievable (Buchholz and Entenmann, 1998).

The physical, mechanical and thermal properties of PLA along with those of PP, PET and PA-6.6 and another biodegradable polymer, poly(hydroxybutyrate)

*Table 13.2* Physical, thermal and mechanical properties of PHB, PLA and some synthetic polymers

| Property | PHB (Doi, 1990) | PLA | PP (Doi, 1990) | PET (Doi, 1990) | PA-6,6 (Doi, 1990) |
|---|---|---|---|---|---|
| Melting temperature (°C) | 180 | 153 (Pluta *et al.*, 2002) | 176 | 267 | 265 |
| Glass-transition temperature (°C) | 4 | 58 (Pluta *et al.*, 2002) | −10 | 69 | 50 |
| Crystallinity (%) | 60–80 | 10–50 (Wang *et al.*, 2005) | 50–70 | 30–50 | 40–60 |
| Density (g/cm$^3$) | 1.25 | 1.26 | 0.91 | 1.39 | 1.14 |
| Tensile modulus (GPa) | 3.5 | 3.8 (Huang, 2005) | 1.7 | 2.9 | 2.8 |
| Tensile strength (MPa) | 40 | 53 (Suyatma *et al.*, 2004) | 38 | 70 | 83 |
| Extension to break (%) | 6 | 4 (Suyatma *et al.*, 2004) | 400 | 100 | 60 |

(PHB), for comparison, are shown in Table 13.2. The melting temperatures of PHB and PLA are near to those of PP, while those of PET and PA-6,6 are substantially higher. While the glass-transition temperature of PHB is comparable with that of PP, PLA has a relatively high $T_g$ in the same region as PET and PA-6.6, which partly accounts for its low biodegradation rate and the higher temperatures required to compost it effectively. PLA has a very slow rate of crystallisation, which allows greater control over its degree of crystallinity (Wang *et al.*, 2005). Its crystallinity is highly variable compared with the other polymers in the table and much lower values can be obtained. The degree of crystallinity of PLA is also highly dependent on the proportions of D- and L-lactic acid in the polymer, and higher crystallinities are obtained with the more optically pure polymers.

Comparing the mechanical properties in Table 13.2, it can be seen that the tensile moduli of PHB and PLA are considerably higher than those of the synthetic polymers, which makes these polymers suitable where stiffness is required as less material would be needed for a component. The tensile strength values of PLA and PHB are closer to that of PP than either PET or PA-6.6 and are therefore weaker. The brittleness of both biopolymers is reflected in the very low extension to break values (<6%) compared with PP, PET and PA-6.6.

PLA is now an established polymer in biomedical applications by virtue of its biocompatibility and biodegradability. Both properties are due in part to its initial hydrolysis product being lactic acid, which is rapidly transformed into pyruvic acid and ultimately to carbon dioxide and water through common metabolic pathways. The realisation of the potential of PLA as a 'green' packaging polymer came somewhat later, when the focus of research switched from highly durable materials to those that would disappear in a convenient period of time.

As poly(lactic acid) is a polyester, it is unsurprising that the main mode of depolymerisation is hydrolysis, wherein the ester bond is cleaved to give a chain terminated with a hydroxyl group and another terminated with a carboxyl group. Although this process can occur in the presence of water alone, it is greatly accelerated by acid or alkaline conditions or the presence of catalysts such as enzymes. Of these, the esterases are the most significant and living cells produce a great number and variety.

It is known that the hydrolysis of the poly(lactic acid)/poly(glycolic acid) (PLA/GA) polymers is catalysed by carboxyl groups. As these are produced in increasing number as hydrolysis proceeds, the process is in effect autocatalytic (Pitt *et al.*, 1981). It is not known if this is the case with pure PLA polymers. Another feature observed with the PLA/GA copolymers is the heterogeneous degradation of larger samples, where the interior degrades faster than the surface (Vert *et al.*, 1995). In addition, larger samples degrade faster than very small or very thin samples (Vert *et al.*, 1995) and this has important implications both in biomedical applications and in optimising the biodegradation properties of packaging materials.

Currently PLA is marketed as a 'green' plastic, most notably by the Cargill Dow consortium under the brand name NatureWorks. The main target application is food packaging and in 2006 the US retail chain Walmart announced that all clam-shell packaging used in its stores would be PLA. However, PLA does have other properties that make it attractive to the packaging industry, such as high optical clarity and high stiffness relative to cellophane, which would allow 'downgauging' of 25%, and, although its high glass-transition temperature is a hindrance to its compostability, it also makes it suitable for sealed package applications (Leaversuch 2001; 2002).

### 13.2.2 Poly(hydroxyalkanoate)s

In 1925, a microbiologist named Lemoigne working at the Pasteur Institute in Paris isolated a new polymer (Lemoigne, 1925) in a chloroform extract of a *Bacillus megaterium* culture and showed that it was a polyester of 3-hydroxybutyric acid (Lemoigne 1926; 1927), thereby giving one of the earliest accounts of a poly(hydroxyalkanoate).

The term poly(hydroxyalkanoate)s is generally used to describe polymers derived from 3-hydroxyalkanoic acids (*beta*-hydroxyalkanoic acids), distinguishing them from poly(lactic acid), which is an *alpha*-hydroxyalkanoic acid. PHA polymers have the generic structure given in Fig. 13.4.

As shown in Fig. 13.4, the length of the alkyl side chain may vary from 1 to 4 depending on the parent acid. For example, poly(3-hydroxybutyrate) gives a polymer in which the side chain is a methyl group. The polymer of 3-hydroxyvaleric

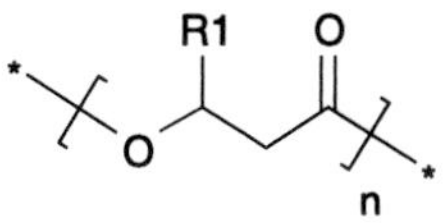

*13.4* General structure of poly(hydroxyalkanoate)s. R1 = $CH_3$, $C_2H_5$, $C_3H_7$, etc.

acid has an ethyl side group. There is a tendency in the literature to use trivial names for the acids rather than the IUPAC systematic nomenclature; thus poly(3-hydroxybutyrate) is used in preference to poly(3-hydroxybutanoate) and poly(3-hydroxyvalerate) rather than poly(3-hydroxypentanoate).

There are two important structural differences between PLA and the poly(hydroxyalkanoate) polymers. The first is that the ester linkages in PHA polymers are separated by two carbon atoms as opposed to one in the case of PLA. The second is the possibility of extended alkyl side chains in the PHA polymers. These confer plasticity on the polymer and offer the possibility of tailoring the polymer properties by varying the side chain length. The molecular weight of microbial P(3HB) is in the range $10^5$–$10^6$ daltons and the polymer is usually more than 50% crystalline with a well-defined melt temperature around 180 °C.

P(3HB) is produced and stored as granules of diameter 0.3 to 1.0 μm in the cytoplasm of cells of many prokaryotic microorganisms (bacteria and cyanobacteria) either as a carbon and energy reserve (when carbon is freely available but other nutrients are limited) or as a sink for reducing equivalents when oxygen is limited (Doi, 1990). Stored PHA may comprise between 30 to 80% of the dry weight of the cell (Doi, 1990). In the environment, microbes produce a range of copolyesters of 3-hydroxyalkanoic acids. These have been found in sewage sludge (Wallen and Rohwedder, 1974), marine sediments (Findlay and White, 1983) and both marine and freshwater cyanobacteria (Capon *et al.*, 1983).

### *Production*

A good example of a commercially produced PHA copolymer is *Biopol*, which is a copolymer of 3-hydroxybutyrate and 3-hydroxyvalerate, P(3HB-*co*-3HV). This is produced on an industrial scale by the bacterium *Alcaligenes eutrophus* growing on propionic acid and glucose in a controlled fermentation process developed by Holmes at ICI (Holmes *et al.*, 1981). Since the initial patent in 1981, many variations of the process have been developed to produce alternative copolymers or to produce PHAs from alternative substrates. For example, one account (Skraly and Peoples, 2005) describes transgenic organisms capable of producing homopolymers of 3-hydroxypropionate or 3-hydroxyvalerate as well as their copolymers by the enzyme-catalysed conversion of diols such as 1,2-propanediol, 1,3-propanediol and, more interestingly, glycerol, which is a by-product of the biodiesel process. The PHA polymers are readily recovered and industrially useful as polymers or as starting materials for a range of chemical intermediates including 1,3-propanediol, 3-hydroxypropionaldehyde, acrylics, malonic acid, esters and amines (Skraly and Peoples, 2005). The genetic technologies available for improving PHA yields from fermentation processes were reviewed in 2003 by Reddy *et al.* (2003).

Another patented method involves first subjecting spent vegetable oil to a short anaerobic treatment to convert the oils to organic acids followed by separation of the acids from the sludge and fermenting them with a hydrogen bacterium to produce polyhydroxyalkanoic acids (Kazunari and Yoshito, 2000). This procedure is a means of disposing of a problematic waste by converting it to a useful product.

Very recently, researchers have reported the principle of producing styrene monomer by pyrolysis of waste polystyrene and using this as a substrate for poly(hydroxyalkanoate) production by *Pseudomonas putida* (Booth, 2006). They did not, however, offer any comparison with the simple alternative of using the regenerated styrene for further polystyrene production.

*Limitations*

The onset of thermal degradation of PHA is close to its melt temperature, which can impose limits on its commercial application. This has led investigators to study other poly(hydroxyalkanoate)s such as poly(3-hydroxybutyrate-co-3-hydroxyvalerate) (PHBV) in the search for a more thermally stable poly(hydroxyalkanoate). Another limitation is the slow crystallisation rate, which, for example, can make it difficult to form films.

*Applications*

A distinguishing feature of poly(hydroxyalkanoate)s is their biocompatibility, particularly of the short-chain polymers (3-HB) and PHBV (Gogolewski *et al.*, 1993). This, coupled with their thermoplasticity, enables the manufacture of articles such as bone implants that do not cause inflammatory reactions with the associated risk of rejection (Saito *et al.*, 1991).

Early commercial applications of PHAs included packaging films for bags and paper coatings as well as a variety of disposable items such as razors and hygiene products as well as cosmetic packaging. PHA is also a useful feedstock for producing pure *R*-3-hydroxybutyric acid, used in the production of the anti-glaucoma drug Trusopt (Reddy *et al.*, 2003).

*Biodegradability*

PHAs are storage polymers that provide a rapidly available intracellular carbon source; thus they are readily hydrolysed by endo- and exoesterases. The main enzyme involved is PHB depolymerase, which is an extracellular endohydrolase identified in organisms such as *Bacillus megaterium*, an extremely common soil bacterium (Reddy *et al.*, 2003). As these enzymes are produced extracellularly by many soil organisms, they are rapidly biodegraded in the environment, with up to 85% degradation in seven weeks being reported (Fletcher, 1993; Johnstone, 1990). A sample of PHA degraded within 254 days in a freshwater lake where the temperature remained less than 6 °C (Johnstone, 1990).

## 13.2.3 Starch

In the context of biodegradable nanocomposites, the role of starch is chiefly as a matrix polymer, whereas cellulose, in the form of fibrils or nanocrystals, plays the part of a reinforcing nanofiller. Starch is both a feedstock for biopolymer production and is itself a biopolymer, which can be processed to make nanocomposites.

In 2000 the world production of starch was estimated at 48.5 million tonnes, which was derived from maize (39.4 million tonnes), potatoes (2.6 million tonnes) and wheat (4.1 million tonnes) with other sources such as cassava and sago accounting for 2.5 million tonnes. Climate and geography dictate the principal source: in the US 24.6 out of 24.9 million tonnes were produced from maize, whereas in Europe 2.8 out of 8.4 million tonnes were derived from wheat.

Starch is a polymer of alpha-glucose, which exists in two forms: the linear amylose (Fig. 13.5) and the highly branched amylopectin (Fig. 13.6). In native starch, the proportions of the two molecules vary according to the plant of origin, with amylose comprising 15–30% and amylopectin 70–85% of the starch (BeMiller and Whistler, 2009). Amylose tends to exist in a tightly bound helical structure, which renders it largely insoluble in cold water, whereas the looser structure of amylopectin renders it more water soluble and digestible. The process of plasticisation opens up the helical amylose structure to give a material suitable

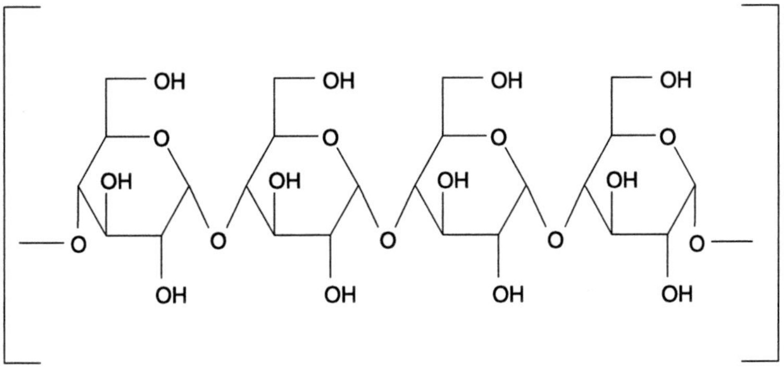

*13.5* Chemical structure of amylose.

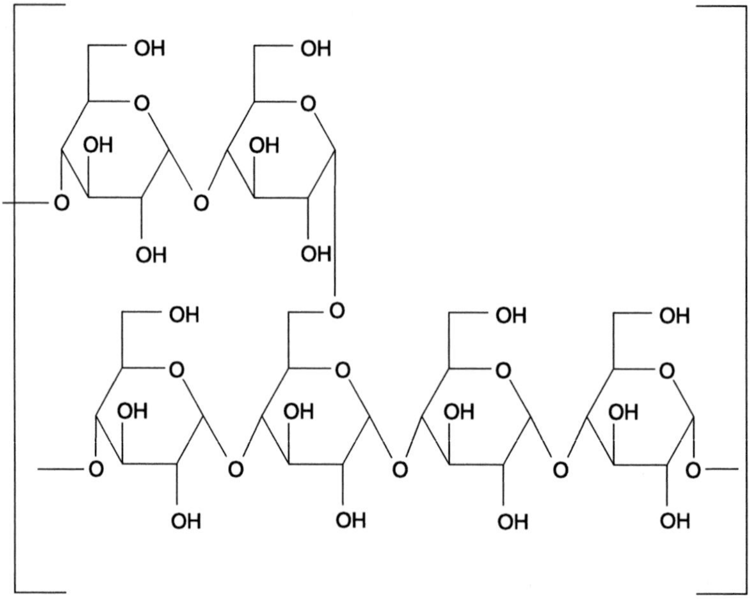

13.6 Chemical structure of amylopectin.

for moulding and extrusion (Olkku and Rha, 1978). The molecular weight of amylose varies between 40 000 and 340 000 (250 to 2000 anhydroglucose units) while the molecular weight of the branched amylopectin may reach 80 000 000 (BeMiller and Whistler, 2009), which accounts for the high viscosity of plasticised starch. The proportions of amylose and amylopectin strongly influence the structure and physical characteristics of starch (Yoo and Jane, 2002) and in particular its behaviour in industrial processing and cooking. Consequently, much plant breeding and genetic research has been devoted to manipulating the relative amounts of these components. 'Waxy rice', for example, has little or no amylose, whereas the amylose content of maize can vary from 28% to 70% depending on the plant variety. The amylose content of wheat starch is typically 25% and that of potato is 20%. The production of processed food, such as potato crisps, generates a large amount of process water rich in so-called 'grey starch' and there is much interest in its potential as a starch feedstock for bio-derived materials. The variability in the structure and physical properties of the starch are so dependent on the plant from which it is obtained that the source should be considered the prime variable. Accordingly, the literature contains extensive reviews devoted to the properties of individual starches from different sources, such as maize (Gallant and Bouchet, 1986), wheat flour (Olkku and Rha, 1978) and peas (Ratnayake *et al.*, 2002), while tuber-derived starches have been extensively reviewed as a group (Hoover, 2001).

### 13.2.4 Chitin

Chitin (Fig. 13.7) is the aminopolysaccharide equivalent of cellulose in terms of its natural abundance and role as structural biopolymer. It is a linear polymer of 2-acetamido-2-deoxy-$\beta$-D-glucose, although it is still frequently referred to by its trivial name, *N*-acetylglucosamine, and thus is an amino analogue of cellulose (Rinaudo, 2006). Purified chitin is an inelastic white material, which is insoluble in water. Deacetylation of chitin, by 5% sodium hydroxide for example, gives chitosan (Fig. 13.8), although complete deacetylation is difficult to achieve. Chitosan is soluble in acetic acid. Chitin itself is soluble in trichloroacetic acid (TCA) and other similar polar protic strong solvents, a property which provides a useful route to fibre spinning and further chemical modification.

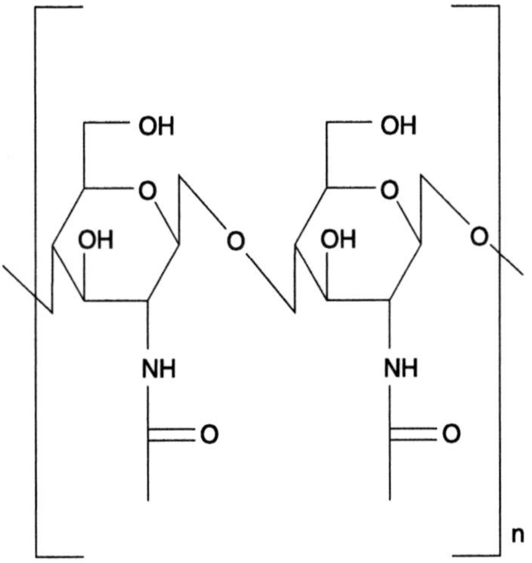

*13.7* Structure of chitin.

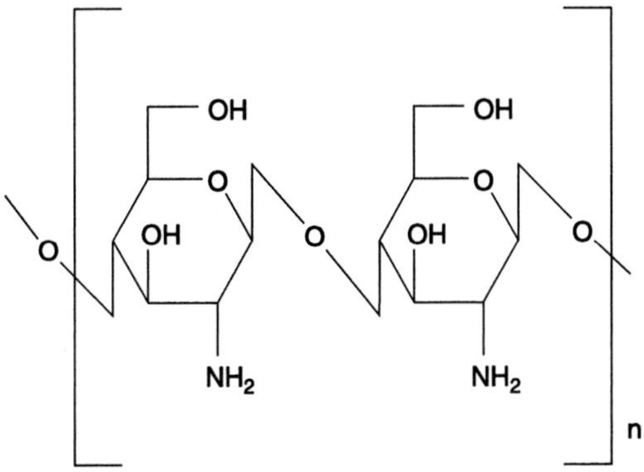

*13.8* Structure of chitosan.

The chemistry of chitin has been reviewed in detail (Pillai *et al.*, 2009), as have its biomedical applications (Jayakumar *et al.*, 2010; 2011), which arise from its adsorptive properties, non-toxicity, biocompatibility and biodegradability. Its adsorptive property also gives it the ability to form hydrogels with polymers such as poly(ethylene glycol) (PEG) and gelatin.

Chitin is a major component of invertebrate exoskeletons such as arthropod and crustacean shells as well as fungal cell walls. It is obtained mainly from marine sources as a by-product of harvesting organisms such as shrimp, prawn, lobster and other shellfish. The rate of production is such that in some areas it is a marine pollutant, for which reason its exploitation as a polymer feedstock has both economic and environmental benefits, if it is diverted from disposal to utilisation. The structure of the invertebrate exoskeleton is in itself an example of a naturally occurring nanocomposite and has been discussed in detail (Raabe *et al.*, 2005).

Like cellulose, chitin is capable of acting as both the matrix polymer and the nanofiller in nanocomposites. For example, chitin-reinforced PCL has been explored as a route to a biodegradable nanocomposite (Morin and Dufresne, 2002) and chitin whiskers have been investigated as a nanofiller for natural rubber (Nair *et al.*, 2003). Studies on chitin nanocomposites tend to focus on biomedical applications such as drug delivery from chitosan organically modified rectorite films (Wang *et al.*, 2007).

### 13.2.5 Polycaprolactone (PCL)

Polycaprolactone is a semicrystalline linear polyester produced by ring-opening polymerisation of *epsilon*-caprolactone, which is commonly derived from fossil carbon. It has a much lower glass-transition temperature ($T_g = -60\,°C$) than other biodegradable polymers, which assists its biodegradability despite its high degree of crystallinity, typically 50%. The melting point ($T_m = 60\,°C$) is also rather low. Like poly(butylene succinate) (PBS), PCL is often used in biodegradable polymer blends such as PCL/PHB (Lovera *et al.*, 2007), PCL/starch (Averous *et al.*, 2000) and PCL/PLA (Liu *et al.*, 2000). As PCL and PLA are both synthesised by ring-opening polymerisation, PCL/PLA multiblock copolymers can readily be produced, to give biodegradable thermoplastic elastomers (Cohn and Hotovely Salomon, 2005).

### 13.2.6 Poly(butylene succinate) (PBS)

Of the synthetic biodegradable polymers available, poly(butylene succinate), see Fig. 13.9, has attracted much attention because it is comparable with poly(propylene) in terms of good thermal resistance and melt processability as well as its chemical resistance, whilst remaining biodegradable (Ray *et al.*, 2005). Although it was originally derived from fossil carbon, there is considerable

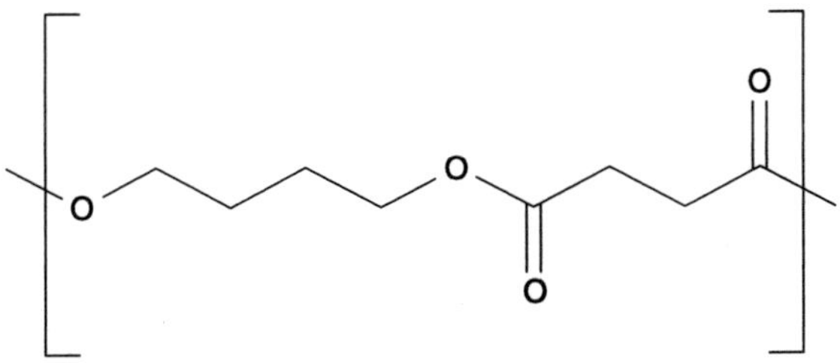

13.9 Structure of poly(butylene succinate).

interest in developing it as a bio-sourced polymer, with the succinic acid moiety being produced by bacteria such as *E. coli* or commercially developed strains such as *Basfi succiniproducens* from glucose or glycerol feedstock. In Europe, the main producer is BASF, although another example of commercially available PBS is Bionolle, manufactured by Showa Highpolymer. PBS is suitable for film forming, sheet extrusion, injection moulding and moulded foam products, and the range of applications for which is it currently marketed include: mulching film, compost bags and household goods. It is also used in some civil engineering and construction applications. It has good fibre-forming properties and is therefore suitable for spun-fibre applications such as textiles. According to the manufacturers, PBS is degradable in a variety of media including compost, soil, fresh and salt water and activated sludge. Breakdown in as little as 12 weeks in suitable aerobic conditions has been reported (Mohanty *et al*., 2003a).

A related polymer is poly(butylene succinate)-co-adipate (PBSA). This is a random copolymer, which is synthesised by the polycondensation of butane-4-diol in the presence of adipic and succinic acids. PBSA has lower crystallinity than PBS and also greater flexibility in the polymer chains (Ray *et al*., 2008).

PBS is often blended with other polymers to give materials with improved properties such as impact strength. Examples of PBS blends with other biodegradable polymers include: PBS/PLA (Chen *et al*., 2005; Harada *et al*., 2007; Park and Im, 2002) and PBS/PHA (Qiu *et al*., 2003).

## 13.3 Methods of production of biodegradable polymer nanocomposites

### 13.3.1 Melt processing

Although *in situ* polymerisation is a feasible technique (Paul *et al*., 2005b), the technique most favoured would appear to be melt extrusion (Ray *et al*., 2002a; 2003d; 2003b; Fujimoto *et al*., 2003b) or variations on melt compounding (Pluta *et al*., 2002; Nam *et al*., 2003; Cabedo *et al*., 2005; Paul *et al*., 2005b). The conditions used for melt compounding and extrusion of PLA/clay nanocomposites are summarised in Table 13.3. The favoured processing temperature is 180–190 °C,

*Table 13.3* Processing conditions employed in the production of PLA/clay nanocomposites by melt intercalation

| Equipment | Temperature (°C) | Speed (rpm) | Time (min) | Reference |
|---|---|---|---|---|
| Counter-rotating mixer | 180–195 | 60 | 10 | Pluta *et al.*, 2002 |
| Twin-screw extruder | 190 | | | Ray *et al.*, 2002 |
| Twin-screw extruder | 190 | | | Maiti *et al.*, 2003 |
| Counter-rotating mixer | 190 | | | Nam *et al.*, 2003 |
| Counter-rotating mixer | 180–190 | 20 | 4 | Paul *et al.*, 2003 |
| | | 60 | 3 | |
| Counter-rotating mixer | 190 | | | Pham Hoai *et al.*, 2005 |
| Counter-rotating mixer | 150 | 60 | 8 | Cabedo *et al.*, 2005 |
| Counter-rotating mixer | 150 | 20 | 4 | Feijoo *et al.*, 2005 |
| | | 60 | 4 | |

although a lower temperature of 150 °C has been used (Cabedo *et al.*, 2005). In the melt-compounding process, the modal speed was 60 rpm, with some researchers favouring an initial phase at a slower speed, although reasons for this were not given. The total processing time ranged from three minutes to 10 minutes (Pluta *et al.*, 2002).

There is considerable variation in the PLA used in nanocomposite production, with five companies providing the entire stock (Table 13.4). Most researchers reported the quality of the PLA by noting its weight-average molecular weight, polydispersity and D-lactide content. The molecular weights of the PLA preparations ranged from 110 000 to 325 000; over a threefold variation. The polydispersity was more conservative, varying from 1.45 to 2.0. The purity was not reported by several workers, although it is known that the proportions of D- and L-isomers have a great influence on the physical properties of the polymer. However, from the known reports of purity it can be seen that the D-lactide content varied from zero to 12% (Paul *et al.*, 2003; Cabedo *et al.*, 2005).

The chemical nature of the organoclay also varied among the reports, although all the nanoclays were based on montmorillonite (MMT) with a cation exchange capacity in excess of 90 me/100 g. The counterion was nearly always a quaternary ammonium compound, i.e. a nitrogen atom is covalently bonded to four alkyl groups so that the 'lone pair' is also shared and the nitrogen centre acquires a positive charge. By using long-chain alkyl groups, the organoclay is more compatible with the polymer and so more easily wetted. The choice of alkylammonium compound is dictated partly by the commercial need for a low-cost compound. Thus, although Ray *et al.* (2002) concentrated on an octadecylammonium organoclay of high purity, most of the remaining researchers, as can be seen in Table 13.5, elected to use organoclays based on the cheaper hydrogenated tallow alkyls (HT), which are of animal origin. However, the purity of HT is somewhat variable, depending on its origin. A typical analysis for HT is: 65% C18; 30% C16 and 5% C14 (Chen and Yoon, 2005).

*Table 13.4* Grades of polylactide used on production and evaluation of PLA/clay nanocomposites

| Manufacturer | Weight-average molecular weight | Polydispersity | Purity (% D-lactide) | Reference |
|---|---|---|---|---|
| Shimadzu | 200 000 | – | – | Ogata *et al.*, 1997 |
| Cargill Dow (USA) | 166 000 | 2.0 | 4.1 | Pluta and Galeski, 2002 |
| Unitika (JP) | 187 000 | 1.76 | 1.1–1.7 | Ray *et al.*, 2002 |
| Shimadzu Co. (JP) | 218 000 | 1.55 | – | Lee *et al.*, 2003a |
| Polysciences Inc. | 325 000[b] | – | – | Krikorian and Pochan, 2003 |
| Unknown | 159 000 | 1.86 | – | Nam *et al.*, 2003 |
| Galactic SA | 155 000 | 1.9 | 0 | Paul *et al.*, 2003 |
| Unitika (JP) | 159 000 | 1.86 | <1 | Pham Hoai *et al.*, 2005 |
| Shimadzu Co. (JP) | 110 000 | 1.45 | – | Lee *et al.*, 2005b |
| Galactic (B) | – | – | 12 | Cabedo *et al.*, 2005; Feijoo *et al.*, 2005 |
| Cargill Dow | 161 500[a] | 1.9 | 30 | Paul *et al.*, 2005a |

Notes: [a] Calculated from Mn and the polydispersity.

[b] The value is for the viscosity-average molecular weight.

– indicates that data was not available.

## 13.3.2 Solvent casting

Solvent casting is a suitable technique at the early stages of investigation of polymer nanocomposites because only small quantities of polymer need be used and the absence of heat treatment avoids thermal degradation of the polymer or nanofiller should they be susceptible. Chloroform remains the solvent of choice for many workers because it can be used equally successfully with PLA, PCL, PHA and PHBV and is also suitable for suspending cellulose microfibrils (Sanchez-Garcia *et al.*, 2008). Biopolymer-layered silicate nanocomposites have also been made using chloroform both to dissolve PLA and to suspend the nanoclay (Krikorian and Pochan, 2003; McLauchlin and Thomas, 2009).

Others have favoured dimethylformamide (DMF) as a solvent for PHBV-cellulose nanocomposites (Ten *et al.*, 2010). In all these studies, the dispersal method was sonication, with or without cooling. Ten *et al.* also reported a 17% reduction in the weight-average molecular weight of PHBV due to the sonication technique employed (Ten *et al.*, 2010). It should be noted that, unlike PLA and

*Table 13.5* Organo-montmorillonites used in production and evaluation of PLA/clay nanocomposites

| Source | CEC (me/100 g) | Organic counterion | Reference |
|---|---|---|---|
| Southern Clay Products Cloisite 25A | 92 | Dimethyl-2-ethylhexyl (hydrogenated tallow alkyl) ammonium | Pluta *et al.*, 2002<br>Paul *et al.*, 2003; 2005a<br>Lee *et al.*, 2003a |
| Southern Clay Cloisite 30B | 90 | Bis-(2-hydroxyethyl)methyl (hydrogenated tallow alkyl) | Paul *et al.*, 2003; 2005b<br>Krikorian and Pochan, 2003 |
| Southern Clay Products Cloisite 15A | 125 | Dimethyl-di(hydrogenated tallow alkyl) | Krikorian and Pochan, 2003 |
| Southern Clay Products Cloisite 20A | 95 | Dimethyl di(hydrogenated tallow alkyl)ammonium | Paul *et al.*, 2003; 2005b |
| Nanocor | 110 | Octadecylammonium 119 | Ray *et al.*, 2003a; 2003b; 2003d<br>Fujimoto *et al.*, 2003b |
| Kunimine | 109 | Dimethyldioctadecylammonium | Nam *et al.*, 2003 |
| Investigator's laboratory | – | Dimethyl-hydrogenated tallow ammonium | Feijoo *et al.*, 2005 |
| Investigator's laboratory | – | Dimethyl-benzyl-dihydrogenated tallow ammonium | Feijoo *et al.*, 2005 |

PCL, PHB itself is not soluble in chloroform at room temperature and requires heating to 70 °C for effective dissolution (Yun *et al.*, 2008).

While organic solvents are required for casting PLA films, some poly(hydroxyalkanoate) derivatives can be cast by alternative methods, such as casting from a latex of bacterial poly($\beta$-hydroxyoctanoate) (PHO) obtained from *Pseudomonas oleovorans*, which can be cast into amorphous, thermoplastic elastomeric films (Dubief *et al.*, 1999).

## 13.4 Properties of biodegradable polymer nanocomposites

### 13.4.1 Mechanical properties

Nanocomposites can be characterised either by tensile testing or by dynamic mechanical analysis. An example of the latter is given in Table 13.6. These materials were prepared by twin-screw extrusion (Ray *et al.*, 2002a).

*Table 13.6* Storage modulus of PLA-organo-montmorillonite exfoliated nanocomposites prepared by twin-screw extrusion

| | Storage modulus, *G′* (GPa) | | | |
|---|---|---|---|---|
| Sample (%$C_{18}$-MMT) | −20 °C | 40 °C | 100 °C | 145 °C |
| PLA | 1.74 | 1.60 | 0.13 | 0.06 |
| PLACN1 (3% clay) | 2.32 | 2.07 | 0.16 | 0.09 |
| PLACN2 (5% clay) | 2.90 | 2.65 | 0.25 | 0.10 |
| PLACN3 (7% clay) | 4.14 | 3.82 | 0.27 | 0.19 |

Source: Ray *et al.*, 2002a.

At all temperatures, the storage modulus increased with increasing organoclay loading. The increases were more marked at temperatures below the $T_g$ of the materials, so at 40 °C a clay loading of 7% increased the storage modulus from 1.60 GPa to 3.82 GPa, indicating that the polymer had become more elastic. The authors also noted increases in *G′* above the glass-transition temperature, from 0.06 GPa to 0.19 GPa, and attributed this to the reinforcement effect of the clay particles, which restricted the movement of the polymer chains.

In another study, PLA/MMT nanocomposites were prepared by solvent casting from a chloroform suspension (Lee *et al.*, 2003b). These samples also contained ammonium carbonate as a blowing agent and sodium chloride particles, which were leached out post-forming to produce scaffolds such as would be used in biomedical applications. Exfoliation of the layer silicate was confirmed by X-ray diffraction analysis and the disappearance of the peak at $2\theta = 4.56°$ due to interlayer spacing was observed. The authors reported a $T_g$ of 63 °C for pristine PLA decreasing to 53 °C for the nanocomposite containing 5% organically modified montmorillonite (o-MMT), although the $T_g$ of the 8% o-MMT nanocomposite was 58 °C. The recrystallisation temperature decreased from 111 °C to 90 °C and this was attributed to the clay platelets acting as nucleating sites and promoting crystallisation.

The mechanical properties were assessed by tensile testing and showed an increase in tensile modulus from 121 MPa for pristine PLA to 170 MPa for the 8% o-MMT nanocomposite. The tensile strength decreased from 48 MPa to 23 MPa. The increase in modulus was ascribed to mechanical reinforcement of the polymer by the clay platelets. In another study (Ogata *et al.*, 1997), it was concluded that intercalation did not take place in solvent-cast composites although a twofold increase in Young's modulus at 10% clay loading was observed. This increase was instead ascribed to a 'superstructure' forming in the blend film in the thickness direction.

The storage modulus of solvent cast nanocomposites has also been assessed elsewhere (Krikorian and Pochan, 2003). Here the conclusions were that exfoliated clay loading up to 15% gave a 61% increase in storage modulus at temperatures 'around body temperature', which is assumed to be from 36 °C to 38 °C.

PHA-cellulose nanocomposites based on a latex of bacterial poly($\beta$-hydroxyoctanoate) obtained from *Pseudomonas oleovorans* have been cast into amorphous, thermoplastic elastomeric films (Dubief *et al.*, 1999). Nanocomposite materials were prepared by stirring mixtures of the latex and colloidal suspensions of either hydrolysed starch or cellulose whiskers. After mixing, the preparations were either cast and evaporated or freeze-dried and moulded. The cellulose used was in the form of tunicin whiskers consisting of parallelepiped rods with the length ranging from 100 nm to several micrometres (average value around 1 μm) for widths of the order of 10–20 nm. The aspect ratio of these whiskers as estimated from transmission electron microscopy was around 67. The properties of the nanocomposites were assessed by determining the variation of storage modulus with temperature and were strongly related to the aspect ratio of the filler and to geometric and mechanical percolation effects. The authors also noted that specific polymer/filler interactions and geometrical constraints imposed by the particle size of the latex influenced the mechanical reinforcement effect of the cellulose whiskers.

In an accompanying publication (Dufresne *et al.*, 1999), the same group showed that the high mechanical properties of the composite were the result of mechanical percolation created when the tunicin whiskers were present at 1% by volume or greater, the so-called 'percolation threshold'. However, rather than through a direct interaction such as interparticular hydrogen bonding, the network was created through transcrystalline layers of polymer forming between the whiskers, the deleterious consequence of which was that the mechanical properties of the semicrystalline medium-chain-length PHA composite decreased disastrously when the melt temperature of the matrix was reached. The phenomena of transcrystallisation and mechanical percolation are therefore important considerations in the development of this type of nanocomposite.

This group also prepared a nanocomposite based on tunicin whiskers and plasticised starch using glycerol as plasticiser (Angles *et al.*, 2000). The product contained four components (starch, cellulose, glycerol and water) and examination showed it to be a heterogeneous system containing glycerol- and amylopectin-rich domains with the plasticisers (water and glycerol) accumulating in the cellulose/amylopectin interface. They observed that transcrystallisation of amylopectin had occurred, similar to that encountered with the PHA composites, and noted that this phenomenon increased with increasing water content. The mechanical properties, such as tensile modulus, were highly dependent on the water content and temperature, which was especially significant because at room temperature the glycerol-plasticised starch matrix was in its glass-rubber transition zone. This made the composite highly sensitive to fluctuations in temperature (Angles *et al.*, 2000).

Although early work on biodegradable nanocomposites focused on PLA, PBS is attracting increasing attention due to its ease of processing (Ray *et al.*, 2005;

2006a; 2006b; Guangxin and Yoon, 2005; Ray and Bousmina, 2006; Shih *et al.*, 2007; 2008; Makhatha *et al.*, 2008).

The intercalation and exfoliation behaviour of PBSA-organoclay nanocomposites has also been studied in detail by combining melt rheology studies with scanning transmission electron microscopy (STEM) (Bandyopadhyay *et al.*, 2010). It was found that increasing the organoclay loading up to 4% suppressed the viscous behaviour of pure PBSA, which became more gel-like at 6% clay content. At a clay content of 5% the behaviour was between the liquid and gel states. STEM studies indicated a highly delaminated nanocomposite microstructure at low clay contents, which became more flocculated with increasing clay content until a stacked–intercalated structure was produced at higher clay content. The authors reported a good agreement between the observed structures and the melt-state rheological behaviour. Other studies have reported similar trends and noted the shear-thinning behaviour of the nanocomposites (Eslami *et al.*, 2010; Bhatia *et al.*, 2009; 2010).

## 13.4.2 Thermal properties

The glass-transition temperature ($T_g$) of pure PLA is little affected by the incorporation of organoclays. For example, one study (Pluta *et al.*, 2002) reported a $T_g$ of 58.4 °C for annealed PLA and 58.2 °C for an annealed PLA/o-MMT nanocomposite. However, crystallising the samples isothermally at 120 °C both lowered $T_g$ and produced greater differences between the materials, so that the $T_g$ of crystallised PLA was 53 °C and that of the crystallised nanocomposite was measurably higher at 55 °C. They noted that this increase was for polystyrene nanocomposites and is ascribed to the restricted segmental motions at the interface of the organic and inorganic phases of the material. Other studies (Ray *et al.*, 2003d) indicate that the $T_g$ values of unfilled PLA and PLA containing 4% organoclay are effectively the same (60 °C). It is more likely that the difference in $T_g$ recorded by these two authors is due to the different D-lactic acid contents, which were >4% and <2%, respectively. It has also been shown that the type of nanoclay had minimal effect on both $T_g$ and melting temperature (Paul *et al.*, 2003). However, in this study, $T_g$ was reduced from 55 °C to 15 °C by addition of 20 wt% PEG as plasticiser. It has also been reported that the melting point of PLA (177 °C) did not change with clay loading (Krikorian and Pochan, 2003).

It has been reported that the maximal thermal stability of PBSA/clay nanocomposites is obtained at 3% clay content (Bhatia *et al.*, 2009). However, PBS/multi-walled carbon nanotube nanocomposites have a decomposition temperature 10 °C higher than PBS (Song and Qiu, 2009).

## 13.4.3 Barrier properties

It is in food packaging that the barrier properties of polymer thin films become highly relevant. In general, it has been found that incorporating up to 5% nanofiller by weight can greatly reduce the permeability to oxygen and water (Lange and Wyser, 2003). While it is generally accepted that the improvements in barrier properties arise because of the creation of a so-called 'tortuous path', which slows down the passage of gas or molecules, not all reported experiments show this. For example, melt-processed PBS-organoclay nanocomposites did not show greatly improved $O_2$ barrier properties over the pure PBS (Ray *et al.*, 2006a). The authors suggested that the high degree of crystallisation caused by the nucleating effect of the organoclay compromised the barrier property in this case. However, a 50% improvement in water barrier and a 48% improvement in oxygen barrier were reported for PLA-layered silica nanocomposites that were prepared by twin-screw extrusion and converted to blown films (Thellen *et al.*, 2005).

As an example of the improvements in barrier properties achievable, the water permeability coefficient of PLA was reduced from $2.30 \times 10^{-14}$ to $1.03 \times 10^{-14}$ by addition of 10% food grade mica by the solvent-casting technique, while the oxygen permeability coefficient was reduced from $2.77 \times 10^{-18}$ to $1.09 \times 10^{-18}$. The water vapour permeability coefficient of PHBV was reduced from $1.27 \times 10^{-14}$ to $0.60 \times 10^{-14}$ and the oxygen permeability increased from $1.44 \times 10^{-18}$ to $2.33 \times 10^{-18}$. The water vapour permeability coefficient of PCL was reduced from $3.39 \times 10^{-14}$ to $1.26 \times 10^{-14}$ and the oxygen permeability decreased from $7.06 \times 10^{-18}$ to $3.67 \times 10^{-18}$ (Sanchez-Garcia and Lagaron, 2010). Thus reductions of 50% or more in permeability are achievable in biodegradable nanocomposites.

### 13.4.4 Biodegradability

The biodegradation of PLA is sometimes compared to that of plasticised starch, which is also a candidate biopolymer. However, unlike starch, PLA requires a higher temperature than starch for biodegradation to occur and does not lend itself to home composting where lower temperatures frequently apply, compared with the more controlled conditions of municipal composting. However, a greatly enhanced biodegradation of PLA with 4% clay loading relative to neat PLA has been reported (Ray *et al.*, 2003a). This was attributed to the greater amorphous nature of the PLA in the nanocomposite (40%), compared with the neat polymer (36%), as it is known that amorphous polymers are more readily degraded by microorganisms. A similar conclusion was reached with PBS/layer silicate nanocomposites where the increased biodegradation was ascribed to reduced crystallinity (Han *et al.*, 2008). The degree of biodegradation appears to depend greatly on the organoclay used in the manufacture of the nanocomposite (Maiti *et al.*, 2003).

A study of hydrolytic stability (Paul *et al.*, 2005a) was carried out by immersing specimens in a pH 7.4 buffer at 37 °C for five and a half months. The authors do not stipulate that sterile conditions were used, although no infection of the samples was reported. A microcomposite prepared with $Na^+$/montmorillonite degraded within two months whereas neat PLA and a nanocomposite based on Cloisite 25A were still structurally unaltered after five months. The rapid degradation of the microcomposite was attributed to the hydrophilicity of the sodium clay used.

## 13.5 Applications of biodegradable polymer nanocomposites

### 13.5.1 Biomedical applications

Research in this field is very extensive, so this section offers a brief introduction and directs the reader to suitable reviews of the literature (Jayakumar *et al.*, 2010; Katti, 2004). When originally discovered, PLA was largely ignored due to its biodegradability, as, at the time, the durable properties of other polymers were more popular. Interest in PLA resumed in the 1960s because of its biocompatibility (Kulkarni *et al.*, 1966). As chitin also has this property (Rinaudo, 2006; Jayakumar *et al.*, 2010), research into biomedical applications of these two polymers comprises the bulk of research carried out to date. PLA also has the property of being absorbed by living tissue, which makes it suitable as a temporary 'scaffold' for regeneration of tissue and bone. As with the food sector, legislation restricts the nanofillers that can be used in nanocomposites for biomedical applications, so that the main filler used is hydroxyapatite, which is itself biocompatible. Thus PLA/hydroxyapatite and collagen/hydroxyapatite nanocomposites are used in many orthopaedic applications. Another extensive research field is that of using PLA-based nanocomposites as controlled drug release agents (Chen *et al.*, 2007).

The salient properties of biodegradable nanocomposites for biomedical applications are the extent and speed of absorption by the living tissue (Gogolewski *et al.*, 1993; Katti, 2004; Li *et al.*, 2004; Weir *et al.*, 2004; Cui *et al.*, 1998) and their mechanical properties, particularly at body temperature (Katti, 2004; Li *et al.*, 2004; Lee *et al.*, 2005a).

### 13.5.2 Packaging

Nanocomposite technology affords a practical solution to the problems often associated with biodegradable and bio-derived polymers for food use, namely weak mechanical properties and poor barrier properties (Sanchez-Garcia and Lagaron, 2010; Sorrentino *et al.*, 2007; Lagaron *et al.*, 2005; Ray *et al.*, 2003c). The alternative, polymer blending, is limited to blending only with other biodegradable polymers if biodegradability is to be maintained. Biodegradable nanocomposites often have better water and oxygen barrier properties as well as greater solvent resistance and improved mechanical properties (Ray *et al.*, 2002a; Sanchez-Garcia *et al.*, 2008; Makhatha *et al.*, 2008; Sanchez-Garcia and Lagaron, 2010; Rhim and Ng, 2007).

While these aspects deal with the shelf-life of the food product, nanocomposite technology can play a role in the end of life of the packaging, particularly where composting or anaerobic digestion is involved, and, in doing so, demonstrates clear advantages over polymer blends. Whereas blending with non-biodegradable polymers can compromise the biodegradability of the whole composite, using nanofillers can result in controlled and enhanced biodegradability (Ray *et al.*, 2002a; 2003c; Fujimoto *et al.*, 2003a). Post-consumer food packaging is usually contaminated with food waste, either remnants of the food that it protected or commingled with uneaten or discarded food. Although recycling is a possible option, the principle of disposing food waste and food packaging by either composting or anaerobic digestion is increasingly seen as a viable method, particularly if energy recovery in the form of biogas can be achieved at the same time (WRAP, 2009).

It has been pointed out that many of the additives, fillers and organic clay modifiers used in nanocomposite manufacture are not approved for food contact (Sorrentino *et al.*, 2007), although one viewpoint is that there is no reason to believe that there is any immediate risk by using substances in approved lists (Lagaron *et al.*, 2005). A comprehensive review of the legislation on food technology in 2005 did not reveal any legislation regarding nanocomposites or nanoparticles (Arvanitoyannis *et al.*, 2005). However, the impacts and implications of nanotechnology in food applications were reviewed in 2009 (Bouwmeester *et al.*, 2009). The potential for consumer exposure to and ingestion of nanoparticles was assessed with respect to the applications (barrier, strength and anti-microbial). It was concluded that, as the nanoparticles were to be directly incorporated into the polymer, there was potential for exposure to the food but not directly to the consumer. Various reports on the subject of nanoparticles in packaging call for new protocols for evaluating the migration of nanoparticles (Dainelli *et al.*, 2008). The safety requirements for 'active and intelligent packaging', which includes biodegradable packaging, are stated in EU regulations 1935/2004/EC and 450/2009/EC.

In summary, this area is still very much under development and, for the time being, the 'precautionary principle' still very much applies in biodegradable nanocomposite research.

### 13.5.3 Automotive applications

The steadily increasing price of oil has an impact on all industries, not least the automotive industry, which has responded by seeking ways to make vehicles more fuel-efficient by substituting non-structural metal body parts with polymer nanocomposite materials (Presting and König, 2003; Hussain *et al.*, 2006; Salonitis *et al.*, 2010; Mohanty *et al.*, 2003b; Moore, 2003; Leaversuch, 2001). In this application, the concern is not so much for biodegradability of the material, but for its sustainability in the context of rising and fluctuating oil prices. The manufacturer Toyota was one of the 'early adopters' of the new technology, although many other manufacturers are following suit and any list of names would rapidly be out of date.

Research in automotive applications has tended to focus on PLA, PBS and cellulose esters as the matrix polymer (Moore, 2003; Harris and Lee, 2008; Lee *et al.*, 2009), while the nanofillers are either organoclays (Ray *et al.*, 2006a) or plant fibres (Baldwin *et al.*, 2003; Lanzillotta *et al.*, 2002; Oksman *et al.*, 2003; Mohanty *et al.*, 2005) such as kenaf (Lee *et al.*, 2009; Huda *et al.*, 2008), hemp and cellulose fibres (Huda *et al.*, 2004a; 2004b; Samir *et al.*, 2004; Eichhorn *et al.*, 2001).

The three sectors, food packaging, automotive applications and biomedical, are all sectors which have a risk-averse approach to new product development. Consequently it is to be expected that progress in these applications is likely to be slow but steady (Hussain *et al.*, 2006).

## 13.6 References


Angles, M.N., Vignon, M.R., Dufresne, A., EMBRAPA, UNESP and USP 2000, 'Processing and characterisation of plasticised starch 1 tunicin whiskers nanocomposite materials', *Natural Polymers and Composites.*, eds L.H.C. Mattoso, A. Leao and E. Frollini, 14–17 May, p. 206.

Arvanitoyannis, I.S., Choreftaki, S. and Tserkezou, P. 2005, 'An update of EU legislation (directives and regulations) on food-related issues (safety, hygiene, packaging, technology, GMOs, additives, radiation, labelling): presentation and comments', *International Journal of Food Science and Technology*, vol. 40, no. 10, pp. 1021–1112.

Averous, L., Moro, L., Dole, P. and Fringant, C. 2000, 'Properties of thermoplastic blends: Starch-polycaprolactone', *Polymer*, vol. 41, no. 11, pp. 4157–4167.

Baldwin, K.A., Lee, E.C., Flanigan, C.M. and SPE, E.D. 2003, 'Use of natural fibre reinforced composites in the automotive industry', *GPEC 2003: Plastics Impact on the Environment*, SPE, Brookfield, CT, 26–27 February., pp. 87.

Bandyopadhyay, J., Maity, A., Khatua, B.B. and Ray, S.S. 2010, 'Thermal and rheological properties of biodegradable poly(butylene succinate)-co-adipate nanocomposites', *Journal of Nanoscience and Nanotechnology*, vol. 10, no. 7, pp. 4184–4195.

BeMiller, J. and Whistler, R.L. 2009, *Starch: Chemistry and Technology*, 3rd edn, Academic Press, London.

Bhatia, A., Gupta, R.K., Bhattacharya, S.N. and Choi, H.J. 2009, 'An investigation of melt rheology and thermal stability of poly(lactic acid)/poly(butylene succinate) nanocomposites', *Journal of Applied Polymer Science*, vol. 114, no. 5, pp. 2837–2847.

Bhatia, A., Gupta, R.K., Bhattacharya, S.N. and Choi, H.J. 2010, 'Effect of clay on thermal, mechanical and gas barrier properties of biodegradable poly(lactic acid)/poly(butylene succinate)(PLA/PBS) nanocomposites', *International Polymer Processing*, vol. 25, no. 1, pp. 5–14.

Booth, B. 2006, Polystyrene to Biodegradable PHA Plastics. *Environmental Science and Technology*, vol. 40, no. 7, 2074–2075.

Bouwmeester, H., Dekkers, S., Noordam, M.Y., Hagens, W.I., Bulder, A.S. *et al.* 2009, 'Review of health safety aspects of nanotechnologies in food production', *Regulatory Toxicology and Pharmacology*, vol. 53, no. 1, pp. 52–62.

Buchholz, B. and Entenmann, G. 1998, *Semi-Solid Mixtures of Amorphous Oligomers and Crystalline Polymers Based on Lactic Acid*, VS 5725881 A.

Cabedo, L., Feijoo, J.L., Villanueva, M.P., Lagaron, J.M., Saura, J.J. *et al.* 2005, 'Comparacion entre nanocompuestos biodegradables de poli(acido lactico) (PLA) amorfo con arcillas de distinta naturaleza', *Revista De Plasticos Modernos*, vol. 584, pp. 177–183.

Capon, R.J., Dunlop, R.W., Ghisalberti, E.L. and Jeffries, P.R. 1983, 'Poly-3-hydroxyalkanoates from marine and freshwater cyanobacteria', *Phytochemistry*, vol. 22, no. 5, pp. 1181–1184.

Chen, G.X. and Yoon, J.S. 2005, 'Clay functionalization and organization for delamination of the silicate tactoids in poly(l-lactide) matrix', *Macromolecular Rapid Communications*, vol. 26, no. 11, pp. 899–904.

Chen, G.X., Kim, H.S., Kim, E.S. and Yoon, J.S. 2005, 'Compatibilization-like effect of reactive organoclay on the poly (l-lactide)/poly (butylene succinate) blends', *Polymer*, vol. 46, no. 25, pp. 11829–11836.

Chen, C., Lv, G., Pan, C., Song, M., Wu, C., *et al.* 2007, 'Poly (lactic acid)(PLA) based nanocomposites – A novel way of drug-releasing', *Biomedical Materials*, vol. 2, p. L1.

Cohn, D. and Hotovely Salomon, A. 2005, 'Designing biodegradable multiblock PCL/PLA thermoplastic elastomers', *Biomaterials*, vol. 26, no. 15, pp. 2297–2305.

Cui, F.Z., Du, C., Feng, Q.L., Zhu, X.D. and De Groot, K. 1998, 'Tissue response to nano-hydroxyapatite/collagen composite implants in marrow cavity', *Journal of Biomedical Materials Research*, vol. 42, no. 4, pp. 540–548.

Dainelli, D., Gontard, N., Spyropoulos, D., Zondervan-van den Beuken, E. and Tobback, P. 2008, 'Active and intelligent food packaging: legal aspects and safety concerns', *Trends in Food Science and Technology*, vol. 19, no. 1, pp. S103–S112.

Doi, Y. 1990, *Microbial Polyesters*, VCH Publishers, Inc, US.

Dubief, D., Samain, E. and Dufresne, E. 1999, 'Polysaccharide microcrystals reinforced amorphous poly(beta-hydroxyoctanoate) nanocomposite materials', *Macromolecules*, vol. 32, no. 18, pp. 5765–5771.

Dufresne, A., Kellerhals, M.B. and Witholt, B. 1999, 'Transcrystallization in Mcl-PHAs/cellulose whiskers composites', *Macromolecules*, vol. 32, no. 22, pp. 7396–7401.

Eichhorn, S.J., Baillie, C.A., Zafeiropoulos, N., Mwaikambo, L.Y., Ansell, M.P. *et al.* 2001, 'Current international research into cellulosic fibres and composites', *Journal of Materials Science (USA)*, vol. 36, no. 9, pp. 2107–2131.

Eslami, H., Grmela, M. and Buosmina, M. 2010, 'Linear and nonlinear rheology of polymer/layered silica nanocomposites', *J. Rheology*, vol. 54, no. 3, pp. 539–562.

Feijoo, J.L., Cabedo, L., Gimenez, E., Lagaron, J.M. and Saura, J.J. 2005, 'Development of amorphous PLA-montmorillonite nanocomposites', *Journal of Materials Science*, vol. 40, no. 7, pp. 1785–1788.

Findlay, R.H. and White, D.C. 1983, 'Polymeric beta-hydroxyalkanoates from environmental samples and *Bacillus megaterium*', *Appl. Environ. Microbiol.*, vol. 45, pp. 71–78.

Fletcher, A. 1993, 'PHA as natural, biodegradable polyesters' in *Plastics from Bacteria and for Bacteria*, Springer-Verlag, New York, pp. 77–93.

Fujimoto, Y., Ray, S.S., Okamoto, M., Ogami, A., Yamada, K. *et al.* 2003a, 'Well-controlled biodegradable nanocomposite foams. from microcellular to nanocellular', *Macromolecular Rapid Communications*, vol. 24, no. 7, pp. 457–461.

Fujimoto, Y., Ogami, A., Okamoto, M., Ray, S.S., Ueda, K. *et al.* 2003b, 'New polylactide/layered silicate nanocomposites. 5. Designing of materials with desired properties', *Polymer*, vol. 44, no. 21, pp. 6633–6646.

Fukuzaki, H., Aiba, Y., Yoshida, M., Asano, M. and Kumakura, M. 1989, 'Direct copolymerisation of lactic acid with butyrolactone in the absence of catalysts', *Makromolekulare Chemie*, vol. 190, no. 7, pp. 1553–1559.

Gallant, D. and Bouchet, B. 1986, 'Ultrastructure of maize starch granules. A review', *Food Microstructure*, vol. 5, no. 1, pp. 141–155.

Gerngross, T.U. and Slater, S.C. 2000, 'How green are green plastics?', *Scientific American*, vol. 283, no. 2, pp. 24–29.

Gogolewski, S., Jovanovic, M., Perren, S.M., Dillon, J.G. and Hughes, M.K. 1993, 'Tissue response and *in vivo* degradation of selected polyhydroxyacids: Polylactides (PLA), poly(3-hydroxybutyrate) (PHB) and poly(3-hydroxybutyrate-co-3-hydroxyvalerate) (PHB/VA)', *Journal of Biomedical Materials Research*, vol. 27, no. 9, pp. 1135–1148.

Guangxin, C. and Yoon, J.-S. 2005, 'Nanocomposites of poly((butylene succinate)-co-(butylene adipate)) (PBSA) and twice functionalized organoclay', *Polymer International*, vol. 54, no. 6, pp. 939–945.

Han, S.I., Lim, J.S., Kim, D.K., Kim, M.N. and Im, S.S. 2008, '*In situ* polymerized poly(butylene succinate)/silica nanocomposites: Physical properties and biodegradation', *Polymer Degradation and Stability*, vol. 93, no. 5, pp. 889–895.

Harada, M., Ohya, T., Iida, K., Hayashi, H., Hirano, K. *et al.* 2007, 'Increased impact strength of biodegradable poly (lactic acid)/poly (butylene succinate) blend composites by using isocyanate as a reactive processing agent', *Journal of Applied Polymer Science*, vol. 106, no. 3, pp. 1813–1820.

Harris, A.M. and Lee, E.C. 2008, 'Improving mechanical performance of injection molded PLA by controlling crystallinity', *Journal of Applied Polymer Science*, vol. 107, no. 4, pp. 2246–2255.

Holmes, P.A., Wright, L.F. and Collins, S.H. 1981, *Beta-hydroxybutyrate Polymers*, EP0052459.

Hoover, R. 2001, 'Composition, molecular structure, and physicochemical properties of tuber and root starches: A review', *Carbohydrate Polymers*, vol. 45, no. 3, pp. 253–267.

Huang, S.J. 2005, 'Poly(lactic acid) and copolyesters' in *Handbook of Biodegradable Polymers*, ed. C. Bastioli, Rapra Technology Limited, Shawbury, Shrewsbury, pp. 287–301.

Huda, M.S., Mohanty, A.K., Drzal, L.T., Misra, M., Schut, E. and SPE 2004a, 'Physico-mechanical properties of 'green' composites from polylactic acid (PLA) and cellulose fibers', *GPEC 2004: Plastics – Helping Grow A Greener Environment*, SPE, 2004, Brookfield, Ct, 18–19 February, p. 12.

Huda, M.S., Mohanty, A.K., Misra, M., Drzal, L.T., Schut, E. and SPE 2004b, 'Effect of processing conditions on the physico-mechanical properties of cellulose fiber reinforced poly(lactic acid)', *ANTEC 2004. Proceedings of the 62nd SPE Annual conference in Chicago, IL*, 16–20 May, p. 1614.

Huda, M.S., Drzal, L.T., Mohanty, A.K. and Misra, M. 2008, 'Effect of fiber surface-treatments on the properties of laminated biocomposites from poly(lactic acid) (PLA) and kenaf fibers', *Composites Science and Technology*, vol. 68, no. 2, pp. 424–432.

Hussain, F., Hojjati, M., Okamoto, M. and Gorga, R.E. 2006, 'Review article: polymer-matrix nanocomposites, processing, manufacturing, and application: an overview', *Journal of Composite Materials*, vol. 40, no. 17, pp. 1511.

Jayakumar, R., Menon, D., Manzoor, K., Nair, S.V. and Tamura, H. 2010, 'Biomedical applications of chitin and chitosan based nanomaterials – A short review', *Carbohydrate Polymers*, vol. 82, no. 2, pp. 227–232.

Jayakumar, R., Prabaharan, M., Sudheesh Kumar, P.T., Nair, S.V. and Tamura, H. 2011, 'Biomaterials based on chitin and chitosan in wound dressing applications', *Biotechnology Advances*, vol. 29, no. 3, pp. 322–337.

Johnson, R.M., Mwaikambo, L.Y. and Tucker, N. 2003, *Biopolymers*, Rapra Technology Ltd, London.

Johnstone, B. 1990, 'A throw away answer', *Far Eastern Econ. Rev.*, vol. 147, no. 6, pp. 62–63.

Katti, K.S. 2004, 'Biomaterials in total joint replacement', *Colloids and Surfaces B: Biointerfaces*, vol. 39, no. 3, pp. 133–142.

Kazunari, M. and Yoshito, Y. 2000, *Production of Biodegradable Plastic from Vegetable Oil Waste*, JP2000189183.

Krikorian, V. and Pochan, D.J. 2003, 'Poly (l-lactic acid)/layered silicate nanocomposite: fabrication, characterization, and properties', *Chemistry of Materials*, vol. 15, no. 22, pp. 4317–4324.

Kulkarni, R.K., Pani, K.C., Neuman, C. and Leonard, F. 1966, *Polylactic Acid for Surgical Implants*, United States, AD636716.

Lagaron, J.M., Cabedo, L., Cava, D., Feijoo, J.L., Gavara, R. *et al.* 2005, 'Improving packaged food quality and safety. Part 2: Nanocomposites', *Food Additives and Contaminants*, vol. 22, no. 10, pp. 994–998.

Lange, J. and Wyser, Y. 2003, 'Recent innovations in barrier technologies for plastic packaging – A review', *Packaging Technology and Science*, vol. 16, no. 4, pp. 149–158.

Lanzillotta, C., Pipino, A., Lips, D. and SPE 2002, 'New functional biopolymer natural fiber composites from agricultural resources', *ANTEC 2002. Proceedings of the 60th SPE Annual Technical Conference in San Francisco, CA*, 5–9 May, pp. 5.

Leaversuch, R. 2001, 'Nanocomposites broaden roles in automotive, barrier packaging', *Plastics Technology*, vol. 47, no. 10, pp. 64–69.

Leaversuch, R. 2002, 'Renewable PLA polymer gets 'green light' for packaging uses', *Plastics Technology*, vol. 48, no. 3, pp. 50–55.

Lee, C.H., Hyun, Y.H., Lim, S.T., Choi, H.J. and Jhon, M.S. 2003a, 'Fabrication and viscoelastic properties of biodegradable polymer /organophilic clay nanocomposites', *Journal of Materials Science Letters*, vol. 22, no. 1, pp. 53–55.

Lee, J.H., Park, T.G., Park, H.S., Lee, D.S., Lee, Y.K., *et al.* 2003b, 'Thermal and mechanical characteristics of polylactic acid nanocomposite scaffold', *Biomaterials*, vol. 24, no. 16, pp. 2773–2778.

Lee, L.J., Zeng, C., Cao, X., Han, X., Shen, J. *et al.* 2005a, 'Polymer nanocomposite foams', *Composites Science and Technology*, vol. 65, no. 15–16, pp. 2344–2363.

Lee, Y.H., Lee, J.H., An, I.G., Kim, C., Lee, D.S. *et al.* 2005b, 'Electrospun dual-porosity structure and biodegradation morphology of montmorillonite reinforced PLLA nanocomposite scaffolds', *Biomaterials*, vol. 26, no. 16, pp. 3165–3172.

Lee, B., Kim, H., Lee, S., Kim, H. and Dorgan, J.R. 2009, 'Bio-composites of kenaf fibers in polylactide: Role of improved interfacial adhesion in the carding process', *Composites Science and Technology*, vol. 69, no. 15–16, pp. 2573–2579.

Lemoigne, M. 1925, *Ann. Inst. Pasteur (Paris)*, vol. 39, pp. 144–173.

Lemoigne, M. 1926, *Bull. Soc. Chim. Biol.*, vol. 8, pp. 770–782.

Lemoigne, M. 1927, *Ann. Inst. Pasteur (Paris)*, vol. 41, pp. 148–165.

Li, H.Y., Chen, Y.F. and Xie, Y.S. 2004, 'Nanocomposites of cross-linking polyanhydrides and hydroxyapatite needles: Mechanical and degradable properties', *Materials Letters*, vol. 58, no. 22–23, pp. 2819–2823.

Liu, L., Li, S., Garreau, H. and Vert, M. 2000, 'Selective enzymatic degradations of poly (L-lactide) and poly (ε-caprolactone) blend films', *Biomacromolecules*, vol. 1, no. 3, pp. 350–359.

Lovera, D., Márquez, L., Balsamo, V., Taddei, A., Castelli, C. *et al.* 2007, 'Crystallization, morphology, and enzymatic degradation of polyhydroxybutyrate/polycaprolactone (PHB/PCL) blends', *Macromolecular Chemistry and Physics*, vol. 208, no. 9, pp. 924–937.

Maiti, P., Batt, C.A. and Giannelis, E.P. 2003, 'Biodegradable polyester/layered silicate nanocomposites', *Nanomaterials for Structural Applications*, Materials Research Society, 506 Keystone Drive, Warrendale, PA, 15086, USA, MA; 2–6 December 2002.

Makhatha, M.E., Ray, S.S., Hato, J. and Luy, A.S. 2008, 'Thermal and thermomechanical properties of poly(butylene succinate) nanocomposites', *Journal of Nanoscience and Nanotechnology*, vol. 8, no. 4, pp. 1679–1689.

McLauchlin, A.R. and Thomas, N.L. 2009, 'Preparation and thermal characterisation of poly(lactic acid) nanocomposites prepared from organoclays based on an amphoteric surfactant', *Polymer Degradation and Stability*, vol. 94, no. 5, pp. 868–872.

Mohanty, A.K., Drzal, L.T. and Misra, M. 2003a, 'Nano reinforcements of bio-based polymers – The hope and the reality', *Polymeric Materials Science and Engineering Spring 2003 Meeting; New Orleans, LA, USA*, 23–27 March, American Chemical Society, 1155 16th St, NW, Washington, DC, 20036, USA [mailto:service@acs.org], pp. 60–61.

Mohanty, A.K., Misra, M., Drzal, L.T. and SPE 2003b, 'Emerging green nanocomposites: striving for sustainability in automotive applications', *GPEC 2003: Plastics Impact on the Environment*, SPE, Brookfield, CT, 26–27 February, pp. 69.

Mohanty, A.K., Misra, M. and Drzal, L.T. 2005, *Natural Fibers, Biopolymers, and Biocomposites*, CRC Press, Boca Raton, FL, USA.

Moore, S. 2003, 'Polylactic acid: Autos *au naturale*', *Plastics Today*. Online: www.plasticstoday.com/articles/autos-au-naturale.

Morin, A. and Dufresne, A. 2002, *Nanocomposites of Chitin Whiskers from Riftia Tubes and Poly(caprolactone)*, Macromolecules, vol. 35, no. 9, pp. 2190–2199.

Murphy, R.J. and Bartle, I. 2004, *Summary Report, Biodegradable Polymers and Sustainability: Insight from Life Cycle Assessment*, National Non-Food Crops Centre, UK.

Nair, K.G., Dufresne, A., Gandini, A. and Belgacem, M.N. 2003, *Crab Shell Chitin Whisker Reinforced Natural Rubber Nanocomposites. 2. Mechanical Behavior*, Biomacromolecules, vol. 4, no. 6, 1835–1842.

Nam, J.Y., Ray, S.S. and Okamoto, M. 2003, 'Crystallization behavior and morphology of biodegradable polylactide/layered silicate nanocomposite', *Macromolecules*, vol. 36, no. 19, pp. 7126–7131.

Narayan, R. 2009, 'Fundamental principles and concepts of biodegradability – Sorting through the facts, hypes and claims of biodegradable plastics in the marketplace', *Bioplastics Magazine*, vol. 4, no. 1. Online: www.ciras.iastate.edu/bioindustry/biobasedproducts/Narayan_biodegradable_plastics.pdf

Ogata, N., Jiminez, G., Kawai, H. and Ogihara, T. 1997, 'Structure and thermal/mechanical properties of poly(l-lactide)-clay blend', *Journal of Polymer Science: Polymer Physics Edition*, vol. 35, no. 2, pp. 389–396.

Oksman, K., Skrifvars, M. and Selin, J.-F. 2003, 'Natural fibres as reinforcement in polylactic acid (PLA) composites', *Composites Science and Technology*, vol. 63, no. 9, pp. 1317–1324.

Olkku, J. and Rha, C.K. 1978, 'Gelatinisation of starch and wheat flour starch – A review', *Food Chemistry*, vol. 3, no. 4, pp. 293–317.

Park, J.W. and Im, S.S. 2002, 'Phase behavior and morphology in blends of poly (L-lactic acid) and poly (butylene succinate)', *Journal of Applied Polymer Science*, vol. 86, no. 3, pp. 647–655.

Paul, M.-A., Alexandre, M., Degee, P., Henrist, C., Rulmont, A. *et al.* 2003, 'New nanocomposite materials based on plasticized poly(l-lactide) and organo-modified montmorillonites: Thermal and morphological study', *Polymer*, vol. 44, no. 2, pp. 443–450.

Paul, M.-A., Delcourt, C., Alexandre, M., Degee, P., Monteverde, F. *et al.* 2005a, 'Polylactide/montmorillonite nanocomposites. Study of the hydrolytic degradation', *Polymer Degradation and Stability*, vol. 87, no. 3, pp. 535–542.

Paul, M.-A., Delcourt, C., Alexandre, M., Degee, P., Monteverde, F., *et al.* 2005b, '(Plasticized) polylactide/(organo-)clay nanocomposites by *in situ* intercalative polymerization', *Macromolecular Chemistry and Physics*, vol. 206, no. 4, pp. 484–498.

Pham Hoai, N., Kaneko, M., Ninomiya, N., Fujimori, A. and Masuko, T. 2005, 'Melt intercalation of poly(l-lactide) chains into clay galleries', *Polymer*, vol. 46, no. 18, pp. 7403–7409.

Pillai, C.K.S., Paul, W. and Sharma, C.P. 2009, 'Chitin and chitosan polymers: Chemistry, solubility and fiber formation', *Progress in Polymer Science*, vol. 34, no. 7, pp. 641–678.

Pitt, C.G., Chasalow, F.I., Hibionada, Y.M., Klimas, D.M. and Schindler, A. 1981, 'Aliphatic Polyesters. 1. Degradation of poly(epsilon-caprolactone) *in vivo*', *Journal of Applied Polymer Science*, vol. 26, no. 11, pp. 3779–3787.

Pluta, M. and Galeski, A. 2002, 'Crystalline and supermolecular structure of polylactide in relation to the crystallization method', *Journal of Applied Polymer Science*, vol. 86, no. 6, pp. 1386–1395.

Pluta, M., Galeski, A., Alexandre, M., Paul, M.-A. and Dubois, P. 2002, 'Polylactide/montmorillonite nanocomposites and microcomposites prepared by melt blending: Structure and some physical properties', *Journal of Applied Polymer Science*, vol. 86, no. 6, pp. 1497–1506.

Poirier, Y., Dennis, D.E., Klomparens, K. and Somerville, C. 1992, 'Polyhydroxybutyrate, a biodegradable thermoplastic, produced in transgenic plants', *Science (Washington)*, vol. 256, no. 5056, pp. 520–523.

Presting, H. and König, U. 2003, 'Future nanotechnology developments for automotive applications', *Materials Science and Engineering: C*, vol. 23, no. 6–8, pp. 737–741.

Qiu, Z., Ikehara, T. and Nishi, T. 2003, 'Poly (hydroxybutyrate)/poly (butylene succinate) blends: Miscibility and nonisothermal crystallization', *Polymer*, vol. 44, no. 8, pp. 2503–2508.

Raabe, D., Sachs, C. and Romano, P. 2005, 'The crustacean exoskeleton as an example of a structurally and mechanically graded biological nanocomposite material', *Acta Materialia*, vol. 53, no. 15, pp. 4281–4292.

Ratnayake, W.S., Hoover, R. and Warkentin, T. 2002, 'Pea starch: Composition, structure and properties – A review', *Starch-Stärke*, vol. 54, no. 6, pp. 217–234.

Ray, S.S. and Bousmina, M. 2006, 'Crystallization behavior of poly((butylene succinate)-co-adipate) nanocomposite', *Macromolecular Chemistry and Physics*, vol. 207, no. 14, pp. 1207–1219.

Ray, S.S., Maiti, P., Okamoto, M., Yamada, K. and Ueda, K. 2002a, 'New polylactide/layered silicate nan-ocomposites. 1. Preparation, characterisation and properties', *Macromol.*, vol. 35, pp. 3104–3110.

Ray, S.S., Yamada, K., Ogami, A., Okamoto, M. and Ueda, K. 2002b, 'New polylactide/layered silicate nanocomposite: Nanoscale control over multiple properties', *Macromolecular Rapid Communications*, vol. 23, no. 16, pp. 943–947.

Ray, S.S., Yamada, K., Okamoto, M., Ogami, A. and Ueda, K. 2003a, 'New polylactide/layered silicate nanocomposites. 3. High-performance biodegradable materials', *Chemistry of Materials*, vol. 15, no. 7, pp. 1456–1465.

Ray, S.S., Yamada, K., Okamoto, M. and Ueda, K. 2003b, 'Biodegradable polylactide/montmorillonite nanocomposites', *Journal of Nanoscience and Nanotechnology*, vol. 3, no. 6, pp. 503–510.

Ray, S.S., Yamada, K., Okamoto, M. and Ueda, K. 2003c, 'Control of biodegradability of polylactide via nanocomposite technology', *Macromolecular Materials and Engineering*, vol. 288, no. 4, pp. 203–208.

Ray, S.S., Yamada, K., Okamoto, M. and Ueda, K. 2003d, 'New polylactide-layered silicate nanocomposites 2. Concurrent improvements of material properties, biodegradability and melt rheology', *Polymer*, vol. 44, no. 3, pp. 857–866.

Ray, S.S., Bousmina, M. and Okamoto, K. 2005, 'Structure and properties of nanocomposites based on poly(butylene succinate-co-adipate) and organically modified montmorillonite', *Macromolecular Materials and Engineering*, vol. 290, no. 8, pp. 759–768.

Ray, S.S., Okamoto, K. and Okamoto, M. 2006a, 'Structure and properties of nanocomposites based on poly(butylene succinate) and organically modified montmorillonite', *Journal of Applied Polymer Science*, vol. 102, no. 1, pp. 777–785.

Ray, S.S., Vaudreuil, S., Maazouz, A. and Bousmina, M. 2006b, 'Dispersion of multi-walled carbon nanotubes in biodegradable poly(butylene succinate) matrix', *Journal of Nanoscience and Nanotechnology*, vol. 6, no. 7, pp. 2191–2195.

Ray, S., Bandyopadhyay, J. and Bousmina, M. 2008, 'Influence of degree of intercalation on the crystal growth kinetics of poly(butylene succinate)-co-adipate nanocomposites', *European Polymer Journal*, vol. 44, no. 10, pp. 3133–3145.

Reddy, C.S.K., Ghai, R., Rashmi, R., and Kalia, V.C. 2003, 'Polyhydroxyalkanoates: An overview', *Bioresource Technology*, vol. 87, pp. 137–146.

Rhim, J.W. and Ng, P.K.W. 2007, 'Natural biopolymer-based nanocomposite films for packaging applications', *Critical Reviews in Food Science and Nutrition*, vol. 47, no. 4, pp. 411–433.

Rinaudo, M. 2006, *Chitin and Chitosan: Properties and Applications*. Progress in Polymer Science, vol. 31, no. 7, pp. 603–632

Saito, T., Tomita, K., Juni, K. and Ooba, K. 1991, '*In vivo* and *in vitro* degradation of poly(3-hydroxybutyrate) in rat', *Biomaterials*, vol. 12, no. 3, pp. 309–312.

Salonitis, K., Stavropoulos, P. and Chryssolouris, G. 2010, 'Nanotechnology for the needs of the automotive industry', *International Journal of Nanomanufacturing*, vol. 6, no. 1, pp. 99–110.

Samir, M.A.S.A., Alloin, F., Paillet, M. and Dufresne, A. 2004, 'Tangling effect in fibrillated cellulose reinforced nanocomposites', *Macromolecules*, vol. 37, no. 11, pp. 4313–4316.

Sanchez-Garcia, M.D. and Lagaron, J.M. 2010, 'Novel clay-based nanobiocomposites of biopolyesters with synergistic barrier to UV light, gas, and vapour', *Journal of Applied Polymer Science*, vol. 118, no. 1, pp. 188–199.

Sanchez-Garcia, M.D., Gimenez, E. and Lagaron, J.M. 2008, 'Morphology and barrier properties of solvent cast composites of thermoplastic biopolymers and purified cellulose fibers', *Carbohydrate Polymers*, vol. 71, no. 2, pp. 235–244.

Schmitt, E.E. and Polistina, R.A. 1969, *Polyglycolic acid prosthetic devices*, A61L17/12; A61L27/18; C08G63/06; D01F6/62B edn, A61B17/04, USA.

Shih, Y.F., Wang, T.Y., Jeng, R.J., Wu, J.Y. and Teng, C.C. 2007, 'Biodegradable nanocomposites based on poly(butylene succinate)/organoclay', *Journal of Polymers and the Environment*, vol. 15, no. 2, pp. 151–158.

Shih, Y.F., Chen, L.S. and Jeng, R.J. 2008, 'Preparation and properties of biodegradable PBS/multi-walled carbon nanotube nanocomposites', *Polymer*, vol. 49, no. 21, pp. 4602–4611.

Skraly, F.A. and Peoples, O. 2005, *Polyhydroxyalkanoate Production from Polyols.*

Song, L. and Qiu, Z. 2009, 'Crystallization behavior and thermal property of biodegradable poly(butylene succinate)/functional multi-walled carbon nanotubes nanocomposite', *Polymer Degradation and Stability*, vol. 94, no. 4, pp. 632–637.

Sorrentino, A., Gorrasi, G. and Vittoria, V. 2007, 'Potential perspectives of bio-nanocomposites for food packaging applications', *Trends in Food Science and Technology*, vol. 18, no. 2, pp. 84–95.

Suyatma, N.E., Copinet, A., Tighzert, L. and Coma, V. 2004, 'Mechanical and barrier properties of biodegradable films made from chitosan and poly(lactic acid) blends', *Journal of Polymers and the Environment*, vol. 12, no. 1, pp. 1–6.

Ten, E., Turtle, J., Bahr, D., Jiang, L. and Wolcott, M. 2010, 'Thermal and mechanical properties of poly(3-hydroxybutyrate-co-3-hydroxyvalerate)/cellulose nanowhiskers composites', *Polymer*, vol. 51, no. 12, pp. 2652–2660.

Thellen, C., Orroth, C., Froio, D., Ziegler, D., Lucciarini, J. *et al.* 2005, 'Influence of montmorillonite layered silicate on plasticized poly(l-lactide) blown films', *Polymer*, vol. 46, no. 25, pp. 11716–11727.

Vert, M., Schwarch, G. and Coudane, J. 1995, 'Present and future of PLA polymers' in *Degradable Polymers, Recycling, and Plastics Waste Management*, eds. A. Albertsson and S. Huang, 1st edn, Marcel Dekker Inc, NY, pp. 195–204.

Vink, E.T.H., Rabago, K.R., Glassner, D.A. and Gruber, P.R. 2003, 'Applications of life cycle assessment to NatureWorks™ polylactide (PLA) production', *Polymer Degradation and Stability*, vol. 80, no. 3, pp. 403–419.

Wallen, L.L. and Rohwedder, W.K. 1974, 'Poly-beta-hydroxyalkanoate from activated-sludge', *Environ. Sci. Technol.*, vol. 8, pp. 576–579.

Wang, X., Du, Y., Luo, J., Lin, B. and Kennedy, J.F. 2007, 'Chitosan/organic rectorite nanocomposite films: Structure, characteristic and drug delivery behaviour', *Carbohydrate Polymers*, vol. 69, no. 1, pp. 41–49.

Wang, Y., Ribelles, J.L.G., Sanchez, M.S. and Mano, J.F. 2005, 'Morphological contributions to glass transition in poly(l-lactic acid)', *Macromolecules*, vol. 38, no. 11, pp. 4712–4718.

Weir, N.A., Buchanan, F.J., Orr, J.F. and Dickson, G.R. 2004, 'Degradation of poly-L-lactide. Part 1: *In vitro* and *in vivo* physiological temperature degradation', *Proceedings of the Institution of Mechanical Engineers H, Journal of Engineering in Medicine*, vol. 218, no. 5, pp. 307–319.

WRAP, 2009, *Research Report: Biopolymer Packaging in UK Grocery Market*, WRAP, Banbury, UK.

Yoo, S. and Jane, J. 2002, 'Structural and physical characteristics of waxy and other wheat starches', *Carbohydrate Polymers*, vol. 49, no. 3, pp. 297–305.

Yun, S., Gadd, G., Latella, B., Lo, V., Russell, R. *et al.* 2008, 'Mechanical properties of biodegradable polyhydroxyalkanoates/single wall carbon nanotube nanocomposite films', *Polymer Bulletin*, vol. 61, no. 2, pp. 267–275.

# Part III
## Applications of polymer nanocomposites

# 14
# Polymer nanocomposites in fuel cells

H.-W. RHEE and L.-J. GHIL, Sogang University, South Korea

**Abstract:** Polymer electrolyte membrane fuel cells (PEMFCs) operating at room temperature have two major problems: (a) easy poisoning of the Pt catalyst by CO and (b) direct methanol fuel cells have a large fuel crossover. Considerable attention has been spent on PEMFCs operating at high temperatures. The US Department of Energy set a target conductivity of 0.1 S/cm at 120 °C and 50% relative humidity. In order to meet this criterion, the most promising approach may be to use nanocomposites of sulfonated polymer electrolyte membranes with inorganic compounds such as hygroscopic metal oxides and solid proton conductors such as heteropolyacids, anhydrous solid acids and layered metal phosphates.

**Key words:** polymer electrolyte membrane, polymer nanocomposite, organic–inorganic nanohybrid, hygroscopic metal oxide, solid proton conductor.

## 14.1 Introduction

Perfluorosulfonic acid membranes such as Nafion, Flemion, Dow, or Aciplex-S are the best polymer electrolyte membranes (PEMs) due to high proton conductivity (~0.1 S/cm) in addition to their chemical and electrochemical stability. However, they have to be fully hydrated to utilize high ionic conductivity and are mechanically unstable for operation at temperatures above 100 °C or in fully hydrated states. Also they are too expensive for the commercialization of PEM fuel cells (PEMFCs). Thus, much research has focused on sulfonated ionomers based on aromatic hydrocarbons such as poly(sulfone), poly(imide), poly(aryl ether ether ketone), etc. (Prater, 1994). But they have low conductivity, at least one order of magnitude lower than perfluorinated ionomers, and are very brittle due to very low elongation at break. In addition, there is large volume expansion on water uptake even though some of them are crosslinked.

Sulfonated ionomers have sulfonic acid groups and they need solvents with a high dielectric constant such as water, so that protons can dissociate from the sulfonated anions. The liberated protons transfer through the medium by a vehicle mechanism and a hopping mechanism. However, phosphoric or phosphonic acid can be highly conductive due to self-dissociation and dehydration even in the anhydrous form, which is attractive for high-temperature operation and proton transfer through the hopping mechanism (Gillespie *et al.*, 1965).

According to the Gierke cluster network model, all sulfonic-acid-based PEMs have nanoporous ionic clusters (2–5 nm in diameter) interconnected through narrow and short channels (~1 nm in diameter) due to phase separation between the polymer matrix backbones and the sulfonic acid groups. Therefore, it is very important to keep the ion clusters and the membrane channels fully hydrated to maintain the maximum conductivity of a sulfonated PEM. The drawbacks of PEMFCs operating at room temperature (fully hydrated) are their low CO tolerance and slow reaction kinetics on the cathode side, which requires very expensive Pt catalyst. Therefore, considerable interest has recently focused on PEMFCs operating at a low relative humidity (RH) (and consequently at high temperatures above 100 °C at ambient pressure) (Alberti and Casciola, 2003). The attractive features of PEMFCs under these conditions are as follows:

- Reduced sensitivity of the anode catalyst to CO poisoning: hence there is lower Pt/Ru catalyst loading and preferential oxidation (PROX) is eliminated.
- Reduction in activation overpotentials: leading to a better current density and a more compact stack.
- Easier waste heat dispersion.
- Simplification in water management: there is no requirement to keep the membrane hydrated.

However, under such conditions all sulfonated membranes tend to dehydrate and their proton conductivity drastically reduces, and so the overall performance of a PEMFC reduces. Worldwide, most research is focused on the development of membranes that have improved proton conductivity, water retention or management, and reduced methanol crossover in the temperature range 100–150 °C. There are four different approaches for developing alternative membranes for high temperatures: (1) modification of the perfluorinated ionomers (Karlsson and Jannasch, 2005), (2) functionalization of the aromatic hydrocarbon membranes to improve ionic conductivity (Yin *et al.*, 2005), (3) the use of polymer blends or ionic complexes between basic polymers and acidic polymers (Yang *et al.*, 2001), and (4) nanocomposite membranes based on solid inorganic materials (Arico *et al.*, 2001). Perhaps the nanocomposite approach is the mostly widely used and the most promising; in this approach inorganic nanoparticles of a few hundred nanometers are introduced into electrolyte membranes. Recently, numerous polymer-based nanocomposite membranes have fascinated researchers in the fuel-cell field (especially, PEMFCs) (Karen and Richard, 2007). Nanocomposite membranes could drastically reduce the high permeability for fuels (Kickner *et al.*, 2004; Vladimir *et al.*, 2007; Savadogo, 2004). Therefore, nanocomposites are expected to fulfill the following properties required for commercialization of PEMFCs at high temperatures (100–150 °C) and low relative humidity:

- High proton conductivity (the US Department of Energy target for automotive applications is 0.1 S/cm at 120 °C and 50% RH).
- High resistance to electrons.
- Excellent barrier properties to fuels and oxidants to avoid diffusion and leakage.

- High water retention: the presence of a hygroscopic additive binds a larger amount of water in the membrane, increasing the membrane water content at a given RH.
- Improved long-term chemical and thermomechanical properties such as glass transition temperature ($T_g$) and Young's modulus.
- Minimized cathode flooding problem, which will increase catalyst activity through reduced water transport.
- Cost compatible with commercial requirements.
- Improved electrode performance: an extended reaction zone may be available.

The prerequisites for polymer electrolyte membranes can be effectively satisfied by synergistically combining both organic PEM and inorganic nanomaterials, which could be included in nanoclusters of sulfonated ionomer membranes. A polymer membrane filled with inorganic nanoparticles is schematically shown in Fig. 14.1. These composite materials possess enhanced properties compared with single organic or inorganic materials. A polymer nanocomposite is made of a polymeric matrix in which an inorganic phase or reinforcing compound is incorporated at a scale of less than 100 nm. Sulfonated ionomeric membranes when used as a matrix provide processability, flexibility, and ion channels for conduction. The inorganic compounds may be hygroscopic or proton conductive, in the form of nanoparticles, layered structures, or nanofibers to increase ionic conductivity, mechanical and thermal stability as well as to reduce the permeability for fuels.

If hydrophilic nano-sized inorganic materials are included in the nanopores of sulfonated membranes, they could retain moisture even at high temperatures due to their extremely high surface area-to-volume ratio. When the inorganic phase is grafted with conductive groups such as carboxylic, sulfonic, or phosphonic acid groups, their conductivity can be further enhanced. The hygroscopic characteristics of the nanomaterials means they adsorb water molecules to a greater extent, store them in the voids, and allow operation at a lower humidity. Proton transfer takes place on the surface of the inorganic nanoparticles, which in turn improves the surface chemistry. Interaction between the organic and inorganic components influences electrochemical stability (Kickelbock, 2003). Therefore, polymer nanocomposite materials can be used to make fuel cells with an extended durability at high temperatures and lower humidity (Wu *et al.*, 2006; Kickelbock, 2003).

When inorganic compounds are intrinsically proton conductors at high temperatures, they are an additional source of protons and also act as stepping stones in proton transfer. When a PEM is reinforced by hard inorganic materials, the thermal properties are improved and thermomechanical degradation, such as cracks, tears, punctures, or pinhole blisters, is reduced (Collier *et al.*, 2006). The advantages of mechanically reinforced membranes allow us to use much thinner membranes, which then implies much lower resistance of the membranes when in use. Therefore, polymer nanocomposites in PEMFCs are valuable tools for both the research community and industry professionals.

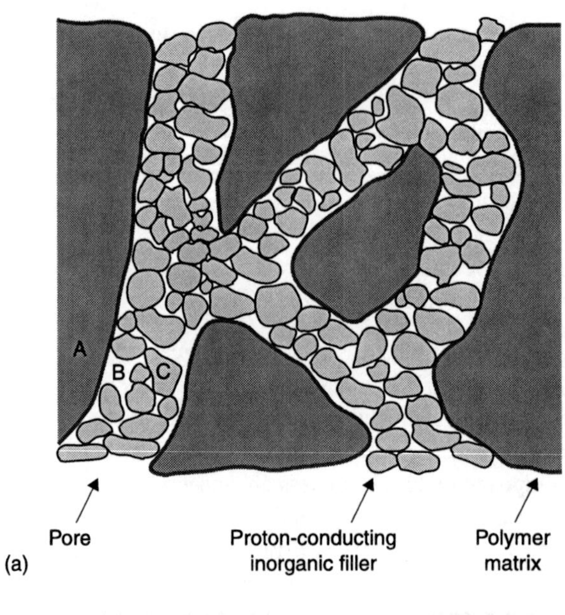


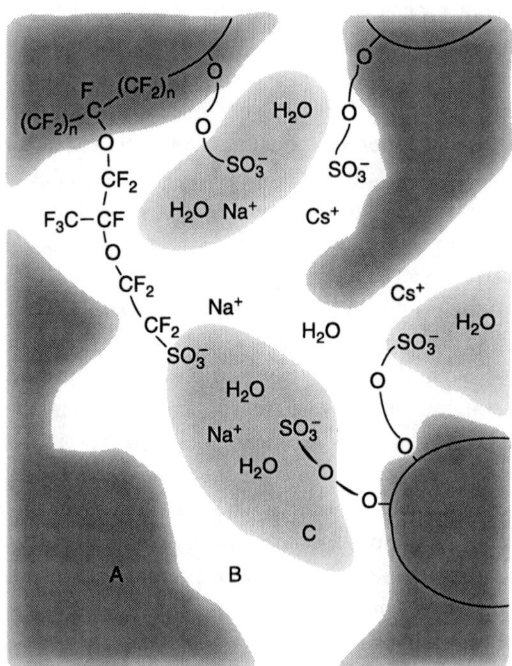


*14.1* (a) Porous polymer membrane filled with proton-conducting inorganic particles. (b) Three-region model for Nafion. (A) hydrophobic region (B) hydrophilic interfacial zone (C) hydrophilic region.

For PEMFCs operating at a high temperature and low RH many types of nanocomposite are available, but only two approaches will be considered in this chapter: (1) the improvement of the characteristics of known membranes by dispersing hygroscopic metal oxides inside the polymeric matrix and (2) membranes obtained by filling the polymeric matrix with inorganic solid proton conductors.

The properties of composite membranes depend not only on the nature of the ionomer and the solid compound used, but also on the amount, homogeneous dispersion, size and orientation of the solid compounds dispersed in the polymeric matrix. In ionomers such as Nafion, where well-separated hydrophilic and hydrophobic domains are present, the distribution of the dispersed compound in the different domains is also important. Due to the high specific surface of nanomaterials, the interfacial interaction with the polymer matrix will become very high and can have a large influence on the properties of the polymer itself. Thus, nanoparticles are expected to modify the original characteristics to a greater extent than large particles.

## 14.2 Using polymer nanocomposites in fuel cells

As mentioned previously, there are two major drawbacks of sulfonated PEMs: (1) their poor conductivity at elevated temperatures and low RH in PEMFCs and (2) large methanol crossover in direct methanol fuel cells (DMFCs). This section will focus on the development of nanocomposite membranes, which are stable in the temperature range 100–150 °C or which are methanol resistant. Many inorganic materials can be used for these nanocomposite membranes and they can be categorized as follows: (a) hygroscopic metal-oxide nanoparticles, which improve water retention at high temperatures and (b) inorganic proton conductors, with reasonable conductivity hardly dependent on humidity. The latter include heteropolyacids, anhydrous solid acids, and layered metal phosphates or their phosphonates in the form of nanoparticles or layered sheets. We will discuss and compare general features of inorganic compounds, preparation techniques, and the properties of nanocomposite membranes with sulfonated polymeric ionomers and finally their application and performance in PEMFCs.

### 14.2.1 Metal oxide-filled nanocomposites

Metal oxides are very hygroscopic and robust compounds and have been widely used as inorganic fillers in polymer composites. The most popular are $SiO_2$, $TiO_2$, and $ZrO_2$, and they have several advantages: (a) they have a large number of hydroxyl groups on the surface, (b) they are easy to prepare by the sol-gel reaction of metal alkoxides under mild conditions, and (c) a wide range of particle size and morphology is available by controlling the process variables (composition, pH, drying, etc.). Therefore, the incorporation of metal oxides within polymer membranes is intended to (a) increase proton conductivity and maintain water retention at high temperatures and low relative humidity, (b) increase resistance to fuel crossover by creating barriers in the flow channels, and (c) improve thermomechanical properties such as rigidity and thermal degradation at high temperatures (Antonucci *et al.*, 1999).

Nanocomposites with a small amount of metal oxides have enhanced water retention and improved membrane performance above 100 °C (Antonucci *et al.*, 1999). This result is believed to be due to increased water adsorption on the hygroscopic metal oxide surfaces, which enhances back-diffusion of water produced at the cathode and reduces the electro-osmotic drag from the anode to the cathode (Hogarth *et al.*, 2005). Arico *et al.* (2001) confirmed the acidity of surface functional groups on the particle surfaces; their strength and number enhanced water retention in membranes in the following order: $SiO_2$ > neutral $Al_2O_3$ > basic $Al_2O_3$ > $ZrO_2$. When the number of particles was above 10%, proton conductivity remarkably decreased and resulted in a brittle membrane (Wang and Saxena, 2001).

There are many studies on Nafion-silica composite membranes with regard to the performance of PEMFCs at temperatures above 100 °C, which helps Nafion's water retention when hydrated. Better PEMFC performance has been observed with a nanocomposite membrane at temperatures above 100 °C, although the membrane showed lower conductivity at lower temperatures. Plausible explanations may be as follows: (a) the replacement of unloaded concentrated water by hydrophilic nanosilica particles, (b) capillary condensation effects, due to smaller empty spaces in $SiO_2$ pores, and (c) improved mechanical properties of the nanocomposite membrane (Wakizoe *et al.*, 1995). In addition, $SiO_2$ nanoparticles might affect membrane permselectivity in two ways: (a) by increasing the membrane fixed charge density proportional to the amount of $SiO_2$, and (b) by increasing the pore density of the organic–inorganic hybrid membrane with incorporated $SiO_2$ nanoparticles, which was supported by water-uptake studies and scanning electron microscope (SEM) images (Sahu *et al.*, 2007).

$TiO_2$ is also a good hydrophilic filler for polymer membranes and its composite matrix has enhanced water absorbent characteristics (Jung *et al.*, 2006) and improved mechanical and thermal properties for a moderate amount (Devrim *et al.*, 2009). When too much $TiO_2$ is added, component mixing decreases, resulting in a brittle membrane (Wang and Saxena, 2001).

$ZrO_2$ is a dioxide metal with hygroscopic properties. Adding this material as an inorganic component could increase membrane stability at high temperatures without disturbing conductivity and the performance of the system. Zirconyl chloride ($ZrOCl_2$) is one of the precursors used as a starting component, and is very useful in producing composite membranes using the impregnation method (Alberti *et al.*, 2007).

Several methods and techniques have been developed to incorporate nano-sized metal oxides into membranes, for example, solution casting, sol-gel processing, and *in situ* impregnation. Solution casting is one of the simplest methods for the preparation of nanocomposite membranes. A mixed solution containing the dissolved polymer and nano-sized inorganic nanofillers is cast as a film followed by the evaporation of the solvent (Gnana *et al.*, 2007). It directly incorporates nano-sized inorganic materials into the polymer matrix when an appropriate amount of nanoparticles is first mixed with the polymer solution under vigorous stirring. For bulk mixing, the inorganic components should be prepared in the form of powders or colloids (Jalani *et al.*, 2005). It is very easy to control the thickness of the composite membranes with solution casting. But a physical interaction between the polymer and the nanomaterial has raised a few questions, such as whether there are homogeneities in the composite membranes and regarding the formation of pores around the metal oxide particles, and the leaching of inorganic materials under repetitive cycles.

The sol-gel method is well established and it is the most convenient process for producing metal oxides from a precursor solution (metal alkoxides or metal chlorides) by hydrolysis and polycondensation reactions. When it is used for the preparation of nanocomposite membranes, it may produce completely transparent and homogeneous membranes, compared with the casting method, which generally results in cloudy membranes due to the larger particle size. The sol-gel process can be used to produce organic–inorganic nanohybrids due to no or negligible phase separation of the nano-sized particles within the membrane matrix (Zoppi *et al.*, 1997; Zoppi and Nunes, 1998). Two different morphologies are schematically compared in Fig. 14.2. In the sol-gel technique, metal oxides are not restricted to the ion channels of the sulfonic acid ionomers and the films become brittle with high contents of metal oxides (> 50%). In addition, the molecular structures of the precursors are very important in controlling the phase separation, flexibility, and ionic conductivity of the resultant membranes (Nunes *et al.*, 2004).

*In situ* impregnation (or the *in situ* sol-gel process) is a versatile and promising technique for preparing low-cost nanocomposites based on the sol-gel reaction. PEM swells in a solvent, to which a a precursor solution (silicon alkoxide) is added, and the precursor starts to migrate into the ion channels or clusters. In other words, the swollen membrane is impregnated by the precursor solution and the sol-gel reaction begins inside the membrane template (Jalani *et al.*, 2005). The *in situ* sol-gel reaction can be catalyzed by an additional acid or an acidified polymer that possesses the pendant $SO_3^-H^+$ group cluster (Kelarakis *et al.*, 2010). Mauritz's group (1998) used the latter to prepare Nafion/$SiO_2$ nanocomposites by infiltration of a Nafion membrane with silicone alkoxides. The advantage of the *in situ* method is that particle size can be controlled by the concentration of precursors; the size and dispersion of particles are of the utmost importance in the final performance of fuel cells. However, this technique sometimes results in significant deposits on the surfaces of the membrane and a concentration gradient across the thickness of the membrane, with a greater $SiO_2$ concentration near the surfaces. This result is due to the difficulty in diffusing the precursor solution

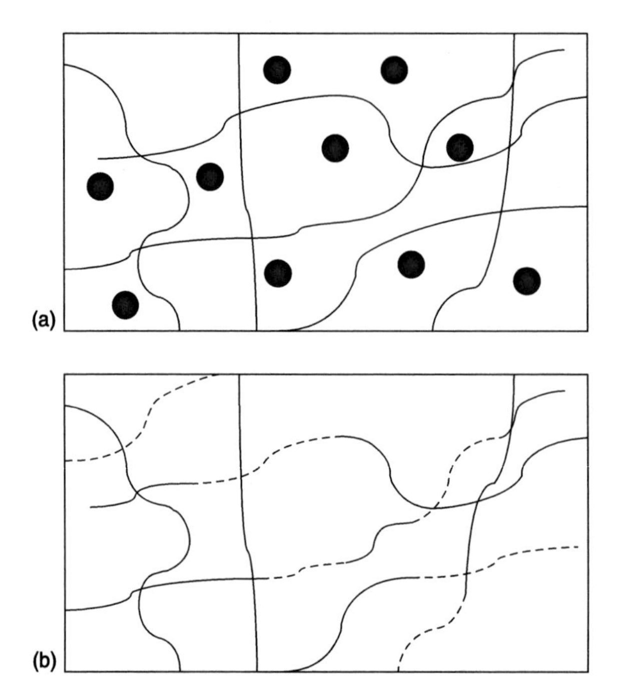


*14.2* Strategies in the preparation of organic–inorganic membranes: (a) isotropic nano-sized inorganic particles and (b) a combination of organic and inorganic monomers, which react to form a common chain or network.

through the narrow channels of Nafion. This gradient could be mitigated by using an acid-catalyzed, pre-hydrolyzed silicone alkoxide solution in alcohol, so that hydrolyzed molecules migrate to the polar clusters at the same time as the membrane is swollen (Klein *et al.*, 2005).

Now we will consider the properties of nanocomposites, including conductivity, permeability, mechanical properties, and water uptake, as well as fuel-cell performance. Table 14.1 compares the water uptake of composite membranes under various conditions. Nanocomposites have a higher water uptake than Nafion on its own due to the water absorbency of the metal oxides in the nanocomposite membranes (Helen *et al.*, 2007). Water uptake is an important property for sulfonated polymer membranes because it is directly related to conductivity, which is almost linearly proportional to the permeability of methanol (Ise *et al.*, 1999). Water uptake by the dried membrane linearly increases with immersion time, which indicates that $SiO_2$ percolates through the swollen membrane with time (Mauritz, 1998). Water uptake also affects the degradation of the membrane–catalyst interface (Zawodzinski *et al.*, 1993).

Datta and co-workers have spent a good deal of effort studying sol-gel synthesized Nafion/$SiO_2$, Nafion/$TiO_2$, and Nafion/$ZrO_2$ nanocomposite

*Table 14.1* Water uptake of Nafion® and metal oxide nanocomposites

| Polymer | Metal oxide | Condition $T_{dry}/T_{wet}$ (°C) | Water uptake (%) | Reference |
|---|---|---|---|---|
| Nafion® | $ZrO_2$ | 100/100 | 28 | Costamagna *et al.*, 2002 |
| | – | | 20 | Costamagna *et al.*, 2002 |
| Nafion® | $TiO_2$ | 80/25 | 29 | Sacca *et al.*, 2006 |
| | – | | 27 | Sacca *et al.*, 2006 |
| Nafion® | $SiO_2$ | 105/25 | 34 | Tiang *et al.*, 2008 |
| | – | | 29.8 | Tiang *et al.*, 2008 |

membranes, in particular their relative performance compared with unmodified Nafion in terms of water uptake, proton conductivity at different relative humidity conditions, fuel-cell performance, and ion exchange measurements (Jalani *et al.*, 2005). At temperatures of about 90 °C and 120 °C, Nafion/$MO_2$ sol-gel membranes were found to display better water uptake at a given RH than Nafion itself. The basic sorption behavior was similar at both temperatures, with water uptake increasing from $SiO_2$ to $TiO_2$ to $ZrO_2$ nanocomposites, which is in order of increasing acid strength. However, at the temperatures of interest, higher proton conductivity has been observed only for the Nafion/$ZrO_2$ sol-gel nanocomposite, with only a 10% enhancement at 40% RH compared with pristine Nafion. Also, the sorption isotherm shape obtained for nanocomposite membranes was similar to that of Nafion, with a sharp increase in the amount of water uptake above a water activity of 0.6. Hence, the basic mechanism of water sorption must be similar for all the oxide-incorporated nanocomposite membranes. The difference is probably due to the change in acidity or active surface area of the membranes.

Figure 14.3 compares the fuel-cell performance of Nafion with that of Nafion/$TiO_2$ nanocomposites. Three composite membranes were prepared using $Ti(OC_4H_9)_4$. The result indicated that $TiO_2$ content increased nearly linearly with membrane thickness under the experimental conditions. The polarization characteristics of all three membrane electrode assemblies (MEAs) of the nanocomposite membranes were improved significantly compared with those of pure Nafion.

Nanocomposite membranes based on sulfonated hydrocarbon have several attributes of interest, including decreased water swelling, reduced permeability towards methanol, and improved morphological stability without compromising proton conductivity for a high degree of sulfonation. For recast sulfonated polyether

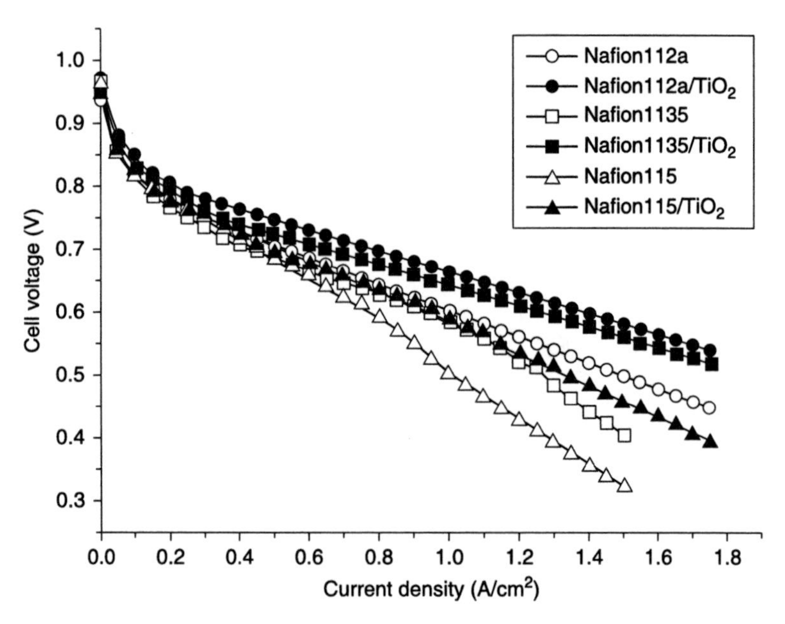


*14.3* Current versus voltage curves for Nafion and Nafion/$TiO_2$ nanocomposites: $T_{cell}$ = 70 °C, $T_{O2}$ = 85 °C, and $T_{H2}$ = 75 °C.

ether ketone (sPEEK)/$SiO_2$ membranes (silica loading up to 20 wt%), the best balance between electrical and mechanical characteristics was accomplished at a silica content of 10 wt%. However, attempts to prepare nanocomposite membranes by *in situ* generation of the silica phase within hydrocarbon membranes were not that successful due to the creation of large particles and auto-aggregation of the silica phase as well as the formation of cavities in the polymer matrix during the hydrolysis of silane (Nunes *et al.*, 2002). It is also very hard to achieve a homogeneous dispersion of $SiO_2$ particles within the hydrocarbon matrix by the sol-gel technique, which results in a very low level of adhesion between the inorganic domains and the polymer matrix in the nanocomposites. However, properly controlled dispersion of the inorganic component at atomic level is feasible either by covalent bonding of the inorganic component to the polymer matrix (Adjemian *et al.*, 2006) or by using organically functionalized silica (Karthikeyan *et al.*, 2006). While adding non-functionalized silica into hydrocarbon membranes impaired the mechanical properties of the composites, nanohybrid membranes or membranes containing organically functionalized silica exhibited similar and higher moduli (10–15 MPa to 15–25 MPa).

Silva's group (2005) prepared sPEEK (degree of sulfonation = 0.87)/$ZrO_2$ nanocomposite membranes and varied a wide range of physical and chemical properties by controlling the inorganic content from 2.5 to 12.5 wt% of $ZrO_2$. The network of $ZrO_2$ within the hydrocarbon polymer led to considerable reduction of water swelling and proton conductivity as well. It also improved morphological stability in DMFC applications in addition to improved barrier properties. MEA tests indicated that nanocomposite membranes with 7.5 wt% $ZrO_2$ presented the highest open-circuit potential due to the balance between methanol crossover and ohmic resistance in comparison with the other studied membranes.

$TiO_2$ has also been used as a filler to alter the transport properties of sPEEK membranes, using either a basic catalyst (pyridine) or a chelating agent (2,4-pentandione) to control the inorganic network features in a sPEEK matrix (Vona *et al.*, 2007). Sol-gel derived sPEEK/$TiO_2$ hybrid membranes (with variations in nano-sized $TiO_2$ content) showed an improvement in water uptake and retention along with a reduction in methanol permeability. Figure 14.4 shows a SEM image of a sPEEK membrane modified with 22 wt% $TiO_2$.

The addition of metal oxides improves various properties of polymer membranes, such as thermomechanical stability, water uptake capacity, and a reduction of methanol crossover, to name a few. But often the improved properties cannot be transformed into the desired level of improvement in cell performance when used in MEAs. This is mainly because of the reduced proton conductivity due to chemical inertness within the sulfonated polymer matrix. Their long-term stability is also still in question.

An exemplary way to prevent the loss of conductivity is to tether the inorganic oxides with proton-conductive organic functional groups such as $-SO_3H$, $-COOH$, etc. Thus, nanocomposite or nanohybrids could be formed from the sol-gel reaction of numerous monomers incorporating a series of organically functional groups that are covalently bound to silica or metal oxides. The inorganic skeleton could be modified by starting from many different silicon or transition metal oxides that predominantly influence the mechanical, optical, or thermal properties.

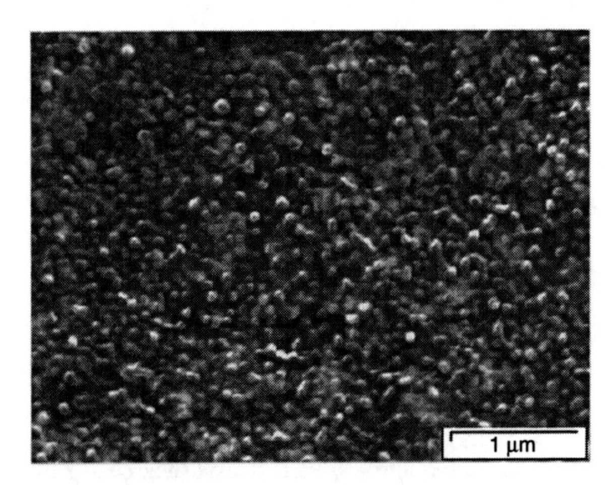


*14.4* SEM image of sPEEK membrane modified with 22 wt% nano-sized $TiO_2$.

On the other hand, the organic part is modified by selecting certain types out of an immense number of available organo-silanes and organic monomers or pre-polymers, which are responsible for the resulting flexibility and processability (Schottner, 2001). These tunable features of organically modified silicates (ORMOSILs) facilitate the preparation of a family of nanocomposite membranes whose structure, morphology, and functionality can be varied.

A significant proton-conductivity enhancement was achieved by using organically modified silane monomers with hydrophilic –Si-OH and proton-conductive $-SO_3H$ functional groups. Sulfonated phenylethylmethoxy silane was developed as an inorganic additive to Nafion (Wang *et al.*, 2002). At 80 °C and 100% relative humidity, the nanocomposite membrane demonstrated conductivity values at least six times higher than commercial Nafion under the same conditions. Hence, there have been many attempts to prepare highly conducting Nafion/bifunctional organosilica nanocomposite membranes with sulfonated diphenyl dimethoxy silane (sDDS) monomers with a $-SO_3H$ functional group (Li *et al.*, 2006) or mercaptopropylmethyldimethoxy silane (MPDS) monomers grafted with a thiol group (R–SH), but these synthetic approaches were not fruitful (Ren *et al.*, 2006). Even though these membranes could not produce proton-conductivity values as high as bifunctionally sulfonated phenylethyl silica-incorporated nanocomposite membranes, their single-cell DMFC performance reached the level of Nafion under the same fuel-cell conditions.

The *in situ* generation of the ORMOSIL phase within Nafion is feasible through the incorporation of organic groups within a silicate structure via the co-polymerization of Si alkoxides (monomer) and organically modified silane monomers. Mauritz was the first to prepare a Nafion/ORMOSIL nanocomposite by *in situ* co-polymerization of tetraethoxysilane (TEOS) and diethoxydimethyl silane (DEDMS) using Nafion as a template (Deng *et al.*, 1995). They later prepared a series of TEOS-$R'_nSi(OR)_{4-n}$ co-monomers. R′ in the co-monomer is the organic group joined to an Si atom via covalent Si–C bonds, which are hydrolytically stable. TEOS acts as a glass network former and inserted $R'_nSi(OR)_{4-n}$ units function either as bridging ($n$ = 2) or end capping ($n$ = 3) network modifiers. Depending upon the co-monomer feed ratio, the ORMOSILs will range from being rigid to flexible or from hydrophilic to hydrophobic; this is the way that the structural topology of their networks is controlled.

The loss of proton conductivity due to the incorporation of inorganic fillers can be avoided by using organically modified silanes as a filler in sPEEK matrices. Diphenyl silanediol (DPSD) is an organically modified alkoxy silane with aromatic phenyl rings. The phenyl group is a highly stable chemical moiety that also plays a prominent role in the structure of polyether ether ketone (PEEK) and allows sulfonation with ease (Fig. 14.5). Sulfonated diphenyl silanediol (sDPSD) can be homogeneously dispersed into a sPEEK matrix (degree of sulfonation as high as 0.9) without any difficulty because of its lack of reactivity towards condensation due to the steric hindrance of the phenyl groups (Licoccia *et al.*,

14.5 Chemical structures of DPSD (left) and sulfonated DPSD (right).

2006). Structural affinities between sDPSD and sPEEK also enhance nanocomposite morphological stability especially by shortening the distance between sulfonic groups, thus helping proton transfer and improving electrical properties. Furthermore, sDPSD is not soluble in water and its presence could be expected to modify the mechanical and solubility properties of sPEEK as well. In effect, sDPSD-incorporated sPEEK membranes are characterized by reduced water swelling, increased thermal stability from 250 °C (pure sPEEK) to 350 °C (nanocomposite), proton-conductivity values as high as 0.1 S/cm at 120 °C, and proton diffusion coefficients higher than those reported for Nafion. Table 14.2 gives an overview of proton conductivity and fuel-cell performance of metal oxide-filled polymer nanocomposites.

In conclusion, nanocomposite membranes based on metal oxides showed better water absorption and better mechanical properties but have not yet established

*Table 14.2* Proton conductivity and fuel-cell performance of metal oxide-filled polymer nanocomposites

| Polymer | Metal oxide | Proton conductivity (S/cm) | Fuel-cell performance ($mA/cm^2$ at 0.6 V) | Reference |
|---|---|---|---|---|
| Nafion® | $SiO_2$ | 0.044 at 25 °C<br>0.099 at 80 °C | | Jalani *et al.*, 2005 |
| Nafion® | $TiO_2$ | 0.023 at 135 °C | 1000 at 80 °C | Jalani *et al.*, 2005 |
| Nafion® | $ZrO_2$ | 0.024 at 135 °C | | Jalani *et al.*, 2005 |
| sPEEK | $SiO_2$ | 0.030 at 100 °C | | Nunes *et al.*, 2002 |
| sPEEK | $ZrO_2$ | 0.005 at 25 °C | | Silva *et al.*, 2005 |
| Nafion® | $SiO_2$-SH | 0.016 at 25 °C | 150 at 70 °C (at 0.4 V, DMFC) | Deng *et al.*, 1995 |
| sPEEK | sDPSD | 0.100 at 120 °C | | Licoccia *et al.*, 2006 |

higher cell performance and long-term stability. Special attention should be paid to the physicochemical characteristics of the various oxides, in particular, to the interfacial chemistry that exists between the oxide particles and the membrane, along with a consideration of how this interface affects elevated-temperature fuel-cell dynamics. The improvement of membrane performance seems to be very little dependent on the nature of the filler, while it is strongly dependent on the size and distribution of the inorganic particles inside the membrane (Alberti *et al.*, 2005). However, it is quite speculative that nanocomposite membranes have met the criteria for fuel-cell performance. Further investigation, therefore needs to be carried out in order to understand the functioning of inorganic fillers in nanocomposite membranes rather than on the beneficial effects of incorporating fillers within a membrane.

### 14.2.2 Heteropolyacid-filled nanocomposites

Metal oxides are passive fillers and may help to retain water or reduce methanol permeability. However, heteropolyacids (HPAs) are highly conductive (~0.19 S/cm) in the hydrated state and are one of the most attractive inorganic proton conductors in crystalline form. They are thermally stable at temperatures higher than 100 °C but are soluble in water and polar solvents. Typical compounds include $H_3PW_{12}O_{40}{\cdot}nH_2O$ (PTA), $H_3PMo_{12}O_{40}{\cdot}nH_2O$ (PMoA), and $H_4SiW_{12}O_{40}{\cdot}nH_2O$ (SiTA), which have different hydrated states (up to 29 water molecules) depending on the environment (Lee *et al.*, 1989; Bardin *et al.*, 1998).

HPAs are nanoparticular acids, 1–100 nm in diameter and composed of primary, secondary, and tertiary structures. The primary structure is the main building block of the HPA and its basic structural unit is the $[PM_{12}O_{40}]^{3-}$ cluster called the Keggin anion, shown in Fig. 14.6. HPAs possess waters of crystallization that bind the Keggin units together in the secondary structure by forming water bridges

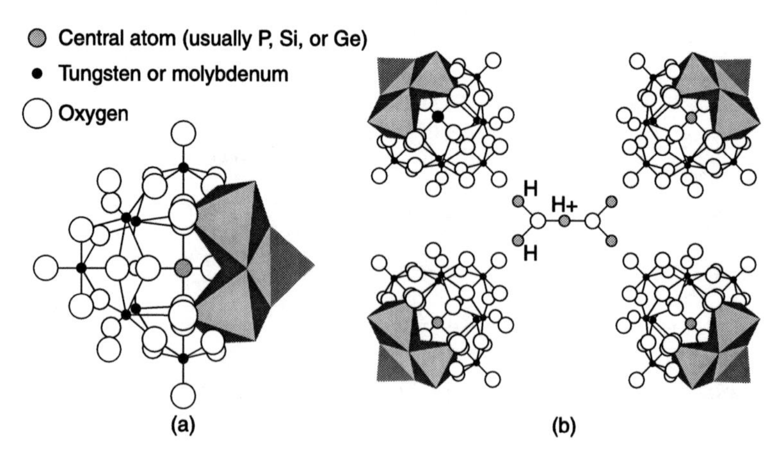


*14.6* The structures of a heteropolyacid: (a) primary (Keggin) structure and (b) secondary (hydrated) structure.

(Okuhara *et al.*, 1996). Tertiary structures can be observed when heavy alkali salts are formed.

In the hydrated phase, the protons reside in the bridging water moieties forming $H_5O_2^+$ and become highly conductive due to strong acidity, and their proton conductivity or acidity is in the following order: $H_3PW_{12}O_{40} > H_4SiW_{12}O_{40}$ and $H_3PW_{12}O_{40} > H_4PMo_{12}O_{40}$ (Okuhara *et al.*, 1996). Protons are most likely located on the bridging oxygen (Lee *et al.*, 1989) and they conduct through self-ionization. Consequently the proton conductivity of HPAs is strictly related to the number of water molecules coordinated to the Keggin unit. The number of water molecules is determined by the relative humidity and temperature. Typically HPAs exist in a series of hydrated phases with the degree of hydration varying from six to 29 molecules of water (waters of hydration) per HPA molecule. For example, the proton conductivities of phosphotungstic acid (PTA) at 29 and six hydrated water molecules decreased from $1.8 \times 10^{-1}$ to $6 \times 10^{-5}$ S/cm, respectively (Pourcelly and Gavach, 1992). Table 14.3 shows the proton conductivity of HPAs for different hydration numbers.

Nafion/HPA nanocomposite membranes can be obtained by simple impregnation of preformed membranes with a HPA solution as well as by mixing a Nafion solution with an appropriate amount of HPA followed by casting (Tazi and Savadogo, 2000). Nafion impregnated with PTA showed more stable electrochemical performance than unmodified Nafion and had a high proton concentration and improved water retention. Similarly substantial increases in water uptake and proton conductivity were accomplished by recast Nafion nanocomposites with phosphomolybdic acid (PMA) and silicotungstic acid (STA)

*Table 14.3* Proton conductivity of heteropolyacids for different hydration numbers

| Structure | Name | Proton conductivity (S/cm) | Hydration number | Reference |
|---|---|---|---|---|
| $H_3PW_{12}O_{40} \bullet nH_2O$ | phosphotungstic acid (PTA) | $2.0 \times 10^{-1}$ | 29 | Pourcelly and Gavach, 1992 |
| | | $7.2 \times 10^{-2}$ | 20 | Pourcelly and Gavach, 1992 |
| | | $2.0 \times 10^{-2}$ | 13 | Pourcelly and Gavach, 1992 |
| $H_3PMo_{12}O_{40} \bullet nH_2O$ | phosphomolybdic acid (PMA) | $2.0 \times 10^{-1}$ | 29 | Pourcelly and Gavach, 1992 |
| | | $2.4 \times 10^{-2}$ | 18 | Pourcelly and Gavach, 1992 |
| | | $1.4 \times 10^{-2}$ | 13 | Pourcelly and Gavach, 1992 |
| $H_4SiW_{12}O_{40} \bullet nH_2O$ | silicotungstic acid (STA) | $7.0 \times 10^{-2}$ | 29 | Okuhara *et al.*, 1996 |

(Ramani *et al.*, 2004). Water uptake in the Nafion/STA composite increased to 60% from the normal value of 27% for Nafion and reached a maximum of 95% for the PMA counterpart. Also the fuel-cell performance of HPA-loaded membranes was far better than Nafion at 80 °C and Nafion/PMA witnessed a maximum current density of 940 mA/cm$^2$ at 0.6 V compared with 640 mA/cm$^2$ for Nafion.

A series of sPEEK nanocomposite membranes were prepared by mixing HPA (60 wt% PTA or PMA) with a polymer solution in dimethyl acetamide (Zaidi *et al.*, 2000). The composite membranes were characterized by a higher glass transition temperature, probably because of the intermolecular interaction between the sulfonic groups and HPA, and by much greater water uptake at room temperature (up to five times for PTA/sPEEK with 80% sulfonation, Fig. 14.7). Proton conductivity was determined over the temperature range 20–150 °C using an open cell. The concomitant water loss over the timescale of the experiment was estimated to be about 65% at 140 °C, which could be the cause of the lower limit of proton conductivity at a given temperature. The nanocomposite membranes were generally more conductive than the pure membrane, but conductivity enhancement decreased with an increasing degree of sulfonation. In all cases, the conductivity dependence on temperature showed a maximum around 120 °C.

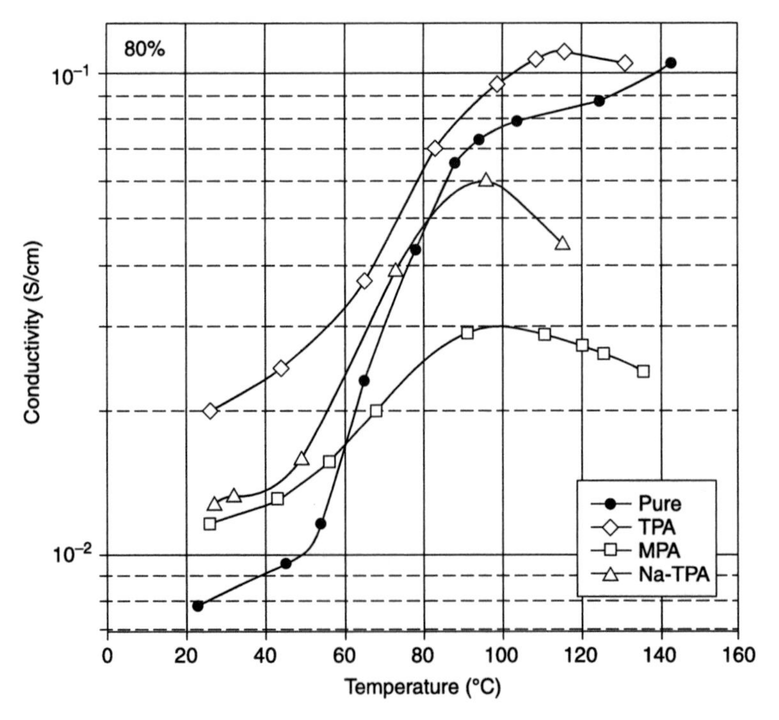


*14.7* Effect of different HPAs on the proton conductivity of an sPEEK (80% DS)/HPA nanocomposite as a function of temperature.

Over the investigated temperature range, the highest conductivities were found for the PTA-based nanocomposite membranes ($\sigma$ from $2 \times 10^{-2}$ to 0.1 S/cm with increasing degree of sulfonation from 70 to 80%).

The extraction of PTA from the composite membranes of HPA/BPSH-40 was carefully examined by using tapping-mode atomic force microscopy (AFM) as shown in Fig. 14.8. A PTA/Nafion composite membrane was used as a control experiment. After immersion in liquid water, PTA/Nafion showed irregular holes (0.2 μm in diameter) on the membrane surface, which was assumed to be due to PTA extraction (Fig. 14.8(a)). In contrast, the HPA/BPSH-40 composite membrane did not show any holes after liquid water treatment, indicative of good retention of the PTA in the composite, as shown in Fig. 14.8(b). This could be partly attributed to the strong hydrogen bonding between BPSH and PTA. In addition, HPA could be controlled to be uniformly dispersed at the nanometer scale, which is definitely another advantage of nanohybrids.

Two major factors limit the performance of nanocomposite membranes based on HPA: (a) the extremely high solubility of HPA additives and (b) their large particle size (30–100 nm) within the membrane matrix, which in turn results in ineffective bridging between the ionic domains. In order to solve these problems organic–inorganic nanohybrids were introduced by several research groups (Staiti *et al.*, 2001; Jung *et al.*, 2006; Boysen *et al.*, 2004). Preparation methods include sol-gel processing of metal oxides such as silica and zirconium oxide, and a synthetic procedure is schematically shown in Fig. 14.9. The synthetic process consists of two main steps: (a) end-capping of the organic polymer with alkoxy silanes through isocyanato coupling and (b) subsequent hydrolysis and condensation of these precursors. The resulting materials are composed of nano-sized silica

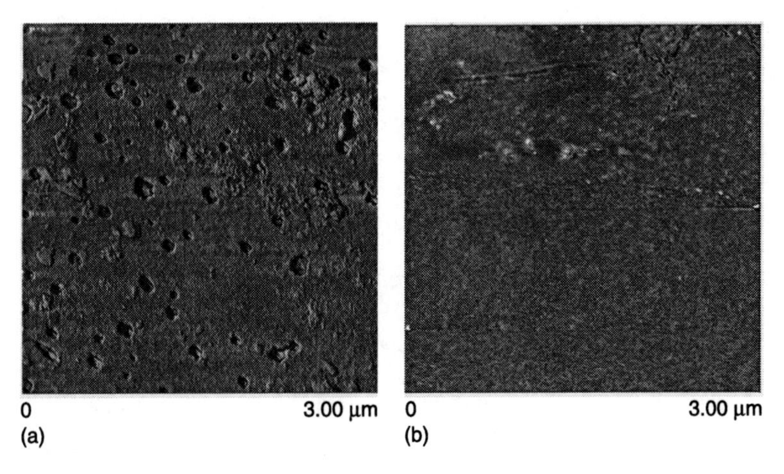


*14.8* Tapping-mode AFM images after immersion of composite membranes in liquid water at 30 °C for 48 h: (a) PTA/Nafion® 117 (4:6) and (b) PTA/BPSH-40 (3:7).

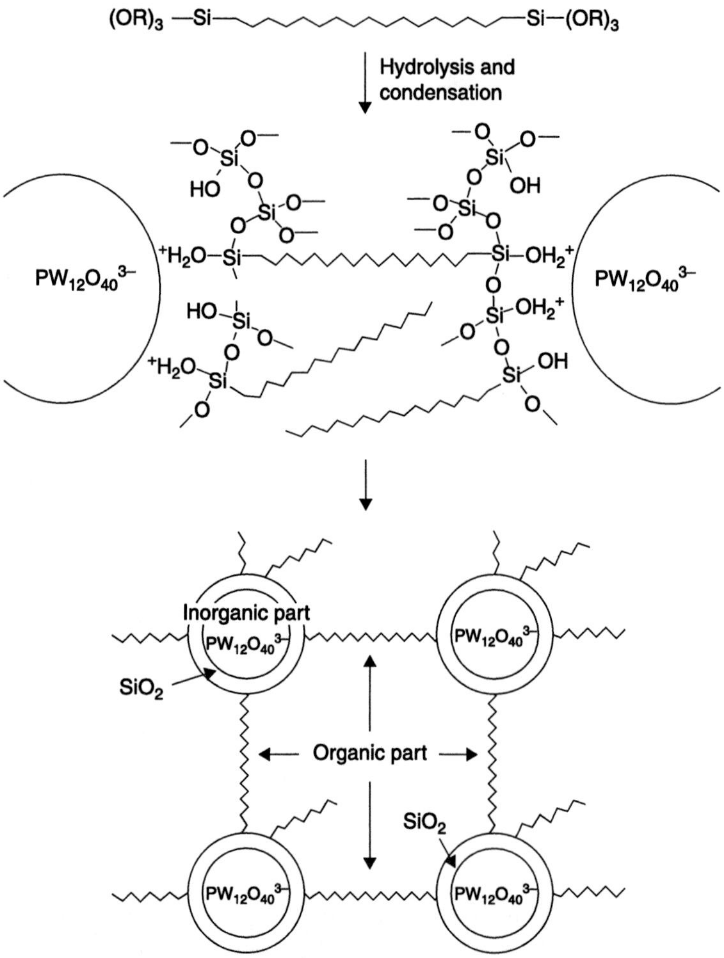

*14.9* Synthetic procedure of PTA/$SiO_2$-modified nanocomposite membrane.

domains interconnected by polymer chains. In the presence of water vapor, these nanohybrids doped with PTA become proton conducting. The PTA is probably bound with the silica (or zirconia) domains by strong coulombic interactions. Accordingly, no PTA leaks out of the hybrid membrane even for doping levels as high as 100 wt% with respect to the hybrid polymer (Staiti *et al.*, 1999, 2001).

A few years later Liu (2005) developed a new type of Nafion composite with PTA immobilized on $SiO_2$ via sol-gel reactions. The silica was first functionalized with aminopropyltriethoxy silane (APTES) to form amine-containing silica materials. PTA was immobilized on the silica by ionic complexation with amino groups in the modified silica. An *in situ* sol-gel reaction was carried out in the presence of Nafion, while an *ex situ* sol-gel reaction was performed without Nafion, which was blended in during the second step. It was speculated that competitive ionic complexion existed between Nafion and PTA with the amino groups on the functionalized silica. The *in situ* Nafion/APTES/PTA composite membrane showed much lower proton conductivity and performance than the Nafion/PTA composite membrane at 80–120 °C and 1 atm. The reason for the low fuel-cell performance of the *in situ* Nafion/APTES/PTA composite membranes was possibly due to residual $NH_2$ poisoning of the Pt catalysts in the MEA.

The *ex situ* method was developed to overcome these problems; a mesoporous silica material (SBA-15) was prepared by Zhao (1998), followed by functionalization with APTES (Feng *et al.*, 1997). PTA was immobilized in the nanopores by ionic complexation with monolayer amino groups at the internal pore surfaces. Composite membranes were obtained by solution casting Nafion with PTA-immobilized SBA-15 at different weight percentages. The nanocomposite had a proton conductivity around 0.16 S/cm (when fully hydrated) and 0.01 S/cm (50% RH) at 120 °C.

Composite membranes based on PTA-impregnated $SiO_2$ and poly(benzimidazole) (PBI) have been prepared and their physicochemical properties were studied. Solution casting yielded membranes with a high tensile strength and a thickness $< 30\,\mu m$. They were chemically stable in boiling water and thermally stable in air up to 400 °C (Staiti *et al.*, 2000). The presence of silica in the composite allowed the membranes to maintain stable proton conductivity at temperatures up to 130 °C under 100% RH. For example, proton conductivity was $10^{-3}$ S/cm at 130 °C and the PBI formed a network to keep the PTA supported on the silica.

As already mentioned, earlier efforts at using HPA itself as a solid electrolyte failed due to its high solubility in water and the strong influence of humidity on proton conductivity (Katsoulis, 1998). But it is interesting to note that the proton conductivity of well-dispersed HPA incorporated in sulfonated polymers is quite promising at temperatures greater than the boiling point of water (Tazi *et al.*, 2000). It was suggested that specific interactions between HPAs and sulfonated polymers could have a significant influence on fuel-cell performance at elevated temperatures (Fig. 14.10).

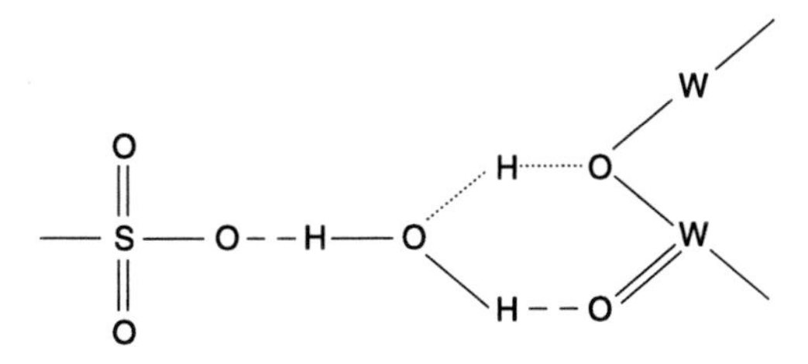


*14.10* Proposed hydrogen-bonding structure in a PTA/BPSH composite membrane.

Honma *et al.* (1999) prepared hybrids of $SiO_2$ and poly(ethylene oxide) (PEO), which incorporated HPAs in the PEO domains. However, there are still problems for the application of HPAs as fillers for membranes. They are soluble and can easily bleed out of the membranes. To overcome the problem of electrolyte dissolution, different approaches have been tried. Ahmed proposed the use of HPAs trapped in zeolites (Ahmed *et al.*, 2005). An approach proposed by other groups is the *in situ* generation of an oxide network by the sol-gel process using alkoxy silanes and the modification of the anion structure of the HPA (Ponce *et al.*, 2003, 2004). In the catalyst literature, the trapping of HPA in a host material, mostly $SiO_2$ networks, has been reported by Misono (1987), Kozhevnikov (1998), and Mizuno and Misono (1998).

Nunes's group (Ponce *et al.*, 2004) investigated an alternative way to avoid the consumption of acid sites. Usually HPA is fixed through covalent bonds or coulombic interactions, which involve acid sites of the polymer matrix. Organosilyl derivatives of HPA, for instance a divacant tungstosilicate $[\gamma\text{-}SiW_{10}O_{36}]^{8-}$, were prepared using 3-glycidoxypropyltrimethoxy silane (GPTS). The introduction of GPTS into the anion structure of HPA enables its attachment to a host material, by an epoxy ring-opening reaction with appropriate functional groups present on the host surface. For instance, the modified HPA was introduced into a sulfonated polymeric matrix, containing an insoluble inorganic filler modified with amino groups. The amino groups reacted with the epoxy groups of the HPA molecules and fixed them without reducing their acidity.

Therefore, HPAs are non-volatile inorganic acids with higher conductivity and thermal stability in their crystalline forms. In the channels of ionomer membranes they function as Brønsted acids and donate protons. They can also solvate protons from the stronger sulfonic acid, which could aid water retention, especially at high temperatures, through a strong and specific interaction with water. Since they are inorganic nanoparticles when in crystalline form, they can reduce fuel crossover, especially in DMFCs, and modify the mechanical properties of membranes when they are uniformly dispersed in the polymer matrices. Table 14.4 gives an overview of proton conductivity and fuel-cell performance for HPA-filled polymer nanocomposites.

### 14.2.3 Anhydrous solid-acid-filled nanocomposites

Superprotonic acid salts are well-known proton conductors with general composition $M_mH_n(XO_4)_{(m+n)/2}$ (M = K, Rb, $NH_4$, or Cs; X = S or Se) and $CsH_2(RO_4)$ (R = P or As). Structurally they are categorized between normal acids and normal salts, and comprise oxyanions, which are linked via hydrogen bonds. They undergo a structural phase change at a certain temperature and bare protons are mobile in anhydrous states due to the existence of a dynamically disordered network of H bonds (Baranov *et al.*, 1989) (Fig. 14.11). Therefore, the superprotonic phases are characterized by fast protonic diffusion and intense

*Table 14.4* Proton conductivity and fuel-cell performance of heteropolyacid-filled nanocomposites

| Polymer | Heteropolyacid | Proton conductivity (S/cm) | Fuel-cell performance at 0.6V ($mA/cm^2$) | Reference |
|---|---|---|---|---|
| Nafion® | PTA | – | 650 at 110 °C | Ramani *et al.*, 2004 |
| Nafion® | PMA | – | 940 at 80 °C | Ramani *et al.*, 2004 |
| Nafion® | STA | 0.020 at 25 °C | 695 at 90 °C | Tazi and Savadogo, 2000 |
| sPEEK | PTA | 0.003 at 25 °C (DS 70)<br>0.017 at 100 °C (DS 70)<br>0.020 at 25 °C (DS 80)<br>0.095 at 100 °C (DS 80) | – | Zaidi *et al.*, 2000 |
| sPEEK | PMA | 0.003 at 25 °C (DS 70)<br>0.011 at 100 °C (DS 70)<br>0.012 at 25 °C (DS 80)<br>0.030 at 100 °C (DS 80) | – | Zaidi *et al.*, 2000 |
| PEO | PTA + $SiO_2$ | 0.004 at 120 °C | – | Staiti *et al.*, 1999 |
| PBI | PTA + $SiO_2$ | 0.002 at 100 °C (RH 100%)<br>0.0003 at 100 °C (RH 40%) | – | Staiti *et al.*, 2001 |
| PBI | STA + $SiO_2$ | 0.002 at 100 °C (RH 100%) | – | Staiti *et al.*, 2001 |
| Nafion® | PTA + $SiO_2$ | – | 800 at 145 °C (at 0.4V, DMFC) | Staiti *et al.*, 2001 |
| Nafion® | STA + $SiO_2$ | – | 600 at 145 °C (at 0.4V, DMFC) | Staiti *et al.*, 2001 |
| Nafion® | PMA + $SiO_2$ | 0.030 at 90 °C | – | Ramani *et al.*, 2004 |
| Nafion® | STA + thiophene | 0.095 at 80 °C | 810 at 80 °C | Tazi and Savadogo, 2000 |

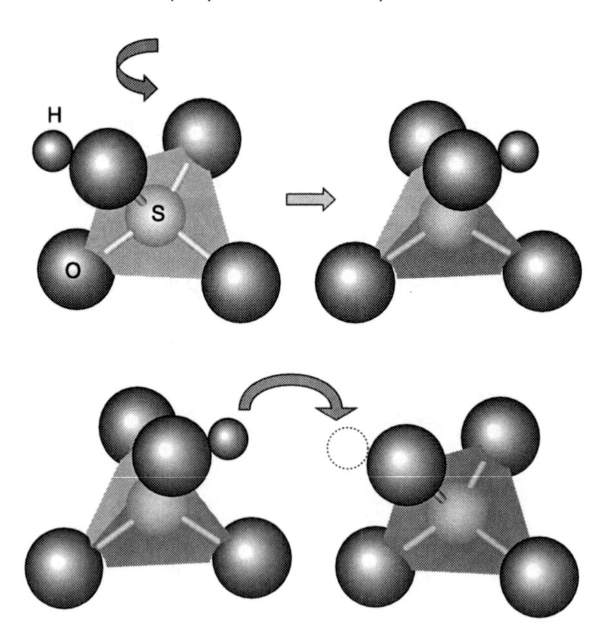


*14.11* Structural phase change of solid acids.

vibrations of the $XO_4$ or $RO_4$ tetrahedra. Thus, proton conductivity is an intrinsic property because of the perfect crystal lattice, and any structural defects decrease proton conductivity. These compounds do not contain water molecules in their structures, so that they have high thermal and electrochemical stability. Their conductivity does not depend on atmospheric humidity and can be enhanced by preparing inorganic composites with metal oxides with a high surface area (Ponomareva *et al.*, 1996; Ponomareva and Lavrova, 1998).

The most well-known compound is $CsHSO_4$, whose structure is shown in Fig. 14.12. It exhibits a high proton conductivity of $10^{-2}$ S/cm at temperatures > 141 °C where a phase transition occurs and it is superprotonic. As shown in Table 14.5 and Fig. 14.13, the superprotonic transition temperature is generally 70–250 °C and these materials have very high conductivity ($10^{-3}$ to $10^{-1}$ S/cm). This makes these materials very promising for applications in fuel cells operating at medium temperatures (100–250 °C). To obtain a lower superprotonic transition temperature and high conductivity new solid compounds were prepared, for example, $CsHSO_4$•$CsH_2PO_4$ and $CsHSO_4$•$CsH_2PO_4$ (Crisholm and Haile, 2000; Haile *et al.*, 1995).

Solid acids are good candidates for proton-conducting membranes due to their high proton conductivity in the anhydrous state, thus eliminating the delicate problem of water management. Nevertheless, they also present serious drawbacks such as brittleness, a narrow temperature range for the superprotonic phase

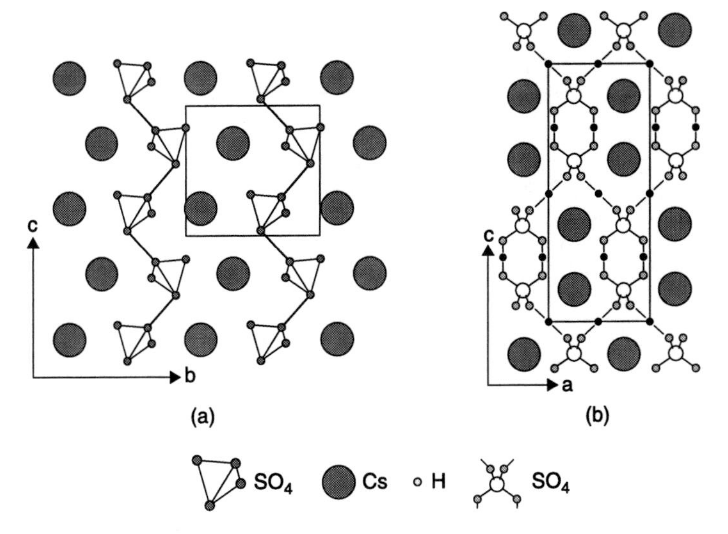


*14.12* Crystal structure of the solid acid $CsHSO_4$: (a) monoclinic phase II and (b) tetragonal phase III, where each oxygen position has half occupancy and the hydrogen atoms are placed in the middle of the disordered hydrogen bones (dashed lines).

*Table 14.5* Different solid acids and their properties

| Compounds | Transition temperature (°C) | Proton conductivity (S/cm) | Structure of superprotonic phase | Reference |
|---|---|---|---|---|
| $KHSO_4$ | 177 | $10^{-1}$ above 177°C | – | Baranov *et al.*, 2005 |
| $RbHSO_4$ | – | ~ $10^{-2}$ | – | Baranov *et al.*, 2005 |
| $RbHSeO_4$ | 172 | ~ $10^{-2}$ | – | Baranov *et al.*, 2005 |
| $CsHSO_4$ | 140 | $4 \times 10^{-2}$ at 200°C | – | Haile *et al.*, 2001 |
| $CsHSeO_4$ | 124 | ~ $10^{-2}$ | – | Haile *et al.*, 2001 |
| $KH_2PO_4$ | 178 | $7 \times 10^{-5}$ at 185°C | Monoclinic | Ortiz *et al.*, 1999 |
| $RbH_2PO_4$ | 71 | $6.8 \times 10^{-2}$ at 340°C | Monoclinic | Ortiz *et al.*, 1999 |
| $CsH_2PO_4$ | 230 | $2.2 \times 10^{-2}$ at 240°C | Cubic | Baranov *et al.*, 1989 |
| $NH_4H_2PO_4$ | 160 | $4 \times 10^{-2}$ at 180°C | Cubic | Baranov *et al.*, 1989 |
| $K_3H(SO_4)_2$ | 205 | $4 \times 10^{-3}$ at 208°C | – | Boysen *et al.*, 2004 |
| $K_3H(SeO_4)_2$ | 117 | ~$10^{-3}$ at 127°C | Rhombohedral | Sinitsyn *et al.*, 2000 |
| $Rb_3H(SeO_4)_2$ | 174 | $2 \times 10^{-4}$ at 186°C | Rhombohedral | Sinitsyn *et al.*, 2000 |
| $Cs_3H(SeO_4)_2$ | 182 | ~ $2 \times 10^{-4}$ at 187°C | – | Boysen *et al.*, 2004 |
| $(NH_4)_3H(SeO_4)_2$ | 27 | $4 \times 10^{-3}$ at 110°C | – | Boysen *et al.*, 2004 |

(between the transition temperature and the melting temperature) (Piao *et al.*, 2009), chemical instability (Baranov *et al.*, 2005), water solubility, and poor mechanical behavior (Boysen *et al.*, 2004). Therefore, their composite materials aim at combining the advantages of the solid acids to eliminate these drawbacks.

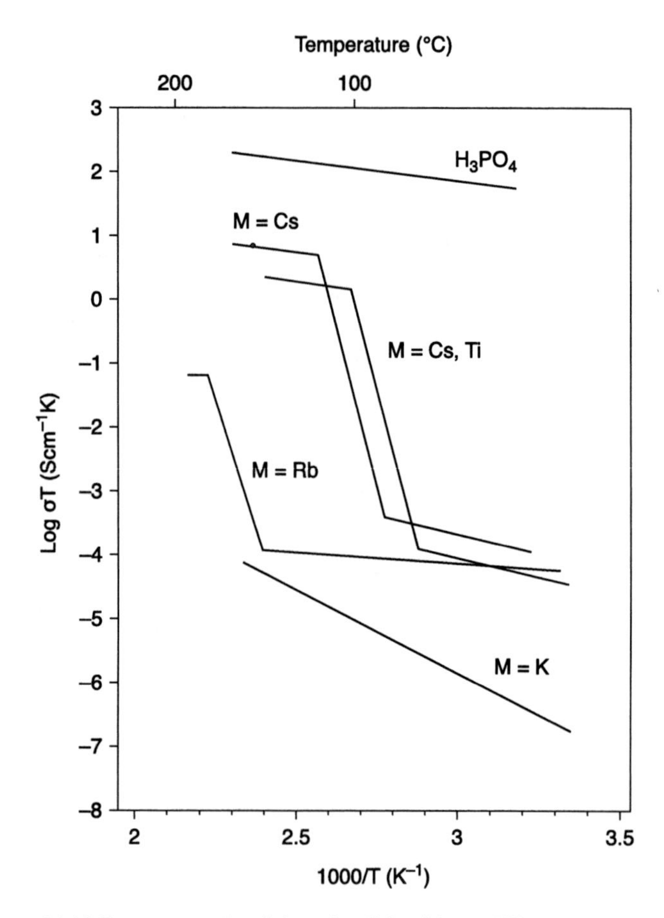


*14.13* Proton conductivity of solid acids at different temperatures.

This type of nanocomposite material has been, however, poorly explored until now, compared with other types of inorganic filler. Only preliminary investigations of polymer composites based on $CsHSO_4$ and poly(vinylidene fluoride) (PVDF) (Boysen *et al.*, 2000) and $CsHSO_4$ with poly(acrylonitrile) could be found in the literature (Andronie *et al.*, 2008).

The first report of the application of $CsHSO_4$ in a fuel cell was in 2001 (Haile *et al.*, 2001). This system was able to operate at 150–160 °C in a $H_2/O_2$ configuration with an open-circuit voltage of 1.1 V and a short-circuit current density of 44 mA/cm$^2$. It was also stable in humid environments but the performance was well short of the performance of equivalent composite polymer membranes operating at 130–140 °C (Costamagna *et al.*, 2002). Later the same groups used a $CsH_2PO_4$ electrolyte membrane in a fuel cell operating at 250 °C. The cell was able to operate stably, and peak and maximum power densities of 48.9 mW/cm$^2$

and 301 mA/cm$^2$ (short circuit), respectively, were attained. This was a very positive result but the conductivity of $CsH_2PO_4$ at low temperatures (< 140 °C) was very low ($10^{-6}$ S/cm). Therefore, it is likely that these fuel cells may have a number of startup issues when used in cyclic applications (like automobiles).

There has been more research on a mix of solid acids with oxide particles. Heterogeneous doping by highly dispersed inert oxides has shown an appreciable increase in proton conductivity for some solid acids, in particular at temperatures lower than the superprotonic temperature (Otomo *et al.*, 2008; Saito *et al.*, 2009). In these composites, a strong surface interaction between the components took place, which led not only to the stabilization of a new phase on the surface of the oxide, but also to a change of the bulk properties of the salts (Lavrova and Ponomareva, 2008). Often the oxide particles used are porous and the effect of their porosity on conductivity has been studied. However, the low-temperature proton conductivity of these composites remains unclear but it is often claimed that the solid acid adopts a highly proton-conducting amorphous phase in the mesopores at lower temperatures (Lavrova and Ponomareva, 2008). Another explanation would be the existence of a metastable phase during cooling of the composites induced by shear elastic forces (Otomo *et al.*, 2008). A final explanation would be that the change in crystalline structure and the enhanced interfacial interaction between the two phases in the composite yields a lower melting point for the solid acid in the composite than for the pure solid acid (Piao *et al.*, 2009).

The dispersion of $SiO_2$ particles into a solid-acid matrix has been observed to improve the mechanical properties of the membrane (Bondarenko *et al.*, 2009). The introduction of $CsH_2PO_4$ into porous anodic alumina membranes yielded a high power output at room temperature and, furthermore, the anodic alumina membrane support guarantees fuel-cell performance with good reproducibility without any external support (Bocchetta *et al.*, 2009).

Anhydrous acids generally suffer from poor mechanical properties and water solubility, as well as extreme ductility and volume expansion with rising temperature. However, when they are incorporated in the channels of a polymer membrane, they could be very useful at high temperatures because their conductivity does not depend on humidity. However, technological challenges still remain before solid acids can be used in fuel cells: (a) the preparation of thin, impermeable solid-acid membranes, (b) the enhancement of electrode performance, and (c) a system design to protect the electrolyte from liquid water during fuel-cell shut-off (Haile *et al.*, 2004).

### 14.2.4 Layered metal phosphate- and phosphonate-filled nanocomposites

Metal phosphates and phosphonates can conduct protons and can be expressed in the form $M^{IV}(RPO_3)_2{\cdot}n\,H_2O$, where M is a tetravalent metal such as Zr, Ti, Ce, Th, or Sn and R is an inorganic or organic group such as –H, –OH, $-CH_3OH$, or $-(CH_2)_n-$. Zirconium phosphate (ZrP), $Zr^{IV}(O_3POH)_2{\bullet}H_2O$, and its phosphonate (ZrPP), $Zr^{IV}(O_3PR)_2{\bullet}nH_2O$, have been the most extensively studied by Alberti's group (Alberti *et al.*, 1996).

They form two types of layered structures, $\alpha$ and $\gamma$ (Fig. 14.14). $\alpha$-ZrP has a pendant OH group, which extends into the interlayer region and forms a hydrogen-bonded network with water. $\gamma$-ZrP, $Zr^{IV}PO_4{\bullet}(O_2P(OH)_2{\bullet}nH_2O$, has the advantage of having an extra water molecule per unit and is more acidic than $\alpha$-ZrP (Alberti *et al.*, 1996).

Layered $\alpha$-ZrP is a surface proton conductor, whose conductivity is four orders of magnitude greater than in the bulk, and its surface area and capacity for water adsorption have a significant influence on proton conductivity. However, crystallinity also plays an important role in its conductivity (Alberti *et al.*, 1984). Isoconductance measurements indicate that proton conductivity varies linearly with the number of surface phosphate groups (Alberti *et al.*, 1978). In addition, the conductivity of $\alpha$-ZrP is highly dependent on hydration, varying by two orders

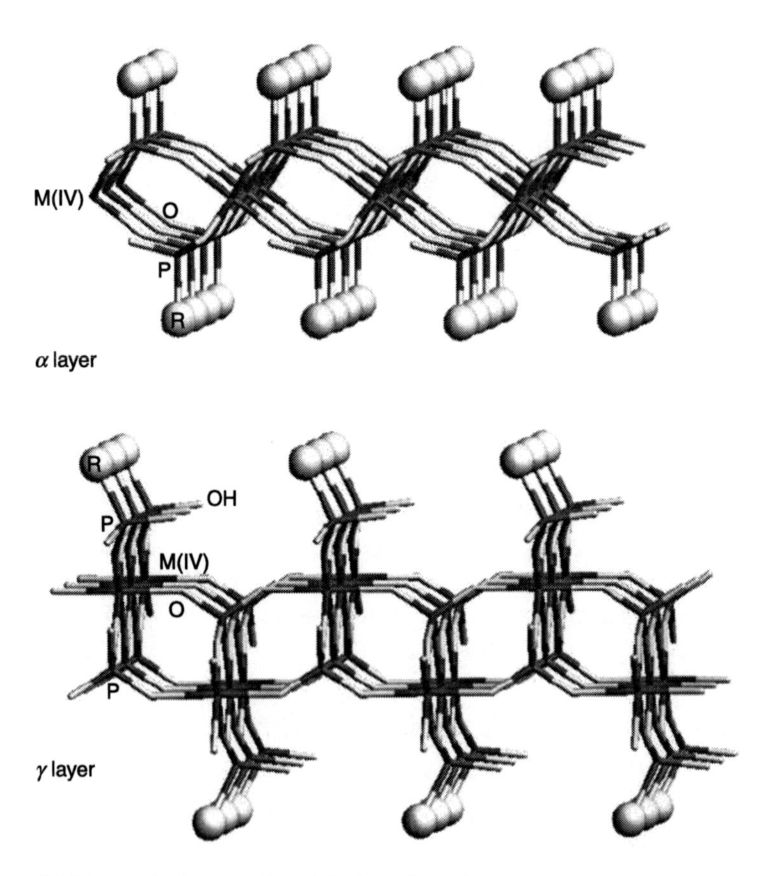


*14.14* $\alpha$ and $\gamma$ layers of metal phosphonates.

of magnitude as the relative humidity increases from 5 to 90% ($\sigma = 10^{-7}$–$10^{-4}$ S/cm) (Casciola and Bianchi, 1985). Recent research has confirmed the dominance of surface transport and demonstrated enhancements that could be made through modification of the P–OH groups ($\sigma = 10^{-5}$–$10^{-4}$ S/cm) (Carriere *et al.*, 2003).

$\alpha$-ZrP was first synthesized by Clearfield and Stynes. Nonetheless, extensive investigation into its use as a proton conductor did not commence until the 1990s (Clearfield, 1990, 1996) due to its low conductivity ($\sigma = 10^{-7}$–$10^{-6}$ S/cm). Costamagna *et al.* (2002) investigated Nafion/zirconium phosphate composite membranes at up to 130 °C and 3 bar. Nafion was impregnated with $\alpha$-ZrP by reacting $ZrOCl_2$ with $H_3PO_4$ at 80 °C in Nafion channels. The results highlighted how the MEA performance of composites could be superior to polymer membranes at high temperatures even though proton conductivity was not significantly increased. Current densities in MEAs were 4–6 times better than for pure Nafion (130 °C at 3 bar, fully hydrated). Concurrently, the results demonstrated that current platinum catalyst technology was compatible with this composite arrangement, which might potentially overcome any development challenges, and that there was a reduction in methanol crossover.

Similar experiments showed that Nafion/ZrP nanocomposite membranes could be produced with a conductivity of 0.64 S/cm compared with 0.40 S/cm for pure Nafion under the same conditions (100% RH, conditioned in boiling water for 1 h). Furthermore, sPEEK composite membranes demonstrated a threefold improvement in conductivity when loaded with 10 wt% $\alpha$-ZrP (Ruffmann *et al.*, 2003). Adjemian *et al.* (2002) incorporated inorganic species (silica and $\alpha$-ZrP) in Nafion channels and they found that silica composite membranes exhibited better robustness, water retention, and better performance even though there were only slight improvements in conductivity.

On the other hand, early efforts using a composite of $\alpha$-ZrP with silica or alumina were made to increase the low conductivity of $\alpha$-ZrP (Glipa *et al.*, 1997) but they failed (Slade *et al.*, 1992). However, more recent efforts with colloidal dispersions of inorganic silicates have been successful (Glipa *et al.*, 1997). Pressed pellets with a surface area of up to 500 m$^2$/g achieved a maximum proton conductivity of $3 \times 10^{-3}$ S/cm for Zr = 25% at 100 °C and 97% RH, one order of magnitude higher than crystalline $\alpha$-ZrP ($\sigma = 1.5 \times 10^{-4}$ S/cm) under the same conditions. Under anhydrous conditions, the maximum conductivity was $8 \times 10^{-8}$ S/cm at 200 °C. This indicates that there is no proton diffusion along the anhydrous surface and only the bulk phosphate groups contribute to conduction.

Layered metal (IV) phosphonates are better proton conductors than their phosphate counterparts. Their structures are $\alpha$-[$M^{IV}(O_3P\text{–}R)_2 \cdot nH_2O$] and $\gamma$-[$M^{IV}PO_4(O_2P(OH)R)_2 \cdot nH_2O$], where organic moieties (R) are bridged through phosphorus atoms to an inorganic two-dimensional matrix (Alberti and Casciola, 1997). When the organic moieties R contain protongenic groups, such as $-PO_3H$, $-SO_3H$, –COOH, or $NH_3^+$, these layered metal (IV) phosphonates become very good proton conductors and exhibit very high conductivity at temperatures up to 200 °C.

Attempts to improve the proton conductivity of solid-acid membranes have included the synthesis of new layered compounds, where Brønsted bases are intercalated in the interlayer region or organic functional groups replace the hydroxyl group of the phosphate (Alberti *et al.*, 1996). A significant improvement was achieved with the intercalation of strongly acidic functional groups of $-SO_3H$ (there was little improvement with the weak acid, –COOH) into the interlayer region.

Zirconium sulfophosphonates $[Zr(O_3PC_6H_4SO_3H)_{0.85}(O_3PC_2H_5)_{1.15}\bullet nH_2O]$ and zirconium alkyl sulfophenylphosphonates $[Zr(O_3PC_6H_4SO_3H)_x(O_3PCH_2OH)_{2-x}\bullet nH_2O]$ have been investigated due to their proton conductivity under different temperature and relative humidity regimes (Alberti *et al.*, 1992; Stein *et al.*, 1996). The best proton conductivity achieved in the anhydrous state was with ethylsulphophenyl phosphonate: $5 \times 10^{-6} \leq \sigma \leq 1.2 \times 10^{-5}$ S/cm at 180 °C and 0% RH (compared with $10^{-6}$ S/cm for microcrystalline $\alpha$-ZrP) (Alberti *et al.*, 1992, 1996). For hydrated results at 20 °C, proton conductivity increased from $10^{-4}$ at 22% RH to $1.6 \times 10^{-2}$ S/cm at 90% RH, while at 100 °C and 90% RH conductivities as high as 0.05 S/cm were reported (comparable with Nafion). Unfortunately zirconium sulfophenylphosphonates, compared with $\alpha$-ZrP, exhibited a greater dependency on relative humidity, especially when the humidity was less than 50%. They are, however, still capable of conducting at 0.01 S/cm at 65% RH and 100 °C (Alberti *et al.*, 1992). One of the significant advantages of zirconium sulfophenylphosphonates is that there was no drop in conductivity as the temperature increased from ambient conditions up to 100 °C (Alberti *et al.*, 2000). Later work has shown the zirconium sulfoarylphosphonates also display high conductivity, although it is more strongly affected by humidity. In this work, zirconium compounds were synthesized with the sulfonic acid attached to a phenyl, benzyl, or fluorinated benzyl group. The highest proton conductivity was $5 \times 10^{-2}$ S/cm at 100 °C and 100% RH and $2 \times 10^{-2}$ S/cm at 150 °C and 100% RH (Zaidi *et al.*, 2000). Table 14.6 shows the proton conductivities of layered zirconium phosphates and phosphonates. As proton conductors the latter have additional desirable attributes, which include:

- Moderate proton conductivity when humidified ($\sim 10^{-2}$ S/cm).
- Brønsted acidity with the ability to donate protons.
- Thermal stability at temperatures above 180 °C.
- Hygroscopic and hydrophilic character.
- Easy synthesis in a manner that is compatible with the chemical and physical limits of the polymer membrane.

Titanium phosphates are also proton conductors. Mesoporous MCM-type zirconium and titanium phosphates were synthesized by the sol–gel route using surfactant templates and they had surface areas of 240–330 m$^2$/g (Rodriguez *et al.*,

*Table 14.6* Proton conductivity of layered metal phosphonates

| Compounds | Proton conductivity (S/cm) | Conditions | Reference |
|---|---|---|---|
| $\alpha$-$Zr(O_3POH)_2 \bullet H_2O$ | $10^{-5}$ to $10^{-6}$ | 20°C, 90% RH | Alberti and Casciola, 1997 |
| $\alpha$-$Zr(O_3POH)_2$ | $1.0 \times 10^{-7}$ | 180°C, 0% RH | Alberti and Casciola, 1997 |
| pellicular $\alpha$-$Zr(O_3POH)_2 \bullet nH_2O$ | $1.0 \times 10^{-4}$ | 20°C, 90% RH | Alberti *et al.*, 1984 |
| $\alpha$-$Zr(O_3PCH_2OH)_{1.27}(O_3PC_6H_4$ | $1.6 \times 10^{-2}$ | 20°C, 90% RH | Alberti *et al.*, 1992 |
| $SO_3H)_{0.73} \bullet nH_2O$($\alpha$-ZrSP) | $8.0 \times 10^{-3}$ | 180°C, 60% RH | Alberti *et al.*, 1992 |
| $\alpha$-$Zr(O_3PCH_2OH)_{1.15}(O_3PC_6H_4SO_3H)_{0.85}$ | $1.2 \times 10^{-4}$ | 25°C, 85% RH | Alberti *et al.*, 1992 |
| $\alpha$-$Zr(O_3PC_6H_4SO_3H)_2 \bullet 3.6H_2O$ | $2.1 \times 10^{-2}$ | 20°C, 90% RH | Alberti *et al.*, 1992 |
| $\gamma$-$Zr(PO_4)(H_2PO_4) \bullet 2H_2O$ | $2.0 \times 10^{-5}$ | 20°C, 90% RH | Anderson and Garrigan, 1984 |
| $\gamma$-$Zr(PO_4)(H_2PO_4)_{0.54}(HO_3PC_6$ | $1.0 \times 10^{-2}$ | 20°C, 90% RH | Anderson and Garrigan, 1984 |
| $H_4SO_3H)_{0.46} \bullet nH_2O$($\gamma$-ZrSP) | $1.3 \times 10^{-3}$ | 20°C, 90% RH | Clearfield, 1990 |
| Zirconium phosphate pyrophosphate | $2.0 \times 10^{-6}$ | 100°C, 20% RH | Clearfield, 1990 |

1999). All conductivities were very low (< $5 \times 10^{-7}$ S/cm). Better results were achieved by intercalating sulfophenylphosphonate groups (Alberti *et al.*, 2001). $Ti(HPO_4)_{1.00}(O_3PC_6H_4SO_3H)_{0.85}(OH)_{0.3} \bullet nH_2O$ generally had a conductivity of an order of magnitude greater than similar zirconium derivatives (0.1 S/cm at 20 °C and 90% RH and < 0.1 S/cm at 100 °C and > 65% RH). For Nafion composite membranes, there was a slight improvement in conductivity together with increases in thermal stability but the humidity dependence remained unchanged.

It is important to note that fully sulfonated compounds are highly deliquescent and hard to recover from solution; this can be circumvented by preparing mixed derivatives where the sulfonated groups are replaced by non-sulfonated groups (e.g. $-O_3PCH_2OH$). It is also of importance to realize that sulfonated materials are likely to have a temperature limit at 200 °C; in some instances the decay of proton conductivity has been attributed to the decomposition of the $-SO_3H$ groups (Alberti and Casciola, 1997). The glass transition limitations of perfluorinated and aromatic sulfonated membranes have been overcome by composite modification. The proton conductivity of nanocomposite membranes mostly surpasses the base polymer membranes at high temperatures. The best results were obtained by maximizing the interfacial surface of dispersed colloidal zirconium phosphate derivatives through exfoliation. This simultaneously affected conductivity, mechanical properties, and fuel crossover.

Increased swelling in polymer membranes leads to increases in proton conductivity (Kreuer, 1997). Therefore, to improve the water retention properties of Nafion above 100 °C, Rhee's group prepared Nafion nanocomposite membranes with layered zirconium sulfophenylphosphonate (ZrSPP) applicable for both DMFCs (Kim *et al.*, 2008) and high-temperature PEMFCs (Kim *et al.*, 2006). A Nafion/ZrSPP nanocomposite membrane for a DMFC tolerated a high degree of methanol crossover while maintaining its essential proton conductivity. Layered ZrSPP was prepared from the precipitation of $Zr^{4+}$ ions and m-sulfophenyl phosphonic acid at a mole ratio of P to Zr = 2.0. The refractive index indicated that the methanol permeability of unmodified Nafion was $2.3 \times 10^{-6}$ $cm^3cm/cm^2s$ at 25 °C while that of Nafion/ZrSPP decreased to $6.5 \times 10^{-7}$ $cm^3cm/cm^2s$. Also Nafion/ZrSPP nanocomposite membranes delivered a constant power output of 104 $mW/cm^2$ at 0.4 V under 1 M methanol. When Nafion/ZrSPP nanocomposite membranes were applied to high-temperature PEMFC, they almost maintained the maximum proton conductivity of Nafion ($\sigma$ = 0.08 S/cm) even at 110 °C when hydrated at 98% RH. Additionally, this membrane had four times better fuel-cell performance than recast Nafion at 100 °C (Fig. 14.15). The results indicated that ZrSPP particles acted as solid proton conductors and were hygroscopic. Table 14.7 summarizes the proton conductivity and fuel-cell performance of layered metal phosphonate-filled polymer nanocomposites.

Therefore, the improved properties of nanoparticular composite membranes are a result of complex interactions between the structure and proton mobility (Damay

*Table 14.7* Proton conductivity and fuel-cell performance of layered metal phosphonate-filled nanocomposite

| Polymer | Layered metal phosphate | Proton conductivity (S/cm) | Fuel-cell performance ($mA/cm^2$ at 0.6 V) | Reference |
|---|---|---|---|---|
| Nafion® | $\alpha$-ZrP | 0.025 at 75 °C | 440 at 75 °C (at 0.37 V, DMFC) | Costamagna *et al.*, 2002 |
| Nafion® | $\alpha$-ZrP | | 1500 at 130 °C (at 0.45 V) | Costamagna *et al.*, 2002 |
| Nafion® | $\alpha$-ZrSPP | 0.040 at 80 °C | | Costamagna *et al.*, 2002 |
| sPEEK | $\alpha$-ZrP | 0.080 at 100 °C (RH 100%) | | Ruffmann *et al.*, 2003 |
| sPEEK | $\alpha$-ZrSPP | 0.040 at 150 °C (RH 100%) | | Ruffmann *et al.*, 2003 |
| Nafion® | ZrSPP | 0.030 at 80 °C (RH 100%) 0.007 at 80 °C (RH 50%) | 600 at 80 °C 250 at 100 °C | Kim *et al.*, 2006 |

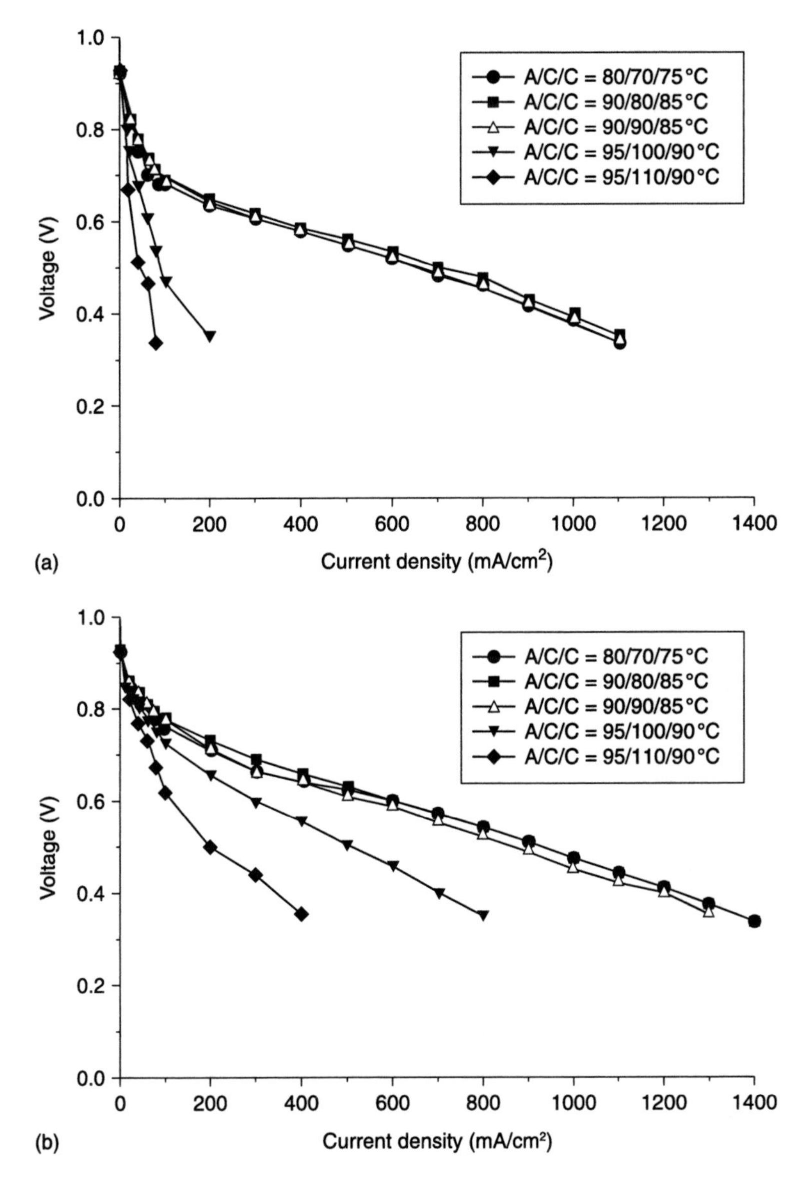


*14.15* Unit cell performance of (a) Nafion® and (b) Nafion®/ZrSPP 20 wt% at different temperatures.

and Klein, 2003). The improved properties might be caused by changes in the thermodynamic properties of the membrane due to the hygroscopic nature of the trapped nanoparticles. Recently, it has been suggested that only the increase in the rigidity of the membrane is responsible for the increase in proton conductivity (Yang *et al.*, 2004). What these studies suggested is that the effects of adding nanoparticles to a polymer membrane arise for the following reasons: (a) thermodynamic changes due to hygroscopicity, (b) changes in capillary forces and the vapor liquid equilibrium as a result of changes in pore properties, (c) surface charge interactions between the composite species, and (d) changes in the morphology of the membrane. This demonstrates that the exact nature of the composite interactions still requires further investigation at a fundamental level in order to understand the key relations, which can help engineer better composite membranes in a controlled manner.

## 14.3 Conclusions and future trends

This chapter presents the status of current research on polymer nanocomposites for PEMFC applications at high temperatures and low RH. Polymer electrolyte membranes require high proton conductivity at low water content, long-term durability under fuel-cell operating conditions, and low fuel permeability. Nanocomposite membranes may be the most promising candidates for solving the problems of the currently used sulfonated membranes including Nafion and sulfonated hydrocarbons.

Four types of nanocomposite membrane have been described: hygroscopic metal oxides, heteropolyacids, anhydrous solid acids, and layered metal phosphates/phosphonates. Hygroscopic and nano-sized inorganic metal oxides within a polymer membrane improve properties such as water uptake capacity and thermomechanical stability, and reduce methanol crossover. However, the improved properties could not be transformed into the desired level of improvement in terms of conductivity and performance in fuel cells.

In crystalline form, heteropolyacids are one of the most attractive inorganic fillers because they are highly conductive ($10^{-3}$ to $10^{-1}$ S/cm) and thermally stable. However, they have two major drawbacks: their extremely high solubility and large particle size (30–100 nm) within the membrane matrix.

Anhydrous solid acids are categorized between normal acids and normal salts. They undergo a structural phase change at a certain temperature and bare protons are mobile in anhydrous states due to the existence of a dynamically disordered network of H bonds. But they generally suffer from poor mechanical properties and water solubility, as well as extreme ductility and volume expansion with rising temperature. However, when they are incorporated into the channels of polymer membranes, they are very promising because their conductivity does not depend on humidity. However, technological challenges still remain before solid acids can be used in fuel cells.

Layered zirconium phosphonates and phosphonates have been widely studied but they become good proton conductors only when humidified. Zirconium phosphonates are better conductors than the corresponding phosphates. Zirconium sulfoarylphosphonates have high conductivity values, comparable with Nafion, although the latter is more strongly affected by humidity. When sulfonic acid was attached to a phenyl, benzyl, or fluorinated benzyl group, the highest proton conductivity was $5 \times 10^{-2}$ S/cm at 100 °C and 100% RH.

Presently, most nanocomposites are based on sulfonic acid moieties in polymer membranes and functionalized inorganic materials, so they always need a solvent such as water to dissociate and conduct protons. More research effort is required to achieve a high degree of functionalization for both organic and inorganic compounds without any deterioration of membrane stability. Phosphoric acids or phosphonic acidic moieties may be good candidates for proton-conducting moieties due to self-ionization and dehydration without a solvent, in addition to being able to produce more complex polymer architectures.

For the nanocomposites discussed, the particle size of the inorganic compounds could not easily be controlled at a nanometer scale, so that the particle size is sometimes over a few hundred nanometers. This prevents good dispersion of the nanoclusters or the ion channels in the sulfonated membranes. It is highly desirable to control particle size to a few nanometers and the dispersion of the inorganic component at a molecular level, either by covalent bonding of the inorganic component to the polymer matrix or by using organic functionalized components.

As previously mentioned, proton conductivity is the most important property but mechanical strength must also be considered. When polymer electrolyte membranes are used in fuel cells, the total resistance of the cell is more important. Therefore, another direction of research on nanocomposite membranes could be the development of membranes with high mechanical strength and high ductility without loss of proton conductivity, which will reduce the thickness and resistance of the membranes.

Finally, there is also a need to study the morphology and structure of nanocomposite membranes as well as stability. Consequently the important parameters in developing optimized nanocomposite membranes may be:

- to identify the appropriate inorganic filler and dispersal conditions for the polymer matrix
- to optimize the concentration of the inorganic materials
- to identify the best synthesis and casting conditions
- to acquire a knowledge of the interactions between the inorganic nanoparticles and the organic polymer matrix.

## 14.4 Acknowledgements

This work was supported by the Global Frontier R&D Program of the Center for Multiscale Energy Systems, funded by the National Research Foundation under the Ministry of Education, Science and Technology, and by the Korean Center for Artificial Photosynthesis (KCAP) located at Sogang University, funded by the Ministry of Education, Science, and Technology (MEST), through the National Research Foundation of Korea (NRF-2011-C1AAA001-2011-0030278).

## 14.5 References

Adjemian, K. T., Lee, S. J., Srinivasan, S., Benziger, J. and Bocarsly, A. B. (2002), 'Silicon oxide Nafion composite membranes for proton-exchange membrane fuel cell operation at 80–140 °C', *J Electrochem Soc*, 149, A256–A261.

Adjemian, K. T., Dominey, T., Krishnan, L., Ota, H., Majsztrik, P., *et al.* (2006), 'Function and characterization of metal oxide-Nafion composite membranes for elevated-temperature $H_2/O_2$ PEM fuel cells', *Chem Mater*, 18, 2238–2248.

Ahmed, I., Zaidi, S. M. J. and Rahman, S. U. (2005), 'Proton conductivity and characterization of novel composite membranes for medium-temperature fuel cells', *Proceedings of the ICOM*, 21–26, 102.

Alberti, G. and Casciola, M. (1997), 'Layered metal phosphonates, a large class of inorgano-organic proton conductors', *Solid State Ionics*, 97, 177–186.

Alberti, G. and Casciola, M. (2003), 'Composite membranes for medium temperature PEM FCs', *Annu Rev Mater Res*, 33, 129–154.

Alberti, G., Casciola, M., Costantino, U., Levi, G. and Ricciardi, G. (1978), 'On the mechanism of diffusion and ionic transport in crystalline insoluble acid salts of tetravalent metals', *J Inorg Nucl Chem*, 40, 533–537.

Alberti, G., Casciola, M., Costantino, U. and Leonardi, M. (1984), 'Conductivity of anhydrous pellicular zirconium phosphate in hydrogen form', *Solid State Ionics*, 14, 289–295.

Alberti, G., Casciola, M., Costantino, U., Peraio, A. and Montoneri, E. (1992), 'Protonic conductivity of layered zirconium phosphonates containing $-SO_3H$ groups. I. Preparation and characterization of a mixed zirconium phosphonate of composition', *Solid State Ionics*, 50, 315–322.

Alberti, G., Casciola, M., Constantino, U. and Vivani, R. (1996), 'Layered and pillared metal(IV) phosphates and phosphonates', *Adv Mater*, 8, 291–303.

Alberti, G., Casciola, M. and Palombari, R. (2000), 'Inorgano-organic proton conducting membranes for fuel cells and sensors at medium temperatures', *J Membr Sci*, 17, 233–239.

Alberti, G., Costantino, U., Casciola, M., Ferroni, S., Massinelli, L., *et al.* (2001), 'Preparation, characterization and proton conductivity of titanium phosphate sulfophenylphosphonate', *Solid State Ionics*, 145, 249–255.

Alberti, G., Casciola, M., Pica, M., Torpanelli, T. and Sganappa, M. (2005), 'New preparation methods for composite membranes for medium temperature fuel cells based on precursor solutions of insoluble inorganic compounds', *Fuel Cells*, 5, 366–374.

Alberti, G., Casciola, M., Capitani, D., Donnadio, A., Narducci, R. *et al.* (2007), 'Novel Nafion-zirconium phosphate nanocomposite membranes with enhanced stability of proton conductivity at medium temperature and high relative humidity', *Electrochimica Acta*, 52, 8125–8132.

Anderson, G. L. and Garrigan, P. L. (1984), In Proceedings of the symposium on molten carbonate fuel cell technology, *Electrochem Soc*, 18, 297.

Andronie, A., Morozan, A., Nastase, C., Nastase, F., Dumitru, A. *et al.* (2008), '$CsHSO_4$/nanooxide polymer membranes for fuel cell', *Funct Nanoscale Mater*, 2, 415–418.

Antonucci, P. L., Aricò, A. S., Creti, P., Ramunni, E. and Antonucci, V. (1999), 'Investigation of a direct methanol fuel cell based on a composite Nafion®-silica electrolyte for high temperature operation', *Solid State Ionics*, 125, 431–437.

Arico, A. S., Srinivasan, S. and Antonucci, V. (2001), 'DMFCs: From fundamental aspects to technology development', *Fuel Cells*, 1(2), 133–161.

Baranov, A. I., Merinov, B. V., Tregubchenko, A. V., Khiznichenko, V. P., Shuvalov, L. A. *et al.* (1989), 'Fast proton transport in crystals with a dynamically disordered hydrogen bond network', *Solid State Ionics*, 36, 279–282.

Baranov, A. I., Grebenev, V. V., Khodan, A. N., Dolbinina, V. V. and Efremova, E. P. (2005), 'Optimization of superprotonic acid salts for fuel cell applications', *Solid State Ionics*, 176, 2871–2874.

Bardin, B. B., Bordawekar, S. V., Neurock, M. and Davis, R. J. (1998), 'Acidity of Keggin-type heteropolycompounds evaluated by catalytic probe reactions, sorption microcalorimetry, and density functional quantum chemical calculations', *J Phys Chem B*, 102, 10817.

Bocchetta, P., Ferraro, R. and Di, Q. F. (2009), 'Advances in anodic alumina membranes thin film fuel cell: $CsH_2PO_4$ pore-filler as proton conductor at room temperature', *J Power Sources*, 187, 49–56.

Bondarenko, A. S., Zhou, W. H. and Bouwmeester, H. J. M. (2009), 'Superprotonic $KH(PO_3H)$-$SiO_2$ composite electrolyte for intermediate temperature fuel cells', *J Power Sources*, 194, 843–846.

Boysen, D. A., Chisholm, R. I., Haile, S. M. and Narayanan, S. R. (2000), 'Polymer solid acid composite membranes for fuel-cell applications', *J Electrochem Soc*, 147, 3610–3613.

Boysen, D. A., Uda, T., Chisholm, C. R. I. and Haile, S. M. (2004), 'High-performance solid acid fuel cells through humidity stabilization', *Science*, 303, 68–70.

Carriere, D., Moreau, M., Lhalil, K., Barboux, P. and Boilot, J. P. (2003), 'Proton conductivity of colloidal nanometric zirconium phosphates', *Solid State Ionics*, 162, 185–190.

Casciola, M. and Bianchi, D. (1985), 'Frequency response of polycrystalline samples of $\alpha$-$Zr(HPO_4)_2 \cdot H_2O$ at different relative humidities', *Solid State Ionics*, 17, 287–293.

Chen, R. H., Yen, C. C., Shern, C. S., Fukami, T. (2006), 'Impedance spectroscopy and dielectric analysis in $KH_2PO_4$ single crystal', *Solid State Ionics*, 177, 2857–2864.

Clearfield, A. (1990), 'Hydrothermal synthesis of selected phosphates and molybdates', *Prog Cryst Growth Charact Mater*, 21, 1–28.

Clearfield, A. (1996), 'Recent advances in metal phosphonate chemistry', *Curr Opin Solid State Mater*, 1, 268–278.

Collier, A., Wang, H., Yuan, X. Z., Zhang, J. and Wilkinson, D. P. (2006), 'Degradation of polymer electrolyte membranes', *Int J Hydrogen Energy*, 31, 1838–1854.

Costamagna, P., Yang, C., Bocarsly, A. B. and Srinivasan, S. (2002), 'Nafion 115/zirconium phosphate composite membranes for operation of PEMFCs above 100 °C', *Electrochimica Acta*, 47, 1023–1033.

Crisholm, C. R. I. and Haile, S. M. (2000), 'Superprotonic behavior of $Cs_2(HSO_4)(H_2PO_4)$ – A new solid acid in the $CsHSO_4$-$CsH_2PO_4$ system', *Solid State Ionics*, 136–137, 229–241.

Damay, F. and Klein, L. C. (2003), 'Transport properties of Nafion composite membranes for proton-exchange membranes fuel cells', *Solid State Ionics*, 162–163, 261–267.

Deng, Q., Moore, R. B. and Mauritz, K. A. (1995), 'Novel Nafion/ORMOSIL hybrids via *in situ* sol-gel reactions', *Chem Mater*, 7, 2259–2268.

Devrim, Y., Erkan, S., Bac, N. and Eroglu, I. (2009), 'Preparation and characterization of sulfonated polysulfone/titanium dioxide composite membranes for proton exchange membrane fuel cells', *International Journal of Hydrogen Energy*, 34, 3467–3475.

Feng, X., Fryxell, G. E., Wang, L. Q., Kim, A. Y., Liu, J. *et al.* (1997), 'Functionalized monolayers on ordered mesoporous supports', *Science*, 276, 923–926.

Gillespie, R. J. and Robinson, E. A. (1965), *Nonaqueous Solvent System*, New York, Academic Press.

Glipa, J., Leloup, M., Jones, D. J. and Roziere, J. (1997), 'Enhancement of the protonic conductivity of $\alpha$-zirconium phosphate by composite formation with alumina or silica', *Solid State Ionics*, 97, 227–232.

Gnana, K. G., Lee, D., Kim, A., Kim, P., Nahm, K. *et al.* (2007), 'Structural and transport properties of porous PVdF-HFP electrolyte membranes modified with an inorganic filler', *Composite Interfaces*, 15, 731–746.

Haile, S. M., Lentz, G., Kreuer, K. D. and Maier, J. (1995), 'Superprotonic conductivity in $Cs_3(HSO_4)_2(H_2PO_4)$', *Solid state Ionics*, 77, 128–134.

Haile, S. M., Boysen, D. A., Chisholm, C. R. I. and Merle, R. B. (2001), 'Solid acids as fuel cell electrolytes', *Nature*, 410, 910–913.

Haile, S. M., Shao, Z. and Kwak, A. (2004), 'Preparation and investigation of anode-supported thin-film fuel cells operated in a single chamber configuration', *Solid State Ionics*, 175, 39–46.

Helen, M., Viswanathan, B. and Srinivasa, S. M. (2007), 'Synthesis and characterization of hybrid membranes based on a-ZrP and silicotungstic acid', *J Membr Sci*, 292, 105.

Hogarth, W. H. J., Costa, J. C. D. and Lu, G. Q. (2005), 'Solid acid membranes for high temperature (> 140 °C) proton exchange membrane fuel cells', *J Power Sources*, 142(1–2), 223–237.

Honma, I., Takeda, Y. and Bae, J. M. (1999), 'Protonic conducting properties of sol-gel derived organic/inorganic nanocomposite membranes doped with acidic functional molecules', *Solid State Ionics*, 120(1–4), 255–264.

Ise, M., Kreuer, K. D. and Maier, J. (1999), 'Electroosmotic drag in polymer electrolyte membrane: An electrophoretic NMR study', *Solid State Ionics*, 125, 213–223.

Jalani, N. H., Dunn, K. and Datta, R. (2005), 'Synthesis and characterization of Nafion-$MO_2$ (M = Zr, Si, Ti) nanocomposite membranes for high temperature PEM fuel cells', *Electrochimica Acta*, 51, 553–560.

Jung, U. H., Park, K. T., Park, E. H. and Kim, S. H. (2006), 'Improvement of low-humidity performance of PEMFC by addition of hydrophilic $SiO_2$ particles to catalyst layer', *J Power Sources*, 159, 529–532.

Karen, I. and Richard, A. (2007), 'Polymer nanocomposites', *MRS Bulletin*, 32, 314–322.

Karlsson, L. E. and Jannasch, P. (2005), 'Polysulfone ionomers for proton-conducting fuel cell membranes – 2. Sulfophenylated polysulfones and polyphenylsulfones', *Electrochimica Acta*, 50, 1939–1946.

Karthikeyan, C. S., Nunes, S. P. and Schulte, K. (2006), 'Permeability and conductivity studies on ionomer-polysilsesquioxane hybrid materials', *Macromol Chem Phys*, 207, 336–341.

Katsoulis, D. E. (1998), 'A survey of applications of polyoxometalates', *Chem Rev*, 98, 359–387.

Kelarakis, A., Alonso, R., Lian, H., Burgaz, E., Estevez, L. *et al.* (2010), *Functional Polymer Nanocomposites for Energy Storage and Conversion*, Washington DC, American Chemical Society.

Kickelbock, G. (2003), 'Concepts for the incorporation of inorganic building blocks into organic polymers on a nanoscale', *Prog Polym Sci*, 28, 83–114.

Kickner, M. A., Ghassemi, H., Kim, Y. S., Einsla, B. R. and McGrath, J. E. (2004), 'Alternative polymer systems for proton exchange membrane (PEMs)', *Chem Rev*, 104, 4587–4612.

Kim, C. S., Krishnan, P. and Park, J. S. (2006), 'Performance of a poly(2,5-benzimidazole) membrane based high temperature PEM fuel cell in the presence of carbon monoxide', *J Pow Sour*, 159, 817–823.

Kim, Y. S., Cho, H. S., Song, M. K., Ghil, L. J., Kang, J. S. *et al.* (2008), 'Characterization of Nafion®/zirconium sulphophenyl phosphate nanocomposite membrane for direct methanol fuel cells', *J Nanosci Nanotech*, 8, 4640–4643.

Klein, L. C., Daiko, Y., Aparicio, M. and Dammy, F. (2005), 'Methods for modifying proton exchange membranes using the sol-gel process', *Polymer*, 46, 4504–4509.

Kozhevnikov, I. V. (1998), 'Catalysis by heteropoly acids and multicomponent polyoxometalates in liquid phase reactions', *Chem Rev*, 98, 171–198.

Kreuer, K. D. (1997), 'On the development of proton conducting materials for technological applications', *Solid State Ionics*, 97, 1–15.

Lavrova, G. V. and Ponomareva, V. G. (2008), 'Intermediate-temperature composite proton electrolyte $CsH_5(PO_4)_2/SiO_2$: Transport properties versus oxide characteristic', *Solid State Ionics*, 179, 1170–1173.

Lee, K. Y., Mizuno, N., Okuhara, T. and Misono, M. (1989), 'Catalysis by heteropoly compounds. XIII. An infrared study of ethanol and diethyl-ether in the pseudoliquid phase of 12-tungstophosphoric acid', *Bull Chem Soc*, 62, 1731–1739.

Li, C., Sun, G., Ren, S., Liu, J., Wang, Q. *et al.* (2006), 'Casting Nafion-sulfonated organosilica nano-composite membranes used in direct methanol fuel cells', *J Membr Sci*, 272, 50–57.

Licoccia, S., Di Vona, M. L., D'Epifanio, A., Marani, D., Vittadello, M. *et al.* (2006), 'ORMOSIL/sPEEK based hybrid composite proton conducting membranes', *J Electrochem Soc*, 153, A1226–A1231.

Liu, Y. (2005), *Organic/inorganic composite membranes for high temperature proton exchange membrane fuel cells*, MS dissertation, University of Connecticut.

Mauritz, K. A. (1998), 'Organic-inorganic hybrid materials: Perfluorinated ionomers as sol-gel polymerization templates for inorganic alkoxides', *Mater Sci Eng*, C6, 121–133.

Misono, M. (1987), 'Heterogeneous catalysis by heteropoly compounds of molybdenum and tungsten', *Catal Rev Sci Eng*, 29, 269.

Mizuno, N. and Misono, M. (1998), 'Heterogeneous catalysis', *Chem Rev*, 98, 199–218.

Nunes, S. P., Ruffmann, B., Rikowski, E., Vetter, S. and Richau, K. (2002), 'Inorganic modification of proton conductive polymer membranes for direct methanol fuel cells', *J Membr Sci*, 203, 215–225.

Nunes, S. P., Ponce, M. L., Prado, L. A. and Silva, V. (2004), 'Membranes for direct methanol fuel cell based on modified heteropolyacids', *Desalination*, 162, 383–391.

Okuhara, T., Mizuno, N. and Misono, M. (1996), 'Catalyst chemistry of heteropoly compounds', *Adv Catal*, 41, 113–252.

Otomo, J., Ishigooka, T., Kitano, T., Takahashi, H. and Nagamoto, H. (2008), 'Phase transition and proton transport characteristics in $CsH_2PO_4/SiO_2$ composites', *Electrochim Acta*, 53, 8186–8195.

Piao, J. H., Liao, S. J. and Liang, Z. X. (2009), 'A novel cesium hydrogen sulfate-zeolite inorganic composite electrolyte membrane for polymer electrolyte membrane fuel cell application', *J Power Sources*, 193, 483–487.

Ponce, M. L., Pardo, L., Ruffmann, B., Richau, K., Mohr, R. *et al.* (2003), 'Reduction of methanol permeability in polyetherketone-heteropolyacid membranes', *J Memb Sci*, 217, 5–15.

Ponce, M. L., Prado, L. A. S. A., Silva, V. and Nunes, S. P. (2004), 'Membranes for direct methanol fuel cell based on modified heteropolyacids', *Desalination*, 162, 383–391.

Ponomareva, V. G. and Lavrova, G. V. (1998), 'Influence of dispersed $TiO_2$ on protonic conductivity of $CsHSO_4$', *Solid State Ionics*, 106, 137–141.

Ponomareva, V. G., Uvarov, N. F., Lavrova, G. V. and Hairetdinov, E. F. (1996), 'Composite protonic solid electrolytes in the $CsHSO_4$-$SiO_2$ system', *Solid State Ionics*, 90, 161–166.

Pourcelly, G. and Gavach, C. (1992), *Proton Conductors: Solids, Membranes and Gels – Materials and Devices*, New York, Cambridge University Press.

Prater, K. B. (1994), 'Polymer electrolyte fuel cells: A review of recent developments', *J Power Sources*, 51, 129–144.

Ramani, V., Kunz, H. R. and Fenton, J. M. (2004), 'Investigation of Nafion®/HPA composite membranes for high temperature/low relative humidity PEMFC operation', *J Membr Sci*, 232, 31–44.

Ren, S., Sun, G., Li, C., Liang, Z., Wu, Z. *et al.* (2006), 'Organic silica/Nafion® composite membrane for direct methanol fuel cells', *J Membr Sci*, 282, 450–455.

Rodriguez, C. R., Jimenez, J. J., Jimenez, L. A., Maireles, T. P., Ramos, B. J. R., *et al.* (1999), 'Proton conductivity of mesoporous MCM type of zirconium and titanium phosphates', *Solid State Ionics*, 125, 407–410.

Ruffmann, B., Silva, H., Schulte, B. and Nunes, S. P. (2003), 'Organic/inorganic composite membranes for application in DMFC', *Solid State Ionics*, 162–163, 269–275.

Sacca, A., Carbone, A., Pedicini, A., Portale, G., Dilario, L. *et al.* (2006), 'Structural and electrochemical investigation on re-cast Nafion membranes for polymer electrolyte fuel cells (PEFCs) application', *J Membr Sci*, 278, 105–113.

Sahu, A. K., Selvarani, G., Pitchumani, S., Sridhar, P. and Shukla, A. K. (2007), 'A sol-gel modified alternative Nafion-silica composite membrane for polymer electrolyte fuel cells', *J Electrochem Soc*, 154, B123–B132.

Saito, J., Yano, H., Miyatake, K., Uchida, M. and Watanabe, M. (2009), 'Durability of a novel sulfonated polyimide membrane in polymer electrolyte fuel cell operation', *Electrochim Acta*, 54, 1076–1082.

Savadogo, O. (2004), 'Emerging membranes for electrochemical systems. Part II. High temperature composite membranes for polymer electrolyte fuel cell (PEFC) applications', *J Power Sources*, 127, 135–161.

Schottner, G. (2001), 'Hybrid sol-gel-derived polymers: Applications of multifunctional materials', *Chem Mater*, 13, 3422–3435.

Silva, V. S., Ruffmann, B., Silva, H., Gallego, Y. A., Mendes, A., Madeira, L. M. and Nunes, S. P. (2005), 'Proton electrolyte membrane properties and direct methanol fuel cell performance', *J Power Sources*, 140, 34–40.

Sinitsyn, V. V., Baranov, A. I. and Ponyatovsky, E. G. (2000) 'Pressure effect on superprotonic phase transition in mixed $[(NH_4)_xRb_{1-x}]_3H(SO_4)_2$ crystals', *Solid State Ionics*, 136–137, 167–171.

Slade, R. C. T., Jinku, H. and Knowles, J. A. (1992), 'Conductivity variations in composites of $\alpha$-zirconium phosphate and fumed silica', *Solid State Ionics*, 50, 287–290.

Staiti, P., Freni, S. and Hocevar, S. (1999), 'Synthesis and characterization of proton-conducting materials containing dodecatungstophosphoric and dodecatungstosilic acid supported on silica', *J Power Sources*, 79, 250–255.

Staiti, P., Minutoli, M. and Hocevar, S. (2000), 'Membrane based on phosphotungstic acid and polybenzimidazole for fuel cell', *J Power Sources*, 90, 231–235.

Staiti, P., Arico, A. S., Baglio, V., Lufrano, F., Passalacqua, E. *et al.* (2001), 'Hybrid Nafion-silica membranes doped with heteropolyacids for application in direct methanol fuel cells', *Solid State Ionics*, 145, 101–107.

Stein, E. W., Clearfield, A. and Subramanian, M. A. (1996), 'Conductivity of group IV metal sulfophosphonates and a new class of interstratified metal amine-sulfophosphonates', *Solid State Ionics*, 83, 113–124.

Tazi, B. and Savadogo, O. (2000), 'Parameters of PEM fuel-cells based on new membranes fabricated from Nafion, silicotungstic acid and thiophene', *Electrochim Acta*, 45, 4329–4339.

Tiang, J., Gao, P., Zhang, Z., *et al.* (2008), 'Preparation and performance evaluation of a Nafion/$TiO_2$ composite membrane for PEMFCs', *J Hydro Ener*, 33, 5686–5690.

Vladimir G., Bluemle, M. J., Castro, E., Yu-Min, T., Zawodzinski, T. A. and Mann, J. A. (2007), 'Characterization of transport properties in gas diffusion layers for proton exchange membrane fuel cells', *J Pow Sour*, 165, 793–802.

Vona, M. L. D., Ahmed, Z., Bellitto, S., Lenci, A., Traversa, E. *et al.* (2007), 'sPEEK-$TiO_2$ nanocomposite hybrid proton conductive membranes via *in situ* mixed sol-gel process', *J Membr Sci*, 296, 156–161.

Wakizoe, M., Velev, O. A. and Srinivasan, S. (1995), 'Analysis of proton exchange membrane fuel cell performance with alternate membranes', *Electrochimica Acta*, 40, 335–344.

Wang, Z. and Saxena, S. K. (2001), 'Raman spectroscopic study on pressure-induced amorphization in nanocrystalline anatase ($TiO_2$)', *Solid State Communications*, 118, 75–78.

Wang, H., Holmberg, B. A., Huang, L., Wang, Z., Mitra, A. *et al.* (2002), 'Nafion-bifunctional silica composite proton conductive membranes', *J Mater Chem*, 12, 834–837.

Wu, C. M., Xu, T. W. and Liu, J. S. (2006), *Charged Hybrid Membranes by the Sol-Gel Approach: Present Status and Future Perspectives*, New York, Nova Science.

Yang, C., Costamagna, P., Srinivasan, S., Benziger, J. and Bocarsly, A. B. (2001), 'Approaches and technical challenges to high temperature operation of proton exchange membrane fuel cells', *J Power Sources*, 103, 1–9.

Yang, C., Srinivasan, S., Bocarsly, A. B., Tulyani, S. and Benziger, J. B. (2004), 'A comparison of physical properties and fuel cell performance of Nafion and zirconium phosphate/Nafion composite membranes', *J Membr Sci*, 237, 145–161.

Yin, Y., Hayashi, S., Yamada, O., Kita, H. and Okamoto, K. I. (2005), 'Branched/crosslinked sulfonated polyimide membranes for polymer electrolyte fuel cells', *Macromol Rapid Commun*, 26, 696–700.

Zaidi, M. J., Mikhailenko, S. D., Robertson, G. P., Guiver, M. D. and Kaliaguine, S. (2000), 'Proton conducting composite membranes from polyether ether ketone and heteropolyacids for fuel cell applications', *J Membr Sci*, 173, 17–34.

Zawodzinski, T. A., Derouin, C., Radzinski, S., Sherman, R. J., Springer, T. E. *et al.* (1993), 'Water uptake by and transport through Nafion 117 membranes', *J Electrochem Soc*, 140, 1981–1985.

Zhao, D., Huo, Q., Feng, J., Chmelk, B. F. and Stucky, G. D. (1998), 'Nonionic triblock and star diblock copolymer and oligomeric surfactant syntheses of highly ordered, hydrothermally stable, mesoporous silica structures', *J Am Chem Soc*, 120, 6024–6036.

Zoppi, R. A. and Nunes, S. P. (1998), 'Electrochemical impedance studies of hybrids of perfluorosulfonic acid ionomer and silicon oxide by sol-gel reaction from solution', *J Electroanal Chem*, 445, 39–45.

Zoppi, R. A., Yoshida, I. V. P. and Nunes, S. P. (1997), 'Hybrids of perfluorosulfonic acid ionomer and silicon oxide by sol-gel reaction from solution: morphology and thermal analysis', *Polymer*, 39, 1309–1315.

# 15
# Polymer nanocomposites for aerospace applications

J. NJUGUNA, K. PIELICHOWSKI and J. FAN,
Cranfield University, UK and Cracow University of Technology, Poland

**Abstract:** Advances in nanotechnology will lead to improvements in capabilities across a spectrum of applications. The uses of polymer nanocomposites in aerospace structures have had a significant effect on aerospace design and applications, primarily by providing safer, faster and eventually cheaper transportation in the future. This chapter reviews key properties of polymer nanocomposites for potential aerospace applications. In particular, the chapter discusses mechanical, field emission, thermal, electrical and optical properties of polymer nanocomposites for aerospace needs.

**Key words:** polymer nanocomposites, aerospace structures, mechanical properties, field emission properties, thermal properties, electrical properties, optical properties.

## 15.1 Introduction

Key to the success of many modern structural components is the tailored behaviour of materials. A relatively inexpensive way of obtaining macroscopically desired responses is to enhance base material properties through the addition of microscopic or nanoscopic matter to manipulate the macrostructures. Accordingly, in many modern engineering designs, materials with highly complex microstructures are now in use. The macroscopic characteristics of a modified base material are an assemblage of different 'pure' components. This newly developed approach offers promising results, including the enhancement of electrical, thermal and mechanical properties through the use of nano-sized organic and inorganic particles. Over the past 15 years, fundamental and applied research has been carried out in the field of polymer nanocomposites.

Many types of nanomaterial (such as carbon nanotubes, nanofibres, $SiO_2$ and montmorillonite) are now available due to the establishment of well-developed manufacturing technologies, such as chemical vapour deposition, ball milling and electrospinning. Through improvements in bulk manufacturing, fibre-reinforced polymer nanocomposites are being used in an increasing number of practical applications (for example, in the manufacture of composite components in aerospace and microelectronics). The improvements that have been identified for high-performance structures and payloads are due to the modification of mechanical, thermal and electrical properties. High-performance structural design criteria impose a number of restrictions on the properties of materials to be used. Lighter, thinner, stronger and cheaper structures are very important goals.

Launching a heavy lift system into low Earth and geosynchronous orbits generally costs €5000–15 000/kg and €28 000/kg, respectively. Because of increasing oil and gas prices, the demand for lightweight materials in the aerospace industry is tremendous. Even in general aviation, fuel costs account for around 50% of the operational costs. Consequently, over the last three decades, the usage of fibre-reinforced polymer (FRP) composites in these applications has increased from less than 5% by structural weight (Boeing 737) to 50% (Boeing 787), contributing over 20% more fuel efficiency.

However, in these conventional structural materials, the fibre orientation is usually in-plane ($x$- and $y$-direction), resulting in fibre-dominated material properties in these directions whereas the matrix dominates in the $z$-direction. Therefore, FRPs are very sensitive to intrinsic damage such as delamination (in particular), matrix cracking and fatigue damage. Several approaches have been adopted to tackle these, which include:

- improving the fracture toughness of the ply interfaces via epoxy/elastomer blends and
- reducing the mismatch of elastic properties (and stress concentrations) at the interfaces between the laminated plies.

These materials also lack other required functional properties such as high electrical and thermal conductivity for electrostatic dissipation and lightning-strike protection. Currently, it is believed that the best route to achieve multifunctional properties in a polymer is to blend it with nanoscale fillers. This is because of the three main characteristics of polymer nanocomposites:

1. reduced nanoscopic confinement of matrix polymer chains;
2. variation in properties of nanoscale inorganic constituents; many studies have reported that the mechanical, conductivity, optical, magnetic, biological and electronic properties of several inorganic nanoparticles significantly change as their size is reduced from the macroscale to the microlevel and nanolevel; and
3. nanoparticle arrangement and creation of a large polymer/particle interfacial area.

## 15.2 Types of fibre-reinforced polymer (FRP) nanocomposites

### 15.2.1 Laminated Carbon nanotubes (CNT)/epoxy FRP nanocomposites

Nanoscale fillers such as carbon nanotubes (CNTs) and carbon nanofibres (CNFs) offer new possibilities for low-weight composites with extraordinary mechanical, electrical and thermal properties. Taking into consideration their high axial

Young's modulus, high aspect ratio, large surface area, low density and excellent thermal and electrical properties, these fillers can be used as modifiers for the polymer matrices of fibre-reinforced polymer composites leading to advanced mechanical behaviour. However, with nanotube-reinforced polymer composites there has only been a moderate strength enhancement, which is significantly below the theoretically predicted potential. To achieve the full potential of nanotubes, there are two critical issues that have to be solved:

- the dispersion of the nanotubes in the polymer matrix,
- the interfacial bonding between the nanotubes and the polymer matrix.

Nevertheless, based on a scaling argument correlating the radius ($r$), fibre strength ($\sigma$) and interface strength ($\tau$) with the energy absorbed per unit cross-sectional area by fibre pull-out (i.e., $G_{pull\text{-}out} \sim r\sigma^2/\tau$), it was shown very recently that improvements in toughness in polymer/CNT nanocomposites cannot be attributed to the nanotube pull-out mechanism, as the pull-out energy significantly decreases when the fibre radius is scaled down to the nanoscale. In line with this argument, many studies have reported reductions in toughness with the incorporation of CNTs, even at low loadings. Further evidence from work with other nanoscale fillers suggests that conventional toughening mechanisms may not transfer to polymer nanocomposites directly.

In general, weakly interacting nanotube bundles and aggregations of nanotubes result in a poor dispersion state that significantly reduces the aspect ratio of the reinforcement. The reason for the weak interfacial bonding behaviour lies in the atomically smooth, non-reactive surface of the nanotubes, which does not efficiently transfer the load from the polymer matrix to the nanotube lattice. To solve this problem, a number of methods have been developed to maximize the benefits of nanotubes in polymer composites, i.e. surfactant-assisted dispersion,[1] sonication at high power,[2] *in situ* polymerization,[3] electric field or magnetic-induced alignment of nanotubes,[4,5] plasma polymerization[6] and surface modifications such as inorganic coating,[7] polymer wrapping[8] and protein functionalization.[9]

One key area where nanocomposites can make a significant impact is in addressing interlaminar toughness in fibre-reinforced composites. The improvement of the interlaminar toughness of fibre-reinforced composites has been the focus of research for a considerable time, since it is directly related to the dynamic as well as the damage-tolerance performance of the composite. The problem has been addressed in various ways: stitching, Z-pinning or interleaving, with a notable increase in toughness while also providing improvements in mechanical properties, such as fatigue life. Other approaches focus on tailoring the matrix or interface properties in order to provide the necessary interlaminar fracture toughness. Matrix toughening may be performed through chemical modification or, more recently, with the incorporation of fillers in the matrix material. Interface modification can also be performed by grafting, which tailors the chemical compatibility between the fibres and the matrix.

Gojny *et al.*[10] investigated the interlaminar shear strength of nano-reinforced FRPs and described an efficient technique (mini-calendering) for dispersing carbon-based nanoparticles in epoxy resins. The application of a mini-calender to disperse carbon nanotubes (and carbon black) proved to be an efficient approach for reaching a good state of dispersion and enabled the manufacture of high volumes of nanocomposites. The resulting nanotube/epoxy composites exhibit a significant increase in fracture toughness as well as an enhancement in stiffness even with low nanotube content. Gojny *et al.*[11] also investigated the influence of CNTs on the interlaminar shear strength of a glass-fibre-reinforced polymer (GFRP) composite. They reported an increase of +19% in interlaminar shear strength with a weight fraction as low as 0.3 wt% of amino-functionalized double-wall CNTs (DWCNT-$NH_2$) in the epoxy matrix, Fig. 15.1.

It has been claimed that the nanometre size of the particles means they can be used as modifiers in fibre-reinforced polymers. Composites have been produced via the resin-transfer-moulding (RTM) process and the particles were not filtered by the glass-fibre bundles. A follow-up review by the same research team reported that the interlaminar shear strength of the nanoparticle-modified GFRP was significantly improved (+16%) whilst adding only 0.3 wt% of CNTs.[12] The interlaminar toughness ($G_{Ic}$ and $G_{IIc}$) was not affected in a comparable manner.

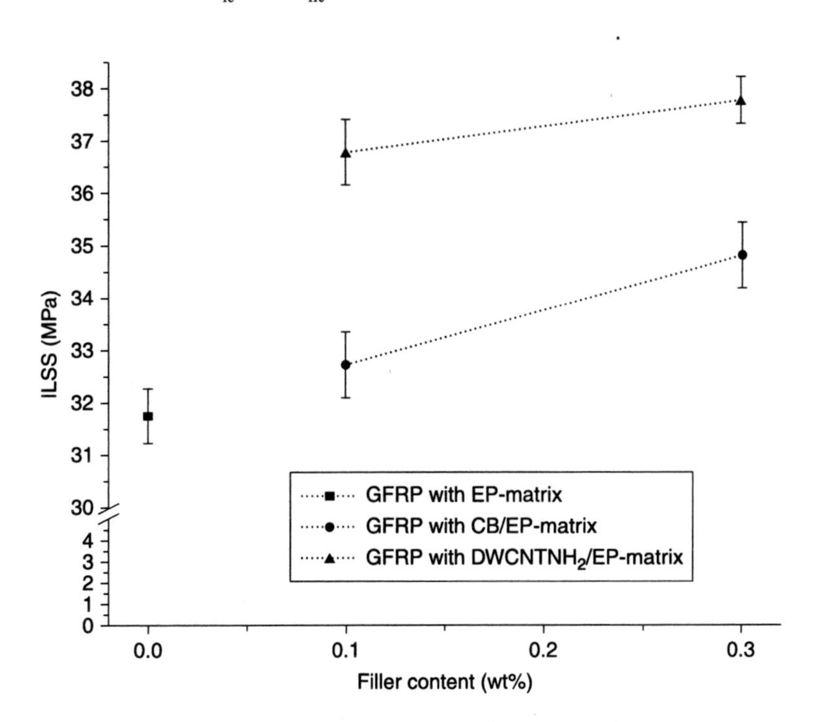


*15.1* Interlaminar shear strength (ILSS) of (nano-reinforced) GFRPs: epoxy (EP), carbon black (CB) and double-wall carbon nanotubes (DWCNT).[11]

The laminates containing CNTs exhibited a relatively high electrical conductivity at very low filler content.

Zhao *et al.*[13] fabricated CNTs and continuous carbon-fibre (T300) reinforced unidirectional epoxy-resin matrix composites. They prepared CNTs by catalytic decomposition of benzene using the floating transition method at 1100–1200 °C. Benzene was used as a carbon source and ferrocene as a catalyst with thiophene. The CNTs used were straight with a diameter of 20–50 nm, internal diameter of 10–30 nm and length of 50–1000 μm. The volume fraction of continuous carbon fibre (first filler) in the composites without second filler (CNT) was 60%. The flexural strength of the composites reached a maximum value of 1780 MPa when the weight per cent of CNT in the epoxy-resin matrix was only 3%. The study concluded that the flexural strength and modulus of the composites increased at first and then decreased with an increase of the CNT content in the epoxy-resin matrix.

Hsiao *et al.*[14] and Meguid and Sun[15] investigated the tensile and shear strength of nanotube-reinforced composite interfaces by single lap shear testing. They observed a significant increase in the interfacial shear strength for epoxies with contents between 1 and 5 wt% of multi-walled nanotubes (MWNTs) when compared with the neat epoxy matrix. In particular, instead of processing and characterizing CNT/polymer composites, Hsiao *et al.*[14] explored the potential of using CNTs to reinforce the adhesives joining two composite structures. In the study, different weight fractions of MWNT were dispersed in epoxy to produce toughened adhesives. The reinforced adhesives were used to bond the graphite fibre/epoxy composite adherends. This experimental study showed that adding 5 wt% MWNT to an epoxy adhesive effectively transferred the shear load from the adhesive to the graphite fibre system in the composite laminates and improved the average shear strength of the adhesion by 46% (±6%). A significant enhancement of the bonding performance was observed as the weight fraction of CNTs was increased. As shown in Fig. 15.2 (left), the 5 wt% MWNT effectively transferred the load to the graphite fibres in the adherends and the resulting failure was in the graphite fibre system. On the other hand, for epoxy adhesives containing no MWNTs (see Fig. 15.2 (right)), failure occurred at the epoxy along the bonding interface and no significant fractures of the graphite fibre were observed. Despite the promising results, the researchers concurred that further experiments involving increasing MWNT weight fractions and more detailed scanning electron microscopy (SEM) observations are required in order to understand and model the role of MWNTs in enhancing adhesion.

Various studies can be found on the incorporation of CNFs in polymeric matrices giving the final mechanical and electrical properties of these materials. As in all cases where nano-sized fillers are involved, the development of high-performance CNF/polymer composites requires a homogeneous dispersion of CNFs in the polymeric matrix because it is crucial to the composite's performance. Early studies by Hussain[16] reported that matrix reinforcement with nanowhiskers can damage the fibres in composite materials. As such, he incorporated microscale and nanoscale $Al_2O_3$ particles in filament-wound carbon-fibre/epoxy composites.

*15.2* Left: SEM image of a fracture surface of the bonding area of 5 wt% MCNT + epoxy; failure of the graphite fibre of the adherends was observed. Right: SEM image of a fracture surface of the bonding area of the epoxy-only case; failure occurred at the epoxy surface of the adherends and no significant graphite fibre fracture was observed.[14]

He observed an increase in modulus, flexural strength, interlaminar shear strength and fracture toughness when the matrix was filled at 10 vol% with alumina particles (25 nm diameter). This effect stemmed largely from the large surface area of the filler and the ability of the particles to mechanically interlock with the fibres. Hybrid reinforced composites consisting of two or more different types of reinforcing fibres have also been studied in polymer matrix composite systems. It has also been reported that hybridization by incorporating whiskers into the matrix causes fibre damage resulting in a decrease in ultimate strength. However, the work claimed that the incorporation of a rigid spherical filler, especially a fine or nano-sized filler, did not cause serious damage to the fibre surfaces.

Mahfuz *et al.*[17] studied the tensile response of carbon-nanoparticle/whisker-reinforced composites and observed a 15–17% improvement in tensile strength and modulus. Iwahori and Ishikawa[18] reported compressive strength improvements in carbon-fibre-reinforced polymer (CFRP) composite laminates by using cup-stacked carbon nanofibres (CSCNFs) dispersed in epoxy as three-phase composites. Iwahori *et al.*[19] went a step further and employed two types of CSCNF with different aspect ratios, i.e. with a fibre length of 500 nm to 1 μm (AR10) and a fibre length of 2.5 to 10.0 μm (AR50), respectively. These two types of CSCNF were dispersed into the epoxy resin. At the first trial stage, a manual fabrication process of the composite plates by impregnation of the diluted compound with the same epoxy into a dry carbon-fibre fabric was employed, followed by hot-press curing. Compression strength improvements of around 15% were attained in the three-phase composites, in comparison with the control case with no CSCNFs. Encouraged by the promising mechanical properties, they also manufactured cup-stacked carbon nanotubes (CSCNTs) dispersed in CFRP fabric to obtain more

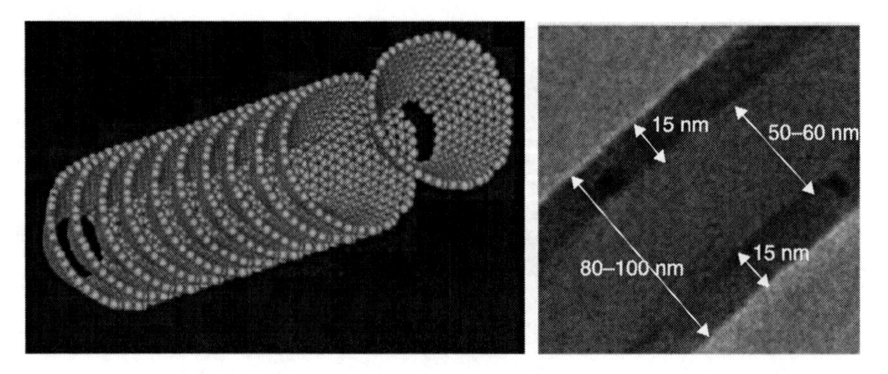


*15.3* Left: Schematic view of cup-stacked carbon nanotube.[19] Right: Typical transmission electron micrograph.[20]

stable mechanical properties than manual fabrication processes. Figure 15.3 shows typical CSCNTs manufactured. They evaluated the mechanical properties of the CSCNT-dispersed CFRP and found an improvement in stiffness and strength (e.g. compressive strength) in two-phase and three-phase nanocomposite materials. The researchers accepted that another key issue in pre-impregnated composite fibre (prepreg) development is the optimization of the aspect ratio of CSCNFs.

Although the details of this process were not disclosed, it is noteworthy that it has one advantage for good dispersion because of the multiple number of edges of graphene sheets on the CSCNF surface. Such edges may help to increase interaction between the CSCNF and polymer. A good dispersion of the CSCNF was suggested in the micrographs for the three-phase composites made through the prepreg route. There was a large improvement of the compression strength for these three-phase composites made using the prepreg method compared with manually impregnated samples. For a T-700 CF UD prepreg sample, the compression strength in the fibre direction improved by 25% in comparison with the control sample (no CSNF). However, the elastic modulus during compression of this composite was not affected as naturally expected. More recently, Yokozeki *et al.*[20] investigated the damage accumulation that occurred in carbon-fibre-reinforced nanocomposite laminates under tensile loading. The nanocomposite laminates used in the study were manufactured from prepregs consisting of traditional carbon fibres and epoxy resin filled with CSCNTs. The thermomechanical properties of the unidirectional carbon-fibre-reinforced nanocomposite laminates were evaluated, and cross-ply laminates were subjected to tension tests to observe the damage accumulation of matrix cracks. As shown in Fig. 15.4, the number of matrix cracks in CSCNT-dispersed CFRP is much less than in conventional CFRP. A clear retardation of matrix-crack accumulation in CSCNT-dispersed CFRP laminates (both 5 wt% and 12 wt%) compared with laminates without CSCNT can

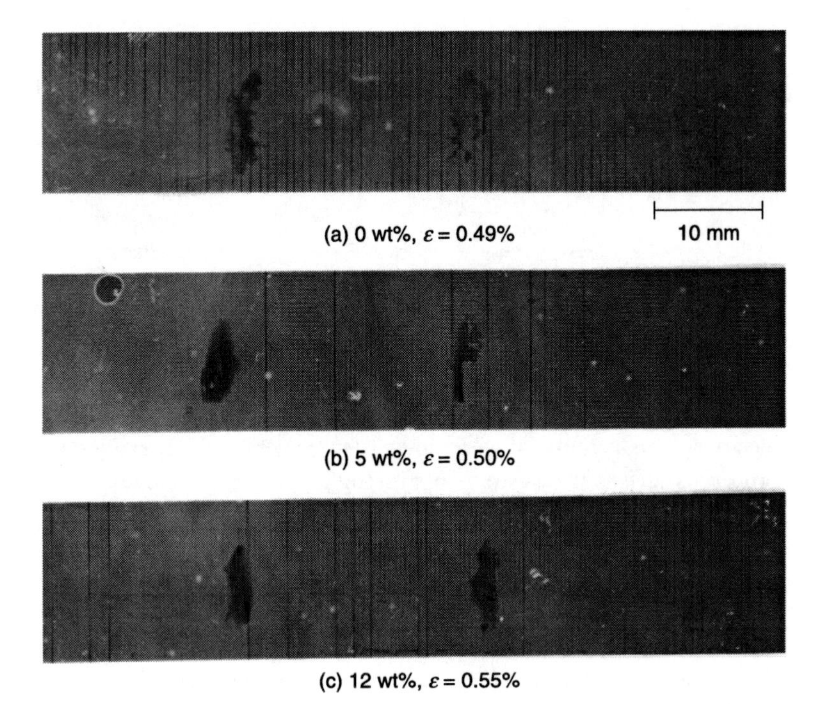


*15.4* (a–c) Comparison of the accumulation of matrix cracks in cross-ply laminates; $\varepsilon$ is the applied strain.[20]

be observed. Fracture toughness associated with matrix cracking was evaluated based on an analytical model using the experimental results. It was suggested that the dispersion of CSCNTs resulted in fracture-toughness improvement and residual thermal strain decrease, which was considered to cause the retardation of matrix-crack formation.

In the work of Wichmann *et al.*,[12] different nanoparticles, such as fumed silica and carbon black, were used to optimize the epoxy matrix system of a glass-fibre-reinforced composite. Their nanometre size enabled their application as particle reinforcement in FRPs produced by a modified RTM, without being filtered by the glass-fibre bundles. Figure 15.5 is a schematic of a modified-RTM device. An electrical field was applied during curing in order to enhance the orientation of the nanofillers in the *z*-direction. The interlaminar shear strengths of the nanoparticle-modified composites were significantly improved (+16%) and an increase in fracture toughness of 42% was observed by adding only 0.3 wt% of CNTs. The interlaminar toughness was not affected in a comparable manner. Only the fumed-silica nanocomposites exhibited a negligible decrease in Young's modulus. However, with 0.5 vol% of epoxy-functionalized fumed-silica nanoparticles, $K_{Ic}$ increased by 55%. The laminates containing CNTs exhibited relatively high

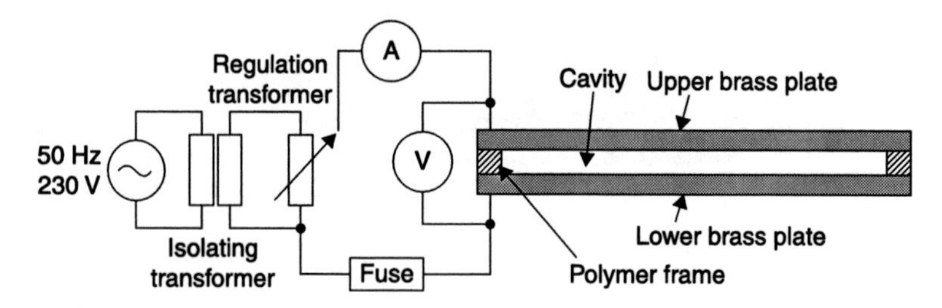


*15.5* Modified RTM-device. The electrical field is applied between the brass plates (*z*-direction).[12]

electrical conductivity at very low filler content, which suggests functional properties such as stress–strain monitoring and damage detection.

It should be acknowledged that traditional fibre-reinforced composite materials with excellent in-plane properties perform poorly when out-of-plane through-thickness properties are important. Composite architectures with fibres designed orthogonal to the two-dimensional (2D) layout in traditional composites could alleviate this weakness in the transverse direction, but efforts so far have only had limited success. Nevertheless, the combination of a nanotube-modified matrix together with conventional fibre reinforcements (e.g. carbon, glass or aramid fibres) could lead to a new generation of multifunctional materials. Besides electrical conductivity, which can be induced by the carbon nanoparticles, an additional *z*-reinforcement can be expected. The fibre orientation in structural components is usually in plane (*x*- and *y*-directions), leading to fibre-dominated material properties in these directions, whereas the *z*-direction remains matrix dominated. With regard to the nanometric size, carbon nanoparticles allow an infiltration between the microscale fibres. The use of CNTs at the reinforcing phase should improve the matrix properties, especially in the *z*-direction, which is equivalent to improving the interlaminar properties.

One of the applications of CNT-reinforced polymers for filament-wound CFRP was demonstrated by Spindler-Ranta and Bakis.[21] An amount of 1 wt% single-walled nanotubes (SWNTs) was added to an epoxy polymer matrix. However, this study concluded that SWNTs did not produce any noticeable effect on the CNT-reinforced composites and filament-wound CFRP rings. In contrast, Veedu *et al.*[22] reported significant improvements in the interlaminar fracture toughness, hardness, delamination resistance, in-plane mechanical properties, damping, thermoelastic behaviour and thermal and electrical conductivities. They presented an approach to three-dimensional(3D) through-the-thickness reinforcement, without altering the 2D stack design, using interlaminar CNT forests, which provide enhanced multifunctional properties along the thickness direction. The CNT forests allowed the fastening of adjacent plies in the 3D composite. They grew MWNTs on the surface of micro-fibre fabric cloth layouts, normal to the fibre lengths, resulting in a 3D effect between plies under loading. These nanotube-coated fabric cloths served as building blocks for multilayered 3D composites, with the nanotube forests providing interlaminar strength and toughness under various loading conditions.

### 15.2.2 Laminated layered silicates/epoxy FRP nanocomposites

In the early 1990s, the Toyota research group synthesized polyamide-6-based clay nanocomposites that demonstrated the first use of nanoclays as a reinforcement for polymer systems.[23] They concluded that nanoclays not only influenced the crystallization process, but that they were also responsible for morphological changes. Recognizing these benefits, many researchers, using a variety of clays and polymeric matrices, have produced nanocomposites with improved properties.[24]

Haque *et al.*,[25] using a similar manufacturing process (i.e. vacuum-assisted resin infusion moulding or VARIM), showed a large improvement of the mechanical properties of their S2-glass-fibre laminates with a very low layered-silicate content. They showed that, by dispersing 1 wt% nanosilicates, S2-glass/epoxy-clay nanocomposites exhibited an improvement of 44%, 24% and 23% in interlaminar shear strength, flexural strength and fracture toughness respectively. Similarly, the nanocomposites exhibit approximately 26 °C higher decomposition temperatures than conventional composites. The increased properties at low loading were attributed to several factors:

- enhanced matrix properties due to lamellar structures,
- synergistic interaction between the matrix, clay and fibres,
- enhanced matrix–fibre adhesion promoted by the clay.

The clays were also presumed to decrease the mismatch in the coefficient of thermal expansion, significantly reducing residual stresses and leading to higher quality laminates. Increased interfacial bonding, matrix agglomeration and coarse morphology were observed on the fractured surface of the low loading nanocomposites. The degradation of these properties at higher clay loadings was believed to be caused by phase-separated structures and also by defects in the cross-linked structures. However, the authors acknowledged that further work was necessary in order to produce clay–epoxy nanocomposites with a fully exfoliated structure.

Similarly, Chowdhury *et al.*[26] employed the VARIM process to manufacture woven carbon-fibre-reinforced polymer matrix composites. They investigated the effects of nanoclay particles on the flexural and thermal properties. Different weight percentages of a surface-modified montmorillonite mineral were dispersed in SC-15 epoxy using sonication. The nanophased epoxy was then used to manufacture 6000 fibre tow-plain weave carbon/epoxy nanocomposites using the

VARIM technique. Flexural test results of thermally post-cured samples indicated a maximum improvement in strength and modulus of about 14% and 9%, respectively. Dynamic mechanical analyses (DMA) of the thermally post-cured samples showed a maximum improvement in the storage modulus of about 52% and an increase in the glass transition temperature of about 13 °C. In terms of mechanical and thermal properties, 2 wt% nanoclay seems to be an optimum loading for carbon/SC-15 epoxy composites. Microstructural studies revealed that nanoclay promotes good adhesion of the fibre and matrix, thereby increasing the mechanical properties.

Lin *et al.*[27] successfully prepared layered silicate/glass-fibre/epoxy hybrid composites using a vacuum-assisted resin transfer moulding (VARTM) process. Figure 15.6 shows a schematic of the experimental set-up for the closed-mould VARTM process. They selected clay and short-length glass fibres to reinforce an epoxy resin. To study the effects of fibre direction on clay distribution in the hybrid composites, unidirectional glass fibres were placed in two directions: parallel and perpendicular to the resin flow direction. The intercalation behaviour of the clay and the morphology of the composites were investigated using X-ray diffraction (XRD) and transmission electron microscopy (TEM). The complementary use of XRD and TEM revealed an intercalated clay structure in the composites. Dispersion of the clay in the composites was also observed using SEM. The clays were dispersed both between the bundles of the glass fibres and within the interstices of the fibre filaments. The mechanical properties of the

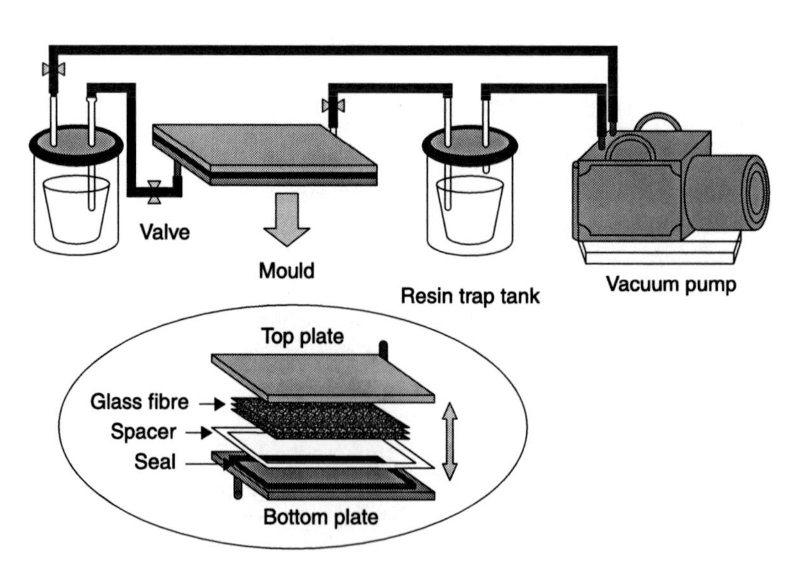


*15.6* Preparation of nanoclay/glass-fibre/epoxy composites using VARTM.[27]

ternary composites were also evaluated. The results indicated that introducing a small amount of organoclay to glass-fibre/epoxy composites enhanced their mechanical and thermal properties, confirming the synergistic effects of glass fibres and clays in the composites.

Aktas *et al.*[28] developed a novel approach for the characterization of nanoclay dispersion in polymeric composites using electron microprobe analysis (EMPA). Dispersion analysis was performed on three sets of centre-gated discs fabricated by RTM. The first set was neat epoxy polymer without reinforcement, whereas the second set comprised 17 vol% randomly oriented chopped glass-fibre preforms. The final set, in addition to the glass-fibre reinforcement, contained 1.7 wt% Cloisite 25A nanoclay. After curing, a sample along the radius of a nanoclay-reinforced disc was analysed with an electron microprobe analyser. The scanning electron micrographs indicated that the nanoclay exists in clusters of various sizes ranging from over 10 μm down to submicrometre scale. Nanoclay clusters larger than 1.5 μm were analysed by digital image processing of scanning electron micrographs taken along the sample radius. The dispersion of nanoclay clusters smaller than 1.5 μm was quantified by compositional analysis via wavelength dispersive spectrometry (WDS). The distribution of nanoclay clusters larger than 1.5 μm was found to be approximately constant along the radius with an average value of 1.4% by volume. Similarly, nanoclay clusters smaller than 1.5 μm were found to be distributed evenly with an average value of 0.41 wt%. In addition, the glass transition temperature improved by 11% with the addition of nanoclay.

Gilbert *et al.*[29,30] and Timmerman *et al.*[31] demonstrated that fracture toughness and mechanical properties are increased by the incorporation of metal and inorganic particles. They developed La PolynanoGrESS (layered polynanomeric graphite epoxy scaled system), which utilizes the nanoparticle effect in an epoxy matrix and scales to a continuous carbon-fibre-reinforced composite system. Typically, Timmerman *et al.*[31] modified the matrices of carbon-fibre/epoxy composites with layered inorganic clays and a traditional filler to determine the effects of particle reinforcement, both microscale and nanoscale, based on the response of these materials to cryogenic cycling. The mechanical properties of the laminates studied were not significantly altered through nanoclay modification of the matrix. The incorporation of a nanoclay reinforcement at a suitable concentration resulted in laminates with microcrack densities lower in response to cryogenic cycling than those seen in unmodified or macro-reinforced materials. Lower nanoclay concentrations resulted in a relatively insignificant reduction in microcracking and higher concentrations displayed a traditional filler effect.

Brunner *et al.*[32] extended the work of Timmerman *et al.*, using epoxy with a relatively small amount of nano-sized filler as a matrix in fibre-reinforced laminates. They focused on investigating whether a nano-modified epoxy matrix yields improved delamination resistance in a fibre-reinforced laminate compared with a laminate with neat epoxy as the matrix material. To start with, neat and nano-modified epoxy specimens without fibre reinforcement were prepared for a comparison of the fracture toughness of the matrix material itself. Additional properties of the neat and nano-modified epoxy were also determined (partly taken from Timmerman *et al.*[31]) and compared. The study reported an improvement in fracture toughness up to about 50% and energy release rates increased by about 20% with the addition of 10 wt% of organosilicate clay.

Several other studies have described property enrichment due to the addition of nanoclay to composite matrices. For instance, Schmidt,[33] Mark[34] and Hussain *et al.*[16] demonstrated the technology of dispersing $Al_2O_3$ particles in a matrix and investigated their effect on the mechanical properties of CFRP. The incorporation of filler particles resulted in higher fracture toughness by significantly improving the toughness of the matrix and crack deviation. Studies on carbon/SiC-epoxy nanocomposites reported a 20–30% improvement in mechanical properties.[35] Mohan *et al.*[36] evaluated the tensile performance of S2-glass/epoxy composites dispersed with alumina nanoparticles up to 1.5% weight fraction and found an increase of 12% in tensile modulus and 8% in tensile strength. Kornmann *et al.*[37] successfully synthesized epoxy-layered silicate nanocomposites based on diglycidyl ether of bisphenol A and an anhydride-curing agent. A manufacturing process using hand lay-up, vacuum bagging and hot pressing was also developed to produce glass-fibre-reinforced laminates with this nanocomposite matrix. Transmission electron microscopy indicated that silicate layers dispersed in the epoxy matrix have long-range order with an interlamellar spacing of about 9 nm. X-ray diffraction analysis confirmed this nanostructure both in the nanocomposites and in a fibre-reinforced composite based on the same matrix. Scanning electron micrographs of the laminate with a nanocomposite matrix showed that nanolayers stacked at the surface of the glass fibre, thus possibly improving the interfacial properties of the fibres. Flexural testing of the laminates showed that the nanolayers improved the modulus and the strength, by 6% and 27% respectively. Dynamic mechanical analyses of the epoxy and nanocomposite plates and their corresponding laminates showed a systematic glass transition temperature decrease of the nanocomposite-based materials. This, the researchers suggested, explained the larger water uptake observed at 50 °C in the plate and the laminate based on a nanocomposite matrix compared with those based on the pristine epoxy.

Karaki *et al.*[38] incorporated layered clay, alumina and titanium dioxide into an epoxy matrix and fabricated continuous carbon-fibre-reinforced polynanomeric matrices to study tension–tension fatigue behaviour. They found that the number of microcracks in each layer depended on the type of particles and their concentration. Wang *et al.*[39] demonstrated that the exfoliated clay with only 2.5 wt% in epoxy showed a significant improvement in fracture toughness and concluded that an increase in the number of microcracks and the fractured surface due to crack deflection resulted in the toughness increase. Siddiqui *et al.*[40] investigated the mechanical properties of nanoclay-dispersed CFRP, and showed that the interlaminar fracture toughness is higher than that of conventional CFRP. Ragosta *et al.*[41] showed that critical stress intensity factors of epoxy/silica nanocomposites increased with an increase of silica content. Work by Seferis and co-workers[42] incorporated nano-sized alumina structures in the matrix and interlayer regions of prepreg-based carbon-fibre/epoxy composites. Subramaniyan *et al.*[43] observed that the addition of 5 wt% of nanoclay increased the elastic modulus of epoxy resin under compression by 20% and that the compressive strength of glass-fibre composites with nanoclay when made by the wet lay-up technique increased by 20–25%. Subramaniyan and Sun[44] showed that polymers can be toughened significantly using a relatively small amount of nanoclay particles based on three-point bending tests of edge-notched specimens with sharp crack tips. A parallel study by the same authors[45] reported that the compressive strength of unidirectional GFRP with nanoclays increased compared with conventional GFRP.

Hackman and Hollaway[46] studied potential applications of clay nanocomposite materials in civil engineering structures. They concluded that there is the possibility to increase the service life of materials subjected to aggressive environments because of the increased durability of glass-fibre and carbon-fibre composites. Liu *et al.*[47] demonstrated the improvement of fracture toughness and the reduction of water diffusivity of epoxy/nanoclay composites. Ogasawara *et al.*[48] investigated the helium gas permeability of silicate-clay (montmorillonite) particle/epoxy nanocomposites. They reported that the incorporation of increasing amounts of montmorillonite particles reduced helium gas permeability. With an increase of montmorillonite loading, gas diffusivity decreased while gas solubility increased. Helium diffusion was found to be in agreement with numerical results based on the Hatta–Taya–Eshelby theory.[49,50] They revealed that the dispersion of nanoscale platelets in a polymer is effective in improving the gas barrier property. The study appreciated the fact that surface-modified clays are amenable for making organic/clay nanocomposites because of the weak bonding force between the layers of montmorillonite.[48] In the study, a typical less viscous epoxy of Epikote 807 base resin was used for better dispersion. It was shown that a loading of the nanoclay of about 4 vol% (about 6 wt%) reduced the diffusion coefficient to 1/10, and that theoretical predictions based on an aspect ratio of 0.001 agreed well with the experimental results.

### 15.2.3 Polyamide FRP nanocomposites

Electrospun nanocomposite fibres have great potential in applications where both a high surface-to-volume ratio and strong mechanical properties are required, such as high-performance filters and fibre-reinforcement materials. Since the mechanical properties of fibres generally improve substantially by decreasing fibre diameter, there is considerable interest in the development of continuous electrospun polymer nanofibres. In this respect, Lincoln *et al.*[51] reported that the degree of crystallinity of polyamide-6 (PA-6) annealed at 205 °C increased substantially with the addition of montmorillonite (MMT). This implied that the silicate layers could act as nucleating agents or growth accelerators. In contrast, a study by Fong *et al.*[52] showed a very similar overall degree of crystallinity for electrospun PA-6 and PA-6/Cloisite-30B nanocomposite fibres containing 7.5 wt% of organically modified MMT (OMMT) layers.

Fornes and Paul[53] found that OMMT layers could serve as nucleating agents at 3% concentration in PA-6/OMMT nanocomposites but retarded the crystallization of PA-6 at a higher concentration of around 7%. In addition, the differences in the molecular weight (MW) of PA-6 and the solvent used for electrospinning were also expected to have different effects on the mobility of PA-6 molecular chains and the interactions between the PA-6 chains and OMMT layers, which may also affect the crystallization behaviour of PA-6 molecules during electrospinning. Li *et al.*[54] manufactured PA-6 fibres and nanocomposite fibres with an average diameter of around 100 nm by electrospinning using 88% aqueous formic acid as the solvent. The addition of OMMT layers in a PA-6 solution increased the solution viscosity significantly and changed the resulting fibre morphology and sizes. TEM images of the nanocomposite fibres and ultra-thin fibre sections and wide-angle X-ray diffraction results showed that the OMMT layers were well exfoliated inside the nanocomposite fibres and oriented along the fibre axial direction. The degree of crystallinity and crystallite size both increased for the nanocomposite fibres and, more significantly, for the fibres electrospun from the 15% nanocomposite solution, which exhibited the finest average fibre size. As a result, the tensile properties of electrospun nanocomposites were greatly improved. Young's modulus and the ultimate strength of electrospun nanocomposite fibrous mats improved to 70% and 30%, respectively, compared with PA-6 electrospun mats. However, the ultimate strength of nanocomposite fibrous mats electrospun from 20% nanocomposite solution decreased by about 20% due to their larger fibres. Young's modulus of PA-6 electrospun single fibres with a diameter around 80 nm was almost double the highest value reported for conventional PA-6 fibres and could be improved by about 100% for electrospun nanocomposite single fibres of similar diameters.

Liang *et al.*[55] described a fibre that consisted of a nano-$Fe_2O_3$-particle/PA-6 nanocomposite. The thermal stability of the composite material was enhanced by about 16 °C (from 440 °C to 456 °C) by the addition of $Fe_2O_3$ nanoparticles with 15.0% content (part per hundred parts of resin). The $Fe_2O_3$-reinforced materials processed by melt spinning displayed an improved tensile modulus compared with similarly processed pure PA-6; the improvements in tensile strength and modulus were about 21% and 112%, respectively. Moreover, this fibre absorbed ultraviolet and visible light.

In another interesting study,[56] a range of polymer matrices, including polyvinyl alcohol, poly (9-vinyl carbazole) and polyamide, were examined. To compare production methods, polymer composite films and fibres were produced. It was found that, by adding various mass fractions of nanofillers, both Young's modulus and hardness increased significantly for both films and fibres. In addition, thermal behaviour was seen to be strongly dependent on the nanofillers added to the polymer matrices. Wu *et al.*[57] prepared carbon-fibre-reinforced and glass-fibre-reinforced PA-6 and PA-6/clay nanocomposites. The fabrication method involved first mechanically mixing PA-6 and PA-6/clay with E-glass short fibres (6-mm long) and carbon fibres (6-mm long), separately. A twin-screw extruder at a rotational speed of 20 rpm extruded the fibres. The temperature profiles of the barrel were 190–210–230–220 °C from the hopper to the die. The extrudate was pelletized, dried and injection moulded into standard test samples for mechanical property tests. The injection-moulding temperature and pressure were 230 °C and 13.5 MPa, respectively. The research found that the tensile strength of PA-6/clay containing 30 wt% glass fibres was 11% higher than that of PA-6 containing 30 wt% glass fibre, while the tensile modulus of the nanocomposite increased by 42%. The flexural strength and flexural modulus of neat PA-6/clay were found to be similar to PA-6 reinforced with 20 wt% glass fibres. It was concluded that the effect of nanoscale clay on toughness was more significant than that of the fibre. The heat distortion temperatures of the PA-6/clay and PA-6 were 112 °C and 62 °C, respectively. Consequently, the heat distortion temperature of the fibre-reinforced PA-6/clay system was almost 20 °C higher than for the fibre-reinforced PA-6 system. The notched Izod impact strength of the composites decreased with the addition of the fibre. Scanning electron microphotographs showed that the wet-out of glass fibre was better than carbon fibre. The study concluded that the mechanical and thermal properties of the PA-6/clay nanocomposites were superior to those of the PA-6 composite in terms of heat distortion temperature, tensile and flexural strength and modulus without sacrificing their impact strength. This was attributed to nanoscale effects and the strong interaction force that existed between the PA-6 matrix and the clay interface.

Regarding short fibres, Akkapeddi[58] prepared PA-6 nanocomposites using chopped glass fibres. A typical experiment used a commercial grade PA-6 with a molecular weight of 30 kg/mol and specially designed functional organo-quaternary ammonium-clay complexes (organoclays) based on MMT or hectorite-type clays. Freshly dried PA-6 (moisture $< 0.05\%$) was blended with 3–5 wt% of a selected organoclay powder and extruded at 260 °C in a single step under high shear mixing conditions. Alternatively, the organoclay was master-batched first into PA-6 (at 25 wt% loading and then re-extruded in a second step with more PA-6 to dilute the clay content to $\leq$ 5 wt%). Conventional chopped glass fibre, 10 μm in diameter and about 3 mm in length, was then added as an optional reinforcement through a downstream feed port at zone 6 of a twin-screw extruder. The glass fibre was compounded with the molten, premixed PA-6 nanocomposite either as a one-step extrusion process or in a second extrusion step. The extrudate was quenched in a water bath and pelletized. The pellets were dried under vacuum at 85 °C and injection moulded into standard ASTM test specimens. As shown in Fig. 15.7, significant improvements in modulus were achievable in both the dry and the moisture-conditioned states for PA-6 nanocomposites compared with

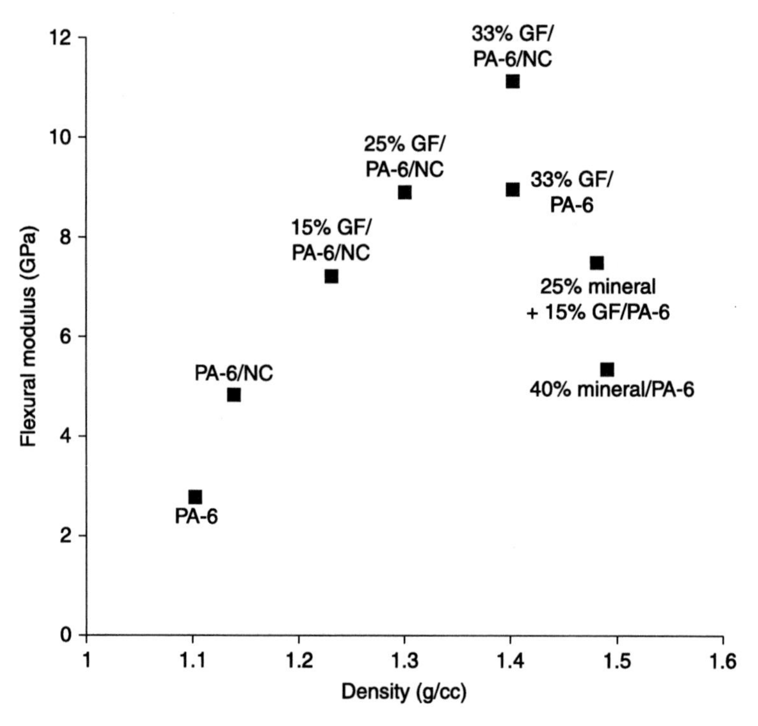


*15.7* Modulus vs. density of glass fibre (GF) for PA-6/nanoclay (PA-6/NC) and PA-6 moulding resins.[58]

standard PA-6, at any given level of glass-fibre reinforcement. In particular, a small amount (3–4 wt%) of nanoscale dispersed layered silicate was capable of replacing up to 40 wt% of a standard mineral filler or 10–15 wt% of glass fibre to give equivalent stiffness at a lower density. In addition, improved moisture resistance, permeation barrier and fast crystallization/mould cycle time contribute to the usefulness of such composites.

Vlasveld *et al.*[59] developed a three-phase thermoplastic composite, consisting of a main reinforcing phase of woven glass or carbon fibres and a PA-6 nanocomposite matrix. The nanocomposites used in this research had moduli that were much higher than unfilled PA-6, including above $T_g$, and moisture-conditioned samples. Flexural tests on commercial PA-6 fibre composites showed a decrease of flexural strength on increasing temperature. The researchers claimed that the strength of the glass-fibre composites was increased by more than 40% at elevated temperatures and the temperature range for a given minimum strength was increased by 40–50 °C. The carbon-fibre composites also showed significant improvements at elevated temperatures, although not at room temperature. Based on flexural tests on PA-6-based glass and carbon-fibre composites over a large temperature range up to near the melting point, it became clear that for these fibre composites it is important to have a reasonably high matrix modulus. Both glass and carbon composites were very sensitive to a decrease of the matrix modulus below values around 1 GPa. At higher moduli, the carbon-fibre composites were more sensitive to the matrix modulus than glass-fibre composites. The modulus of unfilled PA-6 decreased below the (arbitrary) 1 GPa level just above $T_g$. It is noteworthy that the nanocomposites used in this research had moduli that were much higher and stayed above the 1 GPa level up to 160 °C, which was more than 80 °C higher than for unfilled PA-6. The nanocomposites also showed much higher moduli in moisture-conditioned samples. Even in moisture-conditioned samples tested at 80 °C, the modulus was much higher than for the dry unfilled PA-6, which again was well above 1 GPa. DMA measurements indicated that the nanocomposites did not show a change of $T_g$ and that the reduction of the modulus on absorption of moisture was due to the $T_g$ decrease.

Vlasveld *et al.*[60] investigated fibre–matrix adhesion in glass-fibre-reinforced PA-6 silicate nanocomposites. The main reinforcing phase consisted of continuous E-glass fibres, whereas the PA-6-based matrix was a nanocomposite reinforced with platelets of exfoliated layered silicate. Two different types of nanocomposite were used with different degrees of exfoliation of the silicate layers: one with unmodified silicate and one with an organically modified silicate. They developed nanocomposite laminates by the sol-gel and modified-diaphragm methods. The preparation of the PA-6 nanocomposites consisted of melt-compounding Akulon® K122D with Somasif® MEE and Somasif® ME-100 by means of a co-rotating twin-screw extruder at 240 °C. For the Somasif® MEE nanocomposite materials, an 11 wt% MEE master batch was initially compounded. To obtain the various concentrations of the MEE nanocomposite, the master batch was extruded for a second time without dilution for the 11 wt% nanocomposite, or diluted with Akulon® K122D to concentrations of 6.1 and 2.7 wt%. The 2.5 wt% Somasif® ME-100 nanocomposite material was produced by diluting a 10% ME-100 master batch with Akulon® K122D in the extruder. (All percentages are weight percentages of silicate measured with a thermogravimetric analyser (TGA) after heating for 40 min at 800 °C in air.) Two demands in the preparation of the single-fibre fragmentation specimens had to be met: the fibre had to lie straight in the centre of the specimen and the matrix material of the specimen had to be thin enough to be transparent, since the fibre fragments were examined and measured using an optical microscope. A Fontijne hot-plate press heated to 240 °C was used to produce the films for the single-fibre fragmentation test specimen. Single fibres were carefully extracted from a fibre bundle and placed at a distance of approximately 2 cm apart parallel to each other between the PA or nanocomposite films. The hot-plate press was used to melt the polymer films and a pressure of 0.8 N/mm$^2$ was applied for 30 s to provide the necessary bonding with the fibre.

After cooling between cold metal plates, tensile test specimens were prepared. It was observed that the ultimate strength and stiffness increased by adding 1% $SiO_2$ nanoparticles, while little improvement in fatigue behaviour was found. It

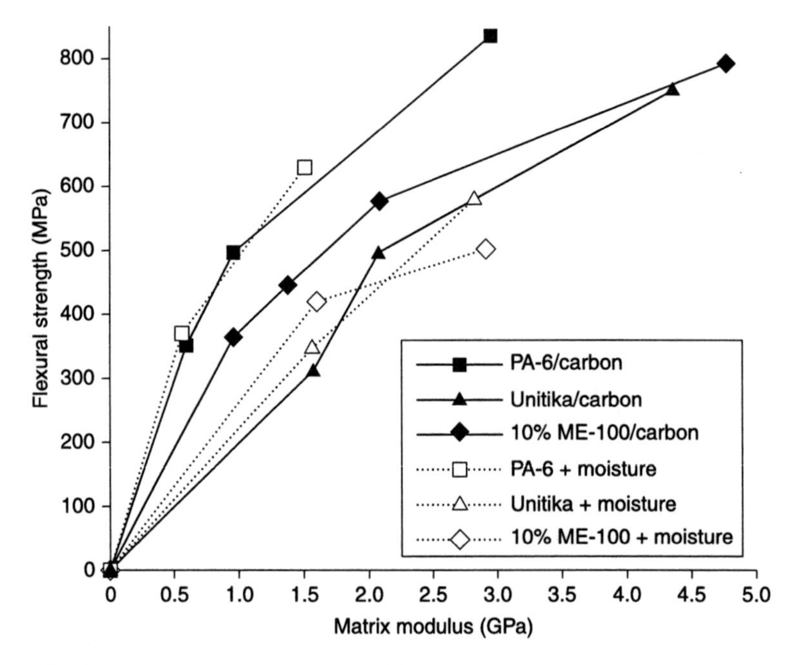


*15.8* Flexural strength of carbon-fibre composites with PA-6, a commercially available PA-6 nanocomposite (Unitika M1030D from Unitika) and a nanocomposite matrix as a function of the matrix modulus (dry and moisture conditioned).[60]

was concluded that the failure mechanism was by interfacial de-bonding and that both the addition of nanoparticles and moisture conditioning had a negative effect on the bonding between the matrix and the glass fibres. In addition, the researchers noted that, in the composites formed, adhesion between the nanocomposites and the carbon fibres (Fig. 15.8) was probably worse than between the unfilled PA-6 and the matrix, reducing the potentially positive influence of the increased matrix modulus.

An assessment of reactively processed anionic polyamide-6 (APA-6) for use as a matrix material in fibre composites was conducted by van Rijswijk *et al.*[61] They also compared it with melt-processed PA-6 and PA-6 nanocomposites. A specially designed lab-scale mixing unit was used to prepare two liquid material formulations at 110 °C under a nitrogen atmosphere: a monomer/activator mixture in tank A and a monomer/initiator mixture in tank B, as shown in Fig. 15.9. After individually degassing both tanks (15 min at 100 mbar), the two material feeds were mixed using a heated (110 °C) static mixer and dispensed (1:1 ratio) into a heated (110 °C) buffer vessel with nitrogen protective environment. A stainless-steel infusion mould was used together with a 3-mm-thick stainless-steel cover plate (not shown) to manufacture neat APA-6 panels (250×250×2 mm). Homogeneous

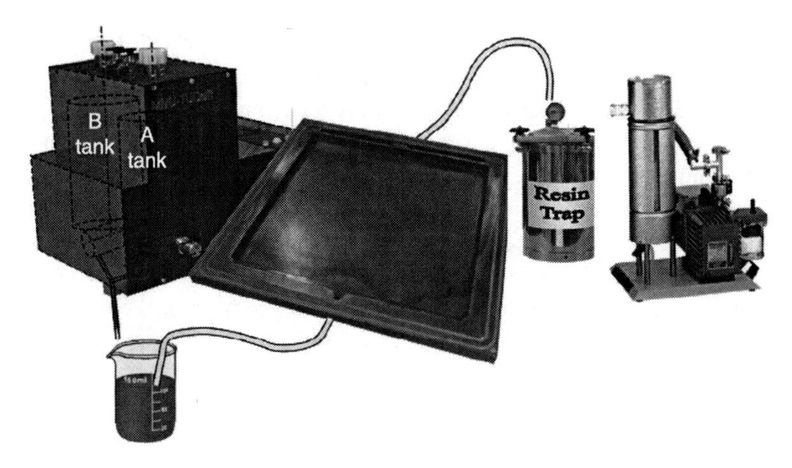


*15.9* Infusion equipment. From left to right: Mini mixing unit (MMU-TU Delft), resin reservoir, stainless-steel infusion mould, resin trap, cold trap and vacuum pump.[61]

heating of the mould was obtained by placing it in a vertically positioned hot flat platen press. Silicon tubes connected the resin inlet of the mould to the buffer vessel and the resin outlet to a vacuum pump. Infusion from bottom to top was necessary to prevent entrapment of air. A pressure-control system was used to precisely set the infusion and curing pressures (absolute pressure in the mould cavity). Loss of control over the pressure in the mould cavity due to solidification of resin in the unheated outlet tube had to be prevented. To avoid this, a buffer cavity was machined in the mould near the outlet to slow down the infusion, hence giving ample time to stop the resin flow before it was able to exit the mould. For each infusion pressure, the infusion time to reach the buffer cavity was determined visually because the steel cover plate had been replaced by a glass one. Additionally, a resin trap and a cold trap were placed directly after the mould to protect the vacuum pump.

The mechanical properties of APA-6 and HPA-6 (Akulon® K222D, low-MW injection-moulding grade hydrolytically polymerized PA-6) nanocomposites were compared with injection-moulded neat HPA-6. As expected, the HPA-6 nanocomposite had the highest modulus over the entire range of temperatures (20–160 °C) and moisture content (0–10 wt%) tested. However, APA-6 came close and had the highest maximum strength due to its characteristic crystal morphology, which was directly linked to the reactive type of processing used. This same morphology, it was claimed, also made APA-6 slightly less ductile compared with melt-processed HPA-6. Compared with the melt-processed HPA-6, APA-6 polymerized at 150 °C and the HPA-6 nanocomposite had a higher modulus at a similar temperature, or a similar modulus at a higher temperature

(40–80 °C increase). It is noteworthy that such an increase in maximum-use temperature, related to the heat distortion temperature, can seriously expand the application possibilities for PA-6 and PA-6 composites. For all PAs, temperature and moisture absorption reduced the modulus and the strength and increased the maximum strain, which was directly related to the glass transition temperature. It was observed that moisture absorption reduced $T_g$ to below the testing temperature. However, the effect of both was in essence the same. Retention of the mechanical properties of APA-6 after conditioning at 70 °C for 500 h and subsequent drying was demonstrated. Conditioning by submersing in water at the same temperature, however, resulted in a brittle material with surface cracks, as is common with most polyamides, which was caused by continued crystallization and the removal of unreacted monomer. Given the fact that submersion at elevated temperatures is usually not an environment in which PA-6 and its composites are applied, this property reduction was therefore not detrimental for the application of these materials. The overall conclusion of this comparative study into the application of polyamides as matrix materials in fibre composites was that both APA-6 and the HPA-6 nanocomposites outperformed melt-processed HPA-6 in terms of modulus and maximum strength. Therefore, the researchers concluded that both 'improved' PAs may be expected to enhance matrix-dominated composite properties like compressive and flexural strength, provided that a strong fibre-to-matrix interphase is obtained.

Another comparative study was conducted by Sandler *et al.*[62] on melt-spun PA-12 fibres reinforced with carbon nanotubes and nanofibres. A range of MWNTs and carbon nanofibres were mixed with a PA-12 matrix using a twin-screw micro-extruder and the resulting blends spun to produce a series of reinforced polymer fibres. The work aimed to compare the dispersion and resulting mechanical properties for nanotubes produced by the electric arc and a variety of chemical vapour deposition techniques. A high quality of dispersion was achieved for all the catalytically grown materials and the greatest improvements in stiffness were observed using aligned, substrate-grown, carbon nanotubes. The use of entangled MWNTs led to the most pronounced increase in yield stress, most likely as a result of increased constraint of the polymer matrix due to the relatively high surface area. The degree of polymer and nanofiller alignment and the morphology of the polymer matrix were assessed using X-ray diffraction and differential scanning calorimetry (DSC). The carbon nanotubes were found to act as nucleation sites under slow-cooling conditions, the effect scaling with effective surface area. Nevertheless, no significant variations in polymer morphology as a function of nanoscale-filler type and loading fraction were observed under the melt-spinning conditions applied. A simple rule-of-mixture evaluation of the nanocomposite stiffness revealed a higher effective modulus for the MWNTs compared with the carbon nanofibres, a result of improved graphitic crystallinity. In addition, this approach allowed a general comparison of the effective nanotube modulus with those of nanoclays as well as common short glass and carbon-fibre fillers in melt-blended PA composites. The experimental results further highlighted the fact that the intrinsic crystalline qualities, as well as the straightness of the embedded nanotubes, were significant factors influencing reinforcement capability.

## 15.2.4 Poly (ether ether ketone) (PEEK) FRP nanocomposites

Jen *et al.*[63] manufactured AS-4/PEEK APC-2 nanocomposite laminates and also studied their mechanical responses. The experimental procedures were as follows: firstly, the nanoparticles were diluted in alcohol (50 ml alcohol: 2 g $SiO_2$) and stirred uniformly. Then, 16 plies of $[0/90]_{4s}$ cross-ply and $[0/\pm45/90]_{2s}$ quasi-isotropic prepregs were cut, $SiO_2$ solution was spread on the prepregs in a temperature-controlled box, and after evaporation of alcohol the nanoparticles were found to weigh in the range 111–148 mg/ply. The next step was to repeat the spreading for 5, 8, 10 and 15 plies, followed by consolidation of the stacked plies in a hot press to form a laminate 2 mm thick. The consolidation process is shown in Fig. 15.10.

Next, the laminates were cut into specimens and tested according to ASTM D3039M. The tensile tests were repeated at 50, 75, 100, 125 and 150 °C. These tests measured stress–strain, strength and stiffness and the obtained data were compared with those for the original APC-2 laminate (no $SiO_2$ nanoparticles) and showed that the optimal content of $SiO_2$ nanoparticles was 1% by total weight.

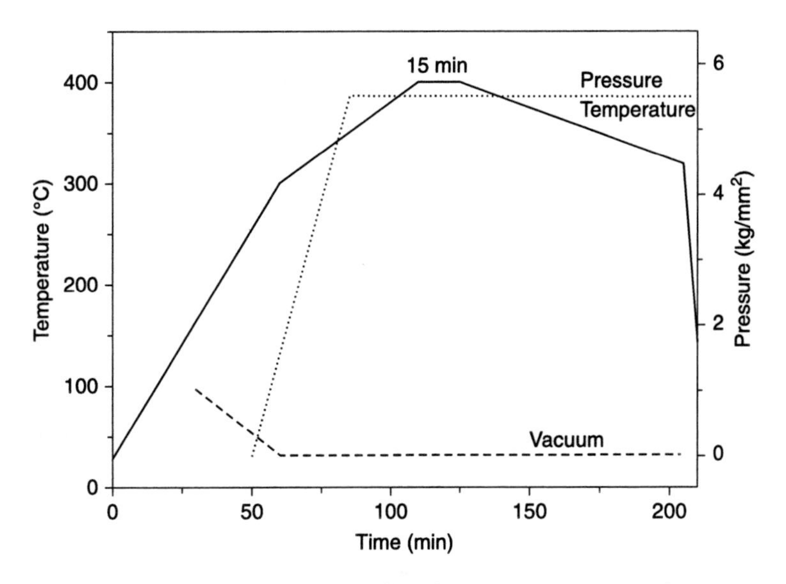


*15.10* Pressure–temperature profile of the curing process of AS-4/PEEK APC-2 nanocomposites.[63]

The ultimate strength increased by about 12.48% and elastic modulus by 19.93% in quasi-isotropic nanolaminates, whilst the improvement for cross-ply nanocomposite laminates was less. At elevated temperatures the ultimate strength decreased slightly below 75 °C and the elastic modulus reduced slightly below 125 °C; however, both properties degraded highly at 150 °C ($\approx T_g$) for the two lay-ups. Finally, after constant stress amplitude tension–tension (T–T) cyclic testing, it was found that both stress-cycle (S–N) curves were very close, being below $10^4$ cycles for cross-ply laminates with or without nanoparticles, and the S–N curve of the nanolaminates was slightly reduced after $10^5$ cycles.

Sandler *et al.*[64] produced poly (ether ether ketone) nanocomposites containing vapour-grown carbon nanofibres using standard polymer processing techniques. Macroscopic PEEK nanocomposite master batches containing up to 15 wt% vapour-grown CNF were prepared using a Berstorff co-rotating twin-screw extruder with a length-to-diameter ratio of 33. The processing temperatures were set to about 380 °C. The strand leaving the extruder was quenched in a water bath, air dried and then regranulated followed by drying at 150 °C for 4 h. Tensile bars according to the ISO 179A standard were manufactured on an Arburg Allrounder 420 injection-moulding machine at processing temperatures of 390 °C, with the mould temperature set to 150 °C. Prior to mechanical testing, all samples were heat treated at 200 °C for 30 min followed by 4 h at 220 °C in an attempt to ensure a similar degree of crystallinity of the polymer matrix. Macroscopic tensile tests were performed at room temperature with a Zwick universal testing machine. The cross-head speed was set to 0.5 mm/min in the 0–0.25% strain range and was then increased to 10 mm/min until specimen fracture occurred. An evaluation of the mechanical composite properties revealed a linear increase in tensile stiffness and strength with nanofibre loading fractions up to 15 wt%, while matrix ductility was maintained up to 10 wt%. Electron microscopy confirmed the homogeneous dispersion and alignment of the nanofibres. An interpretation of the composite performance by short-fibre theory resulted in rather low intrinsic stiffness properties of the vapour-grown carbon nanofibre. Differential scanning calorimetry showed that an interaction between the matrix and the nanoscale filler could occur during processing. However, such changes in polymer morphology due to the presence of nanoscale filler need to be considered when evaluating the mechanical properties of such nanocomposites.

Schmidt's investigation[65] involved multifunctional inorganic–organic composite sol-gel coatings on glass surfaces. The sol-gel process allowed the fabrication of ceramic colloidal particles in the presence of organo-alkoxysilanes carrying various perfluoroalkyl groups and the synthesis of multifunctional transparent inorganic-organic composites. The report claimed that, in addition, these composites can be used as controlled-release systems or designed as graded systems. Using this approach, a coating with a very low surface free energy (with antisoiling properties) and temperature stability up to 350 °C, a controlled-release system for permanent wettability (anti-fogging) and systems containing metal colloids for optical effects were developed. Lin[66] and Wang *et al.*[67] studied the effects of wear and friction by adding SiC nanoparticles to PEEK. The latter studied the effect of the synergism between nanometre SiC and PTFE on the wear of PEEK. Fine powders of PEEK (ICI grade 450P, $\eta = 0.62$) having a diameter of approximately 100 μm were prepared. Nanometre SiC, smaller than 80 nm, was used as a filler. The PTFE powders (25 μm in diameter), nanometre SiC and PEEK were completely mixed ultrasonically and dispersed in alcohol for ~15 min. Then the mixture was dried at 110 °C for 6 h to remove the alcohol and moisture. Finally, the mixture was moulded into block specimens by compression moulding, in which the mixture was heated at a rate of 10 °C $min^{-1}$ to 340 °C, held there for 8 min and then cooled in the mould to 100 °C. After release from the mould, the resultant block specimens were prepared for friction and wear tests. A tribological study found that the incorporation of PTFE into PEEK filled with 3.3 vol% nanometer SiC had a detrimental effect on the tribological properties of the SiC–PTFE–PEEK composite. The morphologies of the worn surfaces and the properties of the transfer films deteriorated, while the load-carrying capacity of the SiC–PTFE–PEEK composite was also adversely affected. The researchers claimed the reason for this was due to $SiF_x$, which formed on the original surface and the worn surface during the compression moulding and sliding friction processes, as a result of a chemical reaction between the nanometre SiC and PTFE. The chemical reaction and the formation of $SiF_x$ dominated the tribological behaviour of the SiC–PTFE–PEEK composites filled with various contents of PTFE and 3.3 vol% nanometre SiC. When the PTFE volume percentage was low then the $SiF_x$ caused friction and the wear of the SiC–PTFE–PEEK composite to rise. However, at high volume percentages the low-friction PTFE dominated the friction and wear behaviour and friction decreased as the percentage of PTFE increased. Chemical reactions and the formation of $SiF_x$ led to changes in the worn-surface morphologies and a detrimental effect on the characteristics of the transfer films.

### 15.2.5 Polyimide polyarylacetylene (PAA) and poly (p-phenylene benzobisoxazole) (PBO) FRP nanocomposites

Ogasawara *et al.*[68] directed their investigations to improving the heat resistance of a relatively new phenylethynyl-terminated imide oligomer (Tri-A PI) by loading of MWNTs. They fabricated MWNT/Tri-A PI composites containing 0, 3.3, 7.7 and 14.3 wt% MWNT using a mechanical blender without any solution (dry condition) for several minutes. The volume fractions of MWNT were calculated to be 2.3, 5.4 and 10.3 vol% from the density of the MWNT (1.9 $g/cm^3$) and the cured polyimide (1.3 $g/cm^3$). Scanning electron micrographs showed the particle size of the imide oligomers to be in the range 0.1–10 μm, and the MWNTs were not dispersed uniformly in the mixture. The loss of aspect ratio during mechanical blending was not significant; therefore MWNTs can be used in various ways; for example, they are suitable for mechanical blending with imide oligomers. The preparation of the nanocomposite involved melt mixing of the MWNT/imide oligomer at 320 °C for 10 min on a steel plate in a hot press and then curing at 370 °C for 1 h under 0.2 MPa of pressure with a PTFE spacer (thickness 1 mm). The resulting composites containing 3.3, 7.7 and 14.3 wt% MWNT exhibited relatively good dispersion at a macroscopic scale. Tensile tests on the composites showed an increase in the elastic modulus and the yield strength, and a decrease in the failure strain. Figure 15.11 shows the effect of MWNT concentration on the Young's modulus of the composites.

Dynamic mechanical analysis showed an increase in the glass transition temperature with incorporation of the carbon nanotubes. The experimental results suggested that the carbon nanotubes were acting as macroscopic cross-links and were further immobilizing the polyimide chains at elevated temperatures. As to the reason why dispersed MWNTs increased the heat distortion temperature, the researchers explained that the dispersed MWNTs impede the molecular motion in the polyimide network at elevated temperatures. The other property improvements in this material are that MWNT showed some potential for controlling electric conductivity and electromagnetic wave absorbability. Although static properties were obtained, discussions were not given and it is evident that more research is required to prove that the suggested phenomenon is a true cause of the higher glass transition temperatures.

There is increasing development in the use of polyarylacetylene (PAA) in advanced heat-resistant composites due to its outstanding heat resistance and excellent ablative properties. Fu *et al.*[69] reviewed the advantages of PAA resin over the state-of-the-art heat-resistant resin. The main potential applications of

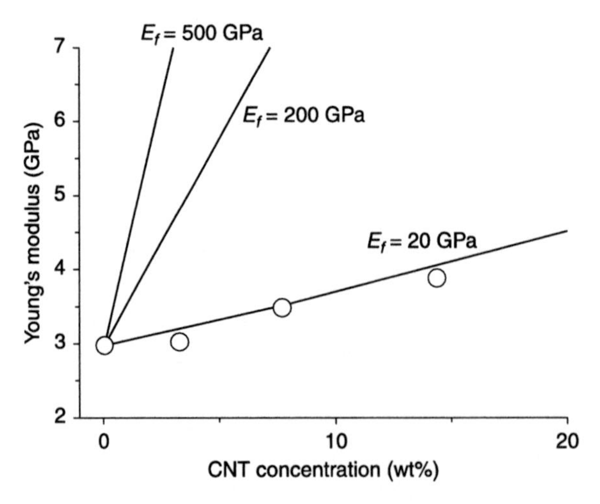


*15.11* Effect of MWNT concentration on Young's modulus of the composites.[68]

PAA resin are in conventional resin matrix composites with ultra-low moisture outgassing characteristics and improved dimensional stability, which are suitable for spacecraft structures, as an ablative insulator for solid rocket motors and as a precursor for carbon–carbon composites. Carbon-fibre-reinforced PAA composites (carbon fibre/PAA) undoubtedly play a very important role in all these fields. Unfortunately, the mechanical properties of the carbon-fibre/PAA material are not yet sufficiently satisfactory to replace widely used heat-resistant composites such as carbon- or graphite-reinforced phenolic resin. The mechanical properties of carbon-fibre-reinforced resin matrix composites depend on the properties of the carbon fibre and the matrix, especially on the effectiveness of the interfacial adhesion between the carbon fibre and the matrix.

PAA has a high content of benzene rings and hence a highly cross-linked network structure, which renders the material brittle. Moreover, the chemically inert characteristics of the carbon-fibre surface lead to weak interfacial adhesion between the fibres and the non-polar PAA resin. To ensure that the material can be used safely in complicated environmental conditions and to exploit the excellent heat-resistant and ablative properties more effectively, it is necessary to improve the mechanical properties of the carbon-fibre/PAA composites. This can be achieved in two ways. One method is to improve the properties of the PAA resin by structural modification or by intermixing with other resins, such as phenolic resin. The other is treatment of the carbon-fibre surface. Treatment of the carbon-fibre surface has been studied for a long time and several methods, such as heat treatment, wet chemical or electrochemical oxidation, plasma treatment, gas-phase oxidation and the high-energy radiation technique have been demonstrated to be effective in the modification of the mechanical interfacial properties of composites based on polar resins such as epoxy. For instance, Zhang *et al.*[70] treated carbon fibres by oxidation–reduction followed by coating with vinyltrimethoxysilane-silsesquioxane (VMS-SSO) to improve the interfacial mechanical properties of the carbon-fibre/PAA composites. The carbon-fibre surface-treatment process is shown in Fig. 15.12.

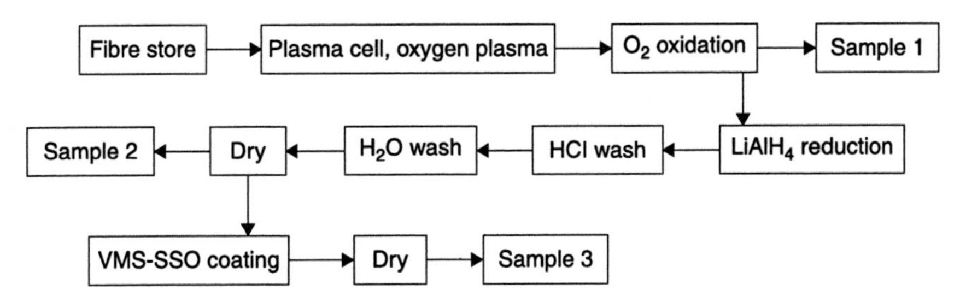


*15.12* Carbon-fibre treatment process. Sample 1 is oxygen plasma oxidation, sample 2 includes $LiAlH_4$ reduction and sample 3 includes coating with VMS-SSO.[70]

Polar functional groups, including carboxyl and hydroxyl, were imported onto the carbon-fibre surface after the oxygen plasma oxidation treatment. The quantity of carboxyl on the carbon-fibre surface decreased and that of hydroxyl increased after $LiAlH_4$ reduction. The $LiAlH_4$ reduction time was decided according to the experimental parameter in Lin.[71] The VMS-SSO coating was grafted onto the carbon-fibre surface by a reaction between the hydroxyl in VMS-SSO and that on the carbon-fibre surface. The VMS-SSO coating concentrations and treatment time were decided according to Zhang *et al.*,[72] who had optimized VMS-SSO coating treatment parameters. The investigation found that the interlaminar shear strength of the carbon-fibre/PAA composites increased by 59.3% after treatment.[70] The conclusion was drawn that carbon-fibre surface oxidation–reduction followed by coating with silsesquioxane is an effective method to improve the interfacial mechanical properties of carbon-fibre/PAA composites. This kind of method could be widely used for different resin matrix composites by changing the functional groups on silsesquioxanes according to that on the resin.

Poly (p-phenylene benzobisoxazole) (PBO), a rigid-rod polymer, is characterized by high tensile strength, high stiffness and high thermal stability. Kumar *et al.*[73] found that PBO/CNT-reinforced fibres exhibited twice the energy-absorbing capability of plain PBO fibres. The nanocomposites were prepared as follows: ~4.3 g (0.02 mol) of 1,4-diaminoresorcinol dihydrochloride, ~4 g (0.02 mol) of terephthaloyl chloride and ~12 g of phosphoric acid (85%) were placed into a 250 ml glass flask, equipped with a mechanical stirrer and a nitrogen inlet/outlet. The resulting mixture was dehydrochlorinated under a nitrogen atmosphere at 65 °C for 16 h and subsequently at 80 °C for 4 h. At this stage, 0.234 g of purified and vacuum-dried HiPco nanotubes was added to the reaction flask. The mixture was heated to 100 °C for 16 h while stirring and then cooled to room temperature. $P_2O_5$ (8.04 g) was added to the mixture to generate poly (phosphoric acid) (77% $P_2O_5$). The mixture was stirred for 2 h at 80 °C and then cooled to room temperature. Further $P_2O_5$ (7.15 g) was then added to the mixture to bring the $P_2O_5$ concentration to 83% and the polymer concentration to 14 wt%. The mixture was heated at 160 °C for 16 h with constant stirring. Stir opalescence was observed during this step. The mixture was finally heated to 190 °C for an additional 4 h while stirring. An aliquot of the polymer solution was precipitated, washed in water and dried under vacuum at 100 °C for 24 h. An intrinsic viscosity of 14 dl/g was determined in methanesulfonic acid at 30 °C. A control polymerization of pure PBO was also carried out under the same conditions without adding SWNT. For PBO/SWNT (90/10) composition, 0.47 g of purified HiPco tubes (SWNT) was added to the mixture. The sequence of steps and polymerization conditions were the same for a PBO/SWNT (95/5) sample. The intrinsic viscosity values of PBO and PBO/SWNT (90/10) were 12 and 14 dl/g, respectively. Single-walled nanotubes were well dispersed during PBO synthesis in polyphosphoric acid (PPA). PBO/SWNT composite fibres were successfully spun from the liquid crystalline solutions using dry-jet wet spinning. The addition of 10 wt% SWNT increased PBO fibre tensile strength by about 50% and reduced shrinkage and high-temperature creep. The existence of SWNT in the spun PBO/SWNT fibres was evidenced by a 1590 $cm^{1}$ Raman peak.

## 15.3 Sandwich structures using polymer nanocomposites

Sandwich composites are used in a wide range of applications from aircraft, ships, ballistic vests and helmets through to racing cars and high-end sports cars. They have a wide range of useful properties, including structural stiffness, crash-energy management and heat shielding, amongst others. These structures, composed of a core of cellular material and outer composite skins, are lightweight and yet offer high resistive stiffness against traction, compressive and bending loads. These properties are utilized to produce functional structures that must sustain high stresses under normal conditions. During severe impact loads in automotive applications, for example, these structures must dissipate impact energy to protect either the rest of the structure or the vehicle's occupants.

Research has shown that damage initiation thresholds and damage size in sandwich composites depend primarily on the properties of the core materials and facings and the relation between them. Much of the early research on sandwich composites under impact focused on a honeycomb core (Nomex, glass thermoplastic or glass-phenolic). A key problem in honeycomb sandwich construction is the low core surface area for bonding. Consequently, expanded foams (often thermoset) are now preferred for achieving reasonably high thermal tolerance, though thermoplastic foams are also used. In turn, the response of foam-core sandwich constructions to impact loading has been studied by many researchers. Accordingly, it is now well understood that the response of foam-core sandwich composites strongly depends on the density and the modulus of the foam.

A possible way of improving the properties of foam materials is through the inclusion of small amounts of nanoparticles (carbon nanotubes and nanofibres, $TiO_2$, nanoclay, etc.) to improve the foam density and modulus properties. Up until now, montmorillonite nanoclays have been the best candidates for foam reinforcement due to ease of processing, enhanced thermal–mechanical properties, wide availability and cost. Likewise, polyurethanes (PU) are core materials of choice due to their tailorable and versatile physical properties, ease of manufacture and their low cost. The use of polyurethanes filled with nanoparticles in constructing either laminates or foams is relatively new. Moreover, the use of nanoparticles in such laminates, or foams in a sandwich composite construction, is in its infancy but has been found to be both realistic and beneficial. For instance, by using less than 5% by weight of nanoclay loadings, significant improvements in foam failure strength and energy absorption have been realized, with over a 50% increase in the impact load-carrying capacity compared with a neat foam sandwich. However, since most current research concentrates on the processing and characterization of nanophased foams and the evaluation of static properties only, dynamic material data on impact failure mechanisms and impact property relations is missing. For the application of nanophased foams in sandwich constructions for ballistic resistance, a proper understanding of their impact behaviour for both high and low-velocity impacts is required.

Therefore, by taking advantage of emerging new materials, nanophased sandwich structures have been fabricated and tested for low-velocity impact resistance, as described in the literature. In a recent development, Njuguna *et al.*[74] fabricated and characterized a series of nanophased hybrid sandwich composites based on polyurethane/montmorillonite (PU/MMT). The polyaddition reaction of the polyol premix with 4,4′-diphenylmethane diisocyanate was applied to obtain nanophased polyurethane foams, which were then used to fabricate sandwich panels. It was found that the incorporation of MMT resulted in a higher number of PU cells with smaller dimensions and higher anisotropy index. The materials obtained exhibited improved parameters in terms of thermal insulation. Importantly, these foams can also be selectively stiffened to meet specific requirements. The results also showed that nanophased sandwich structures are capable of withstanding higher peak loads than those made of neat polyurethane foam cores when subject to low-velocity impacts, despite their lower density than neat PU foams. This is especially significant for multi-impact recurrences within the threshold loads and energies studied. A feasible application for these lightweight structures is as energy-absorbing structures or as inserts in hollow structures.

## 15.4 Properties and applications of polymer nanocomposites

### 15.4.1 Thermal stability and fire retardancy

The commercial importance of polymers has been driving the use of composites in various fields, such as aerospace, automotive, marine, infrastructure and military applications.[75] Performance during use is a key feature of any composite material. Whatever the application, there is a natural concern about the durability of polymeric materials. The deterioration of these materials depends on the duration and the extent of interaction with the environment. Degradation of polymers includes all changes in chemical structure and physical properties due to external chemical or physical stresses caused by chemical reactions, involving bond scissions in the backbone of the macromolecules that lead to materials with characteristics different (usually worse) from those of the starting material (Fig. 15.13). As a consequence of degradation, the resulting smaller fragments do not contribute effectively to the mechanical properties, the article becomes brittle and the life of the material becomes limited. Thus, any polymer or its

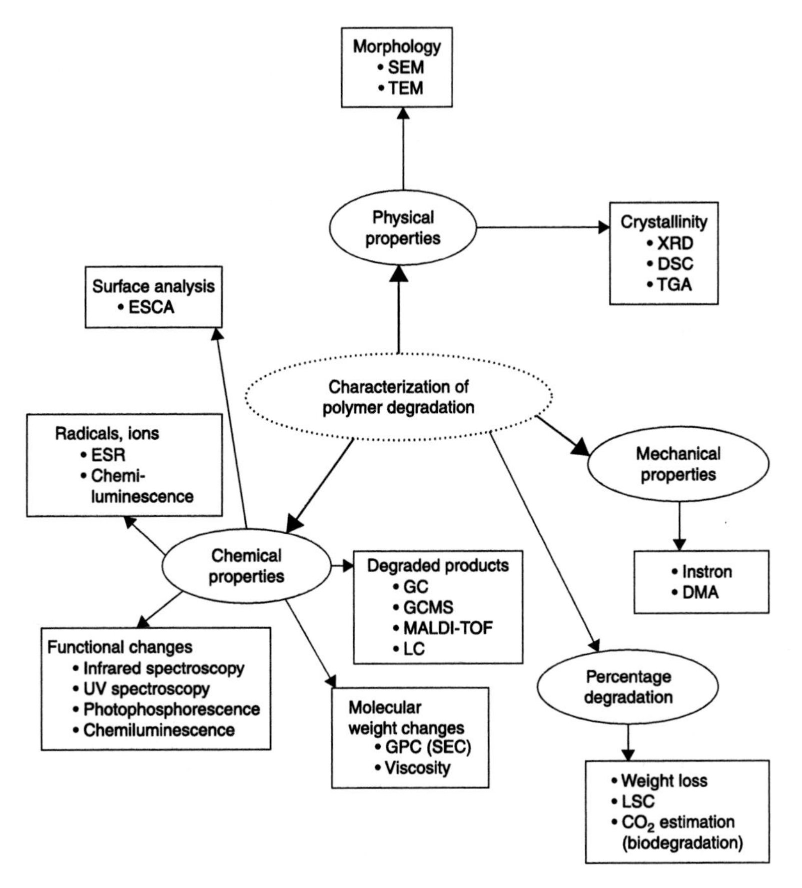


*15.13* Various techniques for the characterization of polymer degradation processes and mechanisms.[75] (ESCA: electron spectroscopy for chemical analysis; ESR: electron spin resonance; GC: gas chromatography; GCMS: gas chromatograph/mass spectroscopy; MALDI-TOF: matrix-assisted laser desorption ionization-time-of-flight mass spectrometry; LC: liquid chromatography; GPC (SEC): gel permeation chromatography (size exclusion chromatography); LSC: laser scanning cytometry.)

(nano)composite that is to be used in outdoor applications must be highly resistant to all environmental conditions.

Research indicates that the modified epoxy nanocomposites possess better flame retardant properties than conventional composites. With the Kissinger method, the activation energies for the thermo-oxidative degradation of epoxy nanocomposites are less than those of the pure epoxy in the first stage of thermo-oxidative degradation. However, in the second stage of thermo-oxidative degradation the activation energies for epoxy nanocomposites are generally higher than those of the pure epoxy. For example, the main mechanism for layered silicate is barrier formation, which influences flame spread in developing fires. Several minor mechanisms are significant, but important fire properties such as flammability or fire load are hardly influenced. Hence combinations with aluminium hydroxide and organo-phosphorus flame retardants need to be evaluated. It has been shown that carbon nanotubes can surpass nanoclays as effective flame-retardant additives if the carbon-based nanoparticles (single- and multi-walled nanotubes as well as carbon nanofibres) form a jammed network structure in the polymer matrix, so that the material as a whole behaves rheologically like a gel.[76]

The thermal degradation of nanocomposites depends on clay loading, structure and the nature of the ambient gas. Recently Leszczyńska *et al.*[77] reviewed the thermal stability of various polymer matrices improved by montmorillonite clay and their influencing factors in detail. For the majority of polymers, due to their hydrophobic character, the clay must be modified with a surfactant in order to make the gallery space sufficiently organophilic to permit it to interact with the polymer. In fact, several factors were found to govern the thermal stability of nanocomposite materials, such as the intrinsic thermal resistance of the polymer matrix, nanofiller content, the chemical constitution of the organic modifier and the chemical character of polar compatibilizers as well as the access of oxygen to the composite material during heating. For the surface modification of clay, the surfactant is usually described as an 'onium' salt, but in fact ammonium salts are most commonly used. The quaternary ammonium ion is nominally chosen to compatibilize the layered silicate with a given polymer resin. However, the molecular structure (length and number of alkyl chains and unsaturation) is also the determining factor of thermal stability of polymer/MMT nanocomposites.

A possible mechanism for the degradation of modifiers in silicates has been given by, among others, Hwu *et al.*[78] and Leszczyńska *et al.*,[77] and research has shown that surfactants degrade between 200 and 500 °C. The amount of surfactant lost during thermogravimetric analysis of various organoclays indicates that surfactants with multiple alkyl tails have greater thermal stability than those with a single alkyl tail. It has been proposed that organic modifiers start decomposing at a temperature of around 200 °C, and the small molecular weight organics are released first while the high-molecular weight organic species are still trapped by the organic layered silicate matrix. With increasing temperature, high-molecular organic polymer chains may exist between the interlayers until the temperature is high enough to lead to their further decomposition. The incorporation of silicate layers with a high-aspect ratio of decomposed/charred material on the clay surface acts as a carbonaceous insulator. Silicate has an excellent barrier property, which prevents permeation of various degraded gaseous products.

The addition of clay enhances performance by acting as a superior insulator and mass transport barrier to the volatile products generated during decomposition. The clay acts as a heat barrier, which could enhance the overall thermal stability of the system as well as assisting in the formation of char during thermal decomposition. In a nanocomposite, the temperature at which volatilization occurs is higher than for a microcomposite. Moreover, thermal oxidation of the polymer is strongly slowed in a nanocomposite with high-char yield, both by a physical barrier effect, enhanced by ablative reassembling of the silicate, and by a chemical catalytic action due to the silicate and the strongly acid sites created by thermal decomposition of the protonated amine silicate modifier.

Polymers that show good fire retardancy on nanocomposite formation exhibit significant intermolecular reactions, such as inter-chain aminolysis or acidolysis, radical recombination and hydrogen abstraction. For polymers that degrade through a radical pathway, the relative stability of the radical is the most important factor in predicting the effect that nanocomposite formation has on the reduction in the peak heat release rate. The more stable the radical produced by the polymer, the better the fire retardancy, as measured by the reduction in the peak heat release rate of the polymer/clay nanocomposite.

CNTs are increasingly finding applications as thermal management materials and are also being considered in potential interface and attachment techniques.[79] Interfaces between materials have a significant effect on the thermal impedance of electronic systems and in practice they can be the dominant factor in achieving effective thermal transfer. The interface materials and processes in question are the methods used to join an electronic device to a thermal transfer medium (e.g. substrate, heat pipe, heat sink), including coatings and bonding techniques. In this respect, they may need to perform the tasks of attachment, stress/strain relief and thermal transfer. The simplest of all interfaces is a dry joint (two surfaces pushed together). In this case, interface thermal resistance can be significant and will depend on the surface materials: their hardness, co-planarity, roughness and the applied pressure to hold the surfaces together. To enhance heat transfer across the interface, thermally conductive materials can be introduced to improve surface coupling and conductivity. Commercialization is being pursued by several companies, e.g. MER Corporation is developing a CNT interface material with Mitsubishi. The material is a film on a 2-μm-thick porous (20% CNT, 80% pores) material, which is filled and laminated with polyethylene or epoxy resin to improve its mechanical properties. This system has similar thermal properties to conventional phase change materials. A hydrocarbon condensed onto the nanotubes improves wetting but may limit thermal conductivity. Nanotube sheets, in particular, have been used to thermally fuse together two polymer sheets in a transparent and seamless fashion.

This concept of a CNT dry adhesive is very attractive in terms of the potential for a high thermal conductivity interface. Zhao *et al.*[80] and Johnson[81] recently reported the development of 'dry adhesive/Velcro', a novel CNT technology for thermal interface materials. MWNT 'adhesives' are electrically and thermally conductive (comparable to commercially available thermal paste). When grown on a surface in an array, CNTs have an extremely large surface-to-volume ratio and bind to each other and to surfaces through van der Waals interactions. When acting collectively, van der Waals forces can provide significant adhesive strength (~12 N/$cm^2$) regardless of the hydrophobicity of the surfaces. Similar work was reported by Xu *et al.*,[82] who developed CNT interfaces for improved thermal and electrical management, which achieved a thermal interface resistance of 20–37 $mm^2$.K/W at a pressure of 0.445 MPa. Additionally, the combination of a CNT array and a phase change material (load 0.35 MPa) produced a minimum resistance of 5.2 $mm^2$.K/W. Hu *et al.*[83] described an exploratory thermal interface structure, made of vertically oriented CNTs directly grown on a silicon substrate, which has been thermally characterized using a 3-omega method. The effective thermal conductivities of the CNT samples, including the effects of voids, were found to be 74 W/m.K to 83 W/m.K in the temperature range of ~22–50 °C, one order higher than the best thermal greases or phase change materials. These results suggest that vertically oriented CNTs can potentially be a promising next-generation thermal interface solution. However, fairly large thermal resistances were observed at the interfaces between the CNT samples and the experimental contact. Minimizing these contact resistances is critical for the application of these materials. Additionally, the potential for a low insertion force, 'flexible' (stress relieving) and reworkable die and substrate attachment system would be of significant benefit for assembly and maintenance. This research has demonstrated the potential of this new technology; however, there are still significant issues in terms of strength, performance and reliability to address prior to these materials being acceptable as an attachment system in a high-performance application. Butt joints give poor thermal resistance and research into devising a low thermal resistance joint is ongoing.

A limiting factor in the use of CNTs is the transfer of heat flux from one nanotube to another in an efficient manner, e.g. in a butt joint. CNTs can be woven into mats to produce a low-density, high-thermal-conductivity material. This can be put into a metal composite by pressure or squeeze casting, or epoxy can be added as a filler to give rigid mats; however, these approaches are still at an early stage. Although expectations of CNTs are very high for their use in composites, there has been some speculation on the results they produce when mixed with polymers. For instance, CNTs are good conductors by themselves but they may not exhibit the same level of conductivity when integrated into other materials. Experiments have shown that thermal conductivity has increased by two or threefold, when it should have been close to 50-fold. The problem is that CNTs vibrate at much higher frequencies than the atoms in the surrounding material, which causes the resistance to be so high that thermal conductivity is limited. Inducing stronger bonds between the nanotube and the other materials might help in solving the problem.

The addition of CNTs to a polymer matrix could increase the glass transition, melting and thermal decomposition temperatures due to their constraint effect on the polymer segments and chains. It is important to improve the thermal endurance of polymer composites.[80] Thermal management is a growing need in aerospace and defence applications alike. As device engineers continue to follow Moore's Law and devices get smaller and more powerful, the management of heat has become a serious issue. However, thermal transport in nanostructures requires an understanding of heat transport beyond that gained at the continuum level and necessitates advances in measurement methods. This arena is one that offers many opportunities for exploration. An increased understanding of interfaces, dispersion and percolation networks and morphology characteristics is expected to provide pervasive and significant advances in the field of nanocomposites. Improved thermal management is essential for meeting market-driven performance, lifetime and cost requirements. This is particularly crucial for the next generation of designs with their need for enhanced functionality, higher volumetric power densities and increased reliability. In many cases, such as all-electric aircraft, military applications and many sensing systems, thermal management solutions also have to perform in harsh environments. The main issues with the current technique are ensuring that the silicon is kept below 350 °C during CNT processing (this is not an issue with SiC devices) and the need to apply a clamping pressure (1–4 atm) to affect a joint. Potential applications include microprocessors and power electronics, such as those used in military appliances.

## 15.4.2 Electronic properties

Despite the excellent dispersion of CNTs and CNFs, the percolation thresholds in various systems are drastically different. Ultra-low percolation thresholds in the range 0.0021–0.0039 wt% and, in contrast, high percolation thresholds in the range 3–5 wt% have been reported in the literature. Major uncertainties are associated with the type and quality of nanotubes, that is, a wide variety of synthesis methods have been employed to obtain nanotubes of different sizes, aspect ratios, crystalline orientation, purity, entanglement and straightness. It has been reported that, when the aspect ratio of CNTs was reduced from 417 to 83–8.3 in epoxy/CNT nanocomposites, the corresponding percolation threshold increased from 0.5–1.5 and > 4 wt%, respectively, indicating that the aspect ratio is a predominant factor.[84]

In contrast, for an aspect ratio of 300, Kim *et al.*[85] reported a percolation threshold of 0.017–0.077 vol% in epoxy/CNT nanocomposites. Even with an aspect ratio of 1000, Allaoui *et al.*[86] obtained a percolation threshold of 0.6 vol%. Moreover, it is also rather interesting to note that, even with the same kind of nanotubes, the percolation threshold varied (from 0.0225 to 10 wt%) depending on the matrix materials. Although the differences can be qualitatively explained based on the type and nature of the matrix/resin and the cross-linking density, concrete knowledge is still lacking. In fact, it is very difficult to control the local cross-linking density as it in turn depends on the nature of fillers, their disentanglement and orientation. With polyvinyl alcohol as the matrix and the same Hyperion nanotubes as fillers, Shaffer and Windle[87] reported a percolation threshold between 5 and 10 wt% for nanotubes. Sandler *et al.*[88] also reported a percolation threshold between 0.0225 and 0.04 wt% in epoxy nanocomposites based on these nanotubes. Although differences in melt viscosity, cross-linking density (and percentage crystallinity for thermoplastics) may qualitatively explain the observed variations with different matrices, proper experimental evidence is still lacking.

In comparison to carbon nanotubes where the graphitic layers are parallel to the nanotube axis, carbon nanofibres, in general, can show a wide range of graphitic layer orientations with respect to the fibre axis. This dramatically affects the percolation threshold of the materials. Nevertheless, it is important to note that the ultra-low percolation thresholds reported in some systems were in fact in samples prepared at a very small scale.

Recently, super-capacitors have attracted much attention because of their high capacitance and potential applications in electronic devices. It has been reported that the performance of super-capacitors with MWNTs deposited with conducting polymers as active materials is greatly enhanced compared with electric double-layer super-capacitors with CNTs, due to the Faraday effect of the conducting polymer. Additionally, polymer/CNT nanocomposites could have many potential applications in electrochemical actuation, electromagnetic interference shielding (EMI), wave absorption, electronic packaging, self-regulating heaters and PTC resistors. Furthermore, synthetic DNA has been proposed for quantum dot and nanotube-based computing systems.[89] The use of DNA as a structural material has opened up many new possibilities for the engineered nanoscale fabrication of computing systems. Beginning with efforts to crystallize DNA to determine its natural structure, the development of synthetic DNA as a structural material for nanoscale fabrication has produced a nanoscale barcode and various 2D and 3D structures.

Nanotube transistors have also been produced using integrated nanotubes, which may lead to large-scale integration. The patterned growth of CNTs on silicon wafers may prove necessary for the integration of nanotubes in aerospace and defence electronics. Also, coiled nanotubes could have even more interesting applications in various areas than their straight counterparts. For example, the conduction of electricity through a coiled nanotube generates an inductive magnetic field, an indication that coiled nanotubes, unlike straight nanotubes, could be of use as electromagnetic nanotransformers or nanoswitches. Recent research and developments have led to the possibility that nanotubes will be useful for downsizing circuit dimensions. Presently, current-induced electromigration causes conventional metal wire interconnects to fail when their diameter becomes too small. The covalently bonded structure of CNT militates against the similar breakdown of nanotube wires and, because of ballistic transport, the intrinsic resistance of the nanotube should essentially vanish. Experimental results have shown that metallic SWNTs can carry up to $10^9$ A/cm$^2$ compared with current densities for normal metals of only $10^5$ A/cm$^2$.

Due to scalability and low power consumption, field emitters are attractive candidates for a wide variety of space applications, particularly where budget is a major consideration. Field emission (FE) using carbon nanotubes has already been demonstrated as ideal for high-voltage low-current electrical power applications such as field-emission electric propulsion (FEEP), colloids, micro-ion thrusters and perhaps even small electrodynamic (ED) tethers.[90] A typical cathode delivers up to 10 mA with gate voltages in the range of 200 to 400 volts and weighs about 20 grams. CNT FE fits well with requirements for micro-satellites, which are considered to be a viable alternative for a variety of applications, and microtechnology, by contributing to a substantial reduction of mass, volume and power requirements for small satellites and satellite sub-systems.

Carbon nanotube technology could have a dramatic breakthrough for magnetic devices, especially magnets, for space and aircraft applications. The basic electronic properties of semiconducting CNTs change when placed in a magnetic field. Nanotube band gaps are comparable with silicon and gallium arsenide, which are currently the mainstays of the computer industry because their narrow band gaps correspond with how much electricity it takes to flip a transistor from 'on to off'. Superconductivity has been reported by both doping a SWNT with caesium, potassium or rubidium and packing small buckyballs inside it.[91] With the possibility of the band gap of carbon nanotubes disappearing altogether in the presence of stronger magnetic fields, CNTs could take over the role of silicon and gallium arsenide, potentially revolutionizing the computer industry. If either the promise of superconductivity with current densities of 106 A/cm$^2$ or higher or the greatly improved strength of a composite material based on carbon nanotube fibres is achieved, there could be significant weight reductions in magnets. Therefore, the most urgent need is to determine the superconducting properties of possible CNTs. That is, we need to know the feasible current density as a function of temperature, magnetic field and strain.

## 15.4.3 Field emission and optical properties

Carbon nanotubes possess the right combination of properties (nanometre-sized diameter, structural integrity, high electrical conductivity and chemical stability) to make good electron emitters. Research on electronic devices has focused primarily on the use of SWNTs and MWNTs as field-emission electron sources for flat panel displays, lamps, gas discharge tubes providing surge protection and X-ray and microwave generators. A potential applied between a nanotube-coated surface and an anode creates a high electric field due to the small radius of the nanofibre tips and the lengths of the nanofibres. The local fields cause electrons to tunnel from a nanotube tip to the tunnel. This process of nanotube-tip electron emission differs from that of bulk metals because it arises from discrete energy states instead of continuous electronic bands and its behaviour depends on the nanotube-tip structure, whether SWNT or MWNT.

The importance of electromagnetic interference (EMI) shielding has also increased in the electronics and communication industries, especially in space and military uses, due to the widespread use of packed highly sensitive electronic devices. Kim *et al.*[92] designed radar-absorbing structures (RASs) with a load-bearing ability in the X-band. Glass/epoxy plain-weave composites with excellent specific stiffness and strength, containing MWNTs to induce dielectric loss, were fabricated. Fabrication involved impregnation of glass/epoxy plain-weave composites by mixing a matrix and MWNTs. The MWNT-filled fabric composites were dried for 5–7 min at 100 °C. Drying times increased with MWNT content. As the viscosity of the premixture increased rapidly above 3.0 wt%, the researchers reported that they found it difficult to maintain the uniformity of MWNTs in the matrix. Specimens were cured at a stabilized pressure of 3 atm and vacuum-bagged in an autoclave initially for 30 min at 80 °C and then for 2 h at 130 °C. Observation of the microstructure of the composites revealed that the uneven distribution of MWNTs could induce high dielectric loss, which was confirmed through a measurement of permittivity.

The optimal design of a two-layered RAS, consisting of MWNT-added glass/epoxy fabric composites, was performed by linking a genetic algorithm with a method for the reflection and transmission of electromagnetic waves in a multilayered RAS. As a result, a two-layered RAS was designed having 90% absorption of electromagnetic (EM) energy over the entire X-band. An RAS fabrication process was proposed based on the non-linearity of the thickness per ply with MWNT content and the number of plies. A comparison between the theoretical and experimental reflection losses confirmed that the process can be used in the fabrication of multilayered RASs. However, the authors commented that further studies directed to broadening the absorbing bandwidth of a RAS, comprising a multilayered RAS and a frequency selective surface, are required. Furthermore, it was reported that GE Plastics has been using CNTs in a poly(phenylene oxide)/polyamide blend for automotive mirror housings for Ford[93] to replace conventional micron-size conducting fillers, which would require loadings as high as 15 wt% to have satisfactory anti-static properties; however, this level of CNT loading would result in poor mechanical properties and a high density of the final composite.

While mechanical properties in this range have been observed for individual SWNTs, those observed for assemblies of these nanotubes in nanotube yarns and sheets are much lower, which restricts the performance of actuators based on them. At high levels of charge injection into a CNT, the predominant cause of actuation is electrostatic, as with dielectric elastomer actuators (DEAs).[94] However, the electrostatic forces are repulsive interactions between like charges injected into the nanotubes, rather than between two electrodes. Charges are injected by applying a voltage between an actuating nanotube electrode and a counter electrode, through an ion-containing solution, as depicted in the upper image of Fig. 15.14 (where the counter electrode is another CNT), leading to charging (Fig. 15.14).

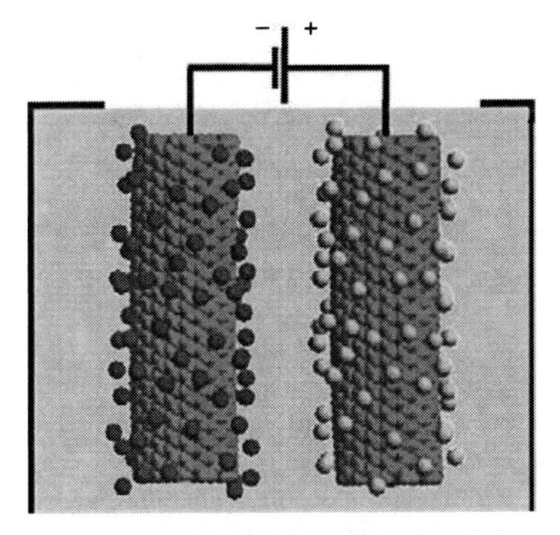

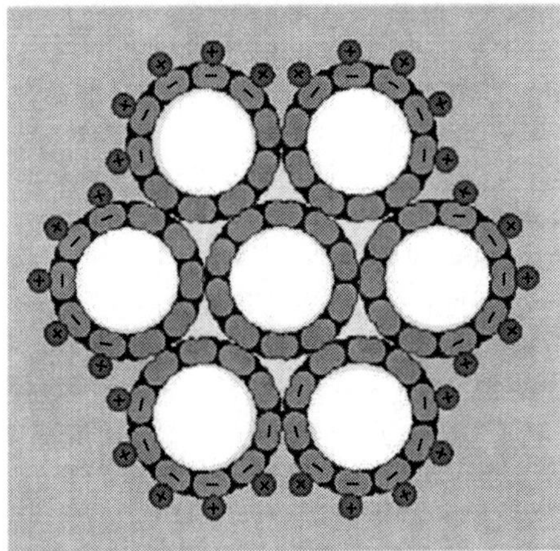

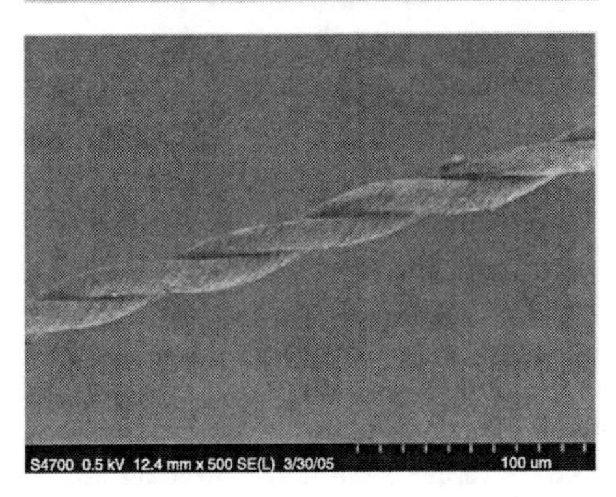


*15.14* Charge injection in a CNT-based electromechanical actuator. Top: An applied potential injects charges into the two nanotube electrodes in solution (left: positive ions, right: negative ions). Middle: Charge injection at the surface of a nanotube bundle. Bottom: A scanning electron micrograph of a twist-spun MWNT yarn, which has a 0.6% actuator stroke at loads of 100 MPa.[94]

Electrostatic repulsive forces between like charges on the CNT work against the stiff carbon–carbon bonds in the nanotubes to elongate and expand the nanotubes, though quantum mechanical effects can predominate over electrostatic forces at low levels of charge injection. Unlike dielectric elastomers, strains are low (<2%) since CNTs are extremely stiff. Actuation is generally achieved in films or yarns (Fig. 15.14 bottom) composed of many nanotubes. The porous nature of the films and fibres enables fast ion transport with response times of <10 ms, with effective strain rates of 19%/s and effective power-to-mass ratios of 270 W/kg (half that of a high revving electric motor). The achievable response rate decreases with increasing nanotube yarn or sheet thickness, increasing interelectrode separation, and decreasing electrolyte ionic conductivity. A high work density combined with good temperature stability (>450 °C in air, >1000 °C in an inert environment) make CNTs prime candidates for situations where weight and temperature are important, such as aerospace and defence applications. Strains are relatively small compared with other polymer actuators (but an order of magnitude larger than found in typical piezoceramics). Strain could be increased by employing electrolytes such as highly purified ionic liquids that can withstand very large potentials without reacting electrochemically. CNT actuators have recently been shown to actuate when used as electrodes in a fuel cell. This possibility is exciting because the energy density of fuel cells is much higher than that of batteries, helping to enable autonomous applications of CNTs and possibly artificial muscle technology.

Research into field effect transistors (FETs) aims to replace the source/drain channel structure with a nanotube. Transistors assembled with carbon nanotubes may or may not work, however, depending on whether the chosen nanotube is semiconducting or metallic, over which the operator has no control. It might be possible to peel back layers from a MWNT to achieve the desired properties, but advances in microlithography are still needed to perfect this reduction method. Recent developments have focused media attention on nanotube nanoelectronic applications. Crossed SWNTs have been used to produce three- and four-terminal electronic devices along with non-volatile memory that functions like an electromechanical relay.

Irradiated by a camera flash, a random network of polyaniline (PANI) nanofibres was turned into a smooth and shiny film due to the highly efficient photothermal conversion and low thermal conductivity of PANI, which demonstrated a versatile new technique for making polymers into potentially useful structures.[95] One of the great advantages of the flash-welding technique is that an area can be selectively welded by using a photomask, differing from other welding techniques like microwave welding. It is used to fabricate polymer films to pre-designed patterns. In addition, the technique can rapidly create asymmetric films, which are widely used in many applications such as separation membranes, chemical sensors and actuators.

Flat panel displays are one of the more lucrative applications of carbon nanotubes but are also one of the most technically complex. Nanotubes have an advantage over liquid crystal displays since they have low power consumption, high brightness, a wide viewing angle, a fast response rate and a wide operating range. In a flat panel display, electric fields direct field-emitted electrons toward an anode where phosphorus produces light. One demonstration used nanotube/epoxy stripes on the cathode glass plate and phosphor-coated indium-tin-oxide (ITO) stripes on the anode plate. The pixels were at the intersections of the cathode and anode stripes. Pulses of ±150 V were switched between the anode and cathode stripes to produce an image. Various types of carbon nanotube are being considered for future field-emission displays (FEDs). Samsung is a leading player in CNT-based FE displays.[96] They have produced several prototype CNT-FEDs, including a 9-in. red–blue–green (RGB) colour FED that can operate at video frame speed. One beauty of the Samsung display is that the CNT-based cathode materials are printable inks.

In addition to flat panel displays, another potentially important application of CNT is in polymer-based light-emitting devices. The advantages of organic light-emitting diodes (OLEDs) based on conjugated polymers are low cost, low operating voltage, excellent processability and flexibility.[97] However, their low quantum efficiency and stability have limited their application and development. Carbon nanotubes are also being considered as electron emitters for high-brightness lighting elements (Fig. 15.15).

Nanotube-based lamps are similar to displays, comprising a nanotube-coated surface opposing a phosphor-coated substrate, but they are less technically challenging and require less investment. With lifetimes expected in excess of 8000 h, they could replace environmentally problematic mercury-based fluorescent lamps used in stadium-style displays. More recently, lighting elements for the three primary colours have been produced and a brightness >10 000 cd/m$^2$ has been demonstrated.[98] The frequency characteristics of the cold cathode element were studied over a range of frequencies. The cut-off frequency in these devices was found to be determined by the vacuum gap capacitance between the cathode and the gate electrode. The devices were repeatedly operated at 50 kHz without any degradation. It is quite feasible to use cold cathode lighting elements to

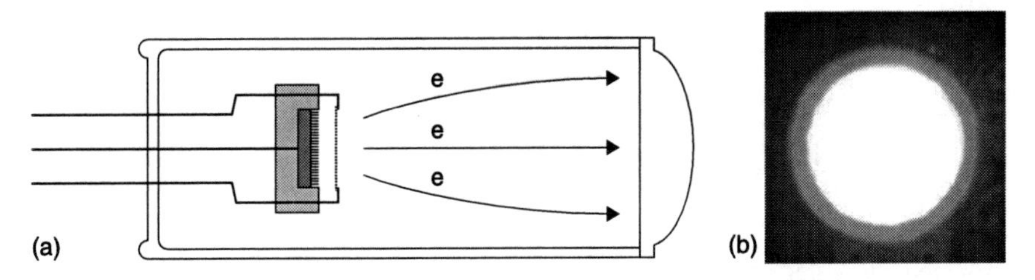


*15.15* (a) Internal arrangement of a CNT-based lighting element. (b) Photograph of a CNT-based lighting element.[98]

assemble a full-colour display with video frame speed for large pixel screens. Nanotube-based gas discharge tubes might also find commercial use in protecting telecommunications networks from power surges. Another application arises if a metal target replaces the phosphorescent screen at the anode. This causes the accelerating voltage to increase, producing X-rays instead of light. The compact geometry of this nanotube-based X-ray generator could lead to use as an X-ray endoscope and for medical exploration.

Non-linear optical organic materials, such as porphyrins, dyes and phthalocyanines, have optical limiting properties, which can be used to control light frequency and intensity in a predictable manner in photonic devices. However, these are narrow-band optical materials. The combination of the unique properties of CNTs with conducting organic polymers (e.g., polyaniline, polypyrrole, polythiophene, poly(3,4-ethylenedioxy thiophene), poly(*p*-phenylene vinylene) and poly(*m*-phenylene vinylene-*co*-2,5-dioctoxy-*p*-phenylene)) makes these materials interesting multifunctional systems with great potential in many applications such as super-capacitors, sensors, advanced transistors, high-resolution printable conductors, electromagnetic absorbers, photovoltaic cells, photodiodes and optical limiting devices. As a result of the optical limiting performance and the good photoconductivity, special attention has been given to CNTs functionalized with polymers such as poly (*N*-vinyl carbazole).

### 15.4.4 Age and durability performance

The study of the degradation and stabilization of polymers is an extremely important area from a scientific and industrial point of view and a better understanding of polymer degradation will help to increase the life of a product.[99] Polymer degradation in broader terms includes biodegradation, pyrolysis, oxidation, mechanical, photo- and catalytic degradation. Because of their chemical structure, polymers are vulnerable to harmful effects in the environment. In the following sections, epoxy and polyamides are considered. It is important to note that little attention has been given to the durability of polymer nanocomposites compared with their preparation techniques and the evaluation of mechanical properties.

#### *Epoxy nanocomposites*

The effects of hydrothermal aging on the thermomechanical properties of high-performance epoxy and its nanocomposites have been reported in the literature.[100] It has been found that the storage modulus and relaxation behaviour are strongly affected by water uptake, while fracture toughness and Young's modulus were less influenced. The dependence of tensile strength and strain at break on water uptake were found to be different in neat epoxy and epoxy–clay systems. Further improvements in flame retardancy (FR) using combinations of nanofiller and traditional FR-additives (e.g. aluminium trihydrate) have been observed. Nanocomposites based on nanofillers and aluminium trihydrate passed the UL 1666 riser test for fire-resistant electrical cables.[101]

There are two factors which have opposite influences on the thermal stability of epoxy–clay nanocomposites. The first is that the addition of clay to epoxy decreases the curing reactivity of epoxy resin. A lower reactivity of the resin generally results in a lower cross-linking density of the cured resin and longer polymer chains among the cross-linking points. A longer polymer chain is less stable thermally than a shorter chain, so nanocomposites with both long and short chains are easier to degrade than pristine epoxy resin. Secondly, silicate layers are good barriers to gases such as oxygen and nitrogen; they can insulate the underlying materials and slow the mass loss rate of decomposition products. Moreover, exfoliated nanocomposites have better barrier properties and thermal stability than intercalated ones. For intercalated nanocomposites (10 wt% clay), the first factor is dominant, whereas for exfoliated nanocomposites (2 wt% clay), the second factor is dominant. Becker *et al.*[102] found that the water uptake (in an aquatic environment) was considerably reduced in epoxy nanocomposites with a particular clay loading percentage.

Jiang *et al.*[103] investigated the resistance to vacuum ultraviolet irradiation of nano-$TiO_2$ modified carbon/epoxy composites. The nano-$TiO_2$ modified composites, including $TiO_2$ + EP648 and M40/$TiO_2$ + EP648, were fabricated so that the nano-$TiO_2$ particles were dispersed in an EP648 epoxy matrix using a high-speed shearing emulsification technique. A jet type of vacuum ultraviolet (VUV) source was used to simulate the VUV spectrum in space, providing various doses of VUV irradiation. Experimental results showed that, compared with the EP648 epoxy and M40/EP648 composite, the specific area mass loss of $TiO_2$ + EP648 decreased by 44% and that of the M40/$TiO_2$ + EP648 composite by 38%. By increasing the dose of VUV irradiation, the internal layer shear strength of the M40/$TiO_2$ + EP648 increased gradually, while that of M40/EP648 showed a decreasing trend. After irradiation, the surface of M40/$TiO_2$ + EP648 changed a little, but that of M40/EP648 was damaged severely. Scanning electron microscopy and atomic force microscopy observations showed that the VUV damage occurred mainly in the epoxy matrix, while the carbon fibres showed good resistance to irradiation.

In another study, Nguyen *et al.*[104] investigated the degradation and potential nanofiller release of amine-cured epoxy nanocomposites containing multi-walled carbon nanotubes (MWCNTs) and nanosilica fillers exposed to UV radiation from 295 nm to 400 nm at 50 °C and 75% RH. During the exposure period, measurements of chemical degradation, mass loss and surface morphological changes were carried out on the samples. In a Fourier transform infrared spectroscopy (FTIR) analysis, the bands at 1508 $cm^{-1}$ and 1714 $cm^{-1}$, representing chain scission and oxidation, respectively, were used to follow various degradation processes of unfilled films and nano-filled epoxy composites exposed to UV radiation (Fig. 15.16). It can be

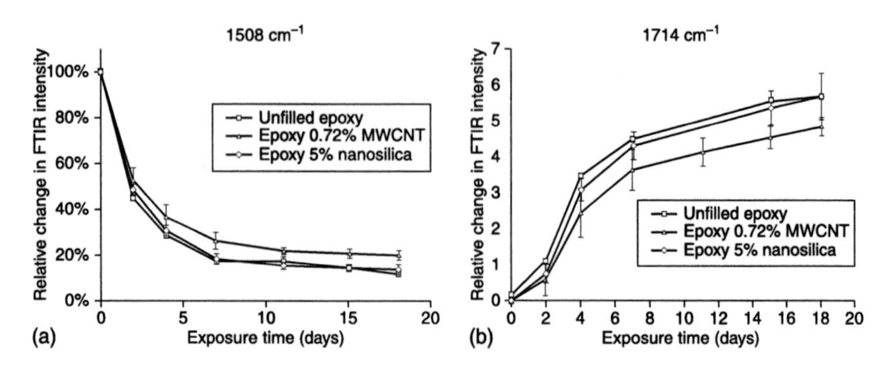


*15.16* (a) Chain scission and (b) oxidation vs. time for unfilled, 0.72% MWCNT filled, and 5% nanosilica-filled amine-cured epoxy samples exposed to UV at 50 °C and 75% RH. Each data point is the average of four specimens and error bars represent one standard deviation.[104]

seen that both unfilled and nano-filled amine-cured epoxy films underwent rapid chemical degradation when exposed to the UV, RH and temperature conditions. The degradation rates of all three materials (at or near the composite surface) reached a plateau in less than 10 days of exposure.

Lee and Lichtenhan[105] reported that the molecular-level reinforcement of cages of polyhedral oligomeric silsesquioxanes (POSS) could significantly retard the physical aging process of epoxy resin in the glassy state. Barral *et al.*[106] used kinetic parameters to predict the lifetime at different temperatures based on thermal degradation. The calculations showed that a POSS/diglycidyl ether of bisphenol A/4,4′-diaminediphenylmethane system can be used at room temperature for a very long time without failure. It must be emphasized that these calculations predicted the expected lifetime of epoxy-resin systems based only on thermal degradation data. Other factors, such as photodegradation, diffusion effects, mechanical and chemical degradation and physical aging, will also affect the expected lifetime. Finally, Ozcelik *et al.*[107] recently studied the thermo-oxidative degradation of graphite/epoxy composite laminates due to exposure to elevated temperatures using weight loss and short beam strength reduction data. Test specimens obtained from 24-ply, unidirectional AS4/3501-6 graphite/epoxy laminates were subjected to 100, 150, 175 and 200 °C for 5000 h (208 days) in air. Predictive differential models for the weight loss and short beam strength reduction were developed using isothermal degradation data only up to 2000 h. The predictive capabilities of both models were demonstrated using the longer term, 5000 h, degradation data. The proposed models were first-order differential expressions, which can be used to predict degradation in an arbitrary, time-dependent temperature environment.

An investigation was undertaken by Barkoula into the environmental degradation of epoxy resin modified with MWCNTs and a carbon-fibre/epoxy composite reinforced with MWCNTs. The samples were exposed to hydrothermal loadings to assess the level of water uptake during loading at various temperatures. Both epoxy/CNT resin and the carbon-fibre/epoxy/CNT composites were tested in this manner. It was found that the modified epoxy resin showed a small increase in weight due to water uptake compared with the neat resin. The increase was thought to be caused by the additional interface between the CNTs and the epoxy, which created a route for water uptake. However, no correlation was found between the extent of CNT modification and the gain in weight.[108] When the composite laminates were tested, they showed a dramatic decrease in water uptake compared with the resin without carbon fibres. This was thought to be due to the introduction of carbon fibres, which do not absorb water. Figure 15.17 shows that there was very little difference between the neat and modified composites in terms of water uptake. The introduction of an interface at the microscale nullified the effect of the nanoscale interface between the CNTs and the epoxy matrix.

*Polyamide nanocomposites*

The presence of MWNTs improves the thermal stability of PA-6 under air obviously, but has little effect on the thermal degradation of PA-6 under a nitrogen atmosphere. A thermal degradation mechanism for PA-6 has been proposed

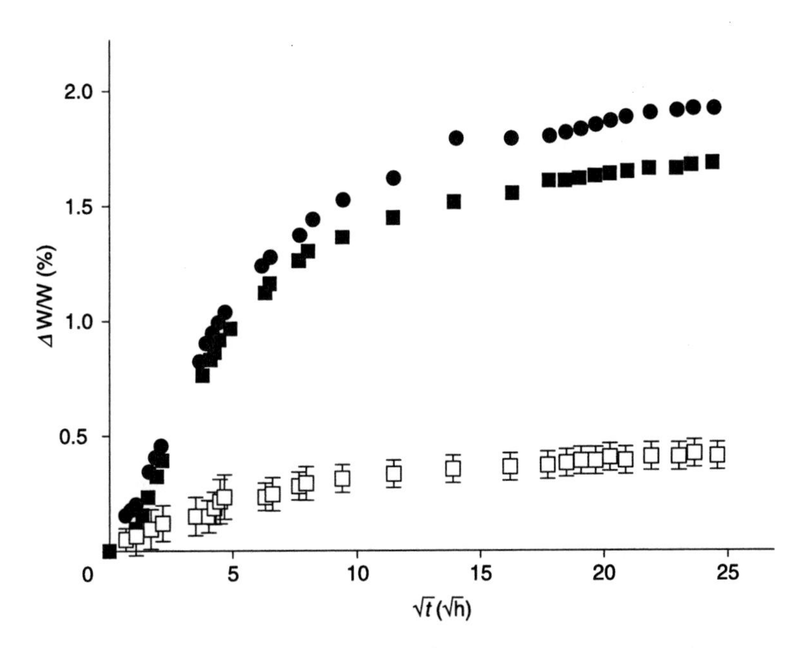


*15.17* Weight gain due to water uptake against time for neat (■) and 0.5% modified (●) epoxy, and neat (□) and modified (○) carbon-fibre/epoxy composites.[108]

by Levchik *et al.*[109] VanderHart *et al.*[110] observed that, in the presence of clay, the $\alpha$-phase of PA-6 transforms into the $\gamma$-phase. Compared with pure PA-6, the PA-6/clay nanocomposite has a higher activation energy, lower thermal decomposition rate constants and better thermal decomposition stability.[111] The activation energy ($E_a$) of PA-6/clay nanocomposites containing 1.2% and 3.5% clay is 174.47 and 309.175 kJ/mol, respectively, and that of pure PA-6 is only 143 kJ/mol. The effect of the modifier on the degradation of the nanocomposite was studied using $^{13}$C NMR. In the presence of a modifier (dihydrogenated-tallow ammonium-ions) the nylon nanocomposite begins to degrade at 240 °C, whereas the virgin polymer does not. They concluded that the organic modifier is less stable. The combination of shear stress and temperature may lead to extensive degradation of the modifier and the extent of clay dispersion may not depend on the modifier.

Davis *et al.*[111] studied the thermal stability of injection-moulded PA-6 nanocomposites using $^{13}$C NMR. Virgin PA-6 and its nanocomposites were injection moulded at 300 °C. PA-6 does not degrade at the processing temperature, whereas there was a significant decrease in the molecular weight of the nanocomposites under the same conditions. It was observed that the degradation might depend upon the percentage of water in the nanocomposites, which might cause hydrolytic cleavage. Fornes *et al.*[112] found the formation of colour with polymer-matrix degradation after twin-screw extrusion of polymer nanocomposites. The researchers reported that the degradation of the nanocomposites depends on the type of nylon-6 materials as well as the chemical structure of the surfactant in the organically modified MMT. Hydroxyethyl groups in the surfactant, as opposed to methyl groups, and tallow substituents, as opposed to hydrogenated-tallow substituents, produce more colours in the nanocomposite, which was related to unsaturation in the alkyl ammonium surfactant causing considerable polymer degradation. The kinetic parameters of PA-6/clay nanocomposite decomposition have also been studied.

Pramoda *et al.*[113] observed that the temperature required for the onset of degradation for PA-6 and 2.5% clay-filled nanocomposites was higher than other compositions (neat polymer, 5% and 7.5%, respectively). Gilman *et al.*[114] proposed that, with the higher loading of clay, the temperatures required for the onset of degradation remain unchanged, which was attributed to agglomeration in the nanocomposites. The presence of organoclay (for PA-6/2.5 wt% clay nanocomposite) increased the activation energy for degradation, $E_a$, compared with neat PA-6 under $N_2$. The major evolved gas products were cyclic monomers, hydrocarbons, $CO_2$, CO, $NH_3$ and $H_2O$ for PA-6 and PA-6/clay nanocomposites. During flammability measurements with a calorimeter in conjunction with an FTIR spectrometer in real time, changes in the condensed phase of PA-6 and a PA-6/clay nanocomposite revealed that the spectra obtained during burning are of sufficiently high quality to show the progression of the material in contact with the probe from a molten polymer to thermal degradation products. The spectral features are consistent with the evolution of caprolactam as a result of depolymerization.

In an intumescent ethylene vinyl acetate (EVA)-based formulation, using a PA-6/clay nanocomposite instead of pure PA-6 (carbonization agent) has been shown to improve the fire properties of the intumescent blend. Using clay as a 'classical' filler gave the same level of FR performance in the first step of combustion as when directly using exfoliated clay in PA-6. But, in the second half of combustion, the clay destabilizes the system and increases flammability. Moreover, kinetic modelling of the degradation of the EVA-based formulations shows that adding clay to the blend gives the same mode of degradation and the same invariant parameters as for the PA-6/clay nanocomposite containing the intumescent blend. The increase in flammability of the standard flammability test paper, K-10, in the second half of combustion shows the advantages of using nanoclay rather than microclay in an intumescent system.[110]

The efficiency of the self-protective coatings, which form during pyrolysis and thermo-oxidative degradation (in the presence of oxygen), of a PA-6/clay nanocomposite has also been investigated.[114] The nanocomposite itself can be protected from fire, flames and oxygen by coating the organosilicon thin films. PA-6 and PA-6/clay nanocomposite (PA-6 nano) substrates were coated by polymerizing A 1,1,3,3-tetramethyldisiloxane (TMDS) monomer doped with oxygen using the cold remote nitrogen plasma (CRNP) process. The thermal degradation of deposits under pyrolytic and thermo-oxidative conditions shows that the residual weight evolution with temperature depends on the chosen atmosphere.

Organically modified clay-reinforced PA-6 was subjected to accelerated heat aging to estimate its long-term thermo-oxidative stability and useful lifetime compared with the virgin material.[115] Changes in molecular weight and thermal and mechanical properties were monitored and considered to be due to the polymer modification encountered during aging. Generally, the strong interaction between the matrix and the clay filler renders the polymer chains, especially those adjacent to silicates, highly restrained mechanically, so that a significant portion of an applied force is transferred to the higher modulus silicates. This mechanism explains the enhancement of tensile modulus that the non-aged clay-reinforced PA-6 exhibited (1320 MPa) compared with the neat polymer (1190 MPa), as shown in Fig. 15.18.

*Enhancing and monitoring structural stability and durability*

In conventional fibre-reinforced polymers (FRPs), fibre-orientation is usually in-plane ($x$- and $y$-directions), resulting in fibre-dominated material properties in these directions, whereas the matrix dominates in the $z$-direction. Therefore, FRPs are very sensitive to intrinsic damage such as delamination (in particular), matrix cracking and fatigue damage. Moreover, their use as structural materials in aerospace applications has increased from less than 5% by structural weight to 50% (mainly due to the need to reduce weight and save fuel), which has

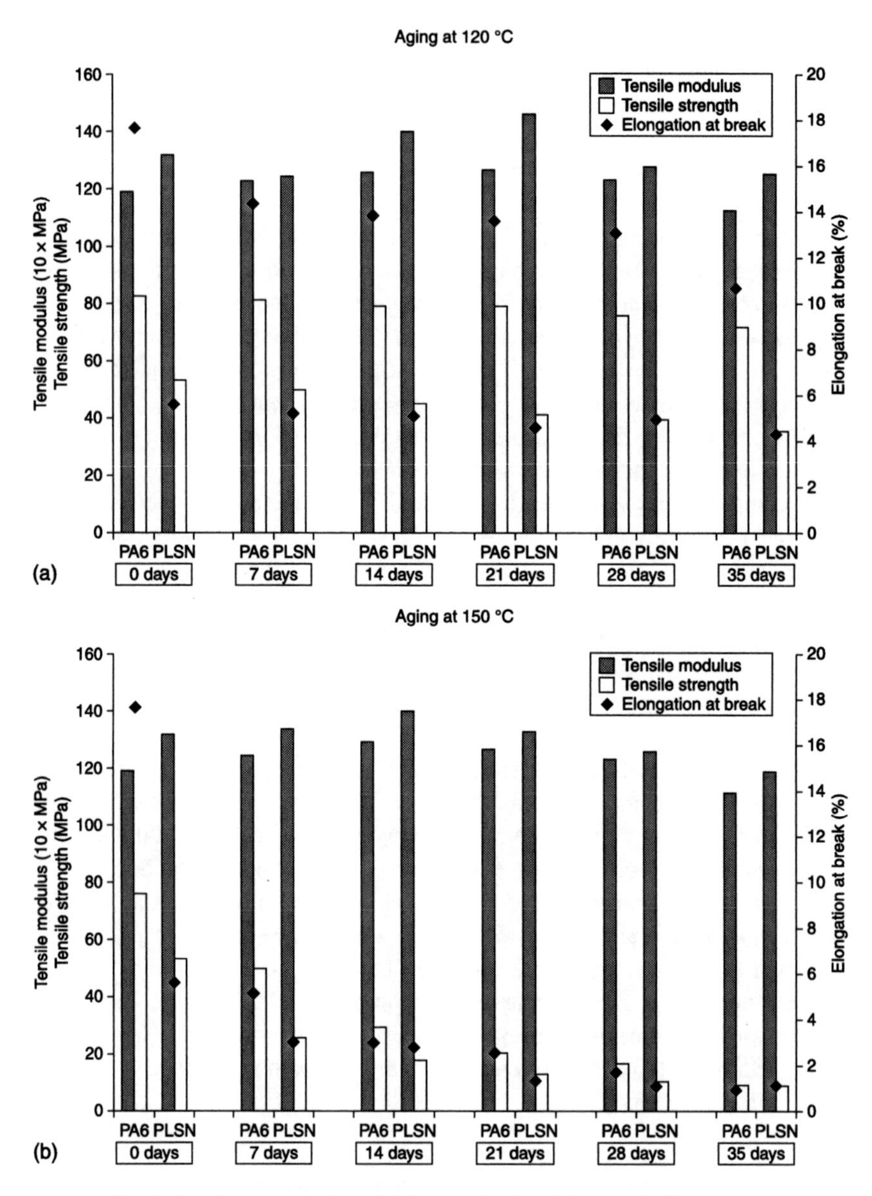


*15.18* Tensile properties of PA-6 and a PA-6 layered silicate nanocomposite (PLSN), oven-aged at (a) 120 °C and (b) 150 °C.[115]

raised many questions relating to vibration and acoustic damping, tribological behaviour and hygroscopic properties. Therefore, for enhanced structural stability, durability and performance, it is necessary to address and accommodate these undesirable issues.

Several approaches have been independently adopted to tackle these problems, which include: (a) using epoxy/elastomer blends at ply interfaces, (b) incorporating interleaved viscoelastic layers, (c) reducing the mismatch in elastic properties (and stress concentrations) at the interfaces between laminated plies and (d) using carbon filaments in place of interleaves. These approaches have their own positives and negatives as these materials are 'functional' rather than 'structural' and their high cost, low strain/stiffness, compatibility/adhesion with FRP laminates, or poor processability always results in a compromise.

A specific area of interest is in tuning an interleaved material to achieve the desired multifunctionalities or to enhance the desired properties compared with FRPs without interleaf layers.

In non-destructive evaluation, damage sensitivity and the reinforcing effects of carbon nanocomposites could be obtained from electrical volume and contact resistivity measurements using acoustic emission techniques. Adding conductive particles to an isolating polymer can result in an electrically conductive composite, if the particle concentration exceeds the percolation threshold, which is the particle volume fraction required to form a conductive network through the bulk polymer. Because of their high aspect ratios, CNTs are very effective fillers in the promotion of electrical conductivity in polymeric matrices. Percolation thresholds as low as 0.0025 wt% have been reported using MWNTs in an epoxy matrix.[116] The percolation threshold for the materials and the dispersion process used in another work was found to be below 0.1 wt%.[11] Conductivities of up to $2 \times 10^{-2}$ S/m were achieved with unfunctionalized CNTs. The functionalization of CNTs with amino groups usually increases the percolation threshold and decreases the maximum achievable conductivity, because the graphitic structure of the nanotubes incurs some defects during the functionalization process.

Furthermore, the average length of a nanotube is reduced, which explains the increasing percolation threshold.[11,117] Park *et al.*[118] explored this concept by applying the electro-micromechanical technique to obtain fibre damage and the reinforcing effect of carbon nanocomposites against content. The sensitivity to fibre damage such as fibre fracture, matrix deformation and fibre tension was highest for 2.0 vol% CNT composites. They suggested that damage sensitivity as measured by electrical resistance might be closely related to the 3D network structure, the percolation structure. For CNT composites, the mechanical properties and apparent modulus indicate that the reinforcing effect increased with CNT content. The researchers confirmed that the apparent modulus, as measured by electro-micromechanical tests, could be used to evaluate the mechanical properties of fibre-reinforced composites. The reinforcing effect on the mechanical properties and apparent modulus was highest for 2.0 vol% CNT, compared with percolation structures with 0.1 and 0.5 vol% CNT. Morphological trends were consistent with damage sensitivity based on electrical properties.

The optimal use of sensor technology is in itself crucial in the design, manufacture, maintenance and correct functioning of a number of aerospace- and defence-related adaptive structures and other strategic equipment. Electrode materials with carbon nanotubes perform better than traditional carbon electrodes, with good conducting ability and high chemical stability. Another possible application could use the inherent multifunctionality of CNT-based materials for designs that are self-sensing – sensor skins capable of probing the environment around a vehicle could be designed so that they are part of the vehicle itself. Their multifunctionality arises because nanotubes can be either metallic or semiconducting based on their chirality. Hence, SWNTs have been used to fabricate several nanoscale devices, such as field-effect transistors and molecular logic devices. Furthermore, simply changing the environment around a nanotube can change its conducting behaviour. This phenomenon has been exploited to create sensors capable of measuring several parameters for a vehicle's structural health (i.e. strain, pressure, temperature, etc.). For aerospace systems, innovative lighter and smaller sensors can be used for enhanced real-time prognostic health monitoring and the diagnosis of aerospace and military structures, such as in strategic or tactical transportation and weapon systems. However, as mentioned above, there are concerns that the current maintenance techniques cannot meet the growing demand for high reliability and readiness. New, advanced, built-in diagnostic techniques have to be developed, which can perform damage diagnosis automatically and without human error; provide advanced warning of structural failure to pilots or operators; minimize unnecessary downtime for scheduled maintenance; maintain reliability and improve the safety of aging structures; reduce maintenance costs and enhance combat readiness.

### 15.4.5 Impact resistance and energy absorption

Damage due to low- and high-velocity impact events weakens the structure of composite materials due to a continuous service load. Furthermore, an impact may generate different types of flaws before full perforation, i.e. sub-surface delamination, matrix cracks, fibre debonding or fracture, indentation and barely visible impact damage (BVID). Over time, these effects can induce variations in the mechanical properties of composite structures (the primary effect of a delamination is to change the local value of the bending stiffness and the transverse-shear stiffness), leading to possible catastrophic failure. It has been reported that the energy-absorption capability and related properties of polymer matrices can be engineered by adding nanoscale fillers. For example, rigid nano-sized particles such as $SiO_2$, $TiO_2$, $CaSiO_3$, $Al_2O_3$ powder, CNTs and clay nanoplatelets have been used, and some important findings are summarized in this section.

Typical fillers for the reinforcement of polymer matrices are particles (e.g. silica or aluminium oxide particles), tubes (e.g. nanofibres or nanotubes) and plates (e.g. nanoclay platelets). Significant enhancement of the impact strength of polymeric nanocomposites has been achieved by adding amino-functionalized

MWCNTs or small amounts of single-walled carbon nanotube (SWCNTs). The sensitivity of FRPs to intrinsic damage (delamination, matrix cracking and fatigue) and their minimal multifunctionality requires a significant effort to improve their performance to meet spacecraft application standards. Currently, however, there are many questions surrounding the incorporation of nanoparticles, like CNTs and CNFs, in FRPs with regard to manufacturing methodology and structural integrity. The toughening of particle-modified semi-crystalline polymers is related to the inter-particle distance, *s*, for any type of added particle. The distance *s* depends on both the concentration, *u*, and the average size, *d*, of the particles.

A number of experiments have shown that fracture toughness improved with the addition of clay nanoplatelets to epoxy when the clay nanoplatelets were not fully exfoliated and intercalated clay nanoplatelets were present. Such very high improvements are not typically observed for composites reinforced with conventional micro-particles. Subramaniyan and Sun[44] reported that core/shell rubber (CSR) nanoparticles, having a soft rubber core and a glassy shell, improved the fracture toughness of an epoxy vinyl ester resin significantly more than MMT nanoclay particles with the same weight fraction. However, hybrid blends of CSR and nanoclay were found to give the best balance of toughness, modulus and strength. The same investigators highlighted that, when the nanoclay particles were used to enhance the polymer matrix in a conventional glass-fibre-reinforced composite, the interlaminar fracture toughness of the composite was less than that of the unreinforced composite. It was suggested that a possible reason for this result was the arrangement of the nanoclay particles along the fibre axis.

Two factors affect the capacity of rigid particles for energy dissipation at high loading rates:

- the ability of the dispersed particles to detach from the matrix and to initiate the local matrix shear in the vicinity of voids and
- the size of the voids.

Therefore, the optimal minimal rigid-particle size for polymer toughening should fulfil two main requirements: to be smaller than the critical size for polymer fracture and also to have a debonding stress that is small compared with the polymer matrix yield stress.

Most of the studies that noted an increase in toughness by incorporating CNTs in polymers (both thermosets and thermoplastics) have attributed it mainly to the nanotube pull-out mechanism and the bridging of cracks in the matrix. The nanotube pull-out mechanism was inspired by conventional polymer/fibre composites where fibre/matrix debonding and fibre pull-out (including work done against sliding friction in pulling out the fibre) govern the extent of energy absorption. Hence, the very high interfacial areas in polymer/nanotube composites are expected to result in drastic improvements in the work necessary for fracture due to nanotube pull-out. To explain and confirm the pull-out mechanism, Barber and co-workers[119] studied the pull-out of individual nanotubes by attaching them to the end of a scanning probe microscope (SPM) tip and pushing them into a liquid epoxy polymer (or liquid melt of polyethylene-butene). After the polymer had solidified, the nanotube was pulled out and the forces were recorded from the deflection of the SPM tip cantilever. Although this provided an estimate of the interfacial strength of individual nanotubes, it is not directly relevant to pull-out toughness measurement, which depends on many factors like the alignment/ orientation and flexible/entanglement nature of the nanotubes. Even by increasing the embedded length of the nanotube in the resin, the nanotube breaks instead of being pulled out from the polymer.

Nanotube-reinforced structures have up to eight times higher tensile strength and advanced energy dissipation mechanisms, so more damping can be achieved with smaller and lighter structural designs. The number of CNT walls and their size affects stress concentration in the composite, and thus short and even round particles are strongest (like diamonds, etc.). However, longer fibres are flexible and may be better for damping. A CNT may act as a nanoscale spring and a crack-trapping material in composites. These damping phenomena could be multiplied when CNTs are dispersed. Orientation and geometry (waviness) of CNT particles may affect the mechanisms of energy dissipation and fracture mechanics. Maximum stiffness is achievable when the longitudinal orientation of CNTs is 90°. Notably, open-end CNTs do not collapse, fail or buckle due to higher stress concentrations, while many researchers have used closed-end CNT-reinforced composites. Thus isolated SWNTs may be desirable for damping applications due to their significant load-bearing ability because of CNT/matrix interactions. Defects of the carbon particles likely limit performance. In future, CNTs may serve as storage containers for sealant or multi-purpose particles of another material.

A Kireitseu *et al.*[120] study on the rotating fan blades of turbine engines represents another feasible aerospace or defence application. The authors considered a large rotating civil engine blade, illustrated in Fig. 15.19, which is typically hollow and usually has stiff rib-like metallic structures in order to increase rigidity and maintain the cross-sectional profile of the blade. They suggested a foam-filled fan where the metal structure or traditional fillers are replaced with a CNT-reinforced syntactic foam and, also, with a CNT-reinforced composite layer on the top. Results of the damping behaviour and impact toughness of the composite sandwiches showed that CNT-reinforced samples have better impact strength and vibration-damping properties over a wide temperature range. Experiments conducted using a vibrating clamped beam with the composite layers indicated up to a 200% increase in the inherent damping level and a 30% increase in the stiffness with some decrease (20–30%) in the density of the composite. The cross-links between the nanotubes and composite layers also served to improve the load transfer within the network, resulting in improved stiffness properties. The critical issues to be considered include: the choice of nanotubes and related matrix

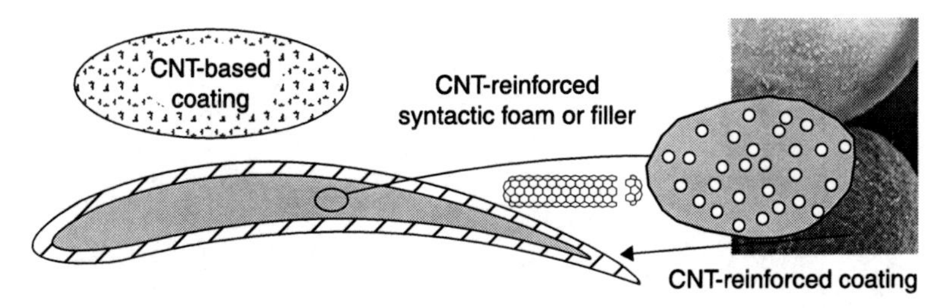


*15.19* Damping material for an aircraft engine blade.[120]

materials for vibration damping; tailoring the nanotube/matrix interface with respect to the matrix; and the orientation, dispersion and bonding of the nanotubes in the matrix. It is anticipated that significant weight, thickness and manufacturing cost reductions could be achieved in this way.

The critical parts in aerospace vehicles depend on both the strength and toughness of the materials they are made of, while the associated technology places strict limitations on the weight of the different components. Notably, matrix toughening may be performed through chemical modification or, more recently, with the incorporation of fillers in the matrix material. An interface can be modified by grafting in order to tailor the chemical compatibility between the fibres and the matrix. In this way, despite their size, nanofillers can have a significant presence in a small zone. In contrast, only a few micro-particles participate in plastic zone deformation. It then follows that nanofillers can lead to better fracture properties of a brittle matrix, resulting in enhancement of the matrix fracture toughness, which can lead to an overall advanced fracture behaviour. Additionally, by involving more nanofibres during delamination the nanofibres can act as reinforcing fibres, thus increasing fracture toughness. These approaches may offer alternatives to the conventional interlaminar toughness improvement of fibre-reinforced composites (especially high-performance, high-temperature composites), which has been addressed in various ways, such as by tailoring the matrix or interface properties, stitching, Z-pinning or interleaving. These approaches also improve mechanical properties, such as fatigue life. Coiled CNTs (Fig. 15.20) are another option for advanced composite reinforcement technology.

Traditional straight carbon nanotubes are recognized as strong reinforcements, which substantially enhance the strength of composites. However, the drawbacks, such as the decrease of fracture toughness and subsequent increase of the brittleness of the composites, may restrict the utilization of these reinforcements for some composite structures. Coiled nanotubes are ideal reinforcements for composite and polymer-based materials. These reinforcements can provide moderate strength improvement as well as enhancing the fracture toughness of the composites without substantially increasing the weight and damaging the nanotube structures.

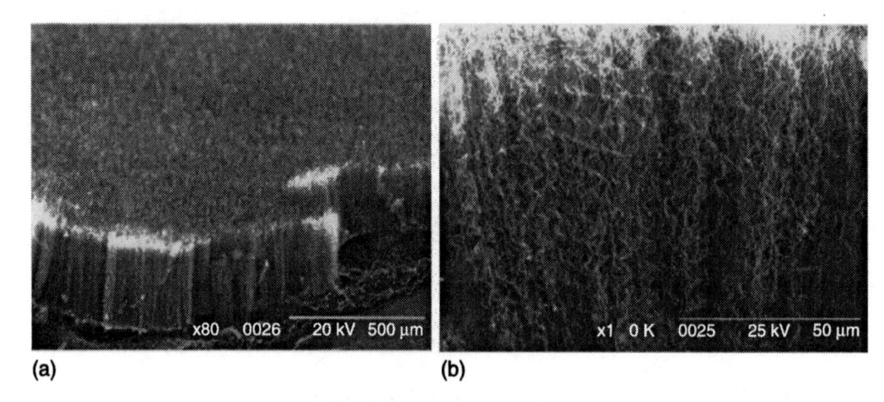


*15.20* (a) Typical SEM image of an aligned coiled nanotube array. (b) Enlarged cross-section of densely packed aligned coiled nanotubes.[121]

## 15.5 Future trends: opportunities and challenges

The use of polymer nanocomposites in structures has several predictable benefits for aerospace structures, primarily by providing safer, faster and eventually cheaper transportation in the future.[122–124] The most obvious of these is significant airframe weight reduction stemming from the low density of polymer nanocomposites and complemented by their high strength and modulus. However, the application to real products is in its infancy and much remains to be achieved.

Experiments on a fully integrated nanotube composite using CNTs have demonstrated a dramatic enhancement of mechanical properties. To demonstrate potential airframe applications, O'Donnell *et al.*[125] conducted a mass analysis study on CNT-reinforced polymer-structured aircraft. The analysis considered a notional Boeing 747-400 and 757-200, an Airbus A320 and an Embraer E145 with a CNT-reinforced polymer (CNRP) as the primary structural material. The entire volume of structural aluminium was replaced with CNT-reinforced polymer, without any modifications to the geometry or design of the airframe. Each airframe modelled saw an average 17.32% weight reduction in the low initial take-off mass category with a minimum mass reduction in the high initial take-off mass category of over 10%. The average fuel saving for all CNRP-structured airframes was 9.8%. Though the probability of CNT-reinforced polymer-structured substitution in existing airframes is unlikely, this type of analysis provides an insight into a small group of benefits when a nano-structured material is applied on the macroscale.

By combining them with conductive polymers, such as PANI, nanotubes can be used for sensitive electrochemical sensors. It has been observed that, with an increase in nanotube concentration, the conductivity of PANI/nanotube films and the current level in the metal/semiconductor devices increases, even at elevated temperatures. Research has also proved that the electrical conductivity increased with the number of nanotubes used in the epoxy-based materials. Volodin *et al.*[126] used coiled nanotubes as a mechanical resonant sensor, which is a significant achievement among many promising applications.

CNTs have tremendous promise in improving the performance of power devices such as thin-film polymeric solar cells, direct-methanol fuel cells, lithium-ion ($Li^+$) batteries, ultra-capacitors and thermionic power supplies.[127] Exceeding the state-of-the-art performance of these devices by using carbon nanotubes, in some cases by merely substituting conventional materials with them, is no longer seen as a proposition but is rapidly becoming reality.

Various advanced composite coatings have been used to improve durability, reliability and performance of many aircraft and armoured-vehicle components. Generally speaking, coatings are significant in many potential aerospace applications. Multifunctional nanocoatings currently in development for aerospace applications include environmentally safe materials for corrosion sensing and protection, and using them on aircraft skins for sensing mechanical damage. Protective coatings for high-performance, high-temperature applications require the development of nanoparticles and nanocomposites, which are a combination of nanoparticles and matrix materials, to form the anti-wear and anticorrosion coatings. Conventional paints are often not only health hazards, but are also labour intensive to develop and to apply. Carbon nanotubes could be used to replace carbon black in powder paints as cost-effective conductive fillers for electrostatic spray painting of vehicular body parts, leading to more durable, corrosion-resistant paints and surfaces. Other applications use the effect of different particles on strength, toughness, thermal transport and various stealth effects. Experimental results[128] have shown that replacing carbon black with CNTs improves skid resistance and reduces tyre abrasion. Other possible applications range from semiconductors, electronic memory, drive products and medical delivery systems to uses in plastics such as vehicle body panels, paint, tyres and as flame retardants in polyethylene and polypropylene.

Single-wall carbon nanotubes have recently been incorporated into poly (3-octylthiophene) (P3OT) to promote exciton dissociation and improve electron transport in a polymeric solar cell.[128] Conducting polymers like P3OT have attracted significant attention in the photovoltaic community since they produce excitons (quantum mechanical particles consisting of bound electron-hole pairs) after optical absorption. When the polymer is placed in a properly structured device with a suitable dopant to promote dissociation of the exciton and conduction of the electron, hole conduction proceeds through the polymer and contributes to an overall photocurrent. SWNTs are an appropriate dopant in these polymers since electron affinity is higher for a SWNT compared with P3OT and electron transport in a metallic SWNT is typical of a ballistic conductor. In addition, the high aspect ratio of a SWNT leads to low doping levels necessary to achieve a sufficient percolation network for high electrical conductivity, while retaining the mechanical properties required for a flexible polymeric device. Clearly, polymer/CNT nanocomposites are an alternative class of organic semiconducting materials, which promise improved performance for organic photovoltaic cells and devices. It is obviously beneficial to design photoconductive devices with a high efficiency in charge carrier generation.

Based on the fact that that the best way to achieve multifunctional properties in a polymer is to blend it with nanoscale fillers, several investigators have reported that the incorporation of nanoparticles in conventional FRPs enhances combinatorial properties at relatively low loadings of filler compared with their microscale and macroscale counterparts. It is thought that nanofillers (particularly high-aspect-ratio CNTs) increase matrix properties, especially in the *z*-direction, by infiltrating between the microscale fibres. Nevertheless, there are many discrepancies and uncertainties in the literature on the mechanical and functional properties and the manufacturing methodologies. It is also necessary to look at these factors from the viewpoints of both binary nanocomposites and ternary hybrids, to give an in-depth understanding and achieve the desired multifunctionality in these materials.[121, 130, 131]

Despite the excellent strength, modulus and electrical and thermal conductivities along with the low density of CNTs, their potential after incorporation into polymers has not been fully realized and the properties of the nanocomposites obtained are often below the predicted values. This is true even after discounting the difficulties associated with disaggregating and controlling their dispersion in polymers (due to their entangled, intertwined networks, high intermolecular van der Waals interactions, differences between single- and multi-walled nanotubes, which affects their surface areas and interfaces, etc.). Most studies to date have reported improvements in stiffness and strength, but toughness results are rather uncertain and require further property optimization studies.

The preparation of nanocomposites of polymers and nanofillers has interested researchers because of their flame retardancy and good mechanical and electrical properties. The dispersion of the fillers is currently difficult and mainly achievable by functionalizing CNTs. Reports on polymer/CNT nanocomposites have mainly focused on the functionalization of CNTs, their preparation and property developments. Unfortunately, or fortunately, the environmental durability of nanocomposites has yet to be studied.

Moreover, it is also important to consider the nanoscopic confinement effect of nanoparticles on the surrounding matrix and the matrix/nanoparticle interface as it has a direct effect on the properties of nanocomposites. Confinement is in turn determined by the size range and dimension of the fillers. For example, size differences between CNTs and CNFs affects their responses under stress conditions. In thermoplastic polymer nanocomposites, it has been noted that with high-aspect-ratio nanoscale fillers (like CNTs and clay) preferential orientation of the polymer lamellae occurs regardless of crystallographic lattice matching (between the polymer chain and the filler crystal).

While with CNFs lattice matching is a prerequisite, it has been revealed that the formation of epitaxially grown transcrystalline regions (due to crystallographic lattice matching) does not guarantee a strong interface between the fillers (where the size or dimensions are not nanoscale) and polymer chains. For example, a TEM micrograph of a polymer/CNF nanocomposite tensile stretched to an extension of 60% clearly revealed partial debonding of the CNFs along the interface. This, in fact, follows conventional rigid-particle toughening of polymers and is based on the interface between the nanoparticles and the polymer matrix. That is, particles must debond at the interface and create a free volume in the material on a submicron level to relieve the constraint adjacent to the crack tip. Similarly, in epoxy/rubber binary blends, cavitation of rubber particles occurs, which releases the high plastic constraint and activates large-scale plastic deformation of the surrounding matrix material. But, with CNTs and other 'nanoscale' fillers, this debonding may be hindered due to the strong tethering junctions with the matrix, ultimately resulting in a brittle failure.

Furthermore, in thermosets, the presence of nanoscale fillers is expected to influence the local cross-linking density (in the vicinity of the fillers) and melt viscosity, which in turn have a drastic effect on the bulk mechanical properties and also the glass transition and heat distortion temperatures of the composites. Hence, an in-depth understanding of the processing and structural properties of these materials is necessary to fully exploit the functional characteristics of CNTs and CNFs.

Researchers have incorporated nanoparticles in conventional structural materials generally using a two-step protocol.[132] Gojny *et al.*[10] used a conventional calendering technique to disperse carbon nanotubes in epoxy resins. This approach utilizes adjacent cylindrical rolls counter-rotating at different velocities to impart high shear stresses (Fig. 15.21).

The mixture must pass through the narrow gap (~5 μm) between the rotating cylinders and so intense uniform shearing is achieved over the entire volume of the material. The suspension of epoxy/nanoparticles was collected and RTM was used to infiltrate the fibre fabric. An electrical field was applied in some cases during curing to enhance orientation of the nanofillers in the *z*-direction. It should be noted that conductivity in the in-plane direction was found to be of the order of $1 \times 10^{-2}$ S/m, while the conductivity in the *z*-direction was $\sim 1 \times 10^{-6}$ S/m (and $\sim 3 \times 10^{-5}$ S/m with an applied electric field during curing), which is still in the antistatic range. The reserachers reported that the interlaminar shear strengths of hybrid materials significantly improved with only 0.3 wt% of CNTs (multi-walled) compared with FRPs without nanoparticles.

A number of studies have revealed that the stiffness, compressive strength and even interfacial/interlaminar shear strength of ternary composites (with CNTs, CNFs or carbon black) were improved compared with binary laminates.[133] Despite lacking any strong evidence, it is thought that the cracks due to fibre rupture or interfacial debonding and even interlaminar delamination could be bridged by

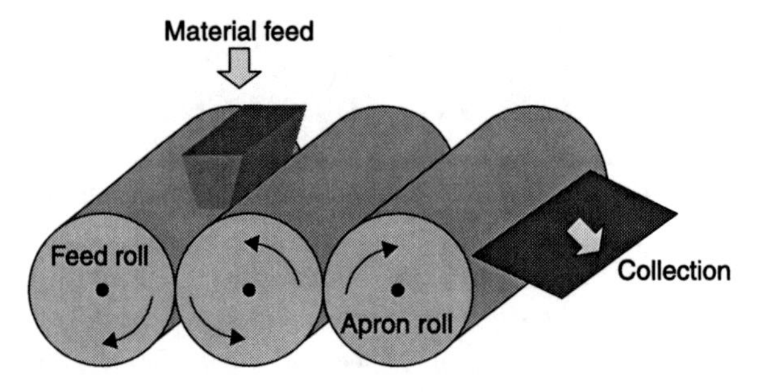


*15.21* Calendering using three-roll milling.[10]

CNTs or CNFs, thus improving the interfacial strength and toughness. Another important factor to consider is the type of resin and hardener. Differences in these have been observed to give dramatic variations in mechanical properties of some hybrid ternary nanocomposites at similar loadings of fillers. A significant number of studies have reported that improved compressive strength, lessening of microcracking and improved interlaminar fracture toughness have been achieved compared with binary conventional FRPs. Although this reinforces the benefits of incorporating nanoparticles in conventional FRPs, the majority of these investigations are preliminary and do not provide an in-depth understanding of the various factors, such as compatibility of the three phases, their interfaces, microstructural variations, nanoparticle network formation, percolation threshold variations or even the mechanisms of damage. But it is of the utmost importance to understand these factors in order to predict the long-term durability of these materials, which is necessary for their successful industrialization.

Overall there is an essential requirement to investigate the durability of these nanocomposites in different environmental conditions to extend the applicability of these hybrid materials. Further, the current limited knowledge of the degradation mechanism of polymer nanocomposites has led to the development of more efficient stabilizers for improving product performance on one hand and, on the other, the development of sensitizers to produce degradable polymers and to preserve the environment. The degradation of polymers has harmful as well as beneficial aspects. If unchecked it can play havoc with the performance of polymer nanocomposites, causing a fire safety hazard or toxicity, but if properly harnessed it can be used for producing new and better materials. It will be worthwhile to focus further studies on nanoparticulates such as clay, carbon nanotubes, metal oxides and metal salts, considering degradation in the environment where they are to be utilized, for example for specific aerospace and military applications. Despite the positive results of much previous research, it has been indicated that nanotubes could be more toxic than other carbon particles or quartz dust when absorbed into lung tissue.[134] Fundamental to the success of nanotechnology will be its perceived safety by the public. Many concerns have been aired concerning the use of manufactured nanoparticles.

## 15.6 Conclusions

Fibre-reinforced composites are a type of engineered material, which exhibits high strength/weight and modulus/weight ratios compared with some metallic materials. In the last two decades, studies have shown the potential improvement in properties and performance of fibre-reinforced polymer matrix materials in which nanoscale and microscale particles have been incorporated. The technology of nanoscale and microscale particle reinforcement can be categorized as inorganic layered clay technology, single-walled and multi-walled carbon nanotube technology, carbon nanofibre technology and metal particle technology. To date, the nanoparticle reinforcement of fibre-reinforced composites has been shown to be a possibility, but much work remains in order to understand how nano-reinforcement results in major changes in material properties. An understanding of these phenomena will facilitate their extension to the reinforcement of more complicated anisotropic structures and advanced polymeric composite systems.

One of the technological drawbacks is that mechanical reinforcement through the application of nanoparticles as a structural element in polymers is more difficult to realize and still remains a challenging task. Efficient exploitation of nanoparticle properties in order to improve material performance is generally related to the degree of dispersion, impregnation in the matrix and interfacial adhesion. The advantage of nanoscale compared with microscale fillers is their enormous surface area, which can act as an interface for stress transfer. The methods published so far on the improvement of the mechanical properties of polymer composites have mainly focused on optimizing the manufacturing processes for the composites, i.e. by using prepared nanoparticles. The property and performance enhancements made possible by nanoparticle reinforcement may be of great use for carbon- or glass-fibre/epoxy composites used in high-performance and high-temperature applications such as aerospace engine components and nacelles, the storage of cryogenic liquids and motorsports. However, precautionary measures should be observed at high temperatures since the structure and properties of these materials can change radically when they are exposed to extreme temperatures, especially in a cyclical fashion.

Better dispersion of the nano-sized, silicate-based filler in matrix systems is expected to yield improved material properties in several areas. Various mechanical properties, specifically improved fracture toughness as well as improved flame-retardant effects, are of interest. A key objective of ongoing worldwide research is investigating whether a nano-modified matrix yields improved delamination resistance in a fibre-reinforced laminate compared with a laminate with a neat polymer as the matrix material. It is necessary, however, to evaluate the damage resistance characteristics of three-phase nanocomposite laminates with regard to their applicability as structural elements. Further, as mentioned above, another complication is that the microstructure of semi-crystalline polymer matrices is influenced not only by the processing history but also by the presence of nanoparticles. The addition of various types of carbon nanotubes, nanofibres or nanoclays to polymers has already been observed to influence the crystallization kinetics and resulting morphology. Such changes in matrix morphology need to be considered when evaluating nanocomposite performance with regard to the intrinsic filler properties. The effects of nanoparticles on such oriented polymer systems, although significant, have not yet been fully established. In addition, as mentioned above, the presence of additives such as colouring pigments has been shown to influence matrix morphology during fibre spinning, and there is the whole technology of nucleating agents, which are deliberately added to influence crystalline microstructure.

Further, through nanoparticle reinforcement, an electrically conductive matrix could provide enhanced possibilities including stress–strain monitoring or damage detection. The application of an electrical field is known to orient the nanoparticles in the in-field-direction, which may result in increased efficiency of the *z*-reinforcement of the laminates. As a further benefit, electrical conductivity in the *z*-direction should increase with this approach. Nanoparticles, especially CNTs, have a high potential in the modification of polymers. They are very effective fillers regarding mechanical properties, especially toughness. Moreover, they allow the introduction of functional properties, which are connected to their electrical conductivity, into polymeric matrices. The electro-micromechanical technique has been studied as an economical non-destructive evaluation method for damage sensing and the characterization of interfacial properties because conductive fibre can act as a sensor itself as well as acting as a reinforcing fibre.

## 15.7 Acknowledgements

Financial support from the Research Councils United Kingdom (RCUK) through Grant No. EP/C509277/1 is acknowledged.

## 15.8 References

1. Gong, X., Liu, J., Baskaran, S., Voise, R. D. and Young, J. S. (2000), 'Surfactant-assisted processing of carbon nanotube/polymer composites', *Chemistry of Materials*, vol. 12, no. 4, pp. 1049–1052.
2. Shaffer, M. S. P. and Windle, A. H. (1999), 'Fabrication and characterization of carbon nanotube/poly(vinyl alcohol) composites', *Advanced Materials*, vol. 11, no. 11, pp. 937–941.
3. Park, C., Ounaies, Z., Watson, K. A., Crooks, R. E., Smith, J., *et al.* (2002), 'Dispersion of single wall carbon nanotubes by *in situ* polymerization under sonication', *Chemical Physics Letters*, vol. 364, no. 3–4, pp. 303–308.

4. Martin, C. A., Sandler, J. K. W., Windle, A. H., Schwarz, M.-K., Bauhofer, W., *et al.* (2005), 'Electric field-induced aligned multi-wall carbon nanotube networks in epoxy composites', *Polymer*, vol. 46, no. 3, pp. 877–886.
5. Shi, D., He, P., Lian, J., Chaud, X., Bud'ko, S. L., *et al.* (2005), 'Magnetic alignment of carbon nanofibers in polymer composites and anisotropy of mechanical properties', *Journal of Applied Physics*, vol. 97, pp. 64312–1.
6. Shi, D., He, P., Lian, J., Chaud, X., Beaugnon, E., *et al.* (2004), 'Plasma coating and magnetic alignment of carbon nano fibers in polymer composites', *JOM*, vol. 56, no. 1, pp. 129–130.
7. Olek, M., Kempa, K., Jurga, S. and Giersig, M. (2005), 'Nanomechanical properties of silica-coated multiwall carbon nanotubes-poly(methyl methacrylate) composites', *Langmuir*, vol. 21, no. 7, pp. 3146–3152.
8. Star, A., Stoddart, J. F., Steuerman, D., Diehl, M., Boukai, A., *et al.* (2001), 'Preparation and properties of polymer-wrapped single-walled carbon nanotubes', *Angewandte Chemie – International Edition*, vol. 40, no. 9, pp. 1721–1725.
9. Bhattacharyya, S., Sinturel, C., Salvetat, J. P. and Saboungi, M.L. (2005), 'Protein-functionalized carbon nanotube-polymer composites', *Appl. Phys. Lett.*, vol. 86, no. 11, pp. 113104.
10. Gojny, F. H., Wichmann, M. H. G., Köpke, U., Fiedler, B. and Schulte, K. (2004), 'Carbon nanotube-reinforced epoxy-composites: enhanced stiffness and fracture toughness at low nanotube content', *Composites Science and Technology*, vol. 64, no. 15, pp. 2363–2371.
11. Gojny, F. H., Wichmann, M. H. G., Fiedler, B., Bauhofer, W. and Schulte, K. (2005), 'Influence of nano-modification on the mechanical and electrical properties of conventional fibre-reinforced composites', *Composites. Part A: Applied Science and Manufacturing*, vol. 36, no. 11, pp. 1525–1535.
12. Wichmann, M. H. G., Sumfleth, J., Gojny, F. H., Quaresimin, M., Fiedler, B. and Schulte, K. (2006), 'Glass-fibre-reinforced composites with enhanced mechanical and electrical properties – Benefits and limitations of a nanoparticle modified matrix', *Engineering Fracture Mechanics*, vol. 73, no. 16, pp. 2346–2359.
13. Zhao, D.L., Qiao, R.-H., Wang C.-Z. and Shen, Z.M. (2006), 'Microstructure and mechanical property of carbon nanotube and continuous carbon fiber reinforced epoxy resin matrix composites', in *Proceedings of the Asian International Conference on Advanced Materials (AICAM 2005)*, 3–5 November 2005, Trans Tech Publications, Beijing, China, pp. 517.
14. Hsiao, K., Alms, J. and Advani, S. G. (2003), 'Use of epoxy/multiwalled carbon nanotubes as adhesives to join graphite fibre reinforced polymer composites', *Nanotechnology*, vol. 14, no. 7, pp. 791–793.
15. Meguid, S. A. and Sun, Y. (2004/6), 'On the tensile and shear strength of nano-reinforced composite interfaces', *Materials & Design*, vol. 25, no. 4, pp. 289–296.
16. Hussain, M., Nakahira, A. and Niihara, K. (1996), 'Mechanical property improvement of carbon fiber reinforced epoxy composites by $Al_2O_3$ filler dispersion', *Materials Letters*, vol. 26, no. 3, pp. 185–191.
17. Mahfuz, H., Adnan, A., Rangari, V. K., Jeelani, S. and Jang, B. Z. (2004), 'Carbon nanoparticles/whiskers reinforced composites and their tensile response', *Composites. Part A: Applied Science and Manufacturing*, vol. 35, no. 5, pp. 519–527.
18. Iwahori, Y. and Ishikawa, T. (2002), 'Mechanical properties of CFRP using dispersed CNF resin', in *Proceedings of the Third Japan–Korea Joint Symposium on Composite Materials*, vol. 191, no. 192.

19. Iwahori, Y., Ishiwata, S., Sumizawa, T. and Ishikawa, T. (2005), 'Mechanical properties improvements in two-phase and three-phase composites using carbon nano-fiber dispersed resin', *Composites. Part A: Applied Science and Manufacturing*, vol. 36, no. 10, pp. 1430–1439.
20. Yokozeki, T., Iwahori, Y. and Ishiwata, S. (2007), 'Matrix cracking behaviors in carbon fiber/epoxy laminates filled with cup-stacked carbon nanotubes (CSCNTs)', *Composites. Part A: Applied Science and Manufacturing*, vol. 38, no. 3, pp. 917–924.
21. Spindler-Ranta, S. and Bakis, C. E. (2002), 'Carbon nanotube reinforcement of a filament winding resin', in *SAMPE 2002*, 12–16 May, Long Beach, CA, p. 13.
22. Veedu, V. P., Cao, A., Li, X., Ma, K., Soldano, C., *et al.* (2006), 'Multifunctional composites using reinforced laminae with carbon-nanotube forests', *Nature Materials*, vol. 5, no. 6, pp. 457–462.
23. Usuki, A., Kawasumi, M., Kojima, Y. and Okada, A. (1993), 'Swelling behavior of montmorillonite cation exchanged for $\omega$-amino acids by $\varepsilon$-caprolactam', *Journal of Materials Research*, vol. 8, pp. 1174–1178.
24. Gao, F. (2004), 'Clay/polymer composites: the story', *Materials Today*, vol. 7, no. 11, pp. 50–55.
25. Haque, A., Shamsuzzoha, M., Hussain, F. and Dean, D. (2003), 'S2-glass/epoxy polymer nanocomposites: manufacturing, structures, thermal and mechanical properties', *Journal of Composite Materials*, vol. 37, no. 20, pp. 1821–1837.
26. Chowdhury, F. H., Hosur, M. V. and Jeelani, S. (2006), 'Studies on the flexural and thermomechanical properties of woven carbon/nanoclay-epoxy laminates', *Materials Science and Engineering: A*, vol. 421, no. 1–2, pp. 298–306.
27. Lin, L., Lee, J., Hong, C., Yoo, G. and Advani, S. G. (2006), 'Preparation and characterization of layered silicate/glass fiber/epoxy hybrid nanocomposites via vacuum-assisted resin transfer molding (VARTM)', *Composites Science and Technology*, vol. 66, no. 13, pp. 2116–2125.
28. Aktas, L., Hamidi, Y. K. and Altan, M. C. (2004), 'Characterisation of nanoclay dispersion in resin transfer moulded glass/nanoclay/epoxy composites', *Plastics, Rubber and Composites*, vol. 33, no. 6, pp. 267–272.
29. Gilbert, E. N., Hayes, B. S. and Seferis, J. C. (2002), 'Metal particle modification of composite matrices for customized density applications', *Polymer Composites*, vol. 23, no. 1, pp. 132–140.
30. Gilbert, E. N., Hayes, B. S. and Seferis, J. C. (2002), 'Variable density composite systems constructed by metal particle modified prepregs', *Journal of Composite Materials*, vol. 36, no. 17, pp. 2045–2060.
31. Timmerman, J. F., Hayes, B. S. and Seferis, J. C. (2002), 'Nanoclay reinforcement effects on the cryogenic microcracking of carbon fiber/epoxy composites', *Composites Science and Technology*, vol. 62, no. 9, pp. 1249–1258.
32. Brunner, A. J., Necola, A., Rees, M., Gasser, P., Kornmann, X., *et al.* (2006), 'The influence of silicate-based nano-filler on the fracture toughness of epoxy resin', *Engineering Fracture Mechanics*, vol. 73, no. 16, pp. 2336–2345.
33. Schmidt, H. (1985), 'New type of non-crystalline solids between inorganic and organic materials', *Journal of Non-Crystalline Solids*, vol. 73, no. 1–3, pp. 681–691.
34. Mark, J. E. (1996), 'Ceramic-reinforced polymers and polymer-modified ceramics', *Polymer Engineering and Science*, vol. 36, no. 24, pp. 2905–2920.
35. Chisholm, N., Mahfuz, H., Rangari, V. K., Ashfaq, A. and Jeelani, S. (2005), 'Fabrication and mechanical characterization of carbon/SiC-epoxy nanocomposites', *Composite Structures*, vol. 67, no. 1, pp. 115–124.

36. Mohan, R. V., Kelkar, A. D. and Akinyede, O. (2005), 'VARTM processing and characterization of composite laminates from epoxy resins dispersed with alumina particles', in *50th International SAMPE Symposium and Exhibition*, 1–5 May 2005, vol. 50, Soc. for the Advancement of Material and Process Engineering, Covina, CA 91724–3748, United States, Long Beach, CA, United States, pp. 2425.
37. Kornmann, X., Rees, M., Thomann, Y., Necola, A., Barbezat, M. and Thomann, R. (2005), 'Epoxy-layered silicate nanocomposites as matrix in glass fibre-reinforced composites', *Composites Science and Technology*, vol. 65, no. 14, pp. 2259–2268.
38. Karaki, T., Killgore, J.P. and Seferis, J.C. (2004), *Characterization of Fatigue Behavior of Polynanomeric Matrix Composites*.
39. Wang, K., Chen, L., Wu, J., Toh, M. L., He, C. and Yee, A. F. (2005), 'Epoxy nanocomposites with highly exfoliated clay: Mechanical properties and fracture mechanisms', *Macromolecules*, vol. 38, no. 3, pp. 788–800.
40. Siddiqui, N. A., Woo, R. S. C., Kim, J., Leung, C. C. K. and Munir, A. (2007), 'Mode I interlaminar fracture behavior and mechanical properties of CFRPs with nanoclay-filled epoxy matrix', *Composites. Part A: Applied Science and Manufacturing*, vol. 38, no. 2, pp. 449–460.
41. Ragosta, G., Abbate, M., Musto, P., Scarinzi, G. and Mascia, L. (2005), 'Epoxy-silica particulate nanocomposites: Chemical interactions, reinforcement and fracture toughness', *Polymer*, vol. 46, no. 23, pp. 10506–10516.
42. Hayes, B. S., Nobelen, M., Dharia, A. K., Seferis, J. C. and Nam, J. (2001), 'Development and analysis of nano-particle modified prepreg matrices', in *33rd International SAMPE Technical Conference – Advancing Affordable Materials Technology*, vol. 33, 5–8 November 2001, Soc. for the Advancement of Material and Process Engineering, Seattle, WA, pp. 1050.
43. Subramaniyan, A. K., Bing, Q., Nakima, D. and Sun, C. T. (2003), In: *CD Proceedings of the 18th Annual Technical Conference of American Society for Composites*, Gainesville, FL, American Society for Composites, Paper 194.
44. Subramaniyan, A. K. and Sun, C. T. (2007), 'Toughening polymeric composites using nanoclay: Crack tip scale effects on fracture toughness', *Composites. Part A: Applied Science and Manufacturing*, vol. 38, no. 1, pp. 34–43.
45. Subramaniyan, A. K. and Sun, C. T. (2006), 'Enhancing compressive strength of unidirectional polymeric composites using nanoclay', *Composites. Part A: Applied Science and Manufacturing*, vol. 37, no. 12, pp. 2257–2268.
46. Hackman, I. and Hollaway, L. (2006), 'Epoxy-layered silicate nanocomposites in civil engineering', *Composites. Part A: Applied Science and Manufacturing*, vol. 37, no. 8, pp. 1161–1170.
47. Liu, W., Hoa, S. V. and Pugh, M. (2005), 'Fracture toughness and water uptake of high-performance epoxy/nanoclay nanocomposites', *Composites Science and Technology*, vol. 65, no. 15–16, pp. 2364–2373.
48. Ogasawara, T., Ishida, Y., Ishikawa, T., Aoki, T. and Ogura, T. (2006), 'Helium gas permeability of montmorillonite/epoxy nanocomposites', *Composites. Part A: Applied Science and Manufacturing*, vol. 37, no. 12, pp. 2236–2240.
49. Hatta, H. and Taya, M. (2006), 'Effective thermal conductivity of a misoriented short fiber composite', *Journal of Applied Physics*, vol. 58, no. 7, pp. 2478–2486.
50. Mori, T. and Tanaka, K. (1973), 'Average stress in matrix and average elastic energy of materials with misfitting inclusions', *Acta Metallurgica et Materialia*, vol. 21, no. 5, pp. 571–574.

51. Lincoln, D. M., Vaia, R. A., Wang, Z. G. and Hsiao, B. S. (2001/2), 'Secondary structure and elevated temperature crystallite morphology of nylon-6/layered silicate nanocomposites', *Polymer*, vol. 42, no. 4, pp. 1621–1631.
52. Fong, H., Liu, W., Wang, C. S. and Vaia, R. A. (2002), 'Generation of electrospun fibers of nylon 6 and nylon 6-montmorillonite nanocomposite', *Polymer*, vol. 43, no. 3, pp. 775–780.
53. Fornes, T. D. and Paul, D. R. (2003), 'Crystallization behavior of nylon 6 nanocomposites', *Polymer*, vol. 44, no. 14, pp. 3945–3961.
54. Li, L., Bellan, L. M., Craighead, H. G. and Frey, M. W. (2006), 'Formation and properties of nylon-6 and nylon-6/montmorillonite composite nanofibers', *Polymer*, vol. 47, no. 17, pp. 6208–6217.
55. Liang, Y., Xia, X., Luo, Y. and Jia, Z. 'Synthesis and performances of Fe2O3/PA-6 nanocomposite fiber', *Materials Letters*, in press, corrected proof.
56. Cadek, M., Le Foulgoc, B., Coleman, J. N., Barron, V., Sandler, J., *et al.* (2002), 'Mechanical and thermal properties of CNT and CNF reinforced polymer composites', in *AIP Conference Proceedings*, no. 63, pp. 562–565.
57. Wu, S., Wang, F., Ma, C. M., Chang, W., Kuo, C., *et al.* (2001), 'Mechanical, thermal and morphological properties of glass fiber and carbon fiber reinforced polyamide-6 and polyamide-6/clay nanocomposites', *Materials Letters*, vol. 49, no. 6, pp. 327–333.
58. Akkapeddi, M. K. (2000), 'Glass fiber reinforced polyamide-6 nanocomposites', *Polymer Composites*, vol. 21, no. 4, pp. 576–585.
59. Vlasveld, D. P. N., Bersee, H. E. N. and Picken, S. J. (2005), 'Nanocomposite matrix for increased fibre composite strength', *Polymer*, vol. 46, no. 23, pp. 10269–10278.
60. Vlasveld, D. P. N., Parlevliet, P. P., Bersee, H. E. N. and Picken, S. J. (2005), 'Fibre–matrix adhesion in glass-fibre reinforced polyamide-6 silicate nanocomposites', *Composites. Part A: Applied Science and Manufacturing*, vol. 36, no. 1, pp. 1–11.
61. van Rijswijk, K., Lindstedt, S., Vlasveld, D. P. N., Bersee, H. E. N. and Beukers, A. (2006), 'Reactive processing of anionic polyamide-6 for application in fiber composites: A comparative study with melt processed polyamides and nanocomposites', *Polymer Testing*, vol. 25, no. 7, pp. 873–887.
62. Sandler, J. K. W., Pegel, S., Cadek, M., Gojny, F., van Es, M., *et al.* (2004), 'A comparative study of melt spun polyamide-12 fibres reinforced with carbon nanotubes and nanofibres', *Polymer*, vol. 45, no. 6, pp. 2001–2015.
63. Jen, M. R., Tseng, Y. and Wu, C. (2005/4), 'Manufacturing and mechanical response of nanocomposite laminates', *Composites Science and Technology*, vol. 65, no. 5, pp. 775–779.
64. Sandler, J., Werner, P., Shaffer, M. S. P., Demchuk, V., Altstädt, V. and Windle, A. H. (2002), 'Carbon-nanofibre-reinforced poly(ether ether ketone) composites', *Composites. Part A: Applied Science and Manufacturing*, vol. 33, no. 8, pp. 1033–1039.
65. Schmidt, H. (1994), 'Multifunctional inorganic-organic composite sol-gel coatings for glass surfaces', *Journal of Non-Crystalline Solids*, vol. 178, pp. 302–312.
66. Lin, J. (1992), *In situ syntheses and phase behavior investigations of inorganic materials in organic polymer solid matrices*, PhD dissertation, Department of Chemistry, Pennsylvania State University.
67. Wang, Q., Xue, Q., Liu, W. and Chen, J. (2000), 'The friction and wear characteristics of nanometer SiC and polytetrafluoroethylene filled polyetheretherketone', *Wear*, vol. 243, no. 1–2, pp. 140–146.

68. Ogasawara, T., Ishida, Y., Ishikawa, T. and Yokota, R. (2004), 'Characterization of multi-walled carbon nanotube/phenylethynyl terminated polyimide composites', *Composites. Part A: Applied Science and Manufacturing*, vol. 35, no. 1, pp. 67–74.
69. Fu, H. J., Huang, Y. D. and Liu, L. (2004), 'Influence of fibre surface oxidation treatment on mechanical interfacial properties of carbon fibre/polyarylacetylene composites', *Materials Science and Technology*, vol. 20, no. 12, pp. 1655–1660.
70. Zhang, X., Huang, Y., Wang, T. and Liu, L. (2007), 'Influence of fibre surface oxidation–reduction followed by silsesquioxane coating treatment on interfacial mechanical properties of carbon fibre/polyarylacetylene composites', *Composites. Part A: Applied Science and Manufacturing*, vol. 38, no. 3, pp. 936–944.
71. Lin, Z., Ye, W., Du, K. and Zeng, H. (2001), 'Homogenization of functional groups on surface of carbon fiber and its surface energy', *Journal of Huaqiao University (Natural Science)*, vol. 22, no. 3, pp. 261–263.
72. Zhang, X., Huang, Y., Wang, T. and Hu, L. (2006), 'Influence of oligomeric silsesquioxane coatings treatment on the interfacial property of CF/PAA composites', *Acta Materiae Compositae Sinica*, vol. 23, no. 1, pp. 105–111.
73. Kumar, S., Dang, T. D., Arnold, F. E., Bhattacharyya, A. R., Min, B. G., *et al.* (2002), 'Synthesis, Structure, and Properties of PBO/SWNT Composites', *Macromolecules*, vol. 35, no. 24, pp. 9039–9043.
74. Njuguna, J., Michalowski, S., Pielichowski, K., Kayvantash, K. and Walton, A. C. (2010), 'Fabrication, characterisation and low-velocity impact on hybrid sandwich composites with polyurethane/layered silicate foam cores', *Polymer Composites*, in press.
75. Pandey, J. K., Raghunatha Reddy, K., Pratheep Kumar, A. and Singh, R. P. (2005), 'An overview on the degradability of polymer nanocomposites', *Polymer Degradation and Stability*, vol. 88, no. 2, pp. 234–250.
76. Kashiwagi, T., Du, F., Douglas, J. F., Winey, K. I., Harris Jr, R. H. and Shields, J. R. (2005), 'Nanoparticle networks reduce the flammability of polymer nanocomposites', *Nature Materials*, vol. 4, no. 12, pp. 928–933.
77. Leszczyńska, A., Njuguna, J., Pielichowski, K. and Banerjee, J. R. (2007), 'Polymer/montmorillonite nanocomposites with improved thermal properties: Part I. Factors influencing thermal stability and mechanisms of thermal stability improvement', *Thermochimica Acta*, vol. 453, no. 2, pp. 75–96.
78. Hwu, J. M., Jiang, G. J., Gao, Z. M., Xie, W. and Pan, W. P. (2002), 'The characterization of organic modified clay and clay-filled PMMA nanocomposite', *Journal of Applied Polymer Science*, vol. 83, no. 8, pp. 1702–1710.
79. Report of a DTI global watch mission (UK), December 2006 (2006), *Developments and trends in thermal management technologies – a mission to the USA*.
80. Zhao, Y., Tong, T., Delzeit, L., Kashani, A., Meyyappan, M. and Majumdar, A. (2006), 'Interfacial energy and strength of multiwalled-carbon-nanotube-based dry adhesive', *Journal of Vacuum Science & Technology B: Microelectronics and Nanometer Structures*, vol. 24, no. 1, pp. 331–335.
81. Johnson, R. C. (2006), 'Carbon-nanotube arrays take heat off chips', *Electronic Engineering Times*, no. 1423, pp. 38.
82. Xu, J. and Fisher, T. S. (2006), 'Enhancement of thermal interface materials with carbon nanotube arrays', *International Journal of Heat and Mass Transfer*, vol. 49, no. 9–10, pp. 1658–1666.

83. Hu, X. J., Padilla, A. A., Xu, J., Fisher, T. S. and Goodson, K. E. (2006), '3-omega measurements of vertically oriented carbon nanotubes on silicon', *Journal of Heat Transfer*, vol. 128, pp. 1109.
84. Bryning, M. B., Islam, M. F., Kikkawa, J. M. and Yodh, A. G. (2005), 'Very low conductivity threshold in bulk isotropic single-walled carbon nanotube–epoxy composites', *Advanced Materials*, vol. 17, no. 9, pp. 1186–1191.
85. Kim, Y. J., Shin, T. S., Choi, H. D., Kwon, J. H., Chung, Y. C. and Yoon, H. G. (2005), 'Electrical conductivity of chemically modified multiwalled carbon nanotube/epoxy composites', *Carbon*, vol. 43, no. 1, pp. 23–30.
86. Allaoui, A., Bai, S., Cheng, H. M. and Bai, J. B. (2002), 'Mechanical and electrical properties of a MWNT/epoxy composite', *Composites Science and Technology*, vol. 62, no. 15, pp. 1993–1998.
87. Shaffer, M. S. P. and Windle, A. H. (1999), 'Fabrication and characterization of carbon nanotube/poly (vinyl alcohol) composites', *Advanced Materials*, vol. 11, no. 11, pp. 937–941.
88. Sandler, J., Shaffer, M. S. P., Prasse, T., Bauhofer, W., Schulte, K. and Windle, A. H. (1999), 'Development of a dispersion process for carbon nanotubes in an epoxy matrix and the resulting electrical properties', *Polymer*, vol. 40, no. 21, pp. 5967–5971.
89. Thiem, C., Drager, S., Flynn, C., Renz, T. and Burns, D. (2004), 'Advanced computer technology for novel information processing paradigms', *Journal of Aerospace Computing, Information, and Communication*, vol. 1, no. 7, pp. 308–317.
90. Spores, R. and Birkan, M. (2002), 'Overview of USAF electric propulsion program', in *38th AIAA/ASME/SAE/ASEE Joint Propulsion Conference and Exhibit*, vol. AIAA-2002-3558, 7–10 July, Indianapolis, Indiana, Defense Technical Information Center.
91. Chapman, J. N., Ruoff, R. S., Chandrasekhar, V. and Dikin, D. (2003), 'Flightweight magnets for space application using carbon nanotubes', in *41st AIAA Aerospace Sciences Meeting & Exhibit*, Reno, NV.
92. Kim, C.-G., Lee, S.-E. and Kang, J.-H. (2006), 'Fabrication and design of multi-layered radar absorbing structures of MWNT-filled glass/epoxy plain-weave composites', *Composite Structures*, vol. 76, no. 4, pp. 397–405.
93. Krishnamoorti, R. and Vaia, R. A. (2001), 'Polymer nanocomposites (synthesis, characterization, and modeling)', *A.C.S. Series*, 804, pp. 15–25.
94. Mirfakhrai, T., Madden, J. D. W. and Baughman, R. H. (2007), 'Polymer artificial muscles', *Materials Today*, vol. 10, no. 4, pp. 30–38.
95. Li, D. and Xia, Y. (2004), 'Welding and patterning in a flash', *Nature Materials*, vol. 3, no. 11, pp. 753–754.
96. Xu, N. S. and Huq, S. E. (2005), 'Novel cold cathode materials and applications', *Materials Science and Engineering R: Reports*, vol. 48, no. 2, pp. 143.
97. Xu, N. S. and Huq, S. E. (2005), 'Novel cold cathode materials and applications', *Materials Science and Engineering: R: Reports*, vol. 48, no. 2–5, pp. 47–189.
98. Chen, J., Ren, H., Ma, R., Li, X., Yang, H. and Gong, Q. (2003), 'Field-induced ionization and Coulomb explosion of $CO_2$ by intense femtosecond laser pulses', *International Journal of Mass Spectrometry*, vol. 228, no. 1, pp. 81–89.
99. Pielichowski, K. and Njuguna, J. (2005), *Thermal Degradation of Polymeric Materials*, RAPRA Technologies Limited, Shawbury, Surrey, UK.

100. Njuguna, J. P. K. (2010), 'Ageing and performance predictions of polymer nanocomposites for exterior defence and aerospace applications', 11–12 March 2010, Hamburg, Germany, Smithers Rapra Technology.
101. Beyer, G. (2005), 'Flame retardancy of polymers by nanocomposites: a new concept', in *High Performance Fillers 2005*, Rapra Technology Ltd., Shrewsbury, UK, pp. 5–7.
102. Becker, O., Varley, R. and Simon, G. (2004), 'Thermal stability and water uptake of high performance epoxy layered silicate nanocomposites', *European Polymer Journal*, vol. 40, no. 1, pp. 187–195.
103. Jiang, L., He, S. and Yang, D. (2003), 'Resistance to vacuum ultraviolet irradiation of nano-$TiO_2$ modified carbon/epoxy composites', *Journal of Materials Research*, vol. 18, no. 3, pp. 654–658.
104. Nguyen, T., Pellegrin, B., Mermet, L., Gu, X., Shapiro, A. and Chin, J. (2009), *Degradation and nanofiller release of polymer nanocomposites exposed to UV*, National Institute of Standards and Technology, Gaithersburg, MD.
105. Lee, A. and Lichtenhan, J. D. (1998), 'Viscoelastic responses of polyhedral oligosilsesquioxane reinforced epoxy systems', *Macromolecules*, vol. 31, no. 15, pp. 4970–4974.
106. Barral, L., Díez, F. J., García-Garabal, S., López, J., Montero, B., *et al.* (2005), 'Thermodegradation kinetics of a hybrid inorganic–organic epoxy system', *European Polymer Journal*, vol. 41, no. 7, pp. 1662–1666.
107. Ozcelik, O., Aktas, L. and Altan, M. C. (2009), 'Thermo-oxidative degradation of graphite/epoxy composite laminates: Modeling and long-term predictions', *Polymer Letters*, vol. 3, no. 12, pp. 797–803.
108. Barkoula, N. M., Paipetis, A., Matikas, T., Vavouliotis, A., Karapappas, P. and Kostopoulos, V. (2009), 'Environmental degradation of carbon nanotube-modified composite laminates: a study of electrical resistivity', *Mechanics of Composite Materials*, vol. 45, no. 1, pp. 21–32.
109. Levchik, S. V., Weil, E. D. and Lewin, M. (1999), 'Thermal decomposition of aliphatic nylons', *Polymer International*, vol. 48, no. 6–7, pp. 532–557.
110. VanderHart, D. L., Asano, A. and Gilman, J. W. (2001), 'Solid-state NMR investigation of paramagnetic nylon-6 clay nanocomposites. 2. Measurement of clay dispersion, crystal stratification, and stability of organic modifiers', *Chemistry of Materials*, vol. 13, no. 10, pp. 3796–3809.
111. Davis, R. D., Gilman, J. W. and VanderHart, D. L. (2003), 'Processing degradation of polyamide 6/montmorillonite clay nanocomposites and clay organic modifier', *Polymer Degradation and Stability*, vol. 79, no. 1, pp. 111–121.
112. Fornes, T. D., Yoon, P. J. and Paul, D. R. (2003), 'Polymer matrix degradation and color formation in melt processed nylon 6/clay nanocomposites', *Polymer*, vol. 44, no. 24, pp. 7545–7556.
113. Pramoda, K. P., Liu, T., Liu, Z., He, C. and Sue, H. J. (2003), 'Thermal degradation behavior of polyamide 6/clay nanocomposites', *Polymer Degradation and Stability*, vol. 81, no. 1, pp. 47–56.
114. Gilman, J. W., Kashiwagi, T. and Lichtenhan, J. D. (1997), 'Nanocomposites: A revolutionary new flame retardant approach', *SAMPE Journal*, vol. 33, no. 4, pp. 40–46.
115. Kiliaris, P., Papaspyrides, C. D. and Pfaendner, R. (2009), 'Influence of accelerated aging on clay-reinforced polyamide 6', *Polymer Degradation and Stability*, vol. 94, no. 3, pp. 389–396.

116. Sandler, J. K. W., Kirk, J. E., Kinloch, I. A., Shaffer, M. S. P. and Windle, A. H. (2003), 'Ultra-low electrical percolation threshold in carbon-nanotube-epoxy composites', *Polymer*, vol. 44, no. 19, pp. 5893–5899.
117. Gojny, F. H., Wichmann, M. H. G., Fiedler, B., Kinloch, I. A., Bauhofer, W., *et al.* (2006), 'Evaluation and identification of electrical and thermal conduction mechanisms in carbon nanotube/epoxy composites', *Polymer*, vol. 47, no. 6, pp. 2036–2045.
118. Park, J.-M., Kim, D.-S., Lee, J.-R. and Kim, T.-W. (2003), 'Nondestructive damage sensitivity and reinforcing effect of carbon nanotube/epoxy composites using electromicromechanical technique', in *Current Trends in Nanoscience – From Materials to Application Symposium A, E-MRS Spring Meeting 2003*, 10–13 June 2003, vol. C23, 12/15, Elsevier, Strasbourg, France, p. 971.
119. Barber, A. H., Cohen, S. R., Eitan, A., Schadler, L. S. and Wagner, H. D. (2006), 'Fracture transitions at a carbon-nanotube/polymer interface', *Advanced Materials*, vol. 18, no. 1, pp. 83–87.
120. Kireitseu, M., Hui, D. and Tomlinson, G. (2008), 'Advanced shock-resistant and vibration damping of nanoparticle-reinforced composite material', *Composites. Part B: Engineering*, vol. 39, no. 1, pp. 128–138.
121. Njuguna, J. Pielichowski, K. and Banerjee, J.R. (2004), 'Recent advances in polymer nanocomposites for aerospace applications', in *Proceedings of The First Seminar on Structure and Properties of Advanced Materials for Environmental Applications, under the auspices of EC Marie Curie Programme (FP 5), Cracow University of Technology, Kraków*, pp. 92–96.
122. Njuguna, J. and Pielichowski, K. (2004), 'Polymer nanocomposites for aerospace applications: Fabrication', *Advanced Engineering Materials*, vol. 6, no. 4, pp. 193–203.
123. Njuguna, J. and Pielichowski, K. (2004), 'Polymer nanocomposites for aerospace applications: Characterisation', *Advanced Engineering Materials*, vol. 6, no. 4, pp. 204–210.
124. Njuguna, J. and Pielichowski, K. (2003), 'Polymer nanocomposites for aerospace applications: Properties', *Advanced Engineering Materials*, vol. 5, no.11, pp. 769–778.
125. O'Donnell, S. E., Sprong, K. R. and Haltli, B. M. (2004), 'Potential impact of carbon nanotube reinforced polymer composite on commercial aircraft performance and economics', in *AIAA 4th Aviation Technology, Integration and Operations (ATIO) Forum*, vol. AIAA-2004-6402, 20–22 September, Chicago, Illinois, USA.
126. Volodin, A., Buntinx, D., Ahlskog, M., Fonseca, A., Nagy, J. B. and VanHaesendonck, C. (2004), 'Coiled carbon nanotubes as self-sensing mechanical resonators', *Nano Letters*, vol. 4, no. 9, pp. 1775–1779.
127. Raffaelle, R. P., Landi, B. J., Harris, J. D., Bailey, S. G. and Hepp, A. F. (2005), 'Carbon nanotubes for power applications', *Materials Science and Engineering B*, vol. 116, no. 3, pp. 233–243.
128. Smith, J. (2005), *Slicing it extra thin.*
129. Landi, B. J., Castro, S. L., Ruf, H. J., Evans, C. M., Bailey, S. G. and Raffaelle, R. P. (2005), 'CdSe quantum dot-single wall carbon nanotube complexes for polymeric solar cells', *Solar Energy Materials and Solar Cells*, vol. 87, no. 1–4, pp. 733–746.
130. Njuguna, J., Pielichowski, K. and Alcock, J. R. (2007), 'Epoxy-based fibre reinforced nanocomposites', *Advanced Engineering Materials*, vol. 9, no. 10, pp. 835.
131. Pielichowski, K., Njuguna, J., Janowski, B. and Pielichowski, J. (2006), 'Polyhedral oligomeric silsesquioxanes (POSS)-containing nanohybrid polymers', in *Supramolecular Polymers Polymeric Betains Oligomers*, vol. 201, pp. 225–296.

132. Njuguna, J., Pielichowski, K. and Desai, D. (2008), 'Nanofiller-reinforced polymer nanocomposites', *Polymers for Advanced Technologies*, vol. 19, no. 8, pp. 947–959.
133. Monteiro-Riviere, N. A., Nemanich, R. J., Inman, A. O., Wang, Y. Y. and Riviere, J. E. (2005), 'Multi-walled carbon nanotube interactions with human epidermal keratinocytes', *Toxicology Letters*, vol. 155, no. 3, pp. 377–384.
134. Bajpai, V., Dai, L. and Ohashi, T. (2004), 'Large-scale synthesis of perpendicularly aligned helical carbon nanotubes', *Journal of the American Chemical Society*, vol. 126, no. 16, pp. 5070–5071.

# 16
# Flame-retardant polymer nanocomposites

J.-M. LOPEZ-CUESTA, Ecole des Mines d'Alès, France

**Abstract:** Nanoparticles can reduce the flammability of polymers because of the reduction in heat release rate, the increase in flame-out and their ability to auto-extinguish. The mechanisms involve physical barrier effects and catalytic processes, which can modify the degradation of polymers and form charred protective layers reinforced by the nanoparticles. The properties conferred by nanoparticles depend on dispersion and the chemical modifications to improve compatibility with polymers. The reduction of total heat released during combustion is achieved by a combination of nanoparticles and flame retardants. Future trends include the development of nanoparticle/polymer hybrids and the optimization of the composition and morphology of the protective barrier.

**Key words:** nanocomposite, flame retardancy, polymer.

## 16.1 Introduction

Most studies on polymer flammability and nanocomposites have examined organo-modified layered silicates (OMLSs). Since the 1990s, research groups in the US at the National Institute of Standards and Technology and Cornell University have shown that dispersed organo-modified clay sheets can reduce the flammability of pristine host polymers.[1–3] Since this pioneering work, many research studies have been carried out on the influence of natural and synthetic layered silicates on flame retardancy, as well as various kinds of nanoparticles, in particular carbon nanotubes (CNTs), nano-oxides, nano-hydroxides, and polyhedral oligomeric silsesquioxane (POSS).

The current research activity of many academic groups focuses mainly on the following topics with the aim of improving the fire performance of nanocomposites:

- the development of new surface and interfacial modifications of nanoparticles;
- the influence of the method of production on nanocomposites;
- the mechanisms of the fire-retardant action;
- the synergistic effects in multicomponent flame-retardant systems containing nanoparticles.

The last of these became particularly prominent when industry tried to implement the use of nanoparticles as flame retardants, particularly with a view to replacing halogenated compounds. As the incorporation of such components is not sufficient to meet fire standards such as minimum limiting oxygen index (LOI) values or V0 class in a vertical UL 94 test, various combinations were investigated.

This chapter first discusses the benefits and advantages of polymer nanocomposites in flame retardancy. Second, methods for improving the fire retardancy of polymers are described, along with a discussion of the fire properties improved by nanoparticles and the mechanisms of the fire-retardant action. The chapter will also focus on new experimental approaches carried out to investigate the fire performance of nanocomposites. In the following section, the incorporation of nanoparticles in flame-retardant polymer materials will be examined, for nanoparticles alone, considering surface and interfacial modifications, and for combinations of nanoparticles with flame retardants. The conclusion will discuss future trends, particularly the potential for specific polymers or oligomers to be grafted onto the surface of nanoparticles as well as generating nanoparticles directly inside a polymer.

## 16.2 Benefits and advantages of polymer nanocomposites for flame retardancy

The development of polymer nanocomposites using nanomaterials is one of the most significant uses of nanotechnology. The ISO standardization committee, TC 229 Nanotechnologies, has considered terminology and nomenclature issues concerning nano-objects and nanocomposites.[4]

Nanocomposites are a category of nanomaterials in which nano-objects are dispersed into a matrix or a phase. Other nanomaterials are mainly nanostructured on their surface, in multilayers, or in the bulk. A nanocomposite is a multiphase solid material in which one of the phases has one, two, or three dimensions of less than 100 nm. Components having at least one dimension less than 100 nm can be incorporated into polymer matrices through various methods and processes, such as melt blending, solvent casting, and *in situ* polymerization. Nanoparticles or nanoadditives can potentially cover a wide range of mainly inorganic materials. On the whole, they are classified into three categories according to their dimensionality (Fig. 16.1):[4]

- one-dimensional (1D) nanoparticles or layered materials, such as organo-modified clays (e.g., montmorillonite), which are characterized by one nanometric dimension;
- two-dimensional (2D) nanoparticles or fibrous materials, such as carbon nanotubes, carbon nanofibres, and sepiolite, which are characterized by elongated structures with two nanometric dimensions;
- three-dimensional (3D) nanoparticles or particulate materials, such as POSS, nanometallic oxides (silica, titania, and alumina), and nanometallic carbides, which are characterized by three nanometric dimensions.

These materials have a huge influence on the properties of the polymer, particularly when compared with the incorporation rate: this scarcely exceeds 10 wt%, firstly because good dispersion in polymers is difficult to achieve, and secondly to

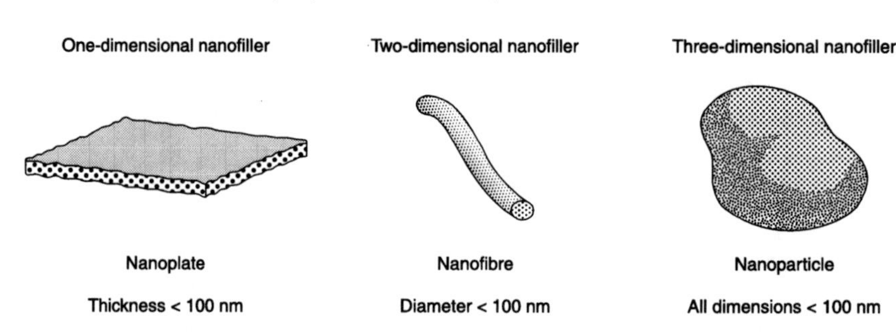


*16.1* Nanoparticles and nano-objects in nanocomposites, from ISO/TS 27687.[4]

maintain the viscosity at a suitable level to allow material processing. Tensile and flexural mechanical properties, such as reinforcement and stiffness, are particularly enhanced, leading to attractive composites with lower density than particulate microcomposites with equivalent properties.[5,6] Moreover, high-aspect-ratio nanoparticles can also give rise to percolation thresholds, leading in turn to outstanding electromagnetic and rheological behavior at an incorporation rate of far less than 1 wt%. The possibility of generating barrier effects with lamellar nanoparticles is of major interest for the reduction of solvent and gas permeability, but also for flame retardancy. This improvement in polymer properties is usually attributed to the huge interfacial area that exists in nanocomposites, which is able to strongly modify the molecular mobility of polymers, and therefore in many cases the glass transition temperature and viscosity in the molten state. A dramatic increase in viscosity is frequently observed as a function of the increase of loading of nanoparticles, particularly for organo-modified layered silicates. This influence limits the acceptable upper values of loading for thermoplastic composites produced in the molten state by processes such as extrusion or injection molding, but also for thermoset composites in which the nanoparticles are mixed with prepolymers and reactive solvents. However, an increase of viscosity conferred by the incorporation of nanoparticles can have a positive effect on flame retardancy, in particular by slowing down the emission of combustible volatiles, and above all by avoiding polymer dripping and by mechanically stabilizing the charred structures. The formation of charred structures is generally observed for polymers that have aromatic groups or that undergo elimination reactions during thermal degradation. Table 16.1 from Lyon and Walters[7] shows the amount of charred residue formed by various common polymers measured using the pyrolysis combustion flow calorimeter (PCFC) at 1 K/s.

The formation of charred structures is often crucial in improving the fire retardancy of polymers. In fact, char formation can act as a barrier structure for mass (volatile combustibles and oxygen) and thermal transfer. In addition to mechanical stabilization, nanoparticles can increase the amount of

*Table 16.1* Char formation of polymers at 800 °C

| Polymer | Char yield (%) |
|---|---|
| Poly(oxymethylene) (POM) | 0 |
| Poly(tetrafluoroethylene) (PTFE) | 0 |
| Poly(methyl methacrylate) (PMMA) | 0 |
| Polypropylene (PP) | 0 |
| Polyethylene(PE) | 0 |
| Polystyrene (PS) | 0 |
| Polyamide 6 (PA6) | 0 |
| Polyethylene terephthalate (PET) | 5.1 |
| Polycarbonate (PC) | 21.7 |
| Polyvinyl chloride (PVC) | 15.3 |
| Poly(vinylidene fluoride) (PVDF) | 7 |

carbonaceous residue through catalytic processes. This indicates that nanoparticles can modify the degradation pathway of polymers. For organo-modified layered silicates, it has been shown by Jiang *et al.*[8] (see Table 16.2) that, when the degradation pathway of a polymer is modified by the presence of organo-modified layered silicates, a significant reduction in the heat release rate (HRR) is observed. In fact, the reduction in HRR for polymer combustion is all the more dramatic since the modification of the degradation pathway is important. For example, in the presence of OMLS, the intra-aminolysis of polyamide chains can become inter-aminolysis, leading to HRR reductions of above 50%.

The reduction in heat release rate and the increase in thermal stability are the main phenomena induced by the presence of nanoparticles in polymers. The increase in thermal stability has been observed for various nanoparticle/polymer combinations[9–11] using thermal analysis (as exemplified for PMMA in Fig. 16.2) and may be due to barrier effects or to a restriction in the motion of the polymer chains, as well as catalytic effects that increase the remaining weight of condensed material at a given temperature. In some cases, the effect is similar to that when an oxidant or inert atmosphere is used for thermogravimetric analysis (TGA), showing that oxygen transfer towards the material is restricted, and hence providing proof for the existence of a barrier effect for volatiles and oxygen diffusion. The improvement in thermal stability conferred by nanoparticles results in a reduced release of combustible volatiles; this in turn means that a material containing nanoparticles has a longer time to ignition when exposed to a heat source.

*Table 16.2* Effects of the incorporation of organo-modified layered silicates on the thermal degradation pathway of various polymers and reduction of peak HRR

| Polymer | Degradation pathway of pristine polymer | Degradation change ascribed to the presence of organo-modified clay | Heat release rate reduction (% HRR) |
|---|---|---|---|
| PA6 | Intra-aminolysis/acidolysis, random scission | Inter-aminolysis/acidolysis, random scission | 50–70 |
| EVA | Chain stripping, disproportionation | Hydrogen abstraction, random scission | 50–70 |
| High impact polystyrene (PS choc) | Beta-scission (chain end and middle) | Recombination, random scission | 40–70 |
| PP | Beta-Scission, disproportionation | Random scission | 20–50 |
| Polyacrylonitrile (PAN) | Cyclization, random scission | No modification | <10 |

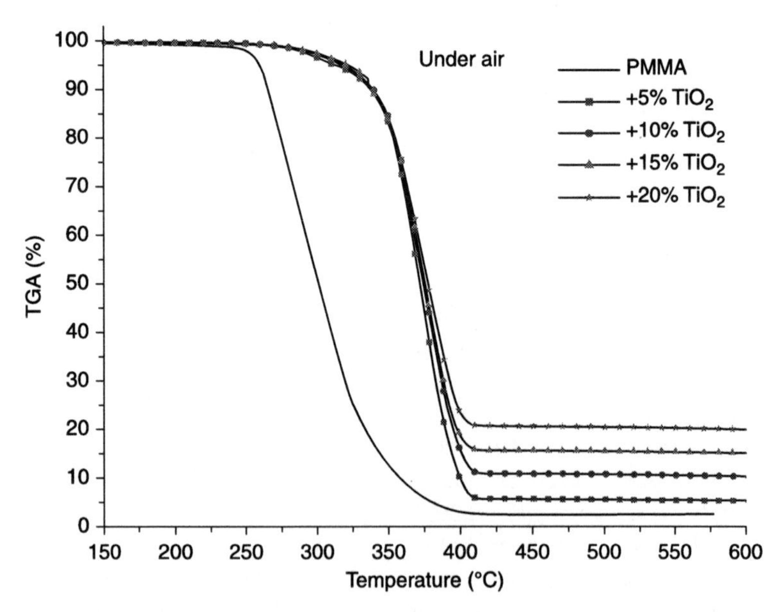


*16.2* Increase of thermal stability of PMMA with titanium oxide nanoparticles.[11]

Various fire properties can characterize a fire-retardant polymer, with the most significant being resistance to ignition, heat release rate, and ability to auto-extinguish. A cone calorimeter (Fig. 16.3) can be used to determine the first of these by applying a selected irradiance produced by a cone heater to a square sample of 10–100 g. The heat release rate of the material measured as a function of time indicates the efficiency of a flame-retardant system. The shape of the curve can also provide information about the formation and perhaps the destruction of a protective layer formed during material combustion.[12] Characteristic parameters obtained from this curve are the time to ignition (TTI), the peak heat release rate (pHRR), and the total heat released (THR). Moreover, the cone calorimeter can be used to measure the mass loss of the sample as a function of time (for a load cell) as well as the rate of smoke release, using the obscuration of a laser beam. For many nanocomposites, especially for organo-modified clays and carbon nanotubes, significant reductions in pHRR are frequently observed, as with organo-modified montmorillonite in ethylene vinyl acetate (EVA) shown by Camino (Fig. 16.4).[13] Reductions in time of flame out are also observed, and are measured using other fire tests such as UL 94V or the epiradiateur test (French standard NFP 92-505). A reduction in TTI is frequently, but not always, observed. Finally, in many cases, the incorporation of nanoparticles does not reduce the THR, which corresponds to the integral of HRR as a function of time. Generally, fire performance increases as a function of the number of nanoparticles, reaches a maximum, and then decreases.

Therefore, although the advantages of nanoparticles in achieving fire-retardant properties such as average heat release rate and peak of heat release rate are

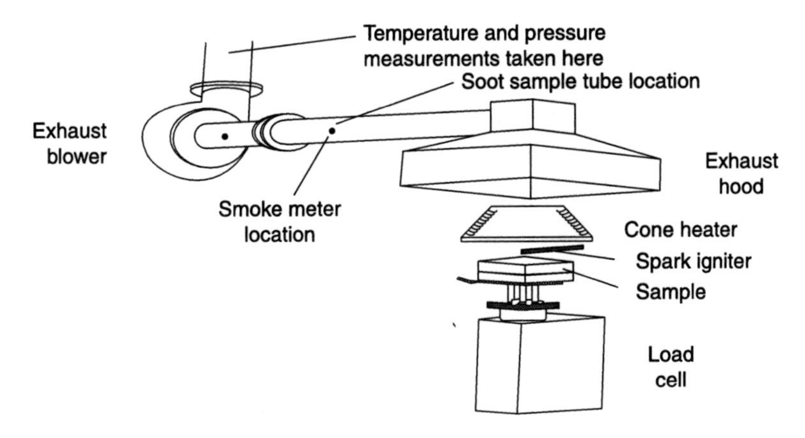


*16.3* Scheme of cone calorimeter.

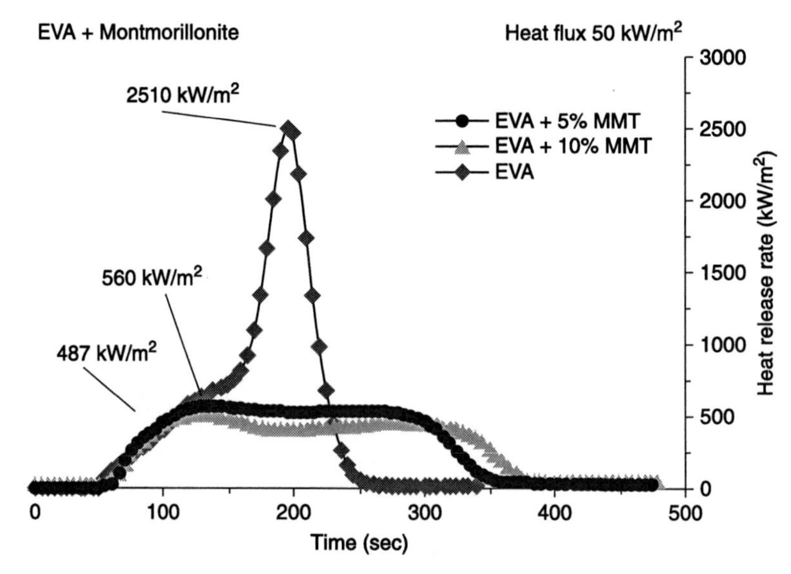


*16.4* Reduction of peak HRR in organo-modified montmorillonite (MMT)/ethylene vinyl acetate (EVA) nanocomposites.

significant, the presence of these components alone is not enough to achieve high levels of flame retardancy in a polymer. The potential of nanoparticles also lies in their synergistic effects with flame retardants (FRs), and they can be used to reduce the global loading of the FRs in polymers. In fact, many flame retardants are used at high percentages (up to 65 wt% for aluminum trihydroxide and up to 40 wt% for chemically reactive flame retardants), which leads to adverse effects on the mechanical properties and processability. The development of high performance flame-retardant systems combining conventional FRs and nanoparticles presupposes that the dispersion of the nanoparticles is conserved, the other components generally being present as fillers or as components that cannot be mixed with the polymer. Moreover, the advantages of nanocomposites regarding fire retardancy are also obvious for thin materials. Since many FRs are present as microparticles (hydroxides, brominated and phosphorated compounds), they are often not suitable for use in thin films and can degrade the properties of functional films. Conversely, well-dispersed nanoparticles can create barrier effects. They could be used in new methods of achieving fire retardancy in miniaturized systems, such as films with micronic thicknesses or thin surface layers. In all cases, the optimization of the use of nanoparticles to improve flame retardancy requires an understanding of their mechanisms of action, which generally involve different kinds of processes, and various physical and chemical phenomena related to, for instance, the surface chemistry of the nanoparticles and their specific surface area. The next section will detail some of these mechanisms, according to the nature of the nanoparticles and polymers.

## 16.3 The role of nanoparticles in improving the flame retardancy of polymers

### 16.3.1 Influence of nanoparticle dispersion and nanocomposite microstructure on fire properties

As mentioned above, the main advantage of nanoparticles in polymers for fire retardancy purposes is the reduction of the heat release rate as measured by a cone calorimeter. Nevertheless, fire-reaction properties also encompass resistance to ignition, the ability to auto-extinguish, and limiting fire propagation. Despite possible modifications of the degradation pathway of the polymers, the presence of nanoparticles does not always have a positive influence on resistance to ignition, regardless of the nature of the nanoparticles and their chemical modifications. For example, some nanoparticles can enhance polymer hydrolysis. Organoclays modified using polar chains can accelerate the thermal degradation of poly($\varepsilon$-caprolactone).[14] For organo-modified clays, the possible decomposition of the organic modifier at a temperature lower than required for polymer processing can trigger ignition. For POSS, the emission of volatile combustibles, such as benzene from trisilanol phenyl POSS incorporated in the polycarbonate, can reduce the time to ignition, as measured by a cone calorimeter, before POSS can contribute to a protective layer able to significantly reduce the heat release rate.[15]

Moreover, in many cases, the presence of a significant number of nanoparticles can substantially increase the thermal conductivity and diffusivity of heat of the polymer, leading to a faster propagation of heat through the polymer (a wicking effect), which then leads to an accelerated release of combustible volatiles. The same phenomena can also affect the ability to auto-extinguish during a UL 94 vertical test or an epiradiateur test, or reduce the rate of burning during a cone calorimeter test, particularly when the main effect of the nanoparticles on fire retardancy is based on the formation of protective barriers resulting from at least one of the following mechanisms:

- the migration of nanoparticles to the surface exposed to heat flux;
- the ablation of the external part of the polymer exposed to the heat source by thermal decomposition;
- the formation of a charred structure ascribed to catalytic processes able to produce polyaromatic species from the degradation products of the polymer.

Overall, the efficiency of these barrier effects is dependent on the irradiance applied and on the thermal energy transferred to the material. Investigations carried out using various OMLS nanocomposites[16] have shown that the reduction in peak HRR increases when increasing the irradiance of the nanocomposites. Consequently, at low irradiance, such as when the polymer begins to ignite, the barrier effects could be ineffective, and in some cases the presence of nanoparticles may even be unfavorable, due to the phenomena mentioned above.

### 16.3.2 Fundamental fire-retardant mechanisms of nanoparticles

As mentioned in the previous section, the formation of a barrier layer is one of the fundamental mechanisms of fire retardancy conferred by the presence of nanoparticles in a polymer. Nevertheless, various kinds of mechanisms can be involved, some of which are connected to the nature of the nanoparticles, such as:

- catalytic effects promoting the formation of charred structures, and reinforced by the nanoparticles;
- purely physical barrier effects due to the particles because of their specific aspect ratio;
- formation of an insulating char structure, able to dissipate incident heat by radiative emission;
- modification of heat diffusivity through the material;
- restriction of macromolecular mobility and an increase in viscosity;
- modification of the degradation pathway of the polymers;
- trapping of radicals formed during combustion;
- formation of new chemical species by reaction with flame retardants or additives (see below).

The majority of studies on the fire retardancy of nanocomposites and the contribution of nanoparticles to fire retardancy have examined layered silicates. Their effectiveness is due to their ability to disperse or to form intercalated structures with the polymers, and also to their specific influence on the degradation pathways of the polymer, which form charred structures from heterogeneous catalytic processes producing polyaromatic compounds. Moreover, the presence of layered silicates contributes to the reinforcement of these charred layers. The nature of the organo-modifiers used to ensure dispersion is also crucial to the influence of the nanoparticles on the degradation pathways of the polymer. The use of alkylammonium modifiers leads to differences in behavior depending on the polar or apolar nature of the alkylammonium chains. Firstly, the type of alkylammonium will govern the dispersion state. Some types can also lead to specific degradation processes in polymers that are sensitive to hydrolysis phenomena such as some polyesters, for example bis(2-hydroxy-ethyl)methyl tallow ammonium modifiers, and this can favor hydrolysis during melt blending. In addition, after the particles have migrated to the surface, thermal decomposition of the organic modifiers by the Hoffman reaction can produce layered silicates with acidic surfaces able to promote dehydrogenation of polymers such as polyolefins. These phenomena eventually result in a charred structure reinforced by the layered silicate aligned parallel to the surface. This kind of structure, initially described by Camino (Fig. 16.5),[13] can be considered analogous to a ceramic char-layered silicate nanocomposite.

The main features of carbon nanotubes that make them interesting for flame retardancy purposes are their outstanding thermal properties, their ability to dissipate heat, and their ability to promote charring. Moreover, other mechanisms connected to their specific surface properties can also be involved, for example radical trapping, as explained in the following.

When nanotubes are relatively well-dispersed in PMMA,[17] a structured layer of nanotubes without any significant cracks or openings forms at the surface of the sample after combustion or gasification performed in a cone calorimeter. This layer covers the entire sample surface of the nanocomposite. Kashiwagi *et al.*[17] have clearly demonstrated the importance of the formation of the protective layer for the reduction in the heat release rate of nanocomposites. This network layer has been considered to act as a heat shield to slow the thermal degradation of PMMA. A test was conducted by the same authors to measure the transmission of a broadband external radiant flux as well as the thermal insulation of this layer. This led to the conclusion that thermal conduction through the network layer appeared to be negligible compared with radiative transfer. The external radiant flux was absorbed by the top layer of the residue and heated the layer almost instantaneously.

Chemical interactions occur between the degradation products of polymers during combustion. Using a cone calorimeter, it was shown in our laboratory that

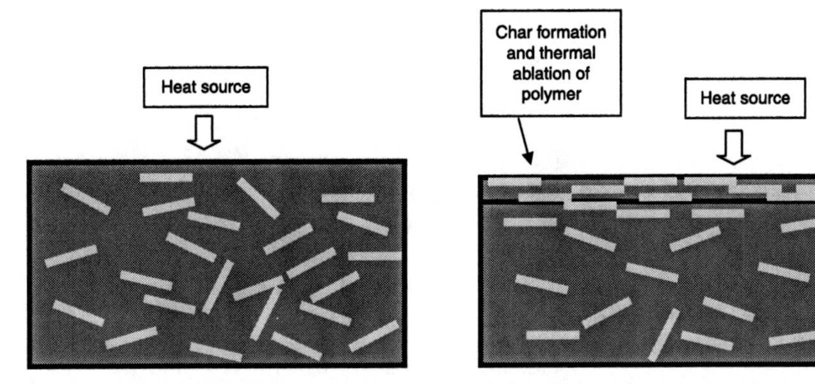


*16.5* Mechanism involving the migration of organo-modified layered silicates and char formation by catalytic processes.

*Table 16.3* Time to ignition (TTI) and maximum peak heat release rate (pHRR) as measured by a cone calorimeter (35 kW/m$^2$) for ternary formulations

| Formulation | TTI (seconds) | Peak HRR (kW/m$^2$) |
|---|---|---|
| EVA | 91 | 707 |
| EVA + 3 wt% MWNT | 84 | 320 |
| EVA + 2 wt% MWNT + 1 wt% crushed MWNT | 107 | 290 |
| EVA + 1 wt% MWNT + 2 wt% crushed MWNT | 155 | 300 |
| EVA + 3 wt% crushed MWNT | 170 | 380 |

CNTs could improve the fire behavior of ethylene vinyl acetate by building a similar protective layer to that observed in PMMA by Peeterbroeck *et al.* (Table 16.3).[18] Moreover, a mechanical crushing of the CNTs incorporated in the same matrix at the same loading entailed a strong increase in the time to ignition, without significant further decrease in the pHRR. One suggested explanation for this unexpected behavior was the occurrence of chemical reactions between the thermo-oxidation products of the EVA and radicals or radical-promoting species located on the crushed multiwalled carbon nanotube (MWNT) surfaces. These species could have formed during crushing in the ball mill. In addition, the presence of a significant amount of radicals or radical promoters on the surface of crushed MWNTs was clearly demonstrated by the chemical reaction with 2,2-diphenyl-1-picrylhydrazyl (DPPH).

The potential that nano-oxides have to improve both the thermal stability and the fire retardancy of polymers has been highlighted in various papers.[11,19–22] It has been shown that the incorporation of $TiO_2$ and $Fe_2O_3$ nano-oxides in PMMA could improve the thermal stability of PMMA by about 50 °C with 5 wt% of nanoparticles.[11] It was hypothesized that this phenomenon could be explained by a restriction of mobility due to adsorption of the polymer on the inorganic surfaces. In fact, a linear relation between the limiting oxygen index and the glass transition temperature was discovered by Laachachi *et al.*[21] (see Fig. 16.6 for $Fe_2O_3$ nanocomposites) for all the nanocomposites investigated, indicating that chain restriction mobility could have an influence on thermal stability for fire, at least at the first stages of polymer degradation. Nevertheless, in another paper[22] the same authors showed that, at a high degradation rate, nanometric $TiO_2$ could catalyze the degradation of PMMA. It can be assumed that the increased stability of the condensed phase, except at high degradation rates, may limit the amount of combustible volatiles released. Hence, the incorporation of both nano-oxides in PMMA led to a decrease in pHRR, as well as a slight increase in TTI in comparison to the corresponding micronic oxides, for which the decrease in pHRR was lower (Table 16.4).

This reduction in TTI when using nanometric rather than micrometric oxide particles has been interpreted in terms of enhanced heat transfer through the material

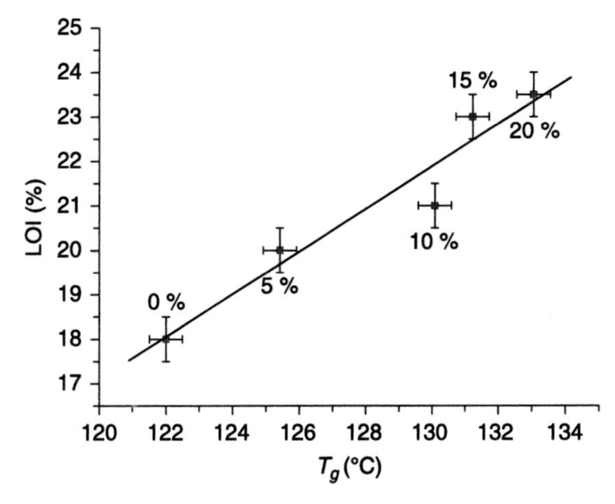


*16.6* Limiting oxygen index (LOI) as a function of glass transition temperature $T_g$ for PMMA/$Fe_2O_3$ nanocomposites at various loadings.

near the surface. This enhancement could be due to an increase of the oxide/polymer contact interface when the particle size decreases. Hence, the temperature of the polymer increases more rapidly, particularly close to the surface; for the polymer there is less time before degradation and ignition. Another consequence of the increase of heat transfer inside the material could be the reduction of both temperature and surface tension gradients in the polymer melt. Thus, there is a slower migration of bubbles corresponding to the release of volatile combustibles, thereby reducing combustible gas flow at the melt/air interface and thus the amount of heat released during combustion.

With regard to TTI, the difference in the results obtained with nanometric $Fe_2O_3$ and $TiO_2$ may be explained by the lower thermal diffusivity of iron oxide compared with titanium dioxide, which leads to an accumulation of heat at the surface exposed to the radiant heat source and, therefore, to a greater rise in surface temperature. This in turns leads to accelerated degradation and ignition. In another

*Table 16.4* Cone calorimetry data for PMMA and its nanocomposites with $TiO_2$ and $Fe_2O_3$ at 35 kW/m$^2$

| Formulation | TTI (seconds) | peak HRR (kW/m$^2$) |
|---|---|---|
| PMMA | 69 | 620 |
| PMMA + 15 wt% nanometric $TiO_2$ | 74 | 440 |
| PMMA + 15 wt% micronic $TiO_2$ | 100 | 510 |
| PMMA + 15 wt% nanometric $Fe_2O_3$ | 88 | 350 |
| PMMA + 15 wt% micronic $Fe_2O_3$ | 97 | 390 |

paper,[23] some of the same authors used laser flash analysis (LFA) to measure the thermal diffusivity of PMMA containing increasing loadings of two nano-oxides and one nano-hydroxide ($TiO_2$, $Al_2O_3$, and boehmite (AlOOH)) up to 15 wt%. It was shown that thermal diffusivity increased as a function of loading and that a correlation between TTI and thermal diffusivity could be established on the basis of the nature of the nano-oxide and its loading in the matrix. The compounds with the highest diffusivity had the longest TTI.

In some cases, nanometric particles can create a purely physical barrier, which strongly depends on the particle size distribution as well as on their aspect ratio. Some comparisons have been made between nanoparticles with the same nature but different aspect ratios, as shown in Fig. 16.7. Laoutid *et al.*[24] incorporated lamellar or fibrous magnesium dihydroxide (MDH) nanoparticles into PMMA to investigate the influence of their shape on fire performance.

Both types of nanoparticle significantly improved the thermal stability of PMMA, under both air and an inert atmosphere. The results obtained using a pyrolysis cone flow calorimeter and a mass-loss cone calorimeter show a significant decrease in pHRR concomitant with charring during combustion. As shown in Table 16.5, lamellar MDH nanoparticles were found to be more efficient for both techniques than fibrous MDH nanoparticles.[24]

In thermoplastic nanocomposites, a pure barrier effect can result either from migration processes as observed with organo-modified layered silicates, or most frequently by the continuous ablation of the polymer during combustion, which leads to a surface layer of residue rich in nanoparticles. In thermosetting polymers containing lamellar nanoparticles, this polymer ablation favors the formation of a protective layer made of an arrangement of nanoparticles parallel to the surface, which is better able to reduce the transfer of volatile combustibles

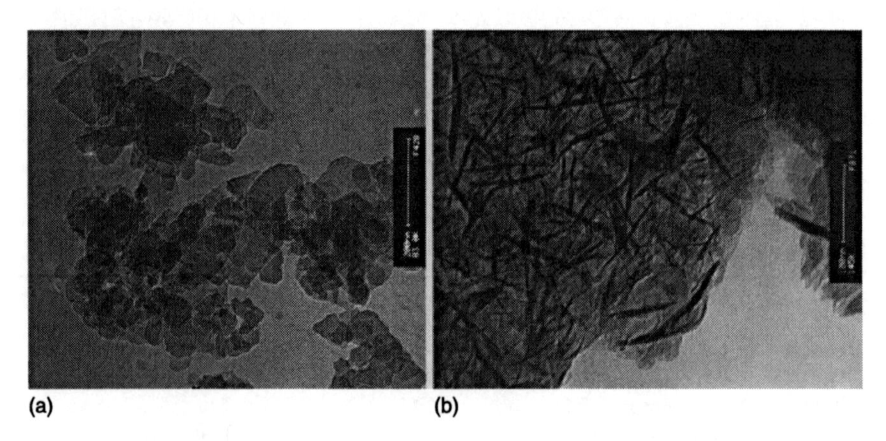


*16.7* Transmission electron micrographs of (a) lamellar MDH and (b) fibrous MDH nanoparticles.

*Table 16.5* Peak heat release rate measured by a cone calorimeter (kW/m$^2$), measured by a PCFC, and heat release capacity measured by a PCFC, for PMMA and PMMA compositions containing lamellar or fibrous MDH nanoparticles

| Formulation | peak HRR (kW/m$^2$) | peak HRR (W/g) | HR capacity (J/g/K) |
|---|---|---|---|
| PMMA | 639 | 419 | 413 |
| PMMA + 5 wt% lamellar MDH | 446 | 287 | 285 |
| PMMA + 10 wt% lamellar MDH | 388 | 220 | 232 |
| PMMA + 20 wt% lamellar MDH | 305 | 204 | 201 |
| PMMA + 5 wt% fibrous MDH | 515 | 358 | 353 |
| PMMA + 10 wt% fibrous MDH | 437 | 277 | 273 |
| PMMA + 20 wt% fibrous MDH | 385 | 224 | 223 |

than particles with isotropic or random orientations. The optimization of the resulting barrier can be achieved by combining two or more kinds of nanometric or submicronic particles. In a recent paper,[25] nanoalumina (AL, median diameter of 13 nm) and submicronic alumina trihydrate (ATH, median diameter of 300 nm) particles were incorporated into an unsaturated polyester resin at various *loadings. The thermal degradation of the composites was studied using thermogravimetric analysis, while their fire behavior was investigated using a cone calorimeter and a pyrolysis combustion flow microcalorimeter. Synergistic effects for thermal stability (determined using TGA in air and a nitrogen atmosphere) and the heat release rate (determined using cone calorimeter testing) were observed for combinations of AL and ATH particles. The best result for fire behavior was obtained for a global loading of 10 wt% with an equal mass ratio for each kind of particle (Fig. 16.8). It may be concluded that this synergistic effect can be attributed to an optimal compacity of the protective layer, due to the particle size distribution of the fillers, formed at the surface of the sample during thermal degradation.

Mass-loss curves also exhibited increased char yield for mixed compounds. The potential for combining particles with different sizes was discussed, as well as the role of water release, with reference to the activation energies of degradation processes. Furthermore, water release from submicronic aluminum trihydroxide did not make a significant contribution to the fire behavior of these mixed compounds. The use of dehydrated ATH instead of pristine ATH at the same percentage led to similar results. Moreover, the replacement of AL by a pyrogenated hydrophilic silica with the same particle size also led to synergistic effects on fire behavior as observed with a cone calorimeter. Consequently, these synergistic effects can be attributed to physical effects resulting from the arrangement of both kinds of mineral particles of very different median size at the surface of the composite during polymer combustion and ablation. The formation of this mineral

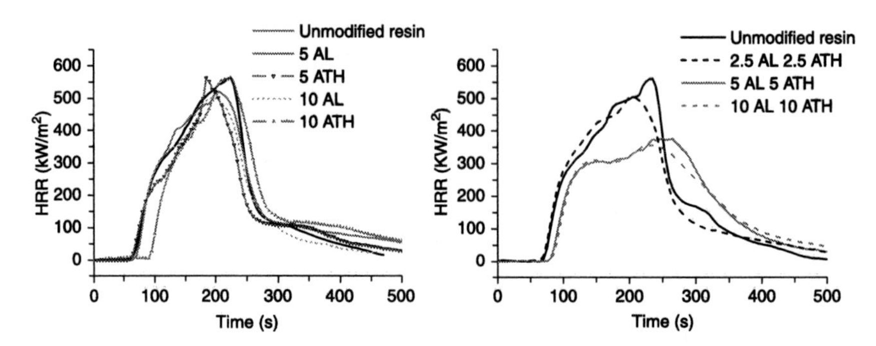


*16.8* Cone calorimeter results of unsaturated polyester resin composites containing AL and ATH, irradiance of 35 kW/m$^2$.

barrier can also promote catalytic effects as a result of the huge specific surface area of oxide nanoparticles.

It can be concluded that some mechanisms involved in the improvement of the fire behavior of thermoplastics containing nanoparticles are also involved in the thermal degradation of thermoset composites.

### 16.3.3 New experimental approaches for investigating the fire performance of nanocomposites

Specific investigations of the fire performance of nanocomposites have led to the development of novel procedures to scrutinize the various mechanisms described in the previous section, and particularly in understanding the constitution of the protective barrier layers. These procedures are enhancements of the techniques usually carried out to study fire behavior, and, in particular, there have been developments in instrumentation. The increased interest in the influence of nanoparticles on fire retardancy has led to multiple combinations of techniques, thereby enriching the standard scientific approaches.

Thus, the complex character of fire-retardant systems incorporating nanoparticles, as well as the diversity of the fire retardancy mechanisms, renders it necessary to establish more connections between the following factors to optimize the composition of these systems:

- the evolution of morphological structure from the starting material to the residue;
- the chemical composition of the condensed phase during polymer degradation;
- the parameters of fire retardancy.

For example, various studies combine X-ray diffraction, IR spectroscopy, X-microanalysis, scanning, transmission electronic microscopy (TEM), and sometimes atomic force microscopy with fire tests such as mass-loss cone calorimetry. The cone calorimeter has become the predominant method for investigating the fire behavior of nanocomposites; new developments include gasification techniques in which the combustion atmosphere can be controlled,[17] as well as the use of temperature sensors able to determine the temperature in the condensed or gaseous phase during a test. Sample residues removed at different stages of degradation can be analyzed. To study the combustion or degradation atmosphere, techniques are usually combined, for example, IR spectroscopy with thermogravimetric analysis (TGA/FTIR), and pyrolysis with gas chromatography and mass spectroscopy (PyGCMS).

Moreover, the need for techniques able to investigate the fire behavior of a very small amount of material has become crucial. In many studies, despite the relatively low number of nanoparticles incorporated in the polymer, the amount of nanocomposite available is not sufficient to perform macroscopic tests such as cone calorimetry, limiting oxygen index, or UL 94V testing. In some cases, the nanoparticles are incorporated into polymers synthesized at a laboratory scale, or have reacted with polymers in microreactors such as micromixers or microextruders. Nevertheless, the potential use of such techniques is limited by the ability to relate the corresponding data to those obtained at a macroscale. In other words, the value of these techniques depends on the ability to determine whether the results are representative of what occurs during the combustion of a piece of material. Recently, the use of TGA, TGA/FTIR, and PyGCMS as tools to understand fire reaction mechanisms using microscale samples has been possible, thanks to the development of a new microcalorimetry technique, pyrolysis combustion flow calorimetry (PCFC). In this technique, developed by Lyon and Walters,[7] 2 mg of material are decomposed in a pyrolysis reactor under an inert atmosphere using a temperature ramp (typically 1 K/s) up to 750 °C. The gases produced are burnt in an 80/20 $N_2/O_2$ mixture. Various parameters are recorded as a function of temperature, such as TTI (s), HRR (W/g), THR (J/g) and also a parameter specific to this technique, heat release capacity (HRC in J/g.K), which corresponds to the ratio between the peak heat release rate (W/g) and the temperature ramp (K/s). If there are several peaks in the HRR versus temperature curve, several HRCs are determined.

The possibility of predicting cone calorimeter or LOI data using PCFC data (THR and HRC) has been investigated by several authors. Yang *et al.*[26] studied the flammability of fibers with and without flame-retardant additives using PCFC. Walters and Lyon[27] attempted to establish connections between PCFC data and the macromolecular structure of polymers according to the van Krevelen approach, based on the contribution of each chemical group to the total heat released.

Morgan *et al.*[28] studied various epoxy-based compounds using PCFC, observing that the best choice based on PCFC results did not correspond to the best choice suggested by cone calorimetry. Hence, the authors judged that the microcalorimeter was not a good screening tool if THR and peak HRR were the only criteria taken into account. The investigation of polycarbonate/acrylonitrile butadiene styrene flame-retardant compounds by Schartel *et al.*[29] has led to correlations between HRC and UL 94 and particularly with LOI. Nevertheless, correlations between PCFC and cone calorimeter data were not achieved. Cogen *et al.*[30] found quite good correlations between HRC (PCFC) and pHRR (cone calorimeter) on the one hand and between THR (PCFC) and THR (cone calorimeter) on the other for halogen-free flame-retardant polyolefin compounds.

In spite of the interesting correlations mentioned above, PCFC cannot be considered as a screening tool for cone calorimeter testing, since the sample characteristics and the test conditions are very different. Indeed, the two techniques appear to be complementary and specific approaches can be adopted, particularly in the study of nanocomposites, which focus on the formation of the protective layer at the surface of the material during combustion. Hence, we have proposed an empirical approach based on experimental data from both techniques, to account for the action of nanoparticles acting as flame-retardant additives.

The method recently proposed by Sonnier *et al.*[31] is based on the assumption that the barrier effects of nanocomposites due to the formation of protective layers during combustion occur in cone calorimetry but not in PCFC. Flame inhibition is not observed in PCFC because combustion is complete. Conversely, fire-retardant mechanisms such as the trapping of radicals or the endothermic release of water, which slow down the degradation of polymeric materials, can be observed in both techniques. Consequently, the relative decrease in pHRR in cone calorimeter testing due to the incorporation of nanoparticles acting as a flame-retardant additive should be higher than (or at least equal to) the relative decrease in HRC (or sumHRC) in PCFC testing. R1 is the ratio between the HRC (or sumHRC) from PCFC for a flame-retardant polymer and the HRC of the pristine polymer (at the same irradiance). R2 is the ratio between the pHRR from cone calorimeter testing of the flame-retardant polymer and that of the pristine polymer (at the same heating rate). In this method, R1 is plotted as a function of R2.

As an example of the use of this method, Fig. 16.9 shows R1 versus R2 for various nanoparticles: nanoboehmite, nanoalumina, and fibrous and lamellar nanomagnesium dihydroxide (MDH), at different loadings in PMMA (5–20 wt%). These data were extracted from the research work of Laoutid *et al.*[24] It appears that all the formulations can be divided into two categories according to the location of the experimental data in the figure. For PMMA/nanoMDH (lamellar and fibrillar), all the experimental points are very close to the line R1 = R2 and follow this line as a function of increasing loading. Even at 20 wt% of nanoMDH, the decrease in pHRR from cone calorimeter results is the same as the decrease in HRC for PCFC results. It can be concluded that no barrier effect is involved with both types of nanoMDH for these loadings. Conversely, for PMMA/

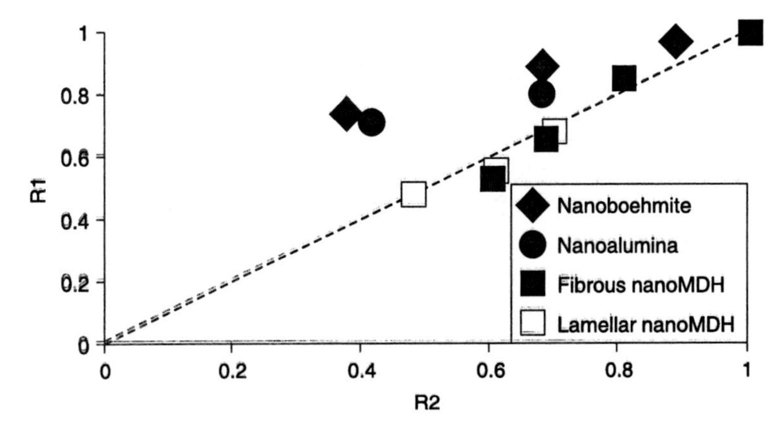


*16.9* R1 versus R2 representation for various PMMA/nanoparticle compositions. For all compositions, the decrease of R1 and R2 was obtained with increased loading up to 20 %wt.

nanoboehmite and PMMA/nanoalumina, the decrease in pHRR from cone calorimetry is stronger than the decrease in HRC obtained by PCFC: R2 < R1. Hence, we could assume that a strong barrier effect does occur for these systems. Observation of the residues after the cone calorimeter tests confirmed this interpretation: the residues formed a cohesive char layer for PMMA/nanoalumina and PMMA/nanoboehmite only. This layer is able to limit heat transfer from the flame to the remaining polymer and the transfer of gases from the pyrolysis zone to the flame.

In conclusion, this approach is representative of new developments to show how nanoparticles form protective barriers. The method can end in two ways: firstly, by investigating the evolution of the chemical structure and composition of the remaining material during degradation at the cone calorimeter; and secondly by using thermogravimetric analyses to complete the HRR evolution measured using PCFC. Moreover, the combination of analytic and complementary fire-testing techniques helps understanding of the mechanisms involved in complex fire-retardant systems that combine nanoparticles with traditional fire retardants. Some of these systems are described in the next section.

## 16.4 Methods of incorporating nanoparticles as flame-retardant components in polymers

### 16.4.1 Processing nanostructures and nanocomposites

Nanoparticles are incorporated in polymers by a variety of methods, giving dispersion at a nanometric scale. Generally, a simple mixing of the initial particles with the polymer is not enough to provide the degree of dispersion required to achieve composites with outstanding properties. Moreover, for layered silicates, the initial particles are micronic and only dispersion in the polymer can lead to a nanocomposite structure. A variety of processes and methods can produce nanocomposites, depending upon the nature of the nanoparticles available or generated, as well as the nature of the polymer. These processes can be classified into three categories: *in situ* polymerization, solvent casting, and melt blending or melt compounding.

#### In situ *polymerization*

*In situ* polymerization can be carried out using a suspension of nanoparticles and monomer, but in most cases the nanoparticles are first incorporated into a solvent for the polymer, to obtain a stable suspension, and then the polymer is added. Some interfacial agents can be added to promote a stable suspension. Polymerization is carried out in solution and afterwards the solvent is removed. This technique was used in pioneering work by Unitika and Toyota[32] and has since been successfully employed for a variety of polymers, with and without tethering of the macromolecular chains to the nanoparticle surface.[6] In general, the thermodynamics is favorable to the conservation of particle dispersion after processing by compression or injection molding, but it requires that the monomer initially disperses the inorganic particles sufficiently well. Nevertheless, the dilution of the polymerized hybrids in the pure polymer can result in the collapse of the nanostructure. This phenomenon has been observed for montmorillonite/poly($\varepsilon$-caprolactone) hybrids diluted in the same polymer.[33]

#### *Solvent casting*

This technique is also called solution intercalation or solvent blending. Its advantage is that the polymer and the nanoparticles can be dissolved and dispersed or exfoliated in the same solvent. Polymer/filler systems do not generally possess favorable thermodynamics for nanocomposite formation. A first critical step is the breaking of agglomerates of primary nanoparticles. The use of ultrasonication is of prime importance and homogeneous mixing and dispersion are crucial, particularly for clays.[34] This technique is especially suitable for thermosetting polymers and polymers that easily swell in solvents, maximizing the mixing between the polymer and the nanoparticles. Moreover, since some polymers have many solvents, it is often relatively easy to select one that is compatible with the nanoparticles and their interfacial modifiers. The second critical step is the conservation of the dispersed/intercalated/exfoliated structure after processing, which is threatened by the unfavorable thermodynamics. Interfacial compatibilizers, such as maleic anhydride grafted polymers, which are also miscible with the solvent, can limit the loss of microstructure.[35]

*Melt compounding*

Melt compounding is the most widely studied method (but is only suitable for thermoplastic nanocomposites), due to its simplicity and the availability of the equipment.[5] The nanocomposite is obtained by mixing the polymer and the nanoparticles in processing equipment, generally a twin-screw extruder or an internal mixer. A two-step process is often used: the first step is to obtain a master batch, and the second is to dilute it. For organo-modified lamellar silicates (OMLSs), the modifier needs to be miscible or compatible with the polymer matrix. Some grafted polymers can improve this compatibility. In addition, nanoparticle modifiers have to be thermally stable with respect to the polymer processing conditions. For lamellar silicates modified with alkylammonium ions, their possible thermal degradation through the Hoffmann reaction limits their use in some engineering polymers. Finally, despite its apparent simplicity, this method often requires accurate control of shearing during the mixing process. The specific design of the screw profiles can determine the final microstructure and properties of the nanocomposites, for example for polypropylene[36] or poly($\varepsilon$-caprolactone).[14]

Methods for preparing nanocomposites depend on thermodynamic interactions or mechanical mixing, or both. The effectiveness and importance of these processing methods differ according to the nature of the polymers and nanoparticles. Thus, the agglomeration phenomena of carbon nanotubes are very specific and significantly different from those of OMLS, and melt processing is less commonly used for nanocomposites reinforced with carbon nanotubes than for OMLS nanocomposites.

Carbon nanotubes exhibit the most outstanding performances as flame-retardant additives compared with other nanoparticles, in terms of the low loading required to entail significant improvements in fire reactions.[37] The reduction of HRR obtained using CNT percentages lower than 1 wt% is usually higher than 50%.[17]

It appears that the reduction in HRR levels off and even begins to increase again for CNT percentages higher than 0.5 wt%. Kashiwagi *et al.*[17] attribute this to the increase of heat conduction through the polymer as a function of the amount of CNT, leading to polymer decomposition, whereas the formation of a thermal shield might be effective for 0.5 wt% CNT.

As mentioned above, the beneficial influence of nanoparticles on fire retardancy is restricted to a limited percentage of nanoparticles, and their advantages are restricted to a reduction in HRR, and not THR. In addition to the heat propagation effect caused by heat conductive fillers, increasing the nanoparticle loading can limit the dispersion of nanoparticles or strongly improve the processability of the polymer, due to its viscosity. For example, OMLS is generally employed at percentages less than or equal to 5 wt%. In contrast to conventional fire retardants, for which the loading is generally high, nanoparticle loading cannot be increased in order to meet fire performance standards; the concept of combining nanoparticles with FRs has emerged as a consequence of this.

### 16.4.2 Synergy of nanoparticles in combination with other flame retardants in polymers

Numerous publications dealing with the combination of nanoparticles (mainly OMLS) and FR compounds have been carried out to generate synergies for a variety of flame-retardant properties.[38] The interest lies particularly in the possibility of reducing the global loading of all FR additives and fillers in the polymer: high loadings lead to degradation of mechanical properties. The growing interest in different kinds of nanoparticles has arisen as a result of the study of combinations of these nanoparticles, mainly with different phosphorus components or metal hydroxides. In particular, the development of intumescent flame-retardant systems containing nanoparticles has been investigated by a number of research groups. Intumescent compounds swell when heated and cause an expansion of the polymeric material with the formation of a stable charred structure.[39] This structure forms a barrier to heat, combustible gases, oxygen, and free radicals during a fire and protects the residual material. Figure 16.10 shows the final charred and expanded residue obtained from a cone calorimeter test performed on an ethylene vinyl acetate copolymer filled with 55 wt% magnesium dihydroxide and 5 wt% of an alkylammonium-modified montmorillonite. The expansion of the charred structure cracks the initial surface layer, which is formed mainly of the minerals since the polymer ablated during combustion.

An intumescent fire-retardant system can be produced from a carbon source, such as a polymer, a polyacid (the most commonly used is ammonium polyphosphate (APP)) and an expansion agent. Due to their widespread use in the cable industry, ethylene vinyl acetate, low-density polyethylene (LDPE) and their blends have been investigated along with polypropylene as potential host polymers for intumescent compounds containing nanoparticles.[40–43] The nanocomposite structure was maintained for all these compounds within the intumescent flame-retardant (IFR) systems. Synergistic effects on HRR values between conventional IFR compounds and OMLS have been observed and polyamide 6 (PA6) has been

*16.10* Intumescent behavior for a cone calorimeter residue for a magnesium dihydroxide/organo-modified montmorillonite/EVA copolymer composition (irradiance of 50 kW/m$^2$).

used as the charring agent. In PP composites,[42] an optimum amount of 4 wt% montmorillonite (OMLS) was shown to maximize the synergistic effect. It was also observed that the amount of maleic anhydride grafted polypropylene (MAPP) used as a compatibilizer in these systems influences the microstructure (exfoliated or intercalated). Moreover, other types of compatibilizing agents have been studied in IFR systems containing OMLS, such as carboxylated polypropylene as a reactive compatibilizer of PP/PA6 blends.[44] The correct selection of compatibilizer is expected to promote good dispersion of the charring polymer, and also to confer good mechanical performance on the fire-retardant polymer blend. Other types of intumescent systems containing nano-oxides have also been developed. Laachachi *et al.*[45,46] used aluminum and titanium nano-oxide with APP intumescent compounds (including melamine phosphate) and also with phosphinate derivatives. APP, melamine phosphate, and $Al_2O_3$ combinations have shown significant synergism for fire retardancy (HRR and char yield) when studied using a cone calorimeter, for a total constant loading of 15 wt% (Fig. 16.11) of phosphorous FR and nanoparticles. These synergistic effects are attributed to the catalytic action of the alumina nanoparticles and to the formation of aluminum phosphate compounds. Nevertheless, for $TiO_2$, the aggregation processes of the nanoparticles did not allow synergism for fire reactions to be achieved. Figure 16.11 shows synergistic effects for peak HRR obtained for the compounds using both nanometric alumina and aluminum phosphinate. HRR and THR values were strongly reduced due to the formation of a charred and vitreous, moderately expanded structure.

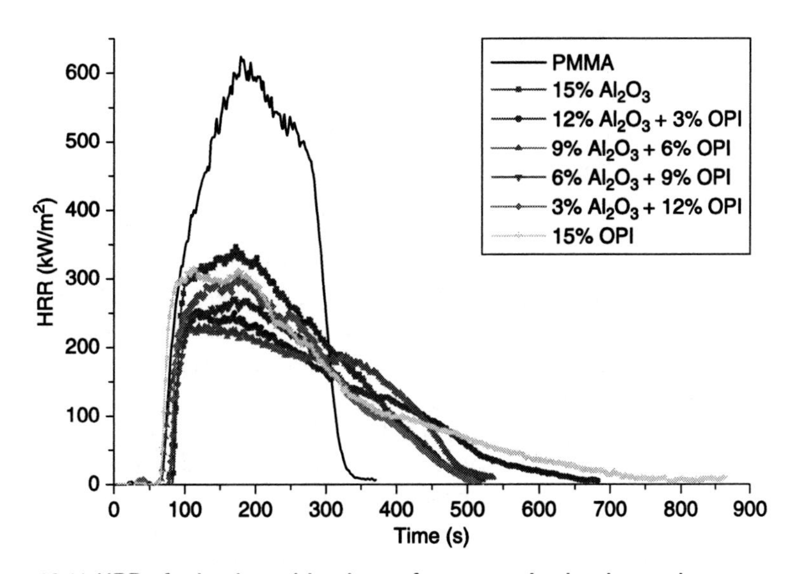


*16.11* HRR of mixed combinations of nanometric alumina and aluminum phosphinate (OP1) and melamine phosphate (AP452) in PMMA.

## 16.5 Practical examples

The main examples of nanoparticles as flame-retardant components are in the cable manufacturing industry dating from 2000. Other uses are in the production of electrical and electronic equipment and public transportation. However, since there is concern in Europe surrounding nanotechnology and nanoparticles, producers generally do not publicize their use in fire-retardant strategies. Cable companies are interested in developing compounds using organo-modified layered silicates mainly with the aim of reducing the loading of metal hydroxides in their EVA and LDPE formulations while maintaining fire performance. Alumina trihydrate (ATH), magnesium dihydroxide (MDH), and hydromagnesite (HM) are the main flame retardants used with loadings up to 65 wt%. Due to these high loadings, mechanical properties such as elongation at break and tensile strength are strongly reduced, whereas density is significantly increased. Hence, cable manufacturers aim to replace a fraction of the metal hydroxides, typically 10%, by a smaller fraction of organo-modified clays, around 5%. The objectives are to:

- take advantage of the barrier structure produced by organo-modified layered silicates,
- reinforce the cohesion of the oxide residue formed after water release.

OMLSs, such as montmorillonite, in combination with magnesium hydroxide can build cohesive layered structures, which can limit the release of volatiles and form expanded and porous residues, similar to intumescent structures with APP. These are called 'mineral intumescences.'[47]

Beyer (Kabelwerk Eupen AG (Belgium))[48] tested the performance of compounds containing nanoparticles in industrial conditions. The compounds were processed using Buss co-kneader mixers. A comparison between EVA/ATH (60% ATH) and EVA/ATH/organoclay (5% organoclay and 60% ATH) showed that, for an irradiance of 50 kW/m$^2$, peak HRR was reduced by 50%. Moreover, the char corresponding to the mixed compound was very rigid and showed very small cracks, in contrast to the EVA/ATH residue.

The same author also investigated EVA/ATH/(CNT/OMLS) compounds.[49] Small insulated wires were produced containing either only OMLS nanoparticles or equal proportions of both kinds of nanoparticle. Small-scale fire testing according to IEC 60332-1 produced very similar results for both compounds, but the compound with CNTs exhibited significantly fewer cracks. This was attributed to the strengthening effect of CNT. The HRR and TTI of both compounds were investigated using a cone calorimeter, with the standard cone holder filled with cable pieces and the wires mounted using a 'single layer design' (parallel and juxtaposed wires within the sample holder). Wires insulated with the CNT compound did not show any increase in HRR for the first 10 min, while the compound containing only organo-modified clays exhibited HRR values close to 80 kW/m$^2$ at 600 s.

Finally, since synergistic effects on various fire properties have been found using combinations of nanoparticles, such as OMLS and silica,[50] for example, which can strongly reinforce the cohesion of the surface layer, complex combinations have also been designed by industrial manufacturers and are currently in use.

## 16.6 Future trends

Future trends in the field will follow from the development of new kinds of nanoparticle, particularly those corresponding to flame-retardant additives such as nano-hydroxides of magnesium and aluminum. In addition, new surface modifications of these nanoparticles are expected to lead to better reactivity with other components of the flame-retardant system in which they are incorporated. Moreover, new methods for producing these materials could be considered, for example the *in situ* synthesis of nano-oxides from precursors during polymer processing.[51]

The majority of studies have been carried out on combinations involving OMLS; this suggests that combinations of OMLS with various flame-retardant intumescent systems have interesting potential. Thanks to the various combinations involving compatibilizers and carbon sources, these systems seem better able to meet industrial demands for fire performance as well as good mechanical properties and excellent processability. Nevertheless, the complexity of such systems will require careful control of the morphology that results from the processing steps.

The amount of char formed and its nature are key parameters in accounting for the FR performance of complex flame-retardant systems containing nanoparticles. Enhancement of the charring activity of IFR systems that contain nanoparticles can be achieved by using nanoparticles alone or through a combination of nanoparticles with strong catalytic activity for carbonization processes.

Combinations of OMLS, nano-oxides or nano-hydroxides and phosphorus FRs seem to have promising potential in this respect. The influence of nano-oxides on thermal stability and their ability to react with phosphorus FRs means they can be included in a number of FR systems, in particular for polymers undergoing hydrolysis reactions. In addition, the advantage of water release by some nanometric particles (nano-hydroxides versus nano-oxides) must also be considered, taking into account their high specific surface area and aspect ratio. These particles can contribute to barrier effects. Hence, innovative FR systems could be designed by optimizing the packing of nano- and microparticles that remain at the surface after polymer ablation resulting from combustion, to maximize the barrier effect.

A constant challenge is to design FR systems that allow maximum effectiveness of fire performance with a minimum of additives incorporated into the polymer. The incorporation of nanoparticles able to produce synergistic effects in FR systems is key to this challenge. One innovative way to fulfill this objective might be to tether the FR agents to the nanoparticles. Some phosphorus compounds, such as oligomers, can be attached by 'grafting from' techniques on some nano-oxides.[52] Consequently, it can be expected that more complex FR systems will be developed with innovative surface modifications of nanoparticles (possibly blended), combined with intumescent systems including various polymer and interfacial agents. These developments will necessarily require control of the morphology during and after processing. Further investigation into the connection between morphology and fire behavior of such complex systems is required.

## 16.7 References

1. E.P. Giannelis, *Adv. Mater.* (1996) 8, 29–35.
2. J.W. Gilman, T. Kashiwagi, J.D Lichtenhan, *J. Sampe* (1997) 33, 40–46.
3. J.W. Gilman, T. Kashiwagi, J. Brown, S. Lomakin, E. Giannelis, *Chem. Mater.* (2000) 12, 1866–1873.
4. ISO/TS 27687 *Nanotechnologies – Terminology and definitions for nano-objects – Nanoparticle, nanofibre and nanoplate*; ISO/TS 11360 *Nanotechnologies – Methodology for the classification and categorization of nanomaterials*.
5. M. Alexandre, P. Dubois, *Mat. Sci. Engineering* (2000) 28, 1–63.
6. S.S. Ray, M. Okamoto, *Prog. Polym. Sci.* (2003) 28, 1539–1641.
7. R.E. Lyon, R.N. Walters, *J. Anal. Appl. Pyrolysis* (2004) 71, 27–46.
8. B.N. Jiang, M. Costache, C. Wilkie, *Polymer* (2005) 46, 10678–10687.
9. J. Zhu, A.B. Morgan, F.J. Lamelas, C.A. Wilkie, *Chem. Mater.* (2001) 13, 3774–3780.
10. B. Lepoittevin, M. Devalckenaere, N. Pantoustier, M. Alexandre, D. Kubies, *et al.*, *Polymer* (2002) 43, 4017–4023.
11. A. Laachachi, E. Leroy, M. Cochez, M. Ferriol, J. M. Lopez-Cuesta, *Pol. Deg. Stab.* (2005) 89, 344–352, A. Laachachi, PhD Dissertation, University of Metz (France) 2005.
12. B. Schartel, R. Hull, *Fire and Mater.* (2007) 31, 5, 327–354.
13. G. Camino, *Nanofire Seminar*, Lyon 2005.
14. S. Labidi, N. Azema, D. Perrin, J.-M. Lopez-Cuesta, *Pol. Deg. Stab.* (2010) 95, 382–388.
15. L. Song, Q. He, Y. Hu, H. Chen, L. Liu, *Pol. Deg. Stab.* (2008) 93, 627–639.
16. B. Schartel, M. Bartholmai, U. Knoll, *Polym. Adv. Technol.* (2006) 17, 772–777.
17. T. Kashiwagi, F. Du, K. I. Winey, K. M. Groth, J. R. Shields, *et al.*, *Polymer* (2005) 46, 471–481.
18. S. Peeterbroeck, F. Laoutid, B. Swoboda, J.-M. Lopez-Cuesta, N. Moreau, *et al.*, *Macromol. Rapid. Commun.* (2007) 28, 260.
19. F. Yang, G. L. Nelson, *Polym. Adv. Technol.* (2006) 17, 320.
20. Y. H. Hu, C. Y. Chen, C. C. Wang, *Polym. Degrad. Stab.* (2004) 84, 545.
21. A. Laachachi, E. Leroy, M. Cochez, M. Ferriol, J. M. Lopez-Cuesta, *Mater. Lett.* (2005) 59, 36.
22. A. Laachachi, M. Ferriol, M. Cochez, D. Ruch, J. M. Lopez-Cuesta, *Polym. Degrad. Stab.* (2008) 93, 1131–1137.
23. B. Friederich, A. Laachachi, M. Ferriol, D. Ruch, M. Cochez, *et al.*, *Polym. Degrad. Stab.* (2010) 95, 1183–1193.

24. F. Laoutid, R. Sonnier, D. François, L. Bonnaud, N. Cinausero, *et al.*, *Polym. Adv. Technol.* (2010) DOI: 10.1002/pat.1661.
25. L. Tibiletti, C. Longuet, L. Ferry, P. Coutelen, A. Mas, *et al.*, *Polym. Degrad. Stab.* (2011) 96, 67–75.
26. C.Q. Yang, Q. He, R.E. Lyon, Y. Hu, *Polym. Deg. Stab.* (2010) 95, 108–115.
27. R.N. Walters, R.E. Lyon, *J. of Appl. Pol. Sci.* (2003) 87, 548–563.
28. A.B. Morgan, M. Galaska, *Polym. Adv. Technol.* (2008) 19, 530–546.
29. B. Schartel, K.H. Pawlowski, R.E. Lyon, *Thermochimica Acta* (2007) 462, 1–14.
30. J.M. Cogen, T.S. Lin, R.E. Lyon, *Fire and Materials* (2009) 33, 33–50.
31. R. Sonnier, L. Ferry, C. Longuet, F. Laoutid, B. Friederich, *Polym. Adv.Technol.* (2011) 22, 1091–1099.
32. Y. Kojima, A. Usuki, M. Kawasumi, A. Okada, Y. Fukushima, *et al.*, *Mater. Research* (1993) 8, 1179–1189.
33. P.B. Messersmith, E.P. Giannelis, *Chem. Mater.* (1993) 5, 1064–1066.
34. A.B. Morgan, J.D. Harris, *Polymer* (2004) 45, 8695–3703.
35. E. Manias, A. Touny, L. Wu, K. Strawhecker, B. Lu, *et al.*, *Chem. Mater.* (2001) 13, 3516–3523.
36. S. Boucard, J. Duchet, J.F. Gérard, P. Prele, S. Gonzalez, *Macrom. Symp.* (2003) 194, 241–246.
37. T. Kashiwagi, 'Progress in flammability studies of nanocomposites with new types of nanoparticles' in *Flame Retardant Polymer Nanocomposites*, A.B. Morgan, C.A. Wilkie, Editors (2007) 285–324, John Wiley and Sons.
38. J.M. Lopez-Cuesta, F. Laoutid, 'Fire retardancy of polymeric materials multicomponent systems' in *Fire Retardancy of Polymeric Materials*, C.A. Wilkie, A.B. Morgan, Editors (2010) CRC Press.
39. M. Le Bras, G. Camino, S. Bourbigot, R. Delobel, *Fire Retardancy of Polymers. The Use of Intumescence* (1998) The Royal Society of Chemistry, Cambridge.
40. S. Bourbigot, M. Le Bras, F. Dabrowski, J. Gilman, T. Kashiwagi, *Fire Mater.* (2000) 24, 201–208.
41. Y. Tang, Y. Hu, Y.S. Wang, Z. Gui, Z. Chen, *et al.*, *Polym. Int.* (2003) 52, 1396–1400.
42. Y. Tang, Y. Hu, B. Li, L. Liu, Z. Wang, *et al.*, *J. Polym. Sci. A Polym. Chem.* (2004) 42, 6161–6173.
43. Y. Tang, Y. Hu, J. Xiao, J. Wang, L. Song, *et al.*, *Polym. Adv. Tech.* (2005) 16, 338–343.
44. Z.L. Ma, W.Y. Zhang, X.Y. Liu, *J. Appl. Polym. Sci.* (2006) 101, 739–746.
45. A. Laachachi, M. Cochez, E. Leroy, P. Gaudon, M. Ferriol, *et al.*, *Polym. Adv. Technol.* (2006) 17, 327–334.
46. A. Laachachi, M. Cochez, E. Leroy, M. Ferriol, J.M. Lopez-Cuesta, *Polym. Deg. Stab.* (2007) 92, 61–69.
47. L. Ferry, P. Gaudon, E. Leroy, J.M. Lopez-Cuesta, 'Intumescence in EVA copolymer filled with magnesium hydroxide and organoclays' in *Fire Retardancy of Polymers* (2005) Royal Society of Chemistry, London.
48. G. Beyer, 'Flame retardant properties of organoclays and CNT and their combinations with alumina trihydrate' in *Flame Retardant Polymer Nanocomposites*, A.B. Morgan, C.A. Wilkie, Editors (2007) 163–190, John Wiley and Sons.
49. G. Beyer, *Fire Mater.* (2005) 29, 61–69.
50. F. Laoutid, L. Ferry, E. Leroy, J.M. Lopez-Cuesta, *Polym. Deg. Stab.* (2006) 91, 2140–2145.

51. P. Van Nieuwenhuyse, V. Bounor-Legare, F. Boisson, P. Cassagnau, Al. Michel, *J. of Non-Crystal. Solids* (2008) 354, 1654–1663.
52. N. Cinausero, N. Azema, J.-M. Lopez-Cuesta, M. Cochez, M. Ferriol, *Pol. Adv. Tech.* (2010). Article on line (www.interscience.wiley.com) DOI: 10.1002/pat.1695.

# 17
# Polymer nanocomposites for optical applications

D. V. SZABÓ, Karlsruhe Institute of Technology (KIT), Germany and T. HANEMANN, Karlsruhe Institute of Technology (KIT), Germany and Albert-Ludwigs-University of Freiburg, Germany

**Abstract:** This chapter reviews the current literature on the optical properties of polymer/nanoparticle composites. The influence of the different nanocomposite formation techniques and the associated nanoparticle agglomeration behaviour on the resulting optical features – such as refractive index, transmittance and fluorescence – is described in detail. For completeness, a review of the linear and non-linear optical properties of polymer-based guest–host systems is included. In addition, some aspects of optical device fabrication using polymer nanocomposites are discussed.

**Key words:** nanocomposite formation, refractive index adjustment, fluorescence, transmittance, device fabrication.

## 17.1 Introduction

In the last three decades a new interdisciplinary branch of science and technology, namely 'microsystems technology' (MST), has been established following the common trend of miniaturization of structures, processing techniques and devices. MST is an enabling technology, allowing the implementation of new applications that could not previously be realized because of technical limitations. MST began with silicon, and since then three additional material classes have been established for use in MST, i.e. polymers, metals and ceramics. Optical devices fabricated using MST have many significant applications in life science; a wide range of material classes are used along with a huge variety of microfabrication technologies. The devices can be classified as follows:

- devices with structural microfeatures: optical fibre connectors with geometric precision in the micrometre range, microspectrometers;
- passive devices: microlenses, waveguides, Y-couplers;
- active devices: phase shifters, Mach–Zehnder interferometers.

Micro-electro-mechanical systems (MEMSs) have mostly been realized using standard silicon processing techniques such as lithography, etching, gas phase deposition techniques and others (Menz *et al.*, 2001). The integration of optical functionalities in MEMS, as for example in micromirror devices (www.dlp.com), leads to micro-opto-electro-mechanical systems (MOEMSs), which have been established in projection devices, tuneable filters and switches (Motamedi, 2005). Polymer optical components developed using these MSTs (e.g. polymer waveguides or free-space optical components like lenses) have been created using different lithographic, thin-film processing or replication methods (Ma *et al.*, 2002; Gale *et al.*, 2005).

In parallel with the developments in MST, nanotechnology has emerged as a key technology of the twenty-first century. Nanotechnology exploits special characteristics of size-related effects. As an example, the optical properties of a device fabricated using nanotechnology depend not only on the properties of the original bulk material, but also on the size and shape of the nanostructures. Hence, researchers now face the challenge of combining MST with nanotechnology to reap the benefits from both approaches. In the field of optical devices, much research effort has been expended, with particular focus on the use of nanoparticle technology for the development of new functional materials with particle-size-dependent optical properties.

In this chapter, the synergy between micro- and nanotechnology with respect to optical materials and devices will be outlined. The incorporation of inorganic nanoparticles or organic dopants in polymers allows new functional materials to be developed with tailored optical properties. The chapter begins with an outline of the optical properties that are of particular interest; this is followed by a discussion of polymer-based nanocomposite formation methods and the associated potential for linear and non-linear optical property tailoring. The chapter ends by considering optical device fabrication using the various different replication techniques that have been established.

## 17.2 Important optical properties of polymer nanocomposites

The design of new devices with passive or active optical functionality is driven by two main considerations:

- the optical, thermomechanical and geometric specifications given by the customer;
- the material properties of the available polymers, fillers or dopants.

The second consideration directly affects the designs that can be realized, and the fabrication method and the performance of the device (Baeumer, 2005). With respect to technological applications, the following optical characteristics are important:

- refractive index;
- good optical transmittance in the range 200–1600 nm;
- polarization;
- non-linear optical properties.

Before describing nanocomposite formation and optical property tailoring, a brief background to these optical properties is given below.

### 17.2.1 Refractive index

The refractive index (RI), $n$, of a material is the ratio of the speed of light $c$ to the phase speed $c_M$ in matter $M$:

$$n = \frac{c}{c_M} \qquad [17.1]$$

The refractive index is mainly influenced by the molecular structure, especially the polarizability $\alpha$ of the number of molecules $N$ per unit volume

$$\frac{n^2 - 1}{n^2 + 2} = \frac{4}{3}\pi N \alpha \qquad [17.2]$$

The wavelength-dependent complex refractive index $n'$ consists of the real part $n$ and the imaginary part $ik'$:

$$n' = n + ik' \qquad [17.3]$$

In Eq. 17.3, $k'$ describes the absorption of light

$$k' = \frac{\alpha' c}{2\omega} \qquad [17.4]$$

where $\alpha'$ is the linear absorption coefficient (cm$^{-1}$), $\omega$ is the frequency of the optical field (s$^{-1}$) and $c$ is the speed of light (cm/s$^{-1}$).

Equation 17.5 below shows that the complex refractive index is closely linked to the permittivity $\varepsilon_r$ and permeability $\mu_r$:

$$n' = \sqrt{\varepsilon_r \mu_r} \qquad [17.5]$$

In the visible range and without absorption a simple correlation between the refractive index and the relative permittivity can be applied

$$n^2 = \varepsilon_r \qquad [17.6]$$

The change of the refractive index with wavelength (dispersion) is quantified by the Abbe number

$$V = \frac{n_D - 1}{n_F - n_c} \qquad [17.7]$$

using three different wavelengths: the D-line (589.2 nm), the F-line (486.1 nm) and the C-line (656.3 nm).

Polymers with different optical axes due to broken symmetry show birefringence $\Delta n$, which can be used for the fabrication of polymer optical polarizers. In addition to birefringence from molecular origins, e.g. the presence of crystalline domains in a semi-crystalline polymer, birefringence can be influenced by the processing technique used, for example by polymer melt processing with subsequent tension freeze-off during cooling or by polymer chain orientation during laminar melt flow through a capillary. A comprehensive review covering the refractive indices of a huge number of polymers can be found in Galiatsatos (2007).

### 17.2.2 Transmittance and absorbance

The optical transmittance of a material in the wavelength region of interest, starting typically at 200 nm in the ultra-violet (UV) range up to 1600 nm in the near-infra-red (NIR) range, is affected by different factors, mostly absorption or scattering. Overtones originating from vibrational transitions of molecular bonds in the infra-red (IR) region cause pronounced absorption in the so-called telecommunication range (1300–1600 nm) in the NIR region. In the visible and UV range, electronic transitions between ground and excited states are responsible for a pronounced loss of transmittance. In particular, electrons fixed in double bonds or in non-binding electron pairs, as in oxygen or nitrogen, are capable of absorbing light in the near-UV and visible regions. The presence of conjugated electrons or intermolecular charge-transfer complexes results in significant absorption with large extinction coefficients; typical representatives are organic molecules like azo or anthraquinone dyes (Zollinger, 1991).

The family of thermoplastic polymers is well suited for optical applications because of the way in which the transmittance of the polymers can be manipulated by the processing technique. Thermoplastic polymers can be subdivided into amorphous and semi-crystalline materials. Amorphous polymers are mainly transparent and colourless while semi-crystalline polymers possess nano- and microcrystalline domains. These crystalline domains normally act as scattering centres and thereby reduce or prevent optical transmittance. Uniaxial or biaxial stretching of these materials enables the polymer chains to be orientated in one or two favoured directions, which therefore destroys the scattering domains.

Scattering can be subdivided into two main types:

- Rayleigh scattering: scattering centres are smaller than the wavelength, the limiting size is one-tenth of the applied wavelength.
- Mie scattering: scattering centres are the same size as the applied wavelength.

Commercial polymers contain a huge number of different organic additives, including UV stabilizers, release agents, plasticizers, moderators, which influence the polymerization reaction, and many more. These guest molecules, dispersed in the polymer matrix (the host), as well as any impurities, affect the optical transmittance of polymer films or bulk materials due to absorption or scattering. The same is true for the polymer ageing or polymer degradation that occurs during polymer-melt-processing methods such as extrusion or moulding.

### 17.2.3 Fluorescence

Fluorescence is a form of luminescence in which the emitted photons are of lower energy than the absorbed ones. This is expressed by Stokes' law: the emitted

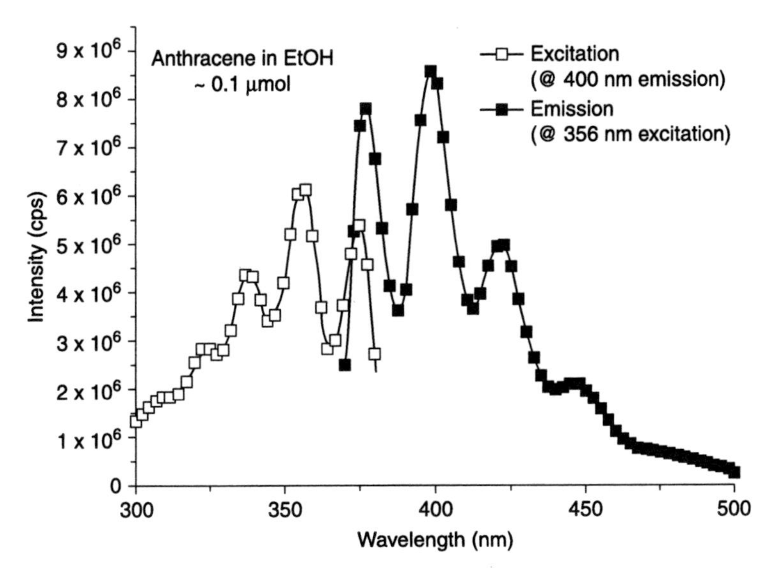


*17.1* Stokes' shift for anthracene dissolved in ethanol (EtOH). Emissions are shifted towards higher wavelengths.

spectrum is shifted to a longer wavelength, and therefore lower energy, than the absorbed radiation (Stokes, 1852); Fig. 17.1 shows the example of anthracene diluted in ethanol. Many organic molecules exhibit fluorescence behaviour, such as 1,4-bis(5-phenyloxazol-2-yl)benzene (POPOP), fluorescein, rhodamine B, anthracene, etc. The common characteristic of these molecules is an aromatic core, containing delocalized electrons in bonding $\pi$-orbitals. The fundamental physics of fluorescence can be found in detail in Lakowicz (1999) and is not described further in this section.

## 17.2.4 Polarization of light

Electromagnetic radiation has a transverse wave form. Polarization describes the direction of oscillation of the electric and magnetic field vectors perpendicular to the wave vector, which describes the propagation direction. In the case of linear polarized light the wave oscillates in a fixed plane perpendicular to the propagation direction. In the case of circular polarization, the plane of oscillation rotates with constant angular velocity. The polarization of light can be actively modulated by non-linear optical devices.

## 17.2.5 Non-linear optical properties

If matter is irradiated with electromagnetic radiation the induced polarization $P$ can be described by a linear correlation with the applied electrical field $E$ and the dielectric susceptibility X

$$P = \varepsilon_0 X(E) E \qquad [17.8]$$

At high electric fields, e.g. generated by laser light, the dielectric susceptibility depends on the independently applied electric fields $E$ following a Taylor series

$$X = X^{(1)} + X^{(2)}E + X^{(3)}E^2 + \ldots \qquad [17.9]$$

where $X^{(1)}$, $X^{(2)}$ and $X^{(3)}$ are the linear, quadratic and cubic components of the dielectric susceptibility tensors. The polarization behaviour can then be described more accurately by

$$P_i(E_j E_k E_l) = X_{ij}^{(1)} E_j + X_{ijk}^{(2)} E_j E_k + X_{ijkl}^{(3)} E_j E_k E_l + \ldots \qquad [17.10]$$

In the case of the typical (linear) optical properties – that is, absorption, refractive index, scattering, etc. – only the linear term $X^{(1)}$ is considered (Hanemann and Haase, 1991; Buckley, 1992).

Typical applications exploiting $X^{(2)}$ electro-optic effects are frequency doubling, as used in quality switches (Q switches) in laser cavities, e.g. Nd:YAG lasers. The Pockels effect describes the generation of birefringence in a suitable material as a function of the strength of an external electrical field. Pockels modulators can therefore be used to change the polarization and, if linear polarized light in combination with crossed polarizers is used, for intensity modulation. Suitable inorganic materials showing $X^{(2)}$ electro-optic properties must possess a non-centrosymmetric crystal structure. In the case of organic dopants, an asymmetric electronic distribution (a push–pull system) in a delocalized electron-rich aromatic structure must be present. These molecules possess large molecular dipole moments. Typical examples are substituted diazo and stilbene chromophores like *N,N*-dimethylaminonitrostilbene, dissolved in or attached to a polymer matrix (Yesodha *et al.*, 2004). When such a guest–host system is heated to slightly above the glass transition temperature and subsequently cooled under an applied electric DC field (>100 V/μm), a non-centrosymmetric structure is realized by the dopant's dipole-moment orientation along the field streamlines.

## 17.3 Polymer/nanoparticle composites

A huge variety of polymer nanocomposites is described in the literature. In parallel with the variety of composite types available, the range of synthesis methods is also very diverse. Therefore, in the following section we will focus on the most important types of polymer nanocomposites and the most relevant synthesis methods when considering polymer nanocomposites with applications in optics.

The optical properties that are of significance in these applications have been discussed above; in addition, the following thermomechanical properties, which relate to fabrication as well as the application of polymer micro-optical components, are of particular importance:

- The heat distortion temperature is the maximum operating temperature of a polymer device under a defined mechanical load, e.g. an optical lens mechanically fixed in a socket, holder or clamp. The heat distortion temperature is below a polymer's glass transition temperature range.
- The thermal expansion coefficient of polymers, with values around $10^{-4}$–$10^{-5}$, is significantly higher than the corresponding values for silicon, ceramics or metals. As a consequence, the combination of polymers with these other material classes leads to internal tensions, which directly affect the optical properties of the composite.
- Polymers containing hydrophilic moieties like carboxylates or amides tend to absorb water up to a level of 10 wt%. This water uptake occurs at the surface of the polymer and causes a swelling of the material accompanied by a change in refractive index and transmittance.

The most widely used optical matrix polymers are polymethylmethacrylate (PMMA) and polycarbonate (PC). These polymers will be discussed below, followed by a description of polymer nanocomposite types and the fabrication methods that are in currently in use.

### 17.3.1 Polymethylmethacrylate (PMMA) and polycarbonate (PC)

With respect to applications in polymer waveguides, free space optics or optical data storage, PMMA and PC are the most important commercially available polymers due to their excellent high transmission characteristics in the visible range. Table 17.1 summarizes the relevant optical and thermomechanical properties of both thermoplastic polymers.

For special applications, like the use of polymer microlenses in micro-invasive surgery, the physical properties of commercially available polymers are not adequate for good device performance. Hence there is a need to tailor the specific properties of the polymer, both the optical and thermomechanical properties,

*Table 17.1* Important properties relevant in applications of the polymers PMMA and PC

| Property | PMMA | PC |
|---|---|---|
| Optical transmittance (visible range, sample thickness 1–2 mm) (%) | 90–92 | 89–90 |
| Refractive index at 589 nm | 1.49–1.51 | 1.57–1.59 |
| Heat distortion temperature (without mechanical load) (°C) | 85–106 | 110–191 |
| Thermal expansion coefficient (range 23–80 °C) (/K) | $8 \times 10^{-5}$ | $7 \times 10^{-5}$ |
| Density (g/cm$^3$) | 1.16–1.19 | 1.14–1.20 |
| Water uptake (%) | 0.3 | 0.2–0.4 |

using nanosized fillers or dopants. In addition, mass fabrication-related material properties, like melt rheology, also have to be adjusted.

## 17.3.2 Polymer nanocomposite types

To illustrate the term 'nanocomposite', the most important types of polymer nanocomposite described in the literature, representing 'bulk' nanocomposites, are shown schematically in Fig. 17.2 (adapted from Vollath, 2008). Using polymers as the matrix, the most common types of classical polymer nanocomposites are zero-dimensional and one-dimensional nanocomposites: composites containing zero-dimensional spherical nanoparticles (upper left image), platelets or sheets (lower image) and composites containing one-dimensional tubes, fibres or rods (upper right image). Spherical nanoparticles are usually inorganic nanoparticles (metals, oxides, sulphides, selenides or nitrides). Platelets or sheets may be clay, graphite or graphene, while tubes, fibres or rods are predominantly carbon nanotubes. The polymer itself may be a standard polymer (e.g. PMMA, PC, epoxy resin or a copolymer) or a polymer with special physical properties (e.g. a conjugated polymer). Usually the zero-dimensional nanocomposites are of interest for classical optical applications. One-dimensional nanocomposites with tubes, fibres or rods

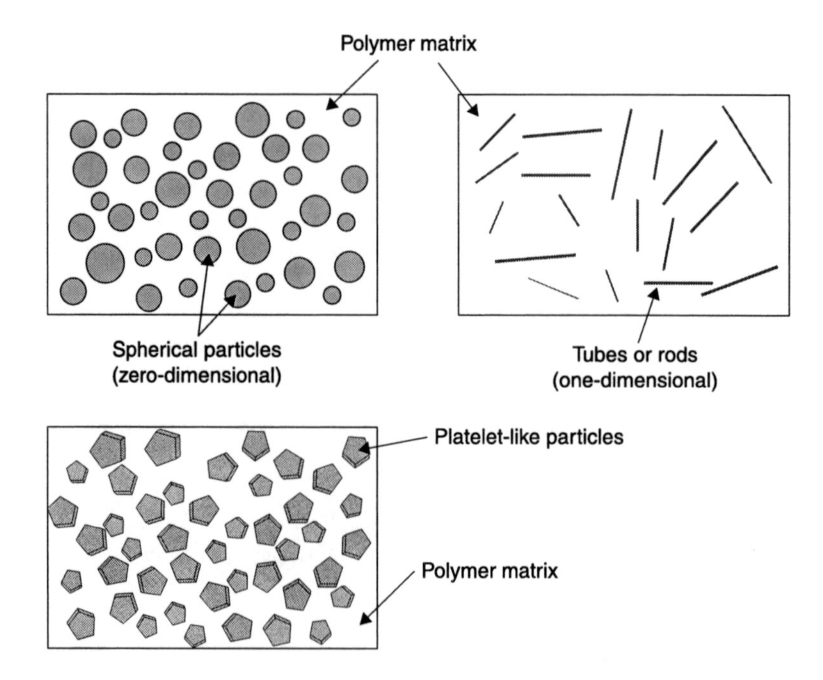


*17.2* Different bulk polymer/nanoparticle composite types. Adapted from Vollath, 2008.

are mainly of interest for electrical conductivity (Coleman *et al.*, 1998; Sandler *et al.*, 2003; Choi *et al.*, 2005; Murphy *et al.*, 2006; Winey *et al.*, 2007; Sahoo *et al.*, 2010) and therefore are not included in this chapter. Synergistic effects with respect to physical properties can be found when combining spherical nanoparticles and tubes in a composite (Sumfleth *et al.*, 2008). The nomenclature for this class of composites usually follows the format of 'polymer–nanoparticle', but, unfortunately, there are discrepancies in the literature. Therefore, the nomenclature used in the sections below follows the form used in the cited references.

In addition to the nanocomposites discussed above, special types of nanocomposites with interesting optical properties have been described in the literature, for example core/shell nanoparticles, polymer-grafted nanoparticles, nanocomposite particles, rods with grafted nanoparticles, and rods made of composites. These special types are illustrated in Fig. 17.3. In these cases, the spherical inorganic particles may be metals, oxides or semiconductors. Core/shell nanoparticles in this context contain an inorganic core and an organic coating, as described by Vollath *et al.* (2004). The organic coating may be a polymerized

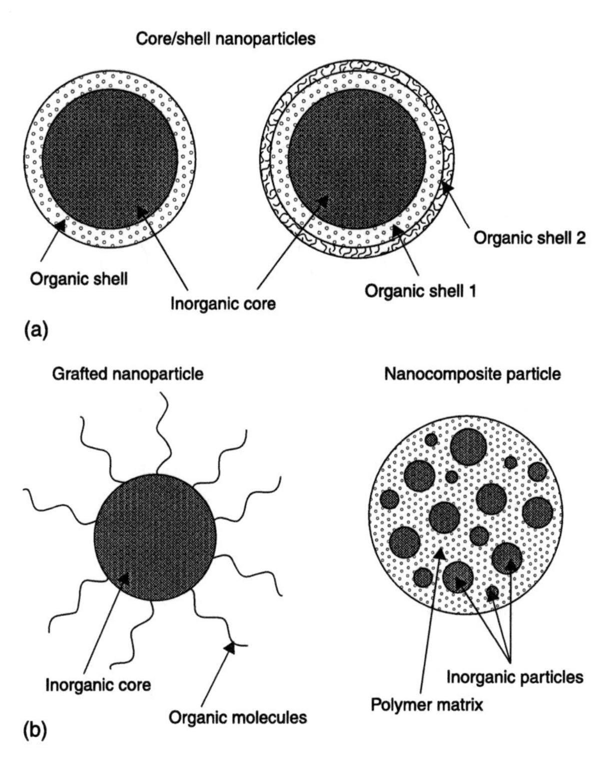


*17.3* Different special types of nanocomposites. (a) Core/shell nanoparticles. (b) Grafted nanoparticles and nanocomposite particles. (*Continued*)

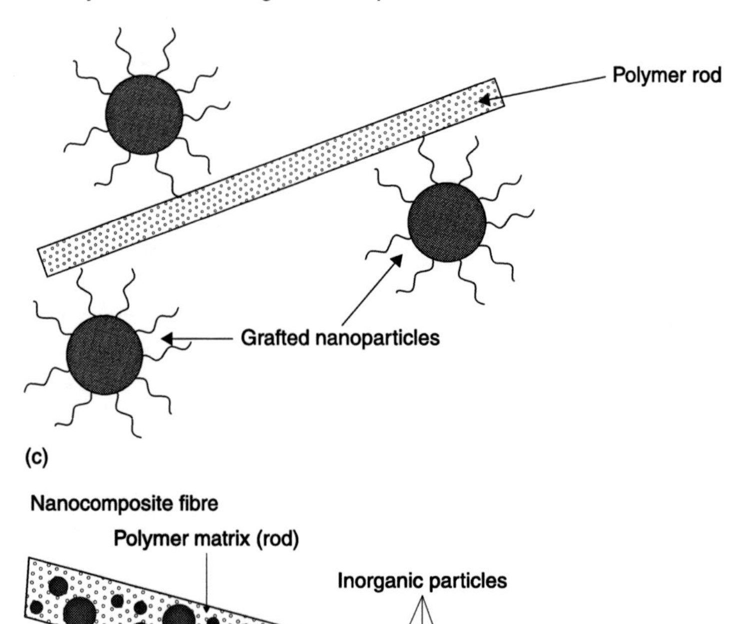


*17.3* (*Continued*) (c) Polymer nanorod with grafted nanoparticles. (d) Nanocomposite fibre.

monomer, an organic dye, a dendrimer or a surfactant. These particles are therefore also called 'grafted nanoparticles', where organic molecules are bonded to the particle surface (Vollath and Szabó, 2004; Vollath *et al.*, 2004; Wang *et al.*, 2006b). Core/shell nanoparticles with two organic layers (Vollath and Szabó, 2006) are a very special variation. For core/shell nanoparticles the nomenclature usually follows the format of 'core/shell'.

Nanocomposite particles, also called microsphere composites, in this context are particles (sometimes micron-sized), made of a polymer and inorganic filler nanoparticles. The inorganic particles themselves may additionally be surface-functionalized or grafted (Tang and Dong, 2009), and the nanocomposite particles may also contain an additional organic shell or encapsulation. Similar statements regarding the range of possible constituent parts of the nanocomposites can be made for rods grafted with nanoparticles and also the nanocomposite rods referred to above. This shows that a combination of several basic elements can lead to the fabrication of a wide variety of polymer nanocomposites, with a similarly wide range of applications.

### 17.3.3 *Ex situ* composite formation

*Ex situ* processes are methods where nanoparticles, synthesized in an external synthesis step, are added or mixed to a monomer or resin (an organic solution), usually followed by a polymerization step. Due to the many different compounding techniques – such as dispersing, blending and extrusion – that are well established in engineering and, furthermore, the broad range of commercial nanoparticles that is available in the size range around 50–100 nm, the *ex situ* process is the most frequently applied composite formation method. In the simplest case, the nanoparticles are used as produced or delivered, but this may result in agglomeration-related problems. Therefore, much research effort needs to be invested in the development of dispersing technologies (Ritzhaupt-Kleissl *et al.*, 2005; Krishnamoorti, 2007).

As mentioned above, *ex situ* methods have been frequently reported in the literature, and several examples are given in the following. Musikhin *et al.* (2002) used *ex situ* methods to generate luminescent polymer–dielectric nanocrystal composites using commercial $Al_2O_3$, $Y_2O_3$, ZnO and $SnO_2/Sb_2O_3/Sb_2O_5$ nanoparticles, respectively. In more elaborate set-ups, the nanoparticles were first surface functionalized and then added to the organic solution (Sangermano *et al.*, 2007). This method was also used by Tang and Dong (2009) to synthesize styrene polymer/ZnO nanocomposite latex. Mahdavian *et al.* (2009) encapsulated commercial $Al_2O_3$ nanoparticles by simultaneous use of an emulsifier with styrene/methylmethacrylate (MMA) using sonification and subsequent miniemulsion polymerization. Nanocomposites consisting of epoxy thermosets and $Al_2O_3$ have been prepared by simple mixing at elevated temperatures (Omrani *et al.*, 2009; Chen *et al.*, 2009a).

Cannillo *et al.* (2006) attached spherical $SiO_2$ nanoparticles (100–200 nm) chemically to polycaprolactone via grafting with a solid load of 1.0 and 2.5 wt%. Rong *et al.* (2005) applied a surface functionalization to commercial $TiO_2$, and then performed a free-radical polymerization of styrene to generate a nanocomposite. Alternative approaches use commercial nanoparticles, applying coupling agents and finally blending the particles with polymer powder (Pan *et al.*, 2008). These methods lead to 'bulk' composite materials. Wang *et al.* (2006a) combined the mixing of commercial nanoparticles in a monomer with polymerization using electron irradiation to obtain $TiO_2$/polymer nanoparticle composites. Another *ex situ* method is the coating of nanoparticles with a polymer by a subsequent polymerization treatment (Chen and Somasundaran, 1998). Very frequently, plasma polymerization processes are used to generate core/shell-type nanoparticles; here also, externally produced nanoparticles are used. Shi *et al.* (2002a, 2002b) combined a fluidized bed reactor with classical plasma polymerization to generate polymer-coated ZnO and $Al_2O_3$, respectively.

Despite the fact that a huge number of different *ex situ* blending and compounding methods are described in the literature, these methods typically suffer due to the presence of nanoparticle agglomerates, which have a direct effect on optical transmittance.

### 17.3.4 *In situ* composite formation

*In situ* composite formation processes are more sophisticated than *ex situ* processes. They usually lead to homogeneous nanoparticle distributions in the matrix, but the resulting materials are not currently ready for the mass market. The various techniques are usually divided into chemical and physical methods.

#### *Chemical* in situ *methods*

These methods use chemical reactions in a liquid environment to generate nanocomposites. The result may be either nanocomposite particles or bulk nanocomposite materials. Very comprehensive reviews on the variety of chemical synthesis methods are given by Caseri (2006) and Althues *et al.* (2007).

A common feature of the chemical *in situ* synthesis of nanocomposites is that (functionalized) nanoparticles are synthesized in a first step, mostly as a sol or dispersed in a solution, followed by a second step in which a monomer or resin is added and brought to polymerization. Much work has been carried out to develop these chemical *in situ* methods and a selection of the literature is presented in the following. Guan *et al.* (2009) synthesized transparent polymer nanocomposites containing ZnS using a one-pot route via *in situ* bulk polymerization. Gonsalves *et al.* (1997) synthesized AlN nanoparticles with a sol-gel method, and then applied an effective solution-mixing method to generate a homogeneous dispersion of AlN nanoparticles in polyimide. Copolymer/GaN nanocomposites were synthesized by *in situ* thermal decomposition of a precursor incorporated into a copolymer (Yang *et al.*, 1999). Xiong *et al.* (2004) prepared polymer/$TiO_2$ nanocomposites by mixing (3-methacryloxypropyl) trimethoxysilane (MPMS)-capped acrylic resins with sol-gel-synthesized $TiO_2$. Quantum dot/polymer nanocomposites were synthesized by polymerization in a microemulsion after synthesis of the nanoparticles by thermal decomposition of a precursor (Esteves *et al.*, 2005; Peres *et al.*, 2005). Althues *et al.* (2006) applied a two-step process to synthesize ZnO in a colloidal suspension, which was finally photopolymerized. A wet-chemical method followed by *in situ* polymerization of a monomer was applied by Cheng *et al.* (2008) in the synthesis of ZnS-containing nanocomposites. The *in situ* generation of $SiO_2$ nanoparticles in an organic solvent, which contained dissolved PMMA, via sol-gel techniques, led to PMMA/nanosilica hybrids after solvent evaporation and drying (Li *et al.*, 2004).

Chemical routes based on sol-gel processes and subsequent *in situ* polymerization are commonly used for the synthesis of hybrid nanocomposite particles and bulk nanocomposites. The main advantages of chemical *in situ* methods are the homogeneous particle size and the presence of isolated particles. One possible drawback may be residues of the chemical precursors.

*Physical* in situ *methods*

These methods are mainly gas-phase methods whereby *in situ* functionalized or encapsulated nanoparticles are synthesized to create hybrid core/shell nanoparticles (Fig. 17.3). A common feature of the methods is that they apply energy to transform chemical compounds (the volatile precursor and gas) into inorganic nanoparticles, and in a subsequent coating step organic compounds are grafted onto the nanoparticle surfaces for coating, encapsulation or surface functionalization. A versatile approach in the gas-phase synthesis of hybrid core/shell nanoparticles is the application of microwaves to generate a plasma. This approach was developed by Vollath and Szabó (1999, 2006) and was used to prepare several materials. Schallehn *et al.* (2003) and Suffner *et al.* (2007) applied chemical vapour synthesis (CVS) for the *in situ* polymer coating of $Al_2O_3$ and $SiO_2$ nanoparticles, respectively. If, instead of a microwave plasma, a traditional hot-wall reactor is used to synthesize inorganic nanoparticles, the coating step is performed in a subsequent radiofrequency (RF)-plasma reactor. Similar processes have been used for the synthesis of ceramic/polymer core/shell nanoparticles and also for the production of metal/polymer core/shell nanoparticles.

Physical gas-phase methods, especially when using microwaves as the energy source, are well suited for the synthesis of extremely small (<10 nm) nanoparticles with a narrow particle size distribution. They enable the *in situ* coating and functionalization of individual nanoparticles with different organic compounds. To date, however, this technology has only been developed on a laboratory scale.

## 17.4 Tailoring optical properties in the near infra-red, visible and ultra-violet ranges

With respect to optical applications, the adjustment of the refractive index of a material is of particular importance. Modification of the refractive index with simultaneous preservation of the transmittance is one of the challenges for particle/matrix nanocomposites, and therefore is reported quite frequently in the literature. Further topics of interest are the fluorescence of nanocomposites, polarization and non-linear optical properties. These properties will be described in more detail in the following sections.

### 17.4.1 Refractive index

To adjust the refractive index of a polymer nanocomposite, three different strategies can be pursued:

1. The chemical synthesis of side chain or block polymers (Fréchet, 2005).
2. The dispersion of nanosized inorganic fillers in a polymer matrix (*in situ* and *ex situ* techniques). As a rule of thumb, scattering and hence a loss of optical transmittance can only be avoided if the particles are smaller than one-tenth of the required optical wavelength, as derived from Rayleigh's scattering law (Miller and Friedmann, 1996). Specific dispersion techniques have to be applied to overcome the agglomeration of nanoparticles, which is often in the micrometre range and results in a significant degradation of optical properties.
3. The addition of organic dopants containing a large number of $\pi$-electrons to polymers increases the refractive index due to a pronounced molecular polarizability following the simple Eqs 17.2 and 17.6.

In a composite, the total refractive index is a function of the refractive indices of the polymer, the filler and the related volume fractions:

$$n_{\text{composite}} = n_{\text{matrix}} c_{\text{filler}} + n_{\text{filler}} c_{\text{filler}} \qquad [17.11]$$

The value changes linearly with the filler concentration, $c_{\text{filler}}$ (concentrations given in vol%). Thus, the density of the selected filler, as well as the difference in refractive indices of the components, plays an important role in achieving the desired effect. For small filler concentrations and almost identical density values for the polymer and the filler or dopant, Eq. 17.11 simplifies to

$$n_{\text{composite}} = n_{\text{matrix}} + n_{\text{filler}} c_{\text{filler}} \qquad [17.12]$$

where concentrations are given in wt%. These equations clearly show that the incorporation of nanoparticles with large differences in refractive index compared with the pristine polymer will significantly modify the refractive index of the resulting composite.

Organic polymers usually have refractive indices in the range 1.3 to 1.7 (see Table 17.2), whereas the refractive index of inorganic compounds may be around 1.5 to approximately 4.0. Table 17.3 shows the densities and refractive indices of the most commonly used inorganic filler materials that have been described in the literature, based on data taken from Kingery *et al.* (1976) and Lide (1992–1993).

In Eqs 17.11 and 17.12, the measurable optical property usually refers to 'bulk' nanocomposites (Fig. 17.2). In the ideal case – assuming monomodal spherical nanoparticles without inter-particular friction, neglecting van der Waals forces between the particles as well as agglomeration, and with a cubic primitive arrangement of the nanoparticles in the polymer matrix – a maximum degree of filling of 52 vol% is feasible. For a dense packing arrangement with a cubic face-centred arrangement of nanoparticles, a maximum degree of filling of 74 vol% can be obtained. In reality, the degree of filling will always be significantly lower. The large specific surface area of the filler causes the formation of an interfacial polymer layer (sometimes also called a 'shell', but not to be confused with 'core/shell' particles) attached to the particle core (Schadler *et al.*, 2007). Consequently, one

*Table 17.2* Densities and refractive indices of organic materials frequently used as a polymer matrix for tuning the refractive index

| Material | Density (g/cm$^3$)[a] | Refractive index $n$ |
|---|---|---|
| PMMA | 1.19 | 1.48899 at 633 nm[b] |
| PC | 1.2 | 1.58018 at 633 nm[b] |
| Polyamide | 1.0–1.15 | 1.53[c] |
| Polyethylene | 0.91–0.965 | 1.51[d] |
| Epoxy resin | 0.75–1.0 | 1.55–1.60[e] |
| Polystyrene | 1.05[f] | 1.5872 at 633 nm[b] |

Sources:

[a]http://www.tfmconsultants.com/specific_density2.html
[b]http://refractiveindex.info/
[c]http://www.goodfellow.com/E/Polyamide-Nylon-6,-6.html
[d]http://www.sdplastics.com/polyeth.html
[e]Galiatsatos, 2007
[f]http://www.viaoptic.de/main/de/download/materialuebersicht.htm

*Table 17.3* Densities and refractive indices of inorganic materials frequently used as fillers for tuning the refractive index of polymers

| Material | Density (g/cm$^3$)[a] | Refractive index $n$[a] | n[b] |
|---|---|---|---|
| $SiO_2$ (amorphous) | 2.17–2.20 | 1.41–1.46 | – |
| $SiO_2$ (quartz) | 2.635–2.660 | 1.544, 1.553 | 1.55 |
| $Al_2O_3$ (hexagonal) | 3.965 | 1.768, 1.760 | – |
| $Al_2O_3$ (gamma) | 3.5–3.9 | 1.7 | – |
| $Al_2O_3$ (corundum) | 3.97 | 1.765 | 1.76 |
| $TiO_2$ (rutile) | 4.26 | 2.616, 2.903 | 2.71 |
| $TiO_2$ (anatase) | 3.84 | 2.554, 2.493 | – |
| $ZrO_2$ (monoclinic) | 5.89 | 2.12, 2.19, 2.20 | – |
| ZnO | 5.606 | 2.008, 2.029 | – |
| PbS | 7.5 | 3.921 | 3.912 |
| ZnS (cubic) | 4.102 | 2.368 | – |

Sources:

[a]Lide, 1992–1993.
[b]Kingery, 1976.

should, in this case, refer to 'core-shell' particles dispersed in a polymer matrix. The presence of this shell will always reduce the maximum degree of filling of the nanoparticles in a polymer matrix. Assuming 5 nm particles, a 0.5 nm thick interfacial layer and a cubic primitive arrangement of the particles, a maximum degree of filling of 30 vol% is achievable; with a cubic face-centred arrangement, a 43 vol% maximum degree of filling is attainable. If the nanoparticle size is increased to 50 nm, the influence of the interfacial layer is drastically reduced.

When using nanoparticles as fillers, in addition to the influence of the interfacial layer, other side effects may occur. Several reported observations in the literature have clearly demonstrated the particle-size dependency of phase stability and also physical properties, such as refractive index or band gap. Zhang and Banfield (1998) analysed the phase stability of nanocrystalline $TiO_2$. They found anatase to be more stable than rutile when the particle size decreased below around 14 nm. As can be seen from Table 17.3, bulk anatase is characterized by a refractive index of 2.554 or 2.493, whereas rutile is characterized by a refractive index of 2.616 or 2.903 for bulk. For amorphous thin $TiO_2$ films, for example, a refractive index of 2.51 (at 550 nm) has been reported (Zhang *et al.*, 2007). Thus, depending on the particle size and phase of the $TiO_2$ used, the influence of the resulting composite on the refractive index will change. Size-dependent refractive indices have been reported for narrow band-gap semiconducting nanoparticles such as PbS (Kyprianidou-Leodidou *et al.*, 1994). With particles larger than 25 nm, the refractive index of PbS at different wavelengths was more or less independent of particle size and near the bulk values. For PbS particles with diameters below 25 nm, the refractive indices decreased significantly with size. At first glance, this is confusing, but a detailed knowledge of the nanoparticles in use will allow target-oriented tailoring of the desired properties.

$TiO_2$ is the most often used nanofiller material. However, different research groups use different particle sizes and phases, so that the results cannot always be compared directly. Figure 17.4 shows the increase in refractive indices for various polymer/$TiO_2$ and $TiO_2$/polymer systems, plotted against weight fraction (a) and volume fraction (b). Data are taken from Nussbaumer *et al.* (2003), Chau *et al.* (2007, 2008), Nakayama and Hayashi (2007), Liu *et al.* (2008), Rao and Chen (2008), Chang *et al.* (2009), Elim *et al.* (2009) and Imai *et al.* (2009). The two graphs show that, with increasing weight or volume fraction of $TiO_2$ nanoparticles, an increase in the refractive indices of the composites can be measured. The observed absolute values differ, however, sometimes even significantly. Differences in values can be explained by the different phases used as fillers, by the different polymer systems used as the matrix, by the different wavelengths used for determination or by the different particle sizes used by the authors. Chau *et al.* (2007) observed the highest increase in refractive index of around 0.9. As these authors did not indicate the refractive index of the epoxy resin that was used, a value of 1.54 was assumed. However, when applying Eq. 17.11 to their data, the resultant values are not realistic. In addition, Rao and Chen (2008) were able to obtain a large absolute increase of the refractive index of around 0.4. Interestingly, these authors were able to introduce a significantly higher volume fraction of $TiO_2$ nanoparticles smaller than 10 nm than is theoretically possible; these data are also, therefore, questionable.

$ZrO_2$ is a material which is also used quite often to adjust the refractive index of polymers. Figure 17.5 compares published data taken from Böhm *et al.* (2004), Sangermano *et al.* (2008), Imai *et al.* (2009) and Xu *et al.* (2009) as a function of weight fraction. As the nanoparticles used possess a higher refractive index than

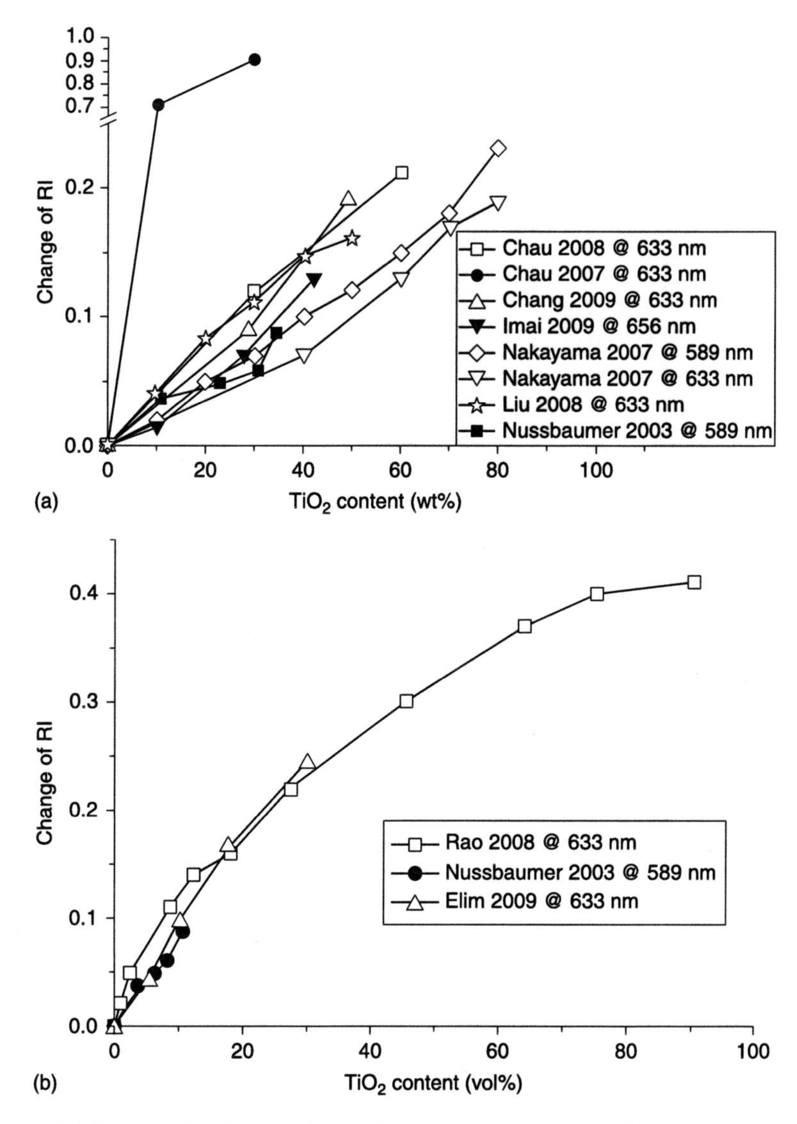


*17.4* Change of refractive index for different polymer/$TiO_2$ systems as a function of inorganic particle content. (a) Dependence of weight fraction (wt%). (b) Dependence of volume fraction (vol%).

the applied organic phases, here also an increase in the refractive index is observed with increasing weight fraction. The absolute increase is lower than for $TiO_2$ nanoparticles, since the refractive index of $ZrO_2$ is lower.

A very strong increase in refractive index was observed when semiconducting nanoparticles with an inherently high refractive index, like ZnS or PbS, are used as a filler, as shown in Fig. 17.6 where the data are taken from Cheng *et al.* (2008),

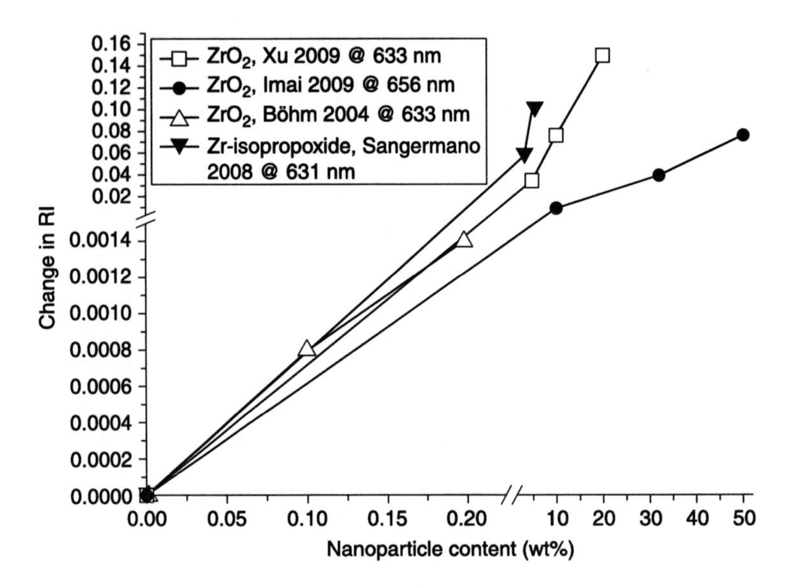


*17.5* Change of refractive index for different polymer/$ZrO_2$ systems as a function of inorganic particle content.

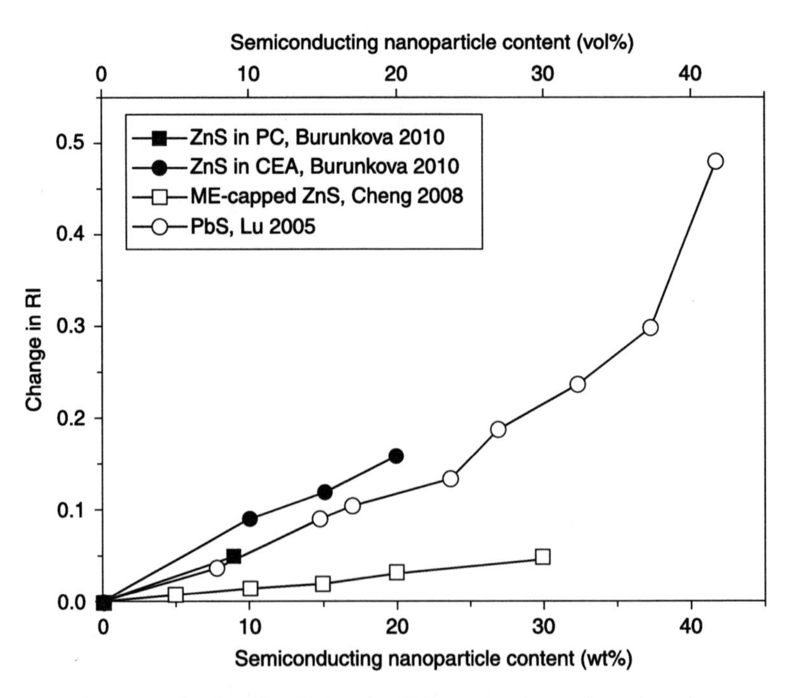


*17.6* Change of refractive index for different polymer/semiconductor nanoparticle systems as a function of inorganic particle content. Filled symbols show the dependence of volume fraction (upper *x*-axis), open symbols show the dependence of weight fraction (lower *x*-axis). (CEA: 2-carboxyethyl acrylate, ME: 2-mercaptoethanol).

Lu *et al.*, (2005) and Burunkova *et al.* (2010). The increase in refractive index using ZnS nanoparticles is of the same order of magnitude as realistic values for $TiO_2$-filled composites. In contrast, the data from Cheng *et al.* (2008) appear quite low. This is because their weight fraction is for functionalized capped nanoparticles. Consequently, the effective ZnS content is lower.

The use of nanoparticles with a lower refractive index than the organic matrix results in a reduced refractive index of the composite. This is shown in Fig. 17.7; here data are taken from Ritzhaupt-Kleissl *et al.* (2005) and Yu *et al.* (2010). An accurate comparison of data is difficult because Yu *et al.* (2010) also changed the composition of the monomer at the different weight fractions. The data from Ritzhaupt-Kleissl *et al.* (2005) show that the applied homogenization method has a significant influence on the refractive index change.

In summary, it can be seen that many research groups have found complementary results concerning the influence of different nanoparticles on the refractive index of polymer/nanoparticle nanocomposites. High degrees of filling have been achieved using chemical *in situ* methods to synthesize nanocomposites. In contrast, *ex situ* methods using commercial nanoparticles only permit low degrees of filling. However, some of the data are questionable insofar as they report either

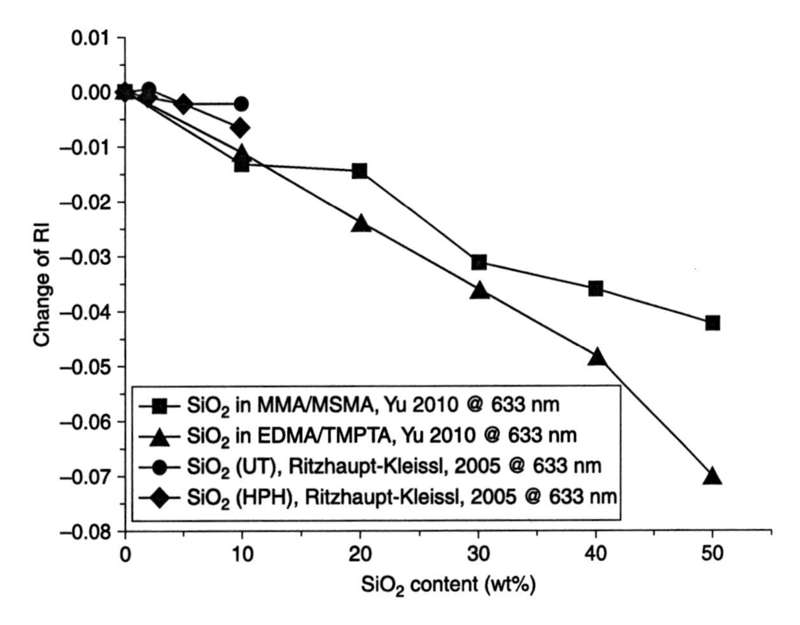


*17.7* Change of refractive index for different polymer/$SiO_2$ systems as a function of inorganic particle content. (UT: UltraTurrax; HPH: high-pressure homogenizer).

implausible filling degrees or non-reliable refractive indices of the composites. Data comparison is sometimes therefore challenging for the following reasons:

- the refractive indices were determined using different wavelengths;
- in the case of $TiO_2$, in particular, not all authors give information about the phase (amorphous, anatase, rutile or other) that was used;
- different units (i.e. wt% or vol%) were used for the filler content;
- surface-modified nanoparticles as well as pristine nanoparticles were used to modify the refractive index;
- different particle sizes were used;
- different processes were used for the synthesis of the composites;
- different homogenization methods were used;
- the influence of any precursor residuals is unclear.

### 17.4.2 Transmittance

In addition to the refractive index, the optical transmittance in the visible (consumer electronics) and the NIR range (optical data transmission) is of particular interest. The main criteria concern particle or agglomerate sizes: they must be below one-tenth of the wavelength used in order to avoid the influence of Rayleigh scattering ($\lambda/10$ criterion). In the case of polymer/nanoparticle composites, primary particle agglomeration causes the optical transmittance of a device to drop significantly, especially at larger sample thicknesses, according to the Lambert–Beer law. Therefore, the published data have to be carefully analysed with respect to the measured sample thickness. A thin layer of some 100 nm thickness can exhibit excellent optical transmittance in the visible range, but this transmittance will drop almost to zero for the technical dimensions of some millimetres that are used in the ASTM standard for transmittance measurements, according to the Lambert–Beer law. Ritzhaupt-Kleissl *et al.* (2006) showed the influence of particle treatment on optical properties: an *ex situ* silanization of nanosized $Al_2O_3$, dispersion in a MMA resin and subsequent polymerization yielded better optical transmittance values in the NIR range than untreated $Al_2O_3$ – when both high-speed stirring and high-pressure homogenization were applied.

In summary, the requirements for high transmittance are particles or agglomerates smaller than one-tenth of the applied wavelength and the use of appropriate homogenization techniques.

### 17.4.3 Fluorescence

In polymer nanocomposites (either nanoparticles embedded in an organic matrix or organic functionalized inorganic nanoparticles) fluorescence may be due either to the polymer matrix, e.g. conjugated polyphenylene vinylene (PPV)-based polymers, or to the inorganic particles or dopants (e.g. quantum dots, rare-earth elements). Additionally, fluorescence may be generated by the interface between the particles and the polymer. Five types of polymer nanocomposite fluorescence are described in detail below.

*Fluorescence of bulk composites using a conjugated polymer matrix with filler nanoparticles*

This type of fluorescence is not described as frequently in the literature as polymer matrix composites with fluorescent filler nanoparticles. Musikhin *et al.* (2002) studied luminescent nanocomposites made from dielectric oxides ($Y_2O_3$, $Al_2O_3$, ZnO, $SnO_2/Sb_2O_3/Sb_2O_3$) and the conjugated polymers PPV or poly[2-(6-cyano-6′-methylheptyloxy)-1,4-phenylene] (CN-PPP). They observed a noticeable shift of emission maxima to higher energies and a significant broadening of the emission, compared with the pristine conjugated polymers. The PPV-based nanocomposites additionally lost the vibronic structure of the pure polymer. The spectra of CN-PPP-based nanocomposites underwent much weaker changes in red shift and peak broadening. Yang *et al.* (2004) dispersed $TiO_2$ nanoparticles in PPV. They observed a shift of fluorescence to higher wavelengths, compared with the pristine conjugated polymer. Kim *et al.* (2009) developed complex poly-[2-methoxy-5-(2-ethylhexyloxy)-1,4-phenylenevinylene] (MEH-PPV) nanoparticles adsorbed with Au nanoparticles with diameters of about 70–180 nm. Photoluminescence increased, compared with the pure conjugated polymer nanoparticles. The vibronic structure of the PPV spectrum was not lost in the nanocomposite. Chen and Green (2010) used grafted Au nanoparticles, which they incorporated into MEH-PPV. They observed the effect of the grafted Au nanoparticles on the fluorescence of MEH-PPV: the emission was red-shifted by 8 nm and the intensity of the second emission peak was decreased. Jetson *et al.* (2010) modified the surface of ZnO nanoparticles in MEH-PPV/ZnO nanocomposites. Their results indicated that surface-modified ZnO nanoparticles were more effective at quenching the emission of MEH-PPV compared with unmodified ZnO nanoparticles, due to charge transfer. Time-resolved fluorescence had different mechanisms for unmodified and modified ZnO. Tang *et al.* (2010) hybridized the cationic conjugated polymer poly [9,9′-bis(6″-(*N,N,N*-trimethylammonium)-hexyl) fluorene-2,7-ylenevinylene-*co*-alt-1,4-phenylene dibromide] (PFV) with Ag–$SiO_2$ nanoparticles and observed an enhanced fluorescence intensity. They claimed that their material had the potential for applications in selective biological sensing and imaging.

*Fluorescence of bulk composites using a polymer matrix with semiconducting (fluorescent) filler nanoparticles*

There is increasing research interest in the incorporation of semiconducting nanoparticles into polymer matrices due to potential applications as light-emitting devices, non-linear optical devices or even biological labels. The most prominent composites with this type of fluorescence are either quantum dots (GaN, ZnS, CdS, CdSe or PbS) or ZnO nanoparticles, mostly embedded in methacryl-based polymers or polyvinyl alcohol (PVA). The whole spectrum of colours can be emitted. Yang *et al.* (1999) demonstrated a strong blue luminescence with an emission maximum near 426 nm in a copolymer/GaN sample, whereas the pristine polymer showed only weak fluorescence. Khanna *et al.* (2005) showed yellow and orange light emission in CdS nanoparticles embedded in PVA. CdS embedded in a poly(ethylene oxide) (PEO) matrix exhibited a fluorescence maximum around 630 nm (Yang *et al.*, 2003). Green-emitting CdSe/poly(butyl acrylate) (PBA) nanocomposites were prepared by Peres *et al.* (2005). Tamborra *et al.* (2008) demonstrated the fluorescence emission and intensity effects of the particle size of the filler and concentration in a PMMA matrix using CdS and CdSe–ZnS nanoparticles. They were able to vary the emission maxima between red, blue and green. Luminescent composites using ZnS and PMMA were reported by Chen *et al.* (2009b). Thin PVA films filled with CdSe–ZnS quantum dots exhibited a fluorescence similar to that of the pristine quantum dots, whereas pure PVA did not show fluorescence (Suo *et al.*, 2010). Wang *et al.* (2011) developed nanocomposites using CdS nanoparticles and polylactic acid (PLA) as the polymer, to produce fluorescent films with good biocompatibility. CdS/polymer bulk composites with poly(*N*-methylolacrylamide) (PNMA) were developed by Fang *et al.* (2010). Their material exhibited fluorescence at 530 nm. The fluorescence that appeared at 350 nm was attributed to the fluorescence of the polymer. In addition, metal nanoparticles like Au show fluorescence if the particle size is below 2 nm (Carotenuto *et al.*, 2009). Au nanoparticles in a styrene matrix, excited with UV light, exhibited red fluorescence. An overview of the literature data is given in Table 17.4. It is evident that, due to the different polymers used, and the different excitation wavelengths used, even similar materials are difficult to compare.

IR-luminescent PbS/polystyrene emits in both the visible and NIR regions (Lim *et al.*, 2004). Figure 17.8 shows the calculated band gap edges as well as the measured NIR emission maxima versus PbS particle size. Lim *et al.* (2004) claim that the calculated and measured values were in quite good agreement.

A broad variety of nanocomposites can be developed using ZnO as filler nanoparticles. Nanocomposite films made of ZnO nanoparticles embedded in a PMMA matrix exhibited a pronounced UV-fluorescence emission, compared with the blank PMMA film (Du *et al.*, 2006), with two emission maxima at 334 and 346 nm. Sato *et al.* (2008) observed a strong fluorescence emission at 407 nm for their PMMA/ZnO nanocomposites. Yang *et al.* (2008) found a broad emission spectrum from a ZnO/epoxy nanocomposite in the range from 400 to 600 nm, with an emission maximum occurring at 442 nm. The fluorescence intensity

*Table 17.4* Fluorescence of quantum dot/polymer nanocomposites

| Quantum dot | Matrix | Excitation wavelength | Emission peaks | Reference |
|---|---|---|---|---|
| GaN, 40 nm | copolymer | 360 nm | 406 nm, 426 nm, 450 nm | Yang *et al.*, 1999 |
| CdS | PEO | 420 nm | 630 nm | Yang *et al.*, 2003 |
| 2% CdS | PVA | n.a. | 446 nm, 550 nm | Khanna *et al.*, 2005 |
| 5% CdS | PVA | n.a. | 570 nm | Khanna *et al.*, 2005 |
| CdSe | PBA | 325 nm | 2.29 eV | Peres *et al.*, 2005 |
| CdS, 3 nm | PMMA | 325 nm | 445 nm | Tamborra *et al.*, 2008 |
| CdS, 4 nm | PMMA | 325 nm | 480 nm | Tamborra *et al.*, 2008 |
| CdS, 7 nm | PMMA | 325 nm | 500 nm | Tamborra *et al.*, 2008 |
| CsSe@ZnS | PMMA | 400 nm | 615 nm | Tamborra *et al.*, 2008 |
| Au, 1.7 nm | PS | 285 nm | 633 nm | Carotenuto *et al.*, 2009 |
| ZnS (PMAA capped) | PMMA | 300 nm | 414 nm | Chen *et al.*, 2009b |
| CdSe–ZnS | PVA | 350 nm | 574 nm | Suo *et al.*, 2010 |
| CdS | PNMA | 302 nm | (350 nm), 530 nm | Fang *et al.*, 2010 |
| CdS, 5 nm | PLA | 370 nm | 410 nm, 640 nm | Wang *et al.*, 2011 |

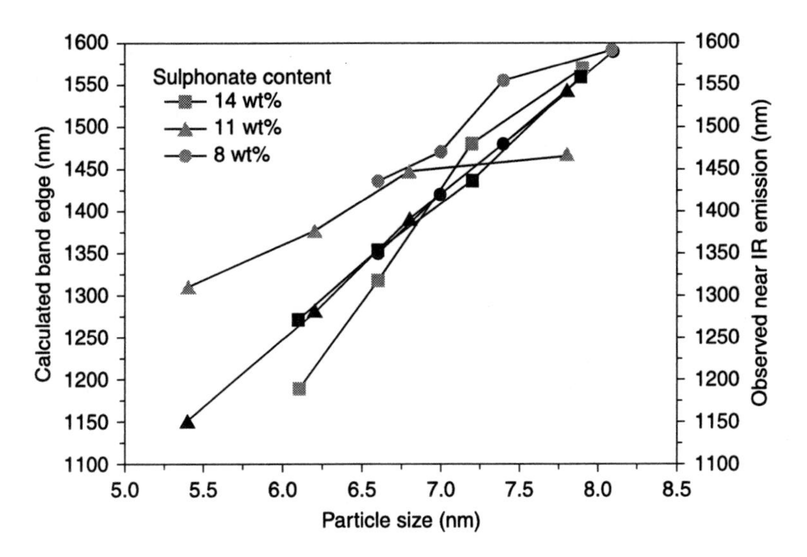


*17.8* Comparison of calculated band edge (black symbols, left *y*-axis) and observed near IR emission (grey symbols, right *y*-axis) of PbS nanoparticles in polystyrene as a function of particle size (Lim *et al.*, 2004).

increased with increasing ZnO content. Sun and Sue (2009) were able to tune the UV emission of ZnO quantum dots in transparent PMMA by adding exfoliated ZnP nanoplatelets. Althues *et al.* (2009) observed a blue shift of the UV emission from 541 to 534 nm with decreasing ZnO particle size from 10.4 to 6.2 nm in a poly(butanediolmonoacrylate) (PBDMA) matrix. An enhanced room-temperature excitonic luminescence was observed in ZnO/PMMA nanocomposites (Paramo *et al.*, 2010). An overview of the excitation and emission values measured in different ZnO/polymer systems is given in Table 17.5. As with previous examples, the differences in the polymers used and the different excitation conditions do not allow a simple comparison of the data.

*Fluorescence of nanocomposites containing rare-earth elements*

One very challenging application of fluorescent nanocomposites is the realization of polymer light-emitting diodes (LEDs) or lasers. A promising approach is the fabrication of polymer waveguides containing inorganic–organic hybrid materials like rare-earth complexes. A PMMA-based planar optical waveguide, containing 2.5 wt% $Eu(DBM)_3$ phen (where DBM is dibenzoylmethane and phen is 1,10-phenanthroline), showed an intense fluorescence emission at 613 nm after pumping by a 457 nm Ar ion laser (Sun *et al.*, 2004). A different europium complex, dissolved in an inorganic–organic hybrid network, produced multi-wavelength photoluminescence in the UV/visible range (Ferreira *et al.*, 2010b). The emission wavelengths can be significantly influenced by the presence of polymer-modified zirconium-propoxide nanoclusters. The addition of an erbium complex to a methacrylate copolymer led, after fabrication of a channel waveguide, to photoluminescence in the NIR range at 1540 nm after excitation at 980 nm (Bo *et al.*, 2009). Luminescent silica-coated $Y_2O_3:Eu^{3+}$ nanopowders, with particle sizes around 10 nm, synthesized via sol-gel methods, showed photoluminescence around 610–620 nm (the excitation wavelength was 260 nm) (Carrillo Romo *et al.*, 2010). A similar approach was described by Macedo and co-workers using silica-coated $Gd_2O_3:Eu^{3+}$, grafted with an organosilane methacrylate and added to styrene prior to radical polymerization (Macedo *et al.*, 2010). Depending on the excitation wavelength, different photoemission spectra in the visible range could be measured.

A comprehensive overview of the different inorganic–organic hybrid materials with the potential to realize passive and optical devices has been published by Ferreira and co-workers (Ferreira *et al.*, 2010a). New photoactive composites based on ethyl vinyl acetate (EVA) and $SrAl_2O_4$:Eu,Dy or $Sr_4Al_{14}O_{25}$:Eu,Dy were developed by Mishra *et al.* (2010). The phosphor nanoparticles were dispersed uniformly in the EVA matrix. After excitation with sunlight (320 nm) phosphorescence at 520 nm was observed for the EVA/$SrAl_2O_4$:Eu,Dy composite, and at 495 nm for the EVA/$Sr_4Al_{14}O_{25}$:Eu,Dy composite, respectively. The intensity was reduced, compared with the pure phosphor.

*Table 17.5* Fluorescence of ZnO/polymer nanocomposites

| ZnO diameter | Matrix | Excitation wavelength | Emission peaks | Comments | Reference |
|---|---|---|---|---|---|
| 5–6 nm | PMMA | 237 nm | 334 nm, 346 nm | | Du *et al.*, 2006 |
| n.a. | PMMA | n.a. | 407 nm | | Sato *et al.*, 2008 |
| n.a. | Poly(2-(carbazol-9-yl) ethyl methacrylate) | n.a. | 352 nm, 366 nm | | Sato *et al.*, 2008 |
| 3.1 nm | Epoxy | 370 nm | 442 nm | | Yang *et al.*, 2008 |
| 5.0 nm | PMMA | 320 nm | 365–387 nm | For varying ZnO content and 0.5% ZnP | Sun and Sue, 2009 |
| 6.2 nm | PBDMA | n.a. | 534 nm | | Althues *et al.*, 2009 |
| 7.6 nm | PBDMA | n.a. | 541 nm | | Althues *et al.*, 2009 |
| 10.4 nm | PBDMA | n.a. | 541 nm | | Althues *et al.*, 2009 |
| 20–50 nm | PMMA | 325 nm | 3.3 eV, 3.1 eV, 2.4 eV | in diols | Paramo *et al.*, 2010 |
| rods | PMMA | 325 nm | 3.3 eV, 3.1 eV, 2.4 eV | in diethyleneglycol | Paramo *et al.*, 2010 |
| n.a. | PNIPAM | 325 nm | 376 nm | | John *et al.*, 2010 |
| 12 nm | no | 485 nm | 550 nm | | Jetson *et al.*, 2010 |
| 9 nm (ME-capped) | no | | 380 nm, 550 nm | weak | Jetson *et al.*, 2010 |
| 9 nm (ME-capped) | MEH-PPV in tetrahydrofuran | | 560 nm, 600 nm | weak | Jetson *et al.*, 2010 |

Note: PNIPAM: poly(*N*-isopropylacrylamide).

*Fluorescence of core/shell nanoparticles*

Experimental evidence for fluorescence generated by the nature of the interface in core/shell nanoparticles was presented by Vollath *et al.* (2004) and Wang *et al.* (2006b). In PMMA-functionalized nanoparticles (core/shell nanoparticles), carboxylic groups at the interface between the inorganic nanoparticles and the polymer shell produce the luminescence. A similar model, also using carboxylic groups at the interface, was proposed by Xiong (2010). With decreasing particle size, the surface area increases and therefore the number of carboxylic groups in the interface increases, resulting in a further rise in intensity. The fluorescence can be varied significantly by using different organic coating compounds and to some extent by varying the nature of the core. Figure 17.9 shows the fluorescence spectra of $ZrO_2$ nanoparticles coated with different organic molecules. Nanocomposites containing fluorescent molecules are characterized by the fluorescence of the organic molecules (Vollath and Szabó, 2004, 2006; Sagmeister *et al.*, 2010). Figure 17.10 shows the fluorescence of core/shell nanocomposites, with different ceramic cores and coated with PMMA. Nanocomposites containing isolating $HfO_2$ and $ZrO_2$ show the strongest fluorescence intensity, whereas the

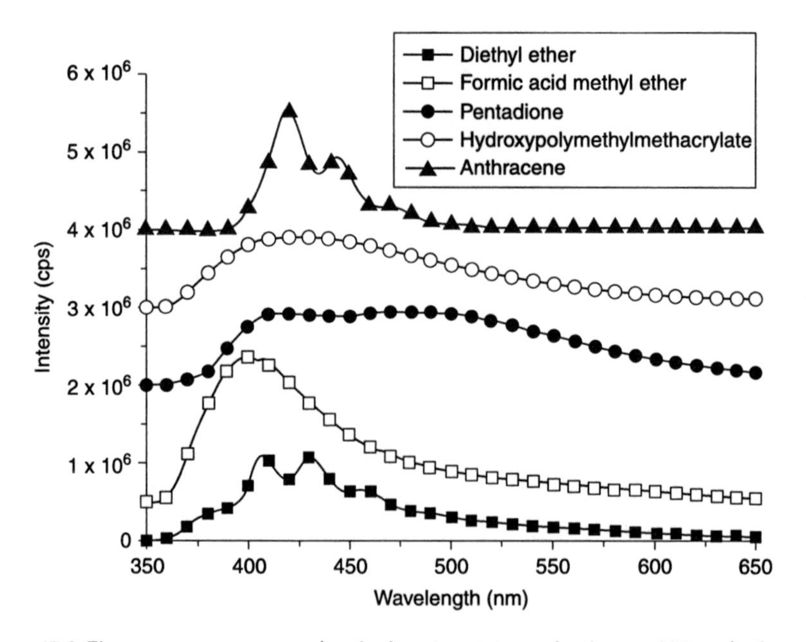


*17.9* Fluorescence spectra (emission spectra, excitation at 325 nm) of core/shell nanoparticles with a $ZrO_2$ core and different organic coatings. The emissions change with the organic coating. The data are stacked for better visibility: the intensity of the anthracene-containing composite is reduced by a factor of 10 for better comparison with the other data.

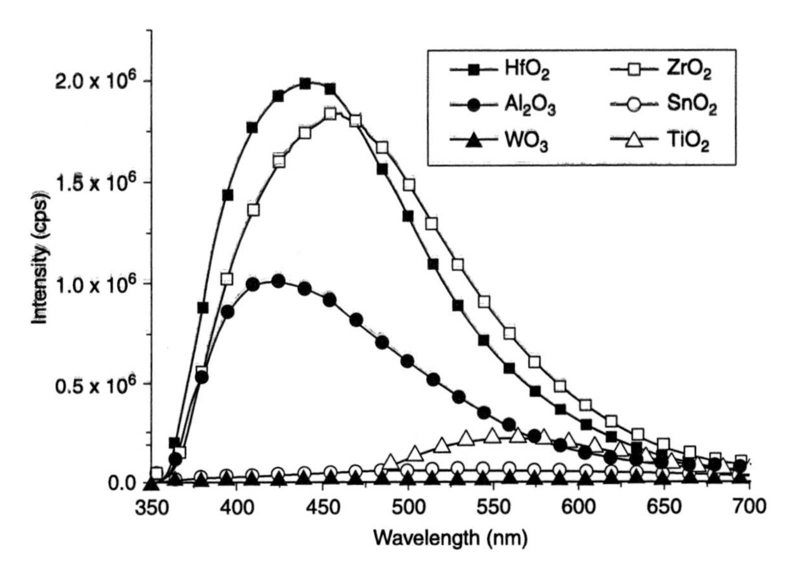


*17.10* Fluorescence spectra (emission spectra, excitation at 325 nm) of core/shell nanoparticles with an inorganic core and coated with PMMA. The shape of the emission spectra is similar, but intensity and maxima change with the type of inorganic core. For $TiO_2$ the excitation was 470 nm.

semiconducting $WO_3$- and $SnO_2$-containing nanocomposites do not show significant fluorescence as powders. Similar spectra, with broad emission ranges, have been observed for several oxide/polymer core/shell nanoparticles (Vollath and Szabó, 2004; Wang *et al.*, 2006a). In addition, other types of core/shell nanoparticles exhibit fluorescence. John *et al.* (2010) used poly(*N*-isopropylacrylamide) (PNIPAM) hydrogel and incorporated ZnO nanoparticles, forming hybrid ZnO/PNIPAM nanoparticles. These particles exhibited fluorescence at 376 nm, when excited at 325 nm. Hybrid ZnO/PNIPAM nanoparticles with amine functional groups can be conjugated with bio-molecules. Thus, this material has potential for biomedical applications. Govindaiah *et al.* (2010) developed bifunctional nanoparticles by functionalizing iron-oxide nanoparticles with a dye, yielding a luminescent, magnetic core/shell nanoparticle, which also has potential for applications in biomedicine or bio-separation.

*Bifunctional nanocomposites*

The development of bifunctional nanocomposites is quite recent. In these nanocomposites, two functions are combined in one composite or particle. The most prominent combination of properties is fluorescence with magnetism. Such systems are of interest for applications in biology, medicine and diagnostics. The systems can be realized using core/shell nanoparticles, where by the particle is designed as a core/shell-1/shell-2 particle. Thus, the synthesis of magnetic and simultaneously fluorescent nanoparticles is possible. By correct selection of the organic dye, the luminescence of, for example, $\gamma$-$Fe_2O_3$ nanoparticles can be adjusted. One can obtain bifunctional nanocomposite particles with superparamagnetic and fluorescing properties, depending on particle size and the organic coating selected (Vollath and Szabó, 2006; Vollath, 2010). Other developments incorporate fluorescent quantum dots as well as superparamagnetic nanocrystals in polymer beads or spheres (Tu *et al.*, 2008; Zhang *et al.*, 2008, 2011; Cho *et al.*, 2010; Evans *et al.*, 2010; Govindaiah *et al.*, 2010), as shown schematically in Fig. 17.3(b). These latter developments, in particular, target applications in biology, biomedicine and bio-analysis.

In summary, the fluorescence of a polymer/nanoparticle nanocomposite can be influenced by the polymer, by the nanoparticles or by the interface between the nanoparticles and the polymer. Therefore, combining the possibilities of polymer chemistry, nanoparticles and interfacial chemistry offers a large toolbox for the design of application-specific fluorescent nanocomposites. In particular, the combination of fluorescent and magnetic properties in one particle/bead/sphere is of particular interest due to potential applications in the biomedical and biological sciences.

## 17.5 Linear and non-linear optical properties of guest–host systems

As described above, the addition of nanoparticles, which are mostly agglomerated, to a polymer matrix reduces the optical transmittance due to scattering even at small nanoparticle contents. Hence it is difficult to achieve significant refractive index increases (according to Eq. 17.12) and excellent optical transmittance. In order to avoid pronounced optical losses, light scattering has to be strictly avoided; therefore polymer/nanoparticle composites cannot be used. Chemical solutions of small organic molecules, such as dyes, in polymers are technically named guest–host systems. The guests (the organic molecules) are randomly distributed at a molecular level in the host (the polymer matrix). Typical geometric dimensions of suitable molecules are below 1 nm; as a result no scattering occurs. As a positive side effect, stable concentrations of up to 20 wt% can be achieved prior to precipitation of the guest. The processing is simple: the guest molecules are dissolved in a monomer/polymer mixture, and the resulting guest–host system can then be polymerized, either thermally or photochemically, using suitable initiators for a solid thermoplastic or thermoset polymer. The coextrusion of a thermoplastic polymer with a dopant is more elaborate and can cause chemical degradation of the guest molecule due to the high melt processing temperature.

For customization of the refractive index, easy-to-polarize organic molecules (see Eqs 17.2 and 17.6), such as electron-rich aromatics like phenanthrene or benzoquinoline (Fig. 17.11), are suitable guest molecules. Their addition to

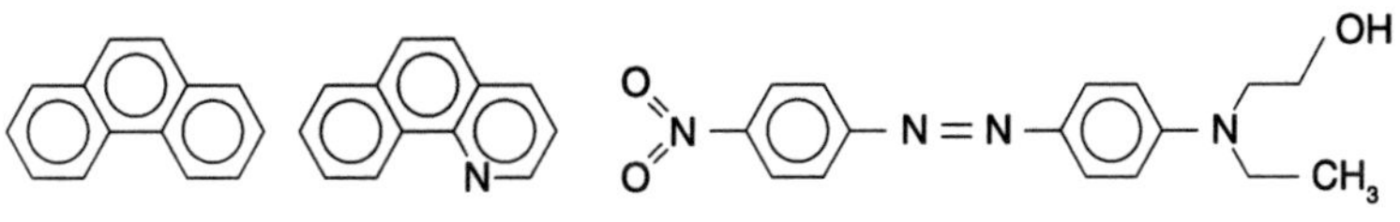

*17.11* Chemical structures of small, electron-rich organic molecules. Left: Phenanthrene. Middle: Benzoquinoline. Right: Disperse Red 1 (DR1).

polymers can give a pronounced increase in refractive index (for PMMA from 1.49 to 1.58, for unsaturated polyester from 1.57 to 1.61, for epoxide from 1.56 to 1.62, all measured at 589 nm) (Hanemann *et al.*, 2008). Due to a high guest molecule concentration of up to 45 wt%, a pronounced yellowing of thicker samples (2 mm), affecting the transmittance below 550 nm, was observed. Another drawback is that the dopants act as a plasticizer and cause a pronounced decrease of the glass transition temperatures and as a consequence a drop in a device's maximum operating temperature (Hanemann *et al.*, 2007). Copolymerization with a crosslinker can compensate for the plasticizing effect; if the crosslinker contains an aromatic moiety a further refractive index increase can also be achieved (Hanemann and Honnef, 2010).

Asymmetric substituted azo dyes, like the commercial dye Disperse Red 1 (DR1), have huge permanent dipole moments. Polymer-based guest–host systems containing these organic dyes can be used in the realization of active micro-optical devices, e.g. a Pockels modulator, which exploit the non-linear optical properties of the mixture.

A Pockels modulator changes the polarization angle of incoming linear polarized light; the angle depends on the strength of an applied external electric field. As an example, Fig. 17.12 shows the change of the polarization angle of incoming linear polarized light (1550 nm) with an applied external modulation field in a PMMA/DR1 single-mode waveguide ($5 \times 5\,\mu m^2$). After the threshold of 4 V/μm, there is a significant change in the polarization angle, which confirms the functionality of the Pockels device (Hanemann *et al.*, 2008).

## 17.6 Replication and optical device fabrication

In the last few years, three main production techniques have been established for the realization of small and micro-optical components (Gale *et al.*, 2005):

- UV embossing or UV moulding;
- hot embossing;
- injection moulding.

For UV embossing or UV moulding, a low-viscosity curable reactive polymer-based resin (an epoxide, a methacrylate or others) containing additives is placed or injected in a mould cavity and solidified by photopolymerization under ambient

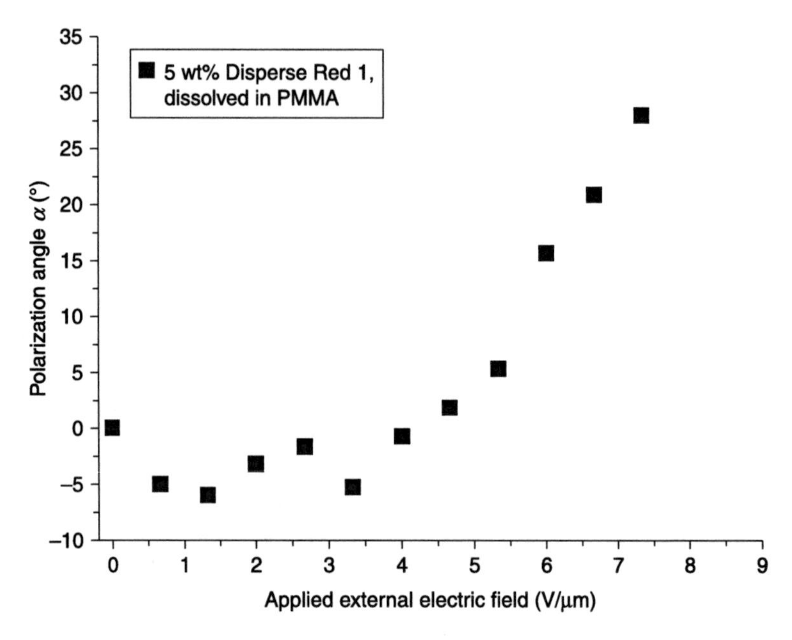


*17.12* Change of polarization angle of incoming 1550-nm radiation with an applied external voltage for PMMA doped with DR1 (Hanemann *et al.*, 2008).

conditions to realize the final component (e.g. a microlens) (Piotter *et al.*, 2008). The additives can be used, for example, to tune the refractive index. An established company using this technique is Heptagon (www.hptg.com). In contrast to hot embossing, only a single-sided structure is possible, because either the substrate or the mould carrying the surface relief has to be highly transparent. The different variants of UV embossing can be used for rapid prototyping (Hanemann *et al.*, 2006) or for fabricating small-scale series.

Semi-finished polymer sheets can be shaped by hot embossing. The polymer is heated to a temperature higher than the glass transition temperature. A shaped mould is pressed into the sheet and a shaped counter tool enables both sides to be shaped. A comprehensive review of hot embossing is Heckele and Schomburg (2004), and quite recently a textbook, *Hot Embossing*, was published by Elsevier (Worgull, 2009). Due to the short flow paths, hot embossing can produce parts with almost no remaining internal stress, which could negatively affect the optical properties of the device. Hot embossing is suitable for small- and mid-scale fabrication runs. Polymer waveguides with a PMMA cladding and a refractive-index-adjusted core, consisting of PMMA and 0.5 wt% nanosized alumina, have been fabricated by hot embossing. The refractive index increase, induced by the nanosized alumina, was measured to be 0.0021 (at 1550 nm), and the specific transmittance was around 96%/mm waveguide length (Ritzhaupt-Kleissl, 2006).

Finally, the different variants of injection moulding are the most important production methods for the mass fabrication of optical components (D'Amore *et al.*, 2004; Mayer, 2007). Polymer granules are heated to a temperature significantly higher than the glass transition temperature. The resulting polymer melt is then injected into a shaped cavity, which is the negative of the desired part. Typical optical components are lenses, prisms, microspectrometers, CD/DVD/Blu-ray optical data storage discs, waveguide structures such as beam splitters and others.

UV moulding and injection moulding have been used for the fabrication of simple optical devices like prisms using polymer/nanofiller composites. Different nanosized ceramic fillers (hydrophilic and hydrophobic aerosils, alumina, zirconia and titania) were dispersed in an unsaturated polyester of MMA-based reactive resins and cured in a simple prism mould or in a reaction moulding set-up suitable for photocuring (Ritzhaupt-Kleissl *et al.*, 2005, 2006). The solid test specimens were characterized with respect to their optical properties such as refractive index and transmittance in the NIR and visible regions. Due to particle agglomeration, the addition of increasing nanoparticle amounts caused a reduction in transmittance, eventually giving a translucent state; photocuring was still possible even at higher nanofiller contents of up to 10 wt%. The applied dispersing method chosen had a pronounced effect on the resulting transmittance, and this could be traced back to the different shear forces generated during dispersion (Ritzhaupt-Kleissl *et al.*, 2005).

The coextrusion of PMMA with nanosized silica and alumina enabled the formation of polymer-based composites suitable for injection moulding (Ritzhaupt-Kleissl *et al.*, 2005). Simple test structures have been produced; unfortunately, increasing the nanofiller content reduced the optical transmittance significantly. The addition of nanosized alumina to a high-performance polycarbonate (APEC 2097, glass transition temperature 202 °C, from Bayer MaterialScience) led to a decomposition of the polymer during compounding via coextrusion. Besides a pronounced discoloration, a significant drop of the glass transition temperature and a change of the molecular mass distribution, measured by size exclusion chromatography, could be detected even at low solid loadings of around 5 wt% (Hanemann *et al.*, 2009).

## 17.7 Conclusion and future trends

The incorporation of inorganic nanoparticles into a polymer matrix allows the optical properties, such as refractive index and fluorescence, of the resulting composite to be tailored. A major drawback, however, is that the nanoparticles often reduce the transmittance properties of samples thicker than 500 μm due to particle agglomeration; this feature currently reduces the likelihood of suitable applications being developed. Only composites containing 'real' nanoparticles significantly smaller than one-tenth of the relevant wavelength do not show this optical scattering. At present only the different *in situ* composite formation techniques can fulfil this requirement. The use of nanosized inorganic–organic hybrid materials or organic molecules as guest dopants extends the range of potential applications to active optical devices like phase and polarization shifters or all-organic light sources. One fundamental benefit of using polymer-based nanocomposites is the use of established polymer processing techniques suitable for low-cost mass fabrication.

In the near future polymer/nanoparticle composites will play a prominent role in the realization of very challenging passive and active optical devices like lenses or prisms as well as tuneable waveguides or modulators, full-organic LEDs and lasers or organic solar cells with improved efficiency.

## 17.8 References

Althues, H, Henle, J and Kaskel, S (2007), 'Functional inorganic nanofillers for transparent polymers', *Chem Soc Rev*, 36, 1454–1465.

Althues, H, Pötschke, P, Kim, G M and Kaskel, S (2009), 'Structure and mechanical properties of transparent ZnO/PBDMA nanocomposites', *J Nanosci Nanotech*, 9, 2739–2745.

Althues, H, Simon, P, Philipp, F and Kaskel, S (2006), 'Integration of zinc oxide nanoparticles into transparent poly(butanediolmonoacrylate) via photopolymerisation', *J Nanosci Nanotech*, 6, 409–413.

Baeumer, S (ed.) (2005), *Handbook of Plastic Optics*. Weinheim, Wiley-VCH.

Bo, S, Hu, J, Liu, X and Zhen, Z (2009), 'Optical properties of ErFOD-doped polymers and fabrication of channel waveguides', *Opt Commun*, 282, 2465–2469.

Böhm, J, Hausselt, J, Henzi, P, Litfin, K and Hanemann, T (2004), 'Tuning the refractive index of polymers for polymer waveguides using nanoscaled ceramics or organic dyes', *Adv Eng Mater*, 6, 52–57.

Buckley, A (1992), 'Polymers for nonlinear optics', *Adv Mater*, 4, 153–158.

Burunkova, J E, Denisyuk, I Y and Williams, T R (2010), 'Filled polymers with high nanoparticles concentration – Synthesis and properties', *J Appl Polym Sci*, 116, 1857–1866.

Cannillo, V, Bondioli, F, Lusvarghi, L, Montorsi, M, Avella, M, *et al.* (2006), 'Modeling of ceramic particles filled polymer-matrix nanocomposites', *Comp Sci Technol*, 66, 1030–1037.

Carotenuto, G, Longo, A and Hison, C L (2009), 'Tuned linear optical properties of gold-polymer nanocomposites', *J Mater Chem*, 19, 5744–5750.

Carrillo Romo, F, García Murillo, A, López Torres, D, Cayetano Castro, N, Romero, V H, *et al.* (2010), 'Structural and luminescence characterization of silica coated $Y_2O_3$:$Eu^{3+}$ nanopowders', *Opt Mater*, 32, 1471–1479.

Caseri, W R (2006), 'Nanocomposites of polymers and inorganic particles: Preparation, structure and properties', *Mater Sci Technol*, 22, 807–817.

Chang, W L, Su, H W and Chen, W C (2009), 'Synthesis and properties of photosensitive polyimide-nanocrystalline titania optical thin films', *Europ Polym J*, 45, 2749–2759.

Chau, J L H, Lin, Y M, Li, A K, Su, W F, Chang, K S, *et al.* (2007), 'Transparent high refractive index nanocomposite thin films', *Mater Lett*, 61, 2908–2910.

Chau, J L H, Tung, C T, Lin, Y M and Li, A K (2008), 'Preparation and optical properties of titania/epoxy nanocomposite coatings', *Mater Lett*, 62, 3416–3418.

Chen, C-H, Jian, J-Y and Yen, F-S (2009a), 'Preparation and characterization of epoxy/γ-aluminum oxide nanocomposites', *Compos Pt A – Appl Sci Manuf*, 40, 463–468.

Chen, L, Wang, C, Li, Q, Yang, S, Hou, L, *et al.* (2009b), '*In situ* synthesis of transparent fluorescent ZnS-polymer nanocomposite hybrids through catalytic chain transfer polymerization technique', *J Mater Sci*, 44, 3413–3419.

Chen, T Y and Somasundaran, P (1998), 'Preparation of novel core-shell nanocomposite particles by controlled polymer bridging', *J Am Ceram Soc*, 81, 140–144.

Chen, X C and Green, P F (2010), 'Control of morphology and its effects on the optical properties of polymer nanocomposites', *Langmuir*, 26, 3659–3665.

Cheng, Y R, Lu, C, Lin, Z, Liu, Y F, Guan, C, *et al.* (2008), 'Preparation and properties of transparent bulk polymer nanocomposites with high nanophase contents', *J Mater Chem*, 18, 4062–4068.

Cho, H-S, Dong, Z, Pauletti, G M, Zhang, J, Xu, H, *et al.* (2010), 'Fluorescent, superparamagnetic nanospheres for drug storage, targeting, and imaging: A multifunctional nanocarrier system for cancer diagnosis and treatment', *ACS Nano*, 4, 5398–5404.

Choi, Y K, Sugimoto, K, Song, S M, Gotoh, Y, Ohkoshi, Y, *et al.* (2005), 'Mechanical and physical properties of epoxy composites reinforced by vapor grown carbon nanofibers', *Carbon*, 43, 2199–2208.

Coleman, J N, Curran, S, Dalton, A B, Davey, A P, McCarthy, B, *et al.* (1998), 'Percolation-dominated conductivity in a conjugated-polymer-carbon-nanotube composite', *Phys Rev B*, 58, R7492–R7495.

D'Amore, A, Gabriel, M, Haese, W, Schift, H and Kaiser, W (2004), 'Concentration of information', *Kunststoffe Plast Europe*, 4–7.

Du, X-W, Fu, Y-S, Sun, J, Han, X and Liu, J (2006), 'Complete UV emission of ZnO nanoparticles in a PMMA matrix', *Semicond Sci Technol*, 21, 1202–1206.

Elim, H I, Cai, B, Kurata, Y, Sugihara, O, Kaino, T, *et al.* (2009), 'Refractive index control and Rayleigh scattering properties of transparent $TiO_2$ nanohybrid polymer', *J Phys Chem B*, 113, 10143–10148.

Esteves, A C C, Barros-Timmons, A, Monteiro, T and Trindade, T (2005), 'Polymer encapsulation of CdE (E = S, Se) quantum dot ensembles via *in situ* radical polymerization in miniemulsion', *J Nanosci Nanotech*, 5, 766–771.

Evans, C W, Raston, C L and Iyer, K S (2010), 'Nanosized luminescent superparamagnetic hybrids', *Green Chemistry*, 12, 1175–1179.

Fang, Y, Chen, L, Wang, C-F and Chen, S (2010), 'Facile synthesis of fluorescent quantum dot-polymer nanocomposites via frontal polymerization', *J Polym Sci Pol Chem*, 48, 2170–2177.

Ferreira, R A S, André, P S and Carlos, L D (2010a), 'Organic-inorganic hybrid materials towards passive and active architectures for the next generation of optical networks', *Opt Mater*, 32, 1397–1409.

Ferreira, R A S, Oliveira, D C, Maia, L Q, Vicente, C M S, André, P S, *et al.* (2010b), 'Enhanced photoluminescence features of $Eu3^+$-modified di-ureasil-zirconium oxocluster organic-inorganic hybrids', *Opt Mater*, 32, 1587–1591.

Fréchet, J M J (2005), 'Functional polymers: From plastic electronics to polymer-assisted therapeutics', *Prog Polym Sci*, 30, 844–857.

Gale, M T, Gimkiewicz, C, Obi, S, Schnieper, M, Söchtig, J, *et al.* (2005), 'Replication technology for optical microsystems', *Opt Lasers Eng*, 43, 373–386.

Galiatsatos, V (2007), Refractive index, stress-optical coefficient, and optical configuration parameter of polymers. In: Mark, J E (ed.) *Physical Properties of Polymers Handbook*. 2nd ed. New York, Springer Science + Business Media.

Gonsalves, K E, Chen, X H and Baraton, M I (1997), 'Mechanistic investigation of the preparation of polymer/ceramic nanocomposites', *Nanostruct Mater*, 9, 181–184.

Govindaiah, P, Park, T-J, Jung, Y, Lee, S, Ryu, D, *et al.* (2010), 'Luminescent iron oxide nanoparticles prepared by one-pot aphen-functionalization', *Macromol Res*, 18, 1109–1114.

Guan, C, Lu, C L, Cheng, Y R, Song, S Y and Yang, B (2009), 'A facile one-pot route to transparent polymer nanocomposites with high ZnS nanophase contents via *in situ* bulk polymerization', *J Mater Chem*, 19, 617–621.

Hanemann, T, Bauer, W, Knitter, R and Woias, P (2006), Rapid prototyping and rapid tooling techniques for the manufacturing of silicon, polymer, metal and ceramic microdevices. In: Leondes, C T (ed.) *MEMS/NEMS Handbook.* Berlin, Springer.

Hanemann, T, Böhm, J, Honnef, K, Ritzhaupt-Kleissl, E and Hausselt, J (2007), 'Polymer/phenanthrene-derivative host-guest systems: Rheological, optical and thermal properties', *Macromol Mater Eng*, 292, 285–294.

Hanemann, T, Böhm, J, Müller, C and Ritzhaupt-Kleissl, E (2008), Refractive index modification of polymers using nanosized dopants. In: Thienpoint, H, van Daele, P, Mohr, J and Taghizadeh, M R (eds.) *Micro-Optics 2008*. Strasbourg, France, SPIE, 69920D-10.

Hanemann, T and Haase, W (1991), Nichtlineare optische Eigenschaften gepolter Polymerfilme. In: *Electronic Displays*, Karlsruhe, pp. 130–135.

Hanemann, T, Hausselt, J and Ritzhaupt-Kleissl, E (2009), 'Compounding, micro injection moulding and characterisation of polycarbonate-nanosized alumina-composites for application in microoptics', *Microsyst Technol*, 15, 421–427.

Hanemann, T and Honnef, K (2010), Polymer-dopant-systems: Tailoring of optical and thermomechanical properties. In: D'Amore, A D, Acierno, D and Grassia, L (eds.) *5th International Conference on Times of Polymers (TOP) and Composites 2010*, Ischia, IT. American Institute of Physics (AIP), Melville, USA, 13–15.

Heckele, M and Schomburg, W K (2004), 'Review on micro molding of thermoplastic polymers', *J Micromech Microeng*, 14, R1.

Imai, Y, Terahara, A, Hakuta, Y, Matsui, K, Hayashi, H, *et al.* (2009), 'Transparent poly(bisphenol A carbonate)-based nanocomposites with high refractive index nanoparticles', *Europ Polym J*, 45, 630–638.

Jetson, R, Yin, K, Donovan, K and Zhu, Z (2010), 'Effects of surface modification on the fluorescence properties of conjugated polymer/ZnO nanocomposites', *Mater Chem Phys*, 124, 417–421.

John, S, Marpu, S, Li, J, Omary, M, Hu, Z, *et al.* (2010), 'Hybrid zinc oxide nanoparticles for biophotonics', *J Nanosci Nanotech*, 10, 1707–1712.

Khanna, P K, Gokhale, R R, Subbarao, V V V S, Singh, N, Jun, K W, *et al.* (2005), 'Synthesis and optical properties of CdS/PVA nanocomposites', *Mater Chem Phys*, 94, 454–459.

Kim, M S, Park, D H, Cho, E H, Kim, K H, Park, Q H, *et al.* (2009), 'Complex nanoparticle of light-emitting MEH-PPV with Au: Enhanced luminescence', *ACS Nano*, 3, 1329–1334.

Kingery, W D, Bowen, H K and Uhlmann, D R (1976), *Introduction to Ceramics*. New York, Chichester, Brisbane, Toronto, Singapore, John Wiley & Sons.

Krishnamoorti, R (2007), 'Strategies for dispersing nanoparticles in polymers', *MRS Bulletin*, 32, 341–347.

Kyprianidou-Leodidou, T, Caseri, W and Suter, U W (1994), 'Size variation of PbS particles in high-refractive-index nanocomposites', *J Phys Chem*, 98, 8992–8997.

Lakowicz, J R (1999), *Principles of Fluorescence Spectroscopy*. New York, Kluwer Academic Press / Plenum Publishers.

Li, C, Wu, J, Zhao, J, Zhao, D and Fan, Q (2004), 'Effect of inorganic phase on polymeric relaxation dynamics in PMMA/silica hybrids studied by dielectric analysis', *Europ Polym J*, 40, 1807–1814.

Lide, D R (1992–1993), *Handbook of Chemistry and Physics*. Boca Raton, Ann Arbor, London, Tokyo, CRC Press.

Lim, W P, Low, H Y and Chin, W S (2004), 'IR-luminescent PbS-polystyrene nanocomposites prepared from random ionomers in solution', *J Phys Chem B*, 108, 13093–13099.

Liu, Y F, Lu, C L, Li, M J, Zhang, L and Yang, B (2008), 'High refractive index organic-inorganic hybrid coatings with $TiO_2$ nanocrystals', *Colloid Surf A-Physicochem Eng Asp*, 328, 67–72.

Lu, C, Guan, C, Liu, Y, Cheng, Y and Yang, B (2005), 'PbS/Polymer nanocomposite optical materials with high refractive index', *Chem Mater*, 17, 2448–2454.

Ma, H, Jen, A K Y and Dalton, L R (2002), 'Polymer-based optical waveguides: Materials, processing, and devices', *Adv Mater*, 14, 1339–1365.

Macedo, A G, Martins, M A, Fernandes, S E M, Barros-Timmons, A, Trindade, T, *et al.* (2010), 'Luminescent $SiO_2$-coated $Gd_2O_3$:$Eu^{3+}$ nanorods/poly(styrene) nanocomposites by *in situ* polymerization', *Opt Mater*, 32, 1622–1628.

Mahdavian, A R, Sarrafi, Y and Shabankareh, M (2009), 'Nanocomposite particles with core-shell morphology III: Preparation and characterization of nano $Al_2O_3$-poly(styrene-methyl methacrylate) particles via miniemulsion polymerization', *Polym Bull*, 63, 329–340.

Mayer, R (2007), 'Precision injection molding', *Optik and Photonik*, 4, 46–51.

Menz, W, Mohr, J and Paul, O (2001), *Microsystem Technologies*. Weinheim, Wiley-VCH.

Miller, J L and Friedmann, E (1996), *Photonic Rules of Thumb*. New York, McGraw-Hill.

Mishra, S B, Mishra, A K, Luyt, A S, Revaprasadu, N, Hillie, K T, *et al.* (2010), 'Ethyl vinyl acetate copolymer-$SrAl_2O_4$:Eu,Dy and $Sr_4Al_{14}O_{25}$:Eu,Dy phosphor-based composites: Preparation and material properties', *J Appl Polym Sci*, 115, 579–587.

Motamedi, M E (ed.) (2005), *MOEMS: Micro-opto-electro-mechanical-systems*. Bellingham, SPIE.

Murphy, R, Nicolosi, V, Hernandez, Y, McCarthy, D, Rickard, D, *et al.* (2006), 'Observation of extremely low percolation threshold in $Mo_6S_{4.5}I_{4.5}$ nanowire/polymer composites', *Scr Mater*, 54, 417–420.

Musikhin, S, Bakueva, L, Sargent, E H and Shik, A (2002), 'Luminescent properties and electronic structure of conjugated polymer-dielectric nanocrystal composites', *J Appl Phys*, 91, 6679–6683.

Nakayama, N and Hayashi, T (2007), 'Preparation and characterization of $TiO_2$ and polymer nanocomposite films with high refractive index', *J Appl Polym Sci*, 105, 3662–3672.

Nussbaumer, R J, Caseri, W R, Smith, P and Tervoort, T (2003), 'Polymer-$TiO_2$ nanocomposites: A route towards visually transparent broadband UV filters and high refractive index materials', *Macromol Mater Eng*, 288, 44–49.

Omrani, A, Simon, L C and Rostami, A A (2009), 'The effects of alumina nanoparticle on the properties of an epoxy resin system', *Mater Chem Phys*, 114, 145–150.

Pan, G, Guo, Q, Tian, A and He, Z (2008), 'Mechanical behaviors of $Al_2O_3$ nanoparticles reinforced polyetheretherketone', *Mater Sci Eng A*, 492, 383–391.

Paramo, J A, Strzhemechny, Y M, Anzlovar, A, Zigon, M and Orel, Z C (2010), 'Enhanced room temperature excitonic luminescence in ZnO/polymethyl methacrylate nanocomposites prepared by bulk polymerization', *J Appl Phys*, 108, 023517.

Peres, M, Costa, L C, Neves, A, Soares, M J, Monteiro, T, *et al.* (2005), 'A green-emitting CdSe/poly(butyl acrylate) nanocomposite', *Nanotechnology*, 16, 1969–1973.

Piotter, V, Bauer, W, Hanemann, T, Heckele, M and Müller, C (2008), 'Replication technologies for HARM devices: Status and perspectives', *Microsyst Technol*, 14, 1599–1605.

Rao, Y Q and Chen, S (2008), 'Molecular composites comprising $TiO_2$ and their optical properties', *Macromolecules*, 41, 4838–4844.

Ritzhaupt-Kleissl, E (2006), *Transparente Polymer-Nanokomposite für Anwendungen in der Mikrooptik*. PhD thesis, University of Freiburg, Germany.

Ritzhaupt-Kleissl, E, Böhm, J, Hausselt, J and Hanemann, T (2005), 'Process chain for tailoring the refractive index of thermoplastic optical materials using ceramic nanoparticles', *Adv Eng Mater*, 7, 540–545.

Ritzhaupt-Kleissl, E, Böhm, J, Hausselt, J and Hanemann, T (2006), 'Thermoplastic polymer nanocomposites for applications in optical devices', *Mater Sci Eng C – Biomimetic Supramol Syst*, 26, 1067–1071.

Rong, Y, Chen, H-Z, Wu, G and Wang, M (2005), 'Preparation and characterization of titanium dioxide nanoparticle/polystyrene composites via radical polymerization', *Mater Chem Phys*, 91, 370–374.

Sagmeister, M, Brossmann, U, List, E, Ochs, R, Szabó, D V, *et al.* (2010), 'Synthesis and optical properties of organic semiconductor/zirconia nanocomposites', *J Nanopart Res*, 12, 2541–2551.

Sahoo, N G, Rana, S, Cho, J W, Li, L and Chan, S H (2010), 'Polymer nanocomposites based on functionalized carbon nanotubes', *Prog Polym Sci*, 35, 837–867.

Sandler, J K W, Kirk, J E, Kinloch, I A, Shaffer, M S P and Windle, A H (2003), 'Ultra-low electrical percolation threshold in carbon-nanotube-epoxy composites', *Polymer*, 44, 5893–5899.

Sangermano, M, Priola, A, Kortaberria, G, Jimeno, A, Garcia, I, *et al.* (2007), 'Photopolymerization of epoxy coatings containing iron-oxide nanoparticles', *Macromol Mater Eng*, 292, 956–961.

Sangermano, M, Voit, B, Sordo, F, Eichhorn, K J and Rizza, G (2008), 'High refractive index transparent coatings obtained via UV/thermal dual-cure process', *Polymer*, 49, 2018–2022.

Sato, M, Kawata, A, Morito, S, Sato, Y and Yamaguchi, I (2008), 'Preparation and properties of polymer/zinc oxide nanocomposites using functionalized zinc oxide quantum dots', *Europ Polym J*, 44, 3430–3438.

Schadler, L S, Kumar, S K, Benicewicz, B C, Lewis, S L and Harton, S E (2007), 'Designed interfaces in polymer nanocomposites: A fundamental viewpoint', *MRS Bulletin*, 32, 335–340.

Schallehn, M, Winterer, M, Weirich, T E, Keiderling, U and Hahn, H (2003), '*In-situ* preparation of polymer-coated alumina nanopowders by chemical vapor synthesis', *Chem Vap Depos*, 9, 40–44.

Shi, D L, He, P, Lian, J, Wang, L M and van Ooij, W J (2002a), 'Plasma deposition and characterization of acrylic acid thin film on ZnO nanoparticles', *J Mater Res*, 17, 2555–2560.

Shi, D L, He, P, Wang, S X, van Ooij, W J, Wang, L M, *et al.* (2002b), 'Interfacial particle bonding via an ultrathin polymer film on $Al_2O_3$ nanoparticles by plasma polymerization', *J Mater Res*, 17, 981–990.

Stokes, G G (1852), 'On the change of refrangibility of light', *Philos T R Soc*, 142, 463–562.

Suffner, J, Schechner, G, Sieger, H and Hahn, H (2007), '*In situ* coating of silica nanoparticles with acrylate-based polymers', *Chem Vap Depos*, 13, 459–464.

Sumfleth, J, de Almeida Prado, L A S, Sriyai, M and Schulte, K (2008), 'Titania-doped multi-walled carbon nanotubes epoxy composites: Enhanced dispersion and synergistic effects in multiphase nanocomposites', *Polymer*, 49, 5105–5112.

Sun, D Z and Sue, H J (2009), 'Tunable ultraviolet emission of ZnO quantum dots in transparent poly(methyl methacrylate)', *Appl Phys Lett*, 94, 253106.

Sun, X, Liang, H, Ming, H, Zhang, Q, Yang, J, *et al.* (2004), 'The investigation on $Eu^{3+}$ high-doped PMMA planar optical waveguide by using scanning near-field optical microscopy', *Opt Commun*, 240, 75–80.

Suo, B, Su, X, Wu, J, Chen, D, Wang, A, *et al.* (2010), 'Poly (vinyl alcohol) thin film filled with CdSe-ZnS quantum dots: Fabrication, characterization and optical properties', *Mater Chem Phys*, 119, 237–242.

Tamborra, M, Striccoli, M, Curri, M L and Agostiano, A (2008), 'Hybrid nanocomposites based on luminescent colloidal nanocrystals in poly(methyl methacrylate): Spectroscopical and morphological studies', *J Nanosci Nanotech*, 8, 628–634.

Tang, E J and Dong, S Y (2009), 'Preparation of styrene polymer/ZnO nanocomposite latex via miniemulsion polymerization and its antibacterial property', *Colloid Polym Sci*, 287, 1025–1032.

Tang, F, He, F, Cheng, H and Li, L (2010), 'Self-assembly of conjugated polymer-Ag@ $SiO_2$ hybrid fluorescent nanoparticles for application to cellular imaging', *Langmuir*, 26, 11774–11778.

Tu, C, Yang, Y and Gao, M (2008), 'Preparations of bifunctional polymeric beads simultaneously incorporated with fluorescent quantum dots and magnetic nanocrystals', *Nanotechnology*, 19, 105601.

Vollath, D (2008), *Nanomaterials: An Introduction to Synthesis, Properties and Applications*. Weinheim, Wiley-VCH.

Vollath, D (2010), 'Bifunctional nanocomposites with magnetic and luminescence properties', *Adv Mater*, 22, 4410–4415.

Vollath, D and Szabó, D V (1999), 'Coated nanoparticles: A new way to improved nanocomposites', *J Nanopart Res*, 1, 235–242.

Vollath, D and Szabó, D V (2004), 'Synthesis and properties of nanocomposites', *Adv Eng Mater*, 6, 117–127.

Vollath, D and Szabó, D V (2006), 'The microwave plasma process – A versatile process to synthesise nanoparticulate materials', *J Nanopart Res*, 8, 417–428.

Vollath, D, Szabó, D V and Schlabach, S (2004), 'Oxide/polymer nanocomposites as new luminescent materials', *J Nanopart Res*, 6, 181–191.

Wang, C-F, Xie, H-Y, Cheng, Y-P, Chen, L, Hu, M Z, *et al.* (2011), 'Chemical synthesis and optical properties of CdS-poly(lactic acid) nanocomposites and their transparent fluorescent films', *Colloid Polym Sci*, 289, 395–400.

Wang, Z, Zu, X, Xiang, X and Yu, H (2006a), 'Photoluminescence from $TiO_2$/PMMA nanocomposite prepared by $\gamma$ radiation', *J Nanopart Res*, 8, 137–139.

Wang, Z G, Zu, X T, Zhu, S, Xiang, X, Fang, L M, *et al.* (2006b), 'Origin of luminescence from PMMA functionalized nanoparticles', *Phys Lett A*, 350, 252–257.

Winey, K I, Kashiwagi, T and Mu, M F (2007), 'Improving electrical conductivity and thermal properties of polymers by the addition of carbon nanotubes as fillers', *MRS Bulletin*, 32, 348–353.

Worgull, M (2009), *Hot Embossing*. Elsevier/William Andrew Applied Science Publishers.

Xiong, H-M (2010), 'Photoluminescent ZnO nanoparticles modified by polymers', *J Mater Chem*, 20, 4251–4262.

Xiong, M, Zhou, S, Wu, L, Wang, B and Yang, L (2004), 'Sol-gel derived organic-inorganic hybrid from trialkoxysilane-capped acrylic resin and titania: Effects of preparation conditions on the structure and properties', *Polymer*, 45, 8127–8138.

Xu, K, Zhou, S X and Wu, L M (2009), 'Effect of highly dispersible zirconia nanoparticles on the properties of UV-curable poly(urethane-acrylate) coatings', *J Mater Sci*, 44, 1613–1621.

Yang, B D, Yoon, K H and Chung, K W (2004), 'Dispersion effect of nanoparticles on the conjugated polymer-inorganic nanocomposites', *Mater Chem Phys*, 83, 334–339.

Yang, Y, Chen, H L and Bao, X M (2003), 'Synthesis and optical properties of CdS semiconductor nanocrystallites encapsulated in a poly (ethylene oxide) matrix', *J Cryst Growth*, 252, 251–256.

Yang, Y, Leppert, V J, Risbud, S H, Twamley, B, Power, P P, *et al.* (1999), 'Blue luminescence from amorphous GaN nanoparticles synthesized *in situ* in a polymer', *Appl Phys Lett*, 74, 2262–2264.

Yang, Y, Li, Y-Q, Fu, S-Y and Xiao, H-M (2008), 'Transparent and light-emitting epoxy nanocomposites containing ZnO quantum dots as encapsulating materials for solid state lighting', *J Phys Chem C*, 112, 10553–10558.

Yesodha, S K, Sadashiva Pillai, C K and Tsutsumi, N (2004), 'Stable polymeric materials for nonlinear optics: A review based on azobenzene systems', *Prog Polym Sci*, 29, 45–74.

Yu, Y Y, Chien, W C and Chen, S Y (2010), 'Preparation and optical properties of organic/inorganic nanocomposite materials by UV curing process', *Mater Des*, 31, 2061–2070.

Zhang, B, Chen, B, Wang, Y, Guo, F, Li, Z, *et al.* (2011), 'Preparation of highly fluorescent magnetic nanoparticles for analytes-enrichment and subsequent biodetection', *J Colloid Interf Sci*, 353, 426–432.

Zhang, B, Cheng, J, Gong, X, Dong, X, Liu, X, *et al.* (2008), 'Facile fabrication of multi-colors high fluorescent/superparamagnetic nanoparticles', *J Colloid Interf Sci*, 322, 485–490.

Zhang, H Z and Banfield, J F (1998), 'Thermodynamic analysis of phase stability of nanocrystalline titania', *J Mater Chem*, 8, 2073–2076.

Zhang, M, Lin, G Q, Dong, C and Wen, L S (2007), 'Amorphous $TiO_2$ films with high refractive index deposited by pulsed bias arc ion plating', *Surf Coat Tech*, 201, 7252–7258.

Zollinger, H (1991), *Color Chemistry*. Weinheim, VCH.

# 18
# Polymer nanocomposite coatings

T.-C. HUANG and J.-M. YEH, Chung Yuan Christian University, Taiwan and C.-Y. LAI, Sipix Technology, Inc., Taiwan

**Abstract:** This chapter discusses the properties of coatings made of polymer nanocomposites based on non-conductive polymers, electroactive polymers and conductive polymers. Anticorrosion resistance and tribology can be efficiently improved when nanoclay or nanoparticles are embedded in the polymer. Polymer nanocomposites made with carbon nanotubes have electrical properties and act as electromagnetic interference shielding. Some coating technologies suitable for polymer nanocomposites are discussed.

**Key words:** coating, anticorrosion properties, tribology, electroactive polymer.

## 18.1 Introduction

Polymer nanocomposites or nanofilled polymers[1] are polymer matrices containing organic or inorganic fillers with a homogeneous nanoscale distribution (normally from 10 to 100 nm in at least one dimension), which are prepared by physical blending or chemical polymerizing technologies. The fillers can be particles, layered materials, fibres or clusters embedded in a wide variety of natural or synthetic polymers. Their distinctive physical and chemical properties, which enhance the performance of the composites, attract much interest even after decades of study. These outstanding properties, due to the fillers, give these polymers a high potential for use in aeronautics, the automotive industry, electronics, medical equipment and consumer goods. Different types of nanofilled polymer, such as powders, bulk and functional thin films, are used widely in industry and academia. Polymer nanocomposite coatings are especially significant because they improve the surface characteristics of substrates for specific purposes. For instance, a polymer nanocomposite, with an inorganic layered filler, coated onto the surface of steel is able to slow corrosion considerably. This protection mechanism can also be used to construct a gas barrier coating,[2,3] since the inorganic layered filler lengthens the penetration pathway for the gas. Other coatings for special functions, such as self-cleaning, temperature resistance, wear resistance and special optics, have appeared in many commercial products. Recently, electroactive coatings based on polymer nanocomposites have been shown to have a much lower resistivity than traditional coatings, leading to new applications as electrochemical sensors,[4] materials with a high dielectric constant,[5] functional membranes[6] and electrochromic materials.[7] Other than their intrinsic material behaviour, the key parameter for defining successful polymer nanocomposite coatings is how easily and efficiently they can be deposited onto substrates.

## 18.2 Coating technologies for polymer nanocomposites

A successful coating requires not only the intrinsic properties of the polymer nanocomposite but also workable technologies for depositing the material on different kinds of substrate. 'Workable', in the sense used here, implies conforming to several conditions to ensure good coating quality characteristics such as surface uniformity, adhesion between interphases, thickness control and the non-toxicity of the material. Production capacity is another significant consideration for large-scale manufacturing. Many different coating technologies have been used in research and production. However, not all of them are practical for nanofilled polymers. Polymer chains can be easily damaged by vaporizing processes, which are normally performed at extremely high temperatures or energies. Hence, a low deposition temperature is usually a basic requirement of a suitable coating technique for polymer nanocomposites. In general, suitable coating techniques for polymer nanocomposites can be classified into four main groups: (1) physical vapour deposition (PVD),[8,9] (2) chemical vapour deposition (CVD),[10–13] (3) chemical and electrochemical deposition[14–16] and (4) roll-to-roll (R2R) casting deposition.[17–19] We provide a brief introduction to each deposition technique in this chapter.

### 18.2.1 Physical vapour deposition

PVD first involves the generation of a vapour of the target substance, e.g. a gas mixture plasma, via a high-energy source. Next, the vaporized material is condensed onto the surface of the substrate in a very low-pressure chamber (0.1–10 Pa) to gradually form a thin film on the substrate. The creation of a high-density plasma using a low-pressure gas requires a static magnetic field of ~0.03 T. The field is formed by a configuration of equivalent (or balanced) magnets within the region in front of the sputtering target and traps electrons close to the surface. The magnetic field lines are predominantly concentrated behind the target, while the plasma density near the substrate is about $n_e$ (~$10^{10}$ $cm^{-3}$). Magnetron sputtering or direct current magnetron sputtering involves the application of a direct potential to the substrate to induce and sustain the plasma. A related technique, unbalanced magnetron sputtering (UMS), was developed to deposit alloys and their nitrides and carbides. However, this type of coating technology still relies on relatively high energies and is rarely utilized for nanofilled polymers due to the hazardous conditions.

### 18.2.2 Chemical vapour deposition

CVD is based on the production of a vapour from the solid target material via a heating process and chemical reactions in the vapour phase. Hence, this deposition involves a homogeneous gas phase or heterogeneous chemical reactions, which occur on or in the vicinity of the heated surface, leading to the formation of powders or films, respectively.[20] CVD is typically performed at very high temperatures, up to 1000 °C, in order to activate the chemical reactions in the vapour phase. These extremely high temperatures lead to a breakdown of the chemical structure of the solid targets, resulting in different properties compared with the original material. Therefore, low-temperature CVD at 350–700 °C has been carried out using inorganic and metallo-organic precursors. Even lower activation temperatures of 200–400 °C have been achieved using a plasma, as described below. The two most widely used CVD devices for providing a high-temperature stage are hot-wall CVD (HWCVD) and hot-filament CVD (HFCVD) reactors, due to their low cost compared with other coating techniques. However, HWCVD is more appropriate for the deposition of polymer nanocomposites because it does not require the excessively high filament temperatures of HFCVD. Even so, these coating technologies are rarely used for polymer nanocomposites, except for very special cases like the deposition of a nanodiamond.

Plasma-assisted CVD is an efficient way to reduce the temperature in CVD. It uses a microwave plasma at an extremely low pressure (10–100 Pa), instead of at high temperature, to activate the chemical reactions. The method requires only a low environmental temperature and uses a 2.45-GHz microwave plasma exciting source. This deposition method has been used to deposit silicon dioxide, carbon nitride and cubic boron nitride. Polymers containing fluorine, such as polyfluorohydrocarbon and polyperfluorocarbon,[21,22] have also been deposited on specific surfaces.[23]

### 18.2.3 Chemical and electrochemical deposition

Chemical and electrochemical deposition is an area of considerable interest, both from a fundamental and an applied standpoint; these methods can be used for electroplating and solution analysis. The low ionizing temperature (in most cases, the temperature can be ambient) should have a negligible hazardous effect on the coating targets. The key objective of chemical and electrochemical deposition is to reduce the precursors to active species through either reducing agents or an input voltage applied to the polyelectrolyte matrix. Based on this prerequisite, conductive polymers such as polyaniline (PANI), polythiophene (PTh), polypyrrole and poly(3,4-ethylenedioxythiophene) (PEDOT) have been the most commonly used for the conductive polymer matrix in nanocomposites. The mechanism for this coating technology is rather simple and normally includes two steps: (1) the target precursors are driven into becoming active monomers, including cathodic or free radicals, and (2) the active monomers diffuse to the cathode (the working electron source and target substrate) and gradually accumulate on the surface of the substrate. However, in the electrochemical method, reduction requires an external current and the sites for anodic and cathodic reactions are separate. For chemical deposition, reduction is induced by a reducing agent and the anodic and cathodic reactions occur together on the workpiece.[24] In addition, these reactions only proceed on catalytically active surfaces, i.e. newly coated metallic surfaces should be catalytically active enough to promote redox reactions. Electrochemical deposition has been widely used with conductive polymer nanocomposites to immobilize nanofillers such as heavy metal colloids or specific enzymes, which are inside the conducting polymer matrix. In addition, bioactive thin films may be able to act as biotransistors to convert analogue biosignals into electronic signals.

### 18.2.4 Roll-to-roll processing deposition

R2R (Fig. 18.1) processing deposition is a high production capacity and low-cost industrial coating technology, which has been employed in the manufacture of flat devices such as organic light-emitting diodes (OLEDs), photovoltaics (PV) and electrophoresis displays (EPDs). A simple roll coating is defined by Coyle *et al.* as follows: the fluid flows into the space between a pair of rotating rollers that control both the thickness and the uniformity of the coated film. There are some variations of this definition, like reverse-roll coating,[25] gravure coating,[26] knife-over-roll coating (gap coating) and slot die coating, which have been used industrially for specific purposes. R2R is a 'non-stop' coating process and can be combined with most other deposition techniques. Many practical techniques, for instance R2R casting, R2R sputtering and R2R plasma-assisted CVD, have been used in the manufacture of nanocomposites for photoconversion thin films.

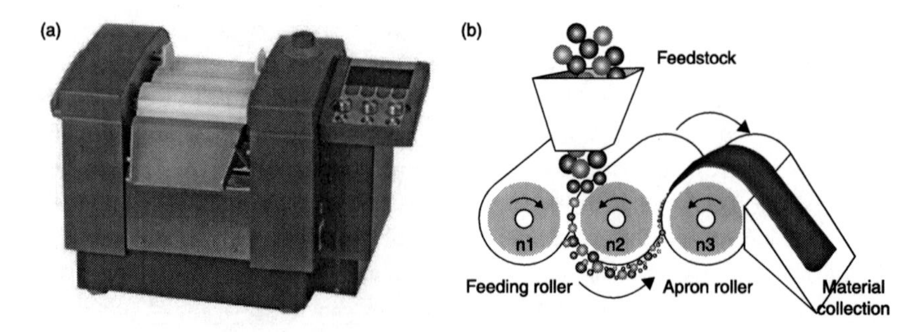


*18.1* (a) Calendering (or three-roll milling) machine used for dispersing particles into a polymer matrix. (b) General configuration and mechanism. After Ma *et al.*[19]

## 18.3 Key properties of polymer nanocomposite coatings

### 18.3.1 Anticorrosion properties

Organic coatings have long been used to protect metals against corrosive environments in both academic and industrial settings. The primary objective of organic coatings is to act as a physical barrier against aggressive species, such as $O_2$ and $H_2O$. In the past decade, scientists have focused on a new type of coating: hybrid organic–inorganic coatings. These coatings combine the flexibility and ease of processing of polymers with the hardness of inorganic materials and have been successfully applied to various substrates. For example, Yeh and co-workers demonstrated that the incorporation of organophilic clay platelets into a polymer matrix, in the form of organic-based coatings, may effectively enhance the corrosion protection of polymers on metallic surfaces.[27–38] This is because of the good dispersion (intercalation and exfoliation, Fig. 18.2) of the plate-like clay platelets in the polymer matrix, which effectively increase the length of the diffusion pathways for oxygen and water. In addition, the dispersion may also decrease the permeability of the coatings.

Apart from clay, another important material with this advantage is $SiO_2$. The incorporation of low-cost $SiO_2$ nanoparticles into polymers can significantly improve thermal properties,[39,40] mechanical properties,[41,42] anticorrosion,[43,44] wear resistance,[45,46] barrier properties[46,47] and the customization of electronic packaging properties.[48] Nanoparticles can significantly alter the mechanical properties of the polymer close to the particle surface due to changes in polymer chain mobility.[46]

Graphene, a monolayer of $sp^2$-hybridized carbon atoms arranged in a two-dimensional lattice, has also attracted tremendous attention as a nanofiller in

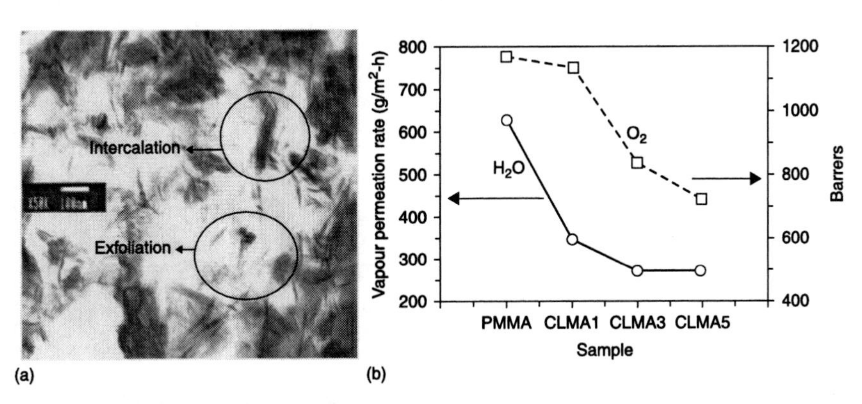


*18.2* (a) Transmission electron micrograph of PMMA/clay nanocomposites. (b) Permeability of $H_2O$ and $O_2$ as a function of clay content in PMMA/clay nanocomposites.[2]

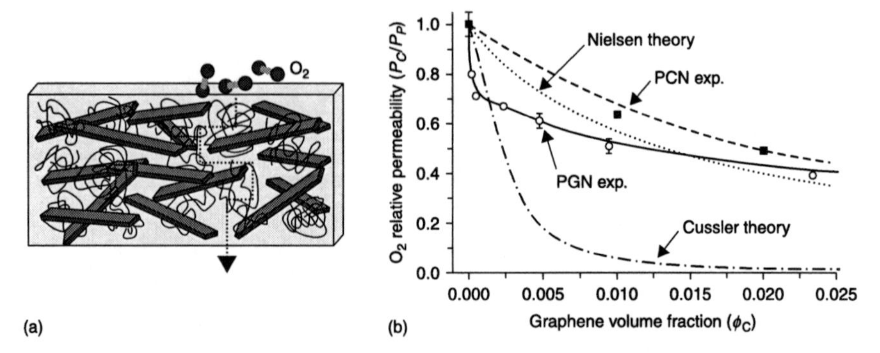


*18.3* (a) A 'tortuous path' of platelets inhibits the diffusion of gases through a polymer composite (Nielsen model). (b) Oxygen permeability of polystyrene/graphene (PGN) and polystyrene/clay (PCN) composites as a function of filler loading, compared with two theoretical models of composite permeability.[63]

recent years due to its exceptional thermal,[49–52] mechanical[53–61] and barrier properties.[62–66] The incorporation of graphene fillers can significantly reduce gas permeation through a polymer composite relative to the neat polymer matrix. A percolating network of platelets can provide a 'tortuous path', which inhibits molecular diffusion through the matrix, thus resulting in significantly reduced permeability (Fig. 18.3). Nguyen *et al.* demonstrated that the incorporation of graphene nanosheets into a polymer matrix improves the barrier capability. At low concentrations (below 0.05 vol%), crumpled graphene sheets are as effective as clay-based nanofillers with loadings that are approximately 25–130 times higher.[63]

Nanofillers can also be loaded into conductive polymers (in addition to non-conductive polymers) to form coating materials, while retaining their barrier capabilities. Conductive polymers such as PANI, polypyrrole, and PTh have also been applied as anticorrosion coatings. PANI in particular is one of the most promising electrode materials due to its simple synthesis and environmental stability.[67] PANI and its derivatives have been extensively studied as anticorrosive coatings for various metals.[68–70] For instance, Ahmad and MacDiarmid[71] and Wei *et al.*[72] evaluated the corrosion protection of PANI by performing a series of electrochemical measurements. Wessling and co-workers reported that conducting polymers form a complex with Fe (passivation oxide layers, Fig 18.4), which improved corrosion protection.[73–75] A possible passivation oxide layer mechanism is shown in Fig. 18.5. The chemical composition of the passivation oxide layers was determined by X-ray photoelectron spectroscopy (XPS). Binding energy plots versus intensity for iron oxide layers on the exterior of the sample and for the oxides right up against the iron surface are shown in Fig. 18.6. These plots indicate that the passive oxide layer is predominately composed of $Fe_2O_3$ on the outside with an $Fe_3O_4$ layer sandwiched between this layer and the nascent steel.

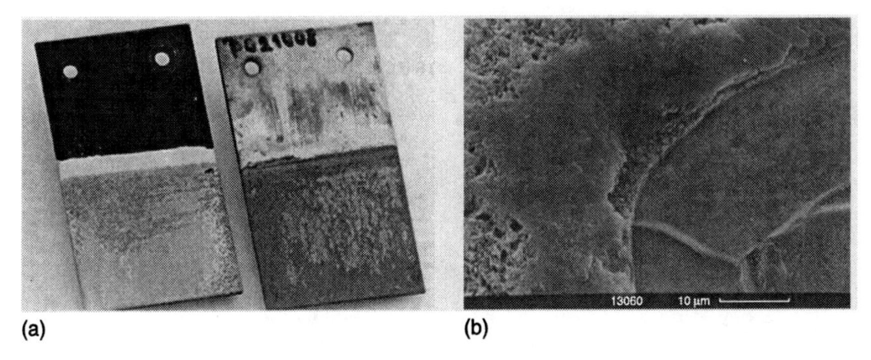


(a) (b)

*18.4* (a) An untreated iron plate (right) exhibits rust after a short time in salt water or after one anodic corrosion current measurement. A passivated plate (left, passivation was performed only on the lower half of the plate and the polyaniline layer was removed after passivation) exhibits no rust even after a corrosion current density measurement. (b) Scanning electron microscopy (SEM) image showing that passivation is achieved with an oxide layer several micrometres thick coating the whole metal surface.[75]

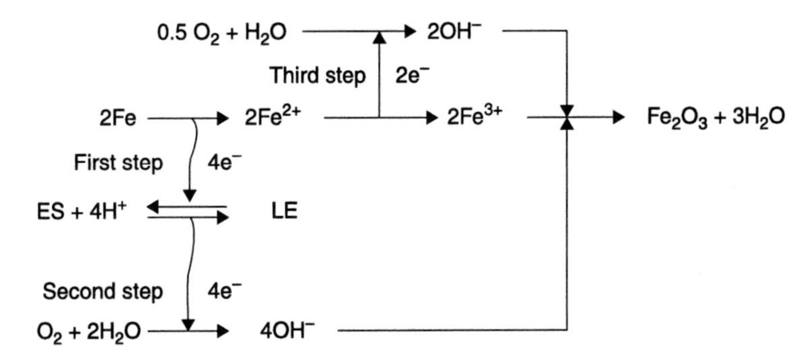


*18.5* Catalytic passivation oxide layer of iron by PANI.[74]

Although PANI has outstanding anticorrosion properties, the adhesion, mechanical properties and barrier effect of the PANI coating can still be improved to enhance the efficiency of corrosion protection. For example, different nanofillers, such as inorganic nanoparticles,[76] graphite,[77] carbon nanotubes[78,79] and nanoclays (layered silicates),[36,37] have been used as additives to strengthen the mechanical and barrier performance of polymers.[80,81] Shabani-Nooshabadi *et al.*[82] synthesized PANI/montmorillonite nanocomposites and coated them on aluminium substrates. The corrosion rate was found to be 190 times lower than that observed for uncoated Al. The redox catalytic and barrier properties affected

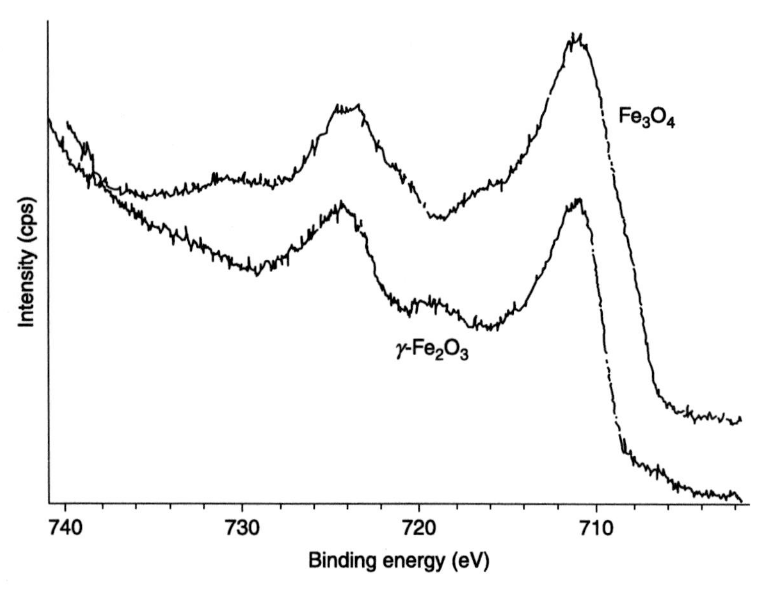


*18.6* XPS analysis of passive iron oxide layers formed in the presence of doped polyaniline.[73]

corrosion performance. Yeh and co-workers[36] synthesized a PANI/clay nanocomposite and coated it onto cold-rolled steel (CRS). The results indicated that a PANI coating containing 1 wt% clay enhanced the corrosion protection of the CRS electrode.

Nanoparticles have been extensively used as stabilizers in the preparation of PANI dispersions and colloids,[83,84] and various hybrid systems consisting of PANI and silica particles have been reported.[85–87] Currently, nanoparticle/PANI hybrid systems have mostly been constructed through the chemical polymerization of aniline in the presence of nanoparticles, or by modifying the nanoparticles with PANI prepared in advance. Radhakrishnan *et al.* used nano-$TiO_2$ as a metal oxide additive in the composite, which gave better dispersion of the formulation as well as better barrier properties in the coatings.[88] Luo *et al.* prepared a $SiO_2$ particle/PANI hybrid system (the $SiO_2$ particles were covered with a PANI nanofilm) on a glassy carbon (GC) electrode surface through a simple electrochemical method for use in biosensor applications.[89]

Nanoparticles ($SiO_2$, $TiO_2$) and nanoclays used as fillers to synthesize PANI nanocomposites can usually improve thermal, mechanical and anticorrosion properties. However, the peak current of the PANI nanocomposite is reduced as a result.[89] This phenomenon may be due to the presence of the non-conductive nanofiller, which partially blocks the electrode surface (Fig. 18.7). A reduction of the electrical conductivity of the PANI/clay nanocomposite was observed in comparison to neat PANI. This is expected because the clay component is not

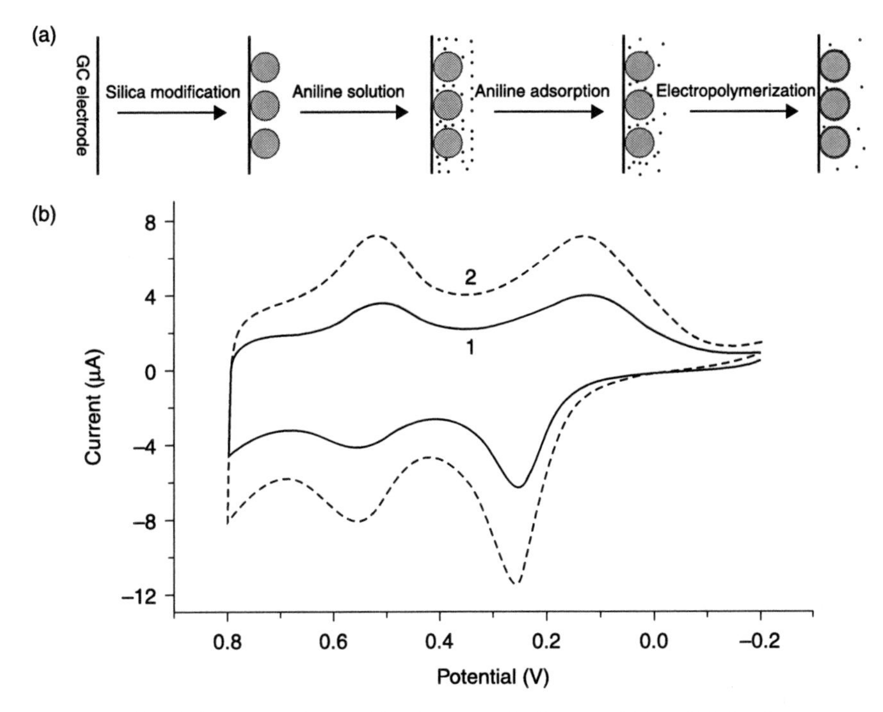


*18.7* (a) Preparation of PANI-covered silica particles. (b) Cyclic voltammograms of (1) PANI/$SiO_2$/GC and (2) PANI/GC electrodes in 1.0M HCl solution.[89]

electrically conductive and because incorporating the clay into the PANI matrix lowers the molecular weight and decreases electrical conductivity (Fig. 18.8).

## 18.3.2 Electrical properties of polymer/carbon nanotube(CNT) materials

Another widely used nanofiller, namely, the carbon nanotube (CNT), has clearly demonstrated its capability as a filler in diverse multifunctional nanocomposites. An enhancement of electrical conductivity by several orders of magnitude at very low percolation thresholds (<0.1 wt%) has been observed for CNTs in polymer matrices. Electrically conducting composites with a volume conductivity higher than $10^{-10}$ S/cm are considered to be an important group of relatively inexpensive materials for many engineering applications (Fig. 18.9),[19,90,91] such as electrically conducting adhesives, antistatic coatings[92] and electromagnetic interference (EMI) shielding for electronic devices.

Ajayan and co-workers[93] first reported the electrochemical oxidation of aniline and the behaviour of PANI films formed on multiwalled CNT (MWNT) electrodes. The results suggest that PANI films on nanotubes are oxidized more readily and

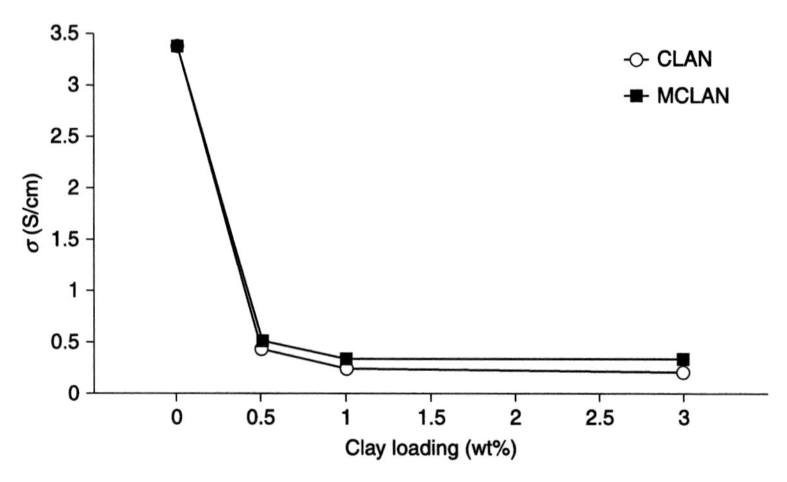


*18.8* Electrical conductivity versus clay loading obtained from four-probe technique measurements.[36]

produce a higher current density during anodic oxidation compared with conventional electrodes such as Pt. This is attributed to the unusual surface topology and relatively large surface area of the nanotube electrodes. The morphology of PANI films formed on nanotube surfaces is granular and nanostructural.

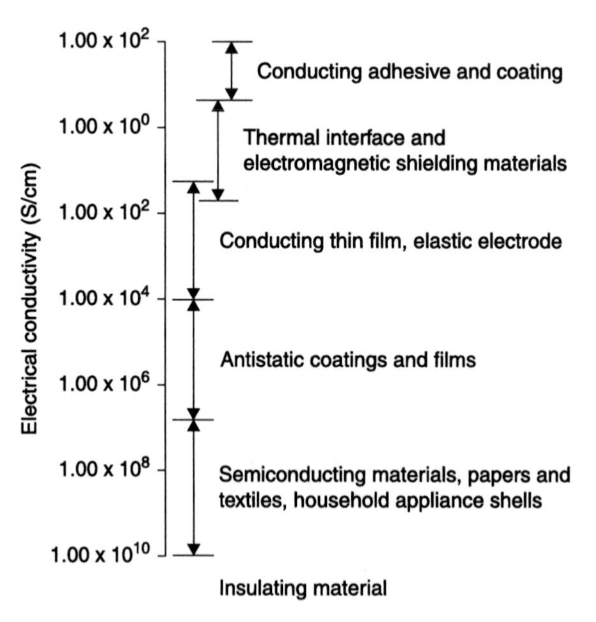


*18.9* Typical applications of conducting composites.[19]

Zengin *et al.*[94] reported that PANI could be doped with CNTs to construct a donor-acceptor (DA) nanofilled thin film. CNTs are relatively good electron acceptors, whereas PANI is a good electron donor. CNTs dispersed well in PANI may serve as electron acceptors and conducting bridges. Conductivity increases from 3.36 S $cm^{-1}$ for HCl-doped PANI to 33.374 S $cm^{-1}$ for doped PANI/CNTs (10 wt% CNT). The Fourier transform infrared (FTIR) spectrum of PANI/CNT composites illustrates several differences from the spectrum of neat PANI (Fig. 18.10). The composite spectrum exhibits an increased quinoid (Q) to benzenoid (B) band (1600–1500 $cm^{-1}$) intensity ratio compared with neat PANI. These data reveal that PANI/CNT is richer in quinoid units than pure PANI. This fact may suggest that PANI/CNT interactions promote or stabilize the quinoid ring structure. The $\pi$-bonded surface of the CNTs might interact strongly with the conjugated structure of PANI, especially through the quinoid ring. The band at 1150 $cm^{-1}$ was described by Macdiarmid and co-workers[95] as an electronic-like band, and it is considered to be a measure of the degree of delocalization of the electrons, and, thus, it is a characteristic peak of the conductivity of PANI. In PANI/CNT composites, the intensity of the signal at 1143 $cm^{-1}$ increased and shifted to 1123 $cm^{-1}$. This increase of the electronic-like absorption peak indicates increased electron delocalization in the composite compared with pure PANI and agrees well with the increased conductivity measurements.

Several methods to modify CNTs have been reported in the literature;[96–101] for instance, physical treatments such as ultrasonication and harsh chemical treatments in strong acids such as sulphuric acid, nitric acid or their mixtures. These

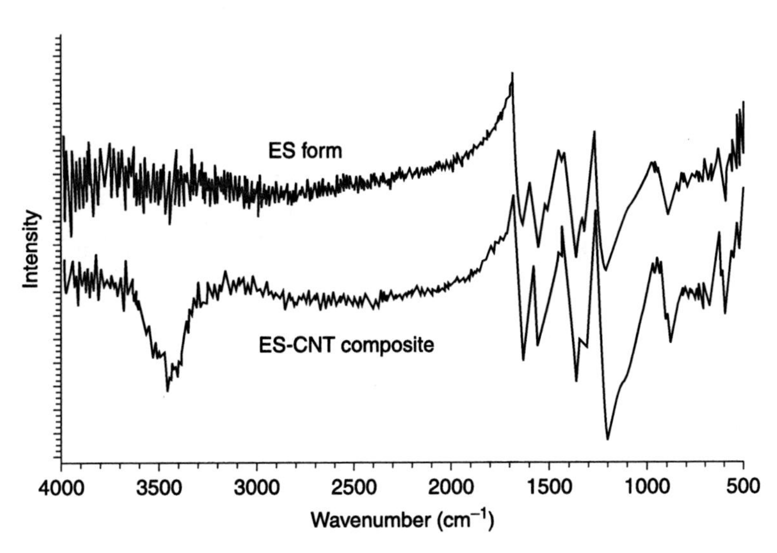


*18.10* FTIR spectra of PANI emeraldine salt (ES) and PANI/CNT ES composites.[94]

treatments work well to improve dispersibility. Unfortunately, these approaches often result in significant damage to the CNT framework, which involves sidewall opening, breaking, and conversion into amorphous carbon.[102,103] The damage to the CNTs definitely diminishes their original outstanding electrical, thermal and physical properties. Recently, the direct Friedel–Crafts acylation of nanofibres and nanotubes has been developed by Baek and co-workers.[104–109] The reaction condition in this approach is known to be a less destructive chemical modification. The CNTs were functionalized with 4-aminobenzoic acid via direct Friedel–Crafts acylation in a mild polyphosphoric acid (PPA)/phosphorous pentoxide ($P_2O_5$) medium. The 4-aminobenzoyl-functionalized CNTs should improve the electrochemical properties of PANI/CNT nanocomposites. Cyclic voltammetry (CV) is generally used to investigate the electrochemical properties of conductive materials. The output current of PANI/CNT nanocomposites was observed to be much larger than that of pure PANI. The results suggest that ion inclusion and exclusion in PANI/CNT nanocomposites were much more effective during the redox process (Fig. 18.11). The results are promising for the possible deposition of high-quality PANI films on nanotubes for future applications in, for example, devices, sensors[110] (in which high currents would improve performance) and nanocomposites for electromagnetic shielding.[64,111]

EMI, also called radio-frequency interference, is a serious issue due to the rapid proliferation of electronics and wireless systems in fields such as navigation and space technology.[112] EMI not only affects the performance of electronic devices

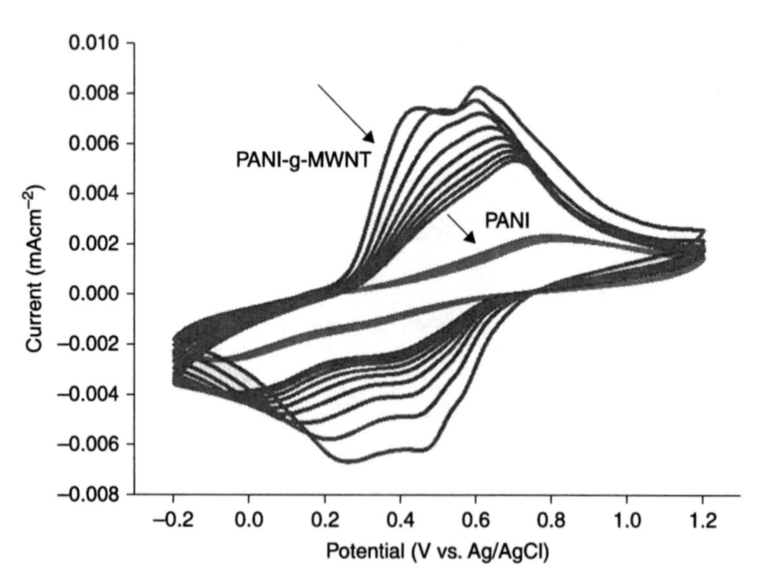


*18.11* Cyclic voltammograms of PANI and PANI-g-MWNT in 1.0 M $H_2SO_4$ aqueous solution. Scan rate is 10 mV $s^{-1}$.[105]

but may also be harmful to life forms, including humans. Therefore, some kind of shielding material must be employed to prevent electromagnetic noise or pollution. Various types of materials have been used for EMI shielding, including metals, carbon materials and conducting polymers. Recently, the EMI shielding and microwave absorption properties of conducting polymers have attracted increased attention due to their good electrical conductivity and processability.[113] The EMI shielding efficiency (SE) of a composite material depends on many factors, including the intrinsic conductivity of the filler, dielectric constant and aspect ratio.[114,115] The small diameter, high aspect ratio, high conductivity and mechanical strength of CNTs, including single-walled CNTs (SWNTs) and MWNTs, make them an excellent option for creating conductive composites for high-performance EMI shielding materials at low filling rates. For example, MWNTs have been added to polymer matrices for EMI shielding materials and tested in the frequency range 8.2–12.4 GHz (X band) with 20 dB for 7% MWNT in polystyrene (PS).[116,117] Joo and co-workers studied the electrical conductivity and EMI properties of MWNTs in poly(methyl methacrylate) (PMMA) containing Fe. They achieved 27 dB for a 40% MWNT loading.[118] The microwave absorption properties of MWNT composites with encapsulated Fe, with different phases and shapes of the included Fe, have also been studied.[119] Recently, Xiang *et al.* synthesized MWNT composites with silica and studied their microwave attenuation in the X band.[120] Grimes *et al.*[121] reported that polymer/SWNT composites possess a high real permittivity component (polarisation, $\varepsilon'$) as well as an imaginary permittivity component (absorption or electric loss, $\varepsilon''$) in the 0.5–2 GHz range. They also found that the permittivity decreased rapidly with increasing frequency.

Eklund and co-workers[122] demonstrated that epoxy/SWNT nanocomposites can be used as effective lightweight EMI shielding materials. The highest EMI shielding effectiveness for epoxy/SWNT composites was found for materials with 15 wt% SWNT. The SE was found to be ~49 dB at 10 MHz, and the shielding effectiveness is around 15–20 dB in the 500 MHz to 1.5 GHz range.

Conducting polymers can be combined with other nanocomponents to enhance EMI shielding. The microwave absorption properties of CNT/conducting-polymer nanocomposites with a core-sheath nanostructure prepared by *in situ* polymerization have been measured.[123] The conductivity of the PANI/CNT nanocomposites was higher than pure PANI and also pure CNTs – an example of the synergistic effect of the two components. PANI/CNT nanocomposites can be used for shielding in the $K_u$ band (12.4–18.0 GHz) as their total shielding effectiveness is in the range of −27.5 to 39.2 dB. The microwave absorption properties of one-dimensional PANI/hexanoic acid (HA)/$TiO_2$ nanocomposites have also been studied.[124] The results showed that the conductivity of the composite significantly decreased after addition of $TiO_2$ and that the PANI/HA/$TiO_2$ nanocomposites synthesized at 0 °C had a maximum reflection loss (RL) of 31 dB (>99.9% power absorption) at 10 GHz. To increase the conductivity and microwave absorption of the above system, SWNTs were also incorporated.[125]

When the SWNT content was 20%, the PANI/HA/$TiO_2$/CNT nanocomposite exhibited an RL of less than −15 dB with a broad bandwidth (4 GHz), while an RL less than −20 dB with a narrow bandwidth (1 GHz) was observed when the SWNT content reached 60%. Although only a few reports on the use of one-dimensional nanostructured conducting polymers for microwave absorption and EMI shielding are available, it is believed that one-dimensional nanocomposites could be important in this field.[64]

### 18.3.3 Electroactive polymer coatings

Although PANI nanocomposites have excellent anticorrosion properties, the poor solubility of PANI in common organic solvents has limited its practical application in many fields. Recently, electroactive polymers incorporating an aniline oligomer have attracted immense research interest because of their advanced properties such as good solubility, mechanical strength and film-forming ability.[126]

A novel route for oxidative coupling polymerization has been developed by Wei, Zhang and co-workers.[127–131] The fabricated copolymers not only contained well-defined conjugated segments, but also provided a better understanding of the structure–property relations and the conducting mechanism of conjugated polymers; knowledge which is limited by the complexity of their molecular structure and their poor solubility in organic solvents. Zagorska and co-workers[132,133] prepared new electroactive polymers combining the electrochromic properties of PTh and oligoaniline. Both the oligoaniline side chains and the PTh main chain could be doped in different ways. Wang and co-workers[134–136] prepared electroactive polymers by including the aniline oligomer in the main and side chains, which exhibited electrochemical properties, a high dielectric constant and electrochromic behaviour.

Albertsson and co-workers[137–140] incorporated the electroactive aniline oligomer into a polymer to obtain biodegradable and electroactive copolymers. The aniline oligomer has a well-defined electroactive structure, good processing properties and high degradability. Wei and co-workers[141,142] incorporated aniline pentamer (AP) into polylactide and obtained copolymers that showed good electroactivity, biodegradability and processing properties. Electroactive polyimide (EPI) and electroactive epoxy (EE) have been extensively studied as electrochemical sensors,[4] functional membranes,[6] electrochromic materials[143] and anticorrosion coatings.[144,145]

EPIs derived from bis-amino-terminated aniline trimer units were prepared by Wei and co-workers.[146] Yeh and co-workers presented the first evaluations of the effect of an amine-capped aniline trimer (AT) on the corrosion protection efficiency of as-prepared EPI.[145] A possible mechanism for the enhanced corrosion protection of EPI coatings on CRS electrodes is: (1) the EPI coatings may act as a physical barrier and (2) the redox catalytic properties of the aniline pentamer units in EPI may induce the formation of a passive metal oxide layer on the CRS electrode, as evidenced by scanning electron microscopy (SEM) and electron spectroscopy for chemical analysis (ESCA) studies (Fig. 18.12). The electrochromic performance of EPI was investigated by studying electrochromic photographs and measuring UV absorption spectra. Across different voltages, the colour of the EPI thin film was found to change from grey (at 0.0 V), to green (at 0.4 V), to blue (at 0.6 V) and to metallic blue (at 1.0 V), as shown in Fig. 18.13.[143]

An electroactive epoxy thermoset (EET) coating, an electroactive polymer, has been studied.[144] A higher concentration of conjugated amino-capped aniline trimer (ACAT) units existed in the as-prepared EET samples, and a higher redox current was observed in the CV studies. Electrochemical corrosion experiments showed that a higher concentration of the ACAT segment in the as-prepared EET led to better corrosion protection of the coated CRS electrode. The mechanism for enhanced corrosion protection is similar to that for the EPI material. Thermogravimetric analysis (TGA), differential scanning calorimetry (DSC) and dynamic mechanical analysis (DMA) showed that incorporating a higher

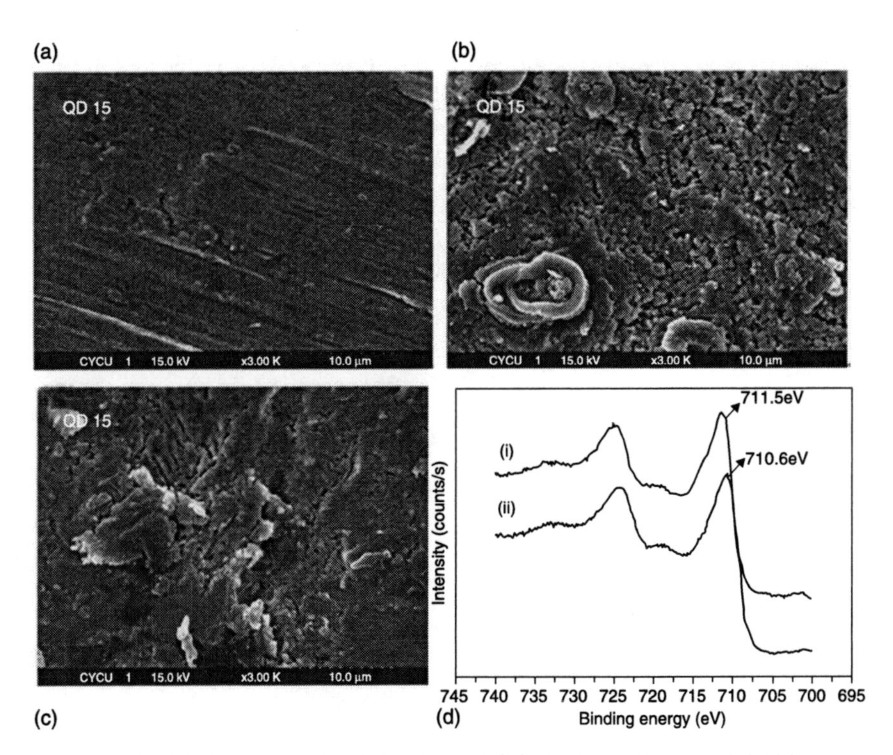


*18.12* SEM images for (a) polished, (b) electroactive copolyimide-coated and (c) electroactive polyimide-coated CRS surfaces. (d) ESCA Fe 2p core level spectra of (i) electroactive copolyimide and (ii) electroactive polyimide.[145]

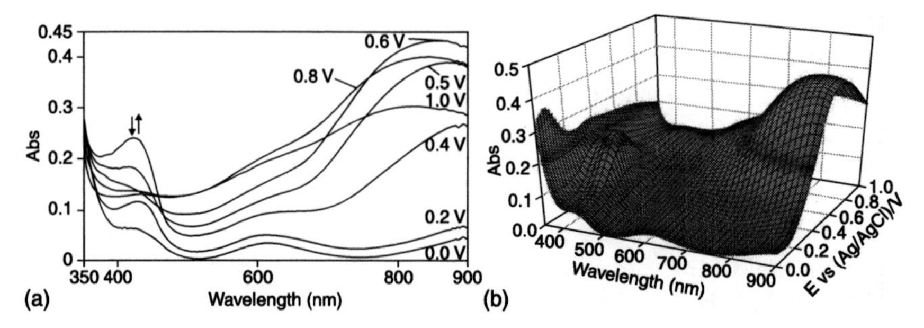


*18.13* (a) Electrochromic behaviour of EPI thin film on glass electrodes coated with indium tin oxide (ITO) at different oxidation potentials. (b) Three-dimensional spectroelectrochemical behaviour of the EPI thin film on an ITO-coated glass electrode between 0.00 and 1.00 V (vs. Ag/AgCl).[143]

percentage of ACAT with a rigid aromatic structure in as-prepared EET led to a significant increase in thermal and mechanical properties (Fig 18.14).

A superhydrophobic electroactive epoxy (SEE) from the surface structure of fresh *Xanthosoma sagittifolium* leaves was prepared and coated on the surface of

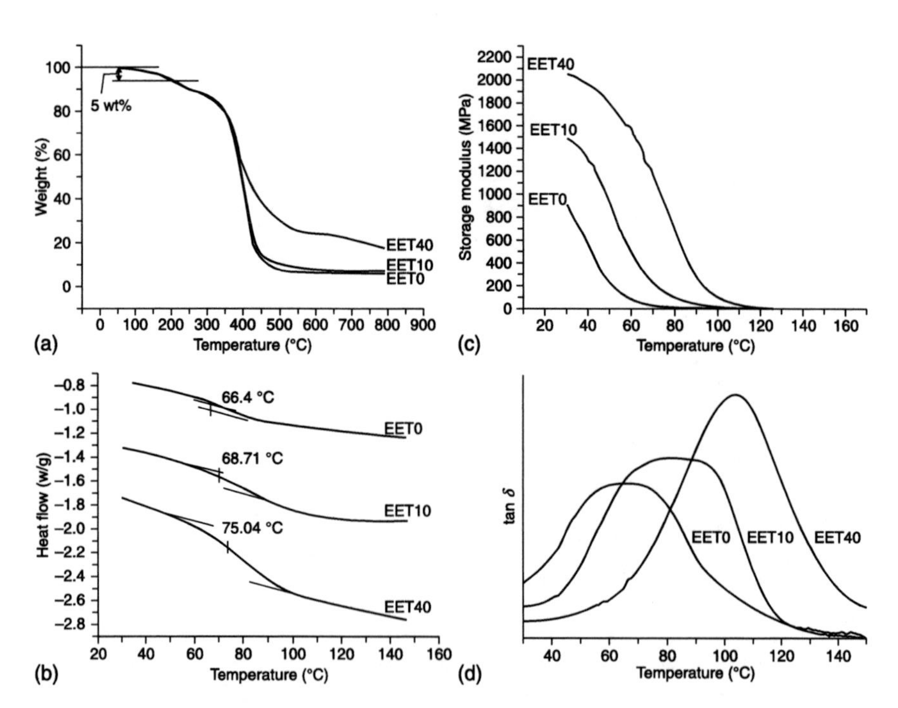


*18.14* (a) TGA curves, (b) DSC curves, (c) storage modulus and (d) tan δ for EET40, EET10 and EET0 membranes.[144]

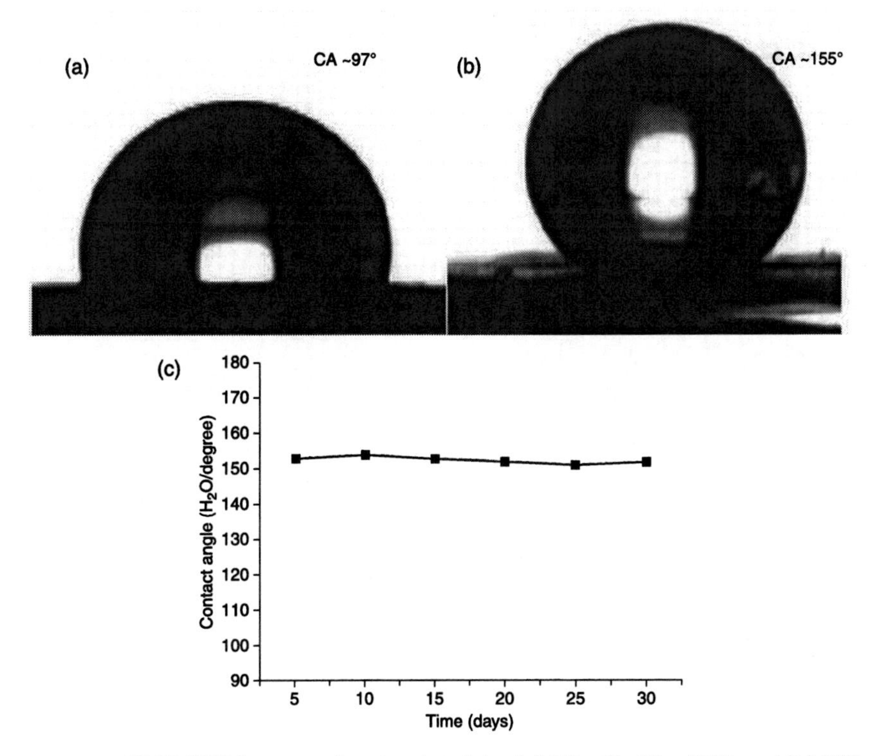


*18.15* CCD images of water droplets: (a) EE with CA of 97° and (b) SEE with CA of 155°. (c) Change in water contact angle with time for SEE.[147]

a CRS electrode using a nanocasting technique for corrosion protection.[147] The electrode was found to have a water contact angle (CA) of 153°, which was significantly higher than smooth EE coated on CRS by spin coating (CA = 97°), as shown in Fig. 18.15.

Nanofillers such as clay, $SiO_2$ and $TiO_2$ also can be included in EPI and EE to capitalize on the properties of these nanoparticles.[148–150] The good dispersion of these nanofillers in the polymer matrix leads to an effective increase in the length of the diffusion pathways for oxygen gas and water vapour and to a decrease in the gas permeability of the coatings, as shown in Fig. 18.16.

## 18.3.4 Friction and wear properties

Both friction and wear belong to the discipline of tribology and surface engineering; friction is the force between two surfaces in contact (the force resists the relative motion of the two objects), and wear is the erosion of material from a solid surface by the action of another solid. Scientists and engineers consider that surface engineering is one of the most crucial branches of technology in present

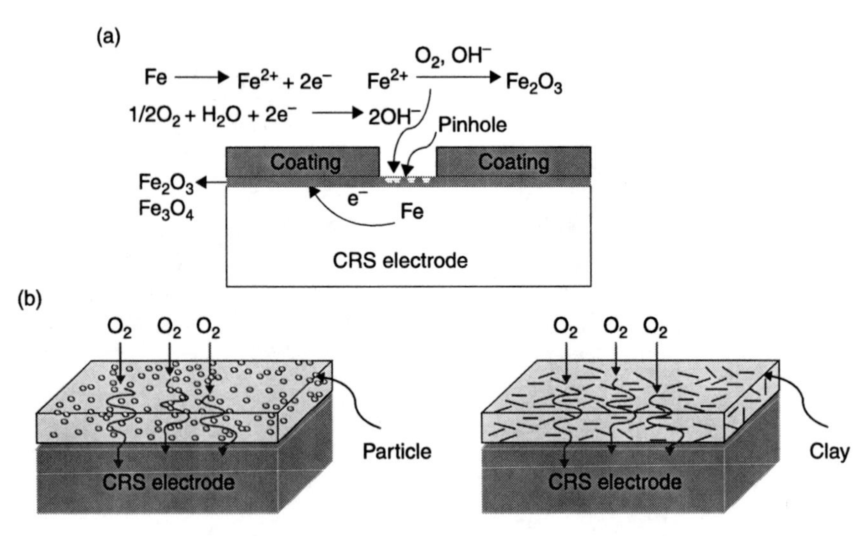


*18.16* (a) Mechanism of CRS passivation by EPI coatings. (b) Diffusion pathway of oxygen gas in polymer nanocomposites.[148–150]

society. Surface engineering is defined as the design of a surface/substrate composite system to achieve performance that could not be achieved by either the surface composition or the substrate alone, through engineering the substrate surface to improve the appearance, to provide protection from environmental damage or to enhance the mechanical or physical performance of the surface.[45] Damage, such as corrosion, friction, wear, heat, radiation, weathering and the like, occurs on the surface of a component or is transferred via the surface into the component. Therefore, surface protection is of considerable significance to modern materials technology, and a wide variety of coatings can improve the wear resistance of a substrate. The addition of a second phase is one of the methods used to improve the tribological properties (such as the coefficient of friction and wear rate) of polymer materials. Factors that have an influence on the friction and wear characteristics of polymer composites are particle size, morphology and the concentration of the filler.

An epoxy/$SiO_2$ composite was prepared by blending and the subsequent addition of a curing agent, and the wear properties were studied. The mixtures obtained were then gently transferred onto a clean, flat, smooth poly(tetrafluoroethylene) (PTFE) sheet, and an aluminium disc was placed on top of the epoxy composite. The PTFE/composite/aluminium sandwich was cured at 23 °C for 168 h and then the thickness of the cured composite was measured. The results showed that spherical silica particles improved the wear resistance of the epoxy matrix even though the filler concentration was relatively low (0.5–4.0 wt%), as shown in Fig. 18.17.[151]

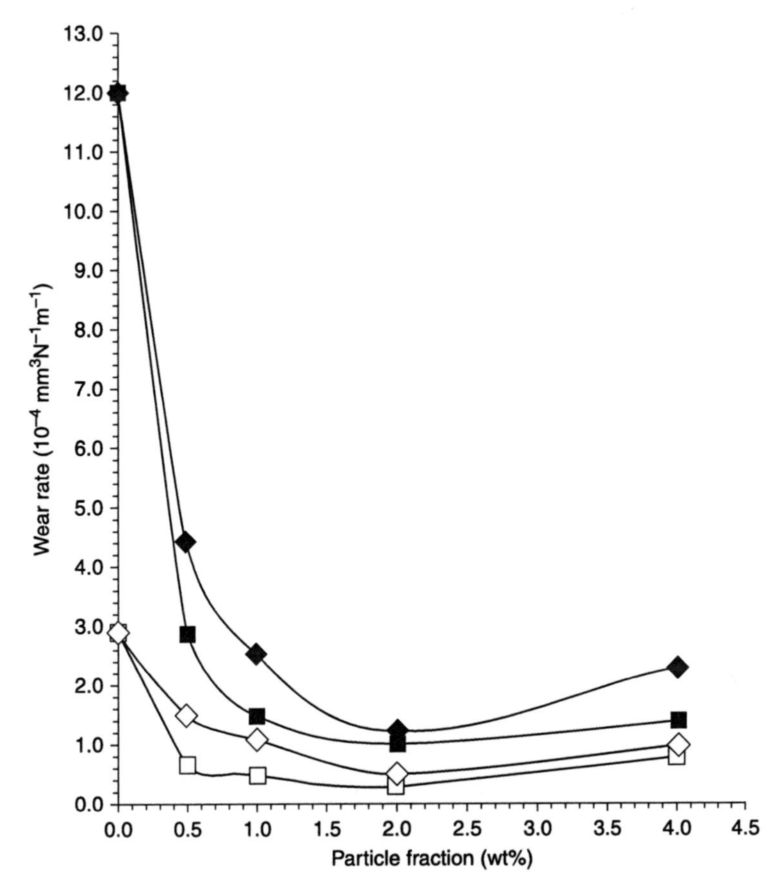


*18.17* Wear rate of epoxy matrix composites filled with silica particles. □: 120 nm, 1N; ◇: 510 nm, 1N; ▪: 120 nm, 2N; ♦: 510 nm, 2 N.[151]

The wear mechanisms of epoxy and particulate-filled epoxy matrix composites under the pin-on-disc condition were studied by Durand *et al.*[152] During wear testing, the unfilled epoxy exhibited brittle behaviour and cracks formed perpendicular to the sliding direction. Therefore, material waves were created and wear debris was produced (Fig. 18.18(a)). With the incorporation of uniformly sized submicron spherical silica particles in the epoxy matrix, the propagation of cracks into the epoxy matrix was hindered by particles on or near the surface layer. This resulted in the formation of finer material waves and, consequently, less debris (Fig. 18.18(b)). Furthermore, hard filler particles in a polymer matrix can reduce the wear rate when the applied stress is less than the critical value.[153] Both of these factors could contribute to a reduction in the wear rate of an epoxy matrix.

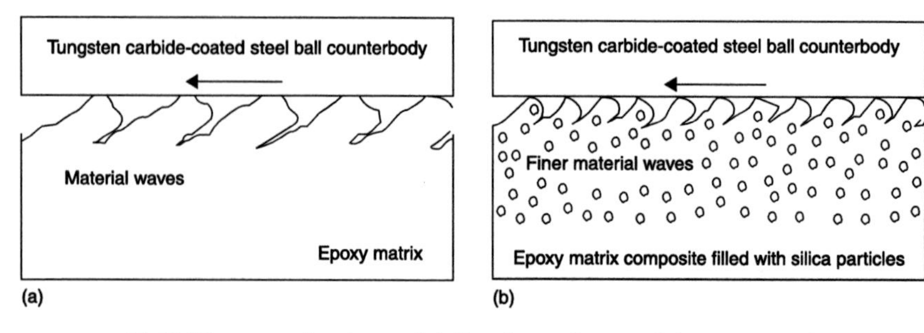


*18.18* Wear mechanisms. (a) Cracks and material waves on the worn surface of an epoxy matrix. (b) Finer cracks and material waves on the worn surface of an epoxy matrix composite filled with silica particles.[151–153]

Wang *et al.*[154–157] incorporated nano-$Si_3N_4$, nano-$SiO_2$, nano-SiC and nano-$ZrO_2$ into poly(ether ether ketone) (PEEK) and the testing results indicated that these nanometre-sized particles could effectively reduce the wear rate of PEEK, which was attributed to a thin, uniform and tenacious transfer film that formed on the surface of the counterpart. Bahadur and co-workers[158,159] investigated the tribological properties of poly(phenylene sulphide) (PPS) filled with nanometre $TiO_2$, ZnO, CuO or SiC, as shown in Table 18.1. An increase in wear resistance and a decrease in the coefficient of friction were observed under specific conditions.

Li *et al.*[160] reported that PTFE filled with nano-ZnO could greatly reduce the wear rate of this polymer. The nano-ZnO antiwear mechanism works by preventing

*Table 18.1* Steady-state wear rates and the coefficients of friction of PPS and its composites filled with different proportions of nanoparticles as filler materials[159]

| Filler | Filler proportion (vol%) | Steady-state wear rate ($nm^3$/km) | Steady-state coefficient of friction |
|---|---|---|---|
| No filler | 0 | 0.324 | 0.43 |
| CuO | 1 | 0.083 | 0.41 |
| | 2 | 0.078 | 0.39 |
| | 3 | 0.161 | 0.37 |
| | 10 | 0.196 | 0.34 |
| $TiO_2$ | 1 | 0.272 | 0.35 |
| | 2 | 0.162 | 0.37 |
| | 3 | 0.445 | 0.45 |
| | 5 | 0.726 | 0.49 |
| ZnO | 1 | 0.922 | 0.39 |
| | 2 | 1.089 | 0.42 |
| | 5 | 1.563 | 0.46 |
| SiC | 2 | 0.628 | 0.41 |
| | 5 | 0.980 | 0.47 |

the destruction of the PTFE's banded structure during the friction process. Sawyer *et al.*[161] explored the friction and wear behaviour of PTFE composites filled with 40-nm $Al_2O_3$, and they found that the friction coefficient of the composites increased slightly compared with an unfilled sample and that the wear resistance increased monotonically with increasing filler concentration. Lai *et al.*[162–164] used nanometre attapulgite and ultrafine diamond to improve the tribological performance of PTFE and analysed the mechanism of the filler action in reducing the wear rate of the PTFE polymer through DSC and SEM. Yu *et al.*[165] compared the friction and wear behaviour of poly(oxymethylene) (POM) filled with micrometre and submicrometre copper particles. The submicrometre copper particles were more effective in lowering the wear and coefficient of friction of the composites. Ng *et al.*[166] dispersed $TiO_2$ nanoparticles in epoxy, and the resultant composites not only appeared to be tougher than the traditional microparticle-filled epoxy but also had a higher scratch resistance. Cai *et al.*[167–168] researched the tribological properties of polyimide (PI) filled with $Al_2O_3$ and CNTs, and they reported that the fillers could effectively enhance the friction resistance and antiwear capacity of the composites. Lai *et al.*[169] showed also that nano-attapulgite particles could improve the tribological behaviour of PI.

Li and co-workers[170] studied the effect of $SiO_2$ particle size on the friction and wear behaviour of PI/$SiO_2$ hybrids synthesized through a sol–gel method. Nanosilica could simultaneously provide PI with strength, toughness, friction and wear resistance. The friction coefficient of the hybrid with 100-nm $SiO_2$ particles was the lowest and approximately 20% lower than that of pure PI. However, the lowest wear rate was recorded for the hybrid with 300-nm $SiO_2$ particles and was approximately 20% lower than that of neat PI. The strength and toughness of PI/$SiO_2$ hybrids are enhanced by approximately 50 and 200%, respectively, compared with pure PI when the size of the silica is less than 300 nm. These results are shown in Fig 18.19.

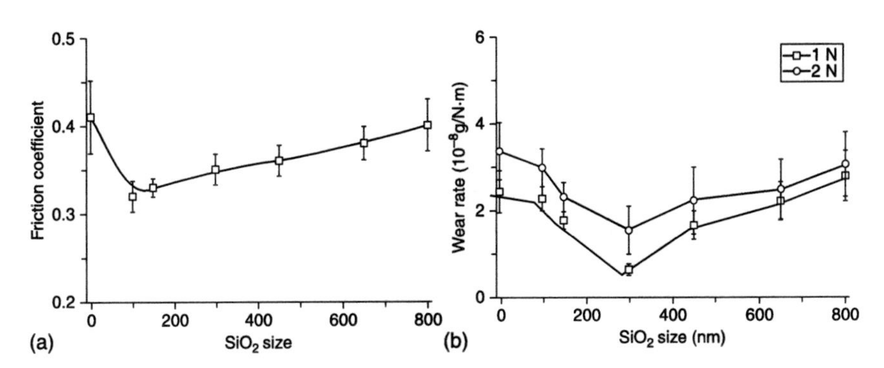


*18.19* (a) Average friction coefficient for samples as a function of the size of the silica in the hybrid films (load: 2 N). (b) Variation of wear rate of PI hybrids with the size of the silica at various applied loads.[170]

Gu and co-workers[171] studied the indentation and scratch behaviour of a $SiO_2$/polycarbonate (PC) composite coating at the microscale and nanoscale. The experimental results showed that hardness and stiffness increased after the addition of nano-$SiO_2$. The scratch tests indicated that the nano-$SiO_2$/PC coating exhibits a smaller scratch depth (Fig. 18.20) and a lower frictional coefficient. The

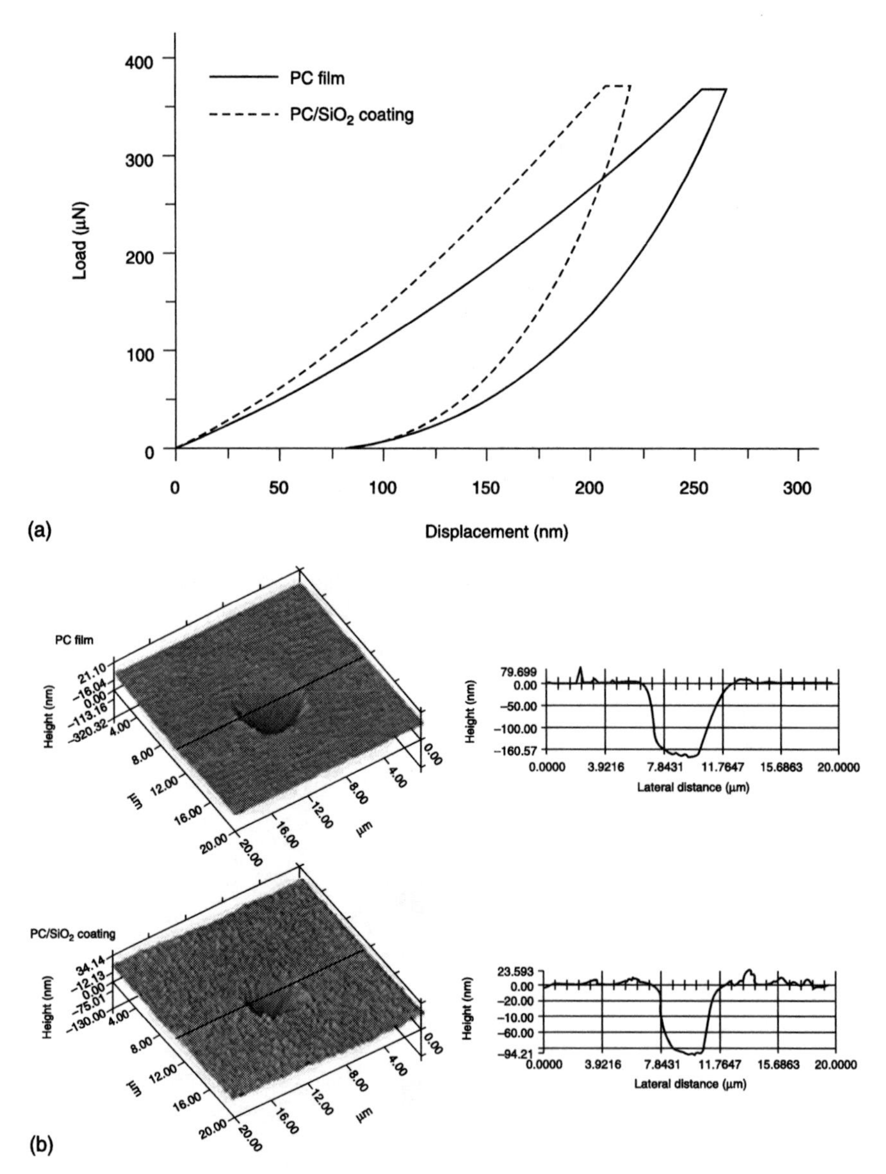


*18.20* (a) Load–displacement curves of PC film and PC/$SiO_2$ coating. (b) Three-dimensional images of the indentations.[171]

addition of nano-$SiO_2$ particles was reported as increasing the hardness and the modulus of the PC film. The elastic recovery of the nanocomposite was also improved. It was suggested that the improvement of these mechanical properties could be ascribed to the excellent performance of the nanoparticles and the increased crystallinity of the nanocomposites.[172–174]

The friction and wear properties of polyester (PE)/clay nanocomposites prepared by a mechanical stirrer was studied by Balasubramanian and co-workers.[175] PE/clay nanocomposites have a slightly higher hardness than pristine PE. The lowest specific wear rate and friction coefficient were obtained for PE containing 3 wt% clay, as shown in Fig. 18.21.

The tribological properties of poly(vinylidene fluoride) (PVDF)/clay nanocomposites were studied by Li and co-workers.[176] Table 18.2 and Fig 18.22 show the crystallinity and mechanical properties of the nanocomposites. The data shows that adding 1–2 wt% clay to PVDF was effective in improving the mechanical and tribological properties of neat PVDF. The nanoclay acts as a reinforcing element for the load bore and thus decreased plastic deformation, which in turn reduced the friction coefficient, while the increased polarity of the material caused by crystal transformation may be the main reason for the lower wear. The PVDF/clay nanocomposite containing 5 wt% nanoclay had the highest

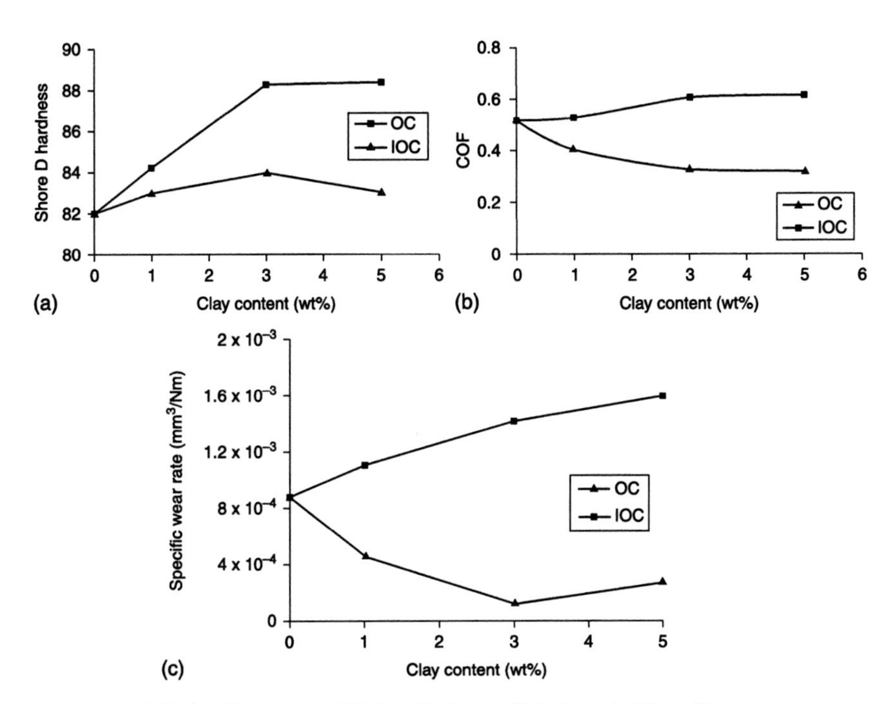


*18.21* (a) Hardness. (b) Coefficient of friction. (c) Specific wear rate. (OC: organo-modified clay; IOC: inorganic clay).[175]

*Table 18.2* Crystallinity and mechanical properties of neat PVDF and its nanocomposites[176]

| | Neat PVDF | +1 wt% nanoclay | +2 wt% nanoclay | +5 wt% nanoclay |
|---|---|---|---|---|
| Crystallinity (%) | 53.0 | 51.2 | 48.9 | 44.2 |
| Hardness (Shore A scale) | 71.9 | 72.4 | 74.8 | 75.6 |
| Tensile strength (MPa) | 55.1 | 51.8 | 53.5 | 52.2 |
| Elongation at break (%) | 48.7 | 49.9 | 53.5 | 43.5 |
| Impact strength (kJ/m$^2$) | 61.0 | 61.7 | 70.6 | 43.4 |

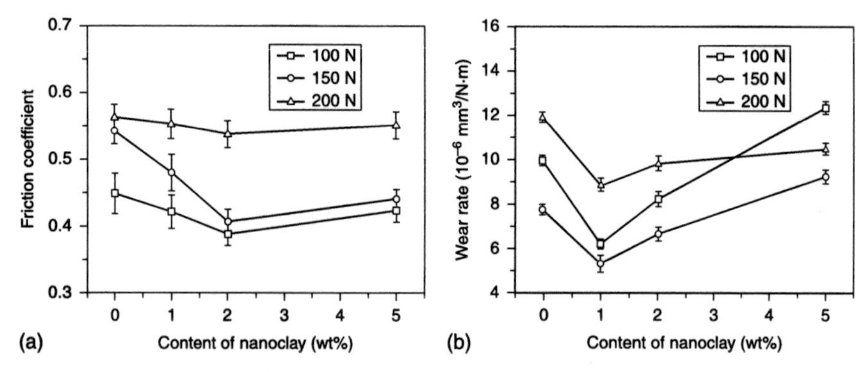


*18.22* Effect of load on (a) the friction coefficient and (b) the wear rate of PVDF and PVDF nanocomposites with different contents of nanoclay.[176]

wear rate. Weak compatibility between the nanoclay and PVDF, and also decreased crystallinity, may be responsible for the higher wear rate.

## 18.4 Future trends

Conductive polymers and non-conductive polymers had been used in many fields in applications such as antistatic coatings, EMI shielding, anticorrosion coatings, wear-resistant coatings etc. Recently, synthetic electroactive polymers have been demonstrated to undergo redox reactions, similarly to conductive polymers, and were also shown to have good solubility and film formation. Because of these advantageous properties, electroactive polymers can be exploited for use in electrochemical sensors, functional membranes, electrochromic materials, anticorrosion coatings and biomaterials. However, the mechanical, thermal and electrical properties have not been studied in much depth since these materials are still a new field of exploration. Their reported behaviour is promising enough for further research to be undertaken in the hope of exploiting their valuable properties.

## 18.5 Acknowledgements

We thank all the publishers (Elsevier Ltd, John Wiley & Sons, Inc. and the American Chemical Society) who give us copyright permission for the figures and tables in this chapter. This work was supported by the National Sciences Council of the Republic of China under Grant number NSC 98-2113-M-033-001-MY3 and NSC 100-2811-M-033-012.

## 18.6 References

1. Manias, E. (2007) 'Nanocomposites: Stiffer by design', *Nat. Mater.*, 6, 9–11.
2. Yeh, J. M., Liou, S. J., Lin, C. Y., Cheng, C. Y., Chang, Y. W. *et al.* (2002) 'Anticorrosively enhanced PMMA/clay nanocomposite materials with quaternary alkylphosphonium salt as an intercalating agent', *Chem. Mater.*, 14, 154–161.
3. Yeh, J. M., Liou, S. J., Lai, C. Y., Wu, P. C. and Tsai, T. Y. (2001) 'Enhancement of corrosion protection effect in polyaniline via the formation of polyaniline-clay nanocomposite materials', *Chem. Mater.*, 13, 1131–1136.
4. Weng, C. J., Jhuo, Y. S., Chen, Y. L., Feng, C. F., Chang, C. H. *et al.* (2011) 'Intrinsically electroactive polyimide microspheres fabricated by electrospraying technology for ascorbic acid detection', *J. Mater. Chem.*, 21, 15666–15672.
5. Chao, D., Jia, X., Liu, H., He, L., Cui, L. *et al.* (2011) 'Novel electroactive poly(arylene ether sulfone) copolymers containing pendant oligoaniline groups: Synthesis and properties', *J. Polym. Sci., Part A: Polym. Chem.*, 49, 1605–1614.
6. Weng, C. J., Jhuo, Y. S., Huang, K. Y., Feng, C. F., Yeh, J. M. *et al.* (2011) 'Mechanically and thermally enhanced intrinsically dopable polyimide membrane with advanced gas separation capabilities', *Macromolecules*, 44, 6067–6076.
7. Jia, X., Chao, D., Liu, H., He, L., Zheng, T. *et al.* (2011) 'Synthesis and properties of novel electroactive poly(amic acid) and polyimide copolymers bearing pendant oligoaniline groups', *Polym. Chem.*, 2, 1300–1306.
8. Jäger, S. and Böttcher, H. (1996) 'Dye composite layers by physical vapor deposition', *Adv. Mater.*, 8, 93–97.
9. Hauert, R., Patscheider, J., Knoblauch, L. and Diserens, M. (1999) 'New coatings by nanostructuring', *Adv. Mater.*, 11, 175–177.
10. Awad, Y., El Khakani, M. A., Brassard, D., Smirani, R., Camiré, N. *et al.* (2010) 'Effect of thermal annealing on the structural and mechanical properties of amorphous silicon carbide films prepared by polymer-source chemical vapor deposition', *Thin Solid Films*, 518, 2738–2744.
11. Wardle, B. L., Saito, D. S., García, E. J., Hart, A. J., de Villoria, R. G. *et al.* (2008) 'Fabrication and characterization of ultrahigh-volume-fraction aligned carbon nanotube–polymer composites', *Adv. Mater.*, 20, 2707–2714.
12. Hart, A. J. and Slocum, A. H. (2006) 'Rapid growth and flow-mediated nucleation of millimeter-scale aligned carbon nanotube structures from a thin-film catalyst', *J. Phys. Chem. B*, 110, 8250–8257.

13. Polini, R., Amar, M., Ahmed, W., Kumashiro, S., Sein, H. *et al.* (2005) 'A study of diamond synthesis by hot filament chemical vapour deposition on nanocomposite coatings', *Thin Solid Films*, 489, 116–121.
14. Ishizawa, H. and Ogino, M. (1996) 'Thin hydroxyapatite layers formed on porous titanium using electrochemical and hydrothermal reaction', *J. Mater. Sci.*, 31, 6279–6284.
15. Helen Annal Therese, G., Vishnu Kamath, P. and N. Subbanna, G. (1998) 'Novel electrosynthetic route to calcium phosphate coatings', *J. Mater. Chem.*, 8, 405–408.
16. Peng, P., Kumar, S., Voelcker, N. H., Szili, E., Smart, R. S. C. *et al.* (2006) 'Thin calcium phosphate coatings on titanium by electrochemical deposition in modified simulated body fluid', *J. Biomed. Mater. Res.*, 76A, 347–355.
17. Park, H. J., Kang, M. G., Ahn, S. H. and Guo, L. J. (2010) 'A facile route to polymer solar cells with optimum morphology readily applicable to a roll-to-roll process without sacrificing high device performances', *Adv. Mater.*, 22, E247–E253.
18. Ahn, S. H. and Guo, L. J. (2008) 'High-speed roll-to-roll nanoimprint lithography on flexible plastic substrates', *Adv. Mater.*, 20, 2044–2049.
19. Ma, P. C., Siddiqui, N. A., Marom, G. and Kim, J. K. (2010) 'Dispersion and functionalization of carbon nanotubes for polymer-based nanocomposites: A review', *Composites Part A*, 41, 1345–1367.
20. Choy, K. L. (2003) 'Chemical vapour deposition of coatings', *Prog. Mater. Sci.*, 48, 57–170.
21. Yasuda, H. and Hsu, T. (1977) 'Some aspects of plasma polymerization investigated by pulsed r.F. Discharge', *J. Polym. Sci., Polym. Chem.*, 15, 81–97.
22. Yasuda, H. and Lamaze, C. E. (1973) 'Polymerization in an electrodeless glow discharge. Ii. Olefinic monomers', *J. Appl. Polym. Sci.*, 17, 1519–1531.
23. Laimer, J. and Matsumoto, S. (1996) 'Pulsed microwave plasma-assisted chemical vapour deposition of diamond', *Int. J. Refract. Met. Hard Mater.*, 14, 179–184.
24. Rao, C. R. K. and Trivedi, D. C. (2005) 'Chemical and electrochemical depositions of platinum group metals and their applications', *Coord. Chem. Rev.*, 249, 613–631.
25. Coyle, D. J., Macosko, C. W. and Scriven, L. E. (1990) 'The fluid dynamics of reverse roll coating', *AIChE J.*, 36, 161–174.
26. Iwasaki, T. (1990) *Gravure Coating Device and Method.* United States patent, 4,948,635, 14 August 1990.
27. Yeh, J. M., Liou, S. J., Lai, M. C., Chang, Y. W., Huang, C. Y. *et al.* (2004) 'Comparative studies of the properties of poly(methyl methacrylate)–clay nanocomposite materials prepared by *in situ* emulsion polymerization and solution dispersion', *J. Appl. Polym. Sci.*, 94, 1936–1946.
28. Yeh, J. M., Chen, C. L., Chen, Y. C., Ma, C. Y., Huang, H. Y. *et al.* (2004) 'Enhanced corrosion prevention effect of polysulfone–clay nanocomposite materials prepared by solution dispersion', *J. Appl. Polym. Sci.*, 92, 631–637.
29. Yeh, J. M., Huang, H. Y., Chen, C. L., Su, W. F. and Yu, Y. H. (2006) 'Siloxane- modified epoxy resin-clay nanocomposite coatings with advanced anticorrosive properties prepared by a solution dispersion approach', *Surf. Coat. Technol.*, 200, 2753–2763.
30. Yeh, J. M., Chen, C. L., Kuo, T. H., Su, W. F., Huang, H. Y. *et al.* (2004) 'Preparation and properties of (BATB-ODPA) polyimide–clay nanocomposite materials', *J. Appl. Polym. Sci.*, 92, 1072–1079.
31. Yu, Y. H., Yeh, J. M., Liou, S. J., Chen, C. L., Liaw, D. J. *et al.* (2004) 'Preparation and properties of polyimide–clay nanocomposite materials for anticorrosion application', *J. Appl. Polym. Sci.*, 92, 3573–3582.

32. Yu, Y. H., Yeh, J. M., Liou, S. J. and Chang, Y. P. (2004) 'Organo-soluble polyimide (TBAPP-OPDA)/clay nanocomposite materials with advanced anticorrosive properties prepared from solution dispersion technique', *Acta Mater.*, 52, 475–486.
33. Huang, T. C., Hsieh, C. F., Yeh, T. C., Lai, C. L., Tsai, M. H. *et al.* (2011) 'Comparative studies on corrosion protection properties of polyimide-silica and polyimide-clay composite materials', *J. Appl. Polym. Sci.*, 119, 548–557.
34. Yeh, J. M., Chen, C. L., Chen, Y. C., Ma, C. Y., Lee, K. R. *et al.* (2002) 'Enhancement of corrosion protection effect of poly(o-ethoxyaniline) via the formation of poly(o-ethoxyaniline)-clay nanocomposite materials', *Polymer*, 43, 2729–2736.
35. Yeh, J. M. and Chin, C. P. (2003) 'Structure and properties of poly(o-methoxyaniline)–clay nanocomposite materials', *J. Appl. Polym. Sci.*, 88, 1072–1080.
36. Chang, K. C., Lai, M. C., Peng, C. W., Chen, Y. T., Yeh, J. M. *et al.* (2006) 'Comparative studies on the corrosion protection effect of DBSA-doped polyaniline prepared from *in situ* emulsion polymerization in the presence of hydrophilic $Na^+$-MMT and organophilic organo-MMT clay platelets', *Electrochim. Acta*, 51, 5645–5653.
37. Chang, K. C., Jang, G. W., Peng, C. W., Lin, C. Y., Shieh, J. C. *et al.* (2007) 'Comparatively electrochemical studies at different operational temperatures for the effect of nanoclay platelets on the anticorrosion efficiency of DBSA-doped polyaniline/$Na^+$-MMT clay nanocomposite coatings', *Electrochim. Acta*, 52, 5191–5200.
38. Yeh, J. M., Chin, C. P. and Chang, S. (2003) 'Enhanced corrosion protection coatings prepared from soluble electronically conductive polypyrrole-clay nanocomposite materials', *J. Appl. Polym. Sci.*, 88, 3264–3272.
39. Chen, C. and Morgan, A. B. (2009) 'Mild processing and characterization of silica epoxy hybrid nanocomposite', *Polymer*, 50, 6265–6273.
40. Cardiano, P., Mineo, P., Sergi, S., Ponterio, R. C., Triscari, M. *et al.* (2003) 'Epoxy-silica polymers as restoration materials. Part II', *Polymer*, 44, 4435–4441.
41. Ochi, M., Takahashi, R. and Terauchi, A. (2001) 'Phase structure and mechanical and adhesion properties of epoxy/silica hybrids', *Polymer*, 42, 5151–5158.
42. Ragosta, G., Abbate, M., Musto, P., Scarinzi, G. and Mascia, L. (2005) 'Epoxy-silica particulate nanocomposites: Chemical interactions, reinforcement and fracture toughness', *Polymer*, 46, 10506–10516.
43. Huang, K. Y., Weng, C. J., Lin, S. Y., Yu, Y. H. *et al.* (2009) 'Preparation and anticorrosive properties of hybrid coatings based on epoxy-silica hybrid materials', *J. Appl. Polym. Sci.*, 112, 1933–1942.
44. Chang, K. C., Lin, H. F., Lin, C. Y., Kuo, T. H., Huang, H. H. *et al.* (2008) 'Effect of amino-modified silica nanoparticles on the corrosion protection properties of epoxy resin-silica hybrid materials', *J. Nanosci. Nanotechnol.*, 8, 3040–3049.
45. Li, Y., Ma, Y., Xie, B., Cao, S. and Wu, Z. (2007) 'Dry friction and wear behavior of flame-sprayed polyamide1010/n-$SiO_2$ composite coatings', *Wear*, 262, 1232–1238.
46. Zou, H., Wu, S. and Shen, J. (2008) 'Polymer/silica nanocomposites: Preparation, characterization, properties, and applications', *Chem. Rev.*, 108, 3893–3957.
47. Petrovicova, E., Knight, R., Schadler, L. S. and Twardowski, T. E. (2000) 'Nylon 11/silica nanocomposite coatings applied by the HVOF process. Ii. Mechanical and barrier properties', *J. Appl. Polym. Sci.*, 78, 2272–2289.
48. Sun, Y., Zhang, Z. and Wong, C. P. (2005) 'Influence of interphase and moisture on the dielectric spectroscopy of epoxy/silica composites', *Polymer*, 46, 2297–2305.
49. Ramanathan, T., Abdala, A. A., Stankovich, S., Dikin, D. A., Herrera Alonso, M. *et al.* (2008) 'Functionalized graphene sheets for polymer nanocomposites', *Nat. Nano.*, 3, 327–331.

50. Verdejo, R., Barroso Bujans, F., Rodriguez Perez, M. A., Antonio de Saja, J. and Lopez Manchado, M. A. (2008) 'Functionalized graphene sheet filled silicone foam nanocomposites', *J. Mater. Chem.*, 18, 2221–2226.
51. Verdejo, R., Saiz Arroyo, C., Carretero Gonzalez, J., Barroso Bujans, F., Rodriguez Perez, M. A. *et al.* (2008) 'Physical properties of silicone foams filled with carbon nanotubes and functionalized graphene sheets', *Eur. Polym. J.*, 44, 2790–2797.
52. Singh, V., Joung, D., Zhai, L., Das, S., Khondaker, S. I. *et al.* (2011) 'Graphene based materials: Past, present and future', *Prog. Mater Sci.*, 56, 1178–1271.
53. Ansari, S. and Giannelis, E. P. (2009) 'Functionalized graphene sheet–poly(vinylidene fluoride) conductive nanocomposites', *J. Polym. Sci., Part B: Polym. Phys.*, 47, 888–897.
54. Jang, J. Y., Kim, M. S., Jeong, H. M. and Shin, C. M. (2009) 'Graphite oxide/ poly(methyl methacrylate) nanocomposites prepared by a novel method utilizing macroazoinitiator', *Compos. Sci. Technol.*, 69, 186–191.
55. Liang, J., Huang, Y., Zhang, L., Wang, Y., Ma, Y. *et al.* (2009) 'Molecular-level dispersion of graphene into poly(vinyl alcohol) and effective reinforcement of their nanocomposites', *Adv. Funct. Mater.*, 19, 2297–2302.
56. Liang, J., Xu, Y., Huang, Y., Zhang, L., Wang, Y. *et al.* (2009) 'Infrared-triggered actuators from graphene-based nanocomposites', *J. Phys. Chem. C*, 113, 9921–9927.
57. Nguyen, D. A., Lee, Y. R., Raghu, A. V., Jeong, H. M., Shin, C. M. *et al.* (2009) 'Morphological and physical properties of a thermoplastic polyurethane reinforced with functionalized graphene sheet', *Polym. Int.*, 58, 412–417.
58. Rafiee, M. A., Rafiee, J., Srivastava, I., Wang, Z., Song, H. *et al.* (2010) 'Fracture and fatigue in graphene nanocomposites', *Small*, 6, 179–183.
59. Verdejo, R., Bernal, M. M., Romasanta, L. J. and Lopez-Manchado, M. A. (2011) 'Graphene filled polymer nanocomposites', *J. Mater. Chem.*, 21, 3301–3310.
60. Stankovich, S., Dikin, D. A., Dommett, G. H. B., Kohlhaas, K. M., Zimney, E. J. *et al.* (2006) 'Graphene-based composite materials', *Nature*, 442, 282–286.
61. Rafiee, M. A., Rafiee, J., Wang, Z., Song, H., Yu, Z. Z. *et al.* (2009) 'Enhanced mechanical properties of nanocomposites at low graphene content', *ACS Nano*, 3, 3884–3890.
62. Zhu, Y., Murali, S., Cai, W., Li, X., Suk, J. W., Potts, J. R. *et al.* (2010) 'Graphene and graphene oxide: Synthesis, properties, and applications', *Adv. Mater.*, 22, 3906–3924.
63. Compton, O. C., Kim, S., Pierre, C., Torkelson, J. M. and Nguyen, S. T. (2010) 'Crumpled graphene nanosheets as highly effective barrier property enhancers', *Adv. Mater.*, 22, 4759–4763.
64. Kim, H., Miura, Y. and Macosko, C. W. (2010) 'Graphene/polyurethane nanocomposites for improved gas barrier and electrical conductivity', *Chem. Mater.*, 22, 3441–3450.
65. Kim, H., Abdala, A. A. and Macosko, C. W. (2010) 'Graphene/polymer nanocomposites', *Macromolecules*, 43, 6515–6530.
66. Potts, J. R., Dreyer, D. R., Bielawski, C. W. and Ruoff, R. S. (2011) 'Graphene-based polymer nanocomposites', *Polymer*, 52, 5–25.
67. Lu, X., Zhang, W., Wang, C., Wen, T. C. and Wei, Y. (2011) 'One-dimensional conducting polymer nanocomposites: Synthesis, properties and applications', *Prog. Polym. Sci.*, 36, 671–712.
68. Martins, N. C. T., Moura e Silva, T., Montemor, M. F., Fernandes, J. C. S. and Ferreira, M. G. S. (2010) 'Polyaniline coatings on aluminium alloy 6061-T6: Electrosynthesis and characterization', *Electrochim. Acta*, 55, 3580–3588.
69. Bagherzadeh, M. R., Mahdavi, F., Ghasemi, M., Shariatpanahi, H. and Faridi, H. R. (2010) 'Using nanoemeraldine salt-polyaniline for preparation of a new anticorrosive water-based epoxy coating', *Prog. Org. Coat.*, 68, 319–322.

70. Sakhri, A., Perrin, F. X., Aragon, E., Lamouric, S. and Benaboura, A. (2010) 'Chlorinated rubber paints for corrosion prevention of mild steel: A comparison between zinc phosphate and polyaniline pigments', *Corros. Sci.*, 52, 901–909.
71. Ahmad, N. and MacDiarmid, A. G. (1996) 'Inhibition of corrosion of steels with the exploitation of conducting polymers', *Synth. Met.*, 78, 103–110.
72. Wei, Y., Wang, J., Jia, X., Yeh, J. M. and Spellane, P. (1995) 'Polyaniline as corrosion protection coatings on cold rolled steel', *Polymer*, 36, 4535–4537.
73. Lu, W. K., Elsenbaumer, R. L. and Wessling, B. (1995) 'Corrosion protection of mild steel by coatings containing polyaniline', *Synth. Met.*, 71, 2163–2166.
74. Wessling, B. (1997) 'Scientific and commercial breakthrough for organic metals', *Synth. Met.*, 85, 1313–1318.
75. Wessling, B. (1994) 'Passivation of metals by coating with polyaniline: Corrosion potential shift and morphological changes', *Adv. Mater.*, 6, 226–228.
76. Yongjun, H. (2005) 'Synthesis of polyaniline/nano-$CeO_2$ composite microspheres via a solid-stabilized emulsion route', *Mater. Chem. Phys.*, 92, 134–137.
77. Li, J., Vaisman, L., Marom, G. and Kim, J. K. (2007) 'Br treated graphite nanoplatelets for improved electrical conductivity of polymer composites', *Carbon*, 45, 744–750.
78. Cong, H., Zhang, J., Radosz, M. and Shen, Y. (2007) 'Carbon nanotube composite membranes of brominated poly(2,6-diphenyl-1,4-phenylene oxide) for gas separation', *J. Membr. Sci.*, 294, 178–185.
79. Geblinger, N., Thiruvengadathan, R. and Regev, O. (2007) 'Preparation and characterization of a double filler polymeric nanocomposite', *Compos. Sci. Technol.*, 67, 895–899.
80. Malinauskas, A., Malinauskiene, J. and Ramanavičius, A. (2005) 'Conducting polymer-based nanostructurized materials: Electrochemical aspects', *Nanotechnology*, 16, R51.
81. Hackman, I. and Hollaway, L. (2006) 'Epoxy-layered silicate nanocomposites in civil engineering', *Composites Part A*, 37, 1161–1170.
82. Shabani-Nooshabadi, M., Ghoreishi, S. M. and Behpour, M. (2011) 'Direct electrosynthesis of polyaniline-montmorillonite nanocomposite coatings on aluminum alloy 3004 and their corrosion protection performance', *Corros. Sci.*, 53, 3035–3042.
83. Riede, A., Helmstedt, M., Riede, V. and Stejskal, J. (1998) 'Polyaniline dispersions. 9. Dynamic light scattering study of particle formation using different stabilizers', *Langmuir*, 14, 6767–6771.
84. Stejskal, J., Kratochvíl, P., Armes, S. P., Lascelles, S. F., Riede, A. *et al.* (1996) 'Polyaniline dispersions. 6. Stabilization by colloidal silica particles', *Macromolecules*, 29, 6814–6819.
85. Liu, P., Liu, W. and Xue, Q. (2004) '*In situ* chemical oxidative graft polymerization of aniline from silica nanoparticles', *Mater. Chem. Phys.*, 87, 109–113.
86. Niu, Z., Yang, Z., Hu, Z., Lu, Y. and Han, C. C. (2003) 'Polyaniline–silica composite conductive capsules and hollow spheres', *Adv. Funct. Mater.*, 13, 949–954.
87. Riede, A., Helmstedt, M., Riede, V., Zemek, J. and Stejskal, J. (2000) '*In situ* polymerized polyaniline films. 2. Dispersion polymerization of aniline in the presence of colloidal silica', *Langmuir*, 16, 6240–6244.
88. Radhakrishnan, S., Siju, C. R., Mahanta, D., Patil, S. and Madras, G. (2009) 'Conducting polyaniline-nano-$TiO_2$ composites for smart corrosion resistant coatings', *Electrochim. Acta*, 54, 1249–1254.
89. Luo, X., Killard, A. J., Morrin, A. and Smyth, M. R. (2007) '*In situ* electropolymerised silica-polyaniline core-shell structures: Electrode modification and enzyme biosensor enhancement', *Electrochim. Acta*, 52, 1865–1870.

90. Grossiord, N., Loos, J., Regev, O. and Koning, C. E. (2006) 'Toolbox for dispersing carbon nanotubes into polymers to get conductive nanocomposites', *Chem. Mater.*, 18, 1089–1099.
91. Zhang, W., Dehghani Sanij, A. and Blackburn, R. (2007) 'Carbon based conductive polymer composites', *J. Mater. Sci.*, 42, 3408–3418.
92. Lou, F. L., Sui, Z. J., Sun, J. T., Li, P., Chen, D. *et al.* (2010) 'Synthesis of carbon nanofibers/mica hybrids for antistatic coatings', *Mater. Lett.*, 64, 711–714.
93. Downs, C., Nugent, J., Ajayan, P. M., Duquette, D. J. and Santhanam, K. S. V. (1999) 'Efficient polymerization of aniline at carbon nanotube electrodes', *Adv. Mater.*, 11, 1028–1031.
94. Zengin, H., Zhou, W., Jin, J., Czerw, R., Smith, D. W., Echegoyen, L. *et al.* (2002) 'Carbon nanotube doped polyaniline', *Adv. Mater.*, 14, 1480–1483.
95. Quillard, S., Louarn, G., Lefrant, S. and Macdiarmid, A. G. (1994) 'Vibrational analysis of polyaniline: A comparative study of leucoemeraldine, emeraldine, and pernigraniline bases', *Phys. Rev. B*, 50, 12496.
96. Bahr, J. L. and Tour, J. M. (2002) 'Covalent chemistry of single-wall carbon nanotubes', *J. Mater. Chem.*, 12, 1952–1958.
97. Gohier, A., Nekelson, F., Helezen, M., Jegou, P., Deniau, G. *et al.* (2011) 'Tunable grafting of functional polymers onto carbon nanotubes using diazonium chemistry in aqueous media', *J. Mater. Chem.*, 21, 4615–4622.
98. Hamon, M. A., Chen, J., Hu, H., Chen, Y., Itkis, M. E. *et al.* (1999) 'Dissolution of single-walled carbon nanotubes', *Adv. Mater.*, 11, 834–840.
99. Hirsch, A. (2002) 'Functionalization of single-walled carbon nanotubes', *Angew. Chem. Int. Ed.*, 41, 1853–1859.
100. Tsang, S. C., Chen, Y. K., Harris, P. J. F. and Green, M. L. H. (1994) 'A simple chemical method of opening and filling carbon nanotubes', *Nature*, 372, 159–162.
101. Liu, J., Rinzler, A. G., Dai, H., Hafner, J. H., Bradley, R. K. *et al.* (1998) 'Fullerene pipes', *Science*, 280, 1253–1256.
102. Kuznetsova, A., Popova, I., Yates, J. T., Bronikowski, M. J., Huffman, C. B. *et al.* (2001) 'Oxygen-containing functional groups on single-wall carbon nanotubes: NEXAFS and vibrational spectroscopic studies', *J. Am. Chem. Soc.*, 123, 10699–10704.
103. Salzmann, C. G., Llewellyn, S. A., Tobias, G., Ward, M. A. H., Huh, Y. *et al.* (2007) 'The role of carboxylated carbonaceous fragments in the functionalization and spectroscopy of a single-walled carbon-nanotube material', *Adv. Mater.*, 19, 883–887.
104. Kumar, N. A., Jeon, I. Y., Sohn, G. J., Jain, R., Kumar, S. *et al.* (2011) 'Highly conducting and flexible few-walled carbon nanotube thin film', *ACS Nano*, 5, 2324–2331.
105. Jeon, I. Y., Kang, S. W., Tan, L. S. and Baek, J. B. (2010) 'Grafting of polyaniline onto the surface of 4-aminobenzoyl-functionalized multiwalled carbon nanotube and its electrochemical properties', *J. Polym. Sci., Part A: Polym. Chem.*, 48, 3103–3112.
106. Choi, E. K., Jeon, I. Y., Bae, S. Y., Lee, H. J., Shin, H. S. *et al.* (2010) 'High-yield exfoliation of three-dimensional graphite into two-dimensional graphene-like sheets', *Chem. Commun.*, 46, 6320–6322.
107. Choi, H. J., Jeon, I. Y., Chang, D. W., Yu, D., Dai, L. *et al.* (2011) 'Preparation and electrocatalytic activity of gold nanoparticles immobilized on the surface of 4-mercaptobenzoyl-functionalized multiwalled carbon nanotubes', *J. Phys. Chem. C*, 115, 1746–1751.
108. Han, S. W., Oh, S. J., Tan, L. S. and Baek, J. B. (2008) 'One-pot purification and functionalization of single-walled carbon nanotubes in less-corrosive poly(phosphoric acid)', *Carbon*, 46, 1841–1849.

109. Lee, H. J., Han, S. W., Kwon, Y. D., Tan, L. S. and Baek, J. B. (2008) 'Functionalization of multi-walled carbon nanotubes with various 4-substituted benzoic acids in mild polyphosphoric acid/phosphorous pentoxide', *Carbon*, 46, 1850–1859.
110. Liao, Y., Zhang, C., Zhang, Y., Strong, V., Tang, J. *et al.* (2011) 'Carbon nanotube/polyaniline composite nanofibers: Facile synthesis and chemosensors', *Nano Lett.*, 11, 954–959.
111. Spitalsky, Z., Tasis, D., Papagelis, K. and Galiotis, C. (2010) 'Carbon nanotube-polymer composites: Chemistry, processing, mechanical and electrical properties', *Prog. Polym. Sci.*, 35, 357–401.
112. Geetha, S., Satheesh Kumar, K. K. and Trivedi, D. C. (2005) 'Polyaniline reinforced conducting e-glass fabric using 4-chloro-3-methyl phenol as secondary dopant for the control of electromagnetic radiations', *Compos. Sci. Technol.*, 65, 973–980.
113. Stafström, S., Brédas, J. L., Epstein, A. J., Woo, H. S., Tanner, D. B. *et al.* (1987) 'Polaron lattice in highly conducting polyaniline: Theoretical and optical studies', *Phys. Rev. Lett.*, 59, 1464–1467.
114. Joo, J. and Lee, C. Y. (2000) 'High frequency electromagnetic interference shielding response of mixtures and multilayer films based on conducting polymers', *J. Appl. Phys.*, 88, 513–518.
115. Chung, D. D. L. (2001) 'Electromagnetic interference shielding effectiveness of carbon materials', *Carbon*, 39, 279–285.
116. Yang, Y., Gupta, M. C., Dudley, K. L. and Lawrence, R. W. (2005) 'Novel carbon nanotube-polystyrene foam composites for electromagnetic interference shielding', *Nano Lett.*, 5, 2131–2134.
117. Yang, Y., Gupta, M. C., Dudley, K. L. and Lawrence, R. W. (2005) 'A comparative study of EMI shielding properties of carbon nanofiber and multi-walled carbon nanotube filled polymer composites', *J. Nanosci. Nanotechnol.*, 5, 927–931.
118. Kim, H., Lee, C., Joo, J., Cho, S., Yoon, H. *et al.* (2004) 'Electrical conductivity and electromagnetic interference shielding of multiwalled carbon nanotube composites containing Fe catalyst', *Appl. Phys. Lett.*, 84, 589–591.
119. Che, R. C., Peng, L. M., Duan, X. F., Chen, Q. and Liang, X. L. (2004) 'Microwave absorption enhancement and complex permittivity and permeability of Fe encapsulated within carbon nanotubes', *Adv. Mater.*, 16, 401–405.
120. Xiang, C., Pan, Y., Liu, X., Sun, X., Shi, X. *et al.* (2005) 'Microwave attenuation of multiwalled carbon nanotube-fused silica composites', *Appl. Phys. Lett.*, 87, 123103–123105.
121. Grimes, C. A., Mungle, C., Kouzoudis, D., Fang, S. and Eklund, P. C. (2000) 'The 500 MHz to 5.50 GHz complex permittivity spectra of single-wall carbon nanotube-loaded polymer composites', *Chem. Phys. Lett.*, 319, 460–464.
122. Li, N., Huang, Y., Du, F., He, X., Lin, X. *et al.* (2006) 'Electromagnetic interference (EMI) shielding of single-walled carbon nanotube epoxy composites', *Nano Lett.*, 6, 1141–1145.
123. Saini, P., Choudhary, V., Singh, B. P., Mathur, R. B. and Dhawan, S. K. (2009) 'Polyaniline-MWCNT nanocomposites for microwave absorption and EMI shielding', *Mater. Chem. Phys.*, 113, 919–926.
124. Phang, S. W., Tadokoro, M., Watanabe, J. and Kuramoto, N. (2008) 'Microwave absorption behaviors of polyaniline nanocomposites containing $TiO_2$ nanoparticles', *Curr. Appl Phys.*, 8, 391–394.
125. Phang, S. W., Tadokoro, M., Watanabe, J. and Kuramoto, N. (2008) 'Synthesis, characterization and microwave absorption property of doped polyaniline

nanocomposites containing $TiO_2$ nanoparticles and carbon nanotubes', *Synth. Met.*, 158, 251–258.
126. Udeh, C. U., Fey, N. and Faul, C. F. J. (2011) 'Functional block-like structures from electroactive tetra(aniline) oligomers', *J. Mater. Chem.*, DOI: 10.1039/c1jm12557e.
127. Chao, D., Lu, X., Chen, J., Zhao, X., Wang, L. *et al.* (2006) 'New method of synthesis of electroactive polyamide with amine-capped aniline pentamer in the main chain', *J. Polym. Sci., Part A: Polym. Chem.*, 44, 477–482.
128. Chao, D., Cui, L., Lu, X., Mao, H., Zhang, W. *et al.* (2007) 'Electroactive polyimide with oligoaniline in the main chain via oxidative coupling polymerization', *Eur. Polym. J.*, 43, 2641–2647.
129. Chao, D., Lu, X., Chen, J., Liu, X., Zhang, W. *et al.* (2006) 'Synthesis and characterization of electroactive polyamide with amine-capped aniline pentamer and ferrocene in the main chain by oxidative coupling polymerization', *Polymer*, 47, 2643–2648.
130. Zhang, J., Chao, D., Cui, L., Liu, X. and Zhang, W. (2009) 'Electroactive polymer with oligoanilines in the main chain: Synthesis, characterization and dielectric properties', *Macromol. Chem. Phys.*, 210, 1739–1745.
131. Chao, D., Ma, X., Liu, Q., Lu, X., Chen, J. *et al.* (2006) 'Synthesis and characterization of electroactive copolymer with phenyl-capped aniline tetramer in the main chain by oxidative coupling polymerization', *Eur. Polym. J.*, 42, 3078–3084.
132. Buga, K., Majkowska, A., Pokrop, R., Zagorska, M., Djurado, D. *et al.* (2005) 'Postpolymerization grafting of aniline tetramer on polythiophene chain: Structural organization of the product and its electrochemical and spectroelectrochemical properties', *Chem. Mater.*, 17, 5754–5762.
133. Dufour, B., Rannou, P., Travers, J. P., Pron, A., Zagórska, M. *et al.* (2002) 'Spectroscopic and spectroelectrochemical properties of a poly(alkylthiophene)-oligoaniline hybrid polymer', *Macromolecules*, 35, 6112–6120.
134. Chao, D., Zhang, J., Liu, X., Lu, X., Wang, C. *et al.* (2010) 'Synthesis of novel poly(amic acid) and polyimide with oligoaniline in the main chain and their thermal, electrochemical, and dielectric properties', *Polymer*, 51, 4518–4524.
135. Cui, L., Chao, D., Zhang, J., Mao, H., Li, Y. *et al.* (2010) 'Synthesis and properties of novel electroactive polyamide containing crown ether in the main chain', *Synth. Met.*, 160, 400–404.
136. He, L., Chao, D., Jia, X., Liu, H., Yao, L. *et al.* (2011) 'Electroactive polymer with oligoanilines in the main chain and azo chromophores in the side chain: Synthesis, characterization and dielectric properties', *J. Mater. Chem.*, 21, 1852–1858.
137. Guo, B., Finne Wistrand, A. and Albertsson, A. C. (2011) 'Universal two-step approach to degradable and electroactive block copolymers and networks from combined ring-opening polymerization and post-functionalization via oxidative coupling reactions', *Macromolecules*, 44, 5227–5236.
138. Guo, B., Finne Wistrand, A. and Albertsson, A. C. (2010) 'Enhanced electrical conductivity by macromolecular architecture: Hyperbranched electroactive and degradable block copolymers based on poly(e-caprolactone) and aniline pentamer', *Macromolecules*, 43, 4472–4480.
139. Guo, B., Finne Wistrand, A. and Albertsson, A. C. (2011) 'Degradable and electroactive hydrogels with tunable electrical conductivity and swelling behavior', *Chem. Mater.*, 23, 1254–1262.

140. Guo, B., Finne Wistrand, A. and Albertsson, A. C. (2010) 'Molecular architecture of electroactive and biodegradable copolymers composed of polylactide and carboxyl-capped aniline trimer', *Biomacromolecules*, 11, 855–863.
141. Huang, L., Hu, J., Lang, L., Wang, X., Zhang, P. *et al.* (2007) 'Synthesis and characterization of electroactive and biodegradable ABA block copolymer of polylactide and aniline pentamer', *Biomaterials*, 28, 1741–1751.
142. Huang, L., Zhuang, X., Hu, J., Lang, L., Zhang, P. *et al.* (2008) 'Synthesis of biodegradable and electroactive multiblock polylactide and aniline pentamer copolymer for tissue engineering applications', *Biomacromolecules*, 9, 850–858.
143. Huang, T. C., Yeh, T. C., Huang, H. Y., Ji, W. F., Chou, Y. C. *et al.* (2011) 'Electrochemical studies on aniline-pentamer-based electroactive polyimide coating: Corrosion protection and electrochromic properties', *Electrochim. Acta*, 56, 10151–10158.
144. Huang, K. Y., Shiu, C. L., Wu, P. S., Wei, Y., Yeh, J. M. *et al.* (2009) 'Effect of amino-capped aniline trimer on corrosion protection and physical properties for electroactive epoxy thermosets', *Electrochim. Acta*, 54, 5400–5407.
145. Huang, K. Y., Jhuo, Y. S., Wu, P. S., Lin, C. H., Yu, Y. H. *et al.* (2009) 'Electrochemical studies for the electroactivity of amine-capped aniline trimer on the anticorrosion effect of as-prepared polyimide coatings', *Eur. Polym. J.*, 45, 485–493.
146. Wang, Z. Y., Yang, C., Gao, J. P., Lin, J., Meng, X. S. *et al.* (1998) 'Electroactive polyimides derived from amino-terminated aniline trimer', *Macromolecules*, 31, 2702–2704.
147. Weng, C. J., Chang, C. H., Peng, C. W., Chen, S. W., Yeh, J. M. *et al.* (2011) 'Advanced anticorrosive coatings prepared from the mimicked *Xanthosoma sagittifolium*-leaf-like electroactive epoxy with synergistic effects of superhydrophobicity and redox catalytic capability', *Chem. Mater.*, 23, 2075–2083.
148. Huang, T. C., Su, Y. A., Yeh, T. C., Huang, H. Y., Wu, C. P. *et al.* (2011) 'Advanced anticorrosive coatings prepared from electroactive epoxy-$SiO_2$ hybrid nanocomposite materials', *Electrochim. Acta*, 56, 6142–6149.
149. Huang, H. Y., Huang, T. C., Yeh, T. C., Tsai, C. Y., Lai, C. L. *et al.* (2011) 'Advanced anticorrosive materials prepared from amine-capped aniline trimer-based electroactive polyimide-clay nanocomposite materials with synergistic effects of redox catalytic capability and gas barrier properties', *Polymer*, 52, 2391–2400.
150. Weng, C. J., Huang, J. Y., Huang, K. Y., Jhuo, Y. S., Tsai, M. H. *et al.* (2010) 'Advanced anticorrosive coatings prepared from electroactive polyimide-$TiO_2$ hybrid nanocomposite materials', *Electrochim. Acta*, 55, 8430–8438.
151. Xing, X. S. and Li, R. K. Y. (2004) 'Wear behavior of epoxy matrix composites filled with uniform sized sub-micron spherical silica particles', *Wear*, 256, 21–26.
152. Durand, J. M., Vardavoulias, M. and Jeandin, M. (1995) 'Role of reinforcing ceramic particles in the wear behaviour of polymer-based model composites', *Wear*, 181–183, 833–839.
153. Shi, M. M. (2000) *Solid lubricating materials*, Beijing, Chemical Industry Press.
154. Wang, Q., Xu, J., Shen, W. and Liu, W. (1996) 'An investigation of the friction and wear properties of nanometer $Si_3N_4$ filled PEEK', *Wear*, 196, 82–86.
155. Wang, Q., Xue, Q. and Shen, W. (1997) 'The friction and wear properties of nanometre $SiO_2$ filled polyetheretherketone', *Tribol. Int.*, 30, 193–197.
156. Wang, Q.-H., Xu, J., Shen, W. and Xue, Q. (1997) 'The effect of nanometer SiC filler on the tribological behavior of PEEK', *Wear*, 209, 316–321.
157. Wang, Q., Xue, Q., Shen, W. and Zhang, J. (1998) 'The friction and wear properties of nanometer $ZrO_2$-filled polyetheretherketone', *J. Appl. Polym. Sci.*, 69, 135–141.

158. Schwartz, C. J. and Bahadur, S. (2000) 'Studies on the tribological behavior and transfer film-counterface bond strength for polyphenylene sulfide filled with nanoscale alumina particles', *Wear*, 237, 261–273.
159. Bahadur, S. and Sunkara, C. (2005) 'Effect of transfer film structure, composition and bonding on the tribological behavior of polyphenylene sulfide filled with nano particles of $TiO_2$, ZnO, CuO and SiC', *Wear*, 258, 1411–1421.
160. Li, F., Hu, K. A., Li, J. L. and Zhao, B. Y. (2001) 'The friction and wear characteristics of nanometer ZnO filled polytetrafluoroethylene', *Wear*, 249, 877–882.
161. Sawyer, W. G., Freudenberg, K. D., Bhimaraj, P. and Schadler, L. S. (2003) 'A study on the friction and wear behavior of PTFE filled with alumina nanoparticles', *Wear*, 254, 573–580.
162. Lai, S. Q., Li, T. S., Liu, X. J. and Lv, R. G. (2004) 'A study on the friction and wear behavior of PTFE filled with acid treated nano-attapulgite', *Macromol. Mater. Eng.*, 289, 916–922.
163. Lai, S. Q., Li, T. S., Liu, X. J., Lv, R. G. and Yue, L. (2006) 'The tribological properties of PTFE filled with thermally treated nano-attapulgite', *Tribol. Int.*, 39, 541–547.
164. Lai, S. Q., Yue, L., Li, T. S. and Hu, Z. M. (2006) 'The friction and wear properties of polytetrafluoroethylene filled with ultrafine diamond', *Wear*, 260, 462–468.
165. Yu, L., Yang, S., Wang, H. and Xue, Q. (2000) 'An investigation of the friction and wear behaviors of micrometer copper particle- and nanometer copper particle-filled polyoxymethylene composites', *J. Appl. Polym. Sci.*, 77, 2404–2410.
166. Ng, C. B., Schadler, L. S. and Siegel, R. W. (1999) 'Synthesis and mechanical properties of $TiO_2$-epoxy nanocomposites', *Nanostruct. Mater.*, 12, 507.
167. Cai, H., Yan, F. and Xue, Q. (2004) 'Investigation of tribological properties of polyimide/carbon nanotube nanocomposites', *Mater. Sci. Eng., A*, 364, 94–100.
168. Cai, H., Yan, F., Xue, Q. and Liu, W. (2003) 'Investigation of tribological properties of $Al_2O_3$-polyimide nanocomposites', *Polym. Test.*, 22, 875–882.
169. Lai, S. Q., Yue, L., Li, T. S., Liu, X. J. and Lv, R. G. (2005) 'An investigation of friction and wear behaviors of polyimide/attapulgite hybrid materials', *Macromol. Mater. Eng.*, 290, 195–201.
170. Lai, S. Q., Li, T. S., Wang, F. D., Li, X. J. and Yue, L. (2007) 'The effect of silica size on the friction and wear behaviors of polyimide/silica hybrids by sol-gel processing', *Wear*, 262, 1048–1055.
171. Wang, Z. Z., Gu, P. and Zhang, Z. (2010) 'Indentation and scratch behavior of nano-$SiO_2$/polycarbonate composite coating at the micro/nano-scale', *Wear*, 269, 21–25.
172. Omrani, A., Simon, L. C. and Rostami, A. A. (2009) 'The effects of alumina nanoparticle on the properties of an epoxy resin system', *Mater. Chem. Phys.*, 114, 145–150.
173. Liu, L., Barber, A. H., Nuriel, S. and Wagner, H. D. (2005) 'Mechanical properties of functionalized single-walled carbon-nanotube/poly(vinyl alcohol) nanocomposites', *Adv. Funct. Mater.*, 15, 975–980.
174. Tjong, S. C. (2006) 'Structural and mechanical properties of polymer nanocomposites', *Mater. Sci. Eng., R*, 53, 73–197.
175. Jawahar, P., Gnanamoorthy, R. and Balasubramanian, M. (2006) 'Tribological behaviour of clay-thermoset polyester nanocomposites', *Wear*, 261, 835–840.
176. Peng, Q. Y., Cong, P. H., Liu, X. J., Liu, T. X., Huang, S. *et al.* (2009) 'The preparation of PVDF/clay nanocomposites and the investigation of their tribological properties', *Wear*, 266, 713–720.

# Index

CPSIA information can be obtained at www.ICGtesting.com
Printed in the USA
LVOW100324251012

304371LV00007B/2/P

9 781845 699406